Foodborne Diseases

Infectious Disease

SERIES EDITOR: *Vassil St. Georgiev*
National Institute of Allergy and Infectious Diseases
National Institutes of Health

Foodborne Diseases

Edited by

Shabbir Simjee, PhD

United States Food and Drug Administration, Laurel, MD;
Current Address: Eli Lilly and Company, Elanco Animal Health,
Basingstoke, Hampshire, United Kingdom

Foreword by

Toni L. Poole, PhD

United States Department of Agriculture,
Agricultural Research Service,
Southern Plains Area Research Center, College Station, TX

HUMANA PRESS ✳ TOTOWA, NEW JERSEY

I dedicate this book to:

My mother Farida Simjee
&
Aunty Attiya Unia

With love, respect and admiration

And to:

Shamim, Asif, Zahra, Suleman, Aaliyah, Fatima (BumBum) and Muhammad

For being a constant source of welcomed, and much appreciated distraction from the mundane routine of life!

© 2007 Humana Press Inc.
999 Riverview Drive, Suite 208
Totowa, New Jersey 07512

www.humanapress.com

The content and opinions expressed in this book are the sole work of the authors and editors, who have warranted due diligence in the creation and issuance of their work. The publisher, editors, and authors are not responsible for errors or omissions or for any consequences arising from the information or opinions presented in this book and make no warranty, express or implied, with respect to its contents.

Due diligence has been taken by the publishers, editors, and authors of this book to assure the accuracy of the information published and to describe generally accepted practices. The contributors herein have carefully checked to ensure that the drug selections and dosages set forth in this text are accurate and in accord with the standards accepted at the time of publication. Notwithstanding, since new research, changes in government regulations, and knowledge from clinical experience relating to drug therapy and drug reactions constantly occur, the reader is advised to check the product information provided by the manufacturer of each drug for any change in dosages or for additional warnings and contraindications. This is of utmost importance when the recommended drug herein is a new or infrequently used drug. It is the responsibility of the treating physician to determine dosages and treatment strategies for individual patients. Further, it is the responsibility of the health care provider to ascertain the Food and Drug Administration status of each drug or device used in their clinical practice. The publishers, editors, and authors are not responsible for errors or omissions or for any consequences from the application of the information presented in this book and make no warranty, express or implied, with respect to the contents in this publication.

This publication is printed on acid-free paper. ∞

ANSI Z39.48-1984 (American Standards Institute) Permanence of Paper for Printed Library Materials.

Production Editor: Christina M. Thomas

Cover design by: Karen Schulz

Cover Illustration: *Toxoplasma gondii* oocysts in sugar fecal float of an infected cat. Image appears in Chapter 12, *Toxoplasma gondii*, by Dolores E. Hill et al. Image supplied by Dolores E. Hill.

For additional copies, pricing for bulk purchases, and/or information about other Humana titles, contact Humana at the above address or at any of the following numbers: Tel: 973-256-1699; Fax: 973-256-8341; E-mail: humana@humanapr.com, or visit our Website: http://humanapress.com

Photocopy Authorization Policy:

Printed in the United States of America. 10 9 8 7 6 5 4 3 2 1

eISBN: 978-1-59745-501-5
Library of Congress Cataloging-in-Publication Data
Foodborne diseases / edited by Shabbir Simjee ; foreword by Toni L. Poole.
       p. ; cm. -- (Infectious disease)
  Includes bibliographical references.
  ISBN 978-1-58829-518-7 (alk. paper)
  1. Foodborne diseases. 2. Food poisoning. 3. Food--Microbiology. I. Simjee, Shabbir. II. Series: Infectious disease (Totowa, N.J.)
  [DNLM: 1. Food Poisoning. 2. Food Microbiology. WC 268 F684 2007]
  QR201.F62F672 2007
  615.9'54--dc22
                          2007004638

Foreword

Many of us have suffered from what we believed may have been a foodborne illness, and have suffered at home without seeking clinical care because we knew it was likely to be over in a day or so. However, foodborne illness may present under unusual circumstances, as I know from having been a victim of a multi-state outbreak of *Salmonella* serotype Agona present in a toasted oat cereal in May of 1998 (JAMA 280 (5):411). This was the first known case of Salmonella contamination of a breakfast cereal; although, *Salmonella* serotype Senftenberg contamination of an infant cereal has been reported in the United Kingdom.

In case of contaminated breakfast cereal, exposure may occur daily or intermittently for several days. By the time I sought clinical advice from a gastroenterologist I felt fine, and after a battery of tests he said, "I wish all of my patients were as healthy as you." It wasn't until I heard a report on CNN later that summer that I realized what had happened. My mother, who shops at the outlet store in Indiana that carried the brand in question, had brought a box of the oat cereal to me during a visit to Tennessee. The box was from the contaminated lot, so it is likely that one undiagnosed clinical case from Tennessee can be added to the MMWR 1998; 47:462-464 report.

This case exemplifies several of the problems that plague food processing, preparation and inter- and intra- country shipping and importation. Exposure and clinical symptoms may differ in ways that neither patients nor clinicians suspect a foodborne pathogen. All of these factors lead to the under reporting of foodborne illnesses.

Today people expect a safe food supply and to meet these goals global cooperation in food pathogen research, monitoring and education are necessary. *Foodborne Diseases* covers several significant foodborne bacterial pathogens, viral infections, toxins and pathogen control strategies in an effort to educate and reduce the risk of foodborne infections in this millennium.

Toni L. Poole, PhD

Contents

Contributors

FRANK M. AARESTUP • *Danish Institute for Food and Veterinary Research, Copenhagen, Denmark*

FRANZ ALLERBERGER, MD • *Austrian Agency for Health and Food Safety (AGES), Vienna, Austria*

HAZEL APPLETON, PhD • *Virus Reference Department, Centre for Infections, Health Protection Agency, London, United Kingdom*

NISHA D. O. ANTOINE, MPH• *Office of Public Health Science, Food Safety and Inspection Service, US Department of Agriculture, Washington, DC*

MALCOLM BANKS, PhD • *Virology Department, Central Veterinary Laboratory, Veterinary Laboratories Agency, Surrey, United Kingdom*

KRISTINA E. BARLOW • *Office of Public Health Science, Food Safety and Inspection Service, US Department of Agriculture, Washington, DC*

ROSS C. BEIER, PhD • *Southern Plains Agriculture Research Center, Agricultural Research Service, US Department of Agriculture, College Station, TX*

JOAN W. BENNETT • *Department of Plant Biology and Pathology, Rutgers University, New Brunswick, NJ*

TODD R. CALLAWAY, MS, PhD • *USDA, ARS, Southern Plains Area Research, College Station, TX*

SHERRI B. DENNIS, PhD • *Center for Food Safety and Nutrition, US Food and Drug Administration, College Park, MD*

CATHERINE M. DENTINGER, FNP, MS • *New York City Department of Health and Mental Hygiene Bureau of Communicable Diseases, New York, NY*

MOSHE S. DREYFUSS, PhD • *Office of Public Health Science, Food Safety and Inspection Service, US Department of Agriculture, Washington, DC*

J. P. DUBEY • *US Department of Agriculture, Animal and Natural Resources Institute, Agricultural Research Service, Animal Parasitic Diseases Laboratory, Beltsville, MD*

DENISE R. EBLEN, PhD • *Office of Public Health Science, Food Safety and Inspection Service, US Department of Agriculture, Washington, DC*

STEVEN L. FOLEY, PhD • *Associate Research Scientist, National Farm Medicine Center, Marshfield Clinic Research Foundation, Marshfield, WI*

MARIA FREDRIKSSON-AHOMAA • *Institute of Hygiene and Technology of Food of Animal Origin, Faculty of Veterinary Medicine, LMU, Oberschleissheim, Germany*

CHRIS I. GALLIMORE • *Enteric Virus Unit, Virus Reference Department, Centre for Infections, Health Protection Agency, London, United Kingdom*

KATHIE A. GRANT, PhD • *Food Safety Microbiology Laboratory, Health Protection Agency Centre for Infections, London, United Kingdom*

JIM GRAY • *Enteric Virus Unit, Virus Reference Department, Centre for Infections, Health Protection Agency, London, United Kingdom*

DOLORES E. HILL • *US Department of Agriculture, Animal and Natural Resources Institute, Agricultural Research Service, Animal Parasitic Diseases Laboratory, Beltsville, MD*

MIREN ITURRIZA-GÓMARA • *Enteric Virus Unit, Virus Reference Department, Centre for Infections, Health Protection Agency, London, United Kingdom*

ELKE JENSEN, PhD • *Office of Public Health Science, Food Safety and Inspection Service, US Department of Agriculture, Washington, DC*

LARS B. JENSEN, PhD, MSc • *National Food Institute, Technical University of Denmark, Copenhagen, Denmark*

JEFFREY JONES • *The Centers for Disease Control, National Center for Infectious Dieseases, Division of Parasitic Diseases, Atlanta, GA*

S. KALE • *Department of Biology, Xavier University, New Orleans, LA*

JIM MCLAUCHLIN, PhD • *Food Safety Microbiology Laboratory, Health Protection Agency Centre for Infections, London, United Kingdom*

JIANGHONG MENG, DVM, PhD, MPVM • *Department of Nutrition and Food Science, University of Maryland, College Park, MD*

MARIANNE D. MILIOTIS, PhD • *Center for Food Safety and Nutrition, US Food and Drug Administration, College Park, MD*

KARA M. MORGAN, PhD • *Office of the Commissioner, US Food and Drug Administration, Rockville, MD*

CELINE A. NADON, PhD • *PulseNet Canada, National Microbiology Laboratory, Public Health Agency of Canada,Winnipeg, Manitoba, Canada*

G. BALAKRISH NAIR • *International Centre for Diarrhoeal Disesase Research, Dhaka, Bangladesh*

ROSELY NICHOLS • *Scottish Parasite Diagnostic Laboratory, Stobhill Hospital, Glasgow, Scotland, United Kingdom*

DAVID J. NISBET, PhD • *USDA, ARS, Southern Plains Area Research Center, College Station, TX*

SIMONA F. OPREA, MD • *William Beaumont Hospital, Royal Oak, MI*

TIM PAGET • *Medway School of Pharmacy, The Universities of Kent and Greenwich at Medway, Kent, UK*

SURESH D. PILLAI, PhD • *Departments of Nutrition and Food Science and Poultry Science, Texas A&M University, College Station, TX*

TONI L. POOLE, MS, PhD • *USDA, ARS, Southern Plains Area Research Center, College Station, TX*

KATRINE M. PRITCHARD • *Management Support Staff, Food Safety and Inspection Service, US Department of Agriculture, Washington, DC*

T. RAMAMURTHY • *National Institute of Cholera and Enteric Diseases, Calcutta, India*

GERRI M. RANSOM, MPH • *Office of Public Health Science, Food Safety and Inspection Service, US Department of Agriculture, Washington, DC*

BONNIE E. ROSE, PhD • *Office of Public Health Science, Food Safety and Inspection Service, US Department of Agriculture, Washington, DC*

MINDI D. RUSSELL• *Food Science Department, Kansas State University, Manhattan, KS*

PARMESH K. SAINI, DVM, PhD • *Office of Public Health Science, Food Safety and Inspection Service, US Department of Agriculture, Washington, DC*

CARL M. SCHROEDER, PhD • *Office of Public Health Science, Food Safety and Inspection Service, US Department of Agriculture, Washington, DC*

HUW V. SMITH • *Director, Scottish Parasite Diagnostic Laboratory, Stobhill Hospital, Glasgow, Scotland, United Kingdom*

CHIRUKANDOTH SREEKUMAR • *US Department of Agriculture, Animal and Natural Resources Institute, Agricultural Research Service, Animal Parasitic Diseases Laboratory, Beltsville, MD*

ELIJAH W. STOMMEL, MD, PhD • *Associate Professor of Medicine, Section of Neurology, Dartmouth-Hitchcock Medical Center, Lebanon, NH*

CHONG GEE TEO, MD, PhD, FRCPATH • *Division of Viral Hepatitis, Centers for Disease Control and Prevention, Atlanta, GA*

DONALD THAYER • *Food Safety Consultant, Lower Gwynedd, PA, Retired Research Leader, Food Safety Intervention Technologies, Eastern Regional Research Center, Agricultural Research Service, USDA, Wyndmoor, PA*

JIUJIANG YU • *Agricultural Research Service, Southern Regional Research Center, New Orleans, LA*

MARCUS ZERVOS, MD • *Chief, Infectious Diseases, Department of Internal Medicine, Henry Ford Hospital, Detroit, MI*

GERALD W. ZIRNSTEIN, PhD • *Office of International Affairs, Food Safety and Inspection Service, US Department of Agriculture, Washington, DC*

Color Plate

The following color illustrations are printed in the insert.

Chapter 3
Fig. 1: Typical colonial growth of *C. botulinum,* together with an atypical lipase negative culture.

Fig. 3: Nagler reaction for the identification of *C. perfringens.* Antisera is spread on the left-hand side of the plate and the right-hand side is untreated. The organisms are (from top to bottom): *C. perfringens* (no reaction, a typical strain) *C. perfringens* (positive reaction), *C. perfringens* (positive reaction), *C. absonum* (negative reaction, lecithinase-positive but not inhibited by the antitoxin), and *C. barattii* (negative reaction, lecithinase positive but not inhibited by the antitoxin).

Chapter 9
Fig. 2: *Cryptosporidium parvum* life cycle. From reference (16). With permission, Elsevier Publishing.

Chapter 15
Fig. 1: Prevalence of *Salmonella* based on the percentage of PF/HACCP "A" *Salmonella* sets, 1998-2002 (28).

Jianghong Meng and Carl M. Schroeder

Abstract

Following a 1982 outbreak of hemorrhagic colitis associated with the consumption of undercooked ground beef, *Escherichia coli*, specifically the serotype O157:H7, burst on the scene as a serious foodborne pathogen. Since then strains of O157:H7 and other enterohemorrhagic *E. coli* (EHEC) have been implicated in numerous outbreaks of foodborne illness throughout the world. A wide range of food vehicles, including meat, fruits, vegetables, and dairy products have been implicated in outbreak, of EHEC infection. Notwithstanding considerable advances in our understanding of the molecular mechanisms involved in *E. coli* pathogenesis, treatment for infected individuals remains for the most part supportive, and thus, prevention is paramount in reducing the incidence of adverse health outcomes associated with EHEC. Continued diligence in implementing strategies aimed at curtailing the presence of EHEC in food, such as the use of good agricultural practices on farm, and public awareness and education campaigns targeting food service personnel and consumers, remains a principal component of protecting public health.

1. INTRODUCTION

In 1885, the then 27-yr-old German pediatrician and bacteriologist Theodor Escherich identified and characterized a bacterium, isolated from the feces of neonates, which he named *Bacterium coli commune (1)*. Renamed in 1911 posthumously to honor its discoverer, *Escherichia coli* has since become a standard component of the microbiological laboratory, having been instrumental in, among other scientific landmarks, the discoveries of competitive inhibition and bacterial genetic recombination *(2,3)*; confirmation of the genetic code *(4)*; and the invention of gene cloning *(5,6)*. Following a 1982 outbreak of infection related to consumption of undercooked ground beef, however, *E. coli*, specifically the serotype O157:H7, emerged as an important cause of foodborne illness *(7)*. Today O157:H7 and other enterohemorrhagic *E. coli* (EHEC) remain a serious public health threat. This review summarizes the ecology, pathogenesis, and control of *E. coli*, paying particular attention to the role of EHEC in foodborne infections.

2. CLASSIFICATION AND IDENTIFICATION

2.1. General Characteristics

Gram-negative and rod-shaped, *E. coli* cells measure approx 1.1–1.5 by 2.0–6.0 µM when living, and 0.4–0.7 by 1.0–3.0 µM when dried and stained. Cells grow singly or in pairs and some are motile by means of peritrichous flagella *(8)*. *E. coli* are facultative anaerobes and grow optimally under aerobic conditions at 37°C, with a mean generation time of approx 20 min in complex media *(9)*. They produce indole, do not ferment citrate,

From: *Infectious Disease: Foodborne Diseases*
Edited by: S. Simjee © Humana Press Inc., Totowa, NJ

are positive for the methyl red test, and negative for oxidase and the Voges–Proskauer reaction *(10)*. Strains that do not readily ferment sucrose or lactose, do not produce gas, and are not motile may be mistaken for *Shigella* spp. *(11)*.

Taxonomically, *E. coli* belongs to the family Enterobacteriaceae. It is one of the six species of the genus *Escherichia* (others include *E. adecarboxylata*, *E. blattae*, *E. fergusonii*, *E. hermanii*, and *E. vulneris*). Serologically, *E. coli* strains are differentiated based on three major cell surface antigens: the O (somatic), H (flagellar), and K (capsular) antigens. At least 167 O antigens, 53 H antigens, and 74 K antigens have been described *(12)*. Characterization of the O and H antigens is sufficient to identify most *E. coli* strains: characterization of the former is done to determine the serogroup, and the latter to determine serotype. The designation O157:H7, for example, is used to describe an *E. coli* strain of serogroup 157 and serotype 7.

Most *E. coli* strains are commensal. At concentrations of approx 10^6 to 10^9 CFU/g in stool *(13)*, and with an estimated mean in vivo generation time of 12 h *(14)*, they help comprise the autochthonous intestinal microbiota of humans. *E. coli*, with the Enterobacteriaceae, benefit their hosts in facilitating competitive exclusion of harmful bacteria and stimulating immune function *(15)*.

2.2. Pathogenic E. coli

Based on their virulence factors (molecules directly involved in pathogenesis but ancillary to normal metabolic functions *[16]*) and the diseases they cause, pathogenic *E. coli* are classified into at least six distinct subgroups or pathotypes (Table 1). These are: (1) diffuse-adhering *E. coli* (DAEC), (2) enteropathogenic *E. coli* (EPEC), (3) enterotoxigenic *E. coli* (ETEC), (4) enteroinvasive *E. coli* (EIEC), (5) entero-aggregative *E. coli* (EAEC), and (6) EHEC. The abbreviation STEC refers to *E. coli* strains that produce Shiga toxin (such strains are also commonly known as verotoxin-producing *E. coli*, VTEC). The abbreviation EHEC, generally speaking, refers to those STEC strains that contain the locus of enterocyte effacement (LEE) (described in Section 7.2.1).

2.2.1. DAEC

DAEC are so named because of their characteristic diffuse adherence pattern to HEp-2 or HeLa cells in culture *(17,18)*. Most DAEC belong to serogroups O1, O2, O21, and O75 *(19)*. Strains of DAEC have been associated with persistent diarrheal illness in children aged between 1 and 5 yr *(20–24)*. Mild diarrhea devoid of fecal leukocytes is a symptom of infection. DAEC-associated diarrhea appears to be age-related, with relative risk of infection increasing with age from 1 to 5 yr *(23)*. Pathogenesis of DAEC is associated with the presence of adhesins, in particular those belonging to the Afa/Dr family *(25)*. Adhesins of this family preferentially bind to the decay-accelerating factor (DAF) protein, which in turn promotes injury to brush border-associated proteins of human intestinal epithelial cells *(26)*. Recent evidence indicates that additional virulence factors, including type III secretion systems *(27)* and autotransporter toxins *(28)*, may be important in facilitating DAEC infection. There have been no recognized outbreaks of foodborne illness related to DAEC.

2.2.2. EPEC

EPEC strains typically cause attaching and effacing (A/E) lesions, possess the EPEC adherence factor (EAF) plasmid, and do not produce Shiga toxin. Serogroups associated

Table 1
Clinical and Epidemiologic Characteristics of the Six *Escherichia coli* Pathotypes

Pathotype	Representative O antigen groups[a]	Associated infections/ syndromes[b]	Common vehicles of infection[c]
Diffuse-adhering *E. coli* (DAEC)	11, 15, 75, 126	Diarrhea (primarily in young children)	None currently recognized
Enteropathogenic *E. coli* (EPEC)	18, 26, 44, 55, 86, 111ab, 114, 119, 125ac, 126, 127, 128ab, 142, 157, 158	Acute nonbloody, watery diarrhea (primarily in infants younger than 2 yr)	Weaning foods and formulas; foods (e.g., rice) contaminated hands; water; fomites (e.g., linens, toys, rubber nipples, and carriages)
Enterotoxigenic *E. coli* (ETEC)	1, 6, 7, 8, 9, 11, 12, 15, 20, 25, 27, 60, 63, 75, 78, 80, 85, 88, 89, 99, 101, 109, 114, 115, 128ac, 139, 148, 149, 153, 159, 166, 167	Traveler's diarrhea; weanling diarrhea; diarrhea ranging from mild and self-limiting to cholera-like purging	Food (e.g., tuna fish, and potato and macaroni salads) and water
Enteroinvasive *E. coli* (EIEC)	11, 28ac, 29, 112, 115, 121, 124, 136, 143, 144, 147, 152, 164, 173	Nonbloody diarrhea and dysentery	Food (e.g., boiled vegetables and cheese), water, and contaminated hands
Enteroaggregative *E. coli* (EAEC)	3, 15, 44, 51, 77, 78, 86, 91, 92, 111, 113, 126, 127, 141, 146	Persistent diarrhea (primarily in infants and children in developing countries)	Food
Enterohemorrhagic *E. coli* (EHEC)	26, 103, 111, 145, 157, approx 50 others	Bloody diarrhea	Foods (particularly ground beef) and water (*see* Table 2)

[a]Data from refs. *(10,19)*.
[b]Data from refs. *(31,120)*.
[c]Data from refs. *(31,19)*.

with human illness include O55, O158, O26, O86, O111ab, O119, O125ac, O126, O127, O128ab, and O142 *(19)*. EPEC strains are a cause of severe diarrhea in infants, especially those in developing countries *(23,29,30)*. In developed countries, nurseries and daycare centers are common sites of EPEC infection. Symptoms of EPEC infection include acute, watery diarrhea, which may be accompanied by vomiting and low-grade fever *(31)*. Pathogenesis follows a three-stage model, which includes (1) localized adherence, (2) signal transduction, and (3) intimate adherence *(32)*. Makino and colleagues have recently described a large (80 persons) outbreak of EPEC infection caused by *E. coli* O157:H45 in rice *(33)*. Barlow et al. have also described an outbreak of infantile diarrhea caused by a novel EPEC serotype, O126:H12, the vehicle of which appears to have been contaminated water *(34)*.

2.2.3. ETEC

ETEC produce heat-stable toxin (ST) and/or heat-labile toxin (LT) *(35)*. Serogroups of ETEC include O6, O8, O15, O20, O25, O27, O63, O78, O85, O115, O128c, O148, O159, and O167 *(19)*. Similar to EPEC, strains of ETEC are an important cause of diarrhea among infants in developing countries, especially among weaning children. Furthermore, ETEC strains account for 30–60% of diarrhea cases in travelers *(36)* and have caused diarrhea among military personnel deployed in developing countries *(37)*. Although humans are the principal reservoir of ETEC, person-to-person transmission is uncommon *(38)*. Infection is characterized by watery diarrhea, ranging from mild and self-limiting to which causes severe, cholera-like purging; and may be accompanied by vomiting and fever *(31)*. Pathogenesis follows from ETEC adhesion to intestinal cells via adhesive fibrilae, from which point production and localized delivery of STs and LTs leads to fluid loss and diarrhea production. Recent years have seen an increase in the number of foodborne outbreaks of ETEC infection in developed countries. Examples include an outbreak caused by *E. coli* O111:B4 among pupils and adults at a school in southern Finland (vehicle unknown) *(39)*, one among Japanese elementary schoolchildren caused by *E. coli* O25:NM in "tuna paste" *(40)*, and one among American partygoers caused by *E. coli* O169:H41 in potato, macaroni, and egg salads *(41)*. Of note, results from the investigation of an outbreak caused by an atypical *E. coli* O39:NM among patrons of a Minnesota restaurant buffet highlight the potential difficulty in classifying diarrheagenic *E. coli* into defined pathotypes *(42)*.

2.2.4. EIEC

EIEC strains invade and multiply within human colonic epithelial cells. Serogroups commonly associated with disease include O28c, O29, O112, O124, O136, O143, O144, O152, O164, and O167 *(19)*. Similar to that of ETEC, humans are the natural reservoir of EIEC. Person-to-person transmission follows the fecal–oral route, with ingestion of contaminated food and water having been implicated in outbreaks of EIEC infection. EIEC infection is clinically indistinguishable from that caused by *Shigella* spp.; both result in bacillary dysentery, or shigellosis, a watery diarrhea containing mucous and trace blood, and often accompanied by severe abdominal cramps and fever *(43)*. The ability of EIEC strains to invade epithelial cells is associated with the presence of an approx 140-MDa plasmid on which reside genes encoding invasion-facilitating outer membrane proteins *(44)*. Bacillary dysentery is often endemic among institutional settings such as prisons, daycare facilities, and nursing homes. An outbreak of infection caused by *E. coli* O166:H15 linked to the consumption of boiled vegetables in Japan provides a recent example of an outbreak of foodborne illness from EIEC *(45)*.

2.2.5. EAEC

EAEC produce a characteristic pattern of aggregative adherence to HEp-2 cells in culture *(18)*. EAEC serotypes include O3, O15, O44, O77, O86, O92, O111, and O127 *(19)*. Strains of these serotypes have been associated with persistent diarrhea in children throughout the developing world, including those living in the Indian subcontinent and in Brazil. Symptoms of infection range from watery diarrhea accompanied by a low-grade fever and occasional vomiting to grossly bloody diarrhea *(31)*. Insights from animal models *(46,47)* and human intestinal biopsy samples *(48)* reveal that EAEC form biofilms in adhering to intestinal mucosal cells. Adherence is largely dependent on the presence of aggregative adherence fimbriae, encoded by an approx 60-MDa plasmid, pAA *(49)*. Following adherence to mucosal cells, EAEC may produce the *E. coli* heat-stable enterotoxin 1 (EAST1). Strains expressing this 38-amino-acid-long toxin have been shown to induce diarrhea in humans *(50)*. In 1993, an EAEC strain of serotype O?:H10 caused an outbreak of infection from which nearly 2700 Japanese students experienced severe diarrhea subsequent to eating school lunches *(51)*. Zhou and colleagues recently described an outbreak of gastroenteritis by *E. coli* isolates that possessed the EAggEC heat-stable enterotoxin (EAST1) gene, but which did not demonstrate entero-aggregative adherence to HEp-2 cells *(45)*.

2.2.6. EHEC

All EHEC strains produce Shiga toxins (named after their discoverer, the Japanese microbiologist Kiyoshi Shiga) which, similar to the Shiga toxin produced by *Shigella dysentariae* type 1, destroy Vero cells (African green monkey kidney cells). *E. coli* O157:H7 is the archetypical EHEC. Having first been described after the 1982 outbreak of illness associated with consumption of undercooked ground beef *(7,52)*, O157:H7 is the leading cause of EHEC infection in the United States and other industrialized countries, including Canada and the United Kingdom. Other serogroups of EHEC implicated in foodborne illness include O26, O103, O111, and O145 *(19)*. Based on the severity of illness EHEC are considered the most serious of the foodborne pathogenic *E. coli*. Therefore, the remainder of this chapter, except in those few instances where explicitly stated otherwise, focuses on EHEC, paying particular attention to their ecology, epidemiology, and pathogenesis related to foodborne illness. Though strains of O157 garner most attention, strains of other EHEC serogroups, especially O111, are being increasingly reported throughout the world *(161)* Also of note, the emergence of sorbitol-fermenting O157 strains will reduce the effectiveness of Sorbitol MacConkey (SMAC) agar for isolation of EHEC.

3. RESERVOIRS OF EHEC

3.1. Cattle

Cattle are the principal reservoirs of EHEC, an observation at least partially attributed to the fact that these animals appear to lack the receptors for *E. coli* O157:H7 Shiga toxins *(53)*. The prevalence of EHEC in cattle nevertheless varies. Zhao and colleagues *(54)* reported that in a survey of 965 dairy calves and 11,881 feedlot cattle in the United States, 31 (3.2%) and 191 (1.6%), respectively, were positive for *E. coli* O157:H7. Elsewhere, in a study of four large-scale (>35,000 animal capacity) beef cattle feedlots in southwest Kansas, 44 of 17,050 (<0.1%) fecal pat samples were positive for *E. coli* O157:H7 *(55)*. Similarly, low prevalences (<2%) of O157:H7 have been described in studies of cattle from Washington *(56)* and Florida *(57)*. Nonculture-based methods may offer increased sensitivity for detection. For instance, using immunomagnetic separation technology, *E. coli* O157:H7 was found in 636 of 4790 (13%) cattle fecal samples *(58)*.

3.2. Food Animals Other Than Cattle

In addition to cattle, EHEC have been recovered from various other food animals, including swine *(3,30,59–64)*, poultry *(30)*, goats, and sheep *(1,36,63,65,66)*. Regarding the latter, Kudva and colleagues have shown that the prevalence of *E. coli* O157:H7 is seasonal, with approximately a third the of sheep in their study positive for O157:H7 in June and none positive in November *(67)*. The same research group has subsequently reported findings suggesting that the colon is the primary gastrointestinal site for EHEC persistence and proliferation in mature ruminants *(68)*.

3.3. Wildlife and Pets

Examples of wildlife and pets in which EHEC have been found include whitetail deer *(22,69–71)*, reindeer *(72)*, boars *(73)*, geese *(29)*, dogs *(74)*, cats *(75)*, and rabbits *(76)*. There is a possibility for EHEC infection following captive animal contact *(76)*. Recent research by Belongia and colleagues suggests a reduced occurrence of clinical EHEC illness among children living in farms, owing in part to repeated EHEC antigenic stimu- lation *(65)*. Although EHEC are found in companion animals *(77)*, there have not been, to the best of our knowledge, reports of EHEC infection from contact with household pets.

3.4. Humans

Persons infected with EHEC may shed the pathogen in their feces for several weeks. For instance, patients with hemorrhagic colitis (HC) or hemolytic uremic syndrome (HUS; *see* Section 5.2) typically shed EHEC for up to 21 days *(2)*. Shedding may in some cases, however, persist longer. For example, in a study of an outbreak of EHEC in a daycare center, Orr et al. found an infected child shed the pathogen for 62 days following the diarrhea onset *(76)*.

4. FOODBORNE OUTBREAKS OF EHEC INFECTION

In 1994, the Council of State and Territorial Epidemiologists designated the infection from *E. coli* O157:H7 as a nationally notifiable disease (meaning health-care providers required to notify local or state health departments in the case of O157:H7 infection). Since that time, the number of annually reported O157:H7 cases are increased from 0.8 to 1.1 per 100,000 population in 2003 *(78)*, an increase of roughly 7%. It is estimated that 85% of EHEC infections are transmitted through food *(79)*. Outbreaks of EHEC infection have been associated with a wide variety of foods (Table 2), specific examples of which are discussed in the following sections.

4.1. Outbreaks Associated With Ground Beef

Ground beef has been frequently associated with outbreaks of foodborne infection caused by EHEC (Table 3), which is perhaps not surprising given the role of cattle as EHEC reservoirs (*see* Section 3.1). The first-recognized outbreak of infection from *E. coli* O157:H7 occurred in the Pacific Northwest United States in 1982 and was associated with the consumption of undercooked ground beef *(7,52)*. Of the 145 outbreaks of O157:H7 infection reported to the US Centers for Disease Control and Prevention (CDC) during the period 1990–1999, ground beef was the confirmed or suspected vehicle in at least 37 (26%) cases *(80)*. Encouragingly, however, infections from *E. coli* O157:H7 through ground beef may be on the decline. Recent data show a reduction in the percent ground beef product samples positive for *E. coli* O157:H7 in the United States over the period 2000–2003 *(81)*. Similarly, during the same period, a decline was observed for the number of O157:H7 clinical infections in the United States *(78)*. Whether these data are

Table 2
Examples of Documented Food- and Waterborne Outbreaks of Infection
From Enterohemorrhagic *Escherichia coli* (EHEC) Worldwide, 1982–2002

Year (Month)	Serotype	Location	Setting	Vehicle	No. of cases (Deaths)	Ref.
1982 (February)	O157:H7	Oregon, US	Community	Ground beef	26	(52)
1982 (May)	O157:H7	Michigan, US	Community	Ground beef	21	(52)
1984 (September)	O157:H7	Nebraska, US	Nursing home	Ground beef	34 (4)	(121)
1985	O157:H7	Canada	Nursing home	Sandwiches	73 (17)	(94)
1987 (June)	O157:H7	Utah, US	Institutions for mentally retarded persons	Ground beef/ person-to-person	51	(122)
1988 (October)	O157:H7	Minnesota, US	School	Precooked ground beef	54	(123)
1989 (December)	O157:H7	Missouri, US	Community	Water	243 (4)	(124)
1990 (July)	O157:H7	North Dakota, US	Community	Roast beef	65	(125)
1991 (November)	O157:H7	Massachusetts, US	Community	Apple cider	23	(85)
1992 (Unknown)	O119:?	France	Community	Goat cheese	>4	(126)
1993 (January)	O157:H7	California, Idaho, Nevada, and Washington, US	Restaurant	Ground beef	732 (4)	(19,127, 128)
1993 (July)	O157:H7	Washington, US	Church picnic	Pea salad	16	(19)
1993 (August)	O157:H7	Oregon, US	Restaurant	Cantaloupe	27	(19)
1994 (February)	O104:H21	Montana, US	Community	Milk	18	(129)
1994 (November)	O157:H7	Washington and California, US	Home	Salami	19	(130)
1995 (February)	O111:NM	Adelaide, Australia	Community	Semidry sausage	>200	(131)
1995 (October)	O157:H7	Kansas, US	Wedding	Fruit salad/ punch	21	(19)
1995 (November)	O157:H7	Oregon, US	Home	Venison jerky	11	(132)
1995 (July)	O157:H7	Montana, US	Community	Lettuce	74	(133)
1995 (September)	O157:H7	Maine, US	Camp	Lettuce	37	(19)
1995 (December)– 1996 (March)	O157:H–	Bavaria, Germany	Throughout Bavaria[a]	Commercial mortadella and teewurst?	28 (3)	(134)

(*Continued*)

Table 2 (*Continued*)

Year (Month)	Serotype	Location	Setting	Vehicle	No. of cases (Deaths)	Ref.
1996	O118:H2	Komatsu, Japan	School	Luncheon (salad?)	126	(83, 135)
1996 (July)	O157:H7	Osaka, Japan	Community	White radish sprouts	7966 (3)	(83)
1996 (October)	O157:H7	California, Washington, and Colorado, US, and British Columbia, Canada	Community	Apple juice	71 (1)	(136)
1996 (November)	O157:H7	Central Scotland	Community	Cooked meat	<501 (21)	(137)
1997 (May)	O157:H–	Scotland	Hospital	Cream cakes	12	(138)
1997 (July)	O157:H7	Michigan, US	Community	Alfalfa sprouts	60	(19)
1997 (July/August)	O26:H11	Southeastern Japan	Daycare center	Prepared foods, vegetables?	32	(139)
1997 (November)	O157:H7	Wisconsin, US	Church banquet	Meatballs, coleslaw	13	(19)
1998 (June)	O157:H7	Wisconsin, US	Community	Cheese curds	63	(140)
1998 (June)	O157:H7	Wyoming, US	Community	Water	114	(141)
1998 (July)	O157:H7	North Carolina, US	Restaurant	Cole slaw	142	(19)
1998 (July)	O157:H7	California, US	Prison	Milk	28	(19)
1999 (March)	O157 PT[b] 21/28	North Cumbria, England	Community	Milk	114	(142)
1999 (May, June, July)	O157:?	Applecross, Scotland	Campsite	Untreated drinking water	6	(143)
1999 (June)	O157 PT[b]2	North Wales	Farm festival	Ice cream, cotton candy	24	(144)
1999 (July)	O111:H8	Texas, US	Cheerleading camp	Lunch, ice, corn, dinner roll	55	(145)
1999 (August)	O157:H7	New York, US	Fair	Well water	900 (2)	(87)
1999 (September)	O157:?	Göteberg, Sweden	Hospital staff party	Lettuce	11	(146)
1999 (November)	O157:H7	California, Nevada, and Arizona, US	Restaurant	Beef tacos	13	(13)
2000 (March and April)	O26:H11	Mecklenburg-West Pomerania, Lower Saxony, Hesse, Germany	Daycare center	Beef ("Seeme-rolle")?	11	(147)
2000 (May)	O157:H7[c]	Walkerton, Ontario	Community	Municipal water supply	2300[c] (7)	(148–150)

(Continued)

Table 2 (*Continued*)

Year (Month)	Serotype	Location	Setting	Vehicle	No. of cases (Deaths)	Ref.
2000 (July)	O157:H7	Wisconsin, US	Restaurant	Watermelon	736 (?)	*(151)*
2001 (July)	O157:H7	Illinois, US	Private home	Ground beef	19 (?)	*(152)*
2001 (November/ December)	O157 PT[b] 21/28	Lancashire, England	Butcher's shop	Cooked meats	30	*(153)*
2002 (June/July)	O157:H7	Colorado, California, Iowa, Michigan, South Dakota, Washington, and Wyoming, US	Community	Ground beef	28	*(74)*
2001 (November)	O157:	Eastern Slovakia		Milk	9	*(154)*
2003–2004 (September/ March)	O157:H-	Copenhagen, Denmark	Community	Organic milk	25	*(159)*
2005 (September/ October)	O157:H7	The Netherlands (locations throughout)	Community	Steak tartare	21	*(157)*
2005 (October)	O157:H7	South Wales, United Kingdom	Schools	Cooked meats	157 (1)	*(158)*
2006 (May/ June)	O157:H7	Edmonton, Alberta	Restaurant	Beef donairs (a specialty food of Middle Eastern origin)	12	*(156)*
2006 (September)	O157:H7	United States (locations throughout)	Community	Spinach	183 (1)	Centers for Diease Control and Prevention[d] *(160)*

[a]Outbreak defined based on hospital admission record review for children admitted with the hemolytic uremic syndrome to the four pediatric hemodialysis centers in Bavaria from 1990 to March 1996.

[b]PT, phage type.

[c]In addition to *E. coli* O157:H7, *Campylobacter jejuni* and *C. coli* were implicated in the Walkerton waterborne outbreak. Although it is not possible to apportion the precise number of the ca. 2300 outbreak cases due to each of these pathogens, insight may be gleaned from the following observations: of the 675 cases for which a stool sample was obtained, 163 (24%) were positive for *E. coli* O157, 97 (14%) were positive for *C. jejuni*, and 7 (1%) were positive for *C. coli*; of the seven persons who died as a result of the outbreak, five developed the hemolytic uremic syndrome (HUS); and stool cultures from the five HUS cases showed that three were infected with *E. coli* O157:H7 and two were infected with *C. jejuni (149,155)*.

[d]Preliminary report.

Table 3
**Food Vehicles Implicated in Outbreaks of Infection With *Escherichia coli*
O157:H7 in United States, 1990–2004**[a]

Rank	Vehicle	No. of outbreaks (% known modes of transmission)
1	Ground beef[b]	63 (39.9)
2	Vegetables/salad bars[c]	33 (20.9)
4	Juice/punch	7 (4.4)
3	Roast beef/steak	9 (5.7)
5	Fruit	5 (3.2)
6	Milk	3 (1.9)
7	Water/Ice	2 (1.3)
7	Veal	2 (1.3)
	Others	34 (21.5)
	Unknown	85

[a]Based on data from the US Centers for Disease Control and Prevention, Foodborne and Diarrheal Diseases Branch, Foodborne Outbreak Response and Surveillance Unit, US Foodborne Disease Outbreaks Annual Summaries, 1990–2004. Available at: http://www.cdc.gov/foodborneoutbreaks/outbreak_data.htm.
[b]Reported as "suspected" in five outbreaks.
[c]Reported as "suspected" in two outbreaks.

reflective of a cause-and-effect relationship is not clear, but they suggest that efforts at reducing *E. coli* O157:H7 in ground beef have a direct impact on public health by curtailing the number of EHEC infections.

4.2. Outbreaks Associated With Produce

Various raw vegetables, including lettuce, alfalfa sprouts, and radish sprouts have been implicated in outbreaks of EHEC infection. Several recent outbreaks underscore the importance of fresh produce as the vehicles for EHEC. In July 1995, for instance, lettuce was implicated in an outbreak of *E. coli* O157:H7 in which 40 Montana residents were clinically confirmed of infection, 13 were hospitalized, and one developed HUS *(82)*. A massive outbreak of EHEC infection occurred in Japan when 7966 cases of illness and 3 deaths ensued from the consumption of white radish sprouts *(83)*. In the fall of 2006 an outbreak of infection with *E. coli* O157:H7 linked to fresh spinach sickened approx 200 people, resulting in approx 100 hospitalizations, 30 cases of HUS, and 1 death.

EHEC contamination of produce may occur through the application of manure-based fertilizers and from spraying and irrigating with contaminated water *(1,74)*. Furthermore, EHEC appear capable of entering lettuce internally via the root system and, as such, avoiding destruction by externally applied sanitizing agents *(84)*. Regardless of the site of contamination, however, an experimental evidence indicates that the practice of washing the produce with water should not be relied on to decrease EHEC contamination significantly *(6)*. Instead, attention should be given to prevent contamination and to control the temperature of cut lettuce and other vegetables.

4.3. Outbreaks Associated With Fresh-Pressed Juice

Concomitant with consumer preference for all-natural-type products, EHEC infection has been increasingly linked to the consumption of fresh-pressed, unpasteurized juices.

Perhaps the best-known examples of EHEC illness from juice have been caused by *E. coli* O157:H7 in a contaminated apple cider. Instances of the contaminated apple cider implicated in O157:H7 infection include a 1991 outbreak in Massachusetts that resulted in 23 illnesses *(85)* and a 1996 outbreak in the western United States that resulted in 66 illnesses and one death *(86)*. Although not a foolproof method, washing and brushing of apples, pasteurization of cider, and the use of preservatives such as sodium benzoate *(13)* can be used to increase the safety of apple cider related to EHEC.

4.4. Outbreaks Associated With Water

The first-documented case of waterborne EHEC infection occurred in 1989 when a community outbreak in Missouri caused 243 cases of illness *(18)*. The outbreak was linked to consumption of unchlorinated water and the number of cases declined shortly after the residents were instructed to boil water and following the chlorination of the water supply. Shortly thereafter, in 1990 at a nursery school in Saitama, Japan, an outbreak of *E. coli* O157:H7 infection related to ingestion of contaminated tap water occurred among 174 children, in which 14 of them developed HUS *(43)*. In 1999, an outbreak of waterborne O157:H7 infection occurred when 900 persons became ill and two died after an ingestion of contaminated well water at a fair in New York *(87)*. In 2000, *E. coli* O157:H7 and *Campylobacter jejuni* contaminated the drinking water supply in Walkerton, Ontario, resulting in 2000 cases of illness and seven deaths *(18)*. The above outbreaks notwithstanding, given the susceptibility of EHEC to conventional water-treatment processes *(17)*, EHEC infection from water consumption should remain minimal in developed countries.

5. CLINICAL CHARACTERISTICS OF EHEC INFECTION

5.1. Spectrum of Disease

Clinical outcomes of infection with EHEC may range from symptom-free carriage to death *(88)*. Infections typically appear as watery diarrhea accompanied by abdominal pain, usually within 3–4 d postingestion. Nausea and vomiting may be accompanying. Low-grade fever occurs in roughly 30% of cases *(88)*. Bloody diarrhea may develop in 24–48 h after appearance of watery diarrhea; and blood leukocytosis has been associated with increased detection of EHEC in stool and a more severe disease course *(74)*. Of those cases with bloody diarrhea, roughly 95% cases resolve within 5–7 d *(88)*.

5.2. Sequellae

The main sequellae from infection with EHEC are HC and HUS. HC manifests as diarrhea characterized by bloody, mucoid stool and is frequently accompanied by fever and severe abdominal pain *(3)*. Individual patient variability in expression of inflammatory mediators is thought to be a key element in determining the progression of severity of the disease in patients *(65)*. The majority of HC cases resolve spontaneously within 7 d *(88)*.

HUS typically occurs in children aged less than 5 yr, following HC *(67)*. Karmali and colleagues were among the first to make the link between Shiga toxin production and HUS, a postdiarrheal clinical condition characterized by acute renal injury, thrombocytopenia, and hemolytic anemia *(89)*. Approx 90% of HUS cases in industrialized countries are the result of EHEC infection *(90)*. Risk factors for progression of *E. coli* O157:H7 infection to HUS include bloody diarrhea, use of antimotility agents, fever, vomiting, elevated serum leukocytes, age of less than 5 yr, and the female sex *(88)*. In a study of German and Austrian pediatric patients, Gerber et al. found that patients with O157:H7 serotypes required dialysis for a longer time and had bloody diarrhea detected more frequently compared to patients with non-O157:H7 serotypes *(91)*. About 3–7% and 20% of sporadic and outbreak cases, respectively, progress to HUS *(77,88,92–94)*. Among the patients with HUS, about 60% of cases resolve, 30% lead to minor sequellae such as proteinuria, 5% lead to severe sequellae such as stroke and chronic renal failure, and 3–5% result in death *(88)*.

5.3. Treatment Options

Patient management for EHEC infections focsses on maintaining fluid and electrolyte balance, control of hypertension, provision of nutritional support, and treatment of severe anemia *(90)*. Diagnosis of HC relies on information concerning patient age, travel history, epidemiological associations, sexual practice, and medical history. Flexible colonoscopy and biopsy may be used to differentiate bacterial HC from other forms of colitis *(95)*. Progression to HUS necessitates prompt hospitalization *(88)*, whereupon further treatment options include dialysis, hemofiltration, packed erythrocyte transfusion, platelet infusions, and, in the cases of utmost severity, renal transplantation *(31)*. Work such as that recently done to construct recombinant bacteria expressing a modified lipopolysaccharide mimicking the Shiga toxin receptor which binds Shiga toxin holds promise related to the future use of probiotics in treating EHEC infection *(1)*.

5.4. Role of Antimicrobials

The usefulness of antimicrobials in preventing the progression of EHEC infection to HC and HUS is not fully resolved. Because antimicrobials may lyse bacterial cell walls, which in turn facilitate the release of Shiga toxins *(96)* and/or increase the expression of Shiga toxins in vivo *(63)*, their use may facilitate the progression of EHEC infections *(97)*. In a retrospective study following an outbreak of EHEC infection in a nursing home, Carter et al. found that the antibiotic therapy during exposure was associated with acquiring a secondary infection, and that the antibiotic therapy after symptom onset was associated with a higher case-fatality rate in the more severe cases *(94)*. Pavia and colleagues, during the investigation of an EHEC outbreak in a home for mentally retarded persons, found that of the eight residents in whom the HUS developed, five had received trimethoprim-sulfamethoxazole, compared with none of the seven residents for whom there were no subsequent complications *(76)*. For further information on the subject of antimicrobial use in EHEC infections, the reader is referred to reviews by Guerrant et al. *(29)*, Paton and Paton *(97)*, and Mead and Griffin *(88)*. Lastly, the use of antimotility agents is contraindicated for EHEC infections *(74)*.

6. PATHOGENESIS OF EHEC

6.1. Susceptible Populations

Those at the extremes of age and those with compromised immune systems are most susceptible to EHEC infection. The highest incidence of *E. coli* O157:H7 occurs in

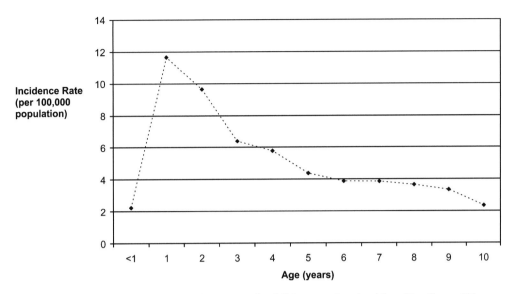

Fig. 1. *Escherichia coli* O157 incidence rates in children aged under 10 yr, Foodborne Diseases Active Surveillance Network (FoodNet), 2002 *(98)*.

children. For example, based on data ascertained by the US Centers for Disease Control and Prevention via the Foodborne Diseases Active Surveillance Network (FoodNet) for the year 1998, the incidence of reported O157:H7 cases was 2.01 in children younger than 1-yr-old, 4.57 in 1–4-yr-olds, and 1.83 in 5–14-yr-olds per 100,000 population. The lowest incidence occurred in persons aged 15 yr or above, ranging from 1.15 to 0.61 reported cases per 100,000 population *(74)*. In 1999, approx 35% of reported cases occurred in 1–10-yr-olds, 17% in 10–20-yr-olds, and 14.1% in persons older than 60 yr *(72)*. Regarding the age group most-at-risk, data for the year 2002 show that the highest incidence rate of *E. coli* O157 occurred in children aged 1 yr, and steadily tapered off thereafter (Fig. 1) *(98)*.

6.2. Infectious Dose

Because of the severity of disease, ethical considerations preclude prospective dose–response studies for EHEC in humans. Nevertheless, based on data gathered during the investigations of outbreaks, retrospective analyses suggest that the infectious dose for STEC may be fewer than 100 cells. In an examination of outbreak data from 1996 among schoolchildren in Japan, Teunis et al. estimated the ingested dose of *E. coli* O157:H7 at 31 CFU among pupils (attack rate of 0.25) and 35 CFU among teachers (attack rate of 0.16) *(99)*.

7. EHEC VIRULENCE FACTORS

7.1. Shiga Toxins

Shiga toxins are named because of their prototype originated from *Shigella dysentariea* type 1 (recall the aforementioned similarity between *E. coli* and *Shigella*). Shiga toxins produced by *E. coli* are characterized into two main groups: Shiga toxin 1 (Stx1), which differs from the Shiga toxin produced by *S. dysentariea* by one amino acid;

and Shiga toxin 2 (Stx2), which shares approx 50% homology at the amino-acid level with Stx1. Together, the toxins in these two groups comprise the Shiga toxin family.

At the molecular level, Shiga toxins consist of A and B subunits, of which the former is an enzymatically active glycohydrolase and the latter mediates substrate-specific binding to intestinal cell wall receptors. The A subunit specifically cleaves a single adenine residue from the eukaryotic 28S rRNA thereby resulting in an inhibition of protein synthesis *(100)*. Shiga toxins also induce macrophage production of tumor necrosis factor α, interleukin-1β, and interleukin-6. Various cells, including tubular epithelial cells, may be the targets for Shiga toxin-mediated apoptosis, which in turn is considered to contribute to the pathogenesis of HUS caused by STEC. Current evidence suggests Stx2 to be more important than Stx1 in the development of HUS *(100)*.

7.2. Adherence Factors

Establishment of infection for many Gram-negative pathogens is dependent on adhesion to host cells. EHEC possess adherence factors which allow them to colonize host sites, such as the small intestine, that are not usually inhabited by *E. coli*. These adhesins typically form fimbriae (pili) or fibrillae. It has recently been demonstrated that expression of thin aggregative fimbriae, termed curli, on the surface of *E. coli* O157:H7 cells results in massive aggregative adherence (compared to localized adherence) in cultured Hep-2 cells *(74)*. Because adhesins are the main determinant in facilitating the bacterial adhesion to host cell tissues, these molecules are potential vaccine candidates for preventing EHEC infection *(101)*.

7.2.1. Locus of Enterocyte Effacement

Bacterial virulence factors such as toxins and adhesins are often encoded by genes located on specific regions of the bacterial genome, termed pathogenicity islands. The locus of enterocyte effacement (LEE) is one such pathogenicity island and is found on the chromosomes of EHEC and EPEC. The LEE encodes a type III secretion system, an adhesin (intimin), an intimin receptor (Tir), a regulator (Ler), and various secreted proteins and chaperones. Proteins encoded by LEE are involved in the adherence of pathogens to intestinal cells, initiation of host signal-transduction pathways, and formation of A/E lesions (*see* Section 2.2.2). The observation that the G+C content of the LEE is ca. 40%, whereas the G+C content of the *E. coli* genome proper is ca. 51%, suggests the EHEC have acquired the LEE by horizontal genetic transfer.

The 43,359-bp sequence of the LEE from an *E. coli* O157:H7 serovar isolated from contaminated hamburger implicated in an outbreak of EHEC infection has been determined *(102)*. This LEE was found to contain 54 open reading frames (ORFs), with genes putatively encoding intimin (*eae*), chaperones (*cesD*), and type III secretion apparatus (multiple different genes). Subsequently, based on systematic mutagenesis of each of its constituent genes and functional characterization of individual deletion mutants, the LEE has been further characterized into astonishing details *(103)*. This same group of researchers have recently reported the evidence indicating multiple *trans*-acting factors interact with LEE *(103,104)*. Further, the characterization of structure and function of LEE and associated molecules promises additional insight into EHEC pathogenesis.

7.2.2. Intimin

Intimin is an outer membrane protein encoded by the *eae* (*E. coli* attaching and effacing) gene. *eae* is located within the LEE pathogenicity island (*see* Section 7.2.1) and encodes a ca. 94 to 97-kDa product. The Eae protein is produced by enteric pathogens that cause A/E lesions (i.e., EHEC, EPEC, *Hafnia alvei*, and *Citrobacter rodentium*). Sequence homology between EPEC and EHEC intimin has been reported to be 86 and 83% at the nucleotide and amino-acid levels, respectively *(76)*. Although EPEC strains produce A/E lesions in both the small and large intestines, EHEC strains produce only in the large intestine. Until recently, intimin was the only factor which had been demonstrated to play a role in intestinal colonization by EHEC; however, recent evidence has begun to shed light on other elements, including OmpA, which are important in EHEC adhesion to host cells *(105)*.

7.2.3. Tir Protein

The translocated intimin receptor (Tir) protein is mainly responsible for the A/E virulence trait associated with EHEC pathogenesis. Similar to *eae*, the gene encoding the Tir protein, *tir*, is located within the LEE and is part of a large quorum-sensing-regulated operon *(106)*, which in addition to Tir encodes its chaperone protein (CesT) and intimin *(107)*. Structurally Tir consists of an extracellular domain containing intimin-specific receptors, a transmembrane domain, and a cytoplasmic domain. Tir is initially transcribed in the bacterial cell as a 78-kDa protein after which it is transported to the host cell and phosphorylated on a tyrosine residue (Tyr 474), thereby increasing to 90-kDa in mass. (It should be noted that the process of tyrosine phosphorylation of Tir has been shown to occur in EHEC serotypes such as O26:NM but not in O157:H7 *[108]*.) At this point Tir interacts with intimin, which in turn facilitates nucleation of actin and other cytoskeletal proteins beneath the adherent bacteria *(109)*, the result of which is the formation A/E lesions on the host cell.

7.3. 60-MDa Plasmid

EHEC strains of serotype O157:H7 typically contain a 60-MDa plasmid, pO157, which, although its precise role in virulence remains clouded, has nevertheless been strongly implicated in the pathogenesis of EHEC infections. pO157 contains 100 ORFs interspersed with seven insertion sequence elements located primarily at the boundaries of putative virulence gene segments *(100)*. Expression of the *toxB* gene on pO157 appears to be required for full epithelial cell adherence of *E. coli* O157:H7 cells *(110)*. Kim et al. *(111)* have recently demonstrated that a homologue of the *msbB* gene located on pO157 encodes an acyltransferase involved in lipid-A biosynthesis. Of note, Brunder and colleagues have reported a novel type of fimbriae in sorbitol-fermenting *E. coli* O157:H(–) strains (important cause of diarrheal disease and HUS in Germany), termed Sfp fimbriae *(112)*. The genes encoding Sfp fimbriae are located on the plasmid pSFO157 found in O157:H(–) strains, which differs markedly from the plasmid pO157, and are not found in classical EHEC.

7.4. Antimicrobial Resistance

The antimicrobial-resistant phenotypes observed most often among *E. coli* have been those for ampicillin, streptomycin, sulfa, and tetracycline. A survey of data from 11 studies published during the past 25 yr showed that among *E. coli* recovered from retail meats, 0–50% were resistant to ampicillin, 0–74% to sulfamethoxazole, and 7–83% to tetracycline *(113)*. Recent years have seen an emergence of *E. coli* strains exhibiting resistance or decreased susceptibility to clinical frontline antimicrobials, including fluoroquinolones and third-generation cephalosporins. Similarly, this has been the case for isolates recovered from food animals *(60,61,114–116)*, retail food *(60,61,113)*, and humans *(76)*. Nevertheless, the extent to which antimicrobial resistance in EHEC portends adverse clinical consequences is not clear (*see* Section 5.4).

7.5. Acid Tolerance

E. coli O157:H7 is, in comparison to most other foodborne pathogens, acid-tolerant. Acid tolerance in *E. coli* is brought about by any one of the three inducible systems, the so-called acid resistant (AR) systems 1, 2, and 3. Its tolerance to acid plays a key role in protecting *E. coli* cells not only from the acidic environments encountered in both the reservoir and the host, but also against the acidic stress associated with various stages of food processing *(117)*. For instance, Lee and Chen *(118)* have demonstrated the protective role of colanic acid in enabling cells of *E. coli* O157:H7 to survive acidic conditions found in foods such as yogurt. Recent data suggest EHEC strains vary in their tolerance to acid, and the strains of serotype O91:H21 (which were shown to survive at pH 2.5 for more than 24 h) may have greater acid-tolerance than that of serotype O157:H7 *(119)*. From a virulence factor point of view, acid tolerance likely facilitates the low infectious dose required for *E. coli* infection *(117)*.

8. SUMMARY AND CONCLUSIONS

Since its advent some 20 yr ago *E. coli* O157:H7 remains an important foodborne pathogen. Moreover, other EHEC, including O26, O103, O111, and O145, have emerged to cause foodborne illness. Although EHEC have been found in a wide range of animals, including deer, rabbits, poultry, swine, and sheep, their animal reservoir must still be considered as cattle. As such, food products derived from cattle, in particular ground beef, remain the important vehicles of EHEC infection; as do fresh produce, water, and other foodstuffs that may be contaminated by cattle manure. In this regard, recent data offer encouragement: from 2000–2003 in the United States, declines have been observed for the number of O157:H7 clinical infections and the percent ground beef product samples positive for *E. coli* O157:H7. Furthermore, the progress aimed at gaining insight into the molecular mechanisms of disease caused by EHEC has been remarkable. Despite these advances, treatment for individuals infected with EHEC remains for the most part supportive. Prevention remains paramount in reducing the incidence of adverse health outcomes associated with EHEC. Continued work is required to further reduce the public health impact of EHEC foodborne infection.

ACKNOWLEDGMENTS

We thank Karl Bettelheim, Peter Feng, Steve Hrudey, Alecia Naugle, Jonathan Rose and M. Jude Smedra for thoughtful review of the manuscript.

REFERENCES

1. Escherich, T. (1885) Die darmbacterien des neugeborenen und saunglings. *Fortschr. Med.* **3**, 515–522; 547–554.
2. Luria, S. E. (1947) Recent advances in bacterial genetics. *Bacteriol. Rev.* **11**, 1–40.
3. Lederberg, J. and Tatum, E. L. (1946) Gene recombination in *Escherichia coli. Nature* **158**, 558.
4. Crick, F. H. and Brenner, S. (1967) The absolute sign of certain phase-shift mutants in bacteriophage T4. *J. Mol. Biol.* **26**, 361–363.
5. Morrow, J. F., Cohen, S. N., Chang, A. C., Boyer, H. W., Goodman, H. M., and Helling, R. B. (1974) Replication and transcription of eukaryotic DNA in *Escherichia coli. Proc. Natl. Acad. Sci. USA* **71**, 1743–1747.
6. Cohen, S. N., Chang, A. C., Boyer, H. W., and Helling, R. B. (1973) Construction of biologically functional bacterial plasmids in vitro. *Proc. Natl Acad. Sci. USA* **70**, 3240–3244.
7. Riley, L. W., Remis, R. S., Helgerson, S. D., et al. (1983) Hemorrhagic colitis associated with a rare *Escherichia coli* serotype. *N. Engl. J. Med.* **308**, 681–685.
8. Orskov, S. (1974) Genus I. *Escherichia.* In: *Bergey's Manual of Determinative Bacteriology* (Buchanan, R. E., Gibbons, N. E., Cowan, S. T., et al., eds.), 8th edn, Williams & Wilkins, Baltimore, MD, pp. 293–296.
9. Martinez-Salas, E., Martin, J. A., and Vicente, M. (1981) Relationship of *Escherichia coli* density to growth rate and cell age. *J. Bacteriol.* **147**, 97–100.
10. Thielman, N. M. and Guerrant, R. L. (1999) *Escherichia coli.* In: *Antimicrobial Therapy and Vaccines* (Yu, V. L., Merigan, T. C. Jr., and Barriere, S. L., eds.), Williams & Wilkins, Baltimore, MD, pp. 188–200.
11. Bettelheim, K. A. (1991) The genus *Escherichia.* In: *The Prokaryotes* (Balows, A., Trüper, H. G., Dworkin, M., Harder, W., and Schleifer, K.-H., eds.), 2nd edn, Springer Verlag, New York, NY, pp. 2697–2735.
12. Lior, H. (1994) Classification of *Escherichia coli.* In Escherichia coli *in Domestic Animals and Humans* (Gyles, C., ed.), CAB International, Wallingford, UK.
13. Conway, P. (1995) Microbial ecology of the human large intestine. In: *Human Colonic Bacteria: Role in Nutrition, Physiology, and Pathology* (Macfarlane, G. T. and Gibson, G. R., eds.), CRC, London, pp. 1–24.
14. Brock, T. D. (1971) Microbial growth rates in nature. *Bacteriol. Rev.* **35**, 39–58.
15. Gibson, G. R. and Roberfroid, M. B. (1995) Dietary modulation of the human colonic microbiota: introducing the concept of prebiotics. *J. Nutr.* **125**, 1401–1412.
16. Donnenberg, M. S. and Whittam, T. S. (2001) Pathogenesis and evolution of virulence in enteropathogenic and enterohemorrhagic *Escherichia coli. J. Clin. Invest.* **107**, 539–548.
17. Scaletsky, I. C., Silva, M. L., and Trabulsi, L. R. (1984) Distinctive patterns of adherence of enteropathogenic *Escherichia coli* to HeLa cells. *Infect. Immun.* **45**, 534–536.
18. Nataro, J. P., Kaper, J. B., Robins-Browne, R., Prado, V., Vial, P., and Levine, M. M. (1987) Patterns of adherence of diarrheagenic *Escherichia coli* to HEp-2 cells. *Pediatr. Infect. Dis. J.* **6**, 829–831.
19. Meng, J., Doyle, M. P., Zhao, T., and Zhao, S. (2001) Enterohemorrhagic *Escherichia coli.* In: *Food Microbiology: Fundamentals and Frontiers* (Doyle, M. P., Beuchat, L. R., and Montville, T. J., eds.), ASM, Washington, DC, pp. 193–213.
20. Scaletsky, I. C., Fabbricotti, S. H., Carvalho, R. L., et al. (2002) Diffusely adherent *Escherichia coli* as a cause of acute diarrhea in young children in Northeast Brazil: a case–control study. *J. Clin. Microbiol.* **40**, 645–648.
21. Poitrineau, P., Forestier, C., Meyer, M., et al. (1995) Retrospective case–control study of diffusely adhering *Escherichia coli* and clinical features in children with diarrhea. *J. Clin. Microbiol.* **33**, 1961–1962.

22. Jallat, C., Livrelli, V., Darfeuille-Michaud, A., Rich, C., and Joly, B. (1993) *Escherichia coli* strains involved in diarrhea in France: high prevalence and heterogeneity of diffusely adhering strains . *J. Clin. Microbiol.* **31,** 2031–2037.

23. Levine, M. M., Ferreccio, C., Prado, V., et al. (1993) Epidemiologic studies of *Escherichia coli* diarrheal infections in a low socioeconomic level peri-urban community in Santiago, Chile. *Am. J. Epidemiol.* **138,** 849–869.

24. Giron, J. A., Jones, T., Millan-Velasco, F., et al. (1991) Diffuse-adhering *Escherichia coli* (DAEC) as a putative cause of diarrhea in Mayan children in Mexico. *J. Infect. Dis.* **163,** 507–513.

25. Peiffer, I., Blanc-Potard, A. B., Bernet-Camard, M. F., Guignot, J., Barbat, A., and Servin, A. L. (2000) Afa/Dr diffusely adhering *Escherichia coli* C1845 infection promotes selective injuries in the junctional domain of polarized human intestinal Caco-2/TC7 cells. *Infect. Immun.* **68,** 3431–3442.

26. Peiffer, I., Bernet-Camard, M. F., Rousset, M., and Servin, A. L. (2001) Impairments in enzyme activity and biosynthesis of brush border-associated hydrolases in human intestinal Caco-2/TC7 cells infected by members of the Afa/Dr family of diffusely adhering *Escherichia coli*. *Cell. Microbiol.* **3,** 341–357.

27. Kyaw, C. M., De Araujo, C. R., Lima, M. R., Gondim, E. G., Brigido, M. M., and Giugliano, L. G. (2003) Evidence for the presence of a type III secretion system in diffusely adhering *Escherichia coli* (DAEC). *Infect. Genet. Evol.* **3,** 111–117.

28. Taddei, C. R., Moreno, A. C., Fernandes Filho, A., Montemor, L. P., and Martinez, M. B. (2003) Prevalence of secreted autotransporter toxin gene among diffusely adhering *Escherichia coli* isolated from stools of children. *FEMS Microbiol. Lett.* **227,** 249–253.

29. Quiroga, M., Oviedo, P., Chinen, I., et al. (2000) Asymptomatic infections by diarrhea-genic *Escherichia coli* in children from Misiones, Argentina, during the first twenty months of their lives. *Rev. Inst. Med. Trop. Sao Paulo* **42,** 9–15.

30. Rosa, A. C., Mariano, A. T., Pereira, A. M., Tibana, A., Gomes, T. A., and Andrade, J. R. (1998) Enteropathogenicity markers in *Escherichia coli* isolated from infants with acute diarrhoea and healthy controls in Rio de Janeiro, Brazil. *J. Med. Microbiol.* **47,** 781–790.

31. Nataro, J. P. and Kaper, J. B. (1998) Diarrheagenic *Escherichia coli*. *Clin. Microbiol. Rev.* **11,** 142–201.

32. Donnenberg, M. S. and Kaper, J. B. (1992) Enteropathogenic *Escherichia coli*. *Infect. Immun.* **60,** 3953–3961.

33. Makino, S., Asakura, H., Shirahata, T., et al. (1999) Molecular epidemiological study of a mass outbreak caused by enteropathogenic *Escherichia coli* O157:H45. *Microbiol. Immunol.* **43,** 381–384.

34. Barlow, R. S., Hirst, R. G., Norton, R. E., Ashhurst-Smith, C., and Bettelheim, K. A. (1999) A novel serotype of enteropathogenic *Escherichia coli* (EPEC) as a major pathogen in an outbreak of infantile diarrhoea. *J. Med. Microbiol.* **48,** 1123–1125.

35. Levine, M. M. (1981) Adhesion of enterotoxigenic *Escherichia coli* in humans and animals. *Ciba. Found. Symp.* **80,** 142–160.

36. Levine, M. M. (1983) Travellers' diarrhoea: prospects for successful immunoprophylaxis. *Scand. J. Gastroenterol. Suppl.* **84,** 121–134.

37. Hyams, K. C., Bourgeois, A. L., Merrell, B. R., et al. (1991) Diarrheal disease during Operation Desert Shield. *N. Engl. J. Med.* **325,** 1423–1428.

38. Levine, M. M., Rennels, M. B., Cisneros, L., Hughes, T. P., Nalin, D. R., and Young, C. R. (1980) Lack of person-to-person transmission of enterotoxigenic *Escherichia coli* despite close contact. *Am. J. Epidemiol.* **111,** 347–355.

39. Viljanen, M. K., Peltola, T., Junnila, S. Y., et al. (1990) Outbreak of diarrhoea due to *Escherichia coli* O111:B4 in schoolchildren and adults: association of Vi antigen-like reactivity. *Lancet* **336,** 831–834.

40. Mitsuda, T., Muto, T., Yamada, M., et al. (1998) Epidemiological study of a food-borne outbreak of enterotoxigenic *Escherichia coli* O25:NM by pulsed-field gel electrophoresis and randomly amplified polymorphic DNA analysis. *J. Clin. Microbiol.* **36,** 652–656.

41. Beatty, M. E., Bopp, C. A., Wells, J. G., Greene, K. D., Puhr, N. D., and Mintz, E. D. (2004) Enterotoxin-producing *Escherichia coli* O169:H41, United States. *Emerg. Infect. Dis.* **10,** 518–521.

42. Hedberg, C. W., Savarino, S. J., Besser, J. M., et al. (1997) An outbreak of foodborne illness caused by *Escherichia coli* O39:NM, an agent not fitting into the existing scheme for classifying diarrheogenic *E. coli. J. Infect. Dis.* **176,** 1625–1628.

43. Day, W. A. and Maurelli, A. T. (2002) *Shigella* and enteroinvasive *Escherichia coli*: paradigms for pathogen evolution and host-parasite interactions. In: Escherichia coli*: Virulence Mechanisms of a Versatile Pathogen* (Donnenberg, M. S., ed.), Academic, San Diego, CA, pp. 209–237.

44. Harris, J. R., Wachsmuth, I. K., Davis, B. R., and Cohen, M. L. (1982) High-molecular-weight plasmid correlates with *Escherichia coli* enteroinvasiveness. *Infect. Immun.* **37,** 1295–1298.

45. Zhou, Z., Ogasawara, J., Nishikawa, Y., et al. (2002) An outbreak of gastroenteritis in Osaka, Japan due to *Escherichia coli* serogroup O166:H15 that had a coding gene for enteroaggregative *E. coli* heat-stable enterotoxin 1 (EAST1). *Epidemiol. Infect.* **128,** 363–371.

46. Vial, P. A., Robins-Browne, R., Lior, H., et al. (1988) Characterization of enteroadherent-aggregative *Escherichia coli*, a putative agent of diarrheal disease. *J. Infect. Dis.* **158,** 70–79.

47. Tzipori, S., Montanaro, J., Robins-Browne, R. M., Vial, P., Gibson, R., and Levine, M. M. (1992) Studies with enteroaggregative *Escherichia coli* in the gnotobiotic piglet gastro-enteritis model. *Infect. Immun.* **60,** 5302–5306.

48. Hicks, S., Candy, D. C., and Phillips, A. D. (1996) Adhesion of enteroaggregative *Escherichia coli* to formalin-fixed intestinal and ureteric epithelia from children. *J. Med. Microbiol.* **44,** 362–371.

49. Nataro, J. P., Deng, Y., Maneval, D. R., German, A. L., Martin, W. C., and Levine, M. M. (1992) Aggregative adherence fimbriae I of enteroaggregative *Escherichia coli* mediate adherence to HEp-2 cells and hemagglutination of human erythrocytes. *Infect. Immun.* **60,** 2297–2304.

50. Menard, L. P. and Dubreuil, J. D. (2002) Enteroaggregative *Escherichia coli* heat-stable enterotoxin 1 (EAST1): a new toxin with an old twist. *Crit. Rev. Microbiol.* **28,** 43–60.

51. Itoh, Y., Nagano, I., Kunishima, M., and Ezaki, T. (1997) Laboratory investigation of enteroaggregative *Escherichia coli* O untypeable:H10 associated with a massive outbreak of gastrointestinal illness. *J. Clin. Microbiol.* **35,** 2546–2550.

52. Wells, J. G., Davis, B. R., Wachsmuth, I. K., et al. (1983) Laboratory investigation of hemorrhagic colitis outbreaks associated with a rare *Escherichia coli* serotype. *J. Clin. Microbiol.* **18,** 512–520.

53. Pruimboom-Brees, I. M., Morgan, T. W., Ackermann, M. R., et al. (2000) Cattle lack vascular receptors for *Escherichia coli* O157:H7 Shiga toxins. *Proc. Natl Acad. Sci. USA* **97,** 10,325–10,329.

54. Zhao, T., Doyle, M. P., Shere, J., and Garber, L. (1995) Prevalence of enterohemorrhagic *Escherichia coli* O157:H7 in a survey of dairy herds. *Appl. Environ. Microbiol.* **61,** 1290–1293.

55. Galland, J. C., Hyatt, D. R., Crupper, S. S., and Acheson, D. W. (2001) Prevalence, antibiotic susceptibility, and diversity of *Escherichia coli* O157:H7 isolates from a longitudinal study of beef cattle feedlots. *Appl. Environ. Microbiol.* **67,** 1619–1627.

56. Hancock, D. D., Besser, T. E., Rice, D. H., Herriott, D. E., and Tarr, P. I. (1997) A longitudinal study of *Escherichia coli* O157 in fourteen cattle herds. *Epidemiol. Infect.* **118,** 193–195.

57. Riley, D. G., Gray, J. T., Loneragan, G. H., Barling, K. S., and Chase, C. C. Jr. (2003) *Escherichia coli* O157:H7 prevalence in fecal samples of cattle from a southeastern beef cow-calf herd. *J. Food Prot.* **66,** 1778–1782.

58. LeJeune, J. T., Besser, T. E., Rice, D. H., Berg, J. L., Stilborn, R. P., and Hancock, D. D. (2004) Longitudinal study of fecal shedding of *Escherichia coli* O157:H7 in feedlot cattle: predominance and persistence of specific clonal types despite massive cattle population turnover. *Appl. Environ. Microbiol.* **70,** 377–384.

59. Stephan, R. and Schumacher, S. (2001) Resistance patterns of non-O157 Shiga toxin-producing *Escherichia coli* (STEC) strains isolated from animals, food and asymptomatic human carriers in Switzerland. *Lett. Appl. Microbiol.* **32,** 114–117.

60. Zhao, S., White, D. G., McDermott, P. F., et al. (2001) Identification and expression of cephamycinase bla(CMY) genes in *Escherichia coli* and *Salmonella* isolates from food animals and ground meat. *Antimicrob. Agents Chemother.* **45,** 3647–3650.

61. Schroeder, C. M., Zhao, C., DebRoy, C., et al. (2002) Antimicrobial resistance of *Escherichia coli* O157 isolated from humans, cattle, swine, and food. *Appl. Environ. Microbiol.* **68,** 576–581.

62. Gallien, P., Klie, H., Lehmann, S., et al. (1994) Detection of verotoxin-producing *E. coli* in field isolates from domestic and agricultural animals in Sachsen-Anhalt. *Berl. Munch. Tierarztl. Wochenschr.* **107,** 331–334.

63. Zhang, X., McDaniel, A. D., Wolf, L. E., Keusch, G. T., Waldor, M. K., and Acheson, D. W. (2000) Quinolone antibiotics induce Shiga toxin-encoding bacteriophages, toxin production, and death in mice. *J. Infect. Dis.* **181,** 664–670.

64. Mizan, S., Lee, M. D., Harmon, B. G., Tkalcic, S., and Maurer, J. J. (2002) Acquisition of antibiotic resistance plasmids by enterohemorrhagic *Escherichia coli* O157:H7 within rumen fluid. *J. Food Prot.* **65,** 1038–1040.

65. Brenner D. J. (1984) Family I. Enterobacteriaceae. In: *Bergey's Manual of Systematic Bacteriology* (Krieg, N. R. and Holt, J. G., eds.), vol. 1, Williams and Wilkins, Baltimore, MD, pp. 408–420.

66. Blanc-Potard, A. B., Tinsley, C., Scaletsky, I., et al. (2002) Representational difference analysis between Afa/Dr diffusely adhering *Escherichia coli* and nonpathogenic *E. coli* K-12. *Infect. Immun.* **70,** 5503–5511.

67. Proulx, F., Seidman, E. G., and Karpman, D. (2001) Pathogenesis of Shiga toxin-associated hemolytic uremic syndrome. *Pediatr. Res.* **50,** 163–171.

68. Grauke, L. J., Kudva, I. T., Yoon, J. W., Hunt, C. W., Williams, C. J., and Hovde, C. J. (2002) Gastrointestinal tract location of *Escherichia coli* O157:H7 in ruminants. *Appl. Environ. Microbiol.* **68,** 2269–2277.

69. Renter, D. G., Sargeant, J. M., Hygnstorm, S. E., Hoffman, J. D., and Gillespie, J. R. (2001) *Escherichia coli* O157:H7 in free-ranging deer in Nebraska. *J. Wildl. Dis.* **37,** 755–760.

70. Fischer, J. R., Zhao, T., Doyle, M. P., et al. (2001) Experimental and field studies of *Escherichia coli* O157:H7 in white-tailed deer. *Appl. Environ. Microbiol.* **67,** 1218–1224.

71. Sargeant, J. M., Hafer, D. J., Gillespie, J. R., Oberst, R. D., and Flood, S. J. (1999) Prevalence of *Escherichia coli* O157:H7 in white-tailed deer sharing rangeland with cattle. *J. Am. Vet. Med. Assoc.* **215,** 792–794.

72. Blanco, J., Blanco, M., Garabal, J. I., and Gonzalez, E. A. (1991) Enterotoxins, colonization factors and serotypes of enterotoxigenic *Escherichia coli* from humans and animals. *Microbiologia* **7,** 57–73.

73. Wahlstrom, H., Tysen, E., Olsson Engvall, E., et al. (2003) Survey of *Campylobacter* species, VTEC O157 and *Salmonella* species in Swedish wildlife. *Vet. Rec.* **153,** 74–80.

74. Barrett, T. J., Kaper, J. B., Jerse, A. E., and Wachsmuth, I. K. (1992) Virulence factors in Shiga-like toxin-producing *Escherichia coli* isolated from humans and cattle. *J. Infect. Dis.* **165,** 979–980.

75. Beutin, L., Geier, D., Zimmermann, S., and Karch, H. (1995) Virulence markers of Shiga-like toxin-producing *Escherichia coli* strains originating from healthy domestic animals of different species. *J. Clin. Microbiol.* **33,** 631–635.

76. Bettelheim, K. A., Goldwater, P. N., Evangelidis, H., Pearce, J. L., and Smith, D. L. (1992) Distribution of toxigenic *Escherichia coli* serotypes in the intestines of infants. *Comp. Immunol. Microbiol. Infect. Dis.* **15,** 65–70.

77. Le Saux, N., Spika, J. S., Friesen, B., et al. (1993) Ground beef consumption in non-commercial settings is a risk factor for sporadic *Escherichia coli* O157:H7 infection in Canada. *J. Infect. Dis.* **167,** 500–502.

78. Anonymous. (2004) Preliminary FoodNet data on the incidence of infection with pathogens commonly transmitted through food—selected sites, United States, 2003. *MMWR Morb. Mortal. Wkly Rep.* **53,** 338–343.

79. Mead, P. S., Slutsker, L., Dietz, V., et al. (1999) Food-related illness and death in the United States. *Emerg. Infect. Dis.* **5,** 607–625.

80. US Centers for Disease Control and Prevention. (2000) US Foodborne Disease Outbreaks. Annual Summaries, 1990–1999. Foodborne and Diarrheal Diseases Branch, Atlanta, GA. Available at http://www.cdc.gov/foodborneoutbreaks/outbreak_data.htm (accessed on 06 February 2007).

81. Holt, K. G., Levine, P., Naugle, A. L., and Eckel, R. (2004) Food Safety and Inspection Service microbiological regulatory testing program for *Escherichia coli* O157:H7 in raw ground beef products, United States, October 1994–September 2003. International Conference on Emerging Infectious Diseases, Atlanta, GA.

82. Ge, B., Zhao, S., Hall, R., and Meng, J. (2002) A PCR-ELISA for detecting Shiga toxin-producing *Escherichia coli. Microbes. Infect.* **4,** 285–290.

83. Michino, H., Araki, K., Minami, S., et al. (1998) Recent outbreaks of infections caused by *Escherichia coli* O157:H7 in Japan. In: Escherichia coli *O157:H7 and other Shiga toxin-producing strains.* ASM, Washington, D.C. 73–82.

84. Solomon, E. B., Yaron, S., and Matthews, K. R. (2002) Transmission of *Escherichia coli* O157:H7 from contaminated manure and irrigation water to lettuce plant tissue and its subsequent internalization. *Appl. Environ. Microbiol.* **68,** 397–400.

85. Besser, R. E., Lett, S. M., Weber, J. T., et al. (1993) An outbreak of diarrhea and hemolytic uremic syndrome from *Escherichia coli* O157:H7 in fresh-pressed apple cider. *JAMA* **269,** 2217–2220.

86. Centers for Disease Control and Prevention. (1997) Outbreaks of *Escherichia coli* O157:H7 infection and cryptosporidiosis associated with drinking unpasteurized apple cider—Connecticut and New York, October 1996. *MMWR Morb. Mortal. Wkly Rep.* **46,** 4–8.

87. Centers for Disease Control and Prevention. (1999) Outbreak of *Escherichia coli* O157:H7 and *Campylobacter* among attendees of the Washington County Fair–New York, 1999. *MMWR Morb. Mortal. Wkly Rep.* **48,** 803–805.

88. Mead, P. S. and Griffin, P. M. (1998) *Escherichia coli* O157:H7. *Lancet* **352,** 1207–1212.

89. Karmali, M. A., Steele, B. T., Petric, M., and Lim, C. (1983) Sporadic cases of haemolytic–uraemic syndrome associated with faecal cytotoxin and cytotoxin-producing *Escherichia coli* in stools. *Lancet* **1,** 619–620.

90. Siegler, R. L. (1995) The hemolytic uremic syndrome. *Pediatr. Clin. North Am.* **42,** 1505–1529.

91. Gerber, A., Karch, H., Allerberger, F., Verweyen, H. M., and Zimmerhackl, L. B. (2002) Clinical course and the role of shiga toxin-producing *Escherichia coli* infection in the hemolytic–uremic syndrome in pediatric patients, 1997–2000, in Germany and Austria: a prospective study. *J. Infect. Dis.* **186,** 493–500.

92. Slutsker, L., Ries, A. A., Greene, K. D., Wells, J. G., Hutwagner, L., and Griffin, P. M. (1997) *Escherichia coli* O157:H7 diarrhea in the United States: clinical and epidemiologic features. *Ann. Intern. Med.* **126,** 505–513.

93. Wall, P. G., McDonnell, R. J., Adak, G. K., Cheasty, T., Smith, H. R., and Rowe, B. (1996) General outbreaks of vero cytotoxin producing *Escherichia coli* O157 in England and Wales from 1992 to 1994. *Commun. Dis. Rep. Rev.* **6,** R26–R33.

94. Carter, A. O., Borczyk, A. A., Carlson, J. A., et al. (1987) A severe outbreak of *Escherichia coli* O157:H7—associated hemorrhagic colitis in a nursing home. *N. Engl. J. Med.* **317,** 1496–1500.

95. Ina, K., Kusugami, K., and Ohta, M. (2003) Bacterial hemorrhagic enterocolitis. *J. Gastroenterol.* **38,** 111–120.

96. Walterspiel, J. N., Ashkenazi, S., Morrow, A. L., and Cleary, T. G. (1992) Effect of sub-inhibitory concentrations of antibiotics on extracellular Shiga-like toxin I. *Infection* **20,** 25–29.

97. Paton, J. C. and Paton, A. W. (1998) Pathogenesis and diagnosis of Shiga toxin-producing *Escherichia coli* infections. *Clin. Microbiol. Rev.* **11,** 450–479.

98. Fullerton, K. E. (2004) Personal communication.

99. Teunis, P., Takumi, K., and Shinagawa, K. (2004) Dose response for infection by *Escherichia coli* O157:H7 from outbreak data. *Risk Anal.* **24,** 401–407.

100. Burland, V., Shao, Y., Perna, N. T., Plunkett, G., Sofia, H. J., and Blattner, F. R. (1998) The complete DNA sequence and analysis of the large virulence plasmid of *Escherichia coli* O157:H7. *Nucleic Acids Res.* **26,** 4196–4204.

101. Smith, H. R., Cheasty, T., and Rowe, B. (1997) Enteroaggregative *Escherichia coli* and outbreaks of gastroenteritis in UK. *Lancet* **350,** 814–815.

102. Guerra, B., Junker, E., Schroeter, A., Malorny, B., Lehmann, S., and Helmuth, R. (2003) Phenotypic and genotypic characterization of antimicrobial resistance in German *Escherichia coli* isolates from cattle, swine and poultry. *J. Antimicrob. Chemother.* **52,** 489–492.

103. Deng, W., Puente, J. L., Gruenheid, S., et al. (2004) Dissecting virulence: systematic and functional analyses of a pathogenicity island. *Proc. Natl Acad. Sci. USA* **101,** 3597–3602.

104. Gruenheid, S., Sekirov, I., Thomas, N. A., et al. (2004) Identification and characterization of NleA, a non-LEE-encoded type III translocated virulence factor of enterohaemorrhagic *Escherichia coli* O157:H7. *Mol. Microbiol.* **51,** 1233–1249.

105. Torres, A. G. and Kaper, J. B. (2003) Multiple elements controlling adherence of enterohemorrhagic *Escherichia coli* O157:H7 to HeLa cells. *Infect. Immun.* **71,** 4985–4995.

106. Sperandio, V., Mellies, J. L., Nguyen, W., Shin, S., and Kaper, J. B. (1999) Quorum sensing controls expression of the type III secretion gene transcription and protein secretion in enterohemorrhagic and enteropathogenic *Escherichia coli*. *Proc. Natl Acad. Sci. USA* **96,** 15,196–15,201.

107. Sanchez-SanMartin, C., Bustamante, V. H., Calva, E., and Puente, J. L. (2001) Transcriptional regulation of the orf19 gene and the tir-cesT-eae operon of enteropathogenic *Escherichia coli*. *J. Bacteriol.* **183,** 2823–2833.

108. DeVinney, R., Stein, M., Reinscheid, D., Abe, A., Ruschkowski, S., and Finlay, B. B. (1999) Enterohemorrhagic *Escherichia coli* O157:H7 produces Tir, which is translocated to the host cell membrane but is not tyrosine phosphorylated. *Infect. Immun.* **67,** 2389–2398.

109. Goosney, D. L., DeVinney, R., Pfuetzner, R. A., Frey, E. A., Strynadka, N. C., and Finlay, B. B. (2000) Enteropathogenic *E. coli* translocated intimin receptor, Tir, interacts directly with α-actinin. *Curr. Biol.* **10,** 735–738.

110. Tatsuno, I., Horie, M., Abe, H., et al. (2001) toxB gene on pO157 of enterohemorrhagic *Escherichia coli* O157:H7 is required for full epithelial cell adherence phenotype. *Infect. Immun.* **69,** 6660–6669.

111. Kim, S. H., Jia, W., Bishop, R. E., and Gyles, C. (2004) An msbB homologue carried in plasmid pO157 encodes an acyltransferase involved in lipid A biosynthesis in *Escherichia coli* O157:H7. *Infect. Immun.* **72,** 1174–1180.

112. Bopp, C. A., Greene, K. D., Downes, F. P., Sowers, E. G., Wells, J. G., and Wachsmuth, I. K. (1987) Unusual verotoxin-producing *Escherichia coli* associated with hemorrhagic colitis. *J. Clin. Microbiol.* **25,** 1486–1489.

113. Schroeder, C. M., White, D. G., and Meng, J. (2004) Retail meat and poultry as a reservoir of antimicrobial-resistant *Escherichia coli*. *Food Microbiol.* **21,** 249–255.

114. van den Bogaard, A. E., London, N., Driessen, C., and Stobberingh, E. E. (2001) Antibiotic resistance of faecal *Escherichia coli* in poultry, poultry farmers and poultry slaughterers. *J. Antimicrob. Chemother.* **47,** 763–771.

115. Schroeder, C. M., Meng, J., Zhao, S., et al. (2002) Antimicrobial resistance of *Escherichia coli* O26, O103, O111, O128, and O145 from animals and humans. *Emerg. Infect. Dis.* **8,** 1409–1414.

116. Meng, J., Zhao, S., Doyle, M. P., and Joseph, S. W. (1998) Antibiotic resistance of *Escherichia coli* O157:H7 and O157:NM isolated from animals, food, and humans. *J. Food Prot.* **61,** 1511–1514.

117. Richard, H. T. and Foster, J. W. (2003) Acid resistance in *Escherichia coli*. *Adv. Appl. Microbiol.* **52,** 167–186.

118. Lee, S. M. and Chen, J. (2004) Survival of *Escherichia coli* O157:H7 in set yogurt as influenced by the production of an exopolysaccharide, colanic acid. *J. Food Prot.* **67,** 252–255.

119. Molina, P. M., Parma, A. E., and Sanz, M. E. (2003) Survival in acidic and alcoholic medium of Shiga toxin-producing *Escherichia coli* O157:H7 and non-O157:H7 isolated in Argentina. *BMC Microbiol.* **3,** 17.

120. Donnenberg, M. S., ed. (2002) Escherichia coli*: Virulence Mechanisms of a Versatile Pathogen*. Academic, London, 417 pp.

121. Ryan, C. A., Tauxe, R. V., Hosek, G. W., et al. (1986) Escherichia coli O157:H7 diarrhea in a nursing home: clinical, epidemiological, and pathological findings. *J. Infect. Dis.* **154,** 631–638.

122. Pavia, A. T., Nichols, C. R., Green, D. P., et al. (1990) Hemolytic–uremic syndrome during an outbreak of *Escherichia coli* O157:H7 infections in institutions for mentally retarded persons: clinical and epidemiologic observations. *J. Pediatr.* **116,** 544–551.

123. Belongia, E. A., MacDonald, K. L., Parham, G. L., et al. (1991) An outbreak of *Escherichia coli* O157:H7 colitis associated with consumption of precooked meat patties. *J. Infect. Dis.* **164,** 338–343.

124. Swerdlow, D. L., Woodruff, B. A., Brady, R. C., et al. (1992) A waterborne outbreak in Missouri of *Escherichia coli* O157:H7 associated with bloody diarrhea and death. *Ann. Intern. Med.* **117,** 812–819.

125. Centers for Disease Control and Prevention. (1991) Foodborne outbreak of gastroenteritis caused by *Escherichia coli* O157:H7—North Dakota, 1990. *MMWR Morb. Mortal. Wkly Rep.* **40,** 265–267.

126. Deschenes, G., Casenave, C., Grimont, F., et al. (1996) Cluster of cases of haemolytic uraemic syndrome due to unpasteurised cheese. *Pediatr. Nephrol.* **10,** 203–205.

127. Bell, B. P., Goldoft, M., Griffin, P. M., et al. (1994) A multistate outbreak of *Escherichia coli* O157:H7-associated bloody diarrhea and hemolytic uremic syndrome from hamburgers. The Washington experience. *JAMA* **272,** 1349–1353.

128. Centers for Disease Control and Prevention. (1993) Preliminary report: foodborne outbreak of *Escherichia coli* O157:H7 infections from hamburgers—western United States, 1993. *MMWR Morb. Mortal. Wkly Rep.* **42,** 85–86.

129. Centers for Disease Control and Prevention. (1995) From the Centers for Disease Control and Prevention. Outbreak of acute gastroenteritis attributable to *Escherichia coli* serotype O104:H21—Helena, Montana, 1994. *JAMA* **274,** 529–530.

130. Centers for Disease Control and Prevention. (1995) From the Centers for Disease Control and Prevention. *Escherichia coli* O157:H7 outbreak linked to commercially distributed dry-cured salami—Washington and California, 1994. *JAMA* **273,** 985–986.

131. Centers for Disease Control and Prevention. (1995) Community outbreak of hemolytic uremic syndrome attributable to *Escherichia coli* O111:NM—South Australia 1995. *MMWR Morb. Mortal. Wkly Rep.* **44,** 550–551, 557–558.

132. Keene, W. E., Sazie, E., Kok, J., et al. (1997) An outbreak of *Escherichia coli* O157:H7 infections traced to jerky made from deer meat. *JAMA* **277,** 1229–1231.

133. Ackers, M. L., Mahon, B. E., Leahy, E., et al. (1998) An outbreak of *Escherichia coli* O157:H7 infections associated with leaf lettuce consumption. *J. Infect. Dis.* **177,** 1588–1593.

134. Ammon, A., Petersen, L. R., and Karch, H. (1999) A large outbreak of hemolytic uremic syndrome caused by an unusual sorbitol-fermenting strain of *Escherichia coli* O157:H–. *J. Infect. Dis.* **179,** 1274–1277.

135. Hashimoto, H., Mizukoshi, K., Nishi, M., et al. (1999) Epidemic of gastrointestinal tract infection including hemorrhagic colitis attributable to Shiga toxin 1-producing *Escherichia coli* O118:H2 at a junior high school in Japan. *Pediatrics* **103,** E2.

136. Centers for Disease Control and Prevention. (1996) From the Centers for Disease Control and Prevention. Outbreak of *Escherichia coli* O157:H7 infections associated with drinking unpasteurized commercial apple juice—British Columbia, California, Colorado, and Washington, October 1996. *JAMA* **276,** 1865.

137. Ahmed, S. and Donaghy, M. (1998) An outbreak of *Escherichia coli* O157:H7 in Central Scotland. In: Escherichia coli *O157:H7 and Other Shiga Toxin-Producing Strains*, ASM, Washington, D.C. 59–65.

138. O'Brien, S. J., Murdoch, P. S., Riley, A. H., et al. (2001) A foodborne outbreak of Vero cytotoxin-producing *Escherichia coli* O157:H-phage type 8 in hospital. *J. Hosp. Infect.* **49,** 167–172.

139. Brook, M. G., Smith, H. R., Bannister, B. A., et al. (1994) Prospective study of verocytotoxin-producing, enteroaggregative and diffusely adherent *Escherichia coli* in different diarrhoeal states. *Epidemiol. Infect.* **112,** 63–67.

140. Centers for Disease Control and Prevention. (2000) From the Centers for Disease Control. Outbreak of *Escherichia coli* O157:H7 infection associated with eating fresh cheese curds—Wisconsin, June 1998. *JAMA* **284,** 2991–2992.

141. Barwick, R. S., Levy, D. A., Craun, G. F., Beach, M. J., and Calderon, R. L. (2000) Surveillance for waterborne-disease outbreaks—United States, 1997–1998. *MMWR CDC Surveill. Summ.* **49,** 1–21.

142. Goh, S., Newman, C., Knowles, M., et al. (2002) *E. coli* O157 phage type 21/28 outbreak in North Cumbria associated with pasteurized milk. *Epidemiol. Infect.* **129,** 451–457.

143. Licence, K., Oates, K. R., Synge, B. A., and Reid, T. M. (2001) An outbreak of *E. coli* O157 infection with evidence of spread from animals to man through contamination of a private water supply. *Epidemiol. Infect.* **126,** 135–138.

144. Payne, C. J., Petrovic, M., Roberts, R. J., et al. (2003) Vero cytotoxin-producing *Escherichia coli O157* gastroenteritis in farm visitors, North Wales. *Emerg. Infect. Dis.* **9,** 526–530.

145. Brooks, J. T., Bergmire-Sweat, D., Kennedy, M., et al. (2004) Outbreak of Shiga toxin-producing *Escherichia coli* O111:H8 infections among attendees of a high school cheerleading camp. *Clin. Infect. Dis.* **38,** 190–198.

146. Welinder-Olsson, C., Stenqvist, K., Badenfors, M., et al. (2004) EHEC outbreak among staff at a children's hospital—use of PCR for verocytotoxin detection and PFGE for epidemiological investigation. *Epidemiol. Infect.* **132,** 43–49.

147. Werber, D., Fruth, A., Liesegang, A., et al. (2002) A multistate outbreak of Shiga toxin-producing *Escherichia coli* O26:H11 infections in Germany, detected by molecular subtyping surveillance. *J. Infect. Dis.* **186,** 419–422.

148. Hrudey, S. E., Payment, P., Huck, P. M., Gillham, R. W., and Hrudey, E. J. (2003) A fatal waterborne disease epidemic in Walkerton, Ontario: comparison with other waterborne outbreaks in the developed world. *Water Sci. Technol.* **47,** 7–14.

149. Hrudey, S. E. 2004. Personal communication.

150. Holme, R. (2003) Drinking water contamination in Walkerton, Ontario: positive resolutions from a tragic event. *Water Sci. Technol.* **47,** 1–6.

151. Centers for Disease Control and Prevention (2002) Summary of notifiable diseases—United States, 2000. *MMWR Morb. Mortal. Wkly Rep.* **49,** i–xxii; 1–100.

152. Centers for Disease Control and Prevention (2003) Summary of notifiable diseases—United States, 2001. *MMWR Morb. Mortal. Wkly Rep.* **50,** i–xxiv; 1–108.

153. Rajpura, A., Lamden, K., Forster, S., et al. (2003) Large outbreak of infection with *Escherichia coli* O157 PT21/28 in Eccleston, Lancashire, due to cross contamination at a butcher's counter. *Commun. Dis. Public Health* **6,** 279–284.

154. Liptakova, A., Siegfried, L., Rosocha, J., Podracka, L., Bogyiova, E., and Kotulova, D. (2004) A family outbreak of haemolytic uraemic syndrome and haemorrhagic colitis caused by verocytotoxigenic *Escherichia coli* O157 from unpasteurised cow's milk in Slovakia. *Clin. Microbiol. Infect.* **10,** 576–578.

155. Bruce-Grey-Owen Sound Health Unit. (2000) The investigative report of the Walkerton outbreak of waterborne gastroenteritis: May–June, 57 pp. Available at http://www.publichealthgreybruce.on.ca/_private/Report/SPReport.htm (accessed on 06 February 2007).

156. Honish, L., Zazulak, I., Mahabeer, R., et al. (2007) Outbreak of *Escherichia coli* O157:H7 gastroenteritis associated with consumption of beef donairs, Edmonton, Alberta, May – June 2006. *Can. Commun. Dis. Rep.* **33,** 14–19.

157. Doorduyn, Y., CM de Jager, C. M., van der Zwaluw, W. K., et al. (2006) Shiga toxin-producing *Escherichia coli* (STEC) O157 outbreak, The Netherlands, September – October 2005. *Euro. Surveill.* **11,** 182–185.

158. Salmon, R., and the Outbreak Control Team. (2005) Outbreak of verotoxin producing *E. coli* O157 infections involving over forty schools in south Wales, September 2005. *Euro. Surveill.* **10,** E051006.1.

159. Jensen, C., Ethelberg, S., Gervelmeyer, A., Nielsen, E. M., Olsen, K. E. P., Molbak, K., and the Outbreak Investigation Team. (2006) First general outbreak of verocytotoxin-producing *Escherichia coli* O157 in Denmark. *Euro. Surveill.* **11,** 55–58.

160. Centers for Disease Control and Prevention. (2006) Ongoing multistate outbreak of *Escherichia coli* serotype O157:H7 infections associated with consumption of fresh spinach–United States, September 2006. *MMWR Morb. Mortal. Weekly Rep.* **55,** 1045–1046.

161. Brooks, J. T., Bergmire-Sweat, D., Kennedy, M., et al. (2004) Outbreak of Shiga toxin-producing *Escherichia coli* O111:H8 infections among attendees of a high school cheerleading camp. *Clin. Infect Dis.* **38,** 190–198.

2
Listeria

Franz Allerberger

1. INTRODUCTION

Listeria monocytogenes, the causative agent of human listeriosis, was discovered in 1927 by E.G.D. Murray and J. Pirie, working independently of each other on outbreaks among laboratory rabbits and guinea pigs *(1)*. *L. monocytogenes* is an aerobic, Gram-positive bacterium that produces severe sepsis, meningoencephalitis, and a wide variety of focal infections in infants and adults. *Listeria* is frequently found in raw and unprocessed food products such as meats, vegetables, dairy, and delicatessen products intended for consumption without further heating. The first-documented case on human listeriosis involved a soldier who suffered from meningitis at the end of World War I *(2)*. Since then listeriosis has emerged as an atypical foodborne illness of major public health concern because of its severity (infections of the central nervous system, septicemia, and abortion), the high case-fatality rate (20–30% of cases), and the long incubation time. Outbreaks associated with contaminated coleslaw, soft cheese, hot dogs, chocolate milk, rice salad, and corn salad have been reported *(3)*. *L. monocytogenes* causes invasive illness mainly in certain well-defined high-risk groups, including immunocompromised persons, pregnant women, neonates, and the elderly. However, listeriosis can occur in otherwise healthy individuals, particularly during an outbreak.

2. CLASSIFICATION AND IDENTIFICATION

2.1. Classification

The genus *Listeria* belongs to the *Clostridium* subbranch, together with *Staphylococcus*, *Streptococcus*, *Lactobacillus*, and *Bronchothrix*. This phylogenetic position of *Listeria* is consistent with its low G+C DNA content (36–42%).

L. monocytogenes is one of the six species in the genus *Listeria*. The other species are *L. seeligeri*, *L. ivanovii*, *L. innocua*, *L. welshimeri*, and *L. grayi (1)*. Two subspecies of *L. ivanovii* have been described: *L. ivanovii* subsp. *ivanovii* and *L. ivanovii* subsp. *londoniensis (4)*. *L. murrayi*, which was a separate species in the genus *Listeria*, is now included in the species *L. grayi (5)*. Based on the results of DNA–DNA hybridization, multilocus enzyme analysis, and 16S rRNA sequencing, the six species in the genus *Listeria* are divided into two lines of descent: (i) *L. monocytogenes* and its closely related species, *L. innocua*, *L. ivanovii*, *L. welshimeri*, and *L. seeligeri*; and (ii) *L. grayi*. All of these species are widespread in the environment, but only *L. monocytogenes* is

From: *Infectious Disease: Foodborne Diseases*
Edited by: S. Simjee © Humana Press Inc., Totowa, NJ

considered to be a significant human and animal pathogen. However, occasional human infections due to *L. ivanovii* and *L. seeligeri* have also been reported *(6)*. *L. ivanovii* is nevertheless mainly responsible for abortion in sheep.

2.2. Identification

The identification of *Listeria* species is based on a limited number of biochemical markers, among which hemolysis is used to differentiate between *L. monocytogenes* and *L. innocua*, the most frequently encountered nonpathogenic *Listeria* species. Hemolysis, a major differential characteristic of *Listeria* species, may be difficult to read on blood agar in some cases (especially for environmental and food isolates). The API-Listeria test (bioMérieux, Marcy-l'Etoile, France) was specifically designed for the genus *Listeria* and consists of 10 biochemical differentiation tests in a microtube format. It includes a patented "DIM," based on the absence or presence of arylamidase, which distinguishes between *L. monocytogenes* and *L. innocua* without additional tests for hemolytic activity *(7)*. Amino-acid peptidase activity against alanyl as well as glycyl (as in the DIM test) produces the same reactions and can be established in a laboratory without using the API test kit *(8,9)*. The optimum growth temperature of the facultative anaerobic *Listeria* spp. is between 30 and 37°C, but growth occurs at 4°C within a few days. Catalase is produced except in a few strains and the oxidase test is negative *(10)*.

The biochemical tests used to differentiate the species are the acid production from D-xylose, L-rhamnose, α-methyl-D-mannoside and D-mannitol. The scheme for biochemical identification of *Listeria* species is shown in Table 1. Without using the DIM test, the assessment of hemolysis is essential in differentiating *L. monocytogenes* and *L. innocua*, the most frequently isolated nonpathogenic *Listeria* species.

Listeria strains are divided into serotypes on the basis of somatic (O) and flagellar (H) antigens *(11,12)*. There are 13 serotypes known for *L. monocytogenes*: 1/2a, 1/2b, 1/2c, 3a, 3b, 3c, 4a, 4ab, 4b, 4c, 4d, 4e, and 7. Serotyping antigens are shared among *L. monocytogenes*, *L. innocua*, *L. seeligeri*, and *L. welshimeri*. Serotyping, although not allowing speciation, serves a useful purpose in confirming the genus diagnosis of *Listeria* and for allowing a first-level subtyping for epidemiological purposes. The introduction of a commercial kit for serotyping *Listeria* (Denka Seiken, Tokyo, Japan) has greatly improved the availability of this method.

3. RESERVOIRS

L. monocytogenes is widespread in nature and has been isolated from soil, dust, food products for humans—both of animal and vegetable origin, feed, water, and sewage, and it can be carried by almost any animal species, including asymptomatic humans. The principal reservoir of the organism is said to be in soil, forage, water, mud, and silage *(13)*. Microbiological surveys have documented that *L. monocytogenes* may be present in a wide range of retail foods *(14,15)*. Unlike most other foodborne pathogens, *Listeria* tends to multiply in refrigerated foods that are contaminated. In the United States, regulations require food companies to guarantee zero *L. monocytogenes* levels in all ready-to-eat products. In France, milk products and other products having undergone heat treatment should be entirely free from *L. monocytogenes*; for raw products, the objective also remains total absence but a level of 100 *L. monocytogenes* per gram is tolerated at the consumption stage. In Austria, health authorities enforce zero

Table 1
Biochemical Differentiation of Species in the Genus *Listeria*

	L. monocytogenes	*L. seeligeri*	*L. ivanovii*	*L. innocua*	*L. welshimeri*	*L. grayi*
D-xylose	–	+	+	–	+	–
L-rhamnose	+	–	–	V	V	V
α-Methyl- D-mannoside	+	–	–	+	+	+
D-mannitol	–	–	–	–	–	+

+, Positive; −, negative; V, variable.

L. monocytogenes levels in milk, milk products, and ready-to-eat products; although for the latter, there is no written law on this subject.

Investigation of several outbreaks has demonstrated that epidemic listeriosis is caused by foodborne transmission of *L. monocytogenes*. H.P.R. Seeliger even dubbed listeriosis a "man-made disease" *(16)*. Outbreaks of listeriosis have been associated with the ingestion of raw milk, soft cheeses, contaminated vegetables, and ready-to-eat meat products such as paté. Also a substantial proportion of sporadic cases of listeriosis results from foodborne transmission. Eating soft cheeses, food purchased from store delicatessen counters, and eating undercooked chicken have been shown to increase the risk of sporadic listeriosis *(14,15)*. In several cases, by tracing a strain of *L. monocytogenes* isolated from a patient to a food item in the patient's refrigerator, and then to the retail source, public health officials were able to provide microbiological confirmation of foodborne transmission of sporadic listeriosis *(15)*.

In neonatal infections, *L. monocytogenes* can be transmitted from mother to child *in utero* or during passage through the infected birth canal. There are rare reports of nosocomial transmission in the nursery attributed to contaminated material or patient-to-patient transmission via health-care workers.

4. FOODBORNE OUTBREAKS

L. monocytogenes is one of the most feared pathogens in the food industry. It has been the cause of several major outbreaks in the United States, Canada, Switzerland, Austria, France, England, and Wales during the past 20 yr *(17–21)*. A feasibility study for a collaborative surveillance of *Listeria* infections in Europe financed by the European Commission revealed a total of 19 outbreaks of invasive listeriosis from 1991 to 2002 *(22)*. A total of 526 outbreak-related cases were reported from nine different countries. In addition, four outbreaks of acute *Listeria* gastroenteritis were reported in Italy in 1993 involving 18 cases and in 1997 involving 1566 cases, in Denmark in 1996 involving three cases, and in Belgium in 2001 involving two cases of acute gastroenteritis and one case of invasive listeriosis. Incriminated foods at the origin of the outbreaks of invasive listeriosis were: a processed meat product (six outbreaks), cheese (five outbreaks), processed fish product (three outbreaks), butter (one outbreak) and undetermined (three outbreaks). Two outbreaks of gastroenteritis were linked to the relative consumption of contaminated rice salad and corn salad, whereas the Belgian outbreak of gastroenteritis and invasive listeriosis was linked to a contaminated frozen cream cake. The origin of one outbreak of gastroenteritis remained unknown. The number of outbreaks reported increased gradually over time: 11 outbreaks in the 5-yr period between 1997 and 2001, compared to seven from 1992 to 1996.

Changes in the way the food is produced and distributed have increased the potential for widespread outbreaks involving many countries as a result of contamination of widely distributed commercial food products. This risk of transnational outbreaks is not hypothetical. From 1987 to 1989, more than 350 cases of listeriosis occurred in England and Wales because of the contaminated imported paté *(23)*. In 2001, an outbreak in Belgium was identified by French investigators because a person had developed symptoms during his vacation in France *(22)*. Recently, an outbreak of 11 cases in France was linked to the consumption of raw sausage spread. The incriminated product, having been exported to Belgium, Germany, and Luxembourg, might have given rise to cases in these countries. In the absence of European surveillance and mechanisms for collaborative outbreak investigation, it is difficult at present to link cases in these countries to the French outbreak *(22)*.

The discovery of *L. monocytogenes* mainly in raw and ready-to-eat meat, poultry, seafood, and dairy products has prompted numerous product recalls leading to large financial losses for the industry and to numerous health scares. These discoveries and the multiple outbreaks that have occurred as a result of food contamination have led to increased regulatory activity, implementation of Hazard Analysis and Critical Control Points (HACCP) programs throughout the food industry and specific recommendations to high-risk groups. Following these measures, a substantial decrease in incidents has been documented in several countries, such as France and the USA where a respective threefold and twofold reduction over the last decade has been attributed to the series of preventive and control measures *(24,25)*.

5. PATHOGENICITY (VIRULENCE FACTORS)

Only two of the six *Listeria* species currently recognized cause listeriosis: *L. monocytogenes* and *L. ivanovii*. Whereas *L. monocytogenes* can infect humans and a wide range of animals, including mammals and birds, *L. ivanovii* is mainly pathogenic for ruminants. After ingestion of *Listeria*-contaminated food, bacteria pass through the stomach and cross the intestinal barrier, presumably via M-cells. They are then transported by lymph or blood to the mesenteric lymph nodes, the spleen, and the liver. *L. monocytogenes* and *L. ivanovii* are facultative intracellular pathogens, which are able to replicate in macrophages and a variety of nonphagocytic cells, such as epithelial and endothelial cells, and in hepatocytes. After entering the cell, *Listeria* escapes early from the phagocytic vacuole, multiplies in the host cell cytosol, and then moves through the cell by induction of actin polymerization. The bacteria then protrude into cytoplasmic evaginations, and these pseudopod-like structures are phagocytosed by the neighboring cells *(26)*.

All major virulence factors of *L. monocytogenes* and *L. ivanovii* are involved in a single process: the cell-to-cell spread. By this function, the pathogens can avoid extracellular environment and can escape humoral effectors of the immune system during their dissemination in the host. These virulence genes form a 9-kb gene cluster, named *Listeria* pathogenicity island 1 (LIPI-1) *(27)*. Besides LIPI-1, a second island of 22 kb, termed LIPI-2, has been described. LIPI-2 is specific for *L. ivanovii* and may play a role in the tropism of this pathogen for ruminants *(28)*.

The *hly* gene encodes the pore-forming listeriolysin O (LLO), a thiol-activated hemolysin, which is able to lyse erythrocytes and other cells in a cholesterol-dependent manner. LLO is an essential virulence factor of *L. monocytogenes*, and its inactivation leads to avirulence. The action of LLO is needed to disrupt the phagocytic vacuole for release of bacteria into the cytoplasm. The genes *mpl*, *actA*, and *plcB* are located downstream from *hly* and are organized in an operon structure. ActA is a surface protein necessary for actin-based motility and for spread from cell to cell. ActA acts as a scaffold to assemble the actin polymerization machinery of the host cell and is also essential for *Listeria* pathogenicity. PlcB is a zinc-dependent broad-substrate-range phospholipase C, which is similar to the clostridial phospholipase C (α-toxin). This enzyme is secreted in an inactive form and then extracellularly modified by the Mpl protease, similar to PlcB, a zinc metalloenzyme. PlcB is, together with *hly*, involved in the disruption of the primary vacuole formed after phagocytosis of extracellular *Listeria* cells. The main function of PlcB is, however, the dissolution of the double-membrane secondary phago-somes formed after cell-to-cell spread.

The genomes of *L. monocytogenes* and *L. innocua* have already been sequenced, and comparison of these genomes with those of other *Listeria* species will presumably lead to the discovery of novel virulence-associated loci *(29)*. Presently, the variability in *L. monocytogenes* with respect to traits of virulence and pathogenicity is far from being understood. Studies on the genetic diversity of strains associated with the outbreaks in Switzerland, California, France, and Denmark showed the presence of a homogeneous genetic background. The suspicion that outbreak strains might constitute a lineage of clones of increased potential to cause listeriosis in humans was first suggested by typing and has now been further substantiated *(30–32)*. Outbreak-associated strains of a parti-cular genetic background are called epidemic clones (Ecs) and monitoring their presence in foodstuffs is of particular concern for veterinary and human public health.

6. CLINICAL CHARACTERISTICS

As *L. monocytogenes* is prevalent in food for human consumption, exposure to this pathogen by the consumption of contaminated food would be considered fairly common. However clinical disease is rare and mainly occurs among the immunocompromised, the pregnant, and the elderly (age ≥ 60 yr). Listeriosis usually manifests itself as meningo-encephalitis and/or septicemia. In Europe, approx 20% of clinical cases are pregnancy-associated (including neonates within the first 3 wk after birth), and the majority of the rest occur in nonpregnant immunocompromised individuals or in the elderly. The median incubation period is estimated to be 3 wk. Outbreak cases have occurred 3–70 d following a single exposure to an implicated product *(13)*.

While listeriosis during pregnancy usually presents with flu-like symptoms which can lead to infection of the fetus causing abortion, premature birth or stillbirth, in non-pregnancy associated cases it mainly manifests as meningoencephalitis and/or septicemia. The onset of meningoencephalitis (which is rare in pregnant women) can be sudden, with fever, intense headache, nausea, vomiting, and signs of meningeal irritation, or may be subacute, particularly in an immunocompromised or an elderly host *(13)*. Rhomben-cephalitis is an unusual form of listeriosis, which involves the brain stem and is similar to circling disease in sheep. *L. monocytogenes* can produce a wide variety of focal infections: conjunctivitis, skin infection, lymphadenitis, hepatic abscess, cholecystitis, peritonitis, splenic abscess, pleuropulmonary infection, joint infection, osteomyelitis, pericarditis, myocarditis, arteritis, and endophthalmitis *(33)*. Endocarditis owing to *L. monocytogenes* is another rare manifestation. Guerro et al. recently reviewed 68 cases of *L. monocytogenes* endocarditis and reported clinical findings very similar to those known for other listeriosis-manifestations: mean age, 65.5 yr; underlying noncardiac conditions, 41.1%; and lethality, 35.3% *(34)*.

In Austria (total population, approx 8 million), 81 human cases of listeriosis (case definition based on isolation of *L. monocytogenes* from normal sterile material), 14 of them pregnancy-associated, were ascertained from 1996 to 2002. Lethality was 31.9% (abortion ×3, stillbirth ×1, and death in a newborn aged 15 d were not counted). Patients (50.7% female) from nonpregnancy associated cases had a mean age of 65.2 yr (median, 66; range 7–93). Age of pregnant women was documented for only 8 of 14 cases: mean 30.5 yr (median, 29–30; range 25–36). Predisposing factors could be shown for 70 of 72 listeriosis patients (study period, 1997–2002): age $\geq$ 60 yr ($n = 19$), pregnancy ($n = 12$) and 39 cases of carcinoma, blood-malignancies, auto-immune diseases (corticosteroid therapy), and status post solid organ transplantation. Overall, 54.2% of the patients (39 out of 72) had an underlying medical disorder. AIDS, often named as a predisposing factor for listeriosis, has not been observed in Austrian listeriosis patients so far.

Diagnosis was confirmed by isolating the infectious agent from cerebrospinal fluid (CSF), blood, amniotic fluid, placenta, meconium, lochia, gastric washings, and other sites of infection. Of the 14 pregnancy-associated cases of listeriosis, primary causative agent determination was carried out from maternal specimens in four cases: blood culture ×1; uterine smear ×1; cervical smear ×2. In nine cases, diagnosis was made from products of conception: neonate gastric secretions ×2; blood culture ×2; blood culture plus CSF culture ×1; ear swab ×1; meconium ×1; amniotic fluid ×1; culture from placenta ×1; and in one case from the simultaneous testing of gastric secretions plus cervical smear. In 61 (91%) of the 67 nonpregnancy-associated patients, the primary isolate was taken from blood ($n = 42$), CSF ($n = 18$), or simultaneously tested blood plus CSF ($n = 1$). Six cases presented as local infections: cholecystitis in a 40-yr-old [serovar (SV) 1/2a]; peritonitis in a 56-yr-old (SV 4b), a 58-yr-old (SV 3a); and in an 80-yr-old (SV 4b), as well as purulent pleuritis in a 65-yr-old (SV 1/2a) and in a 66-yr-old (SV 1/2b).

In healthy adults, exposure to *L. monocytogenes*-contaminated food usually causes only a short period of fecal shedding without any illness. Now it has been recognized that the foodborne transmission of *L. monocytogens* can also cause a self-limiting acute gastroenteritis in immunocompetent persons. From the data available on outbreaks in Italy and in Illinois (USA), it appears that the febrile gastroenteritis in normal hosts requires the ingestion of a high dose of several million bacteria *(35,36)*. Grif et al. studied the incidence of fecal carriage of *L. monocytogenes* in healthy volunteers *(37)*. The PCR results of the subjects indicate an incidence of 5–9 exposures to *L. mono-cytogenes* per person/year. On an average, the incidence of culture-confirmed fecal carriage in healthy adults was two episodes of *L. monocytogenes* carriage per person/ year. Fecal shedding was of short duration (maximum 4 d). The discrepancy between PCR results and the results from conventional culture could be explained by protective host effects. In particular, secretion of gastric acid provides an important protective factor against the passage of potentially pathogenic organisms. Cobb et al. have shown a drastically increased prevalence of *L. monocytogenes* in the feces of patients receiving long-term H2-antagonists compared to the prevalence in patients with normal gastric secretion *(38)*.

7. CHOICE OF TREATMENT

Although optimal therapy of listeriosis has not been verified by randomized clinical studies, penicillin or ampicillin alone, or in combination with gentamicin, are considered the drugs of choice. However, *L. monocytogenes* strains exhibit tolerance to penicillin and ampicillin *(39)*. Although on an average the minimal inhibitory concentration (MIC) of penicillin is usually less than 1 µg/mL, the minimal bactericidal concentration commonly exceeds 10 or more times that amount *(39,40)*. The combination of penicillin with gentamicin shows synergistic activity in vitro and for these reasons has been considered the first-choice therapy of listeriosis in humans *(39)*. However, in vivo synergism has not been uniformly observed and retrospective clinical studies have not consistently shown better results for the combined therapy of listeriosis than simple penicillin monotherapy *(17,34,39,41,42)*. For patients with β-lactam allergy, trimethoprim–sulfamethoxazole or erythromycin may be considered. Vancomycin together with gentamicin may also be a reasonable alternative therapy *(39)*. Antimicrobial suscepti-bilities of *Listeria* have not changed markedly over the past 35 yr *(43,44)*.

L. monocytogenes reproduces in the reticuloendothelial system and survives intra-cellularly after uptake by macrophages *(45)*. This means that the organism cannot be reached by certain antibiotics, a fact which might contribute to the differences between in vitro and in vivo results *(45,46)*. Antimicrobial drugs of questionable value for human listeriosis include cephalosporins, clindamycin, the fluoroquinolones, and aminogylco-sides (when administered individually) *(45,46)*. Cephalosporins have hardly any in vitro effect against *L. monocytogenes*. The reason is the minimal or nonexistent affinity of penicillin-binding proteins 3 and 5 for cephalosporins *(47)*. Despite good in vitro activity, even cephalothin has no effect on experimental listeriosis in mice *(48)*. In addition, cephalothin lacks satisfactory CSF penetration. Reports of therapeutic failures prove that cephalosporins are not indicated in the treatment of listeriosis *(49)*. Clinically effective substances, apart from gentamicin, are only bacteriostatically effective in vitro against *L. monocytogenes*, thus emphasizing the importance of the body's own cellular defense mechanisms *(45,46)*. Because *Listeria* do not produce β-lactamase, addition of β-lactamase-inhibitors in the treatment of listeriosis is ineffective. The MIC values of ampicillin alone and ampicillin combined with sulbactam against *L. monocytogenes* show no relevant difference *(43)*. Kayser et al. reported good in vitro activity of meropenem *(50)*. Another study with this antibiotic showed good activity in experimental meningitis in guinea pigs *(51)*.

8. RESISTANCE

Clinical and Laboratory Standards Institute (CLSI) has not yet provided specific guidelines for the testing of *Listeria*. Usually susceptibility testing is performed according to CLSI guidelines for bacteria that grow aerobically using Mueller–Hinton agar with 5% horse blood *(43)*. For trimethoprim–sulfamethoxazole, the blood is lysed. The pattern of antimicrobial in vitro susceptibility of *L. monocytogenes* has been relatively stable for many years. Usually, the organism is susceptible to penicillin, ampicillin, gentamicin, erythromycin, tetracycline, rifampicin, and chloramphenicol, but only moderately susceptible to quinolones *(39,43)*. Resistant strains are rarely found *(52)*. However, the transfer of genetic elements from *Enterococcus* and *Streptococcus* to *Listeria* determines resistance to macrolides, lincosamide, streptogramin, and tetracycline *(52,53)*. Threlfall et al. reported 1–5% tetracycline resistance in *L. monocytogenes* from humans and food in the UK *(44)*. Charpentier et al. documented in vitro resistance of an *L. monocytogenes* isolate to trimethoprim for the first time *(54)*. In addition, strains resistant to gentamicin, streptomycin, and sulfamethoxazole have been reported *(55,56)*. In 1984 Rapp et al. and in 1986 Pollock et al. reported on ampicillin resistant isolates *(57,58)*. In 1995, Soriano et al. noted an increase in the MIC values of ampicillin for *L. monocytogenes* and suspected that the resistance against ampicillin might possibly be more widespread than had been previously assumed *(59)*. No indication of increased ampicillin resistance was found by other authors *(24)*. In 1995, Charpentier et al. tested 1100 isolates and found all strains were susceptible to ampicillin and gentamicin *(54)*. In vitro testing on recently obtained clinical isolates from France, Norway, Switzerland and Austria, and on reference strains showed that *L. monocytogenes* has retained in vitro susceptibility even 70 yr after the initial reports *(43)*.

9. EPIDEMIOLOGY

9.1. Listeriosis in Animals

In most countries listeriosis in animals is not a notifiable disease. In Europe, listeriosis in animals is notifiable only in Germany, Finland, Sweden, and Norway. Usually, surveillance in animals is based on clinical observations. The "Report on Trends and Sources of Zoonotic Agents in the European Union and in Norway, 2001" lists the following rates of fecal carriage of *L. monocytogenes*: cattle, 6.30% (dairy cattle, 14.05%; calves, 1.86%); pigs, 0.08%; sheep, 3.11%; goats, 4.64–16.13%; horses, 2.07%; fowl, 1.19%; red deer, 8.33%; rabbit (farmed), 0.67%; cats, 0.69%; and dogs, 0.00%.

9.2. Listeria in Food

In Europe, under Council Directive No. 92/46/EEC control of *L. monocytogenes* in milk products by the dairy industry is compulsory. In the "Report on Trends and Sources of Zoonotic Agents in the European Union and in Norway, 2001" several member states reported on investigations of foodstuffs for *Listeria* spp. Sample sizes and definitions used for positive samples were very different. Therefore, results cannot be compared between the countries. The prevalence of *L. monocytogenes* in food, based on detection of the bacterium, were: beef and veal, 0.6–15.4%; pork, 0–40.6%; minced meat, 11.9–18.3%; meat products, 0–10.2%; poultry meat, 2.6–16.7%; poultry meat products, 0–7.6%; milk, 0–2.2%; milk products, 0–4.4%; fish products, 0–13.5%; vegetables, 0–12.5%; and other ready-to-eat food, 0–16.7%.

9.3. Listeriosis in Humans

In Europe, listeriosis in humans is a mandatory notifiable disease in Finland, Italy, Sweden, Denmark, and Norway. In other countries, information is based on laboratory reports. In total, 860 listeriosis cases were reported in the European Union (total population, 370 million) in 2001. This corresponds to an annual incidence of 0.232 cases per 100,000 inhabitants. The share of pregnancy-associated cases ranged between 4 and 24% in the individual countries that provided this information (France, 24%; England and

Wales, 12%; Germany, 10%). Pregnancy-associated cases refer to listeriosis in pregnant women or in the neonates (up to 28 d of life), and the nonpregnancy-associated cases to anyone older (>28 d). It is also standard to classify one pregnancy-associated maternal-fetal unit or mother and neonate pair as one case. In Europe, the age groups mainly affected are elderly persons over 60 yr old and children up to the age of 4 yr. There is no obvious seasonal clustering of listeriosis, which is consistent with the year-round *Listeria* contamination found in ready-to-eat food.

In Austria from 1996 to 2002, the annual incidence of listeriosis averaged 0.158 cases per 100,000 inhabitants. The frequency of incidence in west European countries (14 EU countries, Norway, Iceland, and Switzerland) varied between 0.03 and 0.75 cases per 100,000 inhabitants per year, whereby five countries had more than 0.4 per 100,000 and three further countries had more than 0.6 per 100,000 *(22)*. In the year 2000, the USA registered 0.4 cases per 100,000 (personal comment: Bala Swaminathan, CDC, USA); the FDA claims 1 case per 100,000 per year *(60)*. The incidence of listeriosis in males and females is approximately the same. In Germany in 2001, 0.3 cases per 100,000 population were reported for males and 0.2 cases per 100,000 population for females.

Most human infections are caused by serovars 1/2a, 1/2b and 4b *(13)*. Serotyping results were available for Austrian isolates from 1997 to 2002 ($n = 72$): SV 4b, 56.9%; SV 1/2a, 26.4%; SV 1/2b, 11.1%; SV 1/2c, 4.2%; and SV 3a, 1.4%. Information provided by Denmark shows that most of their cases are caused by serotype 1 followed by serotype 4. In Germany in 2001, five cases were caused by serotype 1/2a, two by SV 4b and one by serotype 1/2b.

The emergence of listeriosis is the result of complex interactions between various factors that reflect changes in social patterns. Swaminathan named the high degree of centralization and consolidation of food production and processing, the increased use of refrigerators as the primary means of preserving food, and changes in food consumption habits (increased consumer demand for convenient food) as main factors *(61)*. In healthy adults, exposure to *L. monocytogenes*-contaminated food usually causes only a short period of fecal shedding without any illness; however, in pregnant women, newborns, the elderly, and adults with weakened immune systems, ingestion can lead to listeriosis.

10. SUMMARY AND CONCLUSIONS

L. monocytogenes has been recognized as a human pathogen for more than 50 yr. It primarily causes abortion, infections of the central nervous system or septicemia, mainly in certain well-defined high-risk groups, including immunocompromised persons and pregnant women. The widespread use of immunosuppressive medications for treatment of malignancy and management of organ transplantation, has expanded the immuno-compromised population at increased risk of listeriosis. Also consumer life styles have changed with less time for food preparation, more ready-to-eat, and take-away foods. Changes in food production and technology have led to the production of foods with long shelf life that are typical "*Listeria* risk foods," because the bacteria have time to multiply, and the food does not undergo a listericidal process such as cooking before consumption. Increased mass production means outbreaks can change from being small and confined to a community or region, to large, affecting hundreds of people. Unlike infection with other common foodborne pathogens, listeriosis is associated with a high case fatality rate of approx 20–30%. Epidemiological investigations during the last 20 yr have shown that epidemic listeriosis is a foodborne disease. Similarly, recent studies have suggested that a substantial proportion of sporadic cases of listeriosis are also caused by consumption of contaminated food. Despite the high contamination rates of certain food with *L. monocytogenes*, listeriosis is a relatively rare disease compared to other common foodborne illnesses such as salmonellosis. However, because of its high case fatality rate, listeriosis ranks among the most frequent causes of foodborne death: in the USA, France, and Austria, it ranks second only after salmonellosis; in England, it ranks fourth *(62–65)*. Therefore, besides the economic consequences, listeriosis remains of great public health concern. In addition, its common potential epidemic source presents a real threat and persists even in countries with a decreasing or low incidence.

REFERENCES

1. Rocourt, J. (1999) The genus *Listeria* and *Listeria monocytogenes*: phylogenetic position, taxonomy, and identification. In: *Listeria, Listeriosis, and Food Safety* (Ryer, E. T. and Marth, E. H., eds.), 2nd edn, Marcel Dekker, New York, pp. 1–20.
2. McLauchlin, J. (1997) The discovery of *Listeria. PHLS Microbiol. Dig.* **14,** 76–78.
3. Schlech, W. F. (2001) Food-borne listeriosis. *Clin. Infect. Dis.* **31,** 770–775.
4. Boerlin, P., Roucourt, J., Grimont, F., Grimont, P. A. D., Jacquet, C., and Piffaretti, J. C. (1992) *Listeria ivanovii* subspecies *londoniensis. Int. J. Syst. Bacteriol.* **15,** 42–46.
5. Rocourt, J., Boerlin, P., Grimont, F., Jacquest, C., and Piffaretti, J. C. (1992) Assignment of *Listeria grayi* and *Listeria murrayi* to a single species, *Listeria grayi*, with a revised description to *Listeria grayi. Int. J. Syst. Bacteriol.* **42,** 69–73.
6. Gilot, P. and Content, J. (2002) Species identification of *Listeria welshimeri* and *Listeria monocytogenes* by PCR assays targeting a gene encoding a fibronectin-binding protein. *J. Clin. Microbiol.* **40,** 698–703.
7. Bille, J., Catimel, B., Bannerman, E., et al. (1992) API-Listeria, a new and promising one-day system to identify *Listeria* isolates. *Appl. Environ. Microbiol.* **58,** 1857–1860.
8. McLauchlin, J. (1997) The identification of *Listeria* species. *Int. J. Food Microbiol.* **38,** 77–81.
9. Clark, A. G. and McLauchlin, J. (1997) A simple colour test based on an alanyl peptidase reaction which differentiates *Listeria monocytogenes* from other *Listeria* species. *J. Clin. Microbiol.* **35,** 2155–2156.
10. Elsner, H.-A., Sobottka, I., Bubert, A., Albrecht, H., Laufs, R., and Mack, D. (1996) Catalase-negative *Listeria monocytogenes* causing lethal sepsis and meningitis in an adult hematologic patient. *Eur. J. Clin. Microbiol. Infect. Dis.* **15,** 965–967.
11. Seeliger, H. P. R. and Jones, D. (1986) Genus *Listeria* Pirie, 1940, 383AL. In: *Bergey's Manual of Systematic Bacteriology* (Sneath, P. H., Mair, N. S., Sharp, M. E., and Holt, J. G., eds.), vol. 2, Williams and Wilkins, Baltimore, MD, pp. 1235–1245.
12. Allerberger, F. (2003) *Listeria*: growth, phenotypic differentiation and molecular microbiology. *FEMS Immunol. Med. Microbiol.* **35,** 183–189.
13. Chin, J., ed. (2000) *Control of Communicable Diseases Manual*, 17th edn. American Public Health Association, Washington, D.C.
14. Schuchat, A., Deaver, K. A., Wenger, J. D., et al. (1992) Role of foods in sporadic listeriosis: I. Case–control study of dietary risk factors. *JAMA* **267,** 2041–2045.
15. Pinner, R. W., Schuchat, A., Swaminathan, B., et al. (1992) Role of foods in sporadic listeriosis: II. Microbiologic and epidemiologic investigation. *JAMA* **267,** 2046–2050.
16. Allerberger, F., Dierich, M. P., Grundmann, H., Hartung, D., Bannerman, E., and Bille, J. (1997) Typing of Austrian *Listeria monocytogenes* isolates by automated laser fluorescence analysis of randomly amplified polymorphic DNA. *Zbl. Bakt.* **286,** 33–40.
17. Allerberger, F. and Guggenbichler, J. P. (1989) Listeriosis in Austria—report of an outbreak in Austria 1986. *Acta Microbiol. Hung.* **36,** 149–152.

18. Büla, C. J., Bille, J., and Glauser, M. P. (1995) An epidemic of food-borne listeriosis in Western Switzerland: description of 57 casing involving adults. *Clin. Infect. Dis.* **20,** 66–72.
19. Jacquet, V., Catimel, B., Brosch, R., et al. (1995) Investigations related to the epidemic strain involved in the French listeriosis outbreak in 1992. *Appl. Environ. Microbiol.* **61,** 2242–2246.
20. Linnan, M. J., Mascola, L., Lou, X. D., et al. (1988) Epidemic listeriosis associated with Mexican-style cheese. *N. Engl. J. Med.* **319,** 823–828.
21. Schlech, W. F., Lavigne, P. M., Bortolussi, R. A., et al. (1983) Epidemic listeriosis—evidence for transmission by food. *N. Engl. J. Med.* **308,** 203–206.
22. De Valk, H., Jacquet, Ch., Goulet, V., et al. (2003) Feasibility study for a collaborative surveillance of *Listeria* infections in Europe. Report to the European Commission, DGSANCO, Paris.
23. McLauchlin, J., Hall, S. M., Velani, S. K., and Gilbert, R. J. (1991) Human listeriosis and paté; a possible association. *BMJ* **303,** 773–775.
24. Goulet, V., deValk, H., Pierre, O., et al. (2001) Effect of prevention measures on incidence of human listeriosis, France, 1987–1997. *Emerg. Infect. Dis.* **7,** 983–989.
25. Tappero, J. W., Schuchat, A., Deaver, K. A., et al. (1995) Reduction in the incidence of human listeriosis in the United States, effectiveness of prevention efforts? *JAMA* **273,** 1118–1122.
26. Schmidt, H. and Hensel, M. (2004) Pathogenicity islands in bacterial pathogenesis. *Clin. Microbiol. Rev.* **17,** 14–56.
27. Vazquez-Boland, J. A., Kuhn, M., Berche, P., et al. (2001) *Listeria* pathogenesis and molecular virulence determinants. *Clin. Microbiol. Rev.* **14,** 586–640.
28. Gonzalez-Zorn, B., Dominguez-Bernal, G., Suarez, M., et al. (2000) SmcL, a novel membrane-damaging virulence factor in *Listeria. Int. J. Med. Microbiol.* **290,** 369–374.
29. Glaser, P., Frangeul, L., Buchrieser, C., et al. (2001) Comparative genomics of *Listeria* species. *Science* **294,** 849–852.
30. Herd, M. and Kocks, C. (2001) Gene fragments distinguishing an epidemic-associated strain from a virulent prototype strain of *Listeria monocytogenes* belong to a distinct functional subset of genes and partially cross-hybridize with other *Listeria* species. *Infect. Immun.* **69,** 3972–3979.
31. Tran, H. L. and Kathariou, S. (2002) Restriction fragment length polymorphisms detected with novel DNA probes differentiate among diverse lineages of serogroup 4 *Listeria monocytogenes* and identify four distinct lineages in serotype 4b. *Appl. Environ. Microbiol.* **68,** 59–64.
32. Wagner, M. and Allerberger, F. (2003) Characterization of *Listeria monocytogenes* recovered from 41 cases of sporadic listeriosis in Austria by serotyping and pulsed-field gel electrophoresis. *FEMS Immunol. Med. Microbiol.* **35,** 227–234.
33. Lorber, B. (1999) *Listeria monocytogenes.* In: *Principles and Practice of Infectious Diseases* (Mandell, G. L., Bennett, J. E., and Dolin, R., eds.), 5th edn, Churchill Livingstone, Philadelphia, pp. 2208–2215.
34. Guerro, M. L. F., Rivas, P., Ràbago, R., Núnez, A., deGórgolas, M., and Martinell, J. (2004) Prosthetic valve endocarditis due to *L. monocytogenes.* Report of two cases and reviews. *Int. J. Infect. Dis.* **8,** 97–102.
35. Aureli, P., Fiorucci, G. C., Caroli, D., et al. (2000) An outbreak of febrile gastroenteritis associated with corn contaminated with *Listeria monocytogenes. N. Engl. J. Med.* **342,** 1236–1241.
36. Dalton, C. B., Austin, C. C., Sobel, J., et al. (1997) An outbreak of gastroenteritis and fever due to *Listeria monocytogenes* in milk. *N. Engl. J. Med.* **336,** 100–105.
37. Grif, K., Patscheider, G., Dierich, M. P., and Allerberger, F. (2003) Incidence of fecal carriage of *Listeria monocytogenes* in three healthy volunteers: A one-year prospective stool survey. *Eur. J. Clin. Microbiol. Infect. Dis.* **22,** 16–20.

38. Cobb, C. A., Curtis, G. D., Bansi, D. S., et al. (1996) Increased prevalence of *Listeria monocytogenes* in the feces of patients receiving long-term H2-antagonists. *Eur. J. Gastroenterol. Hepatol.* **8,** 1071–1074.

39. Hof, H., Nichterlein, T., and Kretschmar, M. (1997) Management of listeriosis. *Clin. Microbiol. Rev.* **10,** 345–357.

40. Allerberger, F. (1987) *Listeria monocytogenes*—microcalorimetric investigations regarding the antibacterial efficiency of chemotherapeutics. *Thermochimica Acta* **119,** 113–119.

41. Hof, H. and Guckel, H. (1987) Lack of synergism of ampicillin and gentamicin in experimental listeriosis. *Infection* **15,** 40–41.

42. Safdar, A. and Armstrong, D. (2003) Listeriosis in patients at a comprehensive cancer center, 1955–1977. *Clin. Infect. Dis.* **37,** 359–364.

43. Heger, W., Dierich, M. P., and Allerberger, F. (1997) In vitro susceptibility of *Listeria monocytogenes*: comparison of E-test with Agar Dilution Test. *Chemotherapy* **43,** 303–310.

44. Threlfall, E. J., Skinner, J. A., and McLauchlin, J. (1998) Antimicrobial resistance in *Listeria monocytogenes* from humans and food in the UK, 1967–96. *Clin. Microbiol. Infect.* **4,** 410–412.

45. Hof, H. (1991) Therapeutic activities of antibiotics in listeriosis. *Infection* **19,** 229–233.

46. Kluge, R. M. (1990) Listeriosis—problems and therapeutic options. *J. Antimicrob. Chemother.* **25,** 887–890.

47. Hakenbeck, R. and Hof, H. (1991) Relatedness of penicillin-binding proteins from various *Listeria* species. *FEMS Microbiol. Lett.* **84,** 191–196.

48. Hof, H. and Waldenmeier, G. (1988) Therapy of experimental listeriosis—evaluation of different antibiotics. *Infection* **16,** 171–174.

49. Allerberger, F. and Dierich, M. P. (1992) Listeriosis and cephalosporins. *Clin. Infect. Dis.* **15,** 177–178.

50. Kayser, F. H., Morenzoni, G., Strässle, A. and Hadorn, K. (1989) Activity of meropenem, against gram-positive bacteria. *J. Antimicrob. Chemother.* **24,** 101–112.

51. Nairn, K., Shepherd, G. L., and Edwards, J. R. (1995) Efficacy of meropenem in experimental meningitis. *J. Antimicrob. Chemother.* **36(A),** 73–84.

52. Charpentier, E. and Courvalin, P. (1999) Antibiotic resistance in *Listeria* spp. *Antimicrob. Agents Chemother.* **43,** 2103–2108.

53. Hadorn, K., Hächler, H., Schaffner, A., and Kayser, F. H. (1993) Genetic characterization of plasmid-encoded multiple antibiotic resistance in a strain of *Listeria monocytogenes* causing endocarditits. *Eur. J. Clin. Microbiol. Infect. Dis.* **12,** 928–937.

54. Charpentier, E., Gerbaud, G., Jacquet, C., Rocourt, J., and Courcalin, P. (1995) Incidence of antibiotic resistance in *Listeria* species. *J. Infect. Dis.* **172,** 277–281.

55. McGowan, A. P., Reeves, D. S., and McLauchlin, J. (1990) Antibiotic resistance in *Listeria monocytogenes*. *Lancet* **336,** 513–514.

56. Tsakaris, A., Papa, A., Douboyas, J., and Antoniadis, A. (1997) Neonatal meningitis due to multi-drug-resistant *Listeria monocytogenes*. *J. Antimicrob. Chemother.* **39,** 553–554.

57. Rapp, M. F., Pershadsingh, H. A., Long, J. W., and Pickens, J. M. (1984) Ampicillin-resistant *Listeria monocytogenes* meningitis in a previously healthy 14-year-old athlete. *Arch. Neurol.* **41,** 1304.

58. Pollock, S. S., Pollock, T. M., and Harrison, M. J. G. (1986) Ampicillin-resistant *Listeria monocytogenes* meningitis. *Arch. Neurol.* **43,** 106.

59. Soriano, F., Zapardiel, J., and Nieto, E. (1995) Antimicrobial susceptibilities of *Corynebacterium* species and other non-spore-forming gram-positive bacilli to 18 antimicrobial agents. *Antimicrob. Agents Chemother.* **39,** 208–214.

60. Anonymous (2004) FDA assesses listeriosis food risks. *ASM News* **70,** 8.

61. Swaminathan, B. (2001) *Listeria monocytogenes*, In: *Food Microbiology: Fundamentals and Frontiers* (Doyle, M. P., Beuchat, L. R., and Montville, T. J., eds.), American Society for Microbiology, Washington, D.C., pp. 383–409.

62. Mead, S.M., Slutsker, L., Dietz, V., et al. (1999) Food-related illness and death in the United States. *Emerg. Infect. Dis.* **5,** 607–625.

63. Vaillant, V., Baron, E., and de Valk, H. (2003) Morbidité et mortalité dus aux maladies infectieuses d'origine alimentaire en France. Rapport de l'inVS: in press.

64. Adak, G. K., Long, S. M., and O'Brien, S. J. (2002) Trends in indigenous food-borne disease and deaths, England and Wales: 1999 to 2000. *Gut* **51,** 832–841.

65. Allerberger, F., Heller, I., Grif, K., and Wagner, M. (2006) Epidemiologie der Listeriose in Österreich. *Wien. Klin. Wochenschr.* in press.

Clostridium botulinum and Clostridium perfringens

Jim McLauchlin and Kathie A. Grant

1. INTRODUCTION

Clostridium is a diverse genus of Gram-positive, endospore-bearing obligate anaerobes that are widespread in the environment. This genus includes more than 100 species, and the overall range in the G+C content (22–55 mol%) reflects the enormous phylogenetic variation encompassed within this group. The principal foodborne pathogens are *Clostridium botulinum* and *Clostridium perfringens* that cause toxin-mediated disease either by preformed toxin (foodborne botulism) or by the formation of toxin in the enteric tract (infant botulism and *C. perfringens* diarrhea). These two bacteria and their foodborne diseases will be discussed here.

2. *CLOSTRIDIUM BOTULINUM*

2.1. Introduction

Botulism is a rare but potentially fatal disease caused by neurotoxins (BoNTs) usually produced by *C. botulinum*. Disease symptoms result from paralysis owing to the inhibition of neurotransmitter release by BoNTs. Human foodborne disease results from either ingestion of preformed BoNT in foods contaminated by *C. botulinum*, or production of BoNT following intestinal colonization by *C. botulinum* (usually in infants). The disease can also be transmitted by nonfoodborne routes including the production of BoNT from *C. botulinum*-infected wounds as well as accidental and deliberate release of BoNT *(1)*. Although botulism has been reported as causing a tooth abscess, and hence was probably foodborne *(2)*, wound botulism is most often associated with the trauma at other sites (especially at injection sites to illegal drug users) and are outside the scope of this work and will not be discussed here.

Foodborne botulism represents a potential national (or international) public health emergency. Prompt diagnosis and early treatment are essential to reduce the considerable morbidity and mortality of this disease. The laboratory investigation of botulism is highly specialized, but is essential for the rapid identification of reservoirs of infection and provides vital information allowing the most appropriate interventions to be implemented. Key reviews on *C. botulinum* can be found in the literature*(1,3–17)*.

From: *Infectious Disease: Foodborne Diseases*
Edited by: S. Simjee © Humana Press Inc., Totowa, NJ

2.2. Classification, Identification, Isolation and Diagnosis

2.2.1. Taxonomy of C. botulinum

The species *C. botulinum* is defined on the basis of a single phenotypic characteristic, i.e., neurotoxin production, and is classified into four distinct taxonomic lineages (designated groups I–IV; *[18]*). Characteristics of these four groups are shown in Table 1. This grouping combines highly diverse phylogenetic organisms that consequently show considerable differences in genotype and phenotype. Indeed the genetic (phylogenetic) differences within *C. botulinum* are greater than that between *Bacillus subtilis* and *Staphylococcus aureus (18)*. This situation is further, and somewhat inconsistently, complicated because there are genetically very closely related *Clostridium* species that do not possess neurotoxins and are, therefore, not named *C. botulinum* (i.e., *Clostridium sporogenes*, *Clostridium novyi*, and un-named organisms) and neurotoxin containing *Clostridium baratii* and *Clostridium butyricum (18)*. Further complications exist becaue there is evidence for instability and horizontal gene transfer of BoNT genes as well as nontoxigenic *C. botulinum* variants containing cryptic or fragmentary copies of BoNTs. It has been proposed *(18)* that the four taxonomic lineages should be reclassified into separate species, and a precedence has been set by a proposal to rename *C. botulinum* Group IV (as well as some *Clostridium subterminale* and *Clostridium hastiforme*) as *Clostridium argentinense (148)*. However, this proposed revision for the classification of *C. botulinum* has not found common use and the taxonomy remains unresolved.

The genome size of both *C. botulinum* groups I and II has been estimated at between 3.6 and 4.1 Mbp by pulsed-field gel electrophoresis analysis *(19,149)*. A complete genome sequence for a single *C. botulinum* Type A (Hall strain, ATCC 3502) is available, and at the time of writing the annotation is not yet complete: the genome size is 3,886,916 bp in size, with a G+C content of 28.2 mol %, and one 16,344 bp plasmid was detected *(20)*.

2.2.2. Neurotoxin Types

BoNTs are proteins that can be divided into seven antigenically distinct types, designated A to G, and are among the most potent toxins known *(1)*. The distribution of BoNTs among the four taxonomic groups is shown in Table 1. Group I includes the proteolytic strains producing BoNTs A, B, and F, either singly or as dual toxin types AB, AF, and BF. Group II includes the nonproteolytic strains producing toxins B, F, and E. Group III organisms produce either type C or D toxins and are generally nonproteolytic. Group IV strains produce type G toxin and differ from other groups in not producing lipase. Important *C. botulinum* for human foodborne diseases are in groups I (BoNTs A, B, and F) and II (BoNTs B, E, and F), with type F being the least common *(1)*. However human infection resulting from type C has been reported *(21)*. Organisms from groups I–III cause diseases among animals especially aquatic birds, horses, and cattle. Group IV organisms producing BoNTG have been isolated from autopsy specimens, although evidence that botulism was the cause of death in these cases was not demonstrated *(22)*. BoNTG has not been associated with any other naturally occurring cases of botulism in humans or any other animals *(9)*. BoNT production, presumably as a result of horizontal gene transfer, has also been detected in *C. baratii* (type F) and *C. butyricum* (type E), both of which have been associated with human diseases *(23–26)*.

Table 1
Characteristics of *C. botulinum* Groups

	Group			
	I	II	III	IV
Neurotoxin types	A, B, F	B, E, F	C, D	G
Human disease	Yes	Yes	No[a]	No
Growth temperatures				
Minimum	10	3.3	15	ND
Optimum	35–40	18–25	40	37
Minimum pH for growth	4.6	5.0	5.0	ND
Minimum a_w for growth	0.94	0.97	ND	ND
Inhibition by NaCl	10%	5%	2%	ND
$D_{100°C}$ of spores (min)	25	<0.1	0.1–0.9	0.8–1.12
$D_{121°C}$ of spores (min)	0.1–0.2	<0.001	ND	ND
Proteolytic activity	Yes	No	No	Yes (weak)
Lipolytic activity	Yes	Yes	Yes	No
Saccharolytic activity	No	Yes	Yes (weak)	No
Related non-neurotoxigenic *Clostridium* species	*C. sporogenes* *C. putrificum*	*C. beijerinkii*	*C. novyi* *C. haemolyticum*	*C. subterminale* *C. histolyticum* *C. linosium*
Location of neurotoxin genes	Chromosomal	Chromosomal	Phage	Plasmid
Overall G+C content (mol%)	26–29	27–29	26–28	28–30

[a]One case of infant botulism due to type C has been reported *(21)*.
ND, not determined.

2.2.3. Detection of BoNTs By Mouse Bioassay

Detection of BoNT production is the cardinal feature for both the classification and identification of *C. botulinum* as well as providing confirmation of a clinical diagnosis of botulism. Traditionally this has relied on the use of an in vivo mouse bioassay to detect BoNT activity *(8,11)*.

Toxin extracts for bioassays are prepared as described here and injected intraperitoneally into mice. The biological action of the toxin is generally observed within 24 h and initially often presents with hyperactivity followed by reduced mobility, ruffled fur, labored breathing, contraction of the abdominal muscles (wasp waist), and total paralysis. Respiratory failure and death follow unless animals are humanely euthanized. Animals are observed initially after injection and then at 2, 4, 8, 12, and 24 h, and then daily for up to 4 d. Those showing signs within 2–6 h usually die within 24 h. The rapidity of symptoms (and time of death) is proportional to the amount of toxin present in the sample. Confirmation of the toxin (together with the determination of the toxin type) is achieved by an absence of the above symptoms in the mice following injection of the inoculum after preincubation with specific antitoxin *(8,11)*.

Extracts from food or feces are generally prepared by homogenization in gelatin-phosphate buffer (pH 6.2) which is centrifuged and filtered (to prevent infection) prior to injection. When low levels of toxin or nonproteolytic strains are suspected, the preparation can be pretreated with trypsin to activate the neurotoxin.

Assays for toxin are applied to direct analysis of clinical material (serum), extracts from clinical material (feces, gastric contents, and vomitus) and food, as well as to the supernatants from enrichment and pure cultures growing in broths. In vitro growth of *C. botulinum* in enrichment cultures is achieved by inoculation of broths with feces, gastric contents, and vomitus as well as food samples. For nonfoodborne infections (i.e., material from infected wounds), additional toxin tests are performed to the extracts from pus or tissue, together with application of cultural methods in these samples.

It should be noted that the detection of toxin by bioassays is highly specialized and is likely to require specific authority (including ethical considerations) for their performance.

2.2.4. Identification and Isolation of C. botulinum

As stated earlier, *C. botulinum* is defined on the basis of a single phenotypic characteristic, i.e., neurotoxin production, and hence more conventional phenotypic reactions such as the fermentation of carbohydrates, cellular fatty-acid analysis, and 16S rDNA gene sequence in not sufficient to provide an unequivocal identification *(18,27,28)*. The identification process, therefore, relies on the detection of neurotoxins (or their genes).

The initial process for culture of *C. botulinum* is the enrichment in prereduced cooked meat medium with or without glucose, chopped-meat glucose-starch medium, or tryptone–peptone glucose yeast extract broth. Trypsin may be added to inactivate bacteriocins and to activate neurotoxin. Replicate inoculated broths are treated either with or without a heat shock (60–80°C for 10–20 min). Lysozyme is also added to assist in the reversal of heat-injured spores, especially for nonproteolytic *C. botulinum* strains.

Because of the different physiologies of the *C. botulinum* groups (Table 1), the optimal incubation temperature for isolation is problematic. Some authors recommend 35°C, and 28°C for which the Group II organisms are suspected (i.e., when examining fish or shellfish; *[11]*). Others recommend 30°C as a compromise to cover all groups, with the proviso that Group I organisms will not be growing optimally *(8)*. Broths should be tested for toxin production using bioassay after 5 d, and an additional 10 d may be necessary to detect the growth from the delayed germination of injured spores. Broths should also be examined microscopically: typical *C. botulinum* in Gram-stained smears have a 'drum stick' or 'tennis racket' appearance with a subterminal spore markedly swelling the vegetative cell.

Broths suspected to contain *C. botulinum* together with foods and clinical samples (especially feces) should be subcultured onto prereduced solid media and incubated in an anaerobic environment. Specific media for *C. botulinum* isolation have been described, which incorporates egg yolk to detect lipase-positive *C. botulinum* colonies *(29)* as well as the selective antimicrobial agents: cycloserine, sulfamethoxazole, and trimethoprim *(29–31)*. Because of the considerable diversity within this species, especially between *C. botulinum* groups I and II, plates with and without added antimicrobial agents should both be inoculated and incubated in parallel. Care should be taken to analyze all suspicious colonies since lipase negative *C. botulinum* occur, although these are rare *(1)*. Typical colonial growth showing a *C. botulinum*, together with an atypical lipase negative culture are shown in Fig 1.

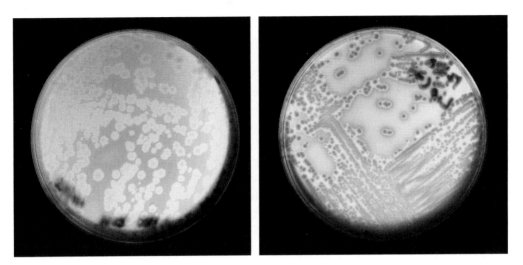

Fig. 1. Typical colonial growth of *C. botulinum*, together with an atypical lipase negative culture. (Please see color insert.)

Classically, *C. botulinum* colonies are usually gray-white on blood containing agars with circular irregular edges, and a clear zone of β-hemolysis. However a large amount of phenotypic variation in colonial appearance should be expected. Purified cultures suspected to be *C. botulinum* should be reinoculated into the broths, and following incubation retested in the bioassay to ensure production of BoNT. Additional methods based on 16S rDNA sequencing and detection of BoNT gene fragments by PCR are increasingly used for identification (*see* Section 2.2.5.). However, because 16S rDNA does not equivocally differentiate *C. botulinum* from its closest relative (e.g., *C. sporogenes* and *C. novyi*), tests for neurotoxin or the presence of neurotoxin genes are still necessary.

2.2.5. Diagnosis of Botulism in Humans

The diagnosis of botulism is based primarily on clinical symptoms and for both foodborne and infant botulism illnesses presents with variable severity, presumably because of the amount of toxin presented to the patient. Common symptoms are diplopia, blurred vision, and bulbar weakness and rapidly progressing symmetric paralysis: a more detailed description of the presentation of this disease is given in Section 2.6. Criteria for a confirmed laboratory diagnosis for foodborne, infant, and wound botulism have been produced (*32*) and these include the detection of botulinum toxin in serum, stool, or patient's food and/or the isolation of *C. botulinum* from stool. Cases are further classified as: confirmed, those with a clinically compatible presentation and occurring among persons who ate the same food as persons who have laboratory-confirmed cases of botulism; and probable, those clinically compatible and with an epidemiologic link (e.g., ingestion of a home-canned food within the previous 48 h) but without laboratory confirmation (*32*). Laboratory confirmation of a clinical diagnosis may not always be possible by the detection of BoNT in patients' body fluids because, an appropriate sample is not always available, the toxin concentration is below the detection limit of the bioassay, or toxin degradation has occurred during transit prior to testing in the laboratory. As stated previously, in 30–40% of patients, the concentration of BoNT is below the detection level for the bioassay *(9)*. Arnon *(4)* summarized results from 21 studies and showed that *C. botulinum* (but not *C. botulinum* neurotoxin) could be recovered from the feces of 1.9% of >1400 individuals feces which were not suspected to have botulism. Thus, the isolation of *C. botulinum* from feces in the absence of this disease symptoms is very rare, and the above case definition with respect to the isolation of the organism is appropriate in almost all circumstances. Difficulties in the laboratory diagnosis of foodborne botulism are illustrated by large outbreaks where despite appropriate testing regimes, none or a very low percentage of laboratory tests provide evidence for the disease *(33,34)*.

All clinical specimens for BoNT detection must be collected prior to the administration of botulinum antitoxin. Suspect foods should also be examined by direct detection of BoNT together with isolation and identification of *C. botulinum* vegetative cells and spores *(8,11)*. In infant botulism, implicated sources are identified by the detection of *C. botulinum* spores in food or in the environment *(8,11)*.

2.2.6. Alternatives to In Vivo Methods

PCR-based assays have been reported for BoNT gene fragments using conventional block-based PCR assays with gel electrophoresis for detection of amplified products *(35–47)*. These assays either detect a single toxin type or suffer from a lack of specificity owing to low melting temperatures of the primers, some requiring subsequent use of hybridization probes to confirm the specific toxin type, hence having shortcomings for the routine identification of cultures or for the detection of the bacterium in enrichment broths. A conventional multiplex PCR assay has been described for BoNT A, B, E and F, and although primers with higher annealing temperatures are used, specific toxin gene detection is still achieved by differential migration of amplified products during agarose gel electrophoresis *(45)*. Advances in nucleic-acid detection technologies have seen a plethora of real-time PCR-based methods for the detection of a wide range of microbial pathogens. Fluorescence-based PCR using hybridization probes allows online monitoring of amplified gene fragments at each cycle of PCR, thus, permitting simultaneous amplification and detection at high sensitivities of pathogen specific nucleic acids within 1–2 h. A real-time PCR assay has been described for a BoNTE *(48)* and for BoNT A, B and E *(49)*. The later method showed that rapid detection of neurotoxin gene fragments could be performed directly on foods, clinical samples, enrichment broths, and bacterial cultures *(49)*. It was also demonstrated that the use of PCR has considerable advantages for the investigation of both foodborne and infant botulism in terms of speed of analysis, nonsubjectivity of testing regimes, reduction in cost and reduction (although not total elimination) in the numbers of animal tests *(49)*.

Considerable efforts have been made in the development and evaluation of immunoassays for the detection of BoNTs, because of the recent increase in interest in bioterrorism agents *(50–54)*. However, despite these efforts, immunoassays have not been found to be suitable alternatives to the mouse for public health applications. This is probably because of both the sensitivity and specificity of the mouse bioassay, and the need to detect relatively small amounts of toxin complex in various matrices. To increase both the sensitivity and specificity of tests, Wictome et al. *(55,56)* described assays based on both the immunoassays as well as the detection of the specific zinc endopeptidase activities of the BoNTs (*see* Section 2.5.). Recent data has shown that the BoNT toxins A–E and F can be unequivocally identified by matrix assisted laser desorption ionization and electrospray mass spectrometry *(57,58)*. The approaches of improved immunoassays, functional in vitro assays, and mass spectrometric approaches have yet to be proven for the detection of these toxins in complex matrices such as serum, feces, and food. However, applications of these technologies to detect other targets suggest that this will be possible in the near future and is likely to be of a level of a similar level of sensitivity to that of the mouse bioassay.

2.2.7. Epidemiological Typing

Despite the realization that *C. botulinum* is widespread in the environment, the use of epidemiological typing systems to track strains of the bacterium through the food chain is relatively underdeveloped, and an association between food sources is assumed on the results of toxin type alone. This is probably adequate where there is a strong epidemiological evidence linking a specific product in a case of food botulism (together with large numbers of the bacterium in incriminated foods); however, typing systems may be invaluable where low levels of the organism are present (as may be the case in infant botulism). From our experience, with a single infant botulism case in the United Kingdom, heterogeneous populations of *C. botulinum* can be present in both the infant's feces and in single packets of implicated food (dried infant-formula feed; *[59]*).

In addition to toxin typing of *C. botulinum*, DNA-based typing methods have been applied to this bacterium and these include pulsed field gel electrophoresis *(19,60,61)*, randomly amplified polymorphic DNA *(62)*, ribotyping *(63,64)*, and amplified fragment length polymorphism analysis *(59)*. These studies indicate a high level of heterogeneity of the bacterium in the environment.

2.3. Reservoirs

C. botulinum is widely distributed in soils and in the sediments of oceans and lakes. However, their types, and the numbers present vary widely. In the Western USA, parts of South America, and China, type A predominates. However, in the Eastern USA, proteolytic type B is most common; and in Europe, nonproteolytic type B. Type E is particularly associated with temperate aquatic environments worldwide (hence, botulism resulting from this type is associated with fish and other seafood), and types C and D in warmer environments. The numbers of the organism present in natural environments also varies widely, some soils being frequently contaminated with >1000 organisms per kg. Examples of the incidence of *C. botulinum* in different soils and sediments are shown in Table 2.

Because of the widespread distribution of this bacterium, foods can contain viable *C. botulinum* spores, but will not cause foodborne botulism unless the organism is able to grow and produce toxin. However, for infant botulism, since viable organisms are required to colonize the gut, *C. botulinum* spores in food (especially, honey and syrup) may act as the reservoir for infection *(65,66)*. In the majority of infant botulism cases, a reservoir has not been identified and other environmental exposures (e.g., spores in dust) have been hypothesized as the reservoir of infection *(65,66)*. A recent case in England identified contaminated dried infant-formula milk as a possible source of infection *(59)*.

Historically, poorly processed canned and bottled foods have been associated with foodborne botulism outbreaks. However, because canning and bottling processes are

Table 2
Occurrence of *C. botulinum* in Soils and Sediments *(7)*

Location	Sample size (g)	% Positive	MPN/kg[a]	BoNT type (% Identified)				
				A	B	C & D	E	F
USA, eastern, soil	10	19	21	12	64	12	12	0
USA, western, soil	10	29	33	62	16	14	8	0
USA, Green Bay, sediment	1	77	1280	0	0	0	100	0
USA, Alaska, soil	1	41	660	0	0	0	100	0
Britain, soil	50	6	2	0	100	0	0	0
Britain, coastal sediment	2	4	18	0	100	0	0	0
Scandinavian, coastal sediment	6	100	>780	0	0	0	100	0
The Netherlands, soil	0.5	94	2500	0	22	46	32	0
Switzerland, soil	12	44	48	28	83	6	0	27
Rome, Italy, soil	7.5	1	2	86	14	0	0	0
Caspian Sea, Iran, sediment	2	17	93	0	8	0	92	0
Sinkiang, China, soil	10	70	25000	47	32	19	2	0
Japan, Hokkaido, soil	5–10	4	4	0	0	0	100	0
Japan, Ishikawa, soil	40–50	56	16	0	0	100	0	0
Brazil, soil	5	35	86	57	7	29	0	7
Paraguay, soil	5	24	10	14	0	14	0	71
South Africa, soil	30	3	1	0	100	0	0	0
Thailand, sediment	10	3	3	0	0	83	17	0
New Zealand, soil	20	55	40	0	0	100	0	0

[a]MPN/kg, highest most probable number/kg of sample.

better controlled by the food industry, home produced products now represent the greatest risk for intoxication. However, swollen or flat cans from any source (flat cans imply loose seams sufficient to allow loss of gas) suggest poor processing and should be considered as likely reservoirs when investigating suspect botulism cases.

2.4. Incidents of Botulism and Risk Factors

2.4.1. Foodborne Incidents

Foodborne outbreaks of botulism occur worldwide and have the potential to cause large morbidity and mortality that require considerable public heath and acute care resources. In addition, the costs for treatment, investigation, food recalls, and any legal actions are likely to be in millions of dollars/pounds even for a relatively small outbreak *(67,68)*. The potential for large outbreaks is illustrated by an outbreak of 91 botulism cases with 18 deaths related to type E and which were associated with traditionally salted fish dish (fesaikh) in Egypt in 1991 *(69)*. Fesaikh is prepared from ungutted fish that, in this outbreak, were suspected to have been salted, warmed (placed in the sun) and then sealed into barrels. This environment was clearly sufficient to allow germination and growth of *C. botulinum* type E, probably in the fish viscera.

As previously stated, foodborne outbreaks are now most often associated with home preservation (e.g., canning, bottling, and preservation in oil) and hence are most common in parts of the United States, central and southern Europe where these preservation practices are more common *(8,70,71)*. In the United States, between 9 and 10 outbreaks of botulism have been reported each year since 1899 *(8,70)*. Home-canned vegetables have traditionally been recognized since the 1950s as the most common cause and were responsible for over 40% of all outbreaks *(8,70)*. Foodborne botulism in an infant has been reported resulting from home-canned infant food *(72)*. Most of the remainder of the cases have been related to readily identifiable poor processing and handling practices as the likely causes allowing the growth of *C. botulinum* (Table 3). The majority of incidents of foodborne botulism in the United States are related to *C. botulinum* type A and to a lesser extent type B, except in Alaska *(8,70)*. Recent changes in 'traditional' fish and marine mammal preservation practices in Alaska have resulted in the highest incidence in this state where 90% of outbreaks are related to type E *(70)*. A list of foods and processing problems for foodborne outbreaks in the United States is shown in Table 3.

All identified foodborne botulism incidents in the United Kingdom are shown in Table 4. The two most recent outbreaks were associated with home-produced products (bottled mushrooms in 1998 and sausage in 2002), both of which were owing to type B and were produced in areas of Europe where botulism is much more common *(150,151)*. The three most recent outbreaks associated with commercially prepared products all had evidence of poor handling and processing. The outbreak associated with tinned salmon in 1978 was caused by a defect in a single can that was contaminated from the factory environment: most probably from factory overalls which were allowed to dry over cans during cooling after being autoclaved *(73)*. The second outbreak was related to a 'shelf-stable' airline meal which was given an unfeasibly long shelf-life which supported the growth of *C. botulinum* type A *(74)*. The third recent outbreak (the largest recorded in the United Kingdom) associated with a commercial product was caused by the consumption of hazelnut yogurt in 1989 *(75)*. This is an unusual vehicle for transmission because the pH of yogurt is normally too low to allow the growth of *C. botulinum*. However, the toxin production occurred in cans of hazelnut conserve which were not sufficiently heated to kill this bacterium and did not contain any other physico-chemical hurdles to prevent the growth of *C. botulinum*. The conserve (together with neurotoxin) was added as a flavoring at the end of the yogurt manufacture, and clearly sufficient toxin retained its biological activity in the acid environment to cause illness.

2.4.2. Infant Botulism

Infant botulism results from the colonization of the immature enteric tract by *C. botulinum* that produces neurotoxin in vivo. As stated previously, risk factors for infant botulism include consumption of honey and corn syrup, age (95% of case occur in the first 6 mo after delivery, the remainder at less than 12 mo), and location (California has the highest rate wordwide). However, the consumption of honey and corn syrup is an identifiable risk factor in <20% of cases *(65)*. Although the disease is orally acquired, it is not clear what proportion is foodborne, or contracted via oral exposure to the environment. The attack rate for infection is likely to be very low and all cases appear sporadic. Almost all cases are owing to *C. botulinum* types A and B *(6)*, with additional

Table 3
Foodborne botulism in the United States, 1990–2000. Adapted from ref. *(7)*

Food type	Incidents	Cases	Comments
Continental USA and Hawaii[a]			
(Total 160 cases: 130 type A, 16 type B, 6 type E, 3 type F, and 5 not known)			
Noncommercial home-canned vegetables	47	70	Likely to be weak acid foods (pH >4.6). Asparagus (9 incidents, 14 cases) and olives (4 cases) were the most common vehicles. Also includes two episodes with garlic in oil which was insufficiently heated to kill *C. botulinum* spores
Noncommercial home-prepared meat products (sausage, pate, beef chili, meatballs, roast beef, hamburger)	7	9	Sausages (3 cases) were the most common vehicle. Failure to refrigerate beef chili after cooking associated with two cases
Noncommercial home prepared other products (salsa, potato salad, bread pudding, soup, apple pie, potatoes, and pickled herring)	14	18	Soup (4 cases) of unspecified type was the most common vehicle. Salsa was responsible for 2 cases and was prepared from raw vegetables and stored at room temperature in an airtight plastic container
Commercial (preserved fish, burrito, and bean dip)	5	10	All outbreaks involved poor handling practices
Restaurant made (cheese sauce and skordalia)	2	25	One outbreak (17 cases) was associated with skordalia which involved potatoes baked in aluminium foil and left at ambient temperature for several days. The second outbreak (8 cases) involved a cheese sauce which was left unrefrigerated
Other and unknown	26	27	One case associated with peyote tea
Alsaka[b]			
(103 cases: 11 type B, 91 type E, and 1 not known)			
Noncommercial preserved fish or marine mammals	49	92	All associated with native foods consisting of whale, beaver, seal, or fish
Unknown	3	11	
Total	**160**	**263**	

[a]Incidence rate <0.1–0.6 cases per million.
[b]Incidence rate 19 cases per million.

very rare cases related to *C. botulinum* type C *(21)*, *C. baratii* (type F) and *C. butyricum* (type E; *[23–26]*).

In the United Kingdom, there have been six cases of infant botulism identified between 1978 and 2001 (Table 5). Infant-formula milk powder was identified as a source of infection in only one case *(59)*. All these six cases have been at the severe end of the spectrum of illness and were ventilated. This raises the possibility of under-recognition in the United Kingdom from the fact that not all cases of infant botulism require ventilation *(3)*.

Table 4
Foodborne Botulism in the United Kingdom 1922–2002

Year	Cases (Deaths)	Home-prepared	Implicated food (Country of origin, if outside the UK)	BoNT type
1922	8 (8)	No	Duck pâté	A
1932	2 (1)	Yes	Rabbit and pigeon broth	NK
1934	1 (0)	Yes	Jugged hare	NK
1935	5 (4)	Yes	Vegetarian nut brawn	A
1935	1 (1)	Yes	Minced meat pie	B
1949	5 (1)	Yes	Macaroni cheese	NK
1955	2 (0)	NK	Pickled fish (Mauritius)	A
1978	4 (2)	No	Canned salmon (USA)	E
1987	1 (0)	No	Rice and vegetable shelf stable airline meal	A
1989	27 (1)	No	Hazelnut yoghurt	B
1998	2 (1)	Yes	Bottled mushrooms (Italy)	B
2002	1 (1)	Yes	Sausage (Poland)	B

NK, not known.
Based on ref. *(140,151)*.

2.5. Pathogenicity and Virulence Factors

2.5.1. Characteristics and Action of Toxins

BoNT is directly responsible for the characteristic clinical symptoms of botulism which arise as a result of blocking of acetylcholine release from peripheral cholinergic sites, particularly neuromuscular junctions leading to typical flaccid paralysis *(14,17)*. All seven BoNTs (A–G) are synthesized as single polypeptide chains and have an approximate molecular weight of 150 kDa. To become biologically active, the single polypeptide is proteolytically cleaved, usually by a botulinum enzyme, to a heavy (~100 kDa) and a light chain (~50 kDa) linked by a single disulfide bond and other noncovalent bonds to give the biologically active toxin (Fig 2.; *[15]*). The BoNT heavy chain contains two functional domains: the C-terminal end, responsible for both receptor binding and internalization; and the N-terminal end, involved in translocation of the light chain across the endocytic membrane into the neuronal cell cytoplasm *(15,16)*. The central region of the light chain shows a high degree of homology conserved in all toxin types and contains the active site of the enzyme, now identified as a zinc endopeptidase, specific for SNARE proteins. SNARE proteins are part of the hosts neurotransmitter release apparatus and when inactivated by the light-chain prevents acetylcholine release.

X-ray crystallography analysis of BoNTA demonstrates the three structurally distinct functional domains, namely, the enzymatic domain corresponding to the light chain, the translocation domain corresponding to the N-terminal part of the heavy chain, and the binding domain composed of two subdomains corresponding to the C-terminal part of the H chain. The enzymatic domain is composed of a mixture of β-sheets and α-helices. The translocation domain, at the N-terminal of the heavy chain, is characterized by two long α-helices that are twisted around each other in coiled-coil-like fashion and a long loop that wraps around the catalytic domain and occludes the active site. The two subdomains comprising the C-terminal of the heavy chain are approximately equivalent

Table 5
Infant Botulism in the United Kingdom

Year	Age (in mo)/Sex	Toxin Type	Laboratory confirmation by analysis of		Ref.
			Feces	Serum	
1978	5.5/F	A	Toxin & organism	ND	*(141)*
1987	4/M	B+F	Toxin & organism	ND	*(142)*
1989	2/F	B	Toxin	ND	*(143)*
1993	4/F	B	Toxin & organism	Toxin	*(144)*
1994	4.5/M	B	Toxin & organism	Toxin	*(145)*
2001	5/F	B	Toxin & organism	ND	*(146)*

M, male; F, female; ND, not detected.

in size and contain predominantly β-sheet structures. The N-terminal subdomain has a so-called "jelly roll fold," whereas the C-terminal subdomain forms a β-trefoil fold and has a β-hairpin at the base of the domain *(15)*.

BoNTs are released from bacterial cells noncovalently complexed with hemagglutinin (HA) and nontoxic, nonhemagglutinin (NTNH) proteins, which, although do not play a role in blocking neurotransmitter release, appear to have a crucial role in protecting BoNT from low pH and enzymic degradation in the human gut *(13)*. BoNT protein complexes dissociate at physiological pH and ionic strength, and thus, intact complex does not reach the peripheral nerve endings and most likely dissociates in the circulation and lymph *(17)*. BoNT complexes vary in size from 300 to 900 kDa depending on toxin type, strain and growth conditions, particularly the availability of iron *(12)*.

Once orally ingested, the BoNT complex is transported across the lumen of the small intestine via intestinal epithelial cells and passes into interstitial fluid and into the circulation. Although transportation across the gut lumen is thought to involve receptor binding, the exact location of the binding domain on the BoNT complex remains controversial with either HA protein or BoNT itself having been proposed as mediating binding *(17)*. On gaining access to the bloodstream and lymph, BoNT is transported to its target site, the neuronal cell. In order to exert its effect, BoNT must leave the vasculature although, at present, it remains to be elucidated if this transportation is by diffusion between the endothelial cells or if an active process of endocytosis and transcytosis is involved *(17)*.

The BoNT mechanism of inhibition of acetylcholine release occurs in four stages: binding, internalization, translocation and intracellular enzymic activity (Fig. 2). Firstly, BoNT binds to the plasma membrane of the neuronal cell followed by internalization by receptor-mediated endocytosis. Then, BoNT is translocated across the endosome membrane, and finally, there is a intracellular endopeptidase action on the SNARE proteins leading to the abolition of endocytosis of acetycholine *(13,17)*. BoNT binds to specific receptors on unmyelinated areas of the presynaptic membrane. Although the exact identity of the BoNT receptor has not been fully elucidated, there is considerable experimental and structural evidence to suggest that binding involves a dual receptor composed of a polysialoganglioside of the G1b series and a specific glycoprotein on the surface of the neuromuscular junction, and it has been proposed for BoNT/B that synaptotagmin II associated with G_{T1b} could be the specific receptor *(13,14,17)*. Different neurotoxins bind to specific receptors as evidenced by the considerable sequence diversity in the binding domain of the heavy chains of the different BoNT serotypes despite their

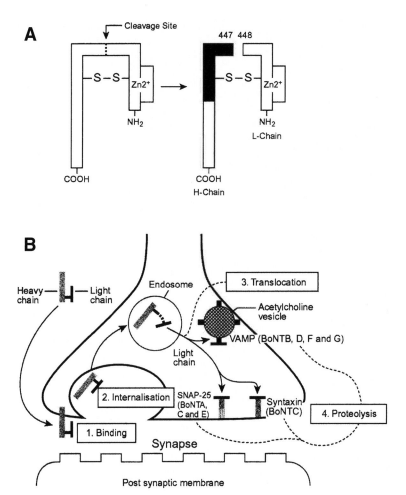

Fig. 2. Schematic representation of the structure and mode of action of *C. botulinum* neurotoxin. (**A**) Cleavage of single neurotoxin peptide to heavy and light chain. The black area (heavy chain) is the translocation domain, and grey the binding domain. (**B**) Schematic mechanism of action of botulinum neurotoxin.

similar structural homology. Receptor specificity together with specificity for the intracellular target accounts for the different sensitivity of animal species to the different serotypes *(13)*. Once bound to its specific receptor, BoNT is internalized by receptor-mediated endocytosis and remains at the nerve terminal where the catalytically active light chain is translocated into the cytosol. The precise mechanism of internalization is not known although it involves clathrin-coated vesicles and is temperature- and energy-dependent *(13,16)*.

During translocation, the light chain of BoNT crosses the endocytic membrane to reach the neuronal cytosol. The process is triggered by acidification of the endocytic vesicle which induces a conformational change in the N-terminal portion (translocation domain) of the heavy chain such that it inserts into the lipid bilayer forming possibly either a channel or cleft which facilitates translocation of the partially unfolded light chain through the membrane *(15,16)*. Although the precise details of the translocation process have yet to be established, at some point both the disulfide bond and noncovalent bonds linking the heavy and light chains are cleaved. Once in the cytoplasm, the light-chain refolds owing to the neutral pH *(13)* and is able to proteolytically cleave highly specific synaptic proteins involved in neurotransmitter release.

The light chain of the BoNT containing metallo-protease activity acts on specific synaptic proteins known as soluble NSF-attachment protein receptors (SNAREs) that are involved in the exocytosis of acetylcholine. SNARE proteins form part of a complex that is involved in fusion of the acetylcholine containing synaptic vesicle with the pre-synaptic membrane. BoNT A and E cleave SNAP-25 (synaptosomal-associated proteins of 25 kDa); BoNTs B, D, F and G cleave synaptobrevin (also known as vesicle-associated membrane or VAMP) and C can act on both the SNAP-25 and syntaxin. Each clostridial neurotoxin recognizes its substrate at specific binding sites (SNARE motifs) and each cleaves a different peptide bond even when the substrate is the same *(17)*. SNARE proteins must be in the uncomplexed form for BoNT action to occur and although cleavage does not prevent formation of the SNARE complex, it results in a nonfunctional complex in which the uncoupling of Ca^{2+} influx and fusion prevents neurotransmitter being released *(13,15)*. The duration of action of the different BoNTs varies significantly with BoNTA having the most persistent action. Duration of BoNT action may be effected by differences in either the persistence of BoNT LC in the cytosol and or persistence of the specific cleaved protein in the SNARE complex. An important factor effecting recovery from BoNT is how quickly and successfully nerve sprouting occurs.

2.5.2. Molecular Genetics of Clostridium botulinum *Neurotoxins*

C. botulinum types A, B, E and F neurotoxin genes are encoded chromosomally, whereas in types C and D the genes are found on a bacteriophage (Table 1). The B*oNT* genes are clustered in close proximity to the nontoxic nonhemagglutinin (*NTNH*) and in most strains with hemagglutanin (HA) genes, forming what is known as the botulinum neurotoxin locus. A positive regulatory gene *bot*R coding a 21–22 kDa protein is associated with the neurotoxin gene locus. *C. botulinum* strains usually produce only one type of neurotoxin and the neurotoxin locus is present in only one copy on the genome. However, in a few strains, two toxin types are synthesized, one usually being produced in excess of the other. These strains contain two BoNT genes. Some *C. botulinum* type A strains have silent *bont* B genes *(76)*, which, although present on the genome with a fully functional BoNTA gene, contain several mutations within the coding region leading to transcription disruption.

The ability of clostridial species other than botulinum to produce neurotoxin, the presence of more than one *bont* gene in some botulinum strains, as well as the similarity between BoNT and tetanus toxin, indicates that bont genes are derived from a common ancestor and are transferable between different clostridial strains; it is likely that this is mediated by mobile genetic elements *(18)*.

2.6. Clinical Characteristics

Early symptoms of foodborne or infant botulism affect the gastrointestinal tract and often involve nausea, abdominal pain, vomiting and diarrhea followed by constipation. Neurological symptoms then follow and are the same, irrespective of the route of entry of botulinum neurotoxin. The classical picture is of descending symmetrical flaccid paralysis, with no fever. Early symptoms include cranial nerve palsies—blurred and double vision (diplopia), difficulty in focusing, drooping eyelids (ptosis), facial weakness, sluggishly reacting or enlarged pupils, difficulty in swallowing (dysphagia), difficulty in speaking (dysphonia) and slurred speech (dysarthria). Weakness in the neck and arms, loss of the gag reflex, weakness in lower limbs, and respiratory paralysis may follow. There may be autonomic signs with dry mouth, fixed or dilated pupils, and gastro-intestinal, urinary, and cardiovascular dysfunction. Altered sensory awareness and fever are not associated with botulism. Deep tendon reflexes may decrease over time in some cases. Initial differential diagnoses from foodborne botulism includes: Guillain-Barré syndrome, myasthenia gravis, spinal/paralytic poliomyelitis, viral or bacterial encephalitis, rabies, cerebrovascular accident, tick paralysis, paralytic shellfish poisoning, diphtheria, chemical intoxications (carbon monoxide, barium, methyl chloride, methyl alcohol, organic phosphorus, and atropine), mushroom poisoning, and reaction to pharmaceutical compounds such as antibiotics (151,8).

For foodborne botulism, the incubation period ranges from 6 h to 10 d, but symptoms are generally present 18–36 h after consumption of contaminated food. In both foodborne and infant botulism there is a spectrum of disease: severity being related to the amount of neurotoxin. If illness is severe, involvement of respiratory muscles occurs, and venti-latory failure and death can occur unless supportive care is provided. Recovery follows regeneration of new neuromuscular connections and 2–8 wk of ventilatory support is common: some patients may require >7 mo before the return of muscular function.

The presentation of infant botulism is very similar to foodborne disease; however, since the infants are unable to complain, symptoms often develop suddenly. In mild cases, weakness, lethargy, and reduced feeding occur. In severe infant botulism cases weak-ened cry, suck, and swallowing are observed together with muscle weakness, diminished gag reflex, and loss of head control. Infants are described as "floppy". In infants, sepsis (especially meningitis), electrolyte-mineral imbalance, metabolic encephalopathy, Reye syndrome, Werdnig–Hoffman disease, congenital myopathy, and Leigh disease should also be considered (8).

Although intestinal colonization by *C. botulinum* is classically associated with infants of <1 yr old, this has also been described, albeit very rarely, in an older infant and among adults (77–82). The diagnosis is supported by prolonged excretion of toxin, *C. botulinum* in the feces and an absence of a specific contaminated food vehicle. A single case has been reported in a 3-yr-old neuroblastoma patient following intense immuno-suppression and antibiotic treatment (83). Specific risk-factors among adults include Chron's disease, gastrointestinal surgery, and prior antibiotic treatment. A cases of *C. baratii* intestinal colonization in an adult has been reported (79). The reservoirs of infec-tion in these cases of adult colonization are equally obscure as for the majority of infant botulism cases, but clearly food and other environmental sources such as dust are the likely candidates.

Similarities exist between infant botulism and sudden infant death syndrome (SIDS) and causal link has been suggested (6). Botulinum neurotoxin (especially type E) was detected in the feces of 12% of SIDS infants (84) although this was not a universal finding (85) and this connection remains controversial.

As noted previously, most cases of botulism are related to BoNT A and BoNT B, BoNTE in which there is an association with marine products, and occasionally BoNT F

(1,6,9). In addition, a single case of infant botulism owing to *C. botulinum* type C *(21)* together with food and infant botulism related to *C. baratii* (type F), and *C. butyricum* (type E; *[23–26]*) have been described. However, there are no recognized differences between the clinical characteristics of botulism and the toxin type or the *Clostridium* species.

2.7. Treatment and Control

The mainstay of treatment of foodborne botulism is the inactivation of toxin in the patient by intravenous administration of equine antitoxin. Early administration is essential because the action is directed towards free toxin which has not yet bound to nerve endings. A single dose is given because the antibodies have a half life of 5–8 d in the patients serum. Treatment is not without risk because up to 9% of patients are hypersensitive to horse serum; an initial skin test prior to administration is advised to establish this. In contrast with foodborne botulism, treatment with equine antitoxin is not recommended in infants because of the possibility of serum sickness. A human derived antitoxin product has been used in the United States *(86)*. Treatment of infant botulism is primarily through meticulous supportive care. Antibiotic treatment has no role in the treatment of foodborne or infant botulism because this may increase toxin release.

Because the administration of antitoxin is the only specific therapy for foodborne botulism, and that this is only effective early in the course of neurological dysfunction *(87)*, the decision for treatment is almost always based on clinical observation, case history, and physical findings; and should not be delayed for the results of laboratory tests.

The control of *C. botulinum* in the food chain relies on the killing of organisms in foods that will support the growth of this bacterium, or the formulation of food ingredients and processes to prevent growth. Although many foods satisfy the nutritional requirements for the growth of *C. botulinum,* not all of these have the necessary anaerobic environments to support growth. Anaerobic requirements are, however, supplied in many canned and bottled vegetables, meat, and fish products, and these should either be subjected to a botulinum 12D 'kill' or have one or more of the following to prevent growth: low pH, low water activity, high salt, high sodium nitrite, and/or other preservatives. Refrigeration will not prevent growth and toxin production unless the temperature is kept below 3°C. Foods processed to prevent spoilage but not refrigerated are the most common vehicles of botulism. Because of the association of infant botulism with honey, and since this food does not undergo a botulinum 'cook,' some countries and manufacturers recommend not feeding honey to infants less than 1 yr old.

Because of the extremely protracted course of both foodborne and infant botulism, the cost of treatment can be considerable. For example, it was estimated that an average cost for hospitalization and treatment of a single infant botulism case was $80,000, with those infants with most protracted illness (>10 mo hospitalization) to be >$635,000 *(6)*.

2.8. Summary and Conclusions

Foodborne botulism results from the ingestion of preformed toxin in contaminated foods which may still be available to cause illness in others and a single case can precede a severe public health emergency. It is, therefore, critical for acute care clinicians to be able to rapidly recognize botulinum cases and to alert public health microbiologists to investigate incidences, identify implicated food sources, and advise those responsible

Table 6
Diseases Caused by *C. perfringens*

C. perfringens type	Diseases in	
	Humans	Other animals
A	Gas gangrene; foodborne, antibiotic-associated, and infectious diarrhea	Enterotoxemia (lambs, cattle, goats, horses, dogs, alpacas, kangaroos, pigs, rabbits, reindeer, and others) Necrotic enteritis (domestic and wild birds) Acute gasrtic dilation (nonhuman primates and others)
B, D, and E	None identified	Enteritis and enterotoxemia (sheep, goats, guinea pigs, rabbits, and others)
C	Necrotizing enteritis (jejunitis)	Enterotoxemia (lambs) Necrotic enteritis (piglets, lambs, calves, foals, and birds)

for food regulation on rapid withdrawal of implicated product and appropriate additional interventions. Identical considerations apply to a deliberate release (bioterrorism) incident if botulinum toxin or *C. botulinum* (or indeed another species of bacteria engineered to produce botulinum neurotoxins) were to be introduced into the food chain. The use of a mouse bioassay to detect active toxin has been the mainstay for laboratory diagnosis and identification of this very diverse group of bacteria; however, molecular biological advances (either for the detection of neurotoxin genes by PCR, detection of toxin by immunoassay, or the detection of biologically active toxins using specific in vitro assays) are likely to become more widespread in the future.

Infant botulism is also a foodborne disease, but occurs via the production of neurotoxin in the immature intestinal tract. The epidemiology is less well-understood than foodborne botulism, but the molecular advances outlined above may also contribute to improved diagnosis and a greater understanding of the disease.

3. CLOSTRIDIUM PERFRINGENS

3.1. Introduction

Clostridium perfringens is probably one of the most widely occurring bacterial pathogens and is ubiquitous in the environment and as part of the normal intestinal flora of humans and other animals *(1)*. This bacterium causes a wide range of diseases in humans, indeed there is hardly a site that has not been reported as infected by *C. perfringens* especially following penetrating wounds *(88)*. The most commonly occurring *C. perfringens* infections are summarized in Table 6.

The pioneering work of Hobbs et al. *(89)* extended earlier observation and firmly established this bacterium as a frequent cause of foodborne diarrheal disease (*see* Section 3.4.1.). Among those diseases which are transmitted to humans via the oral route it is now known that *C. perfringens* (type A) is not only responsible for diarrhea associated with the consumption of contaminated food as originally described by Hobbs et al. *(89)*, but also antibiotic associated and infectious diarrhea. *C. perfringens* (type C) is also a cause of necrotizing enterocolitis (Pig-bel or Darmbrand) in humans which

Table 7
Estimated Illness Resulting From 10 Most Common Foodborne Pathogens in the United States, and England and Wales. Adapted for refs. *(92,93)*

	Numbers of cases		
	Total	Hospitalizations	Deaths
USA (After 1993)			
Total	**76,000,000**	**323,000**	**5200**
Norovirus	23,000,000	50,000	310
Rotavirus	3,900,000	50,000	30
Astrovirus	3,900,000	12,500	10
Campylobacter	2,453,926	13,174	124
Giardia	2,000,000	5,000	10
Salmonella	1,412,498	16,430	582
Shigella	448,240	6,231	70
Cryptosporidium	300,000	1,989	66
C. perfringens	**248,520**	**41**	**7**
Toxoplasma gondii	225,000	5,000	750
England and Wales (After 1992)			
Total	**1,338,772**	**20,759**	**480**
Campylobacter	359,466	16,946	86
C. perfringens	**84,081**	**354**	**89**
Norovirus	57,781	37	9
Yersinia	45,144	216	1
Salmonella	41,616	1,516	119
Astrovirus	17,291	12	4
Bacillus	11,144	27	0
Rotavirus	8,979	46	4
Staphylococcus aureus	2,276	57	0
Cryptosporidium	2,063	39	3
Giardia	1,673	5	0

is also foodborne. This bacterium also causes enteric infections in a very wide range of other animals (Table 6). A more detailed description of the intestinal diseases in humans is given in later sections.

Studies of infectious intestinal disease in the 1990s identified *C. perfringens* as responsible for diarrhea in 4 and 2.3% of patients in the community in England and Wales *(152)* and in Holland *(90)*, respectively. The study in England and Wales also estimated that for each case identified in the laboratory, an additional 186 cases present to a medical practitioner, and 343 cases occur in the community *(91)*. Thus demonstrating that this disease is considerably underdiagnosed. Further analysis in England and Wales *(92)* and in the United States *(93)* estimated *C. perfringens* as the second and the ninth most common foodborne diseases, respectively (Table 7). Adak et al. *(92)* further estimated that *C. perfringens* was responsible for 2% of the total hospitalizations and 18% of all the deaths from preventable foodborne diseases in England and Wales (Table 7).

More detailed reviews about *C. perfringens* and gastrointestinal disease can be found in the following literature *(94–100)*.

3.2. Classification and Identification

3.2.1. Taxonomy

C. perfringens is a Gram-positive, endospore-forming, encapsulated, nonmotile anaerobic bacillus: straight rods with blunt ends occurring singly or in pairs are produced, which are 0.6–2.5 μm wide and 1.3–19 μm long. The bacterium sporulates extremely poorly in laboratory media; but when this occurs, endospores are large oval central or subterminal and distend the cell *(101)*. The organism has a G+C range of 24–27 mol% and on the basis of 16S rDNA sequence, the organism belongs to the *Clostridium* cluster I which is typical for other low G+C species within this genus *(102)*. The bacterium is divided into five different types (A–E) according to the range of 'major' toxins produced (Table 8): the five toxin types can not be distinguished on the basis of cellular and colonial morphology, biochemical reactions, fatty acid, end products of metabolism, or chromosomal arrangement *(101)*. The majority of human infection is related to type A *(1)*. A further range of other 'minor' toxins (gamma, delta, eta, theta, kappa, lambda, mu, nu and neuraminidase) together with the enterotoxin are produced by *C. perfringens* *(97)*. Analysis of the *C. perfringens* genome also identified additional putative toxin genes, including five distinct hyaluronidase genes *(103)*. The role of the enterotoxin and β-toxin in enteric infection are outlined in Section 3.5.

Methods for *C. perfringens* toxin typing were developed in the 1930s and involve the use of specific antitoxin (antisera) in neutralization tests in mice *(97)*. However, these tests are now rarely performed, not least because of ethic issues resulting in the reduction of tests using animals and difficulty in the supply of antitoxin. DNA-based techniques (hybridization and PCR) are now more usually performed *(104,105)*. The presence of toxin genes, however, is not necessarily equivalent to the tests for toxin production: for example, the enterotoxin gene (*cpe*) has been detected in *C. perfringens* type E, but this was shown not to be expressed *(106)*.

C. perfringens is synonymous with *Clostridium welchii*, the latter being more frequently cited in historic UK literature. Precedence was given to *C. perfringens* because this species name was first used.

A complete genome sequence is available for a single *C. perfringens* type A: this strain was originally isolated from a case of gas gangrene and does not contain the enterotoxin gene *(103)*. The genome is 3.03 Mbp in size, has a G+C content of 28.6 mol% and encodes 2660 genes.

3.2.2. Identification

Colonies of *C. perfringens* are circular, semi translucent, smooth, with an entire edge and growth of 1–2 mm in diameter after 24 h. Colonies also appear umbonate with radial striations and a crenated edge, although these are less common. On horse blood agar, colonies show a double zone of hemolysis. The inner clear zone owing to the θ-toxin, the outer more hazy zone owing to the α-toxin *(107)*. Nonhemolytic strains occur.

Failure to grow aerobically, but not anaerobically, on a suitable medium (e.g., blood agar) incubated at 35 or 37°C for 18–24 h is often used as an initial screening for the identification of suspect *C. perfringens* colonies. Clinical laboratories traditionally relied on the Nagler reaction for identification of *C. perfringens* *(153)*. Nagler media contain either a human serum or an egg yolk emulsion, and a dense white opalescence is obtained around *C. perfringens* colonial growth resulting from the activity of the

Table 8
Patterns of Toxin Production for Differentiation of *C. perfringens* Types A–E

C. perfringens Type	Toxin types				
	α	β	ε	ι	Entero[a]
A	+	−	−	−	+
B	+	+	+	−	+
C	+	+	−	−	+
D	+	−	+	−	+
E	+	−	−	+	(+)
Location of toxin	Chromosomal	Plasmid	Plasmid	Plasmid	Chromosomal or plasmid

[a]Not all some strains carry the enterotoxin; enterotoxin in type E cultures is not expressed.

α-toxin is observed, which is inhibited by specific antisera. Nagler plates are prepared and half-treated with the antisera, and suspect colonies are streaked perpendicularly to the antitoxin (Fig. 3). A positive reaction is recognized when the inhibition of the lecithinase (opalescence) occurs in the region of the Nagler plate treated with antitoxin. However, because of the difficulties in obtaining good quality antisera and the availability of alternative methods (*see* final paragraph, this section), the Nagler reaction is now rarely performed. Nagler reaction negative *C. perfringens* have been described which were associated with incidents of food poisoning *(107)*: later analysis by the authors showed that these strains do contain at least a fragment of the α-toxin gene which is not expressed in a biologically active form (Food Safety Microbiology Laboratory, unpublished data).

Identification of *C. perfringens* in both food and clinical laboratories relies on British and European Standard Methods *(108)* which utilize motility nitrate and lactose gelatine media. On lactose gelatine agar, *C. perfringens* ferments lactose (as indicated by a color change of phenol red to yellow resulting from the acidification) and gelatine is liquified when the media is cooled to 5°C for 1 h. In nitrate motility agar slopes, *C. perfringens* reduces nitrate to nitrite, and is nonmotile because the growth occurs only along the site of stab inoculation.

Because all *C. perfringens* produce the α-toxin (as detected in the Nagler reaction) detection of the α-toxin gene is, therefore, a suitable method for identification. A variety of PCR-based procedures have been described for amplification of this gene *(105)*. It is the authors' experience that a simple duplex PCR for both the α-toxins and enterotoxins *(109)* correlates very well with the results of identification using Nagler reactions as well as motility nitrate and lactose gelatine agars (unpublished data). The PCR approach has considerable advantages in terms of the speed of results without the need to manipulate pure cultures and advances in nucleic-acid detection technologies (including real-time PCR procedures) are likely to considerable increase the speed of throughput in the near future.

3.2.3. Selective Isolation

For investigation of food poisoning incidents, samples of feces should be collected as soon as possible and examined for the presence of enterotoxin (*see* Section 3.2.5.) and for the presence of large numbers of the bacterium. *C. perfringens* is readily isolated from stool specimens by plating onto neomycin blood agar followed by overnight

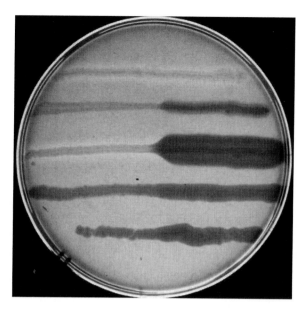

Fig. 3. Nagler reaction for the identification of *C. perfringens*. Antisera is spread on the left-hand side of the plate and the right-hand side is untreated. The organisms are (from top to bottom): *C. perfringens* (no reaction, a typical strain), *C. perfringens* (positive reaction), *C. perfringens* (positive reaction), *C. absonum* (negative reaction, lecithinase-positive but not inhibited by the antitoxin), and *C. barattii* (negative reaction, lecithinase positive but not inhibited by the antitoxin). (Please see color insert.)

anaerobic incubation. Quantification of spores and vegetative cells is important because large numbers of spores (>10^5/g of feces) are usually only found following *C. perfringens* food poisoning. The method outlined in Table 9 given below is that recommended by the Public Health Laboratory Service *(147)*.

Collection and quantitative analysis for the presence of *C. perfringens* in suspect foods is also important for the investigation of food poisoning. The current European Standard Method *(108)* recommends the use of a pour plate technique with tryptose sulfite cycloserine (TSC) agar incubated anaerobically overnight (Table 9). Suspect (sulphite-reducing black) colonies are enumerated and identified. Because the numbers of vegetative *C. perfringens* decline rapidly in some foods (especially those which have been refrigerated) enrichment in cooked meat broths or reinforced clostridial medium broths may be necessary if there is a significant delay before examination.

As outlined above, the conventional methods for isolation of *C. perfringens* from clinical and food material often differ, and it is the authors' experience that these may select different subpopulations of the bacterium. Only enterotoxin containing *C. perfringens* are responsible for food poisoning, but these may be present with large numbers of other nonenterotoxigenic in both food and feces.

3.2.4. Epidemiological Typing

Hobbs et al. *(89)* originally reported the usefulness of serotyping *C. perfringens* using agglutination reactions against specific antisera. This approach was extended by combining schemes from both the United Kingdom with those from the United States and Japan, and over 100 serotypes of *C. perfringens* were recognized *(110)*.

Table 9
Selective Isolation Methods for *C. perfringens* From Feces and Food

I. Examination of feces[a]
   a. Spore count dilution series
      1. Prepare a 1:5 dilution of feces in PBS (minimum 0.1 g feces to 0.5 mL PBS)
      2. Add equal volume of 95% (v:v) ethanol
      3. Leave for 30 min at room temperature
      4. Dilute to 1:1000 in PBS
   b. Vegetative cell count dilution series
      5. Prepare 1:10 and 1:100 dilution of feces in PBS
   c. Enumeration procedure
      6. Inoculate 0.1 mL of each spore count and vegetative cell count dilution series onto
         neomycin (75 mg/L) fastidious anaerobe blood agar
      7. Incubate anaerobically at 35–37°C for 16–24 h and enumerate presumptive
         C. perfringens colonies
II. Examination of foods[b]
      1. Homogenize food in MRD and prepare decimal dilution series
      2. Add 1 mL to an empty Petri dish and pour 10–15 mL of Moulton TSC AGAR, mix well
      3. Allow to solidify and add a 10 mL overlayer of TSC agar
      4. Allow to solidify and incubate anaerobically at 35 or 37°C for 20 h
      5. Enumerate sulphite reducing black colonies with halos (presumptive *C. perfringens*)
         and identify at least five colonies

   Abbreviations: PBS, sterile phosphate-buffered saline; MRD, maximum recovery diluent; TSC, tryptose sulfite cycloserine.
   [a]Public Health Laboratory Service Standard Operating Procedure *(8)*.
   [b]British and European Standard Methods *(108)*.

   However, there are difficulties over the control and production of reagents, stability of types and because >20% of isolates are nontypeable, alternative subtyping methods have been described. Molecular typing techniques for *C. perfringens* have included ribotyping *(111)*, plasmid profiling *(111)*, pulsed field gel electrophoresis (PFGE; Maslanka et al., 1999, Ridell et al., 1998), randomly amplified polymorphic DNA (RAPD; *[111]*) and amplified fragment-length polymorphism (AFLP) analyses (McLauchlin et al., 2000).

   As outlined in later sections, establishing a diagnosis of *C. perfringens* food poisoning can be problematic. Epidemiological typing (together with enterotoxin gene typing), and enterotoxin detection are essential for the investigation of food poisoning outbreaks.

3.2.5. Enterotoxin Detection

   Several biological assays have been described for the detection of enterotoxin. These include the use of rabbit ileal loop, erythemal activity, lethality in mice *(112)*, and the use of vero cells growing in vitro *(113)*. Because of relative insensitivity of these methods and the need to reduce the use of experimental animals, the analysis of cytotoxicity to vero cells growing in vitro is the principal biological assay for enterotoxin.

   Enterotoxin is now almost universally detected using immunoassays. Initial formats for detection included gel-diffusion methods and latex-agglutination tests. These tests have been replaced by other immunoassays of higher sensitivities, which includes commercially available reverse passive latex agglutination (RPLA; *[113]*) and an ELISA test *(114)*. The relative sensitivities of vero cell assays, RPLA, and ELISA have been reported to be 40, 4 and 4 ng/g of feces *(113,115)*; the RPLA has the disadvantage of requiring some subjectivity in reading results and some nonspecific reactions.

Although not strictly a method for enterotoxin detection, PCR has been used for detection of *C. perfringens* enterotoxin gene fragments directly in feces *(109,116)*. Analysis by PCR was of similar sensitivity to the use of immunoassays *(116)*, and from our experience, this can act as a suitable screening test for the presence of enterotoxigenic *C. perfringens* in feces.

3.3. Reservoirs

C. perfringens is ubiquitous in the environment *(95)*. For example, it occurs in soils (commonly at 10^3–10^4/g), in foods (e.g., in at least 50% of raw and frozen meat), is common in dust and the intestinal tracts of humans and animals (human feces, especially in the elderly, commonly contains 10^3–10^6/g). However, there is increasing evidence that this superficial understanding does not reflect a more complex microbiol ecology of subpopulations within this bacterial species. As has already been outlined, *C. perfringens* enterotoxin is required for both the food poisoning and the antibiotic and infectious diarrhea, and not all representatives of this bacterial species encode this gene. Furthermore, analysis of isolates from food poisoning outbreaks in both North America and Europe found that these had the enterotoxin gene chromosomally encoded, whereas the enterotoxin gene was found on a plasmid among unrelated isolates from antibiotic-associated and sporadic diarrhea *(117,118)*. This relationship may not be absolute and a *C. perfringens* strains with a plasmid-encoded enterotoxin has been identified as associated with an outbreak of food poisoning in Japan *(119)*. A comparison of *C. perfringens* with plasmid and chromosomally encoded enterotoxin showed that the former were considerably less heat-resistant than the latter *(120)*. The D values of *C. perfringens* with plasmid and chromosomally encoded enterotoxin for vegetative cells at 55°C in 5–9 min as compared to 12–16 min, respectively, and for spores at 100°C, 0.5–1.9 min as compared to 30–124 min, respectively *(120)*. Furthermore, unrelated *C. perfringens* which produce the enterotoxin were more heat-resistant than those that do not *(121)*: the location of the enterotoxin gene for these isolates was not determined. Recent surveys of foods on retail sale in the United States have shown that >98% of *C. perfringens* do not contain the enterotoxin gene *(122,123)*. The minority of *C. perfringens* from retail foods that did contain the enterotoxin (and, therefore, had the potential to cause food poisoning) all were chromosomally encoded and all had a high heat resistance.

Hence, the recent data have revealed a further level of complexity of the distribution of *C. perfringens* type A. It has been suggested that selection of food poisoning *C. perfringens* may occur in kitchens because of their higher heat resistance. Although this selection may occur, the increased heat resistance appears to be an intrinsic trait of these strains even when directly isolated from food and food components *(157)*. Further work is required to better understand the reservoirs of the *C. perfringens* type A causing enteric infections.

Reservoirs for *C. perfringens* type C are equally intriguing. It is generally believed that *C. perfringens* type C is an obligate parasite of animals' intestinal tracts *(1)*. However when investigating Pig-bel in Papua New Guinea, Walker *(124)* reported that despite readily finding *C. perfringens* type C in the feces of affected individuals, conventional

Table 10
The Principal Finding of the Seminal 1953 Publication From Hobbs et al. *(89)*

A large proportion of food poisoning of unknown origin may be due to heat resistant
  C. perfringens
Almost all outbreaks are caused by meat which has been cooked and allowed to cool slowly
Colic, diarrhea but rarely vomiting occur 8–20 h after consumption
Large numbers of *C. perfringens* are found in the food and feces of patients
Heat resistant *C. perfringens* are commonly found in the feces of animals, and this is
  possibly the source of infection
Results of human volunteer experiments with cooked meat broths showed that:
  diarrhea occured after consumption of broths with viable cultures;
  no diarrhea occured after consumption of uninoculated broths or filtrates of broths containing
  viable cultures

microbiological culture methods of the environment revealed only *C. perfringens* type A. Immunofluorescent assays, however, showed that *C. perfringens* type C was widespread in the feces of unaffected individuals, as well as the soil, in the ovens where meat is cooked. Walker *(124)* further speculated that *C. perfringens* type C is part of the normal fecal flora, or an organism with transient passage through the enteric tract, which in both instances only causes disease under specific circumstances of unusual, increased protein consumption in the diet together with low proteinase activity in the gut (*see* Section 3.4.3.). However, the true reservoirs of both enterotoxigenic *C. perfringens* type A and *C. perfringens* type C appear poorly understood.

3.4. Foodborne Outbreaks

3.4.1. Food Poisoning

The pioneering work of Hobbs et al. *(89)* established *C. perfringens* as a cause of the food poisoning of previously unknown origin. This publication described the disease together with likely food vehicles, established methods for diagnosis, and by human volunteer feeding experiments showing that viable organisms were essential for establishing the diarrhea. The principal finding of this seminal publication is shown in Table 10. The significance of the heat resistance (spore-forming ability) correctly identified by Hobbs et al. *(89)* together with the rapid growth rates at a wide range of temperatures are the features that allow *C. perfringens* to multiply and survive in food and cooking processes. Because of the nutritional requirement of this bacterium (*C. perfringens* requires 13 amino acids to grow), and the need for anaerobic environments, multiplication is particularly associated with protein rich foods, especially those containing meat. Features of *C. perfringens* that favor transmission through food are summarized in Table 11.

C. perfringens food poisoning presents as a profuse diarrheal illness often with abdominal pain 8–24 h after the ingestion of heavily contaminated (usually >10^6 vegetative *C. perfringens* per g or mL) of food. The symptoms usually subside within 10–24 h, and fatalities, although rare, can occur especially in the elderly and immuno-compromised persons. Disease symptoms are caused by the production of enterotoxin (*cpe*) in the intestine from sporulating organism ingested from food *(1)*. Data on >1500 outbreaks of *C. perfringens* food poisoning in the United Kingdom has been collected

Table 11
Properties of *C. perfringens* That Favor Transmission Through Food

Conditions	Properties of *C. perfringens*
Growth temperature	
Range	15–52°C
Optimum	43–45°C
Generation time at 33–49°C	8–20 min
Heat resistance[a]	
Minimum	20 min at 65°C
Maximum	3 h at 100°C
The pH range for growth	6.0–7.0
Minimum a_w for growth	0.95–0.97
Growth in food	Rapid growth in a wide range of highly protein-aceous foods, especially meat based products
Atmosphere for growth	Anaerobic conditions including some modified atmosphere and sous vide products
Distribution	Widespread in the environment, foods and food ingredients, and gastrointestinal tract of animals

[a]Higher heat resistance has been associated with *C. perfringens* strains with a chromosomally encoded enterotoxin gene *(120)*.

([125]; Food Safety Microbiology Laboratory, unpublished data) and are summarized in Table 12. The majority of outbreaks were associated with, meat or poultry, larger scale institutionalized catering, and with poor temperature and time control prior to serving. Most outbreaks were small (less than 50 cases) but 3 and 0.3% involved >200 and >600 cases, respectively (Table 12). Similar (albeit not so extensive) epidemiological data is available for the United States *(126–129)*.

Examples of two typical outbreaks are given here. An explosive outbreak of diarrhea affecting 170 patients occurred in a mental hospital from which three of the patients died. Patients had eaten a roast lamb which had been cooked, stored at room temperature for 20 h and served cold. The same type of *C. perfringens* was recovered from, the lamb dish, patients' feces, and post-mortem samples. The storage of the lamb provided ample time for the growth of the organism which was not killed by thorough reheating of the meat (Food Safety Microbiology Laboratory, unpublished data). In a second outbreak, 36 people were ill with diarrhea following the meal at a public house. Food samples were not available for examination but six large rolled turkey joints had been consumed, cooked to a core temperature of 73°C, cooled, and stored in the refrigerator for up to 36 h. Prior to serving, the meat was reheated with the gravy prepared from meat stock. (Food Safety Microbiology Laboratory, unpublished data). This outbreak occurred during a busy Christmas period when refrigeration was more likely to be poorly controlled. In this outbreak either the meat or the gravy was most likely to have been responsible. Large rolls of meat can be problematic to cook and cool because of their size and the fact that the external contamination becomes located in their centre. It is recommended that the stock is not used to make gravy because it provides ideal conditions for the growth of *C. perfringens* and a potent inoculum for other food items.

Table 12
**Features of 1566 Cases of *C. perfringens* Food Poisoning Outbreaks
in the United Kingdom, 1979–1996**

Number of outbreaks	1566
Number of cases per outbreak	
<50	88.5%
50–99	8.0%
100–299	3.1%
300–499	0.4%
Place where outbreak occurred	
Residential institutions (hospital, old peoples' home, hotel, nursing home, prison, oil rig, ship, and other)	42%
Canteen in nonresidential settings (school, nursery, university, and other)	22%
Commercial catering (restaurant, hotel, wedding, and other functions)	30%
Private house and other	6%
Implicated foods	
Beef	37%
Pork	11%
Lamb	9%
Other meat	16%
Chicken	12%
Turkey	10%
Duck	1%
Soup/stock/gravy	2%
Other	3%
Possible contributing factors (more than one factor may be associated with each outbreak)	
Inadequate:	
Thawing	9%
Cooking	22%
Cooling	62%
Re-heating	49%
Preparation too far in advance	69%
Too large quantities prepared	20%
Use of leftovers food	4%
Cross-contamination	3%

Based on ref. *(125)* and HPA Food Safety Microbiology Laboratory (unpublished).

Outbreaks can be associated with more unusual types of food. An outbreak of 56 cases of diarrhea occurred in an English hotel and was associated with boiled salmon that had been left at room temperature prior to serving cold in the following day *(130)*. The possibility of foods other than meat or fish as vehicles is illustrated by an outbreak in an old peoples' home in Japan where 30 cases of diarrhea occurred after consumption of spinach and fried bean curd dish *(131)*. The food had been cooked, stored at room temperature, and served in the following day after a brief reheat *(131)*. Furthermore, an outbreak with 77 cases of diarrhea occurred in a school, 2–10 h after the consumption of a milk shake. The milk shake was boiled and left for 20 h at room temperature before consumption. *C. perfringens* was isolated from both milk and feces, and is likely to multiply during the slow cooling of large volumes of food or beverages (Food Safety Microbiology Laboratory, unpublished data).

3.4.2. Infectious and Antibiotic-Associated Diarrhea

C. perfringens type A also causes enterotoxin-mediated infectious or antibiotic-associated diarrhea which differs from the classical food-poisoning described by Hobbs et al. *(89)*. Infectious and antibiotic-associated diarrhea is generally more profuse and of longer duration than classical food poisoning: blood and mucus occur in the stools which is also unusual in the classical disease *(132)*. This form of the infection occurs almost exclusively in the elderly, and hospital-acquired cases are often associated with antibiotic treatment *(132)*. The recovery of the bacterium from hospital environments and hands suggests that the disease may be truly infectious. *(132)*. However, carriage of the bacterium is also important in this disease, and although food may also play a role in transmission, this has not been demonstrated.

3.4.3. Necrotizing Enteritis

Necrotizing enteritis resulting from *C. perfringens* type C is very rare in the industrialized world *(133)*. The incubation time is 5–6 h with acute sudden-onset severe lower-abdominal pain, bloody diarrhea, and vomiting. This is in contrast to type A disease in which the diarrhea without vomiting predominates. During necrotizing enteritis, necrotic inflammation of the jejunum and small intestines occurs, and if not treated, it can have a mortality rate of >85%. The disease is mainly owing to the production of β-toxin in the gut, with possible contributions from the δ-toxin and θ-toxins *(96,134)*. The infection is associated with the consumption of foods contaminated with *C. perfringens* type C together with low levels of proteolytic enzymes in the intestines, caused for example by prior low protein (meat) uptake. A further risk-factor is concurrent *Ascaris lumbricoides* (roundworm) infection, because the parasite secretes a powerful trypsin inhibitor *(135)*.

Outbreaks of necrotizing enteritis were detected in Germany just after the World War II. The disease was named Darmbrand ('fire bowels') and was associated with the scarcity of meat and consumption of under-processed home-canned foods, and rabbit. In Papua New Guinea, the disease (named Pig-bel) was much more common and is traditionally associated with feasting. The population has a low-meat diet and consequently low-chymotrypsin secretion. In addition, sweet potatoes are consumed as a staple source of carbohydrate; and because this vegetable contains high concentrations of a trypsin inhibitor, this also acts as a risk factor for the disease *(124)*. During feasts, large amounts of undercooked pork (which contain the bacterium) are consumed and the absence of trypsin and chymotrypsin results in active *C. perfringens* β-toxin in the jejunum and small intestines *(124)*.

3.5. Pathogenicity and Virulence Factors

3.5.1. Enterotoxin

Hobbs et al. *(89)* initially described experiments with the consumption of cooked meat broths by human volunteers. Diarrhea occurred after the consumption of viable cultures of *C. perfringens* but did not occur after the consumption of uninoculated broths or sterile filtrates of broths which contain viable cultures *(89)*. Thus, it was demonstrated that the consumption of viable organisms is essential for the development of the disease. Initial evidence that the enterotoxin is the single factor associated with food poisoning (as well as antibiotic and infectious diarrhea) was shown using purified protein and experimental animal tests *(158)*. Peroral administration of purified enterotoxin to human volunteers with neutralized stomach acid was later shown to generate diarrhea, nausea and abdominal pain indistinguishable from the natural disease *(136)*.

Large amounts of enterotoxin production occur in the *C. perfringens* cytoplasm after activation by factors which also control the onset of sporulation. The protein is not secreted but is released on lysis of the mother cell on completion of sporulation where up to 30% of the total cell protein is enterotoxin. During food poisoning, vegetative cells are consumed in high numbers in contaminated foods and these sporulate (releasing large amounts of enterotoxin) in the small intestines. The bacterium sporulates poorly in most laboratory media: Duncan and Strong *(137)* described a specialist medium to stimulate this process, and most other formulations are derivatives of this original medium *(138)*. The time course for production of enterotoxin in vitro correlates well with the natural disease where spore formation occurs after 3–4 h with maximum numbers after about 7 h. Enterotoxin formation parallels spore formation but decreases after about 10 h when lysis of the mother cells occur *(159)*.

The sequence of the enterotoxin has been investigated by several groups and was found to be highly conserved *(100)*. The enterotoxin comprises a single 35 kDa protein of 319 amino acids which is both heat and pH labile *(100)*. The secondary structure is of 80% β-sheet, and 20% random coil. There is an increased biological activity under limited trypsin and chymotrypsin treatment, which is associated with hydrolysis of the first 25–34 amino acids at the N-terminal: this activation process probably naturally occurs in the intestine during the disease *(100)*. There are no known homologies with any other toxins except for a weak relationship with a *Clostridium botulinum* complex protein *(100)*. The current model of the toxin action involves two domains at either end of the single enterotoxin polypeptide chain *(100)*. The first domain is located at the C-terminal and binds to host cell claudin proteins which are located in the tight junctions of a wide variety of cells, including enterocytes. This process disrupts efficient cell/cell tight junction formation and leads to some cell disintegration and disassociation. The claudin/enterotoxin complex then undergoes a confirmational change and interacts with a second host membrane protein (a binding domain near to the N-terminal is associated with this interaction) and form unique cation permeant pores and allows excess leakage of Ca^{2+} ions from the host cell. The loss of Ca^{2+} ions leads to cytoskeletal collapse and cell lysis.

3.5.2. β-Toxins

There are two β-toxins of *C. perfringens* (designated β1 and β2) which have approx 15% sequence homology. The β1 toxin is associated with necrotizing enteritis causing *C. perfringens* type C strains (as well as type B, Table 8). The β2 toxin also occurs in *C. perfringens* type C strains (with or without the β1 toxin) as well as some *C. perfringens* type A strains *(105)*. The gas gangrene type A strain where there is a whole genome sequence contains the β2 toxin *(103)*. Both the β1 and β2 toxins are cytotoxic and induce hemorrhagic necrosis of the intestinal mucosa in experimental intestinal loop tests; however, the exact mode of action is not understood *(105)*. However, because of similarities with other toxins, it has been suggested that the β-toxins form pores in eukaryotic membranes *(105)*.

3.6. Clinical Characteristics and Diagnosis

Ideally, the diagnosis of *C. perfringens* food poisoning is confirmed by: detection of enterotoxin and high numbers of *C. perfringens* vegetative cells and spores in patient's feces; detection of high numbers of *C. perfringens* vegetative cells in the implicated food; and the demonstration that isolates from the implicated food and patient's feces are the same type and are capable of producing enterotoxin. However, the diagnosis can be problematic because: *C. perfringens* vegetative cells in food decrease, if stored cold or if examination is delayed; spores in feces may germinate, if examination delayed; high counts and mixed *C. perfringens* populations occur in individuals without *C. perfringens* diarrhea, especially in the elderly; not all *C. perfringens* contain the enterotoxin gene and these are phenotypically indistinguishable from those with the enterotoxin gene; *C. perfringens* sporulates poorly in vitro and hence enterotoxin production is difficult to assess; and enterotoxin may not be detected in the feces, especially if examination is delayed >2 d after the onset.

From our experience, the use of PCR to identify enterotoxigenic *C. perfringens* is invaluable when investigating outbreaks. This process allows the distinction of organisms capable of causing disease from those that do not and which may be part of the normal food or fecal flora. Cases of viral diarrhea have been investigated by the authors where large numbers of *C. perfringens* occurred in the feces. PCR was useful in this instance because none of the *C. perfringens* was found to contain the enterotoxin gene. Prior to the availability of PCR, *C. perfringens* "outbreaks" were reported where large numbers of *C. perfringens* but no enterotoxin was detected in patient's feces *(115)*; these outbreaks could have been associated with other microbiological agents.

3.7. Treatment and Control

There is no specific treatment for *C. perfringens* food poisoning, and dehydration should be treated by oral or in more sever cases parenteral fluid replacement. Successful treatment of antibiotic-associated and infectious diarrhea with metronidazole has been reported *(139)*.

Treatment of necrotizing enteritis is related to rapid blood transfusion and fluid replacement together with intestinal decompression, vitamin supplements, and antibiotic treatment. Surgery (resuscitation, typically by removal of 50–200 cm of jejunum or less often the ileum) has been recommended in cases where: toxicity, bleeding, or pain persists; intestinal obstruction and distension persists; and perforation is a suspect. The incidence of Pig-bel in Papua New Guinea was reduced by vaccination with a toxoid of *C. perfringens* type C *(124)*.

Control of *C. perfringens* food poisoning involves the adequate performance of temperature and time control to cooling, storage, and reheating of foods. Since the organism survives many cooking processes, foods should be cooled rapidly (particularly through 55–15°C) as soon as possible after cooking, preferably within 90 min, to minimize the growth of this bacterium. Limiting the sizes of meat joints will greatly facilitate adequate cooling. Precooked foods should be thoroughly reheated to >70°C immediately before consumption to kill vegetative *C. perfringens* cells. Care should also be taken (especially in institutional catering) to avoid the use of repeated reheated ingredients and left-over foods such as soups, stocks, and gravy which could promote the build-up of large numbers of spores. Care should also be taken in general kitchen hygiene by thorough cleaning of all surfaces, cloths and utensils, and the removal of dust and soil from food components.

3.8. Summary and Conclusions

Despite after 50 years since the correct identification and description of control measures of *C. perfringens* in institutional catering by Hobbs et al. *(89)*, this disease remains an important cause of food poisoning. The bacterium also causes infectious and antibiotic associated diarrheal disease as well as necrotising enteritis which is very rare in industrialized countries. The disease is caused by the consumption of foods containing large numbers of the bacterium which produces diarrhoea by the synthesis of enterotoxin in the gut. Diagnosis of *C. perfringens* food poisoning can be problematic and ideally should be confirmed by: detection of enterotoxin and high numbers of *C. perfringens* vegetative cells and spores in patient's feces; detection of high numbers of *C. perfringens* vegetative cells in the implicated food; and the demonstration that isolates from implicated food and patient's feces are the same type and are capable of producing enterotoxin. The bacterium is widespread in the environment, although only a small proportion of *C. perfringens* have the ability to produce enterotoxin. The ecological role of enterotoxin is not known, although recent data has indicated that there are physiological differences associated with this phenotype.

C. perfringens food poisoning can be prevented by adequate temperature and time control together with basic food hygiene. Hobbs et al. *(89)* gave advice on the control of *C. perfringens* that is as relevant today as at the time of writing some 50 yr ago:

> "Outbreaks of this kind should be prevented by cooking meat immediately before consumption, or if this is impossible, by cooling the meat rapidly and keeping it refrigerated until it is required for use."

REFERENCES

1. Hatheway, C. L. and Johnson, E. A. (1998) *Clostridium*: the spore bearing anaerobes. In *Topley and Wilson's Microbiology and Microbial Infections* (Balows, A. and Duerden, B. I. eds.), Vol. 2, 9th edn, Arnold, London, pp. 731–782.
2. Weber, J. T., Goodpasture, H. C., Alexander, H., Werner, S. B., Hatheway, C. L., and Tauxe, R. V. (1993) Wound botulism in a patient with a tooth abscess: case report and review. *Clin. Infect. Dis.* **16,** 635–639.
3. Arnon, S. S. (1980) Infant botulism. *Annu. Rev. Med.* **31,** 541–560.
4. Arnon, S. S. (1985) Infant botulism. In *Clostridia in Gastrointestinal Disease* (Borriello, S. P., ed.), CRC, Boca Raton, FL, pp. 39–57.
5. Sofos, J. N. (1992) Botulism in home processed foods. In: *Handbook of Natural Toxins Vol. 7, Food Poisoning* (Tu, A. T., ed.), Dekker, New York, pp. 171–203.
6. Midura, T. F. (1996) Update, infant botulism. *Clin. Microbiol. Rev.* **9,** 119–125.
7. Dodds, K. L. and Austin, J. W. (1997) *Clostridium botulinum*. In: *Food Microbiology: Fundamentals and Frontiers.* (Doyle, M. P., Beuchat, L. R., and Montville, T. J., eds.), ASM, Washington DC, pp. 288–304.
8. Anon. (1998) *Botulism in the United States, 1899–1996. Handbook for Epidemiologists, Clinicians and Laboratory Workers,* Centres for Disease Control and Prevention, Atlanta, GA.

9. Johnson, E. A. and Goodnough, M. C. (1998) Botulism. In: *Topley and Wilson's Microbiology and Microbial Infections*. (Balows, A. and Duerden, B. I., eds.), Vol. 3, 9th edn, Arnold, London, pp. 723–741.

10. Johnson, E. A. (2000) Neurotoxigenic clostridia. In: *Gram Positive Pathogens* (Fishetti, V. A., ed.), ASM, Washington, pp. 540–548.

11. Solomon, H. M. and Lilly, T. (2001) *Clostridium botulinum*. FDA Bacteriological Analytical Manual Online, 8th edn, Revision A, 1998. Chapter 17. Available at http://www.cfsan.fda.gov/~ebam/bam-17.html#authors (accessed on December 2006).

12. Johnson, E. A. and Bradshaw, M. (2001) *Clostridium botulinum* and its neurotoxins: a metabolic and cellular perspective. *Toxicon*. **39**, 1703–1722.

13. Popoff, M. R. (2002) Molecular biology of clostridial toxins. In: *Pretein Toxins of the Genus* Clostridium *and Vaccination* (Duchesnes, C., Mainil, J., Popoff, M., and Titball, R. eds.), University of Liege, Liege. pp. 25–44.

14. Meunier, F. A., Schiavo, G., and Molgo, J. (2002) Botulinum neurotoxins: from paralysis to recovery of functional neuromuscular transmission. *J. Physiol. Paris*. **96**, 105–113.

15. Turton, K., Chaddock, J. A., and Acharya, K. R. (2002) Botulinum and tetanus neurotoxins: structure, function and therapeutic utility. *Trends Biochem. Sci.* **27**, 552–558.

16. Lalli, G., Bohnert, S., Deinhardt, K., Verastegui, C., and Schiavo, G. (2003) The journey of tetanus and botulinum neurotoxins in neurons. *Trends Microbiol.* **11**, 431–437.

17. Simpson, L. L. (2004) Identification of the major steps in botulinum toxin action. *Annu. Rev. Pharmacol. Toxicol.* **44**, 167–193.

18. Collins, M. D. and East, A. K. (1998) Phylogeny and taxonomy of the food-borne pathogen *Clostridium botulinum* and its neurotoxins. *J. Appl. Microbiol.* **84**, 5–17.

19. Lin, W. J. and Johnson, E. A. (1995) Genome analysis of *Clostridium botulinum* type A by pulsed-field gel electrophoresis. *Appl. Environ. Microbiol.* **61**, 4441–4447.

20. Anon. (2004) *Clostridium botulinum* genome. Available at http://www.sanger.ac.uk/Projects/C_botulinum/ (accessed on August 2004).

21. Oguma, K., Yokota, K., Hayashi, S., et al. (1990) Infant botulism due to *Clostridium botulinum* type C toxin. *Lancet* **336**, 1449–1450.

22. Sonnabend, O., Sonnabend, W., Heinzle, R., Sigrist, T., Dirnhofer, R., and Krech, U. (1981) Isolation of *Clostridium botulinum type G* and identification of type G botulinal toxin in humans: report of five sudden unexpected deaths. *J. Infect. Dis.* **143**, 22–27.

23. Aureli, P., Fenicia, L., Pasolini, B., Gianfranceschi, M., McCroskey, L. M., and Hatheway, C. L. (1986) Two cases of type E infant botulism caused by neurotoxigenic *Clostridium butyricum* in Italy. *J. Infect. Dis.* **154**, 207–211.

24. Gimenez, J. A. and Sugiyama, H. (1988) Comparison of toxins of *Clostridium butyricum* and *Clostridium botulinum* type E. *Infect. Immun.* **56**, 926–929.

25. Hall, J. D., McCroskey, L. M., Pincomb, B. J., and Hatheway, C. L. (1985) Isolation of an organism resembling *Clostridium barati* which produces type F botulinal toxin from an infant with botulism. *J. Clin. Microbiol.* **21**, 654–655.

26. McCroskey, L. M., Hatheway, C. L., Fenicia, L., Pasolini, B., and Aureli, P. (1986) Characterization of an organism that produces type E botulinal toxin but which resembles *Clostridium butyricum* from the feces of an infant with type E botulism. *J. Clin. Microbiol.* **23**, 201–202.

27. Brett, M. M. (1998) Evaluation of the use of the bioMerieux Rapid ID32 A for the identification of *Clostridium botulinum*. *Lett. Appl. Microbiol.* **26**, 81–84.

28. Ghanem, F. M., Ridpath, A. C., Moore, W. E. and Moore, L. V. (1991) Identification of *Clostridium botulinum, Clostridium argentinense*, and related organisms by cellular fatty acid analysis. *J. Clin. Microbiol.* **29**, 1114–1124.

29. Silas, J. C., Carpenter, J. A., Hamdy, M. K. and Harrison, M. A. (1985) Selective and differential medium for detecting *Clostridium botulinum*. *Appl. Environ. Microbiol.* **50**, 1110–1111.

30. Dezfulian, M., McCroskey, L. M., Hatheway, C. L., and Dowell, V. R. (1981) Selective medium for isolation of *Clostridium botulinum* from human feces. *J. Clin. Microbiol.* **13,** 526–531.

31. Mills, D. C., Midura, T. F., and Aron, S. S. (1985) Improved selective medium for the isolation of lipase positive *Clostridium botulinum* from feces of human infants. *J. Clin. Microbiol.* **21,** 947–950.

32. Centers for Disease Control and Prevention. (1997) Case definitions for infectious conditions under public health surveillance. *Morb. Mortal. Wkly Rep.* **46,** RR10 1–55.

33. Centers for Disease Control and Prevention. (1976) Follow up: botulism associated with commercial cherry peppers. *Morb. Mortal. Wkly Rep.* **25,** 148.

34. Terranova, W. A., Breman, J. G., Locey, R. P., et al. (1978) Botulism type B, epidemiological aspects of an extensive outbreak. *Am. J. Epidemiol.* **108,** 150–156.

35. Campbell, K. D., Collins, M. D., and East, A. K. (1993) Gene probes for identification of the botulinal neurotoxin gene and specific identification of neurotoxin types B, E, and F. *J. Clin. Microbiol.* **31,** 2255–2562.

36. Ferreira, J. L., Hamdy, M. K., McCay, S. G., Hemphill, M., Kirma, N., and Baumstark, B. T. (1994) Detection of *Clostridium botulinum* type F using the polymerase chain reaction. *Mol. Cell. Probes* **8,** 365–373.

37. Franciosa, G., Ferreira, J. L., and Hatheway, C. L. (1994) Detection of type A, B, and E botulism neurotoxin genes in *Clostridium botulinum* and other Clostridium species by PCR: evidence of unexpressed type B toxin genes in type A toxigenic organisms. *J. Clin. Microbiol.* **32,** 1911–1917.

38. Fach, P., Gibert, M., Griffais, R., Guillou, J. P., and Popoff, M. R. (1995) PCR and gene probe identification of botulinum neurotoxin A-, B-, E-, F-, and G-producing *Clostridium* spp. and evaluation in food samples. *Appl. Environ. Microbiol.* **61,** 389–392.

39. Takeshi, K., Fujinaga, Y., Inoue, K., et al. (1996) Simple method for detection of *Clostridium botulinum* type A to F neurotoxin genes by ploymerase chain reaction. *Microbiol. Immunol.* **40,** 5–11.

40. Aranda, E., Rodriguez, M. M., Asensio, M. A., and Cordoba, J. J. (1997) Detection of *Clostridium botulinum* types A, B, E and F in foods by PCR and DNA probe. *Lett. Appl. Microbiol.* **25,** 186–190.

41. Franciosa, G., Hatheway, C. L., and Aureli, P. (1998) The detection of a deletion in the type B neurotoxin gene of *Clostridium botulinum* A (B) strains by a two-step PCR. *Lett. Appl. Microbiol.* **26,** 442–446.

42. Hyytia, E., Hielm, S., and Korkeala, H. (1998) Prevalence of *Clostridium botulinum* type E in Finnish fish and fishery products. *Epidemiol. Infect.* **120,** 245–250.

43. Alsallami, A. A. and Kotlowski, R. (2001) Selection of primers for specific detection of *Clostridium botulinum* types B and E neurotoxin genes using PCR method. *Int. J. Food Microbiol.* **69,** 247–253.

44. Dahlenborg, M., Borch, E., and Radstrom, P. (2001) Development of a combined selection and enrichment PCR procedure for *Clostridium botulinum* types B, E, and F and its use to determine prevalence in fecal samples from slaughtered pigs. *Appl. Environ. Microbiol.* **67,** 4781–4788.

45. Lindström, M., Keto, R., Markkula, A., Nevas, M., Hielm, S., and Korkeala, H. (2001) Multiplex PCR assay for detection and identification of *Clostridium botulinum* types A, B, E, and F in food and fecal material. *Appl. Environ. Microbiol.* **67,** 5694–5699.

46. Nevas, M., Hielm, S., Lindström, M., Horn, H., Koivulehto, K., and Korkeala, H. (2002) High prevalence of *Clostridium botulinum* types A and B in honey samples detected by polymerase chain reaction. *Int. J. Food Microbiol.* **30,** 45–52.

47. Szabo, E. A., Pemberton, J. M., and Desmarchelier, P. M. (1993) Detection of the genes encoding botulinum neurotoxin types A to E by the polymerase chain reaction. *Appl. Environ. Microbiol.* **59,** 3011–3020.

48. Kimura, B., Kawasaki, S., Nakano, H., and Fujii, T. (2001) Rapid, quantitative PCR monitoring of growth of *Clostridium botulinum* type E in modified-atmosphere-packaged fish. *Appl. Environ. Microbiol.* **67,** 206–216.

49. Akbulut, D., Grant, K. A., and McLauchlin, J. (2004) Development and application of real-time PCR assays to detect fragments of the *Clostridium botulinum* types A, B and E neurotoxin genes. *Foodborne Pathog. Dis.* **1,** 247–257.

50. Doellgast, G. J., Triscott, M. X., Beard, G. A., and Bottoms, J. D. (1994) Enzyme-linked immunosorbent assay-enzyme-linked coagulation assay for detection of antibodies to *Clostridium botulinum* neurotoxins A, B, and E and solution-phase complexes. *J. Clin. Microbiol.* **32,** 851–853.

51. Szilagyi, M., Rivera, V. R., Neal, D., Merrill, G. A., and Poli, M. A. (2000) Development of sensitive colorimetric capture elisas for *Clostridium botulinum* neurotoxin serotypes A and B. *Toxicon.* **38,** 381–389.

52. Peruski, A. H., Johnson, L. H. III, and Peruski, L. F. Jr. (2002) Rapid and sensitive detection of biological warfare agents using time-resolved fluorescence assays. *J. Immunol. Meth.* **263,** 35–41.

53. Poli, M. A., Rivera, V. R., and Neal, D. (2002) Development of sensitive colorimetric capture ELISAs for *Clostridium botulinum* neurotoxin serotypes E and F. *Toxicon* **40,** 797–802.

54. Ferreira, J. L., Maslanka, S., Johnson, E., and Goodnough, M. (2003) Detection of botulinal neurotoxins A, B, E, and F by amplified enzyme-linked immunosorbent assay: collaborative study. *J. AOAC Int.* **86,** 314–331.

55. Wictome, M., Kirsti, A., Newton, K., et al. (1999) Development of in vitro assays for the detection of botulinum toxins in foods. *Appl. Environ. Microbiol.* **65,** 3787–3792.

56. Wictome, M., Newton, K., Jameson, K., et al. (1999) Development of an in vitro bioassay for *Clostridium botulinum* type B neurotoxin in foods that is more sensitive than the mouse bioassay. *Appl. Environ. Microbiol.* **65,** 3787–3792.

57. van Baar, B. L., Hulst, A. G., de Jong, A. L., and Wils, E. R. (2002) Characterisation of botulinum toxins type A and B, by matrix-assisted laser desorption ionisation and electrospray mass spectrometry. *J. Chromatogr. A.* **970,** 95–115.

58. van Baar, B. L., Hulst, A. G., de Jong, A. L., and Wils, E. R. (2004) Characterisation of botulinum toxins type C, D, E, and F by matrix-assisted laser desorption ionisation and electrospray mass spectrometry. *J. Chromatogr. A.* **1035,** 97–114.

59. Brett, M. M., McLauchlin, J., Harris, A., et al. (2005) A case of infant botulism with a possible link to infant formula milk powder: evidence for the presence of more than one strain of *Clostridium botulinum* in clinical specimens and food. *Emerg. Infect. Dis.* **54,** 769–776.

60. Hielm, S., Bjorkroth, J., Hyytia, E., and Korkeala, H. (1998) Genomic analysis of *Clostridium botulinum* group II by pulsed-field gel electrophoresis. *Appl. Environ. Microbiol.* **64,** 703–708.

61. Hielm, S., Bjorkroth, J., Hyytia, E., and Korkeala, H. (1998) Prevalence of *Clostridium botulinum* in Finnish trout farms: pulsed-field gel electrophoresis typing reveals extensive genetic diversity among type E isolates. *Appl. Environ. Microbiol.* **64,** 4161–4167.

62. Hyytia, E., Bjorkroth, J., Hielm, S., and Korkeala, H. (1999) Characterisation of *Clostridium botulinum* groups I and II by randomly amplified polymorphic DNA analysis and repetitive element sequence-based PCR. *Int. J. Food Microbiol.* **48,** 179–189.

63. Hielm, S., Bjorkroth, J., Hyytia, E., and Korkeala, H. (1999) Ribotyping as an identification tool for *Clostridium botulinum* strains causing human botulism. *Int. J. Food Microbiol.* **47,** 121–131.

64. Skinner, G. E., Gendel, S. M., Fingerhut, G. A., Solomon, H. A., and Ulaszek, J. (2000) Differentiation between types and strains of *Clostridium botulinum* by riboprinting. *J. Food Prot.* **63,** 1347–1352.

65. Arnon, S. S., Midura, T. F., Damus, K., Thompson, B., Wood, R. M., and Chin, J. (1979) Honey and other environmental risk factors for infant botulism. *J. Pediatr.* **94,** 331–336.
66. Crawford, D. and Gorrel, J. G. (2002) Infant botulism and Corn syrup: a case report. *J. Paediatr. Pharmacol. Ther.* **7,** 64–66.
67. Mann, J. M., Lathrop, G. D., and Bannerman, J. A. (1983) Economic impact of a botulism outbreak. Importance of the legal component in food-borne disease. *JAMA.* **249,** 1299–1301.
68. Todd, E. C. (1989) Costs of acute bacterial foodborne disease in Canada and the United States. *Int. J. Food Microbiol.* **9,** 313–326.
69. Weber, J. T., Hibbs, R. G. Jr., Darwish, A., et al. (1993) A massive outbreak of type E botulism associated with traditional salted fish in Cairo. *J. Infect. Dis.* **167,** 451–454.
70. Sobel, J., Tucker, N., Sulka, A., McLaughlin, J., and Maslanka, S. (2004) Foodborne botulism in the United States, 1990–2000. *Emerg. Infect. Dis.* **10(9),** 1606–1611.
71. Squarcione, S., Prete, A., and Vellucci, L. (1999) Botulism surveillance in Italy: 1992–1996. *Eur. J. Epidemiol.* **15,** 917–922.
72. Armada, M., Love, S., Barrett, E., Monroe, J., Peery, D., and Sobel, J. (2003) Foodborne botulism in a 6-mo-old infant caused by home-canned baby food. *Ann. Emerg. Med.* **42,** 226–229.
73. Ball, A. P., Hopkinson, R. B., Farrell, I. D., et al. (1979) Human botulism caused by *Clostridium botulinum* type E: the Birmingham outbreak. *Q. J. Med.* **48,** 473–491.
74. Colebatch, J. G., Wolff, A. H., Gilbert, R. J., et al. (1989) Slow recovery from severe food-borne botulism. *Lancet.* **2,** 1216–1217.
75. O'Mahony, M. O., Mitchell, E., Gilbert, R. J., et al. (1990) An outbreak of foodborne botulism associated with contaminated hazelnut yogurt. *Epidemiol. Infect.* **104,** 389–395.
76. Hutson, R. A., Zhou, Y., Collins, M. D., Johnson, E. A., Hatheway, C. L., and Sugiyama, H. (1996) Genetic characterization of *Clostridium botulinum* type A containing silent type B neurotoxin gene sequences. *J. Biol. Chem.* **271,** 10,786–10,792.
77. Chia, J. K., Clark, J. B., Ryan, C. A., and Pollack, M. (1986) Botulism in an adult associated with food-borne intestinal infection with *Clostridium botulinum.* *N. Engl. J. Med.* **315,** 239–241.
78. McCroskey, L. M. and Hatheway, C. L. (1988) Laboratory findings in four cases of adult botulism suggest colonization of the intestinal tract. *J. Clin. Microbiol.* **26,** 1052–1054.
79. McCroskey, L. M., Hatheway, C. L., Woodruff, B. A., Greenberg, J. A., and Jurgenson, P. (1991) Type F botulism due to neurotoxigenic *Clostridium baratii* from an unknown source in an adult. *J. Clin. Microbiol.* **29,** 2618–2620.
80. Griffin, P. M., Hatheway, C. L., Rosenbaum, R. B., and Sokolow, R. (1997) Endogenous antibody production to botulinum toxin in an adult with intestinal colonization botulism and underlying Crohn's disease. *J. Infect. Dis.* **175,** 633–637.
81. Li, L. Y., Kelkar, P., Exconde, R. E., Day, J., and Parry, G. J. (1999) Adult-onset "infant" botulism: an unusual cause of weakness in the intensive care unit. *Neurology.* **53,** 891.
82. Kobayashi, H., Fujisawa, K., Saito, Y., et al. (2003) A botulism case of a 12-year-old girl caused by intestinal colonization of *Clostridium botulinum* type Ab. *Jpn. J. Infect. Dis.* **56,** 73–74.
83. Shen, W.-P. V., Felsing, N., Lang, D., et al., (1994) Development of infant botulism in a 3 year old female with neurobastoma following autologous bone marrow transplantation: potential use of human botulism immune globulin. *Bone Marrow Transplant* **13,** 345–347.
84. Bohnel, H., Behrens, S., Loch, P., Lube, K., and Gessler, F. (2001) Is there a link between infant botulism and sudden infant death? Bacteriological results obtained in central Germany. *Eur. J. Pediatr.* **160,** 623–628.
85. Byard, R. W., Moore, L., Bourne, A. J., Lawrence, A. J., and Goldwater, P. N. (1992) *Clostridium botulinum* and sudden infant death syndrome: a 10 year prospective study. *J. Paediatr. Child Health.* **28,** 156–157.

86. Frankovich, T. L. and Arnon, S. S. (1991) Clinical trial of botulism immune globulin for infant botulism. *West. J. Med.* **154**, 103.

87. Tackett, C. O., Shandera, X. W., Mann, J. M., Hargrett, N. T., and Blake, P. A. (1984) Equine antitoxin use and other factors that predict the outcome in type A foodborne botulism. *Am. J. Med.* **76**, 794–799.

88. Finegold, S. M. (1977) Anaerobic bacteria in human disease. Academic Press, New York.

89. Hobbs, B. C., Smith, M. E., Oakley, C. L., Warrack, G. H., and Cruickshank, J. C. (1953) *Clostridium welchii* food poisoning. *J. Hyg.* **51**, 75–101.

90. de Wit, M. A., Koopmans, M. P., Kortbeek, L. M., et al. (2001) Sensor, a population-based cohort study on gastroenteritis in the Netherlands: incidence and etiology. *Am. J. Epidemiol.* **154**, 666–674.

91. Anon. (2000) *A Report of the Study of Infectious Intestinal Diseases in England.* The Stationery Office, London.

92. Adak, G. K., Long, S. M., and O'Brien, S. J. (2002) Trends in indigenous foodborne disease and deaths, England and Wales: 1992 to 2000. *Gut.* **51**, 832–841.

93. Mead, P. S., Slutsker, L., Dietz, V., et al. (1999) Food-related illness and death in the United States. *Emerg. Infect. Dis.* **5**, 607–625.

94. Labbé, R. G. (1992) *Clostridium perfringens* gastroenteritis. In: *Handbook of Natural Toxins. Vol. 7, Food Poisoning* (Tu, A. T., ed.), Dekker, New York, pp. 103–119.

95. McClane, B. A. (1997) *Clostridium perfringens.* In: *Food Microbiology: Fundamentals and Frontiers* (Doyle, M. P., Beuchat, L. R., and Montville, T. J., eds.), ASM, Washington, DC, pp. 305–326.

96. Granum, P. E. (1990) *Clostridium perfringens* toxins involved with food poisoning. *Int. J. Food Microbiol.* **10**, 101–112.

97. Hatheway, C. L. (1990) Toxigenic clostridia. *Clin. Microbiol. Rev.* **3**, 66–98.

98. Granum, P. E. and Brynestad, S. (1999) Bacterial toxins as food poisoning. In: *The Comprehensive Sourcebook of Bacterial Protein Toxins* (Alouf, J. E. and Freer, J. H., eds.), 2nd edn, Academic, London, pp. 669–681.

99. Rood, J. I. and Cole, S. T. (1991) Molecular genetics and pathogenesis of *Clostridium perfringens. Microbiol. Rev.* **55**, 621–648.

100. Brynestad, S. and Granum, P. E. (2002) *Clostridium perfringens* and foodborne infections. *Int. J. Food Microbiol.* **74**, 195–202.

101. Cato, E. P., George, W. L., and Finegold, S. M. (1986) Genus *Clostridium.* In: *Bergey's Manual of Systematic Bacteriology* (Sneath, P. H. A., Mair, N. S., Sharpe, M. E., and Holt, J. G., eds.), Vol. 2, Williams & Wilkins, Baltimore, pp. 1141–1200.

102. Collins, M. D., Lawson, P. A., Willems, A., et al. (1994) The phylogeny of the genus *Clostridium*: proposal of five new genera and eleven new species combinations. *Int. J. Syst. Bacteriol.* **44**, 812–826.

103. Shimizu, T., Ohtani, K., Hirakawa, H., et al. (2002) Complete genome sequence of *Clostridium perfringens*, an anaerobic flesh-eater. *Proc. Natl Acad. Sci. USA* **99**, 996–1001.

104. Daube, G., China, B., Simon, P., Hvala, K., and Mainil, J. (1994) Typing of *Clostridium perfringens* by in vitro amplification of toxin genes. *J. Appl. Bacteriol.* **77**, 650–655.

105. Petit, L., Gibert, M., and Popoff, M. R. (1999) *Clostridium perfringens*: toxinotype and genotype. *Trends Microbiol.* **7**, 104–110.

106. Billington, S. J., Wieckowski, E. U., Sarker, M. R., Bueschel, D., Songer, J. G., and McClane, B. A. (1998) *Clostridium perfringens* type E animal enteritis isolates with highly conserved, silent enterotoxin gene sequences. *Infect. Immun.* **66**, 4531–4536.

107. Brett, M. M. (1994) Outbreaks of food-poisoning associated with lecithinase-negative *Clostridium perfringens. J. Med. Microbiol.* **41**, 405–407.

108. Anon. (1999) Microbiology of food and animal feeding stuffs- horizontal method for the enumeration of *Clostridium perfringens* colony count technique (ISO 7937:1997 modified). *BS EN 13401:1999*, British Standards Institute, London.

109. Fach, P. and Popoff, M. R. (1997) Detection of enterotoxigenic *Clostridium perfringens* in food and fecal samples with a duplex PCR and the slide latex agglutination test. *Appl. Environ. Microbiol.* **63(11),** 4232–4236.

110. Stringer, M. F., Turnbull, P. C., and Gilbert, R. J. (1980) Application of serological typing to the investigation of outbreaks of *Clostridium perfringens* food poisoning, 1970–1978. *J. Hyg. (Lond).* **84,** 443–456.

111. Schalch, B., Sperner, B., Eisgruber, H., and Stolle, A. (1999) Molecular methods for the analysis of *Clostridium perfringens* relevant to food hygiene. *FEMS Immunol. Med. Microbiol.* **24,** 281–286.

112. Stark, R. L. and Duncan, C. L. (1971) Biological characteristics of *Clostridium perfringens* type A enterotoxin. *Infect. Immun.* **4,** 89–96.

113. Berry, P. R., Rodhouse, J. C., Hughes, S., Bartholomew, B. A., and Gilbert, R. J. (1988) Evaluation of ELISA, RPLA, and Vero cell assays for detecting *Clostridium perfringens* enterotoxin in faecal specimens. *J. Clin. Pathol.* **41,** 458–461.

114. Forward, L. J., Tompkins, D. S., and Brett, M. M. (2003) Detection of Clostridium difficile cytotoxin and *Clostridium perfringens* enterotoxin in cases of diarrhoea in the community. *J. Med. Microbiol.* **52,** 753–757.

115. Bartholomew, B. A., Stringer, M. F., Watson, G. N., and Gilbert, R. J. (1985) Development and application of an enzyme linked immunosorbent assay for *Clostridium perfringens* type A enterotoxin. *J. Clin. Pathol.* **38,** 222–228.

116. Asha, N. J. and Wilcox, M. H. (2002) Laboratory diagnosis of *Clostridium perfringens* antibiotic-associated diarrhoea. *J. Med. Microbiol.* **51,** 891–894.

117. Collie, R. E. and McClane, B. A. (1998) Evidence that the enterotoxin gene can be episomal in *Clostridium perfringens* isolates associated with non-food-borne human gastrointestinal diseases. *J. Clin. Microbiol.* **36,** 30–36.

118. Sparks, S. G., Carman, R. J., Sarker, M. R., and McClane, B. A. (2001) Genotyping of enterotoxigenic *Clostridium perfringens* fecal isolates associated with antibiotic-associated diarrhea and food poisoning in North America. *J. Clin. Microbiol.* **39,** 883–888.

119. Tanaka, D., Isobe, J., Hosorogi, S., et al. (2003) An outbreak of food-borne gastroenteritis caused by *Clostridium perfringens* carrying the cpe gene on a plasmid. *Jpn. J. Infect. Dis.* **56,** 137–139.

120. Sarker, M. R., Shivers, R. P., Sparks, S. G., Juneja, V. K., and McClane, B. A. (2000) Comparative experiments to examine the effects of heating on vegetative cells and spores of *Clostridium perfringens* isolates carrying plasmid genes versus chromosomal enterotoxin genes. *Appl. Environ. Microbiol.* **66,** 3234–3240.

121. Miwa, N., Masuda, T., Kwamura, A., Terai, K., and Akiyama, M. (2002) Survival and growth of enterotoxin-positive and enterotoxin-negative *Clostridium perfringens* in laboratory media. *Int. J. Food Microbiol.* **72,** 233–238.

122. Lin, Y. T. and Labbe, R. (2003) Enterotoxigenicity and genetic relatedness of *Clostridium perfringens* isolates from retail foods in the United States. *Appl. Environ. Microbiol.* **69,** 1642–1646.

123. Wen, Q. and McClane, B. A. (2004) Detection of enterotoxigenic *Clostridium perfringens* type A isolates in American retail foods. *Appl. Environ. Microbiol.* **70(5),** 2685–2691

124. Walker, P. D. (1985) Pig-bel. In: *Clostridia in Gastrointestinal Disease* (Borriello, S. P., ed.), CRC, Boca Raton, pp. 93–115.

125. Brett, M. M. and Gilbert, R. J. (1997) 1525 outbreaks of *Clostridium perfringens* food poisoning, 1979–1996. *Rev. Med. Microbiol.* **8,** S64–S65.

126. Loewenstein, M. S. (1972) Epidemiology of *Clostridium perfringens* food poisoning. *New Engl. J. Med.* **286,** 1026–1028.

127. Shandera, W. X., Tacket, C. O., and Blake, P. A. (1983) Food poisoning due to *Clostridium perfringens* in the United States. *J. Infect. Dis.* **147,** 167–170.

128. Anon. (1992) Foodborne disease outbreaks, 5-year summary, 1983–1987. *Morb. Mortal. Wkly Rep.* **39(SS-1),** 15–57.

129. Anon. (1996) Surveillance for foodborne-disease outbreaks United States, 1988–1992. *Morb. Mortal. Wkly Rep.* **45(SS-5),** 2–55.

130. Hewitt, J. H., Begg, N., Hewish, J., Rawaf, S., Stringer, M., and Theodore-Gandi, B. (1986) Large outbreaks of *Clostridium perfringens* food poisoning associated with the consumption of boiled salmon. *J. Hyg. (Lond).* **97,** 71–80.

131. Miwa, N., Masuda, T., Terai, K., Kawamura, A., Otani, K., and Miyamoto, H. (1999) Bacteriological investigation of an outbreak of *Clostridium perfringens* food poisoning caused by Japanese food without animal protein. *Int. J. Food Microbiol.* **49,** 103–106.

132. Borriello, S. P. (1995) Clostridial disease of the gut. *Clin. Infect. Dis.* **20,** S242–S250.

133. Severin, W. P. J., de la Fuente, A. A., and Stringer, M. F. (1984) *Clostridium perfringens* type C causing necrotising enteritis. *J. Clin. Pathol.* **34,** 942–944.

134. Jolivet-Reynaud, C., Popoff, M. R., Vinit, M. A., Ravisse, P., Moreau, H., and Alouf, J. E. (1986) Enteropathogenicity of *Clostridium perfringens* beta toxin and other clostridial toxins. *Zbl. Bakt. Microbiol. Hyg. Suppl.* **15,** 145–151.

135. Willis, A. T. and Smith, G. R. (1990) Gas gangrene and other clostridal infections of man and animals. In: *Topley and Wilson's Principals of Bacteriology, Virology and Immunology: Vol. 3, Bacterial Diseases* (Smith, G. R. and Easman, C. S. F., eds.), Arnold, London, pp. 307–329.

136. Skjelkvåle, R. and Uemura, T. (1977) Experimental diarrhoea in human volunteers following oral administration of *Clostridium perfringens* enterotoxin. *J. Appl. Bacteriol.* **43,** 281–286.

137. Duncan, C. L. and Strong, D. H. (1968) Improved medium for sporulation of *Clostridium perfringens*. *Appl. Microbiol.* **16,** 82–89.

138. de Jong, A. E., Beumer, R. R., and Rombouts, F. M. (2002) Optimizing sporulation of *Clostridium perfringens*. *J. Food Prot.* **65,** 1457–1462.

139. Borriello, S. P. and Williams, R. K. (1985) Treatment of *Clostridium perfringens* enterotoxin-associated diarrhoea with metronidazole. *J. Infect.* **10,** 65–67.

140. Brett, M. (1999) Botulism in the United Kingdom. *Euro Surveill.* **4,** 9–11.

141. Turner, H. D., Brett, E. M., Gilbert, R. J., Ghosh, A. C., and Liebeschuetz, H. J. (1978) Infant botulism in England. *Lancet* **i,** 1277–1278.

142. Smith, G. E., Hinde, F., Westmoreland, D., Berry, P. R., and Gilbert, R. J. (1989) Infantile botulism. *Arch. Dis. Child.* **64,** 871–872.

143. Jones, S., Huma, Z., Haugh, C., Young, Y., Starer, F., and Sinclair, L. (1990) Central nervous system involvement in infantile botulism. *Lancet* **335,** 228.

144. Anon. (1993) A case of infant botulism. *Commun. Dis. Rep. Wkly* **3,** 129.

145. Anon. (1994) A case of infant botulism. *Commun. Dis. Rep. Wkly* **4,** 53.

146. Brett, M. M., Hallas, G., and Mpamugo, O. (2004) Wound botulism in the UK and Ireland. *J. Med. Microbiol.* **53,** 555–561.

147. Anon. (1998) Investigation of faeces specimens for bacterial pathogens. *PHLS Standard Operating Procedure B.SOP 30.* PHLS, London.

148. Suen, J. C., Hatheway, C. L., Steigerwalt, A.G., Brenner, D. J. (1988) Genetic confirmation of identities of neurotoxigenic *Clostridium baratii* and *Clostridium butyricum* implicated as agents of infant botulism. *J. Clin. Microbiol.* **26,** 2191–2192.

149. Roberts, E., Wales, J. M., Brett, M. M., and Bradding, P. (1998) Cranial-nerve palsies and vomiting. *Lancet* **352,** 1674.

150. McLauchlin, J., Grant, K. A., Little, C. L. (2006) Foodborne botulism in the UK. *J. Public Health* **28,** 337–342.

151. Werner, S. B., Passaro, D., McGee, J., Schechter, R., Vugia, D. J. (2000) Wound botulism in California, 1951-1998: recent epidemic in heroin injectors. *Clin. Infect. Dis.* **31,** 1018–1024.

152. Tompkins, D., Hudson, M. J., Smith, H. R., et al. (1999) A study of infectious intestinal disease in England: microbiological findings in cases and controls. *Commun. Dis. Public Health* **2,** 108–113.

153. Barrow, G. I. and Feltham, R. K. A. (1991) Cowan and Steel's Manual for the Identification of Medical Bacteria. Cambridge University Press.

154. Maslanka, S. E., Kerr, J. G., Williams, G., et al. (1999). Molecular subtyping of *Clostridium perfringens* by pulsed-field gel electrophoresis to facilitate food-borne-outbreak investigation. *J. Clin. Microbiol.* **37,** 2209–2214.

155. Ridell, J., Bjorkroth, J., Eisgruber, H., Schalch, B., Stolle, A., and Korkeala, H. (1998) Prevalence of the enterotoxin gene and clonality of *Clostridium perfringens* strains associated with food-poisoning outbreaks. *J Food Protect* **61,** 240–243.

156. McLauchlin, J., Ripabelli, G., Brett, M. M., Threlfall, E. J. (2000) Amplified fragment length polymorphism (AFLP) analysis of *Clostridium perfringens* for epidemiological typing. *Int. J. Food Microbiol.* **56,** 21–28.

157. McClane, B. A. (1997) *Clostridium perfringens.* In Doyle, M. P., Beuchat, L. R., Montville, T. J., eds. Food Microbiology, Fundamentals and Frontiers. American Society for Microbiology Press, Washington. pp 305–326.

158. Stark, R. L., Duncan, C. L. (1972) Purification and biochemical properties of *Clostridium perfringens* type A enterotoxin. *Infect. Immun.* **6,** 662–673.

159. Labbé, R. G. (1980) The relationship between sporulation and enterotoxin production in *Clostridium perfringens* type A. *Food Technol.* **34,** 88–90.

Yersinia enterocolitica and Yersinia pseudotuberculosis

Maria Fredriksson-Ahomaa

1. CLASSIFICATION AND IDENTIFICATION

Yersinia enterocolitica and *Yersinia pseudotuberculosis* are included in the genus *Yersinia*. These species were formerly included in the genus *Pasteurella* and later placed into the genus *Yersinia*, named in honor of the French bacteriologist A. J. E. Yersin, a discoverer of the plague bacillus *(1)*. *Y. pseudotuberculosis* was the first species identified in this genus *(2)*. This organism was described in 1889 as a disease in guinea pigs. However, *Y. pseudotuberculosis* has shown to be the ancestor of *Y. pestis*, which was the cause of pandemic plague already during 541–767 AD *(3)*. The *Y. enterocolitica* was identified in 1939 and named by Frederiksen in 1964 *(4)*. The genus *Yersinia* is presently composed of 11 species, three of which can cause disease in humans and animals: *Y. enterocolitica*, *Y. pseudotuberculosis*, and *Y. pestis (5–8)*.

The three pathogenic species, the enteric foodborne pathogens *Y. enterocolitica* and *Y. pseudotuberculosis*, and *Y. pestis* are invasive pathogenic bacteria, which have a common capacity to resist nonspecific immune response and are lymphotrophic *(9)*. These three pathogenic species differ considerably in invasiveness. While *Y. enterocolitica* and *Y. pseudotuberculosis* can cross the gastrointestinal mucosa to infect underlying tissue, *Y. pestis* is injected into the body through an insect bite, and thus, does not require penetrating any body surface on its own *(9)*.

The genus *Yersinia* is classified into the family Enterobacteriaceae, a group of Gram-negative, oxidase-negative and facultatively anaerobic bacteria. All bacteria, belonging to the genus *Yersinia*, are catalase-positive, nonspore-forming rods or coccobacilli of 0.5–0.8 × 1–3 μm in size *(5)*. These bacteria are lactose-negative and have the ability to grow at 0–4°C. *Y. enterocolitica* and *Y. pseudotuberculosis* are urease-positive and can be differentiated from other urease-positive *Yersinia* species on the basis of Voges–Proskauer test and their ability to ferment sorbitol, rhamnose, sucrose, and melibiose (Table 1).

Y. enterocolitica is a heterogeneous species, which can be divided into six biotypes (1A, 1B, 2–5) on the basis of variations in biochemical reactions *(10)*. Subdivision into these six biotypes can be done using the following reactions: pyrazinamidase activity, esculin hydrolysis, tween-esterase activity, indole production, and xylose and trehalose acidification (Table 2). Biotypes 1B and 2–5 include strains that are associated with disease in man and animals when biotype 1A consists of nonpathogenic strains. However, strains

From: *Infectious Disease: Foodborne Diseases*
Edited by: S. Simjee © Humana Press Inc., Totowa, NJ

Table 1
**Biochemical Differentiation of Urea-Positive *Yersinia* Species
After Incubation at 25°C for 18–20 h**

Species	Reaction				
	Voges–Proskauer	Sorbitol	Rhamnose	Sucrose	Melibiose
Y. enterocolitica	+	+	–	+	–
Y. pseudotuberculosis	–	–	+/–	–	+/–
Y. frederiksenii	+	+	+	+	–
Y. intermedia	+	+	+	+	+
Y. kristensenii	–	+	–	–	–
Y. aldovae	+	+	+	–	–
Y. rohdei	–	+	–	+	+/–
Y. mollaretii	–	+	–	+	–
Y. bercovieri	–	+	–	+	–

Table 2
**Biochemical Tests Used for Biotyping *Y. enterocolitica*
and *Y. pseudotuberculosis* Isolates**

Reaction[a]	*Y. enterocolitica*						*Y. pseudotuberculosis*			
	BT[b]1A	BT1B	BT2	BT3	BT4	BT5	BT1	BT2	BT3	BT4
Melibiose	–	–	–	–	–	–	+	–	–	+
Citrate	–	–	–	–	–	–	–	–	+	–
Raffinose	–	–	–	–	–	–	–	–	–	+
Pyrazinamidase	+	–	–	–	–	–	–	–	–	–
Esculin	+	–	–	–	–	–	+	+	+	+
Tween	+	+	–	–	–	–	–	–	–	–
Indole	+	+	+	–	–	–	–	–	–	–
Xylose	+	+	+	+	–	–	+	+	+	+
Trehalose	+	+	+	+	+	–	+	+	+	+

[a]All tests done at 25°C.
[b]BT, biotype.

of biotype 1A have constituted a sizeable fraction of strains from patients with gastroenteritis *(11,12)*. Neubauer et al. *(13,14)* have demonstrated on the basis of the different
DNA–DNA hybridization values and the 16S rRNA gene sequences that *Y. enterocolitica*
should be divided into two subspecies, with one subspecies consisting of strains of biotype
1B, and the other of the remaining strains.

Compared to *Y. enterocolitica* is *Y. pseudotuberculosis* a phenotypically more homogeneous species. Nevertheless, *Y. pseudotuberculosis* can be divided into four biotypes
according to their behavior in citrate, melibiose, and raffinose *(15)* (Table 2). No correlation of pathogenicity of *Y. pseudotuberculosis* with biotype was found in this study.
However, melibiose-positive strains (biotypes 1 and 4) have shown to be more pathogenic
than melibiose-negative strain (biotypes 2 and 3) *(16)*.

Table 3
**Distribution of Pathogenic *Y. enterocolitica* Belonging to Different
Bio-, Sero- and Phage Types**

Biotype	Serotype	Phagetype	Host	Distribution
1B	O:4,32		Man	United States
	O:8	X	Man, pig, wild rodents	Europe, Japan, North America
	O:13		Man, monkey	North America
	O:18		Man	United States
	O:20		Man, rat, monkey	United States
	O:21		Man, rat flea	North America
2 or 3	O:5,27	X	Man, pig, dog, monkey	Australia, Europe, Japan, North America
	O:9	X	Man, pig, cattle, goat, dog, cat, rat	Australia, Canada, Europe, Japan
3	O:1,2,3	I	Chinchilla	Europe, United States
	O:3	II	Man, pig, dog	Japan
			Man, pig, rabbit, rat	Korea
4	O:3	VIII		Europe, Japan
		IXA	Man, pig, dog, cat, rat	South Africa
		IXB		North America
5	O:2,3	II	Hare, goat, sheep, rabbit, monkey	Europe, Australia

Y. enterocolitica and *Y. pseudotuberculosis* can be divided into numerous serotypes on the basis of antigenic variations in cell-wall lipopolysaccharide (LPS). Aleksic and Bockemühl *(17)* have proposed a revised and simplified typing scheme, which includes 20 antigenic factors for *Y. enterocolitica* alone. Serotype O:3 is most frequently isolated from humans in general *(18–22)*. Other common serotypes obtained from humans include serotype, serotype O:9 and serotype O:8. However, several O-antigens, including O:3, O:8, and O:9, have been found in pathogenic and nonpathogenic strains *(23)*. An accurate biochemical characterization is needed before or after serological typing that leads to correct assessment of the relevance of strains especially from foods and the environment, because related species and biotype 1A strains are widely distributed in these samples *(24)*.

The bioserotypes *Y. enterocolitica* differ in their geographical distribution, ecological niches, and pathogenic properties. The vast majority of clinical isolates belong to a relatively few bioserotypes. These bioserotypes have different geographical distributions (Table 3). Strains of biotype 1B serotypes O:4,32; O:8, O:13; O:18; O:20 and O:21 are frequently found associated with human diseases, mostly in the United States and Canada *(25)*, but bioserotype 1B/O:8 has occasionally also been found in Japan and Europe *(26–30)*. Strains that are largely responsible for human yersiniosis in Europe, Japan, Canada, and the United States belong to bioserotype 4/O:3 *(25)*. Bioserotype 3/O:3 has been recovered in Japan *(31)* and China *(32)*; serotypes O:9 and O:5,27 are more widely distributed *(33)*. Non-pathogenic strains of biotype 1A (e.g., serotypes O:5; O:6,30; O:6,31; O:7,8; O:10, O:18; O:46 and nontypable trains) are distributed worldwide and are predominantly isolated from the environment, water, feces, and food *(25)*.

Table 4
Distribution of Serotypes O:1 to O:5 of *Y. pseudotuberculosis* Isolated From Human and Nonhuman Sources

Country	Source	No. of isolates	1a	1b	2a	2b	2c	3	4a	4b	5a	5b	Ref.
Japan	Human	272		30	11	18	6	9	4	11	26	54	*(42)*
										4			
	Pig	195		28		7	10	10		40	2	1	
								7					
	Dog	40		12		2	1	1	2	17	5		
	Cat	19		2		1	1	4		11			
	Rabbit	16		3		1		1		2		9	
	Rat	8		1					7				
	Water	39			2	2	2			12	5	16	
Japan	Human	45		14		1		2		28			*(247)*
	Raccoon dog	23		3		2				17	1		
	Wild mouse	9		2						7			
	Deer	8						1		6	1		
Japan	Water	110		9		5	2	1	14	19	54	6	*(84)*
Italy	Human	5	5										*(127)*
	Pig	8				1		7					
Germany	Human	52	21	10	7	3		7			2	2	*(40)*
	Hare	35	17	8	5	1		4					
	Bird	19	11	4	3			1					
	Sheep	5	2	2				1					
	Cat	5	2	2	1								
	Deer	4		1	1			2					
	Pig	3	1					3					
Belgium	Human	8	4	2				2					*(124)*

Y. enterocolitica can also be divided into phage types. Two schemes (Swedish and French) are used for phage typing of *Y. enterocolitica (34)*. Of these, the French scheme has been used frequently and it recognizes 12 phage types: I–X (including IXa–c). The Swedish scheme recognizes seven phages (A1, A2, B1, B2, C32, C61, and E1) and is used less frequently. Neither of these schemes has produced a large number of distinct epidemiological types because many strains fall into the same phage types. Strains of bioserotype 4/O:3 and phage type VIII predominate in Europe and Japan *(35)*, whereas phage type IXb has been isolated in Canada *(36)* and in the United States *(37)*. *Y. enterocolitica* of bioserotype 1B/O:8 and phage type X, which is a typical North American type, has also sporadically been isolated in Japan *(28,30)*. Because of the need to maintain stocks of biologically active phages and control strains, phage typing is available at only a few laboratories.

A simplified antigenic scheme for serotyping of *Y. pseudotuberculosis* consists of 15 O-serotypes of which serotypes O:1 and O:2 are divided into three subtypes a, b, and c; and serotypes O:4 and O:5 into subtypes a and b *(15,38)*. The serotypes of *Y. pseudotuberculosis* differ in their geographical distribution and ecological niches (Table 4).

Serotypes O:1b and O:3 of *Y. pseudotuberculosis* have been isolated from the patients in Canada *(39)*, whereas serotypes O:1a, O:1b and O:3 have most often been found in humans in Europe *(40,41)*. In the Far East, serotypes O:1b, O:2a, O:2b, O:3, O:4a, O:4b, O:5a and O:5b, of which O:4b and O:5b are dominant, have been isolated from clinical samples *(16,42)*. The relationship between pathogenicity and serological properties of *Y. pseudotuberculosis* is poorly understood *(43)*. However, serotypes O:1 to O:5 have been isolated from humans, whereas serotypes O:6 to O:14 have been isolated from wild animals and from the environment, but not from the clinical samples *(38)*.

2. RESERVOIRS

Y. enterocolitica and *Y. pseudotuberculosis* are widely spread in nature. They have been isolated from many natural sources, such as animals, foods and the environment. Animals have long been suspected to be the reservoirs for *Y. enterocolitica* and *Y. pseudotuberculosis*, and thus, the source of human infections. Numerous studies have been carried out to isolate *Y. enterocolitica* from a variety of animals *(44)*, including wild animals *(45–55)* and farm animals *(45,56–60)*. However, most of *Y. enterocolitica* strains isolated from the animals differ both biochemically and serologically from strains isolated from humans with yersiniosis. Human pathogenic strains of *Y. enterocolitica* have frequently been isolated only from slaughter pigs, which have shown to be a reservoir of serotypes O:3. Serotypes 5,27, O:9, and O:8, which are also common among humans, have sporadically been isolated from pigs *(33)* (Fukushima et al. 1993). In Japan, serotypes O:3, O:8, and O:9 have been isolated from wild rodents, especially from field mice *(61,62)*. Hayashidani et al. *(62)* have suggested that the small rodents living in the wild may be a source of *Y. enterocolitica* O:8 infection for humans in Japan. In this study, restriction endonuclease analysis of virulence plasmid patterns of wild rodents was indistinguishable from the patterns of humans. In France, *Y. enterocolitica* O:9 has frequently been isolated from the stools of cattle and goats in infected herds *(63)*.

The highest prevalence of *Y. enterocolitica*, belonging to serotypes associated with human yersiniosis, has been obtained in pig tonsils, with serotype O:3 being the most common (Table 5). Experimental infection of pigs has shown that *Y. enterocolitica* remains longer *(64,65)*, and the number of isolates is higher *(66)* in tonsils than in feces. *Y. enterocolitica* of bioserotype 4/O:3 has a global distribution among the pig population, but the prevalence does vary between herds in many countries. This herd-wise distribution has been demonstrated by culture methods in Denmark, Norway, Finland, and Canada *(58,67–71)* and by serological tests in Denmark and Norway *(72,73)*. By culture method, 18% *(70)* to 64% *(69)* of the herds have been shown negative for *Y. enterocolitica* 4/O:3. On the contrary, serological investigations have shown that 70–90% of the slaughter herds in Denmark and 63% in Norway are infected with serotype O:3, and that nearly all finishing pigs in infected herds are seropositive *(72,73)*.

Pet animals, such as cats and dogs, have been suspected of being reservoirs for human infections with *Y. enterocolitica*, because of their close contact with humans *(34)*. However, strains of *Y. enterocolitica* 4/O:3 have only occasionally been isolated from dogs and cats *(45,56,58,74–77)*. These strains have mostly been isolated from apparently healthy dogs *(76,77)*. Dogs can asymptomatically carry *Y. enterocolitica* in the pharynx and excrete the organism in feces for several weeks after infection *(78)*.

Table 5
Prevalence of *Y. enterocolitica* Belonging to Serotypes Associated
With Human Disease in Slaughter Pigs

Country	Sample	No. of samples	Prevalence (%) of different serotypes				Ref.
			O:3	O:5,27	O:8	O:9	
Belgium	Tonsils	54	33 (61)				(95)
Canada	Tonsils	202	57 (28)	15 (7)			(230)
	Feces	1420	235 (17)	25 (2)	1 (0.1)	9 (0.6)	(71)
	Tonsils	291	63 (22)	1 (0.3)		2 (0.7)	(65)
	Feces	291	17 (6)				(65)
Chile	Tonsils	100	35 (35)				(248)
China	Feces	510	37 (7)				(32)
Denmark	Tonsils	400	149 (37)				(58)
	Feces	1458	360 (25)				(249)
Finland	Tonsils	185	68 (37)				(250)
	Tonsils	210	122 (56)				(251)
Germany	Tonsils	50	10 (60)				(252)
	Feces	50	5 (10)				(252)
Italy	Tonsils	106	43 (41)			2 (2)	(93)
	Tonsils	150	19 (13)				(253)
	Feces	150	4 (3)	1 (1)			(253)
Japan	Throat	1200	88 (7)	1 (0.1)			(254)
	Feces	1200	86 (7)	1 (0.1)			(254)
Netherlands	Tonsils	86	33 (38)	3 (3)			(91)
	Feces	100	16 (16)			1 (1)	(91)
Norway	Tonsils	461	146 (32)	1 (0.2)			(68)
Spain	Feces	110	6 (5)				(255)
United States	Throat	3375	4 (0.1)	95 (3)			(256)

Several studies have been carried out to isolate *Y. pseudotuberculosis* from farm animals, pets, and wild animals (Table 6). *Y. pseudotuberculosis* is a common inhabitant of the intestine in a wide variety of domestic and wild mammals. Although the host range is broad, the principal reservoir hosts are believed to be rodents, wild birds, and domestic animals (especially pigs and ruminants). Host genetics and age influence the susceptibility to disease. *Y. pseudotuberculosis* has been isolated from 11 species of animals in Japan *(42)*. Of those, monkeys, goats, rabbits, and guinea pigs were diseased, where as cattle, swine, dogs, cats, hares, and rats were become the carriers. Jerrett et al. *(79)* reported that *Y. pseudotuberculosis* is one of the most common infectious cause of death among farmed deer in Australia. *Y. pseudotuberculosis* has frequently been isolated also from raccoon dogs, marten, and fox *(80)*. Frolich et al. *(81)* reported antibodies against *Yersinia* spp. in 163 of 299 (55%) European brown hares by Western blotting in Germany.

Y. pseudotuberculosis is also widely spread in the environment (soil and water) where it can survive for a long time. The environment itself is contaminated from the feces of infected animals, mainly rodents and birds *(82)*. *Y. pseudotuberculosis* has been isolated from fresh water such as river, well, and mountain stream water at a considerable high rate *(42,83,84)*.

Table 6
Prevalence of *Y. pseudotuberculosis* in Animal Sources

Sources	No. of samples	No. of positive samples	%	Country	Reference
Pig	544	14	2.6	Canada	*(36)*
	210	8	4.0	Finland	*(257)*
	217	1	0.5	Italy	*(127)*
	1200	52	4.3	Japan	*(254)*
	585	12	2.1	Japan	*(80)*
	60	1	0.2	Norway	*(258)*
Cattle	2639	185	7	Australia	*(259)*
	618	1	0.2	Japan	*(57)*
	330	1	0.3	New Zealand	*(260)*
Sheep	449	21	4.7	Australia	*(261)*
	66	2	3.0	New Zealand	*(260)*
Dear	153	29	18	Australia	Jerrett et al. 1990
	215	8	3.7	Japan	*(79)*
Dog	252	16	6.3	Japan	Fukushima et al. (1984)
	704	13	1.8	Japan	*(76)*
	176	5	2.8	Japan	*(80)*
Cat	318	4	1.3	Japan	*(262)*
	373	12	3.2	Japan	*(74)*
	724	64	8.8	Saudi Arabia	*(263)*
Hare	139	2	1.4	Japan	*(247)*
	474	6	1.3	Japan	*(80)*
Mouse	669	25	3.6	Japan	*(80)*
	107	9	8.4	China	Zheng et al. *(289)* (1995)
Rat	148	4	2.7	China	Zheng et al. *(289)* (1995)
	301	20	6.6	Saudi Arabia	*(263)*
Rabbits	148	4	2.7	China	Zheng et al. *(289)* (1995)
	70	1	1.4	Italy	*(127)*
Birds	900	1	0.1	Italy	*(127)*
	259	2	0.8	Japan	*(247)*
	108	5	4.6	Japan	*(80)*
	468	3	0.6	Sweden	*(264)*

Food has often been suggested to be the main source of *Y. enterocolitica*, even though pathogenic isolates have seldom been recovered from food samples *(29,85)*. Raw pork products have been widely investigated because of the association between *Y. enterocolitica* 4/O:3 and pigs *(86–94)*. The isolation rate of *Y. enterocolitica* belonging to serotypes associated with human disease has been low in raw pork, except for pig tongues; the most common type isolated was bioserotype 4/O:3 (Table 7). The prevalence of bioserotype 4/O:3 has been exceptionally high in both pig tongues and minced meat in Belgium *(95)*, where head meat including tonsils had been used for minced meat *(96)*. *Y. enterocolitica* strains belonging to bioserotypes associated with human disease have been recovered only a few times from beef, poultry, and milk samples *(94,97,98)*. In these cases, cross-contamination had probably occurred during the process, packaging, or handling.

Table 7
**Isolation Rate of Y. *enterocolitica* Belonging to Serotypes Associated
With Human Disease in Raw Pork**

Sample	No. of samples	Isolation rate (No. of positive samples)				Country	Reference
		O:3	O:5,27	O:8	O:9		
Tongue	302	55 (165)			1 (3)	Belgium	(86)
	37	33 (11)				Canada	(87)
	31	2 (2)		19 (6)		United States	(37)
	47	55 (26)				Norway	(265)
	50	40 (20)				Japan	(266)
	125	6 (8)				Spain	(255)
	29	97 (28)				Belgium	(95)
	40	15 (6)			5 (2)	Netherlands	(91)
	55	25 (14)				Germany	(267)
	86	2 (2)				Italy	(93)
	99	80 (79)				Finland	(268)
	20	75 (15)				Germany	(252)
Offal[a]	34	50 (17)				Finland	(107)
	16	31 (5)				Finland	(268)
	100	46 (46)				Germany	(252)
Pork	91	1 (1)	13 (12)[b]	1 (1)		Canada	(87)
	127	1 (1)				Norway	(89)
	70	31 (22)			4 (3)	Japan	(266)
	267	2 (6)				Denmark	(58)
	50	24 (12)				Belgium	(95)
	400	1 (3)			1 (0.3)	Netherlands	(91)
	67	1 (1)	12 (8)[b]	4 (3)		China	(92)
	48	2 (1)			2 (1)	Germany	(267)
	40	5 (2)	10 (4)		2 (1)	Ireland	(97)
	1278	5 (64)	1 (14)			Japan	(94)
	255	4 (2)				Finland	(269)
	300	2 (6)				Norway	(270)
	120	12 (14)				Germany	(252)

[a]Offal, including liver, heart, and kidney.
[b]Strains belong to serotype O:5.

Inefficient isolation methods can be the reason for low prevalence rates of pathogenic *Y. enterocolitica* (99). The detection rate of pathogenic *Y. enterocolitica* in foods has shown to be clearly higher by PCR than by culturing (Table 8). In a case–control study, *Y. enterocolitica* 4/O:3 infections have been associated with eating raw or undercooked pork in 2 wk before the onset (96). In the United States, *Y. enterocolitica* 4/O:3 infections have been associated with the household preparation of chitterlings (intestines of pigs, which are a traditional holiday dish in the South), particularly among black children (18,100,101).

Most of the *Y. enterocolitica* isolates recovered from the environmental samples, including slaughterhouses, butcher shops, fodder, soil, and water have been nonpathogenic in nature (52,55,58,67,90,102–105). However, strains of bioserotype 4/O:3

Table 8
Detection Rate of Pathogenic *Yersinia enterocolitica* in Natural Samples With PCR

Sample	No. of samples	Detection rate (No. of positive samples)	Reference
Food			
Pig tongues	51	92 (47)	*(269)*
Minced pork	255	25 (63)	*(269)*
Pig offal	34	62 (21)	*(107)*
Pork[b]	300	17 (50)	*(270)*
Pig tongues	15	67 (10)	*(271)*
Ground pork	100	47 (47)	*(271)*
Ground beef	100	31 (31)	*(271)*
Tofu	50	12 (6)	*(271)*
Ground pork	350	38 (133)	*(272)*
Chitterlings	350	79 (278)	*(272)*
Environmental			
Water	105	10 (11)	*(111)*
Slaughterhouse	89	13 (12)	*(107)*

have occasionally been isolated from slaughterhouses *(104,106,107)*, butcher shops *(108,109)*, well-water used as drinking water *(110)*, and sewage water *(58)*. Sandery et al. *(111)* and Waage et al. *(112)* have shown using PCR that pathogenic strains of *Y. enterocolitica* can frequently be detected from the environmental waters.

3. FOODBORNE OUTBREAKS

Yersiniosis has been observed in all continents but seems to be most common in Europe and Japan. Foodborne outbreaks of *Y. enterocolitica* and *Y. pseudotuberculosis* are not common and most of the infections are sporadic. However, foodborne outbreaks of *Y. enterocolitica* and *Y. pseudotuberculosis* may occur and have been reported from many countries (Table 9).

Large outbreaks of *Y. enterocolitica* O:3 gastroenteritis has been documented in Japan and in the former Czechoslovakia *(27,113)*. In all cases, *Y. enterocolitica* O:3 was the causative agent, but the source of infection went undetected. The food handler was expected to be responsible for the two nursery school outbreaks in the former Czechoslovakia *(113)*. In the United States, after 1988, several outbreaks of serotype O:3 infections among infants attributed to chitterlings (a dish made from intestine) have been reported *(18,29,101,114)*. In a case–control study of *Y. enterocolitica* infections among black infants, preparation of chitterlings was significantly ($p < 0.001$) associated with illness *(101)*. In Europe, sporadic infections caused by bioserotype O:3 are common, but outbreaks of disease are uncommon. Because of the long incubation time and the predominance of cases going undiagnosed until 2–3 wk after the infection as arthritic symptoms, the source of infection was not traced in most cases. *Y. enterocolitica* 4/O:3 infection has been associated with consumption of raw or undercooked pork and untreated water in case–control studies *(96,115,116)*.

Several outbreaks of other serotypes have occurred in the United States; six of these were caused by serotype O:8 (Table 7). The outbreaks were associated with sick

Table 9
Foodborne Outbreaks of *Y. enterocolitica* and *Y. pseudotuberculosis*

Year	Location		Organism		No. of Cases	Source	Reference
1970	Norway	Household	YE[a]	O:7,13	2	Well water	*(273)*
1971	Finland	Hospital	YE	O:9	7	Patient	*(274)*
1971	Czechoslovakia	School	YE	O:3	15	Unknown	*(113)*
1972	Japan	School A	YE	O:3	189	Unknown	*(275)*
1972	Japan	School B	YE	O:3	544	Unknown	*(275)*
1972	United States	Household	YE	O:8	15	Dog	*(117)*
1976	United States	School	YE	O:8	228	Chocolate milk	*(118)*
1977	Japan	Kindergarten	YP[b]	O:1b	82	Water	*(42)*
1980	Japan	School	YE	O:3	1051	Unknown	*(27)*
1981	Japan	School	YE	O:3	641	Unknown	*(27)*
1981	Japan	School	YP	O:5a	188	Vegetable juice	*(42)*
1981	Canada	Household	YE	O:21	3	River water	Martin et al. (1992)
1981	United States	Summer camp	YE	O:8	159	Milk powder	*(123)*
1981	United States	Residents	YE	O:8	50	Tofu	*(121)*
1981	Finland	School	YP	O:3	19	Unknown	*(276)*
1981	Japan	School	YP	O:5a	188	Vegetable juice	*(42)*
1982	United States	Brownie troops	YE	O:8	16	Bean sprouts	*(120)*
1982	United States	Residents	YE	O:13	172	Milk	*(122)*
1982	Japan	Mountain area	YP	O:4b	140	Water	*(42)*
1984	Japan	School	YE	O:3	102	Unknown	*(27)*
1984	Japan	Restaurant	YP	O:5a	39	Barbecue	*(277)*
1984	Japan	School	YP	O:3	63	Unknown	*(42)*
1984	Japan	Mountain area	YP	O:4b	11	Water	*(42)*
1985	Japan	School	YP	O:4b	60	Unknown	*(42)*
1986	Belgium	Day-nursery	YE	O:3	21	Unknown	*(181)*
1986	Japan	School	YP	O:4b	549	Unknown	*(42)*
1987	Finland	School	YP	O:1a	34	Unknown	*(180)*
1989	United States	Households	YE	O:3	15	Chitterlings	*(18)*
1995	United States	Residents	YE	O:8	10	Milk	*(98)*
1998	Canada	Households	YP	O:1b	74	Milk	*(126)*
1997	Finland	School	YP	O:3	36	Unknown	*(41)*
1998	Finland	School	YP	O:3	53	Unknown	*(41)*
1998	Finland	Cafeterias	YP	O:3	47	Iceberg lettuce	*(125)*
1999	Finland	School	YP	O:3	22	Iceberg lettuce	*(41)*
2001	Finland	Schools	YP	O:1, O:3	59	Chinese cabbage	*(41)*
2001	United States	Household	YE	O:3	12	Chitterlings	*(101)*
2002	Croatia	Oil tanker	YE	O:3	17	Unknown	*(278)*

[a]YE, *Y. enterocolitica*.
[b]YP, *Y. pseudotuberculosis*.

puppies *(117)*, chocolate milk *(118)*, powdered milk and chow mein *(49)*, tofu *(119)*, bean sprouts *(120)*, and pasteurized milk *(98)*. The source of infection in an inter-familial outbreak of *Y. enterocolitica* O:8 was suggested to be a female dog and its litter from sick puppies. The sequential onset of disease indicated that, once it had been introduced into the households, person-to-person transmission had occurred *(117)*. In the tofu epidemic, the oriental soybean curd was probably packed in untreated spring-water contaminated with *Y. enterocolitica*. Samples of tofu, water supply and asymptomatic employees were positive for *Y. enterocolitica* O:8 *(119)*. Inspection of the tofu plant revealed unsanitary conditions, including poor personal hygiene and unsanitary equipment. In addition, a rare serotype, O:13, has caused a large multistate outbreak in which the pasteurized milk was the common source *(121,122)*. In case–control studies, drinking milk pasteurized by plan A was statistically associated with illness. In a familial outbreak of *Y. enterocolitica* O:21 in Canada, three people in a four-member family were involved. The family admitted of drinking unboiled water obtained directly from the river beside of their home *(123)*.

   Y. pseudotuberculosis infections have been reported from all continents, but the incidence appears to be less frequent than for *Y. enterocolitica*. Most of the infections are sporadic and outbreaks are rare *(124)*. Community outbreaks have mainly occurred in Finland and Japan (Table 9). The source of infection has remained unknown in most of the studies. However, iceberg lettuce was implicated as the vehicle of a widespread foodborne *Y. pseudotuberculosis* outbreak in Finland *(125)*. One more outbreak has also been reported from Canada *(126)*. Using case–control study, homogenized milk, pork, and fruit juice were identified as possible risk factors. *Y. pseudotuberculosis* 1b was isolated from stool specimen and a milk sample; however, the precise cause of this outbreak remained unknown.

   The frequency of *Y. pseudotuberculosis* in stool cultures from patients has shown to be low in Europe *(124,127)*. In Japan, the frequency has shown to be clearly higher. The great differences between Europe and Japan are probably related to geographical differences, but can also be of technical factors including different culturing methods. Since 1970 in Japan, many reports have been made on sporadic infection and outbreaks associated with *Y. pseudotuberculosis* *(42)*. Sporadic infection appears to be concentrated in the 1–16-yr age group, with a peak at 1–2-yr-old children. For instance, Izumi fever, a childhood illness, is caused by *Y. pseudotuberculosis*. The highest incidence of *Y. enterocolitica* and *Y. pseudotuberculosis* infections in Europe has shown to be in young children of 1–5 yr age *(22,40)*.

   The true incidence of *Y. enterocolitica* and *Y. pseudotuberculosis* infections is not known. The results of studies differ depending on the population materials and the diagnostic methods used. The isolation rates of *Y. enterocolitica* from stool cultures of patients with diarrhea in Finland, Canada, Chile, the Netherlands, New Zealand, Italy, and the United States have been reported to be 0.6–2.9% *(19,128–130)*. The infection rate is probably much higher because only the most serious cases are registered. The prevalence of *Y. enterocolitica* O:3/O:9-specific antibodies was relatively high in Finland (19 and 31% by enzyme immunoassay and immunoblotting, respectively) and in Germany (33 and 43%) when healthy blood donors were studied *(131)*. This may indicate a high amount of subclinical *Yersinia* infections in the healthy population.

Table 10
Virulence Determinants of *Y. enterocolitica* and *Y. pseudotuberculosis*

Genomic origin	Determinant	Function	*Y. enterocolitica*	*Y. pseudotuberculosis*
Plasmid				
ysc	Yops	Resistance to phagocytosis	+	+
yadA	YadA	Attachment, invasion	+	(+)
Chromosome				
Inv	Invasin	Attachment, invasion	+	+
ail	Ail	Attachment, invasion	+	–
		Serum resistance	+	+
yst	Yst	Fluid secretion in intestine	+	–
myf	Myf	Attachment	+	–
psa	pH6 antigen	Attachment, serum resistance	–	+
ypm	YPM	Stimulation of T-cells	–	+

4. PATHOGENICITY

Several genes contribute to the virulence of *Y. enterocolitica* and *Y. pseudotuberculosis* (Table 10). All fully virulent *Yersinia* harbor an approx 70-kb virulence plasmid, termed pYV (plasmid for *Yersinia* virulence), which is required for full expression of virulence *(132,133)*. Chromosomal-encoded factors are also needed for virulence, because virulence functions are transferable by the virulence plasmid only to the plasmid-cured strains derived from parental strains *(134,135)*.

Virulence plasmids of pathogenic *Yersinia* are closely related to each other, sharing functional similarities and a high degree of DNA homology *(136)*. The presence of pYV enables virulent *Yersinia* to survive and multiply in lymphoid tissues of their host *(9)*. The pYV codes for an outer membrane protein YadA (*Yersinia* adhesin A), and a set of highly regulated secreted proteins, called Yops (*Yersinia* outer proteins). Yops are secreted by means of a type III secretion system, Ysc. Protein YadA is a major outer membrane protein, which forms a fibrillar matrix on the surfaces of *Y. enterocolitica* and *Y. pseudotuberculosis (137)*, and is only expressed at 37°C *(132)*. YadA is an important virulence factor of *Y. enterocolitica*, but seems to be dispensable for the virulence of *Y. pseudotuberculosis (138)*. YadA plays a protective role in *Y. enterocolitica* with several functions (Table 11).

The *yop* genes located on the pYV code for Yops, which were originally described as *Yersinia* outer membrane proteins because they were detected in the outer membrane fraction of bacterial extracts *(139)*. Today, they are considered secreted proteins *(140)*, which imbue *Yersinia* with the capacity to resist nonspecific immune response *(141)*. Yops protect *Yersinia* from the macrophage by destroying its phagocytic and signaling capacities, and finally, inducing apoptosis. With the type III secretion system (Ysc), extracellularly located *Yersinia* that are in close contact with the eukaryotic cell deliver the toxic bacterial proteins (Yops) into the cytosol of the target cell. Genes specifying the type III machinery (*ysc*) are also located on the pYV. The *yop* and *ysc* genes are

Table 11
Role of YadA in the Virulence of *Y. enterocolitica*

Functions of YadA	References
Serum resistance	*(136)*
Surface hydrophobicity	*(279)*
Autoagglutination	*(280)*
Adhesion to epithelial cells	*(281)*
Hemagglutination	*(137)*
Binding to intestinal brush border membrane ·	*(282)*
Binding to fibronectin	*(283)*
Binding to collagen	*(284)*
Binding to intestinal submucosa	*(285)*
Resistance to killing by polymorphonuclear leukocytes	*(286)*

temperature- and calcium-regulated, being expressed maximally at 37°C in response to the presence of a low calcium concentration *(141)*. All *Yersinia* strains carrying the virulence plasmid exhibit a phenotype known as low-calcium response because it manifests only when pYV-bearing strains are grown at 37°C in media containing a low concentration of Ca^{2+} *(142,143)*. This growth-restriction phenomenon is associated with the massive production of Yops *(9)*.

Invasion of epithelial layers require at least two chromosomal genes, *inv* (invasion) and *ail* (attachment invasion locus) *(144)*. The *inv* codes for Inv, an outer membrane protein, which appears to play a vital role in promoting entry into epithelial cells of the ileum during the initial stages of *Y. enterocolitica* and *Y. pseudotuberculosis* infections *(145,146)*. Inv binds to β1 integrins and thereby mediates invasion into eukaryotic cells in vitro. This gene is found in all *Yersinia* spp.; however, nonpathogenic strains lack functional *inv* homologous sequences *(147)*. Although the Inv protein is maximally synthesized at temperatures below 28°C, the Inv protein is equally well produced at 37°C under acidic conditions *(145)*. The *ail*, in turn, codes for the surface protein Ail, which is produced at 37°C. The Ail also mediates resistance to the bactericidal effect of complement. In contrast to the *inv*, the *ail* is only found in *Y. enterocolitica* bioserotypes associated with the disease *(148)*.

Y. enterocolitica and *Y. pseudotuberculosis* migrate after ingestion through the stomach and the small intestine to the terminal end of ileum. The bacteria bind to the follicle-associated epithelium of the Peyer's patches, which are a part of the gut-associated lymphoid tissue. *Y. enterocolitica* and *Y. pseudotuberculosis* penetrate the intestinal mucosa, through M cells, specialized cells involved in intestinal antigen uptake *(149,150)*. Attachment and invasion of M cells are mediated by chromosomal determinants, Inv and Ail proteins, and plasmid determinant YadA *(144,150)*. After penetration of the intestinal epithelium, enteropathogenic *Yersinia* colonizes the Peyer's patches. The ability to survive and multiply within the lymphoid follicles and other tissue is associated with the presence of virulence plasmid, which is essential for the pathogenesis of *Yersinia (151)*. Antiphagocytic properties of *Yersinia* are mainly mediated by Yop virulon products, which through bacterium–host cell contact subvert phagocyte function and enable survival and extracellular multiplication in host lymphoid tissue *(133)*. Usually the infection is limited to the intestinal area, but sometimes the microorganisms drain into the mesenteric lymph nodes and give rise to a systemic infection.

Both *Y. enterocolitica* and *Y. pseudotuberculosis* express fimbriae that are found on their surface *(152)*. *Y. enterocolitica* synthesizes a fibrillar structure known as Myf (mucoid *Yersinia* factor). The chromosomal locus involves three genes: *myfA*, *myfB*, and *myfC (153)*. Myf may serve as an intestinal colonization factor for *Y. enterocolitica (154)*. The fimbriae in *Y. pseudotuberculosis* are called pH6 antigen, which is synthesized at 37°C under acidic conditions alone. The *psa* genes in the chromosome code for pH6 antigen *(155)*. This antigen induces agglutination of erythrocytes, skin necrosis, upon intra-dermal injection in rabbits, and cytotoxicity for monocytes *(152)*.

Iron is an essential growth factor for the multiplication of almost all bacteria, including *Y. enterocolitica* and *Y. pseudotuberculosis*. The storage, transport, biosynthesis, and regulation of iron are important for the adaptation of *Yersinia* species to different environmental conditions *(133,156)*. The ability to capture iron in vivo is one of the main factors that differentiate high- and low-pathogenic bacteria. *Yersinia* species have developed a variety of ingenious mechanisms for the retrieval of iron, including the ability to bind exogenous siderophores and produce their own siderophores, along with a range of other transport systems *(156)*. One pathway for iron capture is to secrete small molecules called siderophores. These molecules chelate the iron bound to eucaryotic proteins and deliver the metal inside the cytosol of the bacteria. Yersiniabactin is a siderophore, which is synthesized by high-pathogenicity *Yersinia* only. Mouse-lethal serotypes of *Y. enterocolitica*, including serotype O:8, are able to synthesize the yersiniabactin. The yersiniabactin biosynthesis and transport genes are clustered within a region of the chromosome referred to as high-pathogenicity islands (HPI) *(157)*. Low-pathogenicity *Yersinia*, including serotypes O:3 and O:9 of *Y. enterocolitica*, use exogenous siderophores like ferrioxamine, which is commonly used to treat patients with iron over-load, for capturing iron, thus enhancing their ability to disseminate in the host *(152)*. These iron-regulation systems ensure the survival and multiplication of *Yersinia* in an iron-competitive environment.

The chromosome of *Y. enterocolitica*, but not of *Y. pseudotuberculosis*, encodes a heat-stable enterotoxin, Yst *(158)*. It is a 30-amino acid peptide, which resembles closely the heat-stable toxin of enterotoxigenic *Escherichia coli*. The role of Yst in the pathogenesis of *Y. enterocolitica* diarrhea in humans is at present uncertain. Nonpathogenic strains of *Y. enterocolitica* and strains of related species have been found to produce Yst using the infant mouse model *(159)*, and the *yst* gene has been detected in strains of biotype 1A, *Y. kristensenii* and *Y. intermedia (158,160)*. Absence of enterotoxin production in vitro at temperatures exceeding 30°C suggests that this toxin is not produced in the intestinal lumen. However, experimental evidence suggests that Yst may play an important factor involved in *Y. enterocolitica*-associated diarrhea in young rabbits *(161)*. Robins-Brown et al. *(162)* investigated the pathogenicity of *Y. enterocolitica* O:3, including the role of Yst, using gnotobiotic newborn piglet model. There was evidence that enterotoxin may promote intra-intestinal proliferation of *Y. enterocolitica*, thus favoring increased shedding of bacteria and encouraging their spread between hosts.

Urease is produced by all clinical isolates of enteropathogenic *Yersinia*. It is encoded by the urease gene complex (*ure*) in the chromosome *(163)*. This enzyme hydrolyses urea to form carbonic acid and ammonia, leading to an increase in pH. Urease activity may contribute to the virulence by conferring acid tolerance and thereby enhancing bacterial survival in the stomach and other acidic environments *(164)*. The decrease in virulence after intragastric inoculation of urease-negative mutant indicates that the main role of urease is during the initial stage of the bacterial infection, when the bacteria reach the stomach *(165)*.

LPS is a major surface component of the outer membrane of Gram-negative bacteria. In *Yersinia*, the genes directing the biosynthesis of LPS are chromosomally located. The full LPS molecule includes three structurally distinct parts: the lipid A anchored in the outer membrane, an oligosaccharide core, and the O-antigenic polysaccharide (O-antigen) *(166)*. The lipid A part is believed to be responsible for endotoxin activity and to play a central role in sepsis and septic shock owing to Gram-negative bacteria. Skurnik et al. *(167)* have suggested that the outer core provides resistance against defense mechanisms, most probably those involving bactericidal peptides. Although the O-antigen is required for full virulence, its role has yet to be clarified *(168)*. A total absence of O-antigen in *Y. enterocolitica* O:3 has been shown to reduce virulence in the infected mouse model *(169)*.

Y. pseudotuberculosis has been described to produce a superantigen, *Y. pseudotuber-culosis*-derived mitogen (YPM). YPM-producing strains can be separated into three clusters of strains, which produce YPMa, YPMb, or YPMc encoded by the *ypmA*, *ypmB*, and *ypmC*, respectively *(16,170)*. It has been demonstrated that distribution of the *ypm* genes among the strains from diverse geographical areas was in close correlation with the difference in the severity of clinical manifestations between the different areas *(171)*. The frequency of YPMa-producing strains has shown to be higher in the Far East than in Europe, where systemic symptoms in the infected patients were less common compared with the Far East *(171)*. Patients with systemic symptoms such as lympha-denopathy, transient renal dysfunction, and arthritis had shown to have significantly higher titers of anti-YPMa than patients with gastrointestinal tract symptoms alone *(172)*. *Y. pseudotuberculosis* has been divided into six subgroups by Fukushima et al. *(16)*. The presence of three virulence factors—YPM, HPI, and pYV—differed among these groups (Table 12). The pathogenicity of *Y. pseudotuberculosis* depended on the presence of YPMa, HPI, and pYV. The pYV is essential for pathogenicity. The differ-ence in clinical manifestations of *Y. pseudotuberculosis* infection between the Far East and the western countries is related to the heterogeneity in the distribution of YPMa and HPI (Table 12). Most clinical strains were classified into two subgroups: Far East systemic type (group 3: YPMa+/HPI−) and European gastroenteritis type (group 2: YPM−/HPI+). The third group was European low-pathogenicity type (group 3: YPMa−/HPI−).

5. CLINICAL CHARACTERISTICS

The enteric pathogens *Y. enterocolitica* and *Y. pseudotuberculosis* can cause illness ranging from self-limiting enteritis to fatal systematic infection. The major clinical features of infections with these microbes are similar and usually characterized by diarrhea, abdominal pain, and fever (Table 13). Sometimes the symptoms suggest acute appendicitis, and the patient is operated *(173–175)*. The findings may include mesenteric lymphadenitis or terminal ileitis, more rarely true appendicitis is found. However, Lamps et al. *(176)* reported that both *Y. enterocolitica* and *Y. pseudotuberculosis* are important causes of granulomatous appendicitis and *Yersinia* infection may mimic

Table 12
Correlation Between Pathogenicity and Virulence Factors Among
Y. pseudotuberculosis **Strains**

Sub group	Location	Pathogenicity	YPMa	YPMb	YPMc	HPI	pYV
1	Far East	High	+	−	−	+	+
2	Far East, West	High	−	−	−	+	+
3	Far East	High	+	−	−	−	+
4	Far East	NP[a]	−	+	−	−	−
5	Far East, West	Low	−	−	+	+	+
6	Far East, West	High	−	−	−	−	+

[a]NP, Nonpathogenic.

Crohn's disease. Sometimes *Yersinia* infection is suspected on the basis of postinfectious complications such as reactive arthritis, eye inflammation, or erythema nodosum, after an infection, which may passed unnoticed *(175,177–179)*. The clinical manifestations of the infection depend to some extent on the age and physical state of the patient, the presence or absence of underlying medical conditions, and the serotype of the organism *(120)*. The symptom can even be faint and short-lived that yersiniosis is not diagnosed, despite fecal carriage *(19,113,180,181)*.

Gastroenteritis, caused by *Y. enterocolitica*, is the most frequent form of Yersiniosis and occurs most commonly in infants and young children *(19,129,182)*. The incubation period ranges for 1 to 11 days *(120)*. The minimal infective dose for humans has not been determined. The illness is generally mild and self-limiting persisting for 5–14 d, although diarrhea may persist several weeks, occasionally even for several months *(120)*. Sometimes diarrhea may be bloody, and the fever may be high, especially in infants *(101)*. Excretion of the organism in the stools continues for an average of 4 wk after cessation of symptoms, which may contribute to intra-familial spread *(183)*.

Intestinal yersiniosis may mimic acute appendicitis: abdominal pain in right lower quadrant, fever, and no diarrhea *(100,184–186)*. Upon surgery, mesenteric lymphadenitis and true appendicitis may be distinguished by visual examination of the intestine and by light microscopy of the lymph-node biopsy *(180,187)*. Macroscopically there is acute mesenteric lymphadenitis and terminal ileitis, and histologically normal or only mildly inflamed appendix. Unnecessary surgery could be avoided by abdominal ultra sound, computed tomography scanning or magnetic resonance imaging to distinguish between normal appendix and enlarged mesenteric lymph nodes, *(184)*. The acute mesenteric lymphadenitis, caused by either *Y. enterocolitica* or *Y. pseudotuberculosis*, is a common symptom among children and young adolescents *(120,175)*.

Diarrhea and fever are usually milder in adults, but occasionally severe forms of gastrointestinal disease are seen. These include diffuse ulceration and inflammation of large and small bowel *(117)*, peritonitis, toxic megacolon, intestinal perforation, ileocolic intussusception, and mesenteric vein thrombosis leading to intestinal necrosis and cholangitis. Also mild hepatitis and pancreatitis may be symptoms of yersiniosis. *Yersinia* infection may also be asymptomatic and pass unnoticed *(180)*. Postinfection manifestations occur mainly in young adults *(177)*. Less commonly, *Y. enterocolitica* may cause focal infection at extra-intestinal sites with or without bacteremia *(188)*.

Table 13
Clinical Symptoms Associated With *Y. enterocolitica* and *Y. pseudotuberculosis* Infections

Country	No. of patients	Symptoms	Bacteriological findings		Reference
Canada	74	Diarrhea 92%, fever 88%, abdominal pain 76%, fatigue 68%	YP	O:1b	(287)
Denmark	27	Diarrhea 93%, fever 71%, abdominal pain 63%, vomiting 43%	YE	O:3	(21)
Finland	19	Abdominal pain 74%, fever 58%, diarrhea 21%	YP	O:3	(276)
	34	Fever 53%, abdominal pain 38%, nausea 15%, diarrhea 15%,	YP	1a	(180)
	47	Abdominal pain 89%, fever 68%, diarrhea 45%,	YP	O:3	(125)
France	42	Diarrhea 71%, abdominal pain 64%, only fever 29%, arthritis 14%, erythema nodosum 12%	YE	O:9	(63)
Japan	189	Abdominal pain 86%, fever 76%, diarrhea 60%, nausea 24%, vomiting 4%	YE	O:3	(275)
	544	Abdominal pain 64%, fever 50%, diarrhea 32%, nausea 25%, vomiting 11%	YE	O:3	(275)
	39	Fever 100%, abdominal pain 68%, exanthema 68%	YP	O:5a	(277)
The Netherlands	189	Enteritis 63%, extramesenteric form 21%, arthritis and erythema nodosum 23%	YE		(221)
Norway	67	Diarrhea 97%, abdominal pain 77%, fever 54%	YE	O:3	(288)
United States	9	Abdominal pain 100%, Fever 100%, diarrhea 78%, nausea 61%, pharyngitis 22%	YE	O:8	(98)
	10	Diarrhea 100%, fever 80%, vomiting 70%	YE	O:3	(101)

[a]YE, *Y. enterocolitica*.
[b]YP, *Y. pseudotuberculosis*.

Cases of pneumonia, lung abscesses, empyema liver, splenic or renal abscesses, endocarditis, osteomyelitis, septic arthritis, and meningitis have all been reported *(189–195)*. Exudative pharyngitis caused by *Y. enterocolitis* has also been reported *(196,197)*. Rarely, in patients with a predisposing, underlying disease, such as diabetes mellitus and hepatic cirrhosis, or iron overload, or in young infants, infection with *Y. enterocolitica* may cause a septicemia illness *(198,199)*. In Korea, several cases of septicemic form of *Y. pseudotuberculosis* infections have been reported *(200)*. Hosaka et al. *(30)* reported *Y. enterocolitica* O:8 septicemia in an otherwise healthy adult. Sepsis can also occur during blood transfusion *(201)*. One source of *Y. enterocolitica*-contaminated red blood cell concentrate has been reported to be a blood donor with asymptomatic bacteremia *(202,203)*.

Complications of *Yersinia* infection include reactive arthritis (ReA) and occasionally Reiter's syndrome, erythema nodosum, eye inflammations such as uveitis and conjunctivitis, urethritis, balanitis, myocarditis, and glomerulonephritis *(120,177,182,204–206)*. Postinfection complications usually develop within 1 wk to 1 mo after initial infection and may sometimes be the only obvious clinical manifestation of *Yersinia* infection *(177)*. Arthritis and uveitis are associated with the presence of HLA-B27 antigen *(207–211)*. Erythema nodosum is seen in about 10–30% of patients with yersiniosis *(177)*. *Y. pseudotuberculosis* has also been associated with Kawasaki disease (mucocutaneous lymph node syndrome) an important illness of children in Japan and scarlet fever disease *(42)*.

Reactive arthritis triggered by infection with by *Y. enterocolitica* or *Y. pseudotuberculosis* is a frequent cause of acute arthritis predominantly in young and middle-aged patients *(120,206)*. The patients may suffer from arthritis, after 1–3 wk of infection with or without gastroenteritis. Joint symptoms vary from mild arthralgia to severe arthritis. The arthritis is typically an asymmetric oligo- or polyarthritis with sudden onset affecting more often the joints of the lower extremities, such as knee, ankle, and toes, but sometimes also wrist and fingers *(213,214)*. ReA is often migrating and painful. Lower back and muscular pains often coexist. The enteric arthritis is equally common among women and men *(213)*. The mean duration of *Yersinia*-triggered arthritis has been reported to be between 3 and 6 mo. However, a substantial proportion of patients, about 20% in endemic area, experience a chronic course that continues longer than 12 mo *(215)*. The most important factor, which influence prognosis, is HLA-B27. Patients with HLA-B27 have not only a more severe phase of arthritis but also a higher risk for chronic low back pain and sacroilitis, as well as chronic eye inflammation and development of ankylosing spondylitis *(177)*. It has been suggested that *Yersinia* persist after the initial infection in those patients developing arthritis, even through the stool cultures become negative for *Yersinia* shortly after the onset of infection *(179,209)*. Patients with *Yersinia*-triggered arthritis show persistent IgA antibodies against *Yersinia* outer membrane proteins, which is may be the result of chronic stimulation of the gut associated lymphoid tissue by persistent *Yersinia (216)*. Bacteria have not been found in affected joint but degraded LPS of *Yersinia* have been detected in the joint synovium *(214,217)*.

6. CHOICE OF TREATMENT

Yersinia strains are usually susceptible in vitro to co-trimoxazole, aminoglycosides, tetracyclines, chloramphenicol, third-generation cephalosporins, and quinolones *(218–221)*. *Y. enterocolitica* produces frequently β-lactamases and is usually resistant to penicillins and to first-generation cephalosporins *(193,221,222)*. In contrast, *Y. pseudotuberculosis* is susceptible to penicillins because it lacks β-lactamase.

The great majority of *Yersinia* enterocolitis and acute mesenteric adenitis in immunocompetent hosts are self-limiting, and supportive care including fluid and electrolyte replacement is usually sufficient. Antimicrobial therapy has not been proved to be essential or efficient in the treatment of uncomplicated enterocolitis or the pseudoappendicular syndrome. In a placebo-controlled, double blind evaluation of co-trimoxazole therapy for *Y. enterocolitica* enterocolitis in children, treatment failed to shorten the clinical or bacteriologic course of the illness *(183)*. A prospective study has been undertaken to evaluate the incidence, course, effect of treatment and outcome of patients with *Y. enterocolitica* infections *(221)*. A total of 189 patients were followed. The majority of the patients with enteritis recovered without antibiotic treatment. However, the duration of enteritis was not significantly influenced by antibiotic treatment.

Systemic infection, focal extraintestinal infection, and enterocolitis in immunocompromised patients should be treated with antimicrobial therapy. In an experimental model of systemic *Y. enterocolitica* infection in mice, doxycycline, and gentamicin were effective in stopping the proliferation of serotype O:8, where as cefotaxime and imipenem were ineffective. Cover and Aber *(120)* recommended a combination of doxycycline and an aminoglyside for the treatment of patients with bactremia until antimicrobial-susceptibility tests have been performed. Co-trimoxazole, tetracyclines, fluoroquinolones, and aminoglycosides are usually effective in the treatment of septicemic infection and tissue abscesses in liver and spleen *(223)*. Ciprofloxacin has shown to be more appropriate for *Yersinia* osteomyelitis, and co-trimoxazole, chloramphenicol, or third-generation cephalosporin is preferable for central nervous infection. However, fluorochinolones should be considered as a first-line therapy for invasive infection related to *Y. enterocolitica* (193,222). Before, streptomycin or sulphamethoxazole have been recommended for the treatment of severe and acute mesenteric adenitis or enteric fever-like syndrome *(224)*, but these are not any more recommended for the treatment of *Y. enterocolitica* infections. However, the antibiotic therapy should always be based on the results of the antibiogram.

ReA patients are treated symptomatically with nonsteroidal anti-inflammatory drugs or, in severe cases, with steroids, and those with chronic courses are treated with second-line drugs, such as sulfasalazine *(215)*. There was a tendency towards faster remission and relief of pain in patients with ReA receiving ciprofloxacin *(216)*. No effect of ciprofloxacin treatment on arthritis triggered by Yersinia was reported by Sieper et al. *(215)*. Oili et al. *(225)* have described two patients with acute ReA caused by *Y. enterocolitica* who were treated with infliximab with a good response in the acute phase. However, the influence of new drugs on the prognosis of patients with ReA remains to be evaluated.

7. RESISTANCE EPIDEMIOLOGY

A number of antimicrobial agents are active in vitro against *Y. enterocolitica* and *Y. pseudotuberculosis* strains isolated from human *(218,226–228)* and nonhuman sources *(219,224,229,230)*. These include aminoglycosides (e.g., gentamicin, streptomycin, tobramycin, and kanamycin), the third-generation cephalosporins (e.g., ceftriaxone, caftazidime, and cefotaxime), co-trimoxazole, tetracyclines, chloramphenicol, fluoroquinolones (e.g., ciprofloxacin, norfloxacin, and ofloxacin), imipenem, and aztreonam *(227–229,231–233)*. Isolates from human, environmental, and animal sources have essentially equal susceptibility patterns *(234)*.

Yersinia strains are generally resistant to β-lactamase-sensitive penicillins, such as ampicillin, cloxacillin, carbenicillin and ticarcillin, the first-generation narrow-spectrum cephalosporins, erythromycin, clidamycin, and vancomycin *(229,234,235)*. However, some differences in antimicrobial susceptibilities of *Y. enterocolitica* strains to β-lactam antibiotics have been detected among different biotypes (Table 14) and serotypes *(220)*.

Table 14
Antibiotic Susceptibility of Different Biotypes of *Y. enterocolitica*

Antibiotic	Biotype						Reference
	1A	1B	2	3	4	5	
Ampicillin		S[a]	R[b]		R		(234)
Amoxicillin/ clavulanate	R	S	R	R	S	S	(231)
	R			R	S		(226)
		S	R		S		(234)
	R	R	R	R	S		(227)
Carbenicillin		R	S		R		(234)
	R			S	R		(226)
Cefixime	R	S	R	R	S	S	(231)
Cefoxitin	R			R	S	S	(226)
	R	S	R	S	S	S	(231)
Fosfomycin	R	R	S	S	S	S	(231)
Ticarcillin	R	R	R	S	R	R	(231)
	R			S	R		(226)
		R	S		R		(234)

[a]Sensitive.
[b]Resistant.

Most strains of biotype 4 have shown to be sensitive to amoxicillin/clavulanate and to third-generation cephalosporins, but resistant to ampicillin, carbenicillin, ticarcillin, and cephalotin *(226,227,231,234)*. In comparison, most strains of biotype 2 and biotype 3 have shown to be more susceptible to both carbenicillin and ticarcillin, but resistant to amoxicillin–clavulanic acid *(219,226,227,231,234)*. Strains of biotype 1B, on the other hand, have shown to exhibit high rates of suscebtibility to ampicillin and amoxicillin–clavulanic acid, but resistance to carbenicillin, ticarcillin, and cephalothin, whereas strains of biotype 1A have shown to be resistant to amoxicillin–clavulanic acid *(226,231,234)*.

The differences in antimicrobial susceptibility among *Y. enterocolitica* strains of different biotypes are largely attributed to β-lactam resistance, which is mediated by the production of two chromosomally mediated β-lactamases (A and B) *(236,237)*. Lactamase A is a noninducible broad-spectrum enzyme and lactamase B is an inducible cephalosporinase *(238)*. *Y. enterocolitica* strains belonging to bioserotype 4/O:3, phage types VIII (from Europe, Asia, and Brazil) and IXa (from South Africa and Hungary) have shown to be homogenous in expression of β-lactamase, with all strains producing both enzyme A and enzyme B. However, *Y. enterocolitica* 4/O:3/IXb strains isolated in Canada have shown to produce both enzyme A and enzyme B when *Y. enterocolitica* 4/O:3/IXb strains from Australia and New Zealand lacked the cephalosporinase enzyme B *(239)*. These two lactamases account for resistance to ampicillin, cephalothin, carbenicillin, and ticarcillin. Most of *Y. enterocolitica* strains of less-commonly isolated biotypes 2 and 3 lack the enzyme A, which may explain the sensitivity to carbenicillin and ticarcillin.

Emerging antimicrobial resistance phenotypes have been recognized among multiple zoonotic pathogens, including *Y. enterocolitica (240)*. Antimicrobial agents are used in high quantities for therapy, prophylaxis and growth promotion in modern food-animal production *(241)*. Almost all *Y. enterocolitica* strains from pigs examined by Aarestrup et al. *(241)* in Denmark were resistant to ampicillin but susceptible to streptomycin and sulphonamides. However, a limited number of *Y. enterocolitica* strains were resistant nalidixic acid and a single isolate to gentamicin and spectinomycin. In the last few years, clinical isolates that are resistant to chloramphenicol, streptomycin, sulfonamides, co-trimoxazole, and nalidixic acid have been reported *(223,242–244)*. Sanchez et al. have reported the emergence of nalidixic acid-resistant *Y. enterocolitica* around Madrid in Spain *(244)*. Prats et al. *(223)* showed that the rate of resistance among *Y. enterocolitica* O:3 strains has increased up to 90% for streptomycin and sulfonamides, 70% for trimethoprim–sulfamethoxazole, 60% for chloramphenicol, and 5% for nalidixic acid. Kelley et al. *(245)* studied 22 *Y. enterocolitica* strains recovered from poultry litter in the United States and found 4 out of 22 isolates that exhibited multi-resistance to bacitracin, penicillin, streptomycin, and tetracycline. In total, 13 out 22 isolates were resistant to tetracycline. Funk et al. *(246)* reported resistance to oxytetracycline also from slaughter swine in the United States. Tetracyclines are commonly added to swine feeds. Oxytetra-cycline resistance is often plasmid mediated and could be impacted by on-farm antibiotic usage. Resistance to sulfonamides (except trimethoprim–sulfamethoxazole) is widespread *(67,246)*, but Preston et al. *(234)* found a high rate of susceptibility to sulfamethoxazole. Sulfonamides have been used for growth promotion and prevention of diseases in swine. The resistance to sulfonamides among slaughter swine may indicate acquired-resistance from their use in animal feeds *(229)*.

REFERENCES

1. Solomon, T. (1995) Alexander Yersinin and plague bacillus. *J. Trop. Med. Hyg.* **98,** 209–212.
2. Carniel, E. (2003) Evolution of pathogenic *Yersinia*, some lights in the dark. *Adv. Exp. Med. Biol.* **529,** 3–12.
3. Achtman, M., Zurth, K., Morelli, G., Torrea, G., Guiyoule, A., and Carniel, E. (1999) *Yersinia pestis*, the cause of plague, is a recently emerged clone of *Yersinia pseudotuberculosis*. *Proc. Natl Acad. Sci.* **96,** 14,043–14,048.
4. Frederiksen, W. (1964) A study of some *Yersinia pseudotuberculosis*-like bacteria ("*Bacterium enterocoliticum*" and "*Pasteurella X*"). *Scand. Congr. Pathol. Microbiol.* **14,** 103–104.
5. Bercovier, H. and Mollaret, H. H. (1984) Genus XIV. In: *Yersinia: Bergey's Manual of Systematic Bacteriology* (Krieg, N. R., ed.), Vol. 1, Williams & Wilkins, Baltimore, MD, pp. 498–506.
6. Bercovier, H., Steigerwald, A. G., Guiyoule, A., Huntley-Carter, G., and Brenner, D. J. (1984) *Yersinia aldovae* (formerly *Yersinia enterocolitica*-like group X2) a new species of *Enterobacteriaceae* isolated from aquatic ecosystems. *Int. J. Syst. Bacteriol.* **34,** 166–172.
7. Aleksic, S., Steigerwalt, A., Bockemühl, J., Huntley-Carter, G. P., and Brenner, D. J. (1987) *Yersinia rohdei* sp. nov. isolated from human and dog feces and surface water. *Int. J. Syst. Bacteriol.* **37,** 327–332.
8. Wauters, G., Janssens, M., Steigerwalt, A. G., and Brenner, D. J. (1988) *Yersinia mollaretii* sp. nov. and *Yersinia bercovier* sp. nov., formerly called *Yersinia enterocolitica* biogroups 3A and 3B. *Int. J. Syst. Bacteriol.* **38,** 424–429.
9. Cornelis, G. R., Boland, A., Boyd, A. P., et al. (1998) The virulence plasmid of *Yersinia*, an antihost genome. *Microbiol. Mol. Biol. Rev.* **62,** 1315–1352.
10. Wauters, G., Kandolo, K., and Janssens, M. (1987) Revised biogrouping scheme of *Yersinia enterocolitica*. *Contr. Microbiol. Immunol.* **9,** 14–21.

11. Burnens, A. P., Frey, A., and Nicolet, J. (1996) Association between clinical presentation, biogroups and virulence attributes of *Yersinia enterocolitica* strains in human diarrhea disease. *Epidemiol. Infect.* **116,** 27–34.

12. Tennant, S. M., Grant, T. H., and Robins-Brown, R. M. (2003) Pathogenicity of *Yersinia enterocolitica* biotype 1A. *FEMS Immunol. Med. Microbiol.* **38,** 127–137.

13. Neubauer, H., Aleksic, S., Hensel, A., Finke, E. J., and Meyer, H. (2000) *Yersinia enterocolitica* 16S rRNA gene types belong to the same genospecies but form three homology groups. *Int. J. Med. Microbiol.* **290,** 61–64.

14. Neubauer, H., Hensel, A., Aleksic, S., and Mayer, H. (2000) Identification of *Yersinia enterocolitica* within the genus *Yersinia*. *Syst. Appl. Microbiol.* **23,** 58–62.

15. Tsubokura, M. and Aleksic, S. (1995) A simplified antigenic schema for serotyping of *Yersinia pseudotuberculosis*: phenotypic characterisation of reference strains and preparation of O and H factor sera. *Contr. Microbiol. Immunol.* **13,** 99–105.

16. Fukushima, H., Matsuda, Y., Seki, R., et al. (2001) Geographical heterogeneity between Far Eastern and Western Countries in Prevalence of the virulence plasmid, the superantigen *Yersinia pseudotuberculosis*-derived mitogen, and the high-pathogenicity island among *Yersinia pseudotuberculosis* strains. *J. Clin. Microbiol.* **39,** 3541–3547.

17. Aleksic, S. and Bockemühl, J. (1984) Proposed revision of the Wauters et al. antigenic scheme for serotyping of *Yersinia enterocolitica*. *J. Clin. Microbiol.* **20,** 99–102.

18. Lee, L. A., Gerber, A. R., Lonsway, D. R., et al. (1990) *Yersinia enterocolitica* O:3 infection in infants and children associated with household preparation of chitterlings. *N. Engl. J. Med.* **322,** 984–987.

19. Glenn Morris, J., Jr., Prado, V., Ferreccio, C., et al. (1991) *Yersinia enterocolitica* isolated from two cohorts of young children in Santiagi, Chile: incidence of and lack of correlation between illness and proposal virulence factors. *J. Clin. Microbiol.* **29,** 2784–2788.

20. Kontiainen, S., Sivonen, A., and Renkonen, O. V. (1994) Increased yields of pathogenic *Yersinia enterocolitica* strains by cold enrichment. *Scand. J. Infect. Dis.* **26,** 685–691.

21. Munk Petersen, A., Vinther Nielsen, S., Meyer, D., Ganer, P., and Ladefoged, K. (1996) Bacterial gastro-enteritis among hospitalized patients in a Danish County. *Scand. J. Gastroenterol.* **31,** 906–911.

22. Verhaegen, J., Charlier, J., Lemmes, P., Delmee, M., Van Noyen, R., and Wauters, G. (1998) Surveillance of human *Yersinia enterocolitica* infections in Belgium: 1967–1996. *CID* **27,** 59–64.

23. Aleksic, S. (1995) Occurrence of *Yersinia enterocolitica* antigens O:3, O:9 and O:8 in different *Yersinia* species, their corresponding H antigens and origin. *Contr. Microbiol. Immunol.* **13,** 89–92.

24. Wauters, G., Aleksic, S., Charlier, J., and Schulze, G. (1991) Somatic and flagellar antigens of *Yersinia enterocolitica* and related species. *Contr. Microbiol. Immunol.* **12,** 239–243.

25. Bottone, E. J. (1999) *Yersinia enterocolitica*: overview and epidemiologic correlates. *Microb. Infect.* **1,** 323–333.

26. Hoogkamp-Korstanje, J. A. A., DeKonig, J., and Samson, J. P. (1986) Incidence of human infection with *Yersinia enterocolitica* serotypes O3, O8 and O9 and the use of indirect immuno fluorescence in diagnosis. *J. Inf. Dis.* **153,** 138–141.

27. Maruyama, T. (1987) *Yersinia enterocolitica* infection in humans and isolation of the microorganism from pigs in Japan. *Contr. Microbiol. Immun.* **9,** 48–55.

28. Ichinohe, H., Yoshioka, M., Fukushima, H., Kaneko, S., and Maruyama, T. (1991) First isolation of *Y. enterocolitica* serotype O:8 in Japan. *J. Clin. Microbiol.* **29,** 846–847.

29. Ostroff, S. M. (1995) *Yersinia* as an emerging infection: epidemiologic aspects of Yersiniosis. *Contr. Microbiol. Immunol.* **13,** 5–10.

30. Hosaka, S., Uchiayama, M., Ishikawa, M., et al. (1997) *Yersinia enterocolitica* serotype O:8 septicemia in an otherwise healthy adult: analysis of chromosomal DNA pattern by pulsed-field gel electrophoresis. *J. Clin. Microbiol.* **35,** 3346–3347.

31. Fukushima, H., Tsubokura, M., Otsuki, K., and Kawaoka, Y. (1984) Biochemical heterogeneity of serotype O:3 strains of 700 *Yersinia* strains isolated from humans, others mammals, flies, animal feed, and river water. *Curr. Microbiol.* **11,** 149–154.

32. Zheng, X. B. and Xie, C. (1996) Isolation, Characterisation and epidemiology of *Yersinia enterocolitica* from humans and animals. *J. Appl. Bacteriol.* **81,** 681–684.

33. Fukushima, H., Gomyoda, M., and Aleksic, S. (1998) Genetic variation of *Yersinia enterocolitica* serotype O:9 strains detected in samples from western and eastern countries. *Zentralbl. Bacteriol.* **2,** 167–174.

34. Schiemann, D. A. (1989) *Yersinia enterocolitica* and *Yersinia pseudotuberculosis.* In: *Foodborne Bacterial Pathogens* (Doyle, M. P., ed.), Marcel Dekker, New York, pp. 601–672.

35. Kapperud, G. (1991) *Yersinia enterocolitica* in food hygiene. *Int. J. Food Microbiol.* **12,** 53–66.

36. Toma, S. and Deidrick, V. R. (1975) Isolation of *Yersinia enterocolitica* from swine. *J. Clin. Microbiol.* **2,** 478–481.

37. Doyle, M. P., Hugdahl, M. B., and Taylor, S. L. (1981) Isolation of virulent *Yersinia enterocolitica* from porcine tongues. *Appl. Environ. Microbiol.* **42,** 661–666.

38. Fukushima, H. (2003) Molecular epidemiology of *Yersinia pseudotuberculosis. Adv. Exp. Med. Biol.* **529,** 357–358.

39. Toma, S. (1986) Human and nonhuman infections caused by *Yersinia pseudotuberculosis* in Canada from 1962 to 1985. *J. Clin. Microbiol.* **24,** 465–466.

40. Aleksic, S., Bockemühl, J., and Wuthe, H. H. (1995) Epidemiology of *Y. pseudotuberculosis* in Germany, 1983–1993. *Contrib. Microbiol. Immunol.* **13,** 55–58.

41. Hallanvuo, S., Nuorti, P., Nakari, U. M., and Siitonen, A. (2003) Molecular epidemiology of the five recent outbreaks of *Yersinia pseudotuberculosis* in Finland. *Adv. Exp. Med. Biol.* **529,** 309–311.

42. Tsubokura, M., Otsuki, K., Sato, K., et al. (1989) Special features of distributions of *Yersinia pseudotuberculosis* in Japan. *J. Clin. Microbiol.* **27,** 790–791.

43. Nagano, T., Kiyohara, T., Suzuki, K., Tsubokura, M., and Otsuki, K. (1997) Identification of pathogenic strains within serogroups of *Yersinia pseudotuberculosis* and the presence of non-pathogenic strains isolated from animals and the environment. *J. Vet. Med. Sci.* **59,** 153–158.

44. Hurvell, B. (1981) Zoonotic *Yersinia enterocolitica* infection: host range, clinical manifestations, and transmission between animals and man. In: *Yersinia enterocolitica* (Bottone, E. J., ed.), CRC, Boca Raton, FL, pp. 145–159.

45. Ahvonen, P., Thal, E., and Vasenius, H. (1973) Occurrence of *Yersinia enterocolitica* in animals in Finland and Sweden. *Contr. Microbiol. Immunol.* **2,** 135–136.

46. Kaneko, K. I., Hamada, S., Kasai, Y., and Kato, E. (1978) Occurrence of *Yersinia enterocolitica* in house rats. *Appl. Environ. Microbiol.* **36,** 314–318.

47. Kaneko, K. I. and Hashimoto, N. (1981) Occurrence of *Yersinia enterocolitica* in wild animals. *Appl. Environ. Microbiol.* **41,** 635–638.

48. Kato, Y., Ito, K., Kubokura, Y., Maruyama, T., Kaneko, K. I., and Ogawa, M. (1985) Occurrence of *Yersinia enterocolitica* in wild-living birds and Japanese seerows. *Appl. Environ. Microbiol.* **49,** 198–200.

49. Shayegani, M., Stone, W. B., de Forge, I., Root, T., Parsons, L. M., and Maupin, P. (1986) *Yersinia enterocolitica* and related species isolated from wildlife in New York State. *Appl. Environ. Microbiol.* **52,** 420–424.

50. Kaneuchi, C., Shibata, M., Kawasaki, T., Kariu, T., Kanzaki, M., and Maruyama, T. (1989) Occurrence of *Yersinia* spp. in migratory birds, ducks, seagulls, and swallows in Japan. *Jpn. J. Vet. Sci.* **51,** 805–808.

51. Iannibelli, F., Troiano, P., Volterra, L., Nanni, F., and Chiesa, C. (1991) Comparative isolation of *Yersinia* spp. from avian wildlife by different methods. *Contr. Microbiol. Immunol.* **12,** 26–31.

52. Cork, S. C., Marshall, R. B., Madie, P., and Fenwick, S. G. (1995) The role of wild birds and the environment in the epidemiology of Yersiniae in New Zealand. *N. Z. Vet. J.* **43,** 169–174.

53. Suzuki, A., Hayashidani, H., Kaneko, K. I., and Ogawa, M. (1995) Isolation of *Yersinia* from wild animals living in suburbs of Tokyo and Yokohama. *Contr. Microbiol. Immunol.* **13,** 34–45.

54. Wuthe, H.-H., Aleksic, S., and Kwapil, S. (1995) *Yersinia* in the European brown hare of Northern Germany. *Contr. Microbiol. Immunol.* **13,** 51–54.

55. Sulakvelidze, A., Dalakishvili, K., Barry, E., et al. (1996) Analysis of clinical and environmental *Yersinia* isolates in the Republic of Georgia. *J. Clin. Microbiol.* **34,** 2325–2327.

56. Szita, J., Svidró, A., Kubinyi, M., Nyomárkay, I., and Mihályfi, I. (1980) *Yersinia enterocolitica* infections of animals and human contacts. *Acta Microbiol. Acad. Sci. Hung.* **27,** 103–109.

57. Fukushima, H., Saito, K., Tsubokura, M., Otsuki, K., and Kawaoka, Y. (1983) Isolation of *Yersinia* spp. from bovine faces. *J. Clin. Microbiol.* **18,** 981–982.

58. Christensen, S. G. (1987) The *Yersinia enterocolitica* situation in Denmark. *Contr. Microbiol. Immunol.* **9,** 93–97.

59. Fantasia, M., Mingrone, M. G., Martini, A., Boscato, A., and Crotti, D. (1993) Characterization of *Yersinia* species isolated from a kennel and from cattle and pig farms. *Vet. Rec.* **132,** 532–534.

60. Busato, A., Hofer, D., Lentze, T., Gaillard, C., and Burnens, A. (1999) Prevalence and infection risk of zoonotic enteropathogenic bacteria in Swiss cow-calf farms. *Vet. Microbiol.* **69,** 251–263.

61. Iinuma, Y., Hayashidani, H., Kaneko, K., Ogawa, M., and Hamasaki, S. (1992) Isolation of *Yersinia enterocolitica* O8 from free-living small rodents in Japan. *J. Clin. Microbiol.* **30,** 240–242.

62. Hayashidani, H., Ohtomo, Y., Toyokawa, Y., et al. (1995) Potential sources of sporadic human infections with *Yersinia enterocolitica* serovar O:8 in Aomori prefecture, Japan. *J. Clin. Microbiol.* **33,** 1253–1257.

63. Gourdon, F., Beytout, J., Reynaud, A., et al. (1999) Human and animal epidemic of *Yersinia enterocolitica* O:9, 1989–1997, Auvergne, France. *Emerg. Inf. Dis.* **5,** 719–721.

64. Nielsen, B., Heisel, C., and Wingstrand, A. (1996) Time course of the serological response to *Yersinia enterocolitica* O:3 in experimentally infected pigs. *Vet. Microbiol.* **48,** 293–303.

65. Thibodeau, V., Frost, E. H., Chénier, S., and Quessay, S. (1999) Presence of *Yersinia enterocolitica* in tissues of orally inoculated pigs and the tonsils and feces of pigs at slaughter. *Can. J. Vet. Res.* **63,** 96–100.

66. Shiozawa, K., Nishina, T., Miwa, Y., Mori, T., Akahane, S., and Ito, K. (1991) Colonization in the tonsils of swine by *Yersinia enterocolitica*. *Contr. Microbiol. Immunol.* **12,** 63–67.

67. Christensen, S. G. (1980) *Yersinia enterocolitica* in Danish pigs. *J. Appl. Bacteriol.* **48,** 377–382.

68. Nesbakken, T. and Kapperud, G. (1985) *Yersinia enterocolitica* and *Yersinia enterocolitica*-like bacteria in Norwegian slaughter pigs. *Int. J. Food Microbiol.* **1,** 301–309.

69. Asplund, K., Tuovinen, V., Veijalainen, P., and Hirn, J. (1990) The prevalence of *Yersinia enterocolitica* O:3 in Finnish pigs and pork. *Acta Vet. Scand.* **31,** 39–43.

70. Andersen, J. K., Sørensen, R., and Glensbjerg, M. (1991) Aspects of the epidemiology of *Yersinia enterocolitica*: a review. *Int. J. Food Microbiol.* **13,** 231–238.

71. Letellier, A., Messier, S., and Quessy, S. (1999) Prevalence of *Salmonella* spp. and *Yersinia enterocolitica* in finishing swine at Canadian abattoir. *J. Food Prot.* **62,** 22–25.

72. Nielsen, B. and Wegener, H. C. (1997) Public health and pork and pork products: regional perspectives of Denmark. *Rev. Sci. Tech. Off. Int. Epiz.* **16,** 513–524.

73. Skjerve, E., Lium, B., Nielsen, B., and Nesbakken, T. (1998) Control of *Yersinia enterocolitica* in pigs at herd level. *Int. J. Food Microbiol.* **45,** 195–203.

74. Yanagawa, Y., Maruyama, T., and Sakai, S. (1978) Isolation of *Yersinia enterocolitica* and *Yersinia pseudotuberculosis* from apparently healthy dogs and cats. *Microbiol. Immunol.* **22**, 643–646.

75. Pedersen, K. B. and Winblad, S. (1979) Studies on *Yersinia enterocolitica* isolated from swine and dogs. *Acta Path. Microbiol. Scand. Sect. B.* **87**, 137–140.

76. Fukushima, H., Nakamura, R., Iitsuka, S., Tsubokura, M., Otsuki, K., and Kawaoka, Y. (1984) Prospective systematic study of *Yersinia* spp. in dogs. *J. Clin. Microbiol.* **19**, 616–622.

77. Fantasia, M., Mingrone, M. G., Crotti, D., and Boscato, C. (1985) Isolation of *Yersinia enterocolitica* biotype 4 serotype O:3 from canine sources in Italy. *J. Clin. Microbiol.* **22**, 315–324.

78. Fenwick, S. G., Madie, P., and Wilks, C. R. (1994) Duration of carriage and transmission of *Yersinia enterocolitica* biotype 4, serotype O:3 in dogs. *Epidemiol. Infect.* **113**, 471–477.

79. Jerrett, I. V., Slee, K. J., and Robertson, B. I. (1990) Yersiosis in farmed deer. *Aust. Vet. J.* **67**, 212–214.

80. Inoue, M., Nakashima, H., Mori, T., Sakazaki, R., Tamura, K., and Tsubokura, M. (1991) *Yersinia pseudotuberculosis* infection in the mountain area. *Contrib. Microbiol. Immunol.* **12**, 307–310.

81. Frolich, K., Wisser, J., Schmuser, H., et al. (2003) Epizootiologic and ecologic investigations of European brown hares (Lepus europaeus) in selected populations from Schleswig-Holstein, Germany. *J. Wildl. Dis.* **39**, 751–761.

82. Fukushima, H., Gomyoda, M., Shiozawa, K., Kaneko, S., and Tsubokura, M. (1988) *Yersinia pseudotuberculosis* infection contracted through water contaminated by a wild bird. *J. Clin. Microbiol.* **26**, 584–585.

83. Inoue, M., Nakashima, H., Ishida, T., Tsubokura, M., and Sakazaki, R. (1988) Isolation of *Yersinia pseudotuberculosis* from water. *Zentralbl. Bakteirol. Mikrobiol. Hyg. B* **186**, 338–343.

84. Fukushima, H. (1992) Direct isolation of *Yersinia pseudotuberculosis* from fresh water in Japan. *Appl. Environ. Microbiol.* **58**, 2688–2690.

85. De Boer, E. (1995) Isolation of *Yersinia enterocolitica* from foods. *Contr. Microbiol. Immunol.* **13**, 71–73.

86. Wauters, G. (1979) Carriage of *Yersinia enterocolitica* serotype 3 by pigs as a source of human infection. *Contr. Microbiol. Immunol.* **5**, 249–252.

87. Schiemann, D. A. (1980) Isolation of toxigenic *Yersinia enterocolitica* from retail pork products. *J. Food Prot.* **43**, 360–365.

88. Harmon, M. C., Swaminathan, B., and Forrest, J. C. (1984) Isolation of *Yersinia enterocolitica* and related species from porcine samples obtained from an abattoir. *J. Appl. Bacteriol.* **56**, 421–427.

89. Nesbakken, T., Gondrosen, B., and Kapperud, G. (1985) Investigation of *Yersinia enterocolitica*-like bacteria, and thermotolerant campylobacters in Norwegian pork products. *Int. J. Food Microbiol.* **1**, 311–320.

90. De Boer, E., Seldam, W. M., and Oosterom, J. (1986) Characterization of *Yersinia enterocolitica* and related species isolated from foods and porcine tonsils in the Netherlands. *Int. J. Food Microbiol.* **3**, 217–224.

91. De Boer, E. and Nouws, J. F. M. (1991) Slaughter pigs and pork as a source of human pathogenic *Yersinia enterocolitica*. *Int. J. Food Microbiol.* **12**, 375–378.

92. Tsai, S. J. and Chen, L. H. (1991) Occurrence of *Yersinia enterocolitica* in pork products from Northern Taiwan. *Contr. Microbiol. Immunol.* **12**, 56–62.

93. De Giusti, M., de Vito, E., Serra, A., et al. (1995) Occurrence of pathogenic *Yersinia enterocolitica* in slaughtered pigs and pork products. *Contr. Microbiol. Immunol.* **13**, 126–129.

94. Fukushima, H., Hoshina, K., Itowa, H., and Gomyoda, M. (1997) Introduction into Japan of pathogenic *Yersinia* through imported pork, beef and fowl. *Int. J. Food Microbiol.* **35**, 205–212.

95. Wauters, G., Goossens, V., Janssens, M., and Vandepitte, J. (1988) New enrichment method for isolation of pathogenic *Yersinia enterocolitica* serogroup O:3 from pork. *Appl. Environ. Microbiol.* **54,** 851–854.

96. Tauxe, R. V., Wauters, G., Goossens, V., et al. (1987) *Yersinia enterocolitica* infections and pork: the missing link. *Lancet* **i,** 1129–1132.

97. Logue, C. M., Sheridan, J. J., Wauters, G., Mc Dowell, D. A., and Blair, I. S. (1996) *Yersinia* spp. and numbers, with particular reference to *Y. enterocolitica* bio/serotypes, occurring on Irish meat and meat products, and the influence of alkali treatment on their isolation. *Int. J. Food Microbiol.* **33,** 257–274.

98. Ackers, M. L., Schoenfeld, S., Markman, J., Smith, M. G., Nichols, M. A., and DeWitt, W. (2000) An outbreak of *Yersinia enterocolitica* O:8 infections associated with pasteurized milk. *J. Infect. Dis.* **181,** 1834–1837.

99. Fredriksson-Ahomaa, M. and Korkeala, H. (2003) Low occurrence of pathogenic *Yersinia enterocolitica* in clinical, food, and environmental samples: a methodological problem. *Clin. Microbiol. Rev.* **16,** 220–229.

100. Stoddard, J. J., Wechsler, D. S., Nataro, J. P., and Casella, J. F. (1994) *Yersinia enterocolitica* infection in a patient with sickle cell disease after exposure to chitterlings. *Am. J. Pediatr. Hematol. Oncol.* **16,** 153–155.

101. Jones, T. F., Buckingham, S. C., Bopp, C. A., Ribot, E., and Schaffner, W. (2003) From pig to pacifier: chitterling-associated yersiniosis outbreak among black infants. *Emerg. Infect. Dis.* **9,** 1007–1009.

102. Mafu, A. A., Higgins, R., Nadeau, M., and Cousineau, G. (1989) The incidence of *Salmonella, Campylobacter,* and *Yersinia enterocolitica* in swine carcasses and the slaughterhouse environment. *J. Food Prot.* **5,** 642–645.

103. Berzero, R., Volterra, L., and Chiesa, C. (1991) Isolation of Yersiniae from sewage. *Contr. Microbiol. Immunol.* **12,** 40–43.

104. Fransen, N. G., van den Elzen, A. M. G., Urlings, B. A. P., and Bijker, P. G. H. (1996) Pathogenic microorganisms in slaughterhouse sludge—a survey. *Int. J. Food Microbiol.* **33,** 245–256.

105. Sammarco, M. L., Ripabelli, G., Ruberto, A., Iannitto, G., and Grasso, G. M. (1997) Prevalence of Salmonellae, Listeriae, and Yersiniae in the slaughterhouse environment and on work surfaces, equipment, and workers. *J. Food Prot.* **60,** 367–371.

106. Nesbakken, T. (1988) Enumeration of *Yersinia enterocolitica* O:3 from the porcine oral cavity, and its occurrence on cut surface of pig carcasses and the environment in a slaughterhouse. *Int. J. Food Microbiol.* **6,** 287–293.

107. Fredriksson-Ahomaa, M., Korte, T., and Korkeala, H., (2000) Contamination of carcasses, offals and the environment with *yadA*-positive *Yersinia enterocolitica* in a pig slaughterhouse. *J. Food Prot.* **63,** 31–35.

108. Christensen, S. G. (1987) Co-ordination of a nation-wide survey on the presence of *Yersinia enterocolitica* O:3 in the environment of butcher shops. *Contr. Microbiol. Immunol.* **9,** 26–29.

109. Fredriksson-Ahomaa, M., Koch, U., Klemm, C., Bucher, M., and Stolle, A. (2004) Different genotypes of *Yersinia enterocolitica* 4/O:3 strains widely distributed in butcher shops in the Munich area. *Int. J. Food Microbiol.* **95(1),** 89–94.

110. Thompson, J. S. and Gravel, M. J. (1986) Family outbreak of gastroenteritis due to *Yersinia enterocolitica* O:3 from well water. *Can. J. Microbiol.* **32,** 700–701.

111. Sandery, M., Stinear, T., and Kaucner, C. (1996) Detection of pathogenic *Yersinia enterocolitica* in environmental water by PCR. *J. Appl. Bacteriol.* **80,** 327–332.

112. Waage, A. S., Vardund, T., Lund, V., and Kapperud, G. (1999) Detection of low numbers of pathogenic *Yersinia enterocolitica* in environmental water and sewage samples by nested polymerase chain reaction. *J. Appl. Microbiol.* **87,** 814–821.

113. Olsovsky, V., Olsakova, V., Chobot, S., and Sviridov, V. (1975) Mass occurrence of *Yersinia enterocolitica* in two establishments of collective care of children. *J. Hyg. Epid. Microbiol. Immun.* **1**, 22–29.

114. Lee, L. A., Taylor, J., Carter, G. P., Quinn, B., Farmer, J. J., and Tauxe, R. V. (1991) *Yersinia enterocolitica* O:3: an emerging cause of pediatric gastroenteritis in the United States. The *Yersinia enterocolitica* Collaborative Study Group. *J. Infect. Dis.* **163**, 660–663.

115. Ostroff, S. M., Kapperud, G., Huteagner, L. C., et al. (1994) Sources of sporadic *Yersinia enterocolitica* infections in Norway: a prospective case–control study. *Epidemiol. Infect.* **112**, 133–141.

116. Satterthwaite, P., Pritchard, K., Floyd, D., and Law, B. (1999) A case–control study of *Yersinia enterocolitica* infections in Auckland. *Aust. N. Z. Public Health* **23**, 482–485.

117. Gutman, L. T., Ottesen, E. A., Quan, T. J., Noce, P. S., and Katz, S. L. (1973) An inter-familiar outbreak of *Yersinia enterocolitica* enteritis. *N. Engl. J. Med.* **288**, 1372–1377.

118. Black, R. E., Jackson, R. J., Tsai, T., et al. (1978) Epidemic *Yersinia enterocolitica* infection due to contaminated chocolate milk. *N. Engl. J. Med.* **298**, 76–79.

119. Tacket, C. O., Ballard, J., Harris, H., et al. (1985) An outbreak of *Yersinia enterocolitica* infections caused by contaminated tofu (soybean curd). *Am. J. Epidemiol.* **121**, 705–711.

120. Cover, T. L. and Aber, R. C. (1989) *Yersinia enterocolitica*. *N. Engl. J. Med.* **321**, 16–24.

121. Tacket, C. O., Narain, J. P., Sattin, R., et al. (1984) A multistate outbreak of infections caused by *Yersinia enterocolitica* transmitted by pasteurized milk. *JAMA* **251**, 483–486.

122. Toma, S., Wauters, G., McClure, H. M., Morris, G. K., and Weissefeld, A. S. (1984) O:13a,13b, a new pathogenic serotype of *Yersinia enterocolitica*. *J. Clin. Microbiol.* **20**, 843–845.

123. Martin, T., Kasian, G. D., and Stead, S. (1982) Family outbreak of yersiniosis. *J. Clin. Mircobiol.* **16**, 622–626.

124. Van Noyen, R., Selderslaghs, R., Bogaerts, A., Verhaegen, J., and Wauters, G. (1995) *Yersinia pseudotuberculosis* in stool from patients in a regional Belgian hospital. *Contr. Microbiol. Immunol.* **13**, 19–24.

125. Nuorti, J. P., Niskanen, T., Hallanvuo, S., et al. (2004) A widespread outbreak of *Y. pseudotuberculosis* O:3 infections from iceberg lettuce. *J. Infect. Dis.* **189**, 766–774.

126. Nowgesic, E., Fyfe, M., Hockin, J., King, A., Ng, H., and Pacagnella, A. (1999) Outbreak of *Yersinia pseudotuberculosis* in British Columbia, November 1998. *Can. Commun. Disp. Rep.* **25**, 97–100.

127. Chiesa, C., Pacifico, L., Nanni, F., Renzi, A. M., and Ravagnan, G. (1993) *Yersinia pseudotuberculosis* in Italy. Attempt recovery from 37,666 samples. *Microbiol. Immunol.* **37**, 391–394.

128. Fenwick, S. G. and McCarthy, M. D. (1995) *Yersinia enterocolitica* is a common cause of gastroenteritis in Auckland. *N. Z. Med. J.* **14**, 269–271.

129. Stolk-Engelaar, V. M. M. and Hoogkamp-Korstanje, J. A. A. (1996) Clinical presentation and diagnosis of gastrointestinal infections by *Yersinia enterocolitica* in 261 Dutch patients. *Scand. J. Infect. Dis.* **28**, 571–572.

130. Zaidi, A. K. M., Macone, A., and Goldman, D. (1999) Impact of simple screening criteria on utilization of low-yield bacterial stool cultures in a children' s hospital. *Pediatrics* **103**, 1189–1192.

131. Mäki-Ikola, O., Heesemann, J., Toivanen, A., and Granfors, K. (1997) High frequency of *Yersinia* antibodies in healthy populations in Finland and Germany. *Rheumatol. Int.* **16**, 227–229.

132. Portnoy, D. A. and Martinez, R. J. (1985) Role of a plasmid in the pathogenicity of *Yersinia* species. *Curr. Top. Microbiol. Immunol.* **118**, 29–51.

133. Brubaker, R. R. (1991) Factors promoting acute and chronic diseases caused by Yersiniae. *Clin Microbiol. Rev.* **4**, 309–324.

134. Heesemann, J. and Laufs, R. (1983) Construction of a mobilisable *Yersinia enterocolitica* virulence plasmid. *J. Bacteriol.* **155**, 761–767.

135. Heesemann, J., Algermissen, B., and Laufs, R. (1984) Genetically manipulated virulence of *Yersinia enterocolitica*. *Infect. Immun.* **46**, 105–110.

136. Heesemann, J., Keller, C., Morawa, R., Schmidt, N., Siemens, H. J., and Laufs, R. (1983) Plasmids of human strains of *Yersinia enterocolitica*: molecular relatedness and possible importance for pathogenesis. *J. Infect. Dis.* **147**, 107–115.

137. Kapperud, G., Namork, E., Skurnik, M., and Nesbakken, T. (1987) Plasmid-mediated surface fibrillae of *Yersinia pseudotuberculosis* and *Yersinia enterocolitica*: relationship to the outer membrane protein YOP1 and possible importance for pathogenesis. *Infect. Immun.* **55**, 2247–2254.

138. El Tahir, Y. and Skurnik, M. (2001) YadA, the multifaceted *Yersinia* adhesin. *Int. J. Med. Microbiol.* **291**, 209–218.

139. Portnoy, D. A., Moseley, S. L., and Falkow, S. (1981) Characterization of plasmids and plasmid-associated determinants of *Yersinia enterocolitica* pathogenesis. *Infect. Immun.* **31**, 775–782.

140. Michelis, T., Wattiau, P., Brasseur, R., Ruysschaert, J. M., and Cornelis, G. (1990) Secretion of Yop proteins by Yersiniae. *Infect. Immun.* **58**, 2840–2849.

141. Cornelis, G. R. (1998) The *Yersinia* deadly kiss. *J. Bacteriol.* **180**, 5495–5504.

142. Portnoy, D. A., Wolf-Watz, H., Bolin, I., Beeder, A. B., and Falkow, S. (1984) Characterisation of common virulence plasmids in *Yersinia* species and their role in the expression of outer membrane proteins. *Infect. Immun.* **43**, 108–114.

143. Heesemann, J., Gross, U., Schmidt, N., and Laufs, R. (1986) Immunochemical analysis of plasmid-encoded proteins released by enteropathogenic *Yersinia* sp. grown in calcium-deficient media. *Infect. Immun.* **54**, 561–567.

144. Miller, V. L. and Falkow, S. (1988) Evidence for two genetic loci in *Yersinia enterocolitica* that can promote invasion of epithelial cells. *Infect. Immun.* **56**, 1242–1248.

145. Pepe, J. F., Wachtel, M. R., Wagar, E., and Miller, V. L. (1995) Pathogenesis of defined invasion mutants of *Yersinia enterocolitica* in a BALB/c mouse model of infection. *Infect. Immun.* **63**, 4837–4848.

146. Isberg, R. R. and Barnes, P. (2001) Subversion of integrins by enteropathogenic *Yersinia*. *J. Cell Sci.* **114**, 21–28.

147. Pierson, D. E. and Falkow, S. (1990) Non-pathogenic isolates of *Y. enterocolitica* do not contain functional *inv*-homologous sequences. *Infect. Immun.* **58**, 1059–1064.

148. Miller, V. L., Farmer, J. J. III., Hill, W. E., and Falkow, S. (1989) The *ail* locus is found uniquely in *Yersinia enterocolitica* serotypes commonly associated with disease. *Infect Immun.* **57**, 121–131.

149. Autenrieth, I. B. and Firsching, R. (1996) Penetration of M cells and destruction of Peyer's patches by *Yersinia enterocolitica*: an ultrastructural and histological study. *J. Med. Microbiol.* **44**, 285–294.

150. Vazquez-Torres, A. and Fang, F. C. (2000) Cellular routes of invasion by enteropathogens. *Curr. Opin. Microbiol.* **3**, 54–59.

151. Visser, L. G., Hiemstra, P. S., van den Barselaar, M. T., Ballieux, P. A., and van Furth, R. (1996) Role of YadA in resistance to killing of *Yersinia enterocolitica* by antimicrobial polypeptides of human granulocytes. *Infect. Immun.* **64**, 1653–1658.

152. Carniel, E. (1995) Chromosomal virulence factors of *Yersinia*. *Contrib. Microbiol. Immunol.* **13**, 218–224.

153. Iriarte, M., Vanooteghem, J. C., Delor, I., Diaz, R., Knutton, S., and Cornelis, G. (1993) The Myf fibrillae of *Yersinia enterocolitica*. *Mol. Microbiol.* **9**, 507–520.

154. Cornelis, G. (1994) *Yersinia* pathogenicity factors. *Curr. Top. Microbiol. Immunol.* **192**, 246–263.

155. Yang, Y., Merriam, J. J., Mueller, J. P., and Isberg, R. R. (1996) The *psa* locus is responsible for thermoinducible binding of *Yersinia pseudotuberculosis* to cultured cells. *Infect. Immun.* **64,** 2483–2489.

156. Koornhof, H. J., Smego, R. A. Jr., and Nicol, M. (1999) Yersiniosis II: the pathogenesis of *Yersinia* infection. *Eur. J. Clin. Infect. Dis.* **18,** 87–112.

157. Rakin, A., Noelting, C., Schubert, S., and Heesemann, J. (1999) Common and specific characteristics of the high-pathogenicity islands of *Yersinia enterocolitica. Infect. Immun.* **67,** 5265–5274.

158. Delor, I., Kaeckenbeek, A., Wauters, G., and Cornelis, G. R. (1990) Nucleotide sequence of *yst*, the *Yersinia enterocolitica* gene encoding the heat-stable enterotoxin, and prevalence of the gene among pathogenic and non-pathogenic Yersiniae. *Infect. Immun.* **58,** 2983–2988.

159. Kwaga, J. and Iversen, J. O. (1992) Laboratory investigation of virulence among strains of *Yersinia enterocolitica* and related species isolated from pigs and pork products. *Can. J. Microbiol.* **38,** 92–97.

160. Kwaga, J., Iversen, J. O., and Misra, V. (1992) Detection of pathogenic *Yersinia enterocolitica* by polymerase chain reaction and Digoxigenin-labelled polynucleotide probes. *J. Clin. Microbiol.* **30,** 2668–2673.

161. Delor, I. and Cornelis, R. (1992) Role of *Yersinia enterocolitica* Yst toxin in experimental infection of young rabbits. *Infect. Immun.* **60,** 4269–4277.

162. Robins-Browne, R. M., Tzipori, S., Gonis, G., Hayes, J., Withers, M., and Prpic, J. K. (1985) The pathogenesis of *Yersinia enterocolitica* infection in gnotobiotic piglets. *J. Med. Microbiol.* **19,** 297–308.

163. De Koning-Ward, T. F., Ward, A. C., and Robins-Browne, R. M. (1994) Characterization of the urease-encoding gene complex of *Yersinia enterocolitica. Gene* **145,** 25–32.

164. De Koning-Ward, T. F. and Robins-Browne, R. M. (1995) Contribution to urease to acid tolerance in *Yersinia enterocolitica. Infect. Immun.* **63,** 3790–3795.

165. Gripenberg-Lerche, C., Zhang, L., Ahtonen, P., Toivanen, P., and Skurnik, M. (2000) Construction of urease-negative mutants of *Yersinia enterocolitica* serotypes O:3 and O:8 role of urease in virulence and arthritogenicity. *Infect. Immun.* **68,** 942–947.

166. Bruneteau, M. and Minka, S. (2003) Lipopolysaccharides of bacterial pathogens from the genus *Yersinia*: a mini-review. *Biochimie* **85,** 145–152.

167. Skurnik, M., Venho, R., Bengoechea, J. A., and Moriyón, I. (1999) The lipopolysaccharide outer core of *Yersinia enterocolitica* serotype O:3 is required for virulence and plays a role in outer membrane integrity. *Mol. Microbiol.* **31,** 1443–1462.

168. Skurnik, M. and Zhang, L. (1996) Molecular genetics and biochemistry of *Yersinia* lipopolysaccharide. *APMIS* **104,** 849–872.

169. Skurnik, M., Mikkola, P., Toivanen, P., and Tertti, R. (1996) Passive immunization with monoclonal antibodies specific for lipopolysaccharide (LPS) O-side chain protects mice against intravenous *Yersinia enterocolitica* serotype O:3 infection. *APMIS* **104,** 598–602.

170. Donadini, R., Liew, C. W., Kwan, A. H., Mackay, J. P., and Fields, B. A. (2004) Crystal and solution structures of a superantigen from *Yersinia pseudotuberculosis* reveal a jelly-roll fold. *Structure* **12,** 145–156.

171. Yoshino, K., Ramamurthy, T., Nair, G. B., et al. (1995) Geographical heterogeneity between Far East and Europe in prevalence of *ypm* gene encoding the novel superantigen among *Yersinia pseudotuberculosis* strains. *J. Clin. Microbiol.* **33,** 3356–3358.

172. Abe, J., Onimaru, M., Matsumoto, S., et al. (1997) Clinical role for a superantigen in *Yersinia pseudotuberculosis* infection. *J. Clin. Invest.* **99,** 1823–1830.

173. Carniel, E. and Mollaret, H. H. (1990) Yersiniosis. *Comp. Immun. Microbiol. Infect. Dis.* **13,** 51–58.

174. Naktin, J. and Beavis, K. G. (1999) *Yersinia enterocolitica* and *Yersinia pseudotuberculosis. Clin. Lab. Med.* **19,** 523–536.

175. Smego, R. A., Frean, J., and Koornhof, H. J. (1999) Yersiniosis I: microbiological and clinicoepidemiological aspects of plague and non-plague *Yersinia* infections. *Eur. J. Clin. Microbiol. Infect. Dis.* **18,** 1–15.

176. Lamps, L. W., Madhusudha, K. T., Greenson, J. K., et al. (2001) The role of *Yersinia enterocolitica* and *Yersinia pseudotuberculosis* in granulomatous appendicitis: a histologic and molecular study. *Am. J. Surg. Pathol.* **25,** 508–515.

177. Ahvonen, P. (1972) Human yersiniosis in Finland: II. Clinical features. *Ann. Clin. Res.* **4,** 39–48.

178. Sievers, K., Ahvonen, K., and Aho, K. (1972) Epidemiological aspects of *Yersinia* arthritis. *Int. J. Epid.* **1,** 5–46.

179. Toivanen, A., Granfors, K., Lahesmaa-Rantala, R., Leino, R., Ståhlberg, T., and Vuento, R. (1985) Pathogenesis of *Yersinia*-triggered arthritis: immunological, microbiological and clinical aspects. *Immunol. Rev.* **86,** 47–70.

180. Tertti, R., Vuento, R., Mikkola, P., Granfors, K., Mäkelä, A. L., and Toivanen, A. (1989) Clinical manifestation of *Yersinia pseudotuberculosis* infection in children. *Eur. J. Clin. Microbiol. Infect. Dis.* **8,** 587–591.

181. Van Ossel, C. and Wauters, G. (1990) Asymptomatic *Yersinia enterocolitica* infections during an outbreak in a day-nursery. *Eur. J. Clin. Microbiol. Infect. Dis.* **9,** 148.

182. Bottone, E. J. (1997) *Yersinia enterocolitica*: the charisma continues. *Clin. Microbiol. Rev.* **10,** 257–276.

183. Pai, C. H., Gillis, F., Tuomanen, E., and Marks, M. I. (1984) Placebo-controlled double-blind evaluation of trimethoprim-sulfamethoxazole treatment of *Yersinia enterocolitica* gastroenteritis. *J. Pediatr.* **104,** 308–311.

184. Jellou, L., Fremond, B., Dyon, J. F., Orme, R. L., and Babut, J. M. (1997) Mesenteric adenitis caused by *Yersinia pseudotuberculosis* presenting as an abdominal mass. *Eur. J. Pediatr. Surg.* **7,** 180–183.

185. Shorter, N. A., Thompson, M. D., Mooney, D. P., and Modlin, J. F. (1998) Surgical aspects of an outbreak of *Yersinia* enterocolitis. *Pediatr. Surg. Int.* **13,** 2–5.

186. Sakellaris, G., Kakavelakis, K., Stathopoulos, E., Michailidou, H., and Charissis, G. (2004) A palpable right lower abdominal mass due to *Yersinia* mesenteric lymphadenitis. *Pediatr. Surg. Int.* **10,** 155–157.

187. Homewood, R., Gibbons, C. P., Richards, D., Lewis, A., Duane, P. D., and Griffiths, A. P. (2003) Ileitis due to *Yersinia pseudotuberculosis* in Crohn's disease. *J. Infect.* **47,** 328–332.

188. Clarridge, J., Roberts, C., Peters, J., and Musher, D. (1983) Sepsis and empyema caused by *Yersinia enterocolitica*. *J. Clin. Microbiol.* **17,** 936–938.

189. Thirumoorthi, M. C. and Dajani, A. S. (1978) *Yersinia enterocolitica* osteomyelitis in a child. *Am. J. Dis. Child.* **132,** 578–580.

190. Karmali, M. A., Toma, S., Schiemann, D. A., and Ein, S. H. (1982) Infection caused by *Yersinia enterocolitica* serotype O:21. *J. Clin. Microbiol.* **15,** 596–598.

191. Casey, M. F., Gilligan, P. H., and Smiley, M. L. (1987) *Yersinia enterocolitica* meningitis and osteomyelitis: a case report. *Diagn. Microbiol. Infect. Dis.* **8,** 47–50.

192. Giamarellou, H., Antoniadou, A., Kanavos, K., Papaioannou, C., Kanatakis, S., and Papadaki, K. (1995) *Yersinia enterocolitica* endocarditis: case report and literature review. *Eur. J. Clin. Mircobiol. Infect. Dis.* **14,** 126–130.

193. Crowe, M., Ashford, K., and Ispahani, P. (1996) Clinical features and antibiotic treatment of septic arthritis and osteomyelitis due to *Yersinia enterocolitica*. *J. Med. Microbiol.* **45,** 302–309.

194. Bin-Sagheer, S., Myers, J., Lapham, C., and Sarubbi, F. A. (1997) Meningitis caused by *Yersinia enterocolitica*: case report and review. *Infect. Dis. Clin. Pract.* **6,** 198–200.

195. Hopfner, M., Nitsche, R., Rohr, A., Harms, D., Schubert, S., and Folsch, U. R. (2001) *Yersinia enterocolitica* infection with multiple liver abscesses uncovering a primary hemochromatosis. *Scand. J. Gastroenterol.* **36,** 220–224.

196. Tacket, C. O., Davis, B. R., Carter, G. P., Randolph, J. F., and Cohen, M. L. (1983) *Yersinia enterocolitica* pharyngitis. *Ann. Intern. Med.* **99**, 40–42.

197. Rose, F. B., Camp, C. J., and Antes, E. J. (1987) Family outbreak of fatal *Yersinia enterocolitica* pharyngitis. *Am. J. Med.* **82**, 636–637.

198. Kellogg, C. M., Tarakji, E. A., Smith, M., and Brown, P. D. (1995) Bacteremia and suppurative lymphadenitis due to *Yersinia enterocolitica* in a neutropenic patient who prepared chitterlings. *Clin. Infect. Dis.* **21**, 236–237.

199. Deacon, A. G., Hay, A., and Duncan, J. (2003) Septicemia due to *Yersinia pseudotuberculosis*—a case report. *CMI* **9**, 1118–1119.

200. Han, T. H., Paik, I. K., and Kim, S. J. (2003) Molecular relatedness between isolates of *Yersinia pseudotuberculosis* from a patient and an isolate from mountain spring water. *J. Korean Me. Sci.* **18**, 425–428.

201. Mitchell, K. M. T. and Brecher, M. E. (1999) Approaches to the detection of bacterial contamination in cellular blood products. *Transfus. Med. Rev.* **13**, 132–144.

202. Jacobs, J., Jamaer, D., Vandeven, J., Wauters, M., Vermylen, C., and Vandepitte, J. (1989) *Yersinia enterocolitica* in donor blood: a case report and review. *J. Clin. Microbiol.* **27**, 1119–1121.

203. Strobel, E., Heesemann, J., Mayer, G., Peters, J., Müller-Weihrich, S., and Emmerling, P. (2000) Bacterial and serological findings in a further case of transfusion-mediated *Yersinia enterocolitica* sepsis. *J. Clin. Microbiol.* **38**, 2788–2790.

204. Saari, K. M., Mäki, M., Päivänsalo, T., Leino, R., and Toivanen, A. (1986) Acute anterior uveitis and conjunctivitis following *Yersinia* infection in children. *Int. Ophthalmol.* **9**, 237–241.

205. Leirisalo-Repo, M. (1987) *Yersinia* arthritis. Acute clinical picture and long-term prognosis. *Contr. Microbiol. Immunol.* **9**, 145–154.

206. Hannu, T., Mattila, L., Nuorti, J. P., et al. (2003) Reactive arthritis after an outbreak of *Yersinia pseudotuberculosis* serotype O:3 infection. *Ann. Rheum. Dis.* **62**, 866–869.

207. Aho, K., Ahvonen, P., Lassus, A., Sievers, K., and Tiilikainen, A. (1974) HL-A 27 in reactive arthritis. A study of *Yersinia* arthritis and Reiter's disease. *Arthritis Rheum.* **17**, 521–526.

208. Leirisalo-Repo, M. and Suoranta, H. (1988) Ten-year follow study of patients with *Yersinia* arthritis. *Arthritis Rheum.* **31**, 533–537.

209. Toivanen, A. and Toivanen, P. (1988) Pathogenesis of reactive arthritis. In: *Reactive Arthritis* (Toivanen, A. and Toivanen, P., eds.), CRC, Boca Raton, FL, pp. 167–178.

210. Wakefield, D., Stahlberg, T. H., Toivanen, A., Granfors, K., and Tennant, C. (1990) Serologic evidence of *Yersinia* infection in patients with anterior uveitis. *Arch. Ophtalmol.* **108**, 219–212.

211. Careless, D. J., Chiu, B., Rabinovitch, T., Wade, J., and Inman, R. (1997) Immunomagnetic and microbial factors in acute anterior uveitis. *J. Rheumatol.* **24**, 102–108.

212. Ahvonen, P., Sievers, K., and Aho, K. (1969) Arthritis associated with *Y. enterocolitica* infection. *Acta Rheumatol. Scand.* **15**, 232–253.

213. Granfors, K., Jalkanen, S., von Essen, R., et al. (1989) Yersinia antigens in synovial-fluid cells from patients with reactive arthritis. *N. Engl. J. Med.* **320**, 216–221.

214. Sieper, J., Fendler, C., Laitko, S., et al. (1999) No benefit of long-term ciprofloxacin treatment in patients with reactive arthritis and undifferentiated oligoarthritis. *Arthritis Rheum.* **42**, 1386–1396.

215. Hoogkamp-Korstanje, J. A. A., Moesker, H., and Bruyn, G. A. W. (2000) Ciprofloxacin v placebo for treatment of *Yersinia enterocolitica* triggered reactive arthritis. *Ann. Rheum. Dis.* **59**, 914–917.

216. Merilahti-Palo, R., Söderström, K.-O., Lahesmaa-Rantala, R., Granfors, K., and Toivanen, A. (1991) Bacterial antigens in synovial biopsy specimens in yersinia triggered arthritis. *Ann. Rheum. Diss.* **50**, 87–90.

217. Raevuori, M., Harvey, S. M., Pickett, M. J., and Martin, W. J. (1978) *Yersinia enterocolitica*: in vitro antimicrobial susceptibility. *Antimicrob. Agents. Chemother.* **13**, 888–890.

218. Ahmedy, A., Vidon, D. J. M., Delmas, C. L., and Lett, M. C. (1985) Antimicrobial susceptibilities of food-isolated strains of *Yersinia enterocolitica*, *Y. intermedia*, *Y. feredriksenii*, and *Y. kristensenii*. *Antimicrob. Agents Chemoter.* **28,** 351–353.

219. Hornstein, M. J., Jupeau, A. M., Scavizzi, M. R., Philippon, A. M., and Grimont, P. A. D. (1985) In vitro susceptibilities of 126 clinical isolates of *Yersinia enterocolitica* to 21 β-lactam antibiotics. *Antimicrob. Agents Chemother.* **27,** 806–811.

220. Hoogkamp-Korstanje, J. A. (1987) Antibiotics in *Yersinia enterocolitica* infections. *J. Antimicrob. Chemother.* **20,** 123–131.

221. Gayraud, M., Scavizzi, M. R., Mollaret, H. H., Guillevin, L., and Hornstein, M. J. (1993) Antibiotic treatment of *Yersinia enterocolitica* septicemia: a retrospective review of 43 cases. *Clin. Infect. Dis.* **17,** 405–410.

222. Prats, G., Mirelis, G., Llovet, T., Munoz, C., Miro, E., and Navarro, F. (2000) Antibiotic resistance terns in enteropathogenic bacteria isolated in 1985–1987 and 1995–1998 in Barcelona. *Antimicrob. Agents. Chemother.* **44,** 1140–1145.

223. Soriano, F. and Vega, J. (1982) The susceptibility of *Yersinia* to eleven antimicrobial. *J. Antimirob. Chemother.* **10,** 543–547.

224. Oili, K. S., Niinisalo, H., Korpilahde, T., and Virolainen, J. (2003) Treatments of reactive arthritis with infliximab. *Scand. J. Rheumatol.* **32,** 122–124.

225. Pham, J. N., Bell, S. M., and Lanzarone, J. Y. (1991) Biotype and antibiotic sensitivity of 100 clinical isolates of *Yersinia enterocolitica*. *J. Antimicrob. Chemoter.* **28,** 13–18.

226. Stolk-Engelaar, V. M. M., Meis, J. F. G. M., Mulder, J. A., Loeffen, F. L. A., and Hoogkamp-Korstanje, J. A. A. (1995) In-vitro antimicrobial susceptibility of *Yersinia enterocolitica* isolates from stools of patients in the Netherlands from 1982–1991. *J. Antimicrob. Chemother.* **36,** 839–843.

227. Rastawicki, W., Gierczynski, R., Jagielski, M., Kaluzewski, S., and Jeljaszewicz, J. (2000) Susceptibility of Polish strains of *Yersinia enterocolitica* serotype O:3 to antibiotics. *Int. J. Antimicrob. Agents* **13,** 297–300.

228. Kwaga, J. and Iversen, J. O. (1990) In vitro antimicrobial susceptibilities of *Yersinia enterocolitica* and related species isolated from slaughtered pigs and pork products. *Antimicrob. Agents. Chemoter.* **34,** 2423–2425.

229. Hariharan, H., Giles, J. S., Heaney, S. B., Leclerc, S. M., and Schurman, R. D. (1995) Isolation, serotype, and virulence-associated properties of *Yersinia enterocolitica* from the tonsils of slaughter hogs. *Can. J. Vet. Res.* **59,** 161–166.

230. Stock, I. and Wiedemann, B. (1999) An in-vitro study of the antimicrobial susceptibility of *Yersinia enterocolitica* and the definition of a database. *J. Antimicrob. Chemother.* **43,** 37–45.

231. Stock, I. and Wiedemann, B. (1999) Natural antibiotic susceptibility of *Yersinia pseudotuberculosis* strains (Natürliche Antibiotika-Emfindlichkeit von *Yersinia pseudotuberculosis*-Stämmen). *Chemother. J.* **8,** 219–227.

232. Fernandez-Roblas, R., Cabria, F., Esteban, J., Lopez, J. C., Gadea, I., and Soriano, F. (2000) In vitro activity of gemofloxacin (SB265805) compared with 14 other antimicrobials against intestinal pathogens. *J. Antimicrob. Chemoter.* **46,** 1023–1027.

233. Preston, M. A., Brown, S., Borczyk, A. A., Riley, G., and Krishnan, C. (1994) Antimicrobial susceptibility of pathogenic *Yersinia enterocolitica* isolated in Canada from 1972 to 1992. *Antimicrob. Agents Chemother.* **38,** 2121–2124.

234. Hammerberg, S., Sorger, S., and Marks, M. I. (1977) Antimicrobial susceptibilities of *Yersinia enterocolitica* biotype 4, serotype O:3. *Antimicrob. Agents. Chemother.* **11,** 566–568.

235. Stock, I., Heisig, P., and Wiedemann, B. (1999) Expression of β-lactamases in *Yersinia enterocolitica* biovars 2, 4 and 5. *J. Med. Microbiol.* **48,** 1023–1027.

236. Stock, I., Heisig, P., and Wiedemann, B. (2000) β-Lactamase expression in *Yersinia enterocolitica* biovars 1A, 1B and 3. *J. Med. Microbiol.* **49,** 403–408.

237. Pham, J. N., Bell, S. M., Martin, L., and Carniel, E. (2000) The β-lactamases and β-lactam antibiotic susceptibility of *Yersinia enterocolitica*. *J. Antimicrob. Chemoter.* **46,** 951–957.

238. Pham, J. N., Bell, S. M., Hardy, J. M., Martin, L., Guiyoule, A., and Carniel, E. (1995) Susceptibility to β-lactam agents of *Yersinia enterocolitica* biotype 4, serotype 3 isolated in various parts of the world. *J. Med. Microbiol.* **43**, 9–13.

239. White, D. G., Zhao, S., Simjee, S., Wagner, D., and McDermott, P. F. (2002) Antimicrobial resistance of foodborne pathogens. *Microb. Infect.* **4**, 405–412.

240. Aarestrup, F. M., Bager, F., Jensen, N. E., Madsen, M., Meyling, A., and Wegener, H. C. (1998) Resistance to antimicrobial agents used for animal therapy in pathogenic-, zoonotic- and indicator bacteria isolated from different food animals in Denmark: a baseline study for the Danish integrated antimicrobial resistance monitoring program (DANMAP). *APMIS* **106**, 754–770.

241. Trallero, E. P., Zigorraga, C., Cilla, G., Idigoras, P., Lopatequi, C. L., and Solaun, L. (1988). Animal origin of the antibiotic resistance of human pathogenic *Yersinia enterocolitica*. *Scand. J. Infect. Dis.* **20**, 573.

242. Capilla, S., Goni, P., Rubio, M. C., et al. (2003) Epidemiological study of resistance to nalidixic acid and other antibiotics in clinical *Yersinia enterocolitica* O:3 isolates. *J. Clin. Microbiol.* **41**, 4876–4878.

243. Sanchez-Cespede, J., Navia, M. M., Martinez, R., et al. (2003) Clonal dissemination of *Yersinia enterocolitica* strains with various susceptibilities to nalidixic acid. *J. Clin. Microbiol.* **41**, 1769–1771.

244. Kelley, T. R., Pancorbo, O. C., Merka, W. C., and Barnhart, H. M. (1997) Antibiotic resistance of bacterial litter isolates. *Poult. Sci.* **77**, 243–247.

245. Funk, J. A., Troutt, H. F., Davis, S. A., and Fossler, C. P. (2000) In vitro susceptibility of *Yersinia enterocolitica* isolated from the oral cavity of swine. *J. Food Prot.* **63**, 395–399.

246. Fukushima, H. and Gomyoda, M. (1991) Intestinal carriage of *Yersinia pseudotuberculosis* by wild birds and mammals in Japan. *Appl. Environ. Microbiol.* **57**, 1152–1155.

247. Borie, C. F., Jara, M. A., Sánchez, M. L., et al. (1997) Isolation and characterization of *Yersinia enterocolitica* from swine and cattle in Chile. *J. Vet. Med. B.* **44**, 347–354.

248. Andersen, J. K. (1988) Contamination of freshly slaughtered pig carcasses with human pathogenic *Yersinia enterocolitica*. *Int. J. Food Microbiol.* **7**, 193–202.

249. Fredriksson-Ahomaa, M., Björkroth, J., Hielm, S., and Korkeala, H. (2000) Prevalence and characterization of pathogenic *Yersinia enterocolitica* in pig tonsils from different slaughterhouses. *Food Microbiol.* **17**, 93–101.

250. Korte, T., Fredriksson-Ahomaa, M., Niskanen, T., and Korkeala, H. (2004) Low prevalence of pathogenic *Yersinia enterocolitica* in sow tonsils. *Foodborne Pathog. Dis.* **1**, 45–52.

251. Fredriksson-Ahomaa, M., Bucher, M., Hank, C., Stolle, A., and Korkeala, H. (2001) High prevalence of *Yersinia enterocolitica* 4:O3 on pig offal: a slaughtering technique problem. *Syst. Appl. Microbiol.* **24**, 457–463.

252. Bonardi, S., Brindani, F., Pizzin, G., et al. (2003) Detection of *Salmonella* spp., *Yersinia enterocolitica* and verocytotoxin-producing *Escherichia coli* O157 in pigs at slaughter in Italy. *Int. J. Food Microbiol.* **85**, 101–110.

253. Fukushima, H., Maruyama, K., Omori, I., Ito, K., and Iorihara, M. (1989) Role of the contaminated skin of pigs in faecal *Yersinia* contamination of pig carcasses at slaughter. *Fleischwirtschaft* **69**, 409–413.

254. Gurgui Ferrer, M., Mirelis Otero, B., Coll Figa, P., and Prats, G. (1987) *Yersinia enterocolitica* infections and pork. *Lancet* **8**, 334.

255. Funk, J. A., Troutt, H. F., Isaacson, R. E., and Fossler, C. P. (1998) Prevalence of pathogenic *Yersinia enterocolitica* in groups of swine at slaughter. *J. Food Prot.* **61**, 677–682.

256. Niskanen, T., Fredriksson-Ahomaa, M., and Korkeala, H. (2002) *Yersinia pseudotuberculosis* with limited genetic diversity is a common finding in tonsils of fattening pigs. *J. Food Prot.* **65**, 540–545.

257. Nesbakken, T., Nerbrink, E., Røtterud, O. J., and Borch, E. (1994) Reduction of *Yersinia enterocolitica* and *Listeria* spp. on pig carcasses by enclosure of the rectum during slaughter. *Int. J. Food Microbiol.* **23**, 197–208.

258. Slee, K. J., Brightling, P., and Seiler, R. J. (1988) Enteritis in cattle due to *Yersinia pseudotuberculosis* infection. *Aust. Vet. J.* **65,** 271–275.

259. Bullians, J. A. (1987) *Yersinia* species infection of lambs and cull cows at an abattoir. *N. Z. Vet. J.* **35,** 65–67.

260. Slee, K. J. and Skilbeck, N. W. (1992) Epidemiology of *Yersinia pseudotuberculosis* and *Y. enterocolitica* infections in sheep in Australia. *J. Clin. Microbiol.* **30,** 712–715.

261. Fukushima, H., Nakamura, R., Iitsuka, S., Ito, Y., and Saito, K. (1985) Presence of zoonotic pathogens (*Yersinia* app., *Campylobacter jejuni*, *Salmonella* spp., and *Leptospira* spp.) simultaneously in dogs and cats. *Zbl. Bakt. Hyg., I. Abt. Orig.* B **181,** 430–440.

262. Salamah, A. A. (1994) Occurrence of *Yersinia enterocolitica* and *Yersinia pseudotuberculosis* in rodents and cat feces from Riyadh area, Saudi Arabia. *Arab. Gulf J. Sci. Res.* **12,** 546–557.

263. Niskanen, T., Waldenström, J., Fredriksson-Ahomaa, M., Olsen, B., and Korkeala, H. (2003) *virF*-Positive *Yersinia pseudotuberculosis* and *Yersinia enterocolitica* found in migratory birds in Sweden. *Appl. Environ. Microbiol.* **69,** 4670–4675.

264. Nesbakken, T. (1985) Comparison of sampling and isolation procedures for recovery *Yersinia enterocolitica* serotype O:3 from the oral cavity of slaughter pigs. *Acta Vet. Scand.* **26,** 127–135.

265. Shiozawa, K., Akiyama, M., Sahara, K., et al. (1987) Pathogenicity of *Yersinia enterocolitica* biotype 3B and 4, serotype O:3 isolates from pork samples and humans. *Contr. Microbiol. Immunol.* **9,** 30–40.

266. Karib, H. and Seeger, H. (1994) Vorkommen von Yersinien- und Campylobacter-Arten in Lebensmitteln. (Presence of *Yersinia* and *Campylobacter* spp. in foods.) *Fleischwirtschaft* **74,** 1104–1106.

267. Fredriksson-Ahomaa, M., Lyhs, U., Korte, T., and Korkeala, H. (2001c) Prevalence of pathogenic *Yersinia enterocolitica* in food samples at retail level in Finland. *Arch. Lebensmittelhyg.* **52,** 49–72.

268. Fredriksson-Ahomaa, M., Hielm, S., and Korkeala, H. (1999) High prevalence of *yadA*-positive *Yersinia enterocolitica* in pig tongues and minced meat at retail level. *J. Food Prot.* **62,** 123–127.

269. Johannessen, G. S., Kapperud, G., and Kruse, H. (2000) Occurrence of pathogenic *Yersinia enterocolitica* in Norwegian pork products determined by a PCR method and a traditional culturing method. *Int. J. Food Microbiol.* **54,** 75–80.

270. Vishnubhatla, A., Oberst, R. D., Fung, D. Y. C., Wonglumsom, W., Hays, M. P., and Nagaraja, T. G. (2001) Evaluation of a 5′-nuclease (TagMan) Assay for the detection of virulent strains of *Yersinia enterocolitica* in raw meat and tofu samples. *J. Food Prot.* **64,** 355–360.

271. Boyapalle, S., Wesley, I. V., Hurd, H. S., and Reddy, P. G. (2001) Comparison of culture, multiplex, and 5′ nuclease polymerase chain reaction assay for the rapid detection of *Yersinia enterocolitica* in swine and pork products. *J. Food Prot.* **64,** 1352–1361.

272. Lassen, J. (1972) *Yersinia enterocolitica* in drinking water. *Scand. J. Infect. Dis.* **4,** 125–127.

273. Toivanen, P., Toivanen, A., Olkkonen, L., and Aantaa, S. (1973) Hospital outbreak of *Yersinia enterocolitica* infection. *Lancet* **1,** 801–803.

274. Asakawa, Y., Akahane, S., Kagata, N., Noguchi, M., Sakazaki, R., and Tamura, K. (1973) Two community outbreaks of human infection with *Yersinia enterocolitica*. *J. Hyg. Camb.* **71,** 715–723.

275. Tertti, R., Granfors, K., Lehtonen, O. P., et al. (1984) An outbreak of *Yersinia pseudotuberculosis* infection. *J. Infect. Dis.* **149,** 245–250.

276. Nakano, T., Kawaguchi, H., Nakao, K., Maruyama, T., Kamiya, H., and Sakurai, M. (1989) Two outbreaks of *Yersinia pseudotuberculosis* 5a infection in Japan. *Scand. J. Infect. Dis.* **21,** 175–179.

277. Babic-Erceg, A., Klismanic, Z., Erceg, M., Tandara, D., and Smoljanovic, M. (2003) An outbreak of *Y. enterocolitica* O:3 infection on an oil tanker. *Eur. J. Epidemiol.* **18,** 1159–1161.

278. Lachica, R. V., Zink, D. L., and Ferris, W. R. (1984) Association of fibril structure formation with cell surface properties of *Yersinia enterocolitica. Infect. Immun.* **46,** 272–275.

279. Skurnik, M., Böli, I., Heikkinen, H., Piha, S., and Wolf-Watz, H. (1984) Virulence plasmid-associated autoagglutination in *Yersinia* spp. *J. Bacteriol.* **158,** 1033–1036.

280. Heesemann, J. and Grüter, L. (1987) Genetic evidences that outer membrane protein YOP1 of *Yersinia enterocolitica* mediates adherence and phagocytosis resistance to human epithelial cell. *FEMS Microbiol. Lett.* **40,** 37–41.

281. Paerregaard, A., Espersen, F., Jensen, O. M., and Skurnik, M. (1991) Interaction between *Yersinia enterocolitica* and rabbit ileal mucus: growth, adhesion, penetration, and subsequent changes in surface hydrophobicity and ability to adhere to ileal brush border membrane vehicles. *Infect. Immun.* **59,** 253–260.

282. Tertti, R., Skurnik, M., Vartia, T., and Kuusela, P. (1992) Adhesion protein of *Yersinia* species mediates binding of bacteria to fibronectin. *Infect. Immun.* **60,** 3021–3024.

283. Tamm, A., Tarkkanen, A. M., Korhonen, T. K., Kuusela, P., Toivanen, P., and Skurnik, M. (1993) Hydrophobic domains affect the collagen-binding specificity and surface polymerazion as well as the virulence potential of the YadA protein of *Y. enterocolitica. Mol. Microbiol.* **10,** 995–1011.

284. Skurnik, M., El Tahir, Y., Saarinen, M., Jalkanen, S., and Toivanen, P. (1994) YadA mediates specific binding of enteropathogenic *Yersinia enterocolitica* to human intestinal submucosa. *Infect. Immun.* **62,** 1252–1261.

285. Ruckdeschel, K., Roggenkamp, A., Schubert, S., and Heesemann, J. (1996) Differential contribution of *Yersinia enterocolitica* virulence factors to evasion of microbicidal action of neutrophils. *Infect. Immun.* **64,** 724–733.

286. Press, N., Fyfe, M., Bowie, W., and Kelly, M. (2001) Clinical and Microbiological Follow-up of an outbreak of *Yersinia pseudotuberculosis* Serotype 1b. *Scand. J. Infect. Dis.* **33,** 523–526.

287. Ostroff, S. M., Kapperud, G., and Lassen, J. (1992) Clinical features of sporadic *Yersinia enterocolitica* Infections in Norway. *JID* **166,** 812–817.

288. Zheng, X. B., Tsubokura, M., Wang, Y., et al. (1995) *Yersinia pseudotuberculosis* in China *Microbiol. Immunol.* **39,** 821–824.

289. Shayegani, M., Mores, D., De Fargo, J., Root, T., Parsons, L. M., and Maupin, P. S. (1983): Microbiology of a major foodborne outbreak of gastroenteritis caused by *Yersinia enterocolitica* serotype O:8. *J. Clin. Microbiol.* **17,** 35–40.

5

Foodborne Pathogenic Vibrios

T. Ramamurthy and G. Balakrish Nair

1. BACKGROUND

In the past few decades, many foodborne outbreaks associated with consumption of contaminated foods have occurred in both developing and developed countries. Factors that influence the epidemiology of the foodborne diseases include: level of hygiene in handling foods, globalization of the supply and distribution of raw foods, aquaculture practices, farming, introduction of pathogens into new geographical areas, changes in the virulence and environmental resistance of pathogens, decrease in immunity among certain population, and changes in the eating habits.

Members belonging to the genus *Vibrio* of the family Vibrionaceae have acquired increasing significance because of its association with mild-to-severe human diseases. There is a species preponderance of vibrios and their geographic distribution of infection. For example, incidences of some vibrios are related to poor sanitation and hygiene in developing countries, whereas other species are associated with frequent foodborne infections in developed countries. Most of the vibrios reside in aquatic realms and associated fauna and flora as autochthonous microflora. Toxigenic *Vibrio cholerae* is the most important, because this species causes cholera, a devastating disease of global prominence. The other important species is *Vibrio parahaemolyticus*. The recently emerged pandemic strains of this species caused major concerns in both the developed and developing countries. Studies conducted at molecular level have clearly shown that the pandemic strains are of recent origin and are undergoing rapid changes that have not been recorded earlier. Other vibrios of medical importance are *Vibrio vulnificus*, *Vibrio mimicus*, *Vibrio alginolyticus*, *Vibrio fluvialis*, *Vibrio hollisae*, and *Vibrio damsela*. However, the significance of *V. furnissii* and *V. metschnikovii* is not known owing to inadequate number of investigations carried out on these pathogens. With the exception of cholera, *Vibrio* diseases are not designated as reportable.

This chapter reviews the health hazards associated with foods carrying pathogenic vibrios, detection techniques, virulence factors, and the effect of different decontamination methods in eliminating the vibrios.

2. CLASSIFICATION

The family Vibrionaceae consists of seven genera. The genus *Vibrio* currently has 48 species; among them 11 are of clinical importance. Recently, the species *V. damsela* is placed under the genus *Photobacterium*. Species representing the family Vibrionaceae

From: *Infectious Disease: Foodborne Diseases*
Edited by: S. Simjee © Humana Press Inc., Totowa, NJ

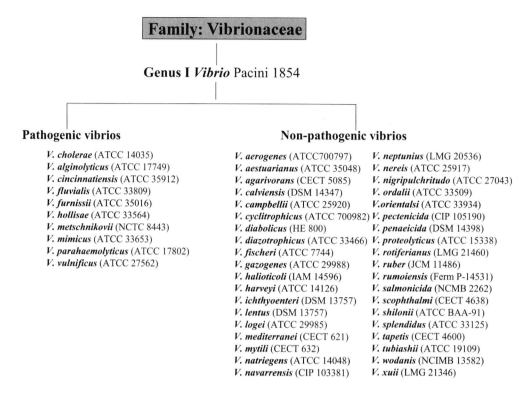

Fig. 1. Classification and names of taxa within the family Vibrionaceae. For each species, the type strain number is given in parentheses. The other six genus in this family includes *Photobacterium, Allomonas, Listonella, Enhydrobacter, Salinivibrio,* and *Enterovibrio.* The species *V. damselae* is now placed under the genus *Photobacterium* and the nomenclature is *Photobacterrium damselae.*

are given in Fig. 1. Members of the genus *Vibrio* are defined as Gram-negative, asporogenous rods that are straight or have a single, rigid curve. They are motile; most have a single polar flagellum, when grown in liquid medium. Sodium ions at varying concentrations stimulate the growth of all species, which are a basic growth requirement for most of the vibrios. Most produce oxidase and catalase, and ferment glucose without producing gas. The G+C content of the genus *Vibrio* is 38–51% compared to the *Photobacterium* (40–44%), *Aeromonas* (57–63%), and *Plesiomonas* (51%).

3. ISOLATION AND IDENTIFICATION

For isolation of vibrios, initial enrichment in alkaline peptone water (pH 8.0) is useful as this medium supports the growth of viable cells. An altered alkaline broth is recommended for specific enrichment of *V. fluvialis* from seafood, and environmental sample *(1).* Majority of the clinically important vibrios can be isolated using common media like blood agar, MacConkey agar, taurocholate–tellurite–gelatin agar, and thiosulfate–citrate–bile salt–sucrose agar (TCBS). The selective nature of TCBS may be inhibitory to some of the vibrios such as *V. hollisae, V. damsela, V. metschnikovii,* and *V. cincinnatiensis.* Use of 1–3% NaCl to the food diluent is necessary to maintain the viability of the halophilic vibrios.

V. parahaemolyticus is known to exist in a viable but nonculturable state especially in foods when incubated at low temperatures and in routine cultural methods. It is difficult to culture cells at this state. The resuscitation of nonculturable cells of *V. parahaemolyticus* is achievable after spreading onto an agar medium supplemented with H_2O_2-degrading compounds, such as catalase or sodium pyruvate *(2)*.

Conventionally, vibrios are identified using biochemical characteristics and specific growth conditions (Table 1). However, it is recommended to use standard strains for quality control to ensure the results. *V. cholerae* and *V. mimicus* are nonhalophils (grows in the absence of NaCl) and are closely related. Because the production of arginine dihydrolase by *V. fluvialis* and *V. hollisae* is often confused with the identification of *Aermonas* spp., salt tolerance test is recommended for confirmation. The vibriostatic compound 2,4-diamino-6,7-diisopropylpteridine (O/129) at concentration of 10 and 150 µg/mL has been used to discriminate between *Aeromonas* and *Vibrio* species. However, because many recent strains of *V. cholerae* have been found resistant to O/129 compound *(3)*, these results should be interpreted with caution and used in conjunction with other biochemical and salt tolerance tests.

Many diagnostic kits are now commercially available for the identification of vibrios of clinical importance (Table 2). An antibody enzyme immuno assay for the detection of *V. cholerae* O1 was developed using the rabbit antiserum, β-D-galactosidase-labeled goat antirabbit immunoglobulin G as the tracer *(4)*. With the detection limit of 4500 cells, this assay can be used for screening *V. cholerae* directly from food homogenates. *V. cincinnatiensis*, a new species, was identified in 1986 from the blood and cerebro-spinal fluid of a patient. On the basis of 5S rRNA comparative sequence analysis, the organism appears to share a recent common ancestry with *V. gazogenes* (98% homo-logy) and close ancestry with *V. mimicus*, *V. fluvialis*, and *V. metschnikovii* *(5)*. Strains belonged to the aerobic biogroup 2 of *V. fluvialis* are now designated as *V. furnissii* *(6)*. Based on the production of gas from the fermentation of carbohydrates, *V. furnissii* can be differentiated from *V. fluvialis*. The distinct features of *V. furnissii* compared to other halophilic vibrios include positive reaction for L-arginine, L-arabinose, maltose, and D-mannitol and negative reaction for L-lysine, L-ornithine, lactose, and Voges–Proskauer *(6)*.

V. cholerae serogroup O1 includes two biotypes—classical and El Tor—each including Inaba and Ogawa, and rarely Hikojima serotypes. Currently, the El Tor biotype is predominant. Vibrios that are bichemically indistinguishable, but do not agglutinate with *V. cholerae* serogroup O1 antiserum (non-O1 strains, formerly known as nonagglutinable vibrios [NAGs] or noncholera vibrios [NCVs] are now included in the species *V. cholerae*). Before 1992, non-O1 strains were recognized to cause sporadic infections and not associated with large epidemics. However, in late 1992, large-scale epidemics of cholera-like infection were reported in India and Bangladesh *(7,8)*. The causative organism was identified as a new serogroup *V. cholerae* O139 (synonym: Bengal), which also elaborates cholera toxin (CT). The clinical and epidemiological picture of illness caused by the O139 serogroup is typical of cholera, and the cases should be reported as cholera *(9)*. The reporting of *V. cholerae* non-O1 infections, other than O139, as cholera is inaccurate and leads to confusion. Apart from O1 and O139 serogroups, 206 O-serogroups are recognized in the current *V. cholerae* serotyping scheme.

Table 1
Biochemical and Growth Characteristics of Clinically Important Vibrios

Test	V. cholerae	V. mimicus	V. parahaemolyticus	V. alginolyticus	V. vulnificus	V. damsela	V. fluvialis	V. hollisae	V. metschnikovii	V. cincinatiensis
Growth in										
TCBS agar	Yellow	Green	Green	Yellow	Green/Yellow	Green	Yellow	Green/NG	Yellow	Yellow
mCPC agar	Purple	NG	NG	Purple	ND	ND	NG	NG	NG	ND
CC agar	Purple	NG	NG	Purple	ND	ND	NG	NG	NG	ND
AGS	K/a	K/A	K/A	K/K	ND	ND	K/K	K/a	K/K	ND
Oxidase	+	+	+	+	+	+	+	+	-	+
Indole[b]	+	+	+	±	+	-	±	+	±	-
Voges–Proskauer[b]	±	-	-	+	-	+	-	-	+	-
NO₃ to NO₂[b]	+	+	+	+	+	+	+	+	-	+
Simmons' citrate	+	+	-	-	±	-	+	-	±	±
ONPG	+	+	-	-	±	-	±	-	±	±
Urea hydrolysis	-	-	±	-	-	-	±	-	-	-
Gelatin hydrolysis	+	±	+	+	±	-	±	-	±	-
Motility	+	+	+	+	+	±	±	-	±	±
Polymyxin B inhibition	±	+	±	±	-	+	+	+	+	+
Arginine dihydrolase[b]	-	-	-	-	-	+	+	-	±	-
Lysin decarboxylase[b]	+	+	+	+	+	±	-	-	±	±
Ornithine decarboxylase[b]	+	+	+	±	±	-	-	-	-	-

Test response[a]

118

Acid production from:								
D-Glucose	+	+	+	+	+	+	+	+
L-Arabinose	+	±	−	−	−	+	+	+
D-Arabitol	−	−	−	−	−	±	−	−
Cellobiose	±	−	−	+	−	±	−	−
Lactose	−	±	±	+	+	±	+	±
Maltose	+	+	+	+	+	+	+	+
D-Mannitol	+	+	±	±	−	±	+	+
Salicin	−	−	−	±	−	−	−	−
Sucrose	+	−	+	+	+	+	−	+
Utilization of:								
L-leucine	−	+	+	−	−	−	−	ND
L-putrescine	−	+	+	−	±	±	+	ND
Ethanol	−	±	±	−	+	+	−	ND
D-glucuronate	±	±	−	+	+	+	−	ND
Growth in NaCl:								
0%	+	+	+	+	+	+	−	−
3%	+	+	+	+	+	+	+	+
6%	±	±	±	+	+	±	+	+
8%	−	−	+	+	−	−	−	−
10%	−	−	−	±	−	−	−	−
Growth at 42°C	+	+	+	+	ND	±	ND	ND
Sensitivity to:								
10 µg of O/129	R	S	R	R	S	R	R	S
150 µg of O/129	S	S	S	S	S	S	S	S

Abbreviations: TCBS, thisulfate–citrate–bile salts–sucrose agar; mCPC, modified polymyxin B-colistin agar; CC agar, cellobiose colistin agar; AGS, arginine glucose slants; ONPG, O-nitrophenyl-β-D-galactopyranoside; NG, no growth; K/A, alkaline slant/acid butt; K/a, alkaline slant/slightly acidic butt; K/K, alkaline slant/alkaline butt; R, resistance; S, susceptible; ND, no data.

[a](+) >90% positive; (±) variable, >50% positive; (∓) variable, <50% positive; (−) <10% positive.
[b]test reaction with 1% NaCl.

119

Table 2
List of Commercial Kits for the Identification of Vibrios/Toxins

Test kit product	Analytical technique	Total time including enrichment culture	Supplier name
CHECK 3 *Vibrio* sp.	Biochemical, visual detection	4–18 h	Contamination Sciences LLC
API 20 E	Biochemical, visual detection	6–18 h	Bio Merieux
Chromogenic *Vibrio*	Chromogenic media for presumptive identification of *V. parahaemolyticus* and *V. vulnificus*	48 h	Biomedix
ISO-GRID method	Membrane filtration with selective and differential culture medium using sucrose fermentation. For *V. parahaemolyticus* count using VSP agar	24 h	QA Life Sciences, Inc.
5898 A	Fatty acid profile based identification of vibrios	26 h	Hewlett-Packard, Microbial ID
Cholera and Bengal Smart	Direct fluorescent assay (DFA) for the detection of *V. cholerae* O1 and O139	5–10 min	New Horizons Diagnostics, Inc.
Toxin Detection kits	*V. parahaemolyticus* heat-stable hemolytic toxin and heat-labile enterotoxin of *V cholerae*	12–18 h	Denka Seiken Co. Ltd
Differentiation kits	Differentiation of classical and other vibrios 0/129 (2,4-diamino-6, 7-diisopropylpteridine) Polymyxin B 50 U disks	12–18 h	Thayer Martin
VET-RPLA TD0920	Reversed passive latex agglutination to detect cholera toxin (1–2 ng/mL)	24 h using bacterial cultures	Oxoid, Inc.

Serotyping system of *V. parahaemolyticus* is based on somatic (O) and capsular (K) antigens that include 13 different O-antigens and 71 different K-antigens *(10)*. Until 1996, *V. parahaemolyticus* infections were associated with sporadic cases of gastroenteritis caused by multiple, diverse serovars. Since 1996, however, an increased incidence of gastroenteritis in many parts of Asia and the United States has been associated with *V. parahaemolyticus* serovar O3:K6 *(11–14)*. Furthermore, other serovars (such as O4:K68, O1:K25, and O1:K untypeble [O1:KUT]), which are virtually identical to O3:K6 by a variety of molecular typing methods *(15,16)* and have been postulated to be the clonal derivatives of the O3:K6.

The API 20E system was shown to be useful for the identification of members of the family Vibrionaceae, especially *V. alginolyticus*, *V. cholerae*, *V. fluvialis*, *V. vulnificus*, and *V. parahaemolyticus (17,18)*. Using commercial bacterial identification kits such as Vitek and API 20E systems, it is difficult to differentiate *Aeromonas schubertii*, *Aeromonas veronii* biotype Veronii from *V. damsela*, and *V. cholerae*, respectively *(19)*. Some of the critical biochemical tests in the Vitek identification need specified reaction time *(20)*.

The cellular fatty acid composition analysis using capillary gas–liquid chromatography (GLC) was designed for the characterization of vibrios and aeromonads. Based on the decision-tree using the GLC profile, the groups *V. alginolyticus–V. parahaemolyticus*; *V. metschnikovii–V. cincinnatiensis*, and *V. cholerae–V. mimicus* could not be differentiated *(21)*. The intracellular trypsin-like activity measured by the fluorescence with fluorogenic substrate benzol-L-arginine-7-aminomythylcoumarine has been suggested for the identification of *V. parahaemolyticus (22)*. However, because of the false-positive results, this assay was not recommended for the direct detection of *V. parahaemolyticus* from seafoods *(23)*. Chemotaxonomic studies carried out with *V. fluvialis* and *V. vulnificus* on the basis of the sugar composition of the polysaccharide portion of their lipopolysaccharides have shown that there is variation in the composition *(24)*. In addition, close similarities in the sugar composition of the same portion were demonstrated between serologically crossreacting non-O1 group *V. cholerae* and *V. fluvialis*, and non-O1 *V. cholerae* and *V. mimicus (24)*.

Monoclonal antibodies (MAbs) raised against an array of clinically important species of vibrios have been shown to be useful for rapidly identifying vibrios. The genus-specific MAbs was very useful to rapidly identify vibrios in clinical specimens, with species-specific MAbs needed to complete the diagnosis *(25)*. For serological testing, MAbs were also raised against *V. cholerae* O1 and O139 serogroups *(26,27)*.

A total of 61 O-serogroups were recognized in the serotyping scheme of *V. fluvialis* and *V. furnissii (28)*. The serogroup O19 of *V. fluvialis* posses the C (Inaba) antigen but not the B (Ogawa) or A antigen of *V. cholerae* O1 *(29)*. Incidence of *V. fluvialis* serogroup O19 was reported in the erstwhile USSR that cross-agglutinated with *V. cholerae* O1 Inaba antiserum *(30)*. Some strains of *V. mimicus* may crossreact with polyclonal antibodies elicited against *V. cholerae* O139. The nature of this crossreactivity resides in the partial structure comprising the galactosyl residue substituted with cyclic phosphate, which is present in the cell-wall-associated polysaccharides *(31)*.

V. vulnificus was first identified and described by CDC in 1976 *(32)*. A species-specific antibody against flagellar (H) antigen expressed by *V. vulnificus* has been used as a rapid and specific diagnostic reagent in the coagglutination test *(33)*. The surface antigenic determinants of the lateral flagella (L-flagella) of *V. alginolyticus* and *V. harveyi* are specific to these two species *(34)*. In *V. fluvialis*, the L-flagellar antigen was found to be species specific and is useful for serological identification *(35)*. Although the H-antigen serotyping was claimed to be species specific *(36)*, its utility is yet to be adopted in the serotyping scheme.

Phage-typing scheme was established for typing *V. fluvialis* strains with a panel of six stable phages that formed 12 different phage types *(37)*. New phage-typing schemes for *V. cholerae* O1 and O139 have been described for routine strain-testing *(38,39)*. However, the applicabilities of phage-typing schemes are on the decline with the advent of more discriminating molecular typing techniques and with the limited availability of phages.

4. IDENTIFICATION OF VIBRIOS BY MOLECULAR TECHNIQUES

Timely resolution of foodborne infections requires rapid and accurate detection of pathogens. Molecular methods have become the vital epidemiological tool in the detection, characterization, and tracking of pathogens. Recently, detection of specific virulence genes or other marker genes is possible by means of polymerase chain reaction (PCR) assay or by DNA probe techniques. Simplex PCR detects a single gene of interest, and multiplex PCR is used for 2–5 target genes in a single reaction. The frequently used PCR assays targeting different vibrios are given in Table 3.

For the detection of *V. cholerae* and *V. mimicus*, the intergenic spacer sequence of 16S–23S rRNA and a hexaplex PCR targeting *ctxA*, *zot*, *ace*, *tcpA*, *ompU*, and *toxR* were useful *(40,41)*. The 16S rRNA gene (rDNA) probe V3VV was found highly specific for the detection of *V. vulnificus (42)*. Heat-stable enterotoxin (STN) is common among *V. cholerae* non-O1, non-O139, and *V. mimicus* strains. A PCR method was established for the detection of *stn* among vibrios *(43)*. Multiplex PCR targeting *ctx* and *hlyA* was developed for the detection of toxigenic *V. cholerae* with a detection limit of 3 CFU/g of food *(44)*. Direct PCR assay for the detection of *ctx* among *V. cholerae* from seafood has also been established with the sensitivity of 1 CFU/10 g *(45)*.

Many PCR assays with high specificity and sensitivity have been developed for the identification of *V. parahaemolyticus (46)*. Among vibrios, the toxin regulatory gene *toxR* seems to vary with each species. Exploiting the *toxR* sequence difference, Kim et al. *(47)* designed a species-specific PCR assay for the identification of *V. parahaemolyticus*. The *gyrB* gene, which is primarily responsible for the control of negative supercoiling of DNA, is universally present in all the bacteria. The primer pair designed from the conserved regions of the *gyrB* sequence of *V. parahaemolyticus* was found to be specific and useful in the routine identification of *V. parahaemolyticus* from food samples *(48)*. Because both pathogenic and nonpathogenic strains of *V. parahaemolyticus* coexist in seafoods, the application of PCR specific for the virulence genes (*tdh* and *trh*) helps in the detection of pathogenic strains of *V. parahaemolyticus (49)*. In seafoods such as oysters, squid, mackerel, and yellow tail, unknown substances inhibited PCR targeting the *tdh* gene. Hence, a DNA-purification step should be adopted before processing the samples for PCR to increase the detection sensitivity *(50)*. PCR assay targeting the thermolabile–hemolysin (*tl*) was reported to be useful for the detection of *V. parahaemolyticus (51)*. The R72H DNA sequence is highly conserved in *V. parahaemolyticus* and can be used for species-specific detection by PCR *(46,52)*. The R72H sequence is composed of a noncoding region and a phosphatidylserine synthetase gene.

For the detection of pandemic strains of *V. parahaemolyticus*, two PCR systems are widely in use. The group-specific PCR targets unique variations in the *toxRS* sequence of pandemic strains *(14)* whereas the *orf8* PCR is useful for the identification of integrated filamentous phage having the open reading frame 8 (ORF8) *(53,54)*. To simplify the assay system, a multiplex PCR assay was designed to distinguish pandemic serogroup strains from other *V. parahaemolyticus* strains *(55)*. However, some recent studies have shown that the ORF8 was absent in considerable number of pandemic strains of *V. parahaemolyticus (16,56,57)*.

Real-time PCR is currently used for the diagnosis of infectious agents. Compared to laborious and resource intensive conventional methods, the real-time PCR is very fast

Table 3
PCR Assay for the Detection of *Vibrio* spp.

Pathogen	Type of PCR (Reference)	Target gene (size, bp)	Primer sequence (5′ to 3′)	PCR condition[a]
V. cholerae	Multiplex (261)	*wbe* O1 (192)	GTT TCA CTG AAC AGA TGG G GGT CAT CTG TAA GTA CAA C	94°C, 1 min 55°C, 1 min
		wbb O139 (449)	AGC CTC TTT ATT ACG GGT GG GTC AAA CCC GAT CGT AAA GG	72°C, 10 min 72°C, 1 min 30 cycles
	Multiplex (262)	*ctxA* (301)	CTC AGA CGG GAT TTG TTA GGC ACG TCT ATC TCT GTA GCC CCT ATT ACG	94°C, 1 min 65°C, 1.5 min
		tcpA classical (617)	CAC GAT AAG AAA ACC GGT CAA GAG ACC AAA TGC AAC GCC GAA TGG AG	72°C, 1.5 min 72°C, 10 min 30 cycles
		tcpA El Tor (471)	GAA GAA GTT TGT AAA AGA AGA ACA C GAA GGA CCT TCT TTC ACG TTG	
	Simplex (263)	*ompW* (588)	CAC CAA GAA GGT GAC TTT ATT GTG GAA CTT ATA ACC ACC CGC G	94°C, 30 s 64°C, 30 s 72°C, 30 s 72°C, 5 min 30 cycles
V. parahaemolyticus	Multiplex (51)	*tlh* (450)	AAA GCG GAT TAT GCA GAA GCA CTG GCT ACT TTC TAG CAT TTT CTC TGC	94°C, 1 min 60°C, 1 min
		trh (500)	TTG GCT TCG ATA TTT TCA GTA TCT CAT AAC AAA CAT ATG CCC ATT TCC G	72°C, 2 min 72° C, 3 min 25 cycles
		tdh (270)	GTA AAG GTC TCT GAC TTT TGG AC TGG AAT AGA ACC TTC ATC TTC ACC	
	Simplex (48)	*gyrB* (285)	CGG CGT GGG TGT TTC GGT AGT	94°C, 1 min

(Continued)

Table 3 (*Continued*)

Pathogen	Type of PCR (Reference)	Target gene (size, bp)	Primer sequence (5′ to 3′)	PCR condition[a]
			TCC GCT TCG GGC TCA TCA ATA	63°C, 1.5 min 72°C, 1.5 min 72°C, 7 min 30 cycles
	Simplex (47)	*toxR* (368)	GTC TTC TGA CGC AAT CGT TG ATA CGA GTG GTT GCT GTC ATG	94°C, 1 min 63°C, 1.5 min 72°C, 2.5 min 72°C, 7 min 25 cycles
	Simplex (14)	*toxRS*[b] (651)	TAA TGA GGT AGA AAC A ACG TAA CGG GCC TAC A	96°C, 1 min 45°C, 2 min 72°C, 3 min 72°C, 7 min 25 cycles
	Simplex (53)	*orf8*[b] (388)	AGT ATT GCT GAA GAG TAC G CTC GAC TTA AAC GAT CCC	96°C, 10 s 55°C, 20 s 72°C, 1 min 72°C, 7 min 30 cycles
V. vulnificus	Simplex (264)	*vvhA* (519)	CCG CGG TAC AGG TTG GCG CA CGC CAC CCA CTT TCG GGC C	94°C, 1 min 62°C, 1 min 72°C, 2 min 72°C, 10 min 25 cycles

[a]PCR conditions include denaturation, annealing, and extension steps.
[b]For specific detection of pandemic strains of *V. parahaemolyticus.*

and highly sensitive. The TaqMan real-time PCR assay targeting the cytolysin and *vvhA* is very sensitive for the detection of viable but nonculturable cells of *V. vulnificus (58)*. SYBR Green-I based real-time PCR with *V. vulnificus*-specific hemolysin gene (*vvh*) specifically detected the target pathogen with minimum direct detection level of 1 pg of DNA or 10^2 CFU/g of oyster meat or 10 mL of water sample *(59)*. A similar assay system was also developed for the detection of *tdh* gene of *V. parahaemolyticus (60)*. Duplex real-time PCR for the detection of foodborne *V. vulnificus*, *V. cholerae*, and *V. parahaemolyticus* in less than 2 h was established targeting the genes encoding cytotoxin hemolysin, *ctx* and *tdh/trh*, respectively *(61)*. Three sets of primers and a TaqMan probe targeting *tdh*, *trh* and *tlh* were successfully used for the detection of *V. parahaemolyticus* in a foodborne outbreak in Florida *(62)*.

Conventional PCR can amplify the target DNA from both viable and nonviable cells of the target pathogen. Detection of viable pathogens from clinical specimens or food samples is desired to ensure that the positive test-results are associated with live pathogens. Positive identification caused by nonviable pathogens may lead to misguided decisions related to the effectiveness of the treatment or whether the food should be discarded. The reverse transcriptase PCR (RT-PCR) is useful for targeting the mRNA from viable cells. Several RT-PCR assay methods are available for the detection of pathogenic vibrios *(63)*.

Polynucleotide and oligonucleotide probes are used for the detection of target sequences in the isolated strain. Use of these probes in a colony hybridization assay helps to screen large number of isolates, which are suspected to carry the virulence genes. The alkaline phosphate and digoxigenin-labeled *tdh* gene probe was useful to detect *V. parahaemolyticus (64)*. 16S rRNA gene sequence analysis was found to be specific for the identification of *V. vulnificus*. In addition, oligonucleotide probes based on the sequence difference in the 23S rRNA have been useful for the *in situ* identification of this species *(65)*. DNA sequences based on the 23S rRNA gene was used as a species-specific probe for the identification of *V. parahaemolyticus (66)*.

5. PHYLOGENETIC APPROACH

Since 1970s, there has been an avalanche of changes, and new insights into the phylogeny of prokaryotic species. The corner stone of natural classification above species level was the recognition that all the traits of an excellent phylogenetic marker are present in ribosomal RNA (rRNA) and its encoding genes. The most powerful and most extensively used phylogenetic marker molecule is the 16S rRNA and the genes coding for this molecule (rDNA). Amplified rRNA genes are sequenced directly by chain-termination method or by automated DNA sequencer. The availability of the 16S rDNA sequence in combination with a database such as ribosomal database project (RDP) and ARBOR (ARB) *(67)*, containing more than 12,000 sequences from more than 90% of all described genera provides an advantage unmatched by any other rapid screening method.

The *hsp*60-gene amplification and sequence comparison was useful, as an alternative target for phylogenetic analysis and species identification of vibrios to complement the conventional methods *(68)*. The sequence comparison of the amplified partial *hsp*60 sequence revealed 71–82% homology among different species of vibrios and 96–100% homology among epidemiologically distinct strains within the same species designation.

In addition, *Aeromonas hydrophila* and *Plesiomonas shigelloides* that are biochemically similar to most of the vibrios can be differentiated by means of *hsp*60 gene sequencing *(68)*. The halophilic *V. parahaemolyticus* is genetically more related to *V. alginolyticus* than *V. vulnificus* (60–70% and 40–50% DNA homology, respectively).

Phylogenetic tree based on the 16S rDNA sequences of clinically important *Vibrio* species showed separated identities (Fig. 2) indicating that the pathogenicity does not evolve with its species divergence *(69)*. In concordance to the GC content of the DNA, *V. alginolyticus* and *V. parahaemolyticus* formed one close group, and *V. mimicus* and *V. cholerae* form a monophyletic grouping indistinguishable from the core *Vibrio* group, which contains mainly the halophilic vibrios.

6. MOLECULAR TYPING

Intraspecies characterization of pathogenic bacteria for epidemiological purpose can now be effectively achieved by molecular typing methods. Enterobacterial repetitive intergenic consensus sequence (ERIC) PCR, restriction fragment length polymorphism (RFLP) in rRNA genes (ribotyping), pulsed-field gel electrophoresis (PFGE), and RFLP analysis of the genetic locus encoding the polar flagellum (Fla locus) or the virulence genes have been used to type the *Vibrio* strains. Arbitrarily primed polymerase chain reaction (AP-PCR) is a simple and valuable tool in molecular epidemiological investigations. *V. mimicus* isolates collected from different geographical locations were not identical but isolates from the same region comprised a single strain *(70)*. However, this assay did not consistently match with serogroups. AP-PCR results showed that *V. vulnificus* strains collected between 1993 and 1999 from patients after consumption of raw oysters in the United States exhibited a high degree of variation in the genomic organization *(71)*. *V. vulnificus* strains isolated from 62 cases of wound infection and bacteremia during 1996–1997 in Israel showed similar pattern by PCR-RFLP *(72)*.

RFLP of the *tdh* gene was shown to be a less discriminating method in the molecular typing of pathogenic *V. parahaemolyticus* strains *(12)* because all the tested strains have two copies of *tdh* as reported earlier *(73)*. Ribotyping determines the number of *rrn* operons in the bacterial genome. For ribotyping, *V. parahaemolyticus* genomic DNA is fragmented either with *Hind*III or *Bgl*I enzymes and hybridization with cDNA probe of *E. coli* 16S and 23S RNA genes *(12,74)*. The *rrn*-mediated recombination may be one of the mechanisms by which pathogenic bacteria may maintain their genomic plasticity, and the possibility of genome rearrangements is greater when there are more *rrn* operons in the genome of an organism *(75)*.

In *V. cholerae*, *Bgl*I restriction enzyme was successfully used to develop a ribotyping scheme *(76)*. Studies conducted in Calcutta showed that the pandemic *V. parahaemolyticus* O3:K6 strains belonged to five different ribotypes with R4 the most common with strains isolated during 1996–1997 *(12)*. This study also indicated that the strains isolated during brief interim period had genetic instability, but over a period of time the clone stabilized as evidenced by the constant prevalence of R4 ribotype. The nonpandemic strains, including "old O3:K6" strains (isolated before 1996) used in this study exhibited heterogeneous ribotype profiles that were distinct from each other, as well as from the R4 pattern exhibited by the pandemic strains. Thus, the new O3:K6 (isolated after 1996) did not seem to have originated from the "old O3:K6" group of strains. Riotyping results of northeastern US *V. parahaemolyticus* outbreak strains showed relatedness

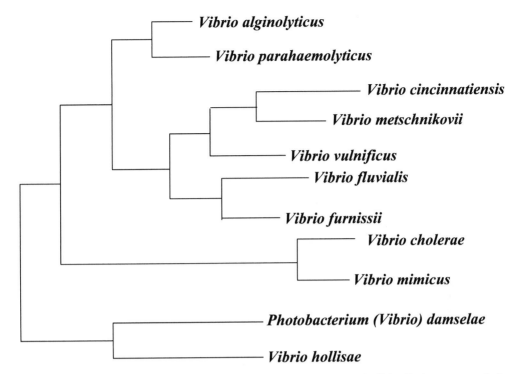

Fig. 2. Schematic representation of phylogenetic relatedness of clinically important vibrios based on 16S rDNA sequences. The genetic distance is not scaled.

with the Asian clone, but was different from the Texas outbreak strains *(77)*. Combination of serotyping and ribotyping showed that the Pacific coast *V. parahaemolyticus* strains appeared to be distinct from that of either the Atlantic coast or the Gulf coast *(78)*.

Among different molecular typing methods, PFGE is highly discriminatory compared with other methods and enables to separate large fragments of digested DNA with a rare cutting enzyme. The *Sfi*I digested chromosomal DNA banding patterns were similar for serovars O3:K6, O4:K68 isolated from seawater and patients in Japan *(79)*. Large number of *V. parahaemolyticus* strains isolated from different outbreaks in Taiwan showed at least 14 PFGE types after *Sfi*I digestion *(80)*. Pandemic strains of O3:K6 isolated during 1996 in Calcutta showed different pulsotypes following *Not*I digestion, compared to the strains isolated before and after this period *(12)*. The same study showed that the strains isolated during the later part exhibited identical or nearly similar pulsotypes, indicating the stability of the strains. *V. parahaemolyticus* strains might show DNA degradation and these strains cannot be used for strain typing by PFGE *(81)*. Prevalence of *V. parahaemolyticus* in imported seafoods from other Asian countries is common in Taiwan *(82)*. However, the isolates were not similar irrespective of the country of origin or kind of seafood when examined by PFGE.

These results reveal prevalence of genetically diverse *V. parahaemolyticus* strains in seafood. Wong et al. *(83)* observed that the strains of O3:K6 which appeared after 1996 formed eight closely related PFGE patterns, with the predominant one (81%) consisting of strains from Taiwan, Korea, Japan, and India. *Ceu*I digestion of chromosomal DNA followed by PFGE is useful for the determination of intra- and interspecies genomic rearrangements and elucidation of the numbers of *rrn* operons in the genomes of prokaryotes. Because *Ceu*I cleaves the DNA only in a 19-bp sequence in the 23 rRNA gene of the *rrn* operon *(84)*, it has also been successfully used to determine the number of *rrn* operon in the genomes of many organisms including *V. cholerae*.

Multilocus sequence typing (MLST) has been used for differentiation of many pathogens *(85,86)* and to detect the role of recombination during bacterial evolution *(87)*. Sequence analysis of amplified *gyrB*, *recA*, *dnaE*, and *gnd* of different serovars of *V. parahaemolyticus* showed that prepandemic strains were highly variable and the pandemic O3:K6 isolates shared two alleles *(88)*. The pandemic clones, defined by a positive GS-PCR for *toxRS* and the presence of ORF8 from the f237 phage, were gene-tically nearly identical at the sequence loci, in contrast to the four distinct serovars: O3:K6, O4:K68, O1:KUT, and O1:K25 *(88)*. These results reconfirm the results of other molecular typing methods that showed multiple serovars occur in a single genetic lineage *(14,53,89)*. The multilocus enzyme electrophoresis (MEE) with four enzyme loci proved to be a powerful molecular tool for the differentiation of *V. mimicus* and *V. cholerae* even when atypical strains were analyzed *(90)*.

7. RESERVOIRS

Aquaculture is being practiced in many countries to augment the need of increasing fish demand. In this closed environment, transmission of fish diseases is high compared to the natural environment. Vibriosis is an economically important fish disease and is responsible for high mortality rates in aquaculture. Some of the fish pathogens can also infect humans. The fish disease causing bacteria, *V. demsela (91)*, *V. vulnificus (92)*, and *V. alginolyticus (93)* were also reported as human pathogens.

V. vulnificus and *V. damsela* are acquired topically from fish through spine puncture or open wounds *(94)*. Apart from seafood, contaminated rice, millet, gruel, vegetables, fruits, poultry, meat, and dairy products have also the potential to transmit cholera *(95)*. The survival dynamics of *V. cholerae* O1 in ceviche differ with serotypes. The survival of *V. cholerae* O1 Inaba decreased in number following a liner or retarded trend, whereas, the other serotype Ogawa followed an accelerated death-trend *(96)*.

Investigations carried out in many countries have shown the presence of human pathogenic vibrios in bivalves, suggesting their potential risk. The species distribution of pathogenic vibrios present in bivalves varies from one geographical region to the other. For example, *V. alginolyticus* and *V. vulnificus* are found more in Italy *(97)*, *V. parahaemolyticus* in Spain *(98)*, *V. alginolyticus* and *V. fluvialis* in the blue crabs of Turkey *(99)*, and mussels from Atlantic coast of Brazil *(100)*. *V. alginolyticus* and *V. parahaemolyticus* were commonly isolated from seafood in Jakarta, Indonesia *(101)*. In Pacific Northwest region, British Columbia coast, the incidence of *V. parahaemolyticus*, *V. fluvialis*, and *V. vulnificus* was high among oysters especially during warmer season than in cold conditions *(102)*. During summer months, the proportion of infections caused by *V. fluvialis*, *V. parahaemolyticus*, *V. costicola*, and *V. damsela* is more in Volga, Russia mainly related to water, consumption of seafood and fresh water fish *(103)*. Even though *V. alginolyticus* and *V. parahaemolyticus* are halophilic organisms, in Calcutta, India, they were isolated from freshwater fishes *(104)*.

The ubiquity and relatively high concentrations of *V. parahaemolyticus*, *V. algino-lyticus*, and other pathogenic vibrios in shellfish are potential public health hazard in

Hong Kong and other subtropical Asian countries *(105)*. In Japan, the incidence of *V. cholerae* non-O1 and *V. mimicus* (serogroup O14) in marine fishes was high during summer *(106)*. An infection of *V. alginolyticus* with acute diarrhea was ascribed to consumption of trout roe in Japan *(107)*. Infection owing to *V. damsela* and *V. hollisae* has generally been associated with the consumption of raw seafood *(108)*. In Japan, the increasing number of *V. parahaemolyticus* infections are linked to the consumption of cultured finfish *(109)*.

There was a strong association between the migration of ridley turtles for hatching eggs in the coast of Costa Rica and the incidence of *V. mimicus* in coastal sand *(110)*. The freshly laid eggs were found contaminated with *V. mimicus*, and some of the isolates (31%) carried the cholera toxin gene (*ctx*). Vibrios from aquaculture ecosystem may also be transmitted directly to humans through handling of fish. A new biotype of *V. vulnificus* caused hundreds of serious infections among persons handling live tilapia produced by aquaculture in Israel *(111)*. High incidence of several vibrios including *V. alginolyticus*, *V. cholerae*, *V. fluvialis*, *V. parahaemolyticus*, and *V. mimicus* in seafood and cultured fishes have been reported in Taiwan *(112)*.

The ability of *V. cholerae* O1 to attach and colonize exoskeleton of edible crustaceans provides a potential means of survival and proliferation in the aquatic environment. In vitro studies clearly showed that *V. cholerae* O1 colonizes and multiplied several folds even at higher saline conditions at 37°C *(113)*. Based on the isolation of *V. cholerae* O1 from oysters and intestinal content of oyster feeding fishes *(114)*, it appears that the pathogen is stable and can be passed into the food chain of aquatic animals.

Historically, it is known that asymptomatic food handlers play a crucial role in the transmission of foodborne diseases. Transmission of *V. cholerae* from food handlers was reported in many countries *(115)*. Cholera epidemic in Latin America caused as many as 1 million cases and death of about 10,000 people mainly related to unchlorinated water supply *(116)*. In addition, cabbage irrigated with raw wastewater contributed to cholera transmission in Trujillo, Peru *(117)*. Formula feeding of infant is a very common practice throughout the world and also an alternative method to prevent mother-to-child infection. Growth of bacterial pathogens in reconstituted infant formula has become a health hazard when contaminated water is used for the rehydration of infant food. Survival studies conducted with common enteric pathogens showed that in tap water *V. cholerae* O1 was the most sensitive at 4°C as well as at 30°C. However, in the reconstituted infant formula, the test pathogen reached 9.2×10^{10} CFU/mL within 24 h at 30°C *(118)*.

8. SPORADIC CASES AND OUTBREAKS BY FOODBORNE VIBRIOS

Foodborne outbreaks by vibrios are generally associated with the consumption of raw or under-cooked seafood. Sporadic incidences of diarrhea caused by non-O1 *V. cholerae* were reported from many countries owing to seafood consumption *(119,120)*. Spread of cholera to Germany through imported fresh fish from Nigeria has been reported *(121)*. Case–control study conducted in Guatemala showed that cholera was associated with the consumption of left-over rice, flavored ice, and street-vended carbonated beverages rather than the suspected drinking municipal tap water *(122)*. In cooked foods, contamination occurs through many sources. Cooked rice rinsed with contaminated drinking water was responsible in an oil rig cholera outbreak *(123)*.

Since the onset of the cholera epidemic in Latin America in 1991, most cases of cholera in the United States have occurred among persons arriving from cholera-affected areas or who have eaten contaminated food brought or imported from these areas. It is estimated that 7974 noncholera *Vibrio* infections and 57 deaths attributed to such infections occur in the United States annually *(124)*, and there was a sharp rise after 1997 *(125)*. The first report on incidence of cholera in United States associated with the food transported from an area with epidemic disease was made during March–April 1991 *(126)*. *V. cholerae* O1 Inaba was isolated from four out of eight patients related to the consumption of crab meat brought from the Ecuador. In December 1994, a cluster of cholera cases occurred among persons in Indiana who had shared a meal of contaminated food brought from El Salvador *(127)*. An outbreak of cholera associated with an imported commercial product occurred in the United States in 1991 *(128)*. Epidemiological investigations revealed detection of El Tor *V. cholerae* O1 among four of the six infected persons who had consumed coconut milk imported from Thailand as well as from the unopened cans of the same brand. Investigation in Thailand of the manufacturing process of the implicated coconut milk showed several sanitary violations that lead to contamination *(128)*.

Outbreaks or epidemics of cholera attributed to raw or uncooked seafood from polluted waters were reported in Guam, Kiribati, Portugal, Italy, and Ecuador. The Louisiana and Texas cases have been traced to eating shellfish from coastal and estuarine waters where a natural reservoir of *V. cholerae*, serotype Inaba, was presumed to exist. These cases were not related to sewage contamination of coastal waters. Detection of carrier status of toxigenic *V. cholerae* was traced in several outbreak investigations. In a village near Chiangmai, Thailand, an outbreak of cholera was reported in 1987 ascribed to the consumption of uncooked pork served to the funeral attendants *(129)*. *V. cholerae* O1 Inaba was isolated from 24 patients in this investigation. In 1988, another outbreak of cholera with 71 culture confirmed cases of *V. cholerae* O1 Ogawa, biotype El Tor occurred in Mae Sot district of Thailand caused by the consumption of raw beef *(130)*. Butchers who carried *V. cholerae* were suspected to be the cause of these two outbreaks.

Outbreaks of cholera in international aircrafts through food contamination were also reported *(131,132)*. Cold seafood salad served in an international aircraft that left Lima, Peru to Los Angeles was responsible for cholera affecting 75 of the 336 passengers, 10 were hospitalized, and one died *(132)*. This was the largest airline-associated outbreak of cholera ever reported and demonstrates the mode of spread of cholera from epidemic area to other parts of the world. In Hong Kong, epidemiological investigation showed linkage of 12 cholera cases with the consumption of seafood in 1994 *(133)*. Microbiological findings demonstrated that contaminated seawater in fish tanks used to keep the shellfishes alive was the most likely vehicle of contamination.

During 1973–1998, 40 outbreaks of *V. parahaemolyticus* infections were reported to the CDC and these outbreaks included more than 1000 illnesses *(134)*. Most of the outbreaks occurred during the warmer months and were attributed to seafood, particularly shellfish. The median attack rate among the persons who consumed the implicated seafood was 56% *(134)*. In the United States, the first outbreak related to the pandemic strain of *V. parahaemolyticus* O3:K6 occurred between May 31 and July 10, 1998, and 416 persons in 13 states had gastroenteritis after the consumption of oysters harvested from Galveston Bay; all 28 stool specimens were culture positive for the pandemic O3:K6 strain *(135)*. During July–August 1998, large outbreaks of *V. parahaemolyticus* infection were reported from California, Oregon, Washington, and Texas *(78,136)*. In subsequent months (July–September), *V. parahaemolyticus* infection spread to Connecticut, New Jersey, and New York *(136)*. However, the O3:K6 serovar was not detected in any of the environmental and food isolates *(78)*. Like the spread of the epidemic strain of *V. cholerae* O1 in 1991 to coastal waters along the Gulf of Mexico, it was suspected that the pandemic *V. parahaemolyticus* O3:K6 was introduced into the US coastal waters by ballast water discharged from ships, which had traveled to Asia *(64,137)*.

Outbreaks of *V. parahaemolyticus* mediated infections were also reported in Canada *(138)* and Spain *(139)*. In the latter, it was because of the consumption of oysters contaminated with nonpandemic strains of *V. parahaemolyticus*. The Canadian outbreak in British Columbia during 1997 might be owing to exceptionally warm ocean-surface water temperatures and heavy rainfall associated with El Nino conditions that appear to be conducive to the growth of *V. parahaemolyticus (138)*. Between November 1997 and April 1998, several human gastroenteritis cases were reported in Antofagasta, Chile *(140)*. This outbreak was associated with the consumption of shellfish, and the etiological agent *V. parahaemolyticus* was isolated from both the affected patients and the shellfish samples. However, in this investigation, tests were not performed to confirm the pandemic *V. parahaemolyticus* strains. In Russia, food poisoning outbreak ascribed to *V. parahaemolyticus* was reported in the city of Vladivostok *(141)*. In Taiwan, *V. parahaemolyticus* was responsible for 61–71% of the total foodborne outbreaks during 1996–1999. Since 1996, the O3:K6 strains have caused numerous outbreaks in Taiwan, accounting for 51–79% of the outbreaks till 1999 *(13,83)*. The emergence and predominance of pandemic strains of *V. parahaemolyticus* was also recorded in Korea, Laos, and Indonesia *(14)*.

Prevalence of additional serovars genetically resembling the pandemic O3:K6 strain of *V. parahaemolyticus* has been recorded in many Asian countries. During 1999, the isolation rate of pandemic *V. parahaemolyticus* in southern Thailand increased to 76% and majority of the strains were O3:K6. The newer serovars such as O1:K25, O1:K41, and O4:K12 were also reported to share all the traits of the pandemic clone except for their somatic and capsular antigens *(57,89)*. Studies conducted in Vietnam showed that the risk factor for *V. parahaemolyticus* related to high socioeconomic status of patients as this group could afford seafood *(142)*. In contrast to *V. cholerae* infections, which tends to occur seasonally, no clear seasonality of the *V. parahaemolyticus* infections were observed in Vietnam. In India, the first outbreak of *V. parahaemolyticus* was reported from Vellore *(143)* owing to nonpandemic strains. The pandemic strains first emerged in India and were associated with the cases of diarrhea admitted to the Infectious Diseases hospital in Calcutta from February 1996 *(11)*.

The first incidence of gastroenteritis involving many cases related to *V. fluvialis* infection was reported in Florida, USA owing to the consumption of contaminated oysters, shrimps, or cooked fish *(144)*. An outbreak of gastroenteritis was reported in the Maharashtra State, India during June 1981 *(145)*. Of the 34 participants who took lunch in a religious ceremony, 14 became ill in 4–6 h after the meal with clinical manifestation of food poisoning. *V. fluvialis* was isolated from nine cases without any other enteric pathogens. The food was suspected for this outbreak, though the left-over food was not available for examination *(145)*.

In Thailand, *Vibrio* bacteremia was mainly owing to non-O1 *V. cholerae* and *V. vulnificus* with symptoms of diarrhea, peritonitis, shock, and skin lesions. Most of the infected patients with these pathogens had cirrhosis *(146)*. In tropical countries, consumption of raw turtle eggs is a common practice. Sporadic cases of *V. mimicus* mediated acute diarrhea have been reported in a hospital-based surveillance at Costa Rica *(147)*. The interesting feature of this study is that all the clinical isolates of *V. mimicus* produced CT. In China, food poisoning of *V. alginolyticus* was first reported in 1989 because of the consumption of salted shrimps *(148)*.

In the United States, 96% of the patients with primary septicemia associated with *V. vulnificus* infection were associated with consumption of raw oysters within 7 d before the onset of the disease *(149)*. *V. vulnificus* was responsible for two fatalities in Florida in 1998 and for a total of 33 deaths in the United States in 1996. During 1996–1997, 62 cases of *V. vulnificus* mediated wound infections, and bacteremia were reported for the first time by handling live cultivated fish from ponds in Israel *(72)*.

9. PATHOGENICITY AND VIRULENCE FACTORS

Overall, the infective dose of vibrios is about 10^6 organisms. The dose can be lowered in the presence of antacids and can be as low as 10^2 of *V. vulnificus* in predisposed patients. The incubation period ranges from 4 to 96 h, usually 12–24 h (mean: 15 h). Vibrios carry many unknown virulence factors, and except for few species, there is no hallmark of any specific gene/virulence factor. Hemolysin is the most widely distributed toxin in the pathogenic vibrios and plays several roles in the infection process. Some strains of *V. fluvialis* produce β-hemolysin and cytotoxin in Hep-2 cell line *(150)*. Vibrios such as *V. fluvialis*, *V. parahaemolyticus*, and *V. alginolyticus* expressed aggregative adherence pattern on Hep-2 cells *(150)* similar to the one described in enteroaggregative *Escherichia coli* *(151)*. In addition to strong lipase and protease activities, most of the human pathogenic vibrios isolated from seafood and aquaculture foods from Taiwan showed cytotoxic activity in Chinese hamster ovary (CHO) cells *(112)*. The protease act not only for processing and activation of protein toxins but also direct extraintestinal infection factors causing edematous or hemorrhagic skin lesions, or disturbance of host defense system. Even though the protease is considered as one of the virulence determinants among vibrios, its prevalence in nonpathogenic vibrios is not known *(152)*. In *V. cholerae* non-O1 and non-O139 and *V. mimicus*, hemagglutination is a common function of lipopolysaccharides (LPS) *(153)*. Although, the immunogenecity of LPS and outer membrane proteins is well documented for important intestinal pathogens, the hemagglutinating properties of cell-surface components are hitherto unrecognized and will definitely contribute toward understanding their role in bacterial adherence. The siderophores of *V. hollisae* and *V. mimicus* have been identified as aerobactin, which is highly prevalent among several strains representing the family Enterobacteriacae *(154)*.

Generally, pathogenic vibrios undergo the viable but nonculturable (VBNC) phenomenon in microcosm studies *(155)*. However, the virulence features of these VBNC strains are recoverable after resuscitation cycle in the murine model *(155)*. In some of the environmental strains of *V. alginolyticus*, virulence genes specific for *V. cholerae* such as *ace*, *zot*, and vibrio pathogenecity island (VPI) were detected *(156)*. Though uncommon, this finding indicates genetic transfer of virulence genes from one species to another in the environment. The ability of *V. damsela* strains to utilize hemoglobin and iron (ferric ammonium citrate) has been demonstrated as a possible marker for the virulence of this species *(157)*. In addition, virulent strains of *V. damsela* were shown to be resistant to bacteriostatic/bactericidal effect exerted by the human serum *(157)*. A heat-labile, protease sensitive cytolysin has also been demonstrated in *V. damsela*, which is antigenically distinct (Mw 57 K) from cytolysins produced by *V. vulnificus*, *V. parahaemolyticus*, and *V. cholerae* O1 El Tor *(158,159)*. This cytolysin showed a strong activity against erythrocytes of mammals, fishes, and on CHO. In addition, it was found that cytolysin was lethal for mice when injected intraperitoneally (mean dose: 1 µg/kg). Two steps of action have been proposed for this toxin, which include a temperature-independent toxin-binding step followed by a temperature-independent membrane-perturbation *(159)*. The gene encoding a novel hemolysin–damselysin, unique to *V. damsela* was cloned and expressed in *Escherichia coli (160)*. The damselysin gene had no homology with hemolysin genes of other vibrios.

The cell-free culture supernatants of *V. fluvialis* strains grown in the presence or absence of lincomycin exhibited several virulence factors including CHO cell elongation, which is a characteristic for *V. cholerae* entrotoxin and heat-labile toxin of *Escherichia coli (161)*. The purified protein from *V. fluvialis* was identified as hemolysin of this species (VFH), which exhibited activity on many mammalian erythrocytes *(162)*. Nucleotide sequence analysis of the *vfh* gene revealed an ORF consisting of 2.2 kb, which encodes a protein of 740 amino acids with molecular weight of 82 kDa. The N-terminal amino acid sequence revealed that the 82-KDa prehemolysin is synthesized in the cytoplasm and secreted into the extracellular environment as the 79-kDa mature hemolysin after cleavage of 25 N-terminal amino acids *(162)*. The hemolysin produced by *V. fluvialis* is heat-labile *(163)*.

The metalloprotease produced by *V. fluvialis* (VFP) belongs to the thermolysin family of toxins *(164)* and was found to have very similar characteristics to *V. vulnificus* protease including hemagglutination, permeability-enhancing and hemorrhagic activities. *V. fluvialis* does not exhibit the mannose-sensitive hemagglutination, which is typical of *V. cholerae (165)*. Although *V. fluvialis* some time causes dysenteric form of the diarrhea, it lacks the invasive property *(166)*. However, this species caused mouse lethality, was positive in the skin permeable factor (SPF), but produced no fluid accumulation (FA) in the suckling mice *(167)*. *V. fluvialis* rarely produces β-hemolysin or adheres to Hep-2 cells *(150)*.

The hallmark of toxigenic *V. cholerae* is the expression of CT. Prevalence of CT producing *V. cholerae* from food and environmental samples are relatively rare. Studies conducted in Taiwan showed about 2% incidence of CT expressing strains of *V. cholerae* non-O1 from the seafood *(112)*. CT produced by *V. cholerae* serogroups O1 and O139 is the main factor that causes diarrhea with severe dehydration by ADP-ribosylation of the α-subunit of the GTP-binding protein, which stimulates adenylate cyclase activity. CT is encoded in the genes *ctxAB*. CT-like toxins were reported in some strains of *V. cholerae* non-O1 and non-O139 strains or *V. mimicus* but not in other vibrios. The filamentous CTXΦ, which encodes CT in epidemic *V. cholerae*, is known to propagate by infecting susceptible strains of *V. cholerae*, which are generally nontoxigenic progenitors, using the toxin coregulated pilus (TCP) as its receptor and, thereby, causing the origination of new toxigenic strains of *V. cholerae (168)*.

Among non-O1 and non-O139 serogroups, most of the CT-producing strains were identified to belong to serogroup O141, and there is a possibility that this serogroup may emerge as potential pathogen in the future *(169)*. A novel hemolysin gene *hlx*, found to be active against sheep, goose, horse, and chicken erythrocytes but not guinea pig or human, was reported among *V. cholerae* and *V. mimicus (170)*. In suckling mice assay (SMA), it was shown that irrespective of the serogroups and elaboration of CT, *V. cholerae* stimulated increased FA in a growth-medium-dependent manner *(171,172)*. However, the FA kinetics was differentiated with neutralization with CT antibody *(171)*.

ToxR-mediated bile resistance is an early step in the evolution of *V. cholerae* as an intestinal pathogen *(173)*. In *V. cholerae* and other vibrios, it was shown that the mutant strains of *toxR* had a reduced minimum bactericidal concentration of bile compared to the wild strains in a ToxT-independent manner *(173)*. Tox-T is the direct transcriptional activator of CT and TCP encoding genes. Unlike *ctx* and *tcp*, which are horizontally acquired genetic elements, the *toxR* is in the ancestral *Vibrio* chromosome.

V. mimicus produces a heat-stable enterotoxin (VM-ST), which is structurally and functionally identical to *V. cholerae* non-O1-ST *(174)*. In addition, *V. mimicus* caused mouse lethality and show a positive test in the SPF assay *(167,175)*. *V. mimicus* elaborates both hemagglutinating protease (agglutination with chicken erythrocytes) as well as a nonprotease hemagglutinin (agglutination with rabbit erythrocytes) *(176)*. The purified protease from *V. mimicus* is a bifunctional molecule capable of mediating proteolysis and hemagglutination, and is immunologically crossreactive with *V. cholerae* protease *(177)*. This protease was found to enhance extraintestinal infections as it induced vascular permeability and form edema when injected into the skin of guinea pig and rats *(178)*. Like *V. fluvialis*, *V. mimicus* do not exhibit mannose-sensitive hemagglutination, which is typical for *V. cholerae (165)*. This pathogen secretes a pore-forming toxin, *V. mimicus* hemolysin (VMH), which causes hemolysis by three sequential steps: binding to erythrocyte membrane, formation of a transmembrane pore, and disruption of the cell membrane *(179)*. The sequence analysis of structural gene *vmhA* encoding the VMH showed 81.6% identity with *V. cholerae* El Tor hemolysin *(180)*. Some of the *V. mimicus* strains may produce thermostable-direct hemolysin (TDH)-like toxin *(181)*, which is distinct compared to *V. parahaemolyticus* TDH *(182)*.

The receptor of CTXΦ, the TCP was also detected in the CTXΦ-positive *V. mimicus* isolates. In vitro studies have shown that the environmental isolates of *V. mimicus* can be infected by CTXΦ to express CT in a TCP-independent way *(183)*. These finding suggest that *V. mimicus* act as an important reservoir for CTXΦ and play a substantial role in the dissemination of toxigenic *V. cholerae* from the environment. The CT epitype B of *V. mimicus* was immunologically indistinguishable from the CT-B produced by the classical biotype strain of *V. cholerae* O1, 569B *(184)*.

Among noncholera vibrios, the frequency of occurrence of *ctx* appears more among *V. mimicus*. Studies conducted in Costa Rica have shown that all the clinical *V. mimicus* isolates produced CT and harbored *ctxA* gene as confirmed by PCR assay *(147)*. In addition, majority of the environmental isolates and some of the clinical isolates are reported to harbor *ctxA (185)*. In Bangladesh, about 10% of the clinical isolates produced CT-like toxin *(186)*. The other virulence determinants for *V. mimicus* includes VM-ST, TDH, and VMH *(185)*. Presence of *zot* gene, which encodes zonula occludens toxin, was detected among *V. mimicus* strains that lacked *ctx (187)*. This indicates a possible of role of *zot* in the toxigenicity of this species. The virulence gene cascade is not uniform in *V. mimicus*, though all the strains have the toxin regulatory gene *toxR (188)*. Among clinical strains of *V. mimicus*, there seems to be significant association of virulence genes and serogroups such as O20, O41 and O115 *(189)*.

The severity of *V. vulnificus* mediated infection depends on a variety of bacterial and host factors. *V. vulnificus* produces a number of enzymes such as hyaluronidase, mucinase, DNases, lipase, and proteases, which may facilitate infection in humans. Because encapsulated *V. vulnificus* strains were commonly isolated from clinical source than the environment, the capsule was considered as one of the virulence factors *(190)*. Acidic capsular polysaccharide (CPS) protects the cells from phagocytosis and other antibacterial responses of the immune system. In addition, this capsule stimulates the release of tumor necrosis factor α and other cytokines and causes septic shock and death. Persons with elevated transferrin-bond iron saturation (>70%) because of hemo-chromatosis, thalassemia, or liver diseases, are at high risk of invasive *V. vulnificus* *(191)*. *V. vulnificus* hemolysin (VVH) is marker for the detection of virulent strains.

Most of the clinical isolates of *V. parahaemolyticus* can be differentiated from the environmental strains by their ability to produce TDH, termed the Kanagawa pheno-menon. This toxin is the best-studied virulence factor of *V. parahaemolyticus*. TDH causes intestinal fluid secretion as well as cytotoxicity in a variety of cell lines *(192)*. The *tdh* gene encodes the production of TDH *(64,73)*. *V. parahaemolyticus* TDH is related (71.5%) to a newly described hemolysin (VC delta TH) of *V. cholerae* O1, which is encoded by the *dth* gene *(193)*. The gene, *tdh* is also present in a high percentage of isolates of *V. hollisae (194,195)*. Phylogenetic analysis of *tdh* genes of *V. hollisae* and other species showed that they are distantly related. These results and the proposed insertion-mediated transfer mechanism suggest that the *tdh* gene is stable in *V. hollisae* and in non-*V. hollisae* species and may have been acquired by recent horizontal gene transfer *(194)*. The *tdh* gene of *V. parahaemolyticus* is flanked by nucleotide sequences identical to the terminally inverted repeat sequences of *IS102* in the chromosome *(196)*. Low content of the G+C (30%) of *tdh* compared to the average G+C content of the *V. parahaemolyticus* chromosome suggest that this locus might be transferred by a mobile genetic element. The amount of TDH produced did not significantly differ between pandemic O3:K6 and nonpandemic O3:K6 strains *(11)*.

Another virulence factor, the TDH-related hemolysin (TRH), encoded by the gene *trh*, has been discovered in clinical strains lacking *tdh (197,198)*. The *trh* shares 69% identity with *tdh2* and the sequence differences were distributed throughout the gene *(199)*. Most of the environmental strains possessing *trh* have been reported in Asia *(83)*. *V. parahaemolyticus* harboring *trh* almost always produces urease, which is atypical for this species *(200–202)*, and the *ure* genes may be part of a pathogenicity island *(203)*. The association between urease production and possession of the *trh* gene is owing to a genetic linkage between the structural gene for urease (*ureC*) and *trh* on the chromo-some of *V. parahaemolyticus (204)*. The urease activity is known to increase gastric-acid tolerance in *Yersinia enterocolitica (205)*, and *Helicobacter pylori (206)*. Presently, there is no in vitro test to detect TRH production. Generally, *V. parahaemolyticus* strains isolated from the environment or foods do not harbor the *tdh* or the *trh* genes.

McCarthy et al. *(207)* demonstrated the prevalence of a thermolabile hemolysin (TLH) in all *V. parahaemolyticus* strains, but not in other species. *V. cholerae* and

V. parahaemolyticus have distinct mechanisms to establish infection. The clinical features of *V. parahaemolyticus* infection include inflammatory diarrhea and in some cases systemic manifestations such as septicemia have been reported. In vitro adherence and cytotoxicity studies with human epithelial cells showed that O3:K6 strains exhibited higher levels of adherence and cytotoxicity to host cells than non-O3:K6 strains suggesting the higher pathogenic potential of pandemic clones. The enhanced adherence and cyto-toxicity may contribute to the unique pathogenic potential of *V. parahaemolyticus* O3:K6 strains *(208)*.

The CPS of *V. parahaemolyticus* is reported to play an important role in the adherence of the pathogen to epithelial cells as evidenced by comparing less CPS producing strains with opaque colonies of the same strain *(209)*. In addition, treatment of *V. parahaemo-lyticus* with anti-CPS but not anti-LPS serum decreases the level of bacterial adherence *(209)*. These results indicate that CPS is one of the factors in bacterial adherence to its target cells.

10. CLINICAL CHARACTERISTICS

Vibrio infections cause three distinct illness syndromes: gastroenteritis, wound or soft tissue infections, that occur by exposure of a wound or broken skin to warm seawater or seafood drippings, and septicemia, an illness characterized by fever or hypotension with isolation of vibrios in the blood culture. Septicemia typically occurs in persons with underlying health problems, such as liver disease, diabetes, cancer, or other immune disorders. In the Gulf coast region (Alabama, Florida, Louisiana, and Texas), raw oyster consumption is an important cause of gastroenteritis among adults without underlying illness *(210)*.

Wound infections are commonly associated with halophilic vibrios such as *V. damsela* *(108,211)*, and *V. vulnificus (72)*. Reports have also shown that *V. damsela* causes aggres-sive infection and necrotizing fascitis *(212,213)* and multiple-system organ failure *(214)*. Thus, a high index of suspicion is needed for the diagnosis of *V. damsela* and concordant infection. Septicemia seems to be a very common symptom for *V. damsela (215,216)*. *V. damsela* is also associated with watery or bloody diarrhea in sporadic cases related to seafood intake. Infection related to *V. hollisae* is characterized by abdominal pain, watery to bloody diarrhea *(108)*. With increasing number of infant gastroenteritis cases, *V. fluvialis* should be considered in the differential diagnosis *(217)*. External otitis caused by *V. alginolyticus* and *V. fluvialis* has been reported with symptoms of itching and seropurulent fluid. The infection may be resulting from prolonged exposure to the seawater *(218)*. Acute suppurative cholangitis caused by *V. fluvialis* has also been reported in Japan *(219)*. Ear infection mediated by *V. parahaemolyticus* and *V. algino-lyticus* seems to be common in Scandinavian area with predisposing conditions such as chronic otitis media, perforation of the tympanic membrane, or ulcus cruris *(220)*.

V. vulnificus, *V. alginolyticus*, *V. parahaemolyticus*, and *V. fluvialis* were reported to cause sight-threatening ocular infections with symptoms such as keratitis, conjunctivitis, and endophthalmitis. Ocular trauma by shellfish from contaminated water is the most common risk factor *(221,222)*. Persons at high risk for *V. vulnificus* infection include those with liver disease, cancer, chronic kidney disease, diabetes, and inflammatory bowl disease. Some vibrios such as *V. fluvialis* and *V. vulnificus* are associated with diarrhea among immunocompromised patients *(223)*. Renal failure appears very common among *V. vulnificus*-infected patients, which can be confused with leptospirosis, scrub typus, malaria, and other forms of sepsis *(224)*.

Fatal gastroenteritis resulting from *V. fluvialis* and nonfatal bacteremia from *V. mimicus* have been reported *(225)*. Leg pain and bullous skin lesions may be a clue to the diagnosis of *V. vulnificus* infection. Febrile patients with cirrhosis and leukemia should be questioned regarding recent seafood ingestion, as they are more susceptible to the infections caused by vibrios *(226)*. Gangrene and endotoxin shock owing to *V. parahaemolyticus* infection has also been reported *(227)*.

Cholera is characterized in its severe form with sudden onset, profuse painless watery stools, occasional vomiting, rapid dehydration, acidosis, circulatory collapse, hypoglycemia in children, and renal failure. Gastric achlorhydria increases the risk of illness. Generally, the breast-fed infants are protected against cholera. Asymptomatic infection is frequent, especially with the El Tor biotype. Cholera occurs significantly more often among persons with 'O' blood group. In severe untreated cases, death may occur within few hours, and the case-fatality rate may exceed 50%; with proper treatment, the case fatality rate is less than 1%. *V. cholerae* non-O1, non-O139 also causes bacteremia because of the consumption of raw calms *(228)*. Cirrhotic patients with *V. cholerae* non-O1, non-O139 bacteremia is rare but highly lethal if treatment measures are not taken in time. Apart from diarrhea, the other symptoms of *V. cholerae* non-O1, non-O139 infection include ascites, fever, abdominal pain, and cellulitis with bullae formation *(229)*.

In infected individuals, *V. parahaemolyticus* can cause chills, fever, nausea, vomiting, watery to bloody diarrhea, abdominal cramps, and, in rare instances, death *(230,231)*. The major inflammatory response resulting from *V. parahaemolyticus* infection at the onset of illness includes edema in laminapropria, congestion in blood vessels, and hemorrhage in the duodenum and rectum *(232)*. Survey conducted in Vietnam showed that 53% of the patients presented with watery stools and 6% reported blood in the stool *(233)*. *V. parahaemolyticus* rarely causes septicemia and fulminant necrotizing fasciitis *(234)*. Generally, soft tissue infections often progress to fatal septicemia. Most of the septicemia patients are immunosuppressed, especially with leukemia and cirrhosis *(235)*. For this reason, patients with immunosuppressed conditions or adrenal insufficiency should avoid raw seafood. *V. parahaemolyticus* elicits a strong immune response so that repeated episodes of gastroenteritis in the same individuals are unlikely *(232)*. A comparative study conducted in Vietnam showed that there is no difference in the expression of virulence between pandemic and nonpandemic strains of *V. parahaemolyticus* isolated from the diarrhea patients as evidenced from the clinical symptom *(57)*.

11. CHOICE OF TREATMENT

Most of the vibrios that cause gastroenteritis are generally susceptible to commonly used antibiotics. *V. vulnificus* wound infections and primary septicemia require antimicrobial treatment to improve the course of illness and to prevent complications. Antimicrobial agents most effective against *V. vulnificus* infections include tetracycline, fluroquinolones, cephalosporins and aminoglycosides *(125)*. Treatment with ceftazidime [2 g iv, three times a day (tid)] and doxycycline (100 mg PO or iv bid) or doxycycline with ciprofloxacin or an aminoglycoside is currently in practice. Taiwan strains of *V. vulnificus* were susceptible to ampicillin and cephalosporin *(236)*. Because the severity of the infection is very high and sometimes lethal, combined therapy with a third-generation cephalosporin or ampicillin and an aminoglycoside was also recommended *(236)*. Surgical treatment includes incision and drainage, fasciotomy, debridement, or amputation of the infected parts *(236)*.

Infection caused by *V. damsela* includes necrotizing fascitis and multisystem/organ failure, for which the medical intervention includes surgical debridement, ventilation support, vasopressors, continuous hemofiltration, and blood-product transfusions. Surgical debridement and amputation of the infected part(s) at the early stage of infection may be the only intervention that saves the lives of the patients affected by *V. damsela (214)*. Ocular infections caused by the vibrios were successfully treated with combination of cefazolin and gentamicin *(222)*. Administration of ciprofloxacin and ceftazidime for external otitis patients infected with *V. alginolyticus* and *V. fluvialis*, respectively succeeded in the remission of the symptom *(218)*. Bacteremia associated with *V. cholerae* non-O1 was effectively treated after antibiotic therapy either cephalothin with gentamicin or ceftriaxone alone *(229)*.

The standard care of cholera patients is to treat mild-to-moderate cholera with oral rehydration salts (ORS) solution or an oral electrolyte rehydration solution, and to treat severe cases with intravenous fluids (Ringer's lactate, WHO–diarrhea treatment solution) and an antimicrobial agent. Prompt restoration of fluids and electrolytes should be the primary goal of treatment. Normal saline solutions should never be used to treat patients with cholera, because it does not contain the electrolytes needed to replace the profound potassium and bicarbonate loss resulting from cholera *(237)*. Antimicrobial therapy has been shown to reduce the magnitude of fluid loss, duration of illness, and duration of excretion of the pathogen. Ciprofloxacin (1 g orally in a single dose) or doxycycline (300 mg orally in a single dose) or tetracycline (500 mg four times a day) are the antibiotics of choice for adults (except pregnant women). Erythromycin for 3 d is recommended for children (10 mg/kg tid) and pregnant women (250 mg four times daily). Till date, trimethoprim–sulfamethoxazole (8 mg/kg trimethoprim and 40 mg/kg sulfamethoxazole daily in two divided doses for 3 d) are the drugs recommended for the treatment of children, while furazolidone (100 mg four times daily for 3 d) is used for the treatment of pregnant women with cholera. Except for tetracycline, other antimicrobial agents are no longer recommended as first-line therapy because of increasing global antimicrobial resistance. Whenever possible, treatment protocols should be based on local antibiogram data. Neither the CDC nor the WHO recommends routine use of cholera vaccine for travelers, because it may create a false sense of security and does not affect cholera severity.

12. RESISTANCE EPIDEMIOLOGY

Antimicrobial agents have been used in aquaculture worldwide to treat infections caused by a variety of bacterial pathogens in cultured fish. Existing molecular characterization of antimicrobial resistance determinants provides further evidence of the transmission of antimicrobial resistance between aquaculture ecosystems and humans. Sulphonamides, tetracycline, amoxicillin, trimethoprim–sulfadimethoxine, and quinolones are used in aquaculture in many countries *(238)*. The use of antimicrobial agents, as with other uses, selects for antimicrobial resistance in the exposed bacterial flora. Horizontal spread of mobile genetic elements from fish pathogens may, therefore, transfer resistance to variety of bacterial species, including bacteria that can infect human *(239)*.

The pathogenic vibrios such as *V. alginolyticus, V. fluvialis, V. damsela, V. metschnikovii,* and *V. vulnificus* isolated from the blue crab (*Callinectes sapidus*) are highly susceptible to doxycycline, tetracycline, and ciprofloxacin *(99)*. Except for colistin, no antibiotic resistance was recorded for *V. cholerae* O1, though this pathogen was frequently detected among diarrheal patients in Indonesia *(240)*. In Guatemala, cholera was mainly transmitted through food contamination and the strains isolated from the patients were resistant to furazolidone, sulfisoxazole, and streptomycin *(122)*.

The clinical strains of *V. mimicus* isolated in India are multidrug resistant *(175)*. In a comparative study, it was shown that the *V. mimicus* strains collected from Okayama in Japan were more resistant to ampicillin than the strains collected at Dhaka, Bangladesh *(241)*. In Bangladesh, all the tested environmental strains of *V. mimicus* were resistant to streptomycin, kanamycin, and co-trimoxazole, whereas the clinical strains remained susceptible to these drugs *(242)*.

In 1991, an epidemic of *V. cholerae* O1 infection affected Latin America. The epidemic strain was susceptible to 12 antimicrobial agents tested except in coastal Ecuador, where the epidemic strain showed multidrug resistance *(243)*. The cholera epidemic in Ecuador began among persons working in shrimp farms. Multidrug resistance was present in noncholera *Vibrio* infections that were pathogenic to shrimps. The resistance may have been transferred to *V. cholerae* O1 from other vibrios *(243)*.

13. DECONTAMINATION OF VIBRIOS FROM FOOD

Several useful methods for the decontamination of vibrios and other human pathogens from foods are available. In vitro techniques tested with pure cultures are useful in its applicability in the food industry. High pressure processing (HHP) technique has shown good potential in reducing bacteria in foods. At 35K psi, for >10 min reduced the count of *V. parahaemolyticus*. When treated with HHP from 200 to 300 MPa for 5–15 min at 25°C, *V. parahaemolyticus, V. cholerae* (toxigenic and nontoxigenic), *V. hollisae,* and *V. mimicus* were inactivated without triggering VBNC state *(244)*. The results of studies involving *V. vulnificus* in oysters revealed that a pressure treatment of 250 MPa for 120 s achieved about 5-log reduction, and *V. parahaemolyticus* required a pressure of 300 MPa for 180 s for a comparable 5-log reduction *(245)*. However, the cells already in a VBNC state appeared to possess greater pressure resistance. Because most of the vibrios may survive at 60°C/140°F for up to 3 min, cooking temperature of seafood must exceed 70°C/158°F for 30 s.

Boiling temperature was very damaging to *V. cholerae* O1 in shrimps with and without carapace, but in shrimps contaminated and stored at –200°C, the test pathogen survived for 26 d *(246)*. In clams, mild heat-treatment for 23 s reduces the large *Vibrio* load including *V. cholerae* O139 *(247)*. *V. cholerae* O1 and non-O1 strains remain viable in hot foods tested at temperatures 50–60°C for 1 h. However, an incubation period of 48 h at 37°C was found to be appropriate for the recovery of *V. cholerae* from hot foods *(248)*. Radiation dose of 3 kGy and exposure of seafood to chlorine or iodophor at 13 ppm for 15 min was one of the guidelines for inactivation of vibrios. Ionizing irradiation dose of 1.41 kGy by ^{60}Co is necessary to eliminate the higher number of toxigenic and nontoxigenic *V. cholerae* viable cells in oysters *(249)*.

Several effective chemicals are in use for the treatment of raw fruits and other foods. For decontamination of *V. cholerae* O1, chlorine was more effective (0.5×10^{-1}) than copper oxychloride, sodium-*O*-phenylphenoate, guazatine (a polyamine mixture), and imazalil (an imidazole), which were effective at the concentration of 10^2 *(250)*. Elimination of *V. cholerae* O1 by lime juice was 2–6 logs greater than the maximum infectious dose and 4–8 logs greater than the minimum infectious dose *(251,252)*. Addition of lime juice to nonacidic foods, salads, beverages, ceviche, and water was shown to prevent infection of *V. cholerae (251,252)*. This measure is particularly important for rural and slum population in the tropics and subtropics. Case–control study during a cholera outbreak in Guinea-Bissau of West Africa, showed that the addition of lime juice in sauce or tomato sauce eaten with stale cooked rice gave a strong protective effect *(253)*.

Studies conducted in Argentina using 20 food items using *V. cholerae* O1 Ogawa showed that the test pathogen survived from 1 d in pasta to 90 d in sterile milk samples *(254)*. Generally, foods with acidity higher than pH 5.5 did not favor the growth of *V. cholerae*. The other factors, such as surface adherence, amino acids, magnesium, could also eventually modify the persistence of *V. cholerae* in foods *(254)*. *V. parahaemolyticus* is very sensitive to butylated hydroxyanisole at 50 ppm, which can be used as preservative of processed food. However, these methods are not popular in industrial practice.

Oysters accumulate greater concentration of *V. cholerae* O1 than *E. coli* and *Salmonella*. In controlled experiments, it was shown that the *V. cholerae* persisted during the depurification process at a higher level *(255)*. Owing to this reason, though effective for other enteric pathogens, the depurification (relaying) is not recommended for most of the vibrios.

14. PREVENTION AND CONTROL

Prevention of contamination at all points of the food chain is preferred over the application of disinfectant after contamination occurs. Prevention of infection can be achieved through application of the principles of food hygiene and the hazard analysis critical control point (HACCP) system. The consumers should avoid eating raw or undercooked oysters, especially during warmer months. Avoidance of cross-contaminating utensils, cutting boards, countertops, and foods with raw shellfish products is mandatory while preparing shellfish. Open wounds are the source for extra-intestinal infections when exposed to seawater or raw shellfish products. Shellfishes should be boiled thoroughly until shells open. Avoid cross-contamination of cooked food with raw seafood, or juices from raw seafood. Wearing gloves when handling raw shellfish is a recommended measure to prevent seafood mediated *Vibrio* infections.

Persons with certain high-risk conditions should especially be advised to avoid raw shellfish. These conditions include liver disease of any type, alcoholism, hemochromatosis (abnormal iron metabolism), diabetes, achlorhydria (reduced acidity in the stomach), hemochronic inflammatory bowel disease, cancer, chemotherapy, immune disorders, and long-term steroid use. Physicians should obtain a travel history and consider cholera in patients with diarrhea who have traveled from cholera endemic countries. Patients with decompensated cirrhosis are susceptible to *Vibrio* bacteremia after ingestion of seafood. A high index of suspicion and early administration of antibiotics may lower the mortality rate *(229)*.

Shellfish harvester should follow the regular environmental monitoring programs particularly during warmer months when human illness increase. In the case of monitoring of *Vibrio*, detection of fecal coliforms as an indicator of fecal pollution may not be helpful in the environmental monitoring activities including in shrimp production areas through aquaculture, because the vibrios are natural inhabitants of coastal waters *(125, 256)*. Compliance of permissible limit recommended by FDA for vibrios (10^3 CFU/g of oyster meat for *V. parahaemolyticus*) should be followed. In several recent incidences, however, *V. parahaemolyticus* outbreaks occurred even though no oyster sample exceeded the permissible limit *(137)*. Toxigenic *V. cholerae* is rarely detected in either imported or domestically produced seafood. The FDA has established the zero tolerance of toxigenic *V. cholerae* to <1 CFU/25 g. Noncholera vibrios are found routinely in seafood but no tolerance limit has been established for their presence. Companies reduce the *Vibrio* levels to <3 most probable number (MPN)/g can label their products as "processed to reduce *V. vulnificus* and/or *V. parahaemolyticus* to undetectable level" *(257)*. The International Commission of Microbiological Specification for Foods (ICMSF) recommended limit for *V. parahaemolyticus* to different seafood are given in Table 4. Interstate Shellfish Sanitation Conference (ISSC) suggest that consumable oyster should not contain more than 3 CFU/g *V. vulnificus* in oyster meat *(258)*.

Various postharvest treatments (PHT) such as mild heat, high pressure, freezing, and irradiation have been developed by the oyster industry to reduce number of organism at least to the permissible limits (Table 5). For industrial elimination *of V. parahaemolyticus* and seafood safety, the following guides are recommended by the ICMSF *(259)*. Shellfishes must be harvested from the approved waters. Cook the food at 50–65°C for 10 min and hold foods at either below 5°C or above 60°C. The pH value of the seafood must be reduced to 4.8 or less. For some enteric pathogens concentrated by shellfish, a period of depuration by holding the shellfish in clean water for several days can remove most of the pathogens. However, vibrios adhere strongly to the shellfish digestive tract and cannot be successfully removed by rinsing or depuration. Another method called relaying, which involves suspending shellfish in water with higher salinity at an offshore location, was found to significantly reduce the number of *V. vulnificus (260)*. In reconstituted infant formula, the enteric pathogens multiply in this medium into several folds *(118)*. Unless refrigerated, the reconstituted infant formula should be consumed soon after preparation to avoid increased risk of illness associated with the raise in number of bacteria that may be introduced through contaminated water.

The development of international epidemiological surveillance system for better understanding of the role of foods as vehicles for infectious diseases is critical. Information generated from such a system is valuable in establishing more meaningful practices and guidelines for prevention and control of foodborne pathogenic vibrios.

15. SUMMARY AND CONCLUSIONS

The isolation of potentially pathogenic vibrios from natural and cultivated seafoods is a risk for human health. With the expansion of aquaculture, medical practitioners can expect more infection caused by vibrios. Clinicians treating patients with diarrhea should include vibrios in the differential diagnosis if the patients or the patient's contacts have taken seafood or exposure to seawater. Diagnosis and treatment may be difficult without proper knowledge of each species, especially in view of antibiotic resistance strains. Seafoods are a significant source of potentially pathogenic vibrios, especially *V. vulnificus*, which is an emerging opportunistic pathogen and responsible for one of the fulminate foodborne diseases in immunocompromised hosts. The spreading

Table 4
**Recommended Limits of *V. parahaemolyticus* in Seafood By the International
Commission on Microbial Specifications for Foods [modified from ICMSF-(265)]**

Product	n^a	c^b	Bacteria (gram or cm^3)	
			m^c	M^d
Fresh and frozen fish and cold-smoked fish	5	2	10^2	10^3
Frozen raw crustaceans	5	1	10^2	10^3
Frozen cooked crustaceans	5	1	10^2	10^3
Cooked, chilled, and frozen crab meat	10	1	10^2	10^3
Fresh and frozen bivalve molluscs	10	1	10^2	10^3

[a]Number of representative sample unit.

[b]Maximum number of acceptable sample units with bacterial counts between *m* and *M*.

[c]Maximum recommended bacterial counts for good quality products.

[d]Maximum recommended bacterial counts for marginally acceptable quality products. Plate counts below *m* are considered good quality. Plate counts between *m* and *M* are considered marginally acceptable quality, but can be accepted if the number of samples does not exceed *c*. Plate counts at or above *M* are considered unacceptable quality.

Table 5
**Comparative Effectiveness of Mitigation Strategies in Reducing Vibrios
From Seafoods (modified from FAO/WHO Codex Alimentarius Commission)**

Mitigation	Effectiveness	Measurement unit	Inactivation (log reduction)
Hydrostatic pressure	+++	200–300 MPa/120–600 s	3.5–6.0
Rapid cooling[a]	+/++	4–6°C	0.8
Irradiation (γ-rays from ^{60}Co)	+++	3 kgy for 15 min	5–6
Pasteurization	+++	50°C for 10–15 min (shell internal temperature)	5[b]
Freezing and thawing[a]	++	−20°C	3–4
Depuration[a]	+/−	None	
Relay at high salinity[a]	++	30–38 ppt, 7–10 d	1–2
Commercial heat-treatment	+++	External temperature 99°C and shell internal temperature 56–69°C for 30 s	5

[a]Considered in the FAO/WHO risk assessment.

[b]4-log reduction in the case of *V. parahaemolyticus* O3:K6 pandemic strains.

of cholera and pandemic strains of *V. parahaemolyticus* from endemic regions to nonendemic countries raised questions regarding food safety, transportation, and food handling. Control, prevention, and risk implied in food import–export are also the subject of recent concern.

With the existing clinical practice, *Vibrio* mediated infections are curable with in short span of time, as most of the pathogens are highly susceptible to many common antimicrobials. In addition, decontamination methods now exist to effectively reduce the number of vibrios without compromising the food-value of raw and finished products. Research attention should be paid for some of the vibrios whose virulence mechanisms are not fully understood. Molecular detection methods should be exploited to develop rapid identification of pathogenic vibrios, to preempt the occurrence of outbreaks.

REFERENCES

1. Nishibuchi, M., Roberts, N. C., Bradford, H. B. Jr., and Seidler, R. J. (1983) Broth medium for enrichment of *Vibrio fluvialis* from the environment. *Appl. Environ. Microbiol.* **46,** 425–429.
2. Mizunoe, Y., Wai, S. N., Ishikawa, T., Takede, A., and Yoshida, S. (2000) Resuscitation of viable but nonculturable cells of *Vibrio parahaemolyticus* induced at low temperature under starvation. *FEMS Microbiol. Lett.* **186,** 115–120.
3. Ramamurthy, T., Pal, A., Pal, S. C., and Nair, G. B. (1992) Taxonomic implication of the emergence of high frequency of occurrence of 2,4-diamino-6,7-diisopropylpteridine-resistant strains of *Vibrio cholerae* from clinical cases of cholera in Calcutta, India. *J. Clin. Microbiol.* **30,** 742–743.
4. Kitagawa, T., Tsutida, Y., Murakami, R., et al. (1992) Detection and quantitative assessment of a *Vibrio cholerae* O1 species in several foods by a novel enzyme immuno assay. *Microbiol. Immunol.* **36,** 13–20.
5. Brayton, P. R., Bode, R. B., Colwell, R. R., et al. (1986) *Vibrio cincinnatiensis* sp. nov., a new human pathogen. *J. Clin. Microbiol.* **23,** 104–108.
6. Brenner, D. J., Hickman-Brenner, F. W., Lee, J. V., et al. (1983) *Vibrio furnissii* (formerly aerobic biogroup of *Vibrio fluvialis*), a new species isolated from human feces and the environment. *J. Clin. Microbiol.* **18,** 816–824.
7. Ramamurthy, T., Garg, S., Sarma, R., et al. (1993) Emergence of a novel strain of *Vibrio cholerae* with epidemic potential in southern and eastern India. *Lancet* **341,** 703–704.
8. Albert, M., Siddique, A. K., Islam, M. S., et al. (1993) Large outbreak of clinical cholera due to *Vibrio cholerae* non-O1 in Bangladesh. *Lancet* **341,** 704.
9. Bhattacharya, S. K., Bhattacharya, M. K., Nair, G. B., et al. (1993) Clinical profile of acute diarrhoea cases infected with the new epidemic strain of *Vibrio cholerae* O139: designation of the disease as cholera. *J. Infect.* **27,** 11–15.
10. Iguchi, T., Kondo, S., and Hisatune, K. (1995) *Vibrio parahaemolyticus* O serotypes from O1 to O13 all produce R-type lipopolysaccharide: SDS PAGE and compositional sugar analysis. *FEMS Microbiol. Lett.* **130,** 287–292.
11. Okuda, J., Ishibashi, M., Hayakawa, E., et al. (1997) Emergence of a unique O3:K6 clone of *Vibrio parahaemolyticus* in Calcutta, India, and isolation of strains from the same clonal group from southeast Asian travelers arriving in Japan. *J. Clin. Microbiol.* **35,** 3150–3155.
12. Bag, P. K., Nandi, S., Bhadra, R. K., et al. (1999) Clonal diversity among the recently emerged strains of *Vibrio parahaemolyticus* O3:K6 associated with pandemic spread. *J. Clin. Microbiol.* **37,** 2354–2357.
13. Chiou, C.-S., Hsu, S.-Y., Chiu, S.-I., Wang, T.-K., and Chao, C.-S. (2000) *Vibrio parahaemolyticus* serovar O3:K6 as cause of unusually high incidence of food-borne disease outbreaks in Taiwan from 1996 to 1999. *J. Clin. Microbiol.* **38,** 4621–4625.
14. Matsumoto, C., Okuda, J., Ishibashi, M., et al. (2000) Pandemic spread of an O3:K6 clone of *Vibrio parahaemolyticus* and emergence of related strains evidenced by arbitrarily primed PCR and *toxRS* sequence analysis. *J. Clin. Microbiol.* **38,** 578–585.
15. Chowdhury, N. R., Chakraborty, S., Eampokalap, B., et al. (2000) Clonal dissemination of *Vibrio parahaemolyticus* displaying similar DNA fingerprint but belonging to two different serovars (O3:K6 and O4:K68) in Thailand and India. *Epidemiol. Infect.* **125,** 17–25.
16. Bhuiyan, N. A., Ansarizzaman, M., Kamruzzaman, M., et al. (2002) Prevalence of the pandemic geneotype of *Vibrio parahaemolyticus* in Dhaka, Bangladesh, and significance of its distribution across different serotypes. *J. Clin. Microbiol.* **40,** 284–286.

17. Overman, T. L., and Overley, J. K. (1986) Feasibility of same-day identification of members of the family Vibrionaceae by the API 20E System. *J. Clin. Microbiol.* **23,** 715–717.

18. Overman, T. L., Kessler, J. F., and Seabolt, J. P. (1985) Comparison of API 20E, API rapid E, and API rapid NFT for identification of members of the family Vibrionaceae. *J. Clin. Microbiol.* **22,** 778–781.

19. Abbott, S. L., Seli, L. S., Catino, M. Jr., Hartley, M. A., and Janda, J. M.. (1998) Misidentification of unusual *Aeromonas* species as members of the genus *Vibrio*: a continuing problem. *J. Clin. Microbiol.* **36,** 1103–1104.

20. Park, T. S., Oh, S. H., Lee, E. Y., et al. (2003) Misidentification of *Aeromonas veronii* biovar sobria as *Vibrio alginolyticus* by the Vitek system. *Lett. Appl. Microbiol.* **37,** 349–353.

21. Urdaci, M. C., Marchand, M., and Grimont, P. A. (1990) Characterization of 22 *Vibrio* species by gas chromatography analysis of their cellular fatty acids. *Res. Microbiol.* **141,** 437–452.

22. Miyamoto, T., Miwa, H., and Hatano, S. (1990) Improved fluorogenic assay for rapid detection of *Vibrio parahaemolyticus* in foods. *Appl. Environ. Microbiol.* **56,** 1480–1484.

23. Venkateshwaran, K., Kurusu, T., Satake, M., and Shinoda, S. (1996) Comparison of a fluorogenic assay with a conventional method for rapid detection of *Vibrio parahaemolyticus* in seafoods. *Appl. Environ. Microbiol.* **62,** 3516–3520.

24. Iguchi, T., Kondo, S., and Hisatsune, K. (1989) Sugar composition of the polysaccharide portion of lippopolysaccharides of *Vibrio fluvialis, Vibrio vulnificus and Vibrio mimicus. Microbiol. Immunol.* **33,** 833–841.

25. Chen, D., Hanna, P. J., Altmann, K., Smith, A., Moon, P., and Hammond, L. S. (1992) Development of monoclonal antibodies that identify *Vibrio* species commonly isolated from infections of humans, fish and shellfish. *Appl. Environ. Microbiol.* **58,** 3694–3700.

26. Ramamurthy, T., Garg, S., and Nair, G. B. (1995) Monoclonal antibodies against Ogawa specific and Ogawa and Inaba common antigenic determinants of *Vibrio cholerae* O1 and their diagnostic utility. *Indian J. Med. Res.* **101,** 10–12.

27. Garg, S., Ramamurthy, T., Mukhopadhyay, A. K., et al. (1994) Production and cross-reactivity patterns of a panel of high affinity monoclonal antibodies to *Vibrio cholerae* O139 Bengal. *FEMS Immunol. Med. Microbiol.* **8,** 293–298.

28. Shimada, T., Arakawa, E., Okitsu, T., et al. (1999) Additional O antigens of *Vibrio fluvialis* and *Vibrio furnissii. Jpn. J. Infect. Dis.* **52,** 124–126.

29. Landersjo, C., Weintraub, A., Ansaruzzaman, M., Albert, M. J., and Widmalm, G. (1998) Structural analysis of the O-antigenic polysaccharide from *Vibrio mimicus* N-1990. *Eur. J. Biochem.* **251,** 986–990.

30. Libinzon, A. E., Levanova, G. F., and Gal'tseva, G. V. (1991) The isolation of *Vibrio fluvialis* on the territory of the USSR. *Zh. Mikrobiol. Epidemiol. Immunobiol.* **2,** 20–23.

31. Landersjo, C., Weintraub, A., Ansaruzzaman, M., Albert, M. J., and Widmalm, G. (1998) Structural analysis of the O-antigenic polysaccharide from *Vibrio mimicus* N-1990. *Eur. J. Biochem.* **251,** 986–990.

32. Hollis, D. G., Weaver, R. E., Baker, C. N., and Thornsberry, C. (1976) Halophilic *Vibrio* sp. isolated from blood cultures. *J. Clin. Microbiol.* **3,** 425.

33. Simonson, J., and Siebeling, R. J. (1986) Rapid serological identification of *Vibrio vulnificus* by anti-H coagglutination. *Appl. Environ. Microbiol.* **52,** 1299–1304.

34. Shinoda, S., Nakahara, N., Uchida, E., and Hiraga, M. (1985) Lateral flagellar antigen of *Vibrio alginolyticus* and *Vibrio harveyi*: existence of serovars common to the two species. *Microbiol. Immunol.* **29,** 173–182.

35. Shinoda, S., Nakahara, N., and Kane, H. (1984) Lateral flagellum of *Vibrio fluvialis*: a species-specific antigen. *Can. J. Microbiol.* **30,** 1525–1529.

36. Tassine, M. G., Siebeling, R. J., Roberts, N. C., and Larson, A. D. (1983) Presumptive identification of *Vibrio* species with H antiserum. *J. Clin. Microbiol.* **18,** 400–407.

37. Suthienkul, O. (1993) Bacteriophage typing of *Vibrio fluvialis. Southeast Asian J. Trop. Med. Pub. Health* **24,** 449–454.

38. Chattopadhyay, D. J., Sarkar, B. L., Ansari, M. Q., et al. (1993) New phage typing scheme for *Vibrio cholerae* O1 El Tor strains. *J. Clin. Microbiol.* **31**, 1579–1585.

39. Chakarabarti, A. K., Ghosh, A. N., Nair, G. B., Niyogi, S. K., Bhattacharya, S. K., and Sarkar, B. L. (2000) Development and evaluation of a phage typing scheme for *Vibrio cholerae* O139. *J. Clin. Microbiol.* **38**, 44–49.

40. Chun, J., Rivera, I. N., and Colwell, R. R. (2002) Analysis of 16S-23S rRNA intergenic spacer of *Vibrio cholerae* and *Vibrio mimicus* for the detection of these species. *Meth. Mol. Biol.* **179**, 171–178.

41. Singh, D. V., Isac, S. R., and Colwell, R. R. (2002) Development of a hexaplex PCR assay for rapid detection of virulence and regulatory genes in *Vibrio cholerae* and *Vibrio mimicus*. *J. Clin. Microbiol.* **40**, 4321–4324.

42. Cerda-Cuellar, M., Jofre, J., and Blanch, A. R. (2000) A selective medium and a specific probe for detection of *Vibrio vulnificus*. *Appl. Environ. Microbiol.* **66**, 855–859.

43. Vicente, A. C., Coelho, A. M., and Salles, C. A. (1997) Detection of *Vibrio cholerae* and *V. mimicus* heat-stable toxin gene sequence by PCR. *J. Med. Microbiol.* **46**, 398–402.

44. Shangkuan, Y. H., Show, Y. S., and Wang, T. M. (1995) Multiplex polymerase chain reaction to detect toxigenic *Vibrio cholerae* and to biotype *Vibrio cholerae* O1. *J. Appl. Bacteriol.* **79**, 264–273.

45. Koch, W. H., Payne, W. L., Wentz, B. A., and Cebula, T. A. (1993) Rapid polymerase chain reaction method of detection of *Vibrio cholerae* in foods. *Appl. Environ. Microbiol.* **59**, 556–560.

46. Robert-Pillot, A., Guenole, A., and Fournier, J.-M. (2002) Usefulness of R72H PCR assay for differentiation between *Vibrio parahaemolyticus* and *Vibrio alginolyticus* species: validation by DNA-DNA hybridization. *FEMS Microbiol. Lett.* **215**, 1–6.

47. Kim, Y. B., Okuda, J., Matsumoto, C., Takahashi, N., Hashimoto, S., and Nishibuchi, M. (1999) Identification of *Vibrio parahaemolyticus* strains at the species level by PCR targeted to the *toxR* gene. *J. Clin. Microbiol.* **37**, 1173–1177.

48. Venkateswaran, K., Dohmoto, N., and Harayama, S. (1998) Cloning and nucleotide sequencing of the *gyrB* gene of *Vibrio parahaemolyticus* and its application in detection of this species in shrimp. *Appl. Environ. Microbiol.* **64**, 681–687.

49. Dileep, V., Kumar, H. S., Kumar, Y., Nishibuchi, M., Karunasagar, I., and Karunasagar, I. (2003) Application of polymerase chain reaction for detection of *Vibrio parahaemolyticus* associated with tropical seafoods and coastal environment. *Lett. Appl. Microbiol.* **36**, 423–427.

50. Hara-Kudo, Y., Sugiyama, K., Nishibuchi, M., et al. (2003) Prevalence of thermostable direct hemolysin-producing *Vibrio parahaemolyticus* O3:K6 in seafood and coastal environment in Japan. *Appl. Environ. Microbiol.* **69**, 3883–3891.

51. Bej, A. K., Patterson, D. P., Brasher, C. W., Vickery, M. C. L., Jones, D. D., and Kaysner, C. (1999) Detection of total and hemolysin-producing *Vibrio parahaemolyticus* in shellfish using multiplex PCR amplification of *tl*, *tdh*, and *trh*. *J. Microbiol. Meth.* **36**, 215–225.

52. Lee, C. Y., Pan, S. F., and Chen, C. H. (1995) Sequence of a cloned pR72H fragment and its use for detection of *Vibrio parahaemolyticus* in shellfish with the PCR. *Appl. Environ. Microbiol.* **61**, 1311–1317.

53. Nasu, H., Iida, T., Sugahara, T., et al. (2000) A filamentous phage associated with recent pandemic *Vibrio parahaemolyticus* O3:K6 strains. *J. Clin. Microbiol.* **38**, 2156–2161.

54. Myers, M. L., Panicker, G. and Bej, A. K. (2003) PCR detection of a newly emerged pandemic *Vibrio parahaemolyticus* O3:K6 pathogen in pure cultures and seeded waters from the Gulf of Mexico. *Appl. Environ. Microbiol.* **69**, 2194–2200.

55. Okura, M., Osawa, R., Iguchi, A., Arakawa, E., Terajima, J., and Watanabe, H. (2003) Genotypic analysis of *Vibrio parahaemolyticus* and development of a pandemic group-specific multiplex PCR assay. *J. Clin. Microbiol.* **41**, 4676–4682.

56. Laohaprertthisan, V., Chowdhury, A., Kongmuang, U., et al. (2003) Prevalence of sero-diversity of the pandemic clone among the clinical strains of *Vibrio parahaemolyticus* isolated from Thailand. *Epidemiol. Infect.* **130,** 395–406.

57. Chowdhury, A., Ishibachi, M., Thiem, V. D., et al. (2004) Emergence and serovar transition of *Vibrio parahaemolyticus* pandemic strains isolated during a diarrhea outbreak in Vietnam between 1997 and 1999. *Microbiol. Immunol.* **48,** 319–327.

58. Cambell, M. S. and Wright, A. C. (2003) Real-time PCR analysis of *Vibrio vulnificus* from oyster. *Appl. Environ. Microbiol.* **69,** 7137–7144.

59. Panicker, G., Myers, M. L., and Bej, A. K. (2004) Rapid detection of *Vibrio vulnificus* in shellfish and Gulf of Mexico water by real-time PCR. *Appl. Environ. Microbiol.* **70,** 498–507.

60. Blackstone G. M., Nordstrom, J. L., Vickery, M. C., Bowen, M. D., Meyer, R. F., and DePaola, A. (2003) Detection of pathogenic *Vibrio parahaemolyticus* in oyster enrichments by real time PCR. *J. Microbiol. Meth.* **53,** 149–155.

61. Fukushima, H., Tsunomori, Y., and Ryotaro, S. (2003) Duplex real-time SYBER green PCR assay for detection of 17 species of food or waterborne pathogens in stools. *J. Clin. Microbiol.* **41,** 5134–5146.

62. Davis, C. R., Heller, L. C., Peak, K. K., et al. (2004) Real-time PCR detection of the thermostable direct hemolysin and thermolabile hemolysin genes in a *Vibrio parahaemolyticus* cultured from mussels and mussel homogenate associated with a food borne outbreak. *J. Food Prot.* **67,** 1005–1008.

63. Bej, A. K., Ng, W. Y., Morgan, S., Jones, D. D., and Mahbubani, M. H. (1996) Detection of viable *Vibrio cholerae* by reverse-transcriptase polymerase chain reaction (RT-PCR). *Mol. Biotech.* **5,** 1–10.

64. McCarthy, S. A., DePaola, A., Kaysner, C. A., Hill, W. E., and Cook, D. W. (2000) Evaluation of nonisotopic DNA hybridization methods for the detection of the *tdh* gene of *Vibrio parahaemolyticus*. *J. Food Prot.* **63,** 1660–1664.

65. Aznar, R., Ludwig, W., Amann, R. I., and Schleifer, K. H. (1994) Sequence determination of rRNA genes of pathogenic *Vibrio* species and whole-cell identification of *Vibrio vulnificus* with rRNA-targeted oligonucleotide probes. *Int. J. Syst. Bacteriol.* **44,** 330–337.

66. Sparagano, O. A., Robertson, P. A., Purdom, I., et al. (2002) PCR and molecular detection for differentiating *Vibrio* species. *Ann. NY Acad. Sci.* **969,** 60–65.

67. Stackebrandt, E. (2000) Classification of the bacteria-Phylogenetic approach. In: *Encyclopedia of Food Microbiology* (Robinson, R. K., Batt, C. A., and Patel, P. D., eds.), Academic, London, pp. 178–183.

68. Kwok, A. Y., Wilson, J. T., Coulthart, M., Ng, L. K., Mutharia, L., and Chow, A. W. (2002) Phylogenetic study and identification of human pathogenic *Vibrio* species based on partial *hsp*60 gene sequences. *Can. J. Microbiol.* **48,** 903–910.

69. Venkateswaren, K. (2000) Standard cultural methods and molecular techniques in foods. In: *Encyclopedia of Food Microbiology* (Robinson, R. K., Batt, C. A., and Patel, P. D., eds.), Vol. 3, Academic, London, pp. 2248–2258.

70. Bi, K, Shi, L., Maehara, Y., Miyoshi, S., Tomochika, K., and Shinoda, S. (2000) Analysis of *Vibrio mimicus* clinical strains by arbitrarily primed polymerase chain reaction. *Microbiol. Immunol.* **44,** 149–153.

71. Vickery, M. C., Harol, N., and Bej, A. K. (2000) Cluster analysis of AP-PCR generated DNA fingerprints of *Vibrio vulnificus* isolates from patients fatally infected after consumption of raw oysters. *Lett. Appl. Microbiol.* **30,** 258–262.

72. Bisharat, N., Agmon, V., Finkelstein, R., et al. (1999) Clinical, epidemiological and microbiological features of *Vibrio vulnificus* biogroup 3 causing outbreaks of wound infection and bacterimia in Israel. Israel *Vibrio* study group. *Lancet* **354,** 1421–1424.

73. Nishibuchi, M. and Kaper, J. B. (1995) Thermostable direct hemolysin gene of *Vibrio parahaemolyticus*: a virulence gene acquired by a marine bacterium. *Infect. Immun.* **64,** 2093–2099.

74. Wong, H. C., Ho, C. Y., Kuo, L. P., Wang, T. K., Lee, C. L., and Shih, D. Y. (1999) Ribotyping of *Vibrio parahaemolyticus* obtained from food poisoning outbreaks in Taiwan. *Microbiol. Immunol.* **43,** 631–636.

75. Krawiec, S. and Riley, M. (1990) Organization of the bacterial genome. *Microbiol. Rev.* **54,** 502–539.

76. Koblavi, S., Grimont, F., and Grimont, P. A. D. (1990) Clonal diversity of *Vibrio cholerae* O1 evidenced by rRNA gene restriction patterns. *Res. Microbiol.* **141,** 645–657.

77. Gendel, S. M., Ulasek, J., Nishibuchi, M., and DePaola, A. (2001) Automated ribotyping differentiates *Vibrio parahaemolyticus* O3:K6 strains associated with the Texas outbreak from other clinical strains. *J. Food Prot.* **64,** 1617–1620.

78. DePaola, A., Ulaszek, J., Kaysner, C. A., et al. (2003) Molecular, serological, and virulence characteristics of *Vibrio parahaemolyticus* isolated from environmental, food, and clinical sources in North America and Asia. *Appl. Environ. Microbiol.* **69,** 3999–4005.

79. Kubota, T. (1999) Analysis of thermostable direct hemolysin-producing *Vibrio parahaemolyticus* by pulsed-field gel electrophoresis. *Nippon Kosu Eisei Zasshi.* **46,** 929–933 (in Japanese).

80. Wong, H. C., Tlu, K., Pan, T. M., Lee, C. L., and Shih, D. Y. C. (1996) Subspecies typing of *Vibrio parahaemolyticus* by pulsed-field gel electrophoresis. *J. Clin. Microbiol.* **34,** 1535–1539.

81. Marshall, S., Clark, C. G., Wang, G., Mulvey, W., Kelly, M. T., and Johnson, W. M. (1999) Comparison of molecular methods for typing *Vibrio parahaemolyticus*. *J. Clin. Microbiol.* **37,** 2473–2478.

82. Wong, H. C., Chen, M. C., Liu, S. H., and Liu, D. P. (2000) Incidence of highly genetically diversified *Vibrio parahaemolyticus* in seafood imported from Asian countries. *Int. J. Food Microbiol.* **15,** 181–188.

83. Wong, H. C., Liu, S. H., Wang, T. K., et al. (2000) Characteristics of *Vibrio parahaemolyticus* O3:K6 from Asia. *Appl. Environ. Microbiol.* **66,** 3981–3986.

84. Liu, S.-L. and Sanderson, K. E. (1995) *Ceu*I reveals conservation of the genome of independent strains of *Salmonella typhimurium*. *J. Bacteriol.* **177,** 3355–3357.

85. Kotestishvili, M., Stine, O. C., Chen, Y., et al. (2003) Multilocus sequence typing has better discriminatory ability for typing *Vibrio cholerae* than does pulsed-field gel electrophoresis, and it clusters epidemic *V. cholerae* serogroups in a distinct genetic cluster. *J. Clin. Microbiol.* **41,** 2191–2196.

86. Kotestishvili, M., Stine, O. C., Kreger, A., Morris, J. G. Jr., and Sulakvelidze, A. (2002) Multilocus sequence typing for characterization of clinical and environmental *Salmonella* strains. *J. Clin. Microbiol.* **40,** 1626–1635.

87. Feil, E. J., Smith, J. M., Enright, M. C., and Spratt, B. G. (2000) Estimating recombinational parameters in *Streptococcus pneumoniae* from mutilocus sequence typing data. *Genetics* **154,** 1439–1450.

88. Chowdhury, N. R., Stine, O. C., Morris, J. G. Jr., and Nair, G. B. (2000) Assessment of evolution of pandemic *Vibrio parahaemolyticus* by multilocus sequence typing. *J. Clin Microbiol.* **42,** 1280–1282.

89. Chowdhury, N. R., Chakraborty, S., Ramamurthy, T., et al. (2000) Molecular evidence of clonal *Vibrio parahaemolyticus* pandemic strains. *Emerg. Infect. Dis.* **6,** 631–636.

90. Vieira, V. V., Teixeira, L. F., Vicente, A. C., Momen, H., and Salles, C. A. (2001) Differentiation of environmental and clinical isolates of *Vibrio mimicus* from *Vibrio cholerae* by multilocus enzyme electrophoresis. *Appl. Environ. Microbiol.* **67,** 2360–2340.

91. Pedersen, K., Dalsgaard, I., and Larsen, J. L. (1997) *Vibrio damsela* associated with fish diseased fish in Denmark. *Appl. Environ. Microbiol.* **63,** 3711–3715.

92. Dalsgaard, I., Hoi, L., Siebeling, R. J., and Dalsgaard, A. (1999) Indole-positive *Vibrio vulnificus* isolated from disease outbreaks on a Danish eel farm. *Dis. Aquat. Organ.* **35,** 187–194.

93. Lee, K. K., Yu, S. R., Chen, F. R., Yang, T. I., and Liu, P. C. (1996) Virulence of *Vibrio alginolyticus* isolated from diseased tiger prawn, *Penaeus monodon*. *Curr. Microbiol.* **32,** 229–231.

94. Lehane, L. and Rawlin, G. T. (2000) Topically acquired bacterial zoonoses from fish: a review. *Med. J. Aust.* **173,** 256–259.
95. Rabbani, G. H. and Greenough, W. B. III. (1999) Food as a vehicle of transmission of cholera. *J. Diarrhoeal Dis. Res.* **17,** 1–9.
96. Torres-Vitela, M. R., Castillo, A., Ibarra-Velazquez, L. M., et al. (2000) Survival of *Vibrio cholerae* O1 in ceviche and its reduction by heat pretreatment of raw ingredients. *J. Food Prot.* **63,** 445–450.
97. Maugeri, T. L., Caccamo, D., and Gugliandolo, C. (2000) Potentially pathogenic vibrios in brackish waters and mussels. *J. Appl. Microbiol.* **89,** 261–166.
98. Garcia Cortes, V. and Antillon, F. (1990) Isolation of entropathogenic *Vibrio* in bivalves and mud from the Nicoya Gulf, Costa Rica. *Rev. Biol. Trop.* **38,** 437–440 (in Spanish).
99. Yalcinkaya, F., Ergin, C., Agalar, C., Kaya, S., and Aksoylar, M. Y. (2003) The presence and antimicrobial susceptibilities of human-pathogen *Vibrio* spp. isolated from blue crab (*Callinectes sapidus*) in Belek tourism coast, Turkey. *Int. J. Environ. Health. Res.* **13,** 95–98.
100. Matte, G. R., Matte, M. H., Sato, M. I., Sanchez, P. S., Rivera, I. G., and Martines, M. T. (1994) Potentially pathogenic vibrios associated with mussels from a tropical region on the Atlantic coast of Brazil. *J. Appl. Bacteriol.* **77,** 281–287.
101. Molitoris, E., Joseph, S. W., Krichevsky, M. I., Sindhuhardja, W., and Colwell, R. R. (1985) Characterization and distribution of *Vibrio alginolyticus* and *Vibrio parahaemolyticus* isolated in Indonesia. *Appl. Environ. Microbiol.* **50,** 1388–1394.
102. Kelly, M. T. and Stroh, E. M. (1988) Occurrence of Vibrionaceae in natural and cultivated oysters populations in the Pacific Northwest. *Diagn. Microbiol. Infect. Dis.* **9,** 1–5.
103. Boiko, A. V. (2000) The etiological structure of acute intestinal infections caused by noncholera vibrios in the Volga delta. *Zh. Mikrobiol. Epidemiol. Immunobiol.* **1,** 15–17.
104. Chatterjee, B. D. and Neogy, K. N. (1972) *Vibrio alginolyticus, Vibrio parahaemolyticus,* and the related biotypes isolated from nonmarine fishes in Calcutta. *Bull. Calcutta Sch. Trop. Med.* **20,** 29–30.
105. Chan, K. Y., Woo, M. L., Lam, L. Y., and French, G. L. (1989) *Vibrio parahaemolyticus* and other halophilic vibrios associated with seafood in Hog Kong. *J. Appl. Bacteriol.* **66,** 57–64.
106. Kodama, H., Hayashi, M., and Gyobu, Y. (1991) Surveys on the contamination of marine fish with non-O1 *Vibrio cholerae* and *Vibrio mimicus* and poisoning cases by these organisms. *Kansenshogaku Zasshi.* **65,** 193–199 (in Japanese).
107. Hiratsuka, M., Saitoh, Y., and Yamane, N. (1980) The isolation of *Vibrio alginolyticus* from a patient with acute entero-colitis. *Tohoku J. Exp. Med.* **132,** 469–472.
108. Morris, J. G. Jr., Miller, H. G., Wilson, R., et al. (1982) Illness caused by *Vibrio damsela* and *Vibrio hollisae. Lancet* **1,** 1294–1297.
109. Ministry of Health and Welfare. (1998) *Vibrio parahaemolyticus,* Japan, 1996–1998. *Infect. Agents Sur. Rep.* **20,** 159–160.
110. Acuna, M. T., Diaz, G., Bolanos, H., et al. (1999) Sources of *Vibrio mimicus* contamination of turtle eggs. *Appl. Environ. Microbiol.* **65,** 336–338.
111. Bisharat, N. and Raz, R. (1996) *Vibrio* infection in Israel due to changes in fish marketing. *Lancet* **348,** 1585–1586.
112. Wong, H. C., Ting, S. H., and Shieh, W. R. (1992) Incidence of toxigenic vibrios in foods available in Taiwan. *J. Appl. Bacteriol.* **73,** 197–202.
113. Castro-Rosas, J. and Escartin, E. F. (2002) Adhesion and colonization of *Vibrio cholerae* O1 on shrimp and crab carapaces. *J. Food Prot.* **65,** 492–498.
114. Morbidity Mortality Weekly Report. (1993) Isolation of *Vibrio cholerae* O1 from oysters-Mobile Bay, 1991–1992. *Morb. Mortal. Wkly. Rep.* **42,** 91–93.
115. Goh, K. T., Lam, S., Kumarapathy, S., and Tan, J. L. (1984) A common source food-borne outbreak of cholera in Singapore. *Int. J. Epidemiol.* **13,** 210–215.

116. Mahon, B. E., Mintz, E., Greene, K., Wells, J. G., and Tauxe, R. V. (1996) Reported cholera in the United States, 1992–1994: a reflection of global changes in cholera epidemiology. *JAMA* **276,** 307–312.
117. Lancet Comment. (1992) Of cabbages and chlorine: cholera and Peru. *Lancet* **340,** 20–21.
118. Wu, F. M., Beuchat, L. R., Doyle, M. P., Mintz, E. D., Wells, J. G., and Swaminathan, B. (2002) Survival and growth of *Shigella flexneri, Salmonella enterica* serovar enteritidis, and *Vibrio cholerae* O1 in reconstituted infant formula. *Am. J. Trop. Med. Hyg.* **66,** 782–786.
119. Piergentili, P., Castellani-Pastoris, M., Fellini, R. D., et al. (1984) Transmission of non O group 1 *Vibrio cholerae* by raw oyster consumption. *Int. J. Epidemiol.* **13,** 3340–3343.
120. Morris, J. R. Jr., Wilson, R., Davis, B. R., et al. 1981. Non-O group 1 *Vibrio cholerae* gastroenteritis in the United States: clinical, epidemiologic, and laboratory characteristics of sporadic cases. *Ann. Intern. Med.* **94,** 656–658.
121. Schurmann, D., Ebert, N., Kampf, D., Baumann, B., Frei, U., and Suttorp, N. (2002) Domestic cholera in Germany associated with fresh fish imported from Nigeria. *Eur. J. Clin. Microbiol. Infect. Dis.* **21,** 827–828.
122. Koo, D., Aragon, A., Moscoso, V., et al. (1996) Epidemic cholera in Guatemala, 1993: transmission of a newly introduced epidemic strain by street vendors. *Epidemiol. Infect.* **116,** 121–126.
123. Johnston, J. M., Martin, D. L., Perdue, J., et al. (1983) Cholera on a Gulf Coast oil rig. *N. Eng. J. Med.* **309,** 523–526.
124. Mead, P., Slutsker, L., Dietz, V., et al. (1999) Food-related illness and death in the United States. *Emerg. Infect. Dis.* **5,** 607–625.
125. Daniels, N. A., Evans, M. C., and Griffin, P. M. (2000) Noncholera vibrios. In: *Emerging Infections* (Scheld, W. M., Craig, W. A., and Hughes, J. M., eds.), Vol. 4, American Society for Microbiology, Washington DC, pp. 137–147.
126. Finelli, L., Swerdlow, D., Mertz, K., Ragazzoni, H., and Spitalny, K. (1992) Outbreak of cholera associated with crab brought from an area with epidemic disease. *J. Infect. Dis.* **166,** 1433–1435.
127. Morbidity Mortality Weekly Report. (1995) Cholera associated with food transported from El Salvador-Indiana, 1994. *Morb. Mortal. Wkly. Rep.* **44,** 385–386.
128. Taylor, J. L., Tuttle, J., Pramukul, T., et al. (1993) An outbreak of cholera in Maryland associated with imported commercial frozen fresh coconut milk. *J. Infect. Dis.* **167,** 1330–1335.
129. Swaddiwudhipong, W., Akarasewi, P., Chayanuyayodhin, T., Kunasol, P., and Foy, H. M. (1990) A cholera outbreak associated with eating uncooked pork in Thailand. *J. Diarrhoeal Dis. Res.* **8,** 94–96.
130. Swaddiwudhipong, W., Jirakanvisun, R., and Rodklai, A. (1992) A common source of food borne outbreak of El Tor cholera following consumption of uncooked beef. *J. Med. Assoc. Thai.* **75,** 413–417.
131. Sutton, R. G. (1974) An outbreak of cholera in Australia due to food served in flight on an international aircraft. *J. Hyg. (London).* **72,** 441–451.
132. Eberhart-Phillips, J., Besser, R. E., Tormey, M. P., et al. (1996) An outbreak of cholera from food served on an international aircraft. *Epidemiol. Infect.* **116,** 9–13.
133. Kam, K. M., Leung, T. H., Ho, Y. Y., Ho, N. K., and Saw, T. A. (1995) Outbreak of *Vibrio cholerae* O1 in Hong Kong related to contaminated fish tank water. *Pub. Health* **109,** 389–395.
134. Daniels, N. A., MacKinnon, L., Bishop, R., et al. (2000) *Vibrio parahaemolyticus* infections in the United States, 1973–1998. *J. Infect. Dis.* **181,** 1661–1666.
135. Daniels, N. A., Ray, B., Easton, A., et al. (2001) Emergence of a new *Vibrio parahaemolyticus* serotypes in raw oyster: A prevention quandary. *JAMA* **284,** 1541–1545.
136. Morbidity Mortality Weekly Report. (1999) Outbreak of *Vibrio parahaemolyticus* infection associated with eating raw oysters and clams harvested from Long Island Sound–Connecticut, New Jersey, and New York. *Morb. Mortal. Wkly. Rep.* **29,** 48–51.

137. Centers for Diseases Control and Prevention. (1999) Outbreak of *Vibrio parahaemolyticus* infections associated with eating raw oysters and clams harvested from Long Island Sound–Connecticut, New Jersey, and New York, 1998. *JAMA* **281,** 603–604.

138. Fyfe, M., Yeung, S. T., Daly, P., Schallie, K., Kelly, M. T., and Buchanan, S. (1997) Outbreak of *Vibrio parahaemolyticus* related to raw oysters in British Columbia. *Can. Commu. Dis. Rep.* **23,** 145–148.

139. Lozano-Leon, A., Torres, J., Osorio, C. R., and Martinez-Urtaza, J. (2003) Identification of *tdh*-positive *Vibrio parahaemolyticus* from an outbreak associated with raw oyster consumption in Spain. *FEMS Microbiol. Lett.* **26(226),** 281–284.

140. Cordova, J. L., Astorga, J., Silva, W., and Riquelme, C. (2002) Characterization by PCR of *Vibrio parahaemolyticus* isolates collected during the 1997–1998 Chilean outbreak. *Biol. Res.* **35,** 433–440.

141. Smolikova, L. M., Lomov, I. M., Khomenko, T. V., et al. (2001) Studies on halophilic vibrios causing a food poisoning outbreak in the city of Vladivostok. *Zh. Mikrobiol. Epidemiol. Immunobiol.* **63,** 3–7.

142. Tuyet, D. T., Thiem, V. D., Seidlein, L. V., et al. (2002) Clinical, epidemiological, and socioeconomic analysis of an outbreak of *Vibrio parahaemolyticus* in Khanh Hoa province, Vietnam. *J. Infect. Dis.* **186,** 1615–1620.

143. Lalitha, M. K., Walter, N. M., Jesudason, M., and Mathan, V. I. (1983) An outbreak of gastroenteritis due to *Vibrio parahaemolyticus* in Vellore. *Indian J. Med. Res.* **78,** 611–605.

144. Klontz, K. C. and Desenclos, J. C. (1990) Clinical and epidemiological features of sporadic infections with *Vibrio fluvialis* in Florida. *J. Diarrhoeal Dis. Res.* **8,** 24–26.

145. Thekdi, R. J., Lakhani, A. G., Rale, V. B., and Panse, M. V. (1990) An outbreak of food poisoning suspected to be caused by *Vibrio fluvialis. J. Diarrh. Dis. Res.* **8,** 163–165.

146. Thamlikitkul, A. (1990) *Vibrio* bacteremia in Siriraj Hospital. *J. Med. Assoc. Thai.* **73,** 136–139.

147. Campos, E., Bolanos, H., Acuna, M. T., et al. (1996) *Vibrio mimicus* diarrhea following ingestion of raw turtle eggs. *Appl. Environ. Microbiol.* **62,** 1141–1144.

148. Ji, S. P. (1989) The first isolation of *Vibrio alginolyticus* from samples which caused food poisoning. *Zhonghua Yu Fang Yi Xue Za Zhi* **23,** 71–73 (in Chinese).

149. Shapiro, R. L., Altekruse, S., Hutwagner, L., et al. (1998) The role of Gulf Coast oysters harvested in warmer months in *Vibrio vulnificus* infection in the United States. *J. Infect. Dis.* **178,** 6752–6759.

150. Scoglio, M. E., Di Pietro, A., Delia, I., Mauro, S., and Lagana, P. (2001) Virulence factors in vibrios and aeromonads. *New. Microbiol.* **24,** 273–280.

151. Nataro, J. P., Kaper, J. B., Robins Browne, R., Prado, V., Vial, P., and Levine, M. M. (1987) Patterns of adherence of diarrheagenic *Escherichia coli* to HEp-2 cells. *Pediatr. Infect. Dis. J.* **6,** 829–831.

152. Alam, M., Miyoshi, S., and Shinoda, S. (1995) Production of antigenically related exocellular proteases mediating hemagglutination by vibrios. *Microbiol. Immunol.* **39,** 67–70.

153. Alam, M., Miyoshi, S. I., Tomochika, K. I., and Shinoda, S. (1996) Purification and characterization of hemagglutinins from *Vibrio mimicus*: a 39-kilodalton major outer membrane protein and lippopolysaccharide. *Infect. Immun.* **64,** 4035–4041.

154. Okujo, N. and Yamamoto, S. (1994) Identification of the siderophores from *Vibrio hollisae* and *Vibrio mimicus* as aerobactin. *FEMS Microbiol. Lett.* **118,** 187–192.

155. Baffone, W., Citterio, B., Vittoria, E., et al. (2003) Retention of virulence in viable but noculturable halophilic *Vibrio* spp. *Int. J. Food Microbiol.* **89,** 31–39.

156. Sechi, L. A., Dupre, I., Deriu, A., Fadda, G., and Zanetti, S. (2000) Distribution of *Vibrio cholerae* virulence genes among different *Vibrio* species isolated in Sardinia, Italy. *J. Appl. Microbiol.* **88,** 475–481.

157. Fouz, B., Toranzo, A. E., Biosca, E. G., Mazoy, R., and Amaro, C. (1994) Role of iron in the pathogenicity of *Vibrio damsela. FEMS Microbiol. Lett.* **15,** 181–188.

158. Kreger, A. S. (1984) Cytolytic activity and virulence of *Vibrio damsela*. *Infect. Immun.* **44,** 326–331.

159. Kothary, M. H. and Kreger, A. S. (1985) Purification and characterization of an extracellular cytolysin produced by *Vibrio damsela*. *Infect. Immun.* **49,** 25–31.

160. Cutter, D. L. and Kreger, A. S. (1990) Cloning and expression of the damselysin gene from *Vibrio damsela*. *Infect. Immun.* **58,** 266–268.

161. Lockwood, D. E., Kreger, A. S., and Richardson, S. H. (1982) Detection of toxins produced by *Vibrio fluvialis*. *Infect. Immun.* **35,** 702–708.

162. Han, J. H., Lee, J. H., Choi, Y. H., Park, J. H., Choi, T. J., and Kong, I. S. (2002) Purification, characterization and molecular cloning of *Vibrio fluvialis* hemolysin. *J. Biochim. Biophy. Acta* **1599,** 106–114.

163. Rahim, Z. and Aziz, K. M. (1996) Factors affecting production of haemolysin by strains of *Vibrio fluvialis*. *J. Diarrhoeal Dis. Res.* **14,** 113–116.

164. Miyoshi, S., Sonoda, Y., Wakiyama, H., et al. (2002) An exocellular thermolysin-like metallo-protease produced by *Vibrio fluvialis*: purification, characterization and gene cloning. *Microb. Pathog.* **33,** 127–134.

165. Myatt, D. C. and Davis, G. H. (1989) Extracellular and surface-bound biological activities of *Vibrio fluvialis*, *Vibrio furnissii* and related species. *Med. Microbiol. Immun. (Berl.)* **178,** 279–287.

166. Carvalho, I. T., Magalhaes, V., Leal, N. C., Melo, V., and Magalhaes, M. (1994) *Vibrio fluvialis* attaches to but does not enter HeLa cells monolayers. *Mem. Inst. Oswaldo Cruz* **89,** 221–223.

167. Rodrigues, D. P., Ribeiro, R. V., Alves, R. M., and Hofer, E. (1993) Evaluation of virulence factors in environmental isolates of *Vibrio* species. *Mem. Inst. Oswaldo Cruz* **88,** 589–592.

168. Waldor, M. K. and Mekalanos, J. J. (1996) Lysogenic conversion by filamentous phage encoding cholera toxin. *Science* **272,** 1910–1914.

169. Yamai, S., Okitsu, T., Shimada, T., and Katsube, Y. (1997) Distribution of serogroups of *Vibrio cholerae* non-O1, non-O139 with specific reference to their ability to produce cholera toxin, and addition of novel serogroups. *Kansenshogaku Zasshi.* **71,** 1037–1045 (in Japanese).

170. Nagamune, K., Yamamoto, K., and Honda, T. (1995) Cloning and sequencing of a novel hemolysis gene of *Vibrio cholerae*. *FEMS Microbiol. Lett.* **128,** 265–269.

171. Nishibuchi, M., Seidler, R. J., Rollins, D. M., and Joseph, S. W. (1983) *Vibrio* factors cause rapid fluid accumulation in suckling mouse. *Infect. Immun.* **40,** 1083–1091.

172. Nishibuchi, M. and Seidler, R. J. (1983) Medium-dependent production of extracellular enterotoxins by non-O1 *Vibrio cholerae*, *Vibrio mimicus*, and *Vibrio fluvialis*. *Appl. Environ. Microbiol.* **45,** 228–231.

173. Provenzano, D., Schuhmacher, D. A., Barker, J. L., and Klose, K. E. (2000) The virulence regulatory protein ToxR mediates enhanced bile resistance in the *Vibrio cholerae* and other pathogenic *Vibrio* species. *Infect. Immun.* **68,** 1491–1497.

174. Arita, M., Honda, T., Miwatani, T., Takeda, T., Takao, T., and Shimonishi, Y. (1991) Purification and characterization of a heat-stable enterotoxin of *Vibrio mimicus*. *FEMS Microbiol. Lett.* **63,** 105–110.

175. Ananthan, S. and Dhamodaran, S. (1996) Toxigenicity and drug sensitivity if *Vibrio mimicus* isolated from patients with diarrhoea. *India J. Med. Res.* **104,** 336–341.

176. Alam, M., Miyoshi, S., Maruo, I., Ogawa, C., and Shinoda, S. (1994) Existence of a novel hemagglutinin having no protease activity in *Vibrio mimicus*. *Microbiol. Immunol.* **38,** 467–470.

177. Chowdhury, M. A., Miyoshi, S., and Shinoda, S. (1990) Purification and characterization of protease produced by *Vibrio mimicus*. *Infect. Immun.* **58,** 4159–4162.

178. Chowdhury, M. A., Miyoshi, S., and Shinoda, S. (1991) Vascular permeability enhancement by *Vibrio mimicus* protease and the mechanisms of action. *Microbiol. Immunol.* **35,** 1049–1058.

179. Miyoshi, S., Sasahara, K., Akamatsu, S., et al. (1997) Purification and characterization of hemolysin produced by *Vibrio mimicus. Infect. Immun.* **65,** 1830–1835.

180. Kim, G. T., Lee, J. Y., Huh, S. H., Yu, J. H., and Kong, I. S. (1997) Nucleotide sequence of the *vmhA* gene encoding hemolysin from *Vibrio mimicus. Biochem. Biophys. Acta* **1360,** 102–104.

181. Uchimura, M., Koiwai, K., Tsuruoka, Y., and Tanaka, H. (1993) High prevalence of thermostable hemolysin (TDH)-like toxin in *Vibrio mimicus* strains isolated from diarrhoeal patients. *Epidemiol. Infect.* **111,** 49–53.

182. Yoshida, H., Honda, T., and Miwatani, T. (1991) Purification and characterization of hemolysin of *Vibrio mimicus* that relates to the thermostable direct hemolysin of *Vibrio parahaemolyticus. FEMS Microbiol. Lett.* **68,** 249–253.

183. Faruque, S. M., Rahman, M. M., Asadulghani, Islam, K. M. N., and Mekalanos, J. J. (1999) Lysogenic conversion of environmental *Vibrio mimicus* strains by CTXΦ. *Infect. Immun.* **67,** 5723–5729.

184. Tamplin, M. L., Jalali, R., Ahmed, M. K., and Colwell, R. R. (1990) Variation in the epitopes of the B subunit if *Vibrio cholerae* non-O1 and *Vibrio mimicus* cholera toxins. *Can. J. Microbiol.* **36,** 409–413.

185. Gyobu, Y., Isobe, J., Kodama, H., Sato, S., and Shimada, T. (1992) Enteropathogenicity and enteropathogenic toxin production of *Vibrio mimicus. Kansenshogaku Zasshi.* **66,** 115–120 (in Japanese).

186. Chowdhury, M. A., Aziz, K. M., Kay, B. A., and Rahim, Z. (1987) Toxin production by *Vibrio mimicus* strains isolated from human and environmental sources in Bangladesh. *J. Clin. Microbiol.* **25,** 2200–2203.

187. Chowdhury, M. A., Hill, R. T., and Colwell, R. R. (1994) A gene for the enterotoxin zonula occludens toxin is present in *Vibrio mimicus* and *Vibrio cholerae* O139. *FEMS Microbiol. Lett.* **119,** 377–380.

188. Bi, K., Miyoshi, S. I., Tomochika, K. I., and Shinoda, S. (2001) Detection of virulence associated genes in clinical strains of *Vibrio mimicus. Microbiol. Immunol.* **45,** 613–616.

189. Shi, L., Miyoshi, S., Hiura, M., Tomochika, K., Shimada, T., and Shinoda, S. (1998) Detection of genes encoding cholera toxin (CT), zonula occuludens toxin (Zot), accessory cholera toxin (Ace), and heat-stable neterotoxin (ST) in *Vibrio mimicus* clinical strains. *Microbiol. Immunol.* **42,** 823–828.

190. Hayat, U., Reddy, G. P., Bush, C. A., Johnson, J. A., Wright, A. C., and Morris, J. G. Jr. (1993) Capsular types of *Vibrio vulnificus*: an analysis of strains form clinical and environmental sources. *J. Infect. Dis.* **168,** 758–762.

191. Hor, L. I., Chang, T. T., and Wang, S. T. (1999) Survival of *Vibrio vulnificus* in whole blood from patients with chronic liver diseases: association with phagocytosis by neutrophils and serum ferritin levels. *J. Infect. Dis.* **179,** 275–278.

192. Raimondi, F., Kao, J. P., Fiorentini, C., et al. (2000) Enterotoxicity and cytotoxicity of *Vibrio parahaemolyticus* thermostable direct hemolysin in in vitro system. *Infect. Immun.* **68,** 3180–3185.

193. Fallariono, A., Attridge, S. R., Manning, P. A., and Focareta, T. (2002) Cloning and characterization of a novel haemolysin in *Vibrio cholerae* O1 that does not directly contribute to the virulence of the organisms. *Microbiology* **148,** 2181–2189.

194. Nishibuchi, M., Janda, J. M., and Ezaki, T. (1996) The thermostable direct hemolysin gene (*tdh*) of *Vibrio hollisae* is dissimilar in prevalence to and phylogenetically distant from the *tdh* genes of other vibrios: implications in the horizontal transfer of the *tdh* gene. *Microbiol. Immunol.* **40,** 59–65.

195. Hickman, F. W., Farmer, J. J. III, Hollis, D. G., et al. (1982) Identification of *Vibrio hollisae* sp. nov. from patients with diarrhea. *J. Clin. Microbiol.* **15,** 395–401.

196. Nishibuchi, M. and Kaper, J. B. (1990) Duplication variation of the thermostable direct haemolysin (*tdh*) gene in *Vibrio parahaemolyticus. Mol. Microbiol.* **4,** 87–99.

197. Honda, S., Goto, I., Minematsu, N., et al. (1987) Gastroenteritis due to Kanagawa negative *Vibrio parahaemolyticus. Lancet* **i,** 331–332.

198. Honda, T., Ni, Y., and Miwatani, T. (1988) Purification and characterization of a hemolysin produced by a clinical isolate of Kanagawa phenomenon-negative *Vibrio parahaemolyticus* and related to the thermostable direct hemolysin. *Infect. Immun.* **56,** 961–965.

199. Nishibuchi, M., Taniguchi, T., Misawa, T., Khaeomanee-iam, V., Honga, T., and Miwatani, T. (1989) Cloning and nucleotide sequencing of the gene (*trh*) encoding the hemolysin related to the thermostable direct hemolysin of *Vibrio parahaemolyticus. Infect. Immun.* **57,** 2691–2697.

200. Okuda, J., Ishibashi, M., Abbott, S. L., Janda, J. M., and Nishibuchi, M. (1997) Analysis of the thermostable direct hemolysin (*tdh*) gene and the *tdh*-related hemolysin (*trh*) genes in urease-positive strains of *Vibrio parahaemolyticus* isolated on the West Coast of the United States. *J. Clin. Microbiol.* **35,** 1965–1971.

201. Osawa, R., Okitsu, T., Morozumi, H., and Yamai, S. (1996) Occurrence of urease-positive *Vibrio parahaemolyticus* in Kanagawa, Japan, with specific reference to presence of thermostable direct hemolysin (TDH) and the TDH-related-hemolysin genes. *Appl. Environ. Microbiol.* **62,** 725–727.

202. Suthienkul, O., Ishibashi, M., Iida, T., et al. (1995) Urease production correlates with possession of the *trh* gene in *Vibrio parahaemolyticus* strains isolated in Thailand. *J. Infect. Dis.* **172,** 1405–1408.

203. Park, K. S., Iida, T., Yamaichi, Y., Oyagi, T., Yamamoto, K., and Honda, T. (2002) Genetic characterization of DNA region containing the *trh* and *ure* genes of *Vibrio parahaemolyticus. Infect. Immun.* **68,** 5742–5748.

204. Iida, T., Suthienkul, O., Park, K. S., et al. (1997) Evidence for genetic linkage of the *ure* and *trh* genes in *Vibrio parahaemolyticus. Emerg. Infect. Dis.* **7,** 477–478.

205. DeKoning-Ward, T. F. and Robins-Browne, R. M. (1995) Contribution of urease to acid tolerance in *Yersinia enterocolitica. Infect. Immun.* **63,** 3790–3795.

206. Turbett, G. R., Høj, P. B., Horne, R., and Mee, B. J. (1992) Purification and characterization of the urease enzymes of *Helicobacter* species from humans and animals. *Infect. Immun.* **60,** 5259–5266.

207. McCarthy, S. A., DePaola, A., Cook, D. W., Keysner, C. A., and Hill, W. E. (1999) Evaluation of alkaline phosphatase and digoxigenin-labelled probes for detection of the thermolabile (*tlh*) gene of *Vibrio parahaemolyticus. Lett. Appl. Microbiol.* **28,** 66–70.

208. Yeung, P. S., Hayes, M. C., DePaola, A., Kaysner, C. A., Kornstein, L., and Boor, K. J. (2002) Comparative phenotypic, molecular, and virulence characterization of *Vibrio parahaemolyticus* O3:K6 isolates. *Appl. Microbiol.* **68,** 2901–2909.

209. Hsieh, Y. C., Liang, S. M., Tsai, W. L., Chen, Y. H., Liu, T. Y., and Liang, C. M. (2003) Study of capsular polysaccharide from *Vibrio parahaemolyticus. Infect. Immun.* **71,** 3329–3336.

210. Levine, W. C. and Griffin, P. M. (1993) *Vibrio* infections on the Gulf Coast: results of first year of regional surveillance. Gulf Coast Vibrio Working Group. *J. Infect. Dis.* **167,** 479–483.

211. Dryden, M., Legarde, M., Gottlieb, T., Brady, L., and Ghosh, H. K. (1989) *Vibrio damsela* wound infections in Australia. *Med. J. Aust.* **151,** 540–541.

212. Yuen, K. Y., Ma, L., Wong, S. S., and Ng, W. F. (1993) Fatal necrotizing fasciitis due to *Vibrio damsela. Scand. J. Infect. Dis.* **25,** 659–661.

213. Tang, W. M. and Wong, J. W. (1999) Necrotizing fasciitis caused by *Vibrio damsela. Orthopedics* **22,** 443–444.

214. Goodell, K. H., Jordan, M. R., Graham, R., Cassidy, C., and Nasraway, S. A. (2004) Rapidly advancing necrotizing fasciitis caused by *Photobacterium (Vibrio) damsela*: a hyperaggressive variant. *Crit. Care. Med.* **32,** 278–281.

215. Perez-Tirse, J., Levine, J. F., and Mecca, M. (1993) *Vibrio damsela.* A cause of fulminant septicemia. *Arch. Intern. Med.* **9,** 1838–1840.

216. Shin, J. H., Shin, M. G., Suh, S. P., Ryang, D. W., Rew, J. S., and Nolte, F. S. (1996) Primary *Vibrio demsela* septicemia. *Clin. Infect. Dis.* **22,** 856–857.
217. Kolb, E. A., Eppes, S. C., and Klein, J. D. (1997) *Vibrio fluvialis*: an under recognized enteric pathogen in infants? *South Med. J.* **90,** 544–545.
218. Garcia-Martos, P., Benjumeda, M., and Delgado, D. (1993) Otis externa caused by *Vibrio alginolyticus*: description of 4 cases. *Acta Otorrinolaringol. Esp.* **44,** 55–57 (in Spanish).
219. Yoshii, Y., Nishino, H., Satake, K., and Umeyama, K. (1987) Isolation of *Vibrio fluvialis*, an unusual pathogen in acute suppurative cholangitis. *Am. J. Gastroentrol.* **82,** 903–905.
220. Hornstrup, M. K. and Gahrn-Hansen, B. (1993) Extraintestinal infections caused by *Vibrio parahaemolyticus* and *Vibrio alginolyticus* in a Danish county. Scand. *J. Infect. Dis.* **25,** 735–740.
221. Lessner, A. M., Webb, R. M., and Rabin, B. (1985) *Vibrio alginolyticus* conjunctivitis. First reported case. *Arch. Opthalmol.* **103,** 229–230.
222. Penland, R. L., Boniuk, M., and Wilhelmus, K. R. (2000) *Vibrio* ocular infections on the U. S. Gulf Coast. *Cornea* **19,** 26–29.
223. Hodge, T. W., Levy, C. S., and Smith, M. A. (1995) Diarrhea associated with *Vibrio fluvialis* infection in a patient with AIDS. *Clin. Infect. Dis.* **21,** 237–238.
224. Lerstloompleephunt, N., Tantawichien, T., and Sitpriji, V. (2000) Renal failure in *Vibrio vulnificus* infection. *Ren. Fail.* **22,** 337–343.
225. Klontz, K. C., Cover, D. E., Hyman, F. N., and Mullen, R. C. (1999) Fatal gastroenteritis due to *Vibrio fluvialis* and nonfatal bacteremia due to *Vibrio mimicus*: unusual *Vibrio* infections in two patients. *Clin. Infect. Dis.* **19,** 541–542.
226. Hally, R. J., Rubin, R. A., Fraimow, H. S., and Hoffman-Terry, M. L. (1995) Fatal *Vibrio parahaemolyticus* septicemia in a patient with cirrhosis: A case report and review of the literature. *Dig. Dis. Sci.* **40,** 1257–1260.
227. Roland, F. P. (1970) Leg gangrene and endotoxin shock due to *Vibrio parahaemolyticus*— an infection acquired in New England coastal waters. *N. Engl. J. Med.* **282,** 1306.
228. Namdari, H., Klaips, C. R., and Hughes, J. L. (2000) A cytotoxin-producing strain of *Vibrio cholerae* non-O1, non-O139 as a cause of cholera and bacteremia after consumption of raw clams. *J. Clin. Microbiol.* **38,** 3518–3519.
229. Lin, C. J., Chiu, C. T., Lin, D. Y., Shee, I. S., and Lien, J. M. (1996) Non-O1 *Vibrio cholerae* bacteremia in patients with cirrhosis: 5-yr experience from a single medical center. *Am. J. Gastroenterol.* **91,** 336–340.
230. Hughes J. M., Boyce, J. M., Aleem, A. R., Wells, J. G., Rahman, A. S., and Curlin, G. T. (1978) *Vibrio parahaemolyticus* enterocolotis in Bangladesh: report of an outbreak. *Am. J. Trop. Med. Hyg.* **27,** 106–112.
231. Blake, P. A., Weaver, R. E., and Hollis, D. G. (1980) Disease of humans (other than cholera) caused by vibrios. *Annu. Rev. Microbiol.* **34,** 341–367.
232. Qadri, F., Alam, M. S., Nishibuchi, M., et al. (2003) Adaptive and inflammatory immune response in patients infected with strains of *Vibrio parahaemolyticus*. *J. Infect. Dis.* **187,** 1085–1096.
233. Tuyet, D. T., Thiem, V. D., Seidlein, L. V., et al. (2002) Clinical, epidemiological, and socioeconomic analysis of an outbreak of *Vibrio parahaemolyticus* in Khanh Hoa province, Vietnam. *J. Infect. Dis.* **186,** 1615–1620.
234. Lim, T. K. and Stebbings, A. E. (1999) Fulminant necrotising fasciitis caused by *Vibrio parahaemolyticus*. *Singapore Med. J.* **40,** 596–597.
235. Ng, T. C., Chiang, P. C., Wu, T. L., and Leu, H. S. (1999) *Vibrio parahaemolyticus* bacteremia: case report. *Changgeng Yi Xue Za Zhi* **22,** 508–514 (in Chinese).
236. Chuang, Y. C., Yuan, C. Y., Liu, C. Y., Lan, C. K., and Huang, A. H. (1992) *Vibrio vulnificus* infection in Taiwan: report of 28 cases and review of clinical manifestation and treatment. *Clin. Infect. Dis.* **15,** 1071–1072.

237. Mahalanabis, D., Molla, A. M., and Sack, D. A. (1992) Clinical management of cholera. In: *Cholera* (Barua, D. and Greenough, W. B., eds.), Plenum, New York, p. 266.

238. Smith, P., Hiney, M. P., and Samuelsen, O. B. (1994) Bacterial resistance to antimicrobial agents used in fish farming: a critical evaluation of method and meaning. *Ann. Rev. Fish Dis.* **4**, 273–313.

239. Aoki, T. (1997) Resistance plasmids and the risk of transfer. In: *Furunculosis: Multidisciplinary Fish Disease Research* (Bernoth, E. M., ed.), Academic, London, pp. 433–440.

240. Lesmana, M., Subekti, D. S., Simanjuntak, P., Punjabi, N. H., Campbell, J. R., and Oyofo, B. A. (2002) Spectrum of *Vibrio* species associated with acute diarrhea in North Jakarta, Indonesia. *Diagn. Mirobiol. Infect. Dis.* **43**, 21–27.

241. Chowdhury, M. A., Yamanaka, H., Miyoshi, S., Aziz, K. M., and Shinoda, S. (1989) Ecology of *Vibrio mimicus* in aquatic environments. *Appl. Environ. Microbiol.* **55**, 2073–2080.

242. Chowdhury, M. A., Aziz, K. M., Rahim, Z., and Kay, B. A. (1986) Antibiotic resistance pattern of *Vibrio mimicus* isolated from human and environmental sources in Bangladesh. *Antimicrob. Agents Chemother.* **30**, 622–623.

243. Weber, J. T., Mintz, E. D., Canizares, R., et al. (1994) Epidemic cholera in Ecuador: multidrug-resistance and transmission by water and seafood. *Epidemiol. Infect.* **112**, 1–11.

244. Berlin, D. L., Herson, D. S., Hicks, D. T., and Hoover, D. G. (1999) Response of pathogenic *Vibrio* species to high hydrostatic pressure. *Appl. Environ. Microbiol.* **65**, 2776–2780.

245. Cook, D. W. (2003) Sensitivity of *Vibrio* species in phosphate-buffered saline and in oysters to high-pressure processing. *J. Food Prot.* **66**, 2276–2282.

246. Nascumento, D. R., Vieira, R. H., Almeida, H. B., Patel, T. R., and Iaria, S. T. (1998) Survival of *Vibrio cholerae* O1 strains in shrimp subjected to freezing and boiling. *J. Food Prot.* **61**, 1317–1320.

247. Liew, W. S., Leisner, J. J., Rusul, G., Radu, S., and Rassip, A. (1998) Survival of *Vibrio* spp. including inoculated *Vibrio cholerae* O139 during heat-treatment of cockles (*Anadara granosa*). *Int. J. Food Microbiol.* **42**, 167–173.

248. Makukutu, C. A. and Guthrie, R. K. (1986) Behavior of *Vibrio cholerae* in hot foods. *Appl. Environ. Microbiol.* **52**, 824–831.

249. de Moraes, I. R., Del Mastro, N. L., Jakabi, M., and Gelli, D. S. (2000) Radiosensitivity of *Vibrio cholerae* O1 incorporated in oysters to ^{60}Co. *Rev. Saude. Pub.* **34**, 29–32 (in Portuguese).

250. de Castillo, M. C., de Allori, C. G., de Gutierrez, R. C., et al. (1998) Action against *Vibrio cholerae* O1 tox$^+$ of chemical products used in the lemon production. *Rev. Latinoam. Microbiol.* **40**, 120–123.

251. Mata, L., Vargas, C., Saborio, D., and Vives, M. (1994) Extinction of *Vibrio cholerae* in acidic substrata: contaminated cabbage and lettuce treated with lime juice. *Rev. Biol. Trop.* **42**, 487–492.

252. Mata, L., Vives, M., and Vicente, G. (1994) Extinction of *Vibrio cholerae* in acidic substrata: contaminated fish marinated with lime juice (ceviche). *Rev. Biol. Trop.* **42**, 479–485.

253. Rodrigues, A., Sandstrom, A., Ca, T., Steinsland, H., Jensen, H., and Aaby, P. (2000) Protection from cholera by adding lime juice to food—results from community and laboratory studies in Guinea-Bissau, West Africa. *Trop. Med. Int. Health* **5**, 418–422.

254. Dobosch, D., Gomez Zavaglia, A., and Kuljich, A. (1995) The role of food in cholera transmission. *Medicina (B. Aires)* **55**, 28–32 (in Spanish).

255. Murphree, R. L. and Tamplin, M. L. (1995) Uptake and retention of *Vibrio cholerae* O1 in the Eastern oyster, *Crassotrea virginica. Appl. Environ. Microbiol.* **61**, 3656–3660.

256. Dalsgaard, A., Huss, H. H., H-Kittikun, A., and Larsen, J. L. (1995) Prevalence of *Vibrio cholerae* and *Salmonella* in a major shrimp production area in Thailand. *Int. J. Food. Microbiol.* **28**, 101–113.

257. DePaola, A. Jr. (2004) Use of zero tolerance for certain vibrios in seafoods. Session 92, IFT Annual Meeting, July 12–16, 2004, Las Vegas.

258. Interstate Shellfish Sanitation Conference. (2000) Issue relating to *Vibrio vulnificus* risk management plan for oyster, ISSC, Colombia, SC.

259. International Commission on Microbiological Specification for Foods. (1996) *Vibrio parahaemolyticus*. In: *Micro-organisms in Foods 5. Microbiological Specification of Food Pathogens*, Blackie Academic and Professional, London, pp. 426–435.

260. Motes, M. L. and DePaola, A. (1996) Offshore suspension relaying to reduce levels of *Vibrio vulnificus* in oysters (*Crassostrea virginica*). *Appl. Environ. Microbiol.* **62,** 3875–3877.

261. Hoshino, K., Yamasaki, S., Mukhopadhyay, A. K., et al. (1998) Development and evaluation of a multiplex PCR assay for rapid detection of toxigenic *Vibrio cholerae* O1 and O139. *FEMS Immunol. Med. Microbiol.* **20,** 201–207.

262. Keasler, S. P. and Hall, R. H. (1993) Detection and biotyping *Vibrio cholerae* O1 with multiplex ploymerase reaction. *Lancet* **341,** 1661.

263. Nandi, B., Nandy, R. K., Mukhopadhyay, S., Nair, G. B., Shimada, T., and Ghose, A. C. (2000) Rapid method for species-specific identification of *Vibrio cholerae* using primers targeted to the gene of outer membrane protein OmpW. *J. Clin. Microbiol.* **38,** 4145–4151.

264. Hill, W. E., Keasler, S. P., Trucksess, M. W., Feng, P., Kaysner, C. A., and Lampel, K. A. (1991) Polymerase chain reaction identification of *Vibrio vulnificus* in artificially contaminated oysters. *Appl. Environ. Microbiol.* **57,** 707–711.

265. ICMSF. (1986) Microorganisms in Foods. 2. Sampling for Microbiological Analysis: Principles and specific Applications. 2nd ed. University of Toranto Press, Buffalo, NY.

6

Enterococcus and its Association with Foodborne Illness

Simona F. Oprea and Marcus J. Zervos

1. INTRODUCTION

Enterococci are an important group of bacteria and their interaction with humans is complex. On one hand, enterococci are part of the normal flora of humans and animals, and some of their strains are used for the manufacturing of foods or as probiotics, whereas others are known to cause serious diseases in humans. With the emergence of enterococci as the third most common cause of nosocomial blood-stream infections *(1)* as well as the alarming rise in enterococci resistance to multiple antimicrobials, more concentrated effort has been invested in the better understanding of this versatile microorganism.

Although they are not classic foodborne pathogens, enterococci have earned their own place, in this book, as organisms that can be acquired from food. They have been associated with foodborne outbreaks because of their presence in foods and their capacity to carry and disseminate resistance genes through the food chain. The role of enterococci in disease raises concerns regarding their use in foods or as probiotics, whereas the imminent crisis of emergent antimicrobial resistance leads to measures that are urgently needed for the judicious use of antimicrobials in agriculture and human medicine.

2. CLASSIFICATION AND IDENTIFICATION

Considered for a long time a major division of the genus *Streptococcus*, enterococci have undergone considerable changes in taxonomy in the last decades, including the recognition of *Enterococcus* as a separate genus in 1984 *(2,3)*.

Enterococci are catalase-negative Gram-positive cocci that occur singly, in pairs, or in short chains. They are facultative anaerobes and very tolerant to extreme temperatures, salinity, and pH; thus, they grow in 6.5% NaCl broth at pH 9.6 and at temperatures ranging from 10 to 45°C, with optimum growth at 35°C *(2)*. Enterococci hydrolyze esculin in the presence of 40% bile-salts. Most enterococci hydrolyze L-pyrrolidonyl-β-naphtylamide (PYR) and all strains hydrolyze leucine-β-naphtylamide by producing leucine aminopeptidase (LAPase). Some species, such as *E. gallinarum*, are motile.

From: *Infectious Disease: Foodborne Diseases*
Edited by: S. Simjee © Humana Press Inc., Totowa, NJ

Current criteria for the inclusion into the *Enterococcus* genus are a combination of DNA–DNA reassociation values, 16S rRNA gene sequencing, whole-cell protein analysis, and conventional phenotypic tests *(2–4)*.

Enterococci belong to the clostridial subdivision of Gram-positive bacteria, together with the other genera of lactic-acid bacteria: *Aerococcus, Carnobacterium, Globicatella, Lactobacillus, Lactococcus, Leuconostoc, Oenococcus, Pediococcus, Streptococcus, Tetragenococcus, Vagococcus,* and *Weissella (5)*. So far 23 distinct *Enterococcus* species have been recognized *(3)* but new species continue to be identified *(6)*. They have been separated into five groups on the basis of acid formation in mannitol and sorbose broths, and hydrolysis of arginine *(3)*. Group I consists of *Enterococcus avium, Enterococcus malodoratus, Enterococcus raffinosus, Enterococcus pseudoavium, Enterococcus saccharolyticus, Enterococcus pallens,* and *Enterococcus gilvus*. They form acid in both mannitol and sorbose broths but do not hydrolyze arginine. Group II consists of *Enterococcus faecalis, Enterococcus faecium, Enterococcus casseliflavus, Enterococcus mundtii,* and *Enterococcus gallinarum*. These five species form acid in mannitol but not sorbose broth and hydrolyze arginine. Group III includes *Enterococcus durans, Enterococcus porcinus, Enterococcus ratti, Enterococcus hirae, Enterococcus dispar,* and mannitol-negative variants of *E. faecalis* and *E. faecium*. The enterococci in Group IV do not produce acid in mannitol or sorbose broths and do not hydrolyze arginine. They are *Enterococcus asini, Enterococcus sulfureus,* and *Enterococcus cecorum*. Finally, Group V consists of the variant strains of *Enterococcus casseliflavus, Enterococcus gallinarum,* and *Enterococcus faecalis* that fail to hydrolyze arginine *(3)*.

Three new species of *Enterococcus* were recently identified from human clinical sources by the analysis of whole-cell protein profiles and DNA–DNA reassociation experiments, in conjunction with conventional physiological tests. These new enterococcal species, provisionally designated *Enterococcus* sp. nov. CDC PNS-E1, *Enterococcus* sp. nov. CDC PNS-E2, and *Enterococcus* sp. nov. CDC PNS-E3, resemble the physiological groups I, II and IV; two were isolated from human blood and one from human brain tissue. Resistance to vancomycin was detected in one of the strains, which is harboring the *vanA* resistance gene *(6)*.

The species most broadly distributed in nature are *E. faecalis* and *E. faecium*. Identification of the different *Enterococcus* species can be done by conventional physiological tests, by commercial systems, or by molecular methods. Correct species identification is important for appropriate antimicrobial treatment, for epidemiologic surveillance, and for the selection of starter strains and labeling of the product to which the starter is added.

3. RESERVOIRS

Enterococci have the capacity to grow and survive in very harsh environments and thus occupy a variety of ecological niches. They can be found in soil, water, food, plants, animals (including mammals, birds, and insects), and humans. The major natural habitat of these organisms appears to be the gastrointestinal tract of humans and other animals were they make up a significant portion of the normal gut flora *(4)*. With the increase in antimicrobial-resistant enterococci worldwide and their occasional implication in serious human diseases, there has been considerable interest in identifying the reservoirs of these organisms and of antimicrobial resistance genes.

3.1. Nonhuman Reservoirs

3.1.1. In Mammals, Birds, and Insects

Enterococci are a natural part of the intestinal flora in most mammals and birds. Insects and reptiles were also found to harbor these organisms. Some of the *Enterococcus* species are host-specific, while others are more broadly distributed.

Exemples of host-specific enterococci are *E. columbae*, specific for pigeons, and *E. asini*, found so far only in donkeys. Several enterococcal species, although identifiable in a number of hosts, appear to have some host-specific properties *(7)*. Few occasions, similar *E. faecium* strains have been found in animals and humans, and in some of these cases strains were epidemiologically related, originating from the farmer and one of his animals *(7)*. Of importance, all of these studies were done on glycopeptide-resistant *E. faecium*.

The most often encountered enterococcal species in the intestines of farm animals are *E. faecalis*, *E. faecium*, *E. hirae*, and *E. durans*. In chickens, *E. faecalis*, found early in life, is later replaced by *E. faecium*, and then by *E. cecorum*. Other species occasionally isolated from chickens are *E. casseliflavus*, *E. gallinarum*, and *E. mundtii*.

3.1.2. In Soil, Plants, and Water

The presence of enterococci in these sources is usually a result of fecal contamination. *E. faecium* and *E. faecalis* have been the species most commonly recovered from water. *E. casseliflavus*, *E. mundtii*, and *E. sulfureus* appear to be plant-associated *(7)*.

3.1.3. In Foods

Enterococci are used as starter and probiotic cultures in foods, and they occur as natural food contaminants *(8)*. They have been ascribed to both beneficial and detrimental roles in foods. In processed meats, enterococci can survive heat processing and cause spoilage, whereas in certain cheeses they contribute to ripening and development of flavor *(5)*. Because of their heat-resistance, enterococci are often the only surviving bacterium (other than spore-formers) in heat-treated, semipreserved, nonsterile foods such as processed meats *(9)*.

Some enterococci produce bacteriocins (called enterocins) that exert anti-*Listeria* activity *(5)* and are also able to inhibit or kill other enterococci, clostridia, bacilli, and staphylococci. On the other hand, enterococci can cause food intoxication through the production of biogenic amines and can serve as a reservoir for virulence traits and antimicrobial resistance *(10,11)*. *E. faecalis* and *E. faecium*, the strains most commonly associated with human infection, are also the strains most commonly found in foods or used as starter cultures.

3.1.4. Probiotics in Animal Feeds

Enterococci are used as probiotics to improve the microbial balance of the intestine and to treat gastroenteritis in humans and animals *(8)*. Their effects when used as probiotics are unclear, but studies have focused on growth-promoting of food animals and diminishing *E. coli* weaning diarrhea *(7)*. The identification of common antimicrobial resistance determinants and antimicrobial resistant enterococcal isolates in humans, food, and food-producing animals, has led to concern about the safety of use of certain enterococcal strains as probiotics *(5,12)*.

3.2. Human Reservoir

Enterococci make up approx 1% of the normal intestinal flora of humans *(13)*, being the predominant Gram-positive cocci in stool with concentrations ranging from 10^5 to 10^7 CFU/g in feces *(14)*. *E. faecalis* is normally found in the stool of 90–100% of animals and humans, whereas *E. faecium* is found in 25%. Small numbers of enterococci also occur in oropharyngeal secretions, the urogenital tract, and on the skin, especially in the perineal area *(2,4)*. The prevalence of the different species appears to vary according to the host and is also influenced by age, diet, underlying diseases, and prior antimicrobial therapy *(2,15)*. The species most commonly isolated from human clinical specimens is *E. faecalis*, followed by *E. feacium*. Infections with *E. casseliflavus* and *E. raffinosus* have also been reported *(2)*. Rarely, other species, including *E. avium*, *E. cecorum*, *E. dispar*, *E. durans*, *E. gilvus*, *E. gallinarum*, *E. hirae*, *E. mundtii*, and *E. pallens*, have also been isolated from human sources *(2)*. Recently, three new enterococcal species, provisionally designated as *Enterococcus* sp. nov. CDC PNS-E1, *Enterococcus* sp. nov. CDC PNS-E2, and *Enterococcus* sp. nov. CDC PNS-E3, were isolated from human blood and human brain tissue *(6)*.

Even though they are commensal organisms, enterococci can act as opportunistic pathogens, particularly in elderly patients with serious underlying diseases and in other immunocompromised patients who have been hospitalized for a prolonged period, use invasive devices, and/or have received broad-spectrum antibiotics *(2)*.

4. FOODBORNE OUTBREAKS

E. faecalis and *E. faecium* are suspected, but unconfirmed causative agents of foodborne illness in its classic sense. They were first associated with foodborne illness in 1926 when two outbreaks of gastroenteritis from cheese were reported *(16)*. Enterococci were implicated by their presence in large numbers in the incriminated foods, and the absence of other pathogens such as *S. aureus* or *Salmonella* spp. *(17)*. On the other hand, it is felt that enterococci can cause food intoxication through the production of biogenic amines, but both of these observations are yet to be confirmed *(10)*. Efforts to prove that enterococci cause foodborne illness, including animal experiments and studies using human volunteers, have yielded contradictory results *(16,17)*. Food intoxication caused by ingestion of biogenic amines determines a number of symptoms including headache, vomiting, increased blood pressure, and even allergic reactions *(10)*. The ability to produce biogenic amines in cheese and fermented sausages has been reported for bacteria of the genus *Enterococcus (10,18–20)*.

Although not considered as classical foodborne pathogens, several concerns exist regarding the presence of enterococci in foods. Enterococcal strains colonizing foods or used intentionally as starters have been repeatedly found to harbor virulence traits and drug-resistance genes and, thus, may not be nonpathogenic. These isolates can enter the human food chain and colonize the human intestine, or transfer virulence traits and antimicrobial-resistance genes to human enterococcal isolates or even to organisms from other genera. The extent to which this occurs is unsettled, and for this reason, enterococci that may be transmitted to humans through the food chain have raised considerable concern.

5. PATHOGENICITY (VIRULENCE FACTORS)

With the exception of *E. faecium* and *E. faecalis*, enterococci are rarely reported to be pathogens in humans *(5)*. In fact, until two decades ago, when they started emerging as important nosocomial pathogens, enterococci were considered rather innocuous. Previously regarded as endogenous pathogens acquired from the patient's own flora, in 1987, person–person transmission was first demonstrated *(31)*. Spread of enterococci among patients and even from an institution to the other has been documented *(31)*.

Enterococci have been associated with high mortality rates (12–68%) in patients with bacteremia *(4,21–31)*, but these studies have nonetheless failed to establish unequivocally the pathogenicity of the causative organism in this setting. In one large study of bloodstream infections, enterococci were the only Gram-positive pathogens independently associated with a high risk of death *(32)*. From the fact that most of the patients are severely debilitated and that enterococci are often part of a polymicrobic bacteremia, it has been difficult to determine the independent contribution of enterococci to morbidity and mortality in these patients.

Even though enterococci are not intrinsically as virulent as other organisms such as *Staphylococcus aureus* and *Streptococcus pyogenes* and lack classic virulence factors, their resistance to a variety of antibiotics is a cause for concern and a contributing factor to the pathogenesis of infection. This property allows them to survive and proliferate in patients receiving antimicrobial therapy, which leads to their ability to cause superinfections in these patients. In addition, enterococci have a high capacity to exchange genetic information, including antimicrobial resistance genes and virulence traits, among themselves and also with organisms of other genera.

Enterococci are able to adhere to left-sided heart valves and renal epithelial cells, properties which possibly enable them to cause endocarditis and urinary tract infections *(4)*. Although they are natural inhabitants of the gastrointestinal tract, they are not known to cause gastroenteritis in humans except for a possible strain of *E. hirae* isolated from a patient with diarrhea that was able to cause diarrhea in suckling rats *(4)*. In intra-abdominal and wound infections, it has been suggested that enterococci act synergistically with other bacteria, but their exact role in those settings still remains to be defined *(4)*.

Several definite and potential virulence factors have been identified in enterococci, but none has been established as having a major or consistent contribution to virulence in humans. The best described are cytolysin *(14,33,34)*, aggregation substance *(34–36)*, pheromones *(14,37)*, lipoteichoic acid *(38,39)*, gelatinase, a metalloendopeptidase *(40,41)*, hyaluronidase *(42)*, and AS-48, a bacteriocin with activity against Gram-positive and Gram-negative bacteria *(43)*. Cytolysin has lytic activity against Gram-positive bacteria and selected eukaryotic cells; in combination with aggregation substance, it was shown to increase mortality in a rabbit-model endocarditis *(14,34)*. Aggregation substance facilitates binding of donor to recipient cells in pheromone mating response, thus facilitating exchange of plasmids carrying virulence traits and antibiotic resistance genes *(14)*.

Other putative virulence factors include the adhesin, called enterococcal surface protein (Esp), and surface carbohydrates *(5,14,44)*.

Most virulence factors were described in *E. faecalis*, and many reside on conjugative plasmids that can spread with ease horizontally between the strains in a natural environment, such as the gastrointestinal tract *(14)*. Virulence traits are less well-characterized in *E. faecium*. As host immunosuppression increases, the requirements for particular virulence traits to enhance the likelihood or severity of the disease decrease; it is, therefore, possible that some fraction of enterococcal disease, often affecting severely debilitated patients, is attributable to ordinary commensal strains without any virulence traits *(45)*.

Enterococci are very much capable of exchanging genetic material between themselves and with other genera, and they generally do so by means of narrow-host pheromone-responsive plasmids, broad host range plasmids, and transposons *(14,46)*. Conjugation experiments have confirmed in vitro vancomycin resistance gene transfer from entero-cocci to *S. aureus (47)* and, more recently and raising serious concern, the same phenomenon occurred in vivo, leading to the emergence of the first three clinical vanco-mycin resistant *S. aureus* (VRSA) isolates *(48–51)*.

Virulence factors such as hemolysin, aggregation substance, Esp, and Gel can also be present in food enterococcal isolates *(8,44)*. In general, the incidence of these virulence traits was lower among *E. faecium* strains than among *E. faecalis* strains and *E. faecium* harbored fewer virulence traits than *E. faecalis (8,44)*. In a comparative study, *E. faecalis* strains causing human infection had more virulence determinants than did food strains, which in turn had more than did starter strains *(8)*. It was also demonstrated that starter strains, added intentionally to certain cheeses and other fermented milk products, can acquire known virulence genes by the natural-conjugation gene-transfer process *(8)*.

6. CLINICAL CHARACTERISTICS

Enterococci are associated with a variety of clinical infections. Considered organisms of low virulence, playing an unclear role in certain infections, especially when isolated in conjunction with more virulent pathogens, enterococci are also capable of causing serious and often life-threatening disease, most notably endocarditis. Most commonly, they have caused urinary tract, bloodstream (including endocarditis), wound, intra-abdominal, and pelvic infections. In addition, enterococci have been suspected, but remain unconfirmed, as causative agents of foodborne illness (subchapter 3). Enterococci have become the second most common agent recovered from nosocomial urinary tract infec-tions and wound infections, and the third leading cause of bacteremia in the United States *(1,2)*. Most clinical isolates of enterococci (80–90%) are *E. faecalis*, more susceptible to vancomycin and penicillins, although strains with high-level resistance to aminoglycosides have emerged, creating serious therapeutic problems for patients with endocarditis *(2,52)*. *E. faecium*, intrinsically more resistant to antibiotics, accounts for 5–10% of enterococcal infections; acquired resistance of this organism to vancomycin, penicillins, and aminoglycosides has rendered *E. faecium* infections very difficult to treat. Rarely, infections are resulting from *E. gallinarum, E. raffinosus, E. casseliflavus, E. avium, E. pseudoavium, E. malodoratus, E. mundtii, E. durans,* or *E. hirae (45)*.

6.1. Urinary Tract Infections

Urinary tract infections are the most common type of clinical disease produced by enterococci. These organisms have been implicated in cystitis, pyelonephritis, prostatitis, and perinephric abscesses. The infections are most often nosocomial, associated with instrumentation or urologic manipulation, and affect men much more commonly than women. In the absence of clinical and laboratory signs or symptoms of infection, isolation of enterococci from urine likely represents colonization of the urinary tract rather then true infection. Bacteremia is a relatively rare complication of enterococcal urinary tract infections *(4)*.

6.2. Bacteremia and Endocarditis

Portals of entry for enterococcal bacteremia include the urinary tract, intra-abdominal/pelvic infections, wounds, infected intravenous or intra-arterial catheters, or the biliary tree. A gastrointestinal source likely accounts for the remaining of the cases, in which a primary infection cannot be identified *(53)*. Unlike in the case of other bacteria, metastatic infections, septic shock, and disseminated intravascular coagulation rarely complicate pure enterococcal bacteremia *(4,52)*. In spite of this fact, mortality rates from enterococcal bacteremia have been reported to be between 12–75% *(52)*. It is likely that these high mortality rates reflect the patient population at risk for developing this complication—older adults with multiple comorbid conditions *(52)*. Nosocomial enterococcal bacteremia is often polymicrobial.

Only about 1 out of 50 cases of enterococcal bacteremia result in endocarditis and the latter is much more commonly seen with community-acquired than with nosocomial bacteremia *(4)*. Enterococci are estimated to account for about 20% of the cases of native-valve endocarditis and for about 6–7% of prosthetic-valve endocarditis *(2)*. The most common clinical picture is that of subacute bacterial endocarditis, with an acute presentation occurring less often. Currently, the greatest challenge in the management of enterococcal endocarditis stems from the increasing resistance of these organisms to multiple antimicrobials.

6.3. Intra-Abdominal, Pelvic, and Soft-Tissue Infections

Enterococci are frequently isolated from intra-abdominal, pelvic, and soft-tissue infections (surgical wounds, decubitus ulcers, and diabetic foot infections). In all these settings, they are usually part of a mixed aerobic and anaerobic flora, and therefore, their individual clinical significance has been difficult to assess. Enterococci can cause peritonitis in patients with nephrotic syndrome or cirrhosis and in patients on chronic ambulatory peritoneal dialysis, in which cases, infections can be monomicrobial.

6.4. Uncommon Infections

Respiratory tract infections owing to enterococci are exceedingly unusual *(4)*. Meningitis, otitis, sinusitis, septic arthritis, osteomelitis, and endophtalmitis, may occur, but are also rare and the significance of isolates from some of these sites should be carefully evaluated before clinical decisions are made *(2,4,52)*. Enterococci have clearly been documented to cause neonatal sepsis characterized by fever, lethargy, and respiratory difficulty accompanied by bacteremia, meningitis, or both *(4)*.

7. CHOICE OF TREATMENT

7.1. Antimicrobial Therapy

For susceptible isolates, penicillin or ampicillin is the treatment of choice. In general, treatment of enterococcal infections depends on whether the organism is susceptible in vitro to β-lactams, aminoglycosides, and glycopeptides, whether the infection is polymicrobial vs monomicrobial, and whether the endocarditis is present *(52)*.

Penicillin or ampicillin remains the antibiotic of choice for urinary-tract infections, peritonitis, and wound infections that do not require bactericidal treatment *(4)*. This appears to apply to bacteremia of noncardiac origin as well, in which the advantage of combination therapy over monotherapy has not been demonstrated *(52)*. Vancomycin (or teicoplanin) is the alternative agent in patients who are allergic to penicillin or for organisms (usually *E. faecium*) with high-level resistance to penicillin *(4)*. Nitrofurantoin and fosfomycin can also be used successfully for treatment of lower urinary-tract infections with enterococci, whereas the utility of fluoroquinolones is much less. Tetracycline and chloramphenicol also demonstrate in vitro activity against some strains of enterococci, but they are bacteriostatic against these organisms, and clinical failures with chloramphenicol have been documented *(4)*.

Skin and soft tissue infections, as well as intra-abdominal and pelvic infections, which are usually polymicrobic, should best be treated with a combination of ampicillin and antibiotics effective against a wide range of anaerobic and aerobic Gram-negative bacilli and staphylococci *(52)*.

β-Lactamase producing enterococci are rare. They remain susceptible to vancomycin (and teicoplanin) and to combinations of β-lactams and β-lactamase inhibitors *(4)*.

For endocarditis caused by susceptible organisms, a combination of a cell-wall active agent (usually penicillin, ampicillin, or vancomycin for penicillin-allergic patients) and an aminoglycoside in synergistic doses (usually gentamicin or streptomycin) remains the standard *(4)*. Therapy should be continued for 4–6 wk *(4,52)*.

Endocarditis caused by enterococci with high-level resistance to both gentamicin and streptomycin is very difficult to treat. Intravenous ampicillin by continuous infusion and extended for 8–12 wk has been recommended *(4)*, but this approach has not been uniformly successful. Early surgical intervention may also be considered.

The therapeutic armamentarium against vancomycin-resistant enterococci (VRE) has been expanded by the recent approval of new agents, like quinupristin/dalfopristin, a streptogramin combination active only against *E. faecium*; linezolid, an oxazolidinone, and daptomycin, a lipopeptide, both active against *E. faecium* and *E. faecalis*. There is, however, very limited experience with treatment of vancomycin-resistant enterococcal endocarditis in general, and with these new agents in particular. In addition, resistance to quinupristin/dalfopristin and linezolid has already been documented among enterococci *(52)*. As with enterococci with high-level resistance to aminoglycosides, early removal of the infected valve might need to be considered.

7.2. Adjunctive Therapy

Given the limitations of antimicrobial therapy, removal of infected foci, such as intravenous catheters, and drainage of abscesses are important adjunctive measures. Early valve replacement may be required in endocarditis caused by multiresistant enterococci.

Infection control measures are essential in limiting the spread of antimicrobial-resistant enterococci. Judicious use of antibiotics, education of health care personnel, use of the microbiology laboratory to quickly identify patients with VRE, and infection control measures that minimize the transmission to other patients are equally important *(52)*.

8. RESISTANCE EPIDEMIOLOGY

The emergence of antimicrobial resistance represents the greatest threat to the treatment of human enterococcal infections. Enterococci are intrinsically resistant to a number of antimicrobial agents normally used to treat infections resulting from Gram-positive organisms. In addition, they have a remarkable ability to acquire new mechanisms of resistance and to transfer resistance determinants, by way of conjugation, among themselves, and between themselves and other organisms.

8.1. Antimicrobial Resistance in Humans

All enterococci have intrinsic (species characteristic) resistance to β-lactam antimicrobial agents, with cephalosporins having much less activity (and no clinical utility against enterococci) compared to penicillins. Among *E. faecium*, there has been a dramatic increase in inherent resistance to penicillin and ampicillin *(4)*. Enterococci also exhibit low-level intrinsic resistance to aminoglycosides, a phenomenon that can be overcome by the addition of a cell-wall active agent resulting in synergistic killing, when treating serious infections. Although trimethoprim–sulfamethoxazole shows activity against enterococci in vitro, it has failed in the therapy of enterococcal infections *(4)*. Low-level intrinsic resistance to clindamycin and lincomycin is also the characteristic for enterococci.

In addition to their intrinsic resistance, enterococci have acquired new mechanisms of resistance to a wide variety of antimicrobial agents, such as β-lactams, aminoglycosides (high-level, MIC > 2000 µg/mL), fluoroquinolones, lincosamides (high-level), macrolides, rifampin, tetracyclines, glycopeptides, and most recently, oxazolidinone and streptogramins *(4,54)*. Most of these resistance mechanisms are mediated by genes encoded on plasmids or transposons, with the exception of linezolid, rifampin, and fluoroquinolones, which have chromosomal mutations at the bases of resistance. For streptogramins, the mechanism of resistance is in part plasmid-mediated and in part unknown *(46–48,53,55,86)*.

Clinical isolates of enterococci, in particular *E. faecium*, have become increasingly resistant to ampicillin. In some hospitals, over 90% of the *E. faecium* isolates are resistant to ampicillin with MIC $\geq$ 32 µg/mL *(54)*. High-level resistance to penicillins is owing predominantly either to the overproduction of a penicillin-binding protein (PBP) with low affinity for penicillins or to mutations that make the low-affinity PBP even less susceptible to inhibition by penicillins *(54)*.

Some, although rare, *E. faecalis* strains produce β-lactamase; importantly, these strains may appear to be susceptible to penicillins in vitro when standard MIC testing is done because they produce the enzyme in small amounts. Therefore, a test for β-lactamase production needs to be used in order to identify these strains. These strains caused outbreaks in the 1980s, but have not been seen in the last 15 yr.

Enterococci with high-level resistance to streptomycin and kanamycin have been relatively common, but high-level resistance to gentamicin has become a clinical problem only since the 1980s *(4)*. Clinical failures and relapses after therapy in patients with endocarditis resulting from *E. faecalis* and *E. faecium* with high-level resistance to all aminoglycosides are increasingly encountered. Some isolates with high-level gentamicin resistance are susceptible in vitro to streptomycin, which can be used in combination with a cell-wall active agent for therapy of endocarditis.

In the late 1980s, VRE were first described, first in Europe, and then in the United States *(55)*. Six types of glycopeptide resistance have already been identified, with three of them being the most common: the VanA phenotype, associated with inducible high-level resistance to both vancomycin and teicoplanin; the VanB phenotype, associated with moderate to high levels of inducible resistance to vancomycin only; and the VanC phenotype, associated with low-level noninducible resistance to vancomycin *(2)*. The VanA and VanB phenotypes are considered the most clinically relevant and are usually associated with *E. faecalis* and *E. faecium* strains. Strains of VRE are often multidrug resistant. Ampicillin and vancomycin resistance are associated with *E. faecium* far more commonly than with other species *(2)*.

The percentage of nosocomial infections caused by VRE increased more than 20-fold during 1989–1993, from 0.3 to 7.9%, and the trend has continued since then, with rates now approaching 20% of all enterococcal isolates *(4,45)*. In a comparison of National Nosocomial Infection Surveillance (NNIS) pathogens from 1994 to 1998 and January–May 1999, there was a 47% increase in VRE *(56)*. In 1997, data obtained from more than 100 clinical laboratories showed that 52% of *E. faecium* and 1.9% of *E. faecalis* isolates were resistant to vancomycin *(4)*.

There is clear evidence for the spread of enterococcal strains between patients, and even the dissemination of such strains from one institution to the other. It still appears that resistant organisms from patients or from the transient carriage on the hands of hospital personnel or environmental surfaces first colonize the gastrointestinal tract (and occasionally the skin) before causing infections in patients *(4,57,58)*. Once colonization occurs, it may persist for months or even years *(4)*.

A multitude of risk factors have been associated with antibiotic-resistant enterococcal colonization or infection. In the acute care setting, colonization with aminoglycoside-resistant enterococci was shown to be associated with intravenous catheters, bladder catheters, prior surgical procedures, and prior antibiotic therapy *(59)*. In the long-term care setting, colonization with high-level aminoglycoside-resistant enterococci was associated with poor functional status, older age, and colonization with methicillin-resistant *S. aureus (60)*. Prior antibiotic therapy with vancomycin, third-generation cephalosporins, and/or agents with anti-anaerobic activity has been noted as a risk factor for colonization or infection with VRE *(61–68)*, as were increased hospital or intensive care unit stay, exposure to other patients with VRE, chronic hemodialysis, extended care facility stay, pressure ulcers, hematological malignancy, and receipt of bone marrow, stem cell or solid organ transplant *(52,66,67,69–76)*. The risk factors and epidemiology of vancomycin-resistant *E. faecalis*, recently analyzed *(67)*, are largely similar to those previously described of vancomycin-resistant *E. faecium*.

Resistance to the most recently introduced anti-Gram-positive agents, oxazolidinones and streptogramins has already emerged among enterococci *(77–79)*. Resistance to quinupristin/dalfopristin, although rare in clinical isolates of *E. faecium*, is very common in isolates recovered from food animals fed the structurally related virginiamycin *(80–82)*. In the case of the most recently approved daptomycin, spontaneous acquisition of resistance in vitro is rare, and it is hoped that this will extrapolate into the clinical setting *(83)*.

It is evident that, with an increase in the consumption of antimicrobial agents by humans and animals, there is a resultant increase in antimicrobial resistance *(12)*.

8.2. Antimicrobial Resistance in Nonhuman Reservoirs

8.2.1. Food Animals

Antimicrobial agents are used in animals for therapy, for prophylaxis of infection during times of stress, such as early weaning, for metaphylactics (treatment of clinically healthy animals belonging to the same flock/pen as sick animals), or for growth promotion *(7)*. The latter usage has been seriously questioned in recent years *(7)*. Often, the same classes of antimicrobials are used for treatment or growth promotion in food animals and for treatment of human infections. Unfortunately, the use of antimicrobial agents in food animals has created a large reservoir of antimicrobial-resistant enterococci and resistance genes, the dissemination of which to humans, may pose a significant threat to human health.

In Europe the rise of VRE was linked to the use of avoparcin (a glycopeptide related to vancomycin) as a growth promoter in food animals. In Europe, VRE was found throughout the community, the food supply, and urban and rural sewage systems. In contrast, in the United States, where avoparcin was never approved for use in animals, VRE is rampant in hospitals, where vancomycin is routinely used, but is rare in the community.

Virginiamycin (a streptogramin) has been used for over two decades in poultry and swine feed in the United States. Concerns exist that the agricultural use of virginiamycin has generated enterococci resistant to streptogramins. Resistance to quinupristin–dalfopristin—a recently introduced streptogramin combination used to treat human infections—has already been documented in clinical isolates from patients treated with this antimicrobial *(84,85)* and is common in isolates recovered from food animals *(85)*. More than half of grocery-store poultry in the United States carry quinupristin–dalfopristin-resistant enterococci *(80,86)* and at least 1% of the US population is colonized with this organism in the intestinal tract *(86,87)*. Rates of *E. faecium* resistance to quinupristin–dalfopristin in stool cultures as high as 7.4% were reported among patients in German hospitals *(88)*.

8.2.2. Foods

Foods, other than meat, have also been found to harbor resistant enterococci. A study carried out in Argentina found high-level resistance to streptomycin and gentamicin in *E. faecium*, and resistance to streptomycin alone in *E. faecalis* from farm lettuce *(89)*. In another study of 24 enterococcal strains isolated from traditional Italian cheeses, one *E. faecium* strain showed *vanA* vancomycin resistance genotype while four strains showed a β-haemolytic reaction on human blood *(11)*. As previously discussed, quinupristin–dalfopristin-resistant *E. faecium* strains are commonly isolated from food animals.

A high prevalence of enterococci was also found in low-microbial load diets processed for safety for patients with hematological malignancy—low-level vancomycin resistance and high-level streptomycin resistance were found in some of the strains, indicating a possible route for the acquisition of antimicrobial-resistant strains by some of the most vulnerable hospital patients *(90)*.

8.3. Transfer of Enterococci and Antimicrobial-Resistant Genes and Virulence Determinants Between Nonhuman Reservoirs and Humans

After the use of avoparcin in livestock as a growth promoter in Europe, introduced in 1974, animals started developing VRE. In the late 1980s, VRE occurrence emerged in humans. Epidemiologic studies examining glycopeptide use in animal husbandry have provided evidence that *vanA*-mediated VRE is now ubiquitous in European communities, the organism readily colonizing intestinal tracts of animals for which avoparcin was used as a feed supplement *(45,91–94)*. Subsequent enteric colonization of humans has been documented *(45,95)*. In contrast, in the United States VRE is largely confined to hospitals, where it is likely thriving under selective pressure from widespread vancomycin use *(61)*.

In the United States, the use of virginiamycin in animal food and the emergence of quinupristin–dalfopristin resistance mirror the avoparcin–VRE situation in Europe. Virginiamycin is used worldwide and has been approved in the United States for use in chickens, turkeys, swine, and cattle, mainly for growth promotion. Virginiamycin has been banned in Europe since 1999, pending further studies on the possible relationship of its use in food animals and resistance to streptogramins used clinically to treat infections in humans. Although rare, resistance to quinupristin/dalfopristin has been documented in human *E. faecium* isolates. Of concern, emergence of resistance during therapy was observed and resulted in clinical failure *(84)*. A study of different ecological sources in Germany discovered rates of resistance to quinupristin/dalfopristin of up to 100% among *E. faecium* isolates from sewage, poultry manure, and pig manure, whereas rates of resistance among broiler chickens, pork meat, human stool samples, and hospitalized patients were 46, 10, 14, and 7.4%, respectively *(88)*. A Centers for Disease Control and Prevention (CDC) multistate surveillance study found a 1–2% rate of resistance to quinupristin/dalfopristin in *E. faecium* isolated from human stool samples *(81)*. Data suggest that quinupristin/dalfopristin resistance is more common in farms that use virginiamycin *(80)*. The rate of virginiamycin resistance decreased after its use was banned in Denmark in 1998 *(96)*. Similarly, the avoparcin ban in 1995 resulted in a significant decrease in the rate of glycopeptide resistance in *E. faecium* from broiler chickens *(96)*.

Gentamicin is commonly used in swine and widely used in chickens and turkeys. A study by Donabedian et al. *(12)* found that when gentamicin-resistant genes are present in the resistant enterococci isolated from animals, they are also present in the enterococci isolated from food products of the same animal species. Molecular analysis of gentamicin-resistant enterococcal strains from humans, food, and farm animals in their study provided evidence of the spread of such isolates from animals to humans over a broad geographical area through the food supply *(12)*.

Molecular screening for virulence determinants in clinical enterococcal isolates, starter strains, and strains used as probiotic cultures in foods, demonstrated the presence of multiple virulence determinants in clinical *E. faecalis*, more so than in food strains, which, in turn had more virulence determinants than did starter strains *(8)*. Of concern, transconjugation in which starter strains acquired additional virulence determinants from medical strains was demonstrated *(8)*. Thus, there is a risk that a safe strain that lacks genes for known virulence factors could acquire such genes by conjugation, once it enters the human gastrointestinal tract.

9. SUMMARY AND CONCLUSIONS

Although part of the normal intestinal flora and once felt to be innocuous, enterococci have proven to have much more complex interactions with the human host, having emerged in recent years as important nosocomial pathogens. Strains with resistance to multiple antimicrobials are on the rise, posing significant therapeutic and epidemiological challenges. Virulence factors have also been described for these organisms traditionally regarded as having low pathogenicity.

The role of enterococci in disease raises valid concerns regarding their safety for use in foods or as probiotics. Should *Enterococcus* strains be selected for use as starter or probiotic cultures, ideally such strains should harbor no virulence determinants and should be sensitive to clinically relevant antibiotics.

In 1994, as the imminent crisis of emergent antimicrobial resistance became more and more recognized, the WHO recommended immediate and drastic reductions in the use of antibiotics in animals, plants, and fish, as well as in human medicine *(97,98)*. Such reductions however, even if they occur, are unlikely, alone, to be efficient in curtailing the emergence of resistant bacteria. In order to achieve this goal, a more comprehensive, multidisciplinary effort is needed, including a better understanding of the epidemiology and pathogenicity of these micro-organisms, judicious use of antimicrobials, effective infection control measures in hospitals, and reduction of resistance in reservoirs such as the environment and animal husbandry.

REFERENCES

1. Jones, R. N., Marshall, S. A., Pfaller, M. A., et al. (1997) Nosocomial enterococcal blood stream infections in the SCOPE Program: antimicrobial resistance, species occurrence, molecular testing results, and laboratory testing accuracy. SCOPE Hospital Study Group. *Diagn. Microbiol. Infect. Di*s. **29,** 95–102.
2. Teixeira, L. M. and Facklam, R. R. (2003) *Enterococcus.* In: *Manual of Clinical Microbiology* (Murray, P. R., Baron, E. J., Jorgensen, J. H., Pfaller, M. A., and Yolker, H. Y., eds.), 8th edn, ASM, Washington, DC, pp. 422–429.
3. Facklam, R. R., Carvalho, M. G. S., and Teixeira, L. M. (2002) History, taxonomy, biochemical characteristics, and antibiotic susceptibility testing of enterococci. In: *The Enterococci. Pathogenesis, Molecular Biology, and Antibiotic Resistance* (Gilmore, M. S., ed.), ASM, Washington, DC, pp. 1–54.
4. Moellering, R. C. Jr. (2000) *Enterococcus* species, *Streptococcus bovis,* and *Leuconostoc* species. In: *Mandell's, Bennett's and Dolin's Principles and Practice of Infectious Diseases* (Mandell, D., Bennett, J. E., and Dolin, R., eds.), 5th edn, Churchill Livingstone, Philadelphia, PA, pp. 2147–2152.
5. Franz, C. M., Holzapfel, W. H., and Stiles, M. E. (1999) Enterococci at the crossroads of food safety? *Int. J. Food Microbiol.* **47,** 1–24.
6. Carvalho, M. D. G., Steigerwalt, A. G., Morey, R. E., Shewmaker, P. L., Teixeira, L. M., and Facklam, R. R. (2004) Characterization of three new enterococcal species, *Enterococcus* sp. Nov. CDC PNS-E1, *Enterococcus* sp. Nov. CDC PNS-E2, and *Enterococcus* sp. Nov. CDC PNS-E3, isolated from human clinical specimens. *J. Clin. Microbiol.* **42,** 1192–1198.
7. Aarestrup, F. M., Butaye, P., and Witte, W. (2002) Nonhuman reservoirs of enterococci. In: *The Enterococci. Pathogenesis, Molecular Biology, and Antibiotic Resistance* (Gilmore, M. S., ed.), ASM, Washington, DC, pp. 55–99.
8. Eaton, T. J. and Gasson, M. J. (2001) Molecular screening of *Enterococcus* virulence determinants and potential for genetic exchange between food and medical isolates. *Appl. Environ. Microbiol.* **67(4),** 1628–1635.
9. International Commission on Microbiological Specifications for Foods. (1978) *Microorganisms in foods. 1. Their significance and methods of enumeration*, 2nd edn. University of Toronto Press, Toronto.
10. Giraffa, G. (2002) Enterococci form foods. *FEMS Microbiol. Rev.* **26(2),** 163–171.
11. Andrighetto, C., Knijff, E., Lombardi, A., et al. (2001) Phenotypic and genetic diversity of enterococci isolated from Italian cheeses. *J. Diary Res.* **68(2),** 303–316.

12. Donabedian, S. M., Thal, L. A., Hershberger, E., et al. (2003) Molecular characterization of gentamycin-resistant *Enterococci* in the United States: evidence of spread form animals to humans through food. *J. Clin. Microbiol.* **41**, 1109–1113.

13. Tannock, G. W. and Cook, G. (2002) Enterococci as members of the intestinal microflora of humans. In: *The Enterococci. Pathogenesis, Molecular Biology, and Antibiotic Resistance* (Gilmore, M. S., ed.), ASM, Washington, DC, pp. 101–132.

14. Jett, B. D., Huycke, M. M., and Gilmore, M. S. (1994) Virulence of enterococci. *Clin. Microbiol. Rev.* **7**, 462–478.

15. Gilmore, M. S. and Ferretti, J. J. (2003) The thin line between gut commensal and pathogen. *Science* **299**, 1999, 2001–2002.

16. Stiles, M. E. (1989) Less recognized or presumptive foodborne pathogenic bacteria. In: *Foodborne Bacterial Pathogens* (Doyle, M. P., ed.), Marcel Dekker, New York, pp. 674–735.

17. Riemann, H. and Bryan, F. L. (eds.) (1979) *Foodborne Infections and Intoxications*, 2nd edn, Academic, New York.

18. Giraffa, G., Carminati, D., and Neviani, E. (1997) Enterococci isolated from diary products: a review of risks and potential technological use. *J. Food Prot.* **60**, 732–738.

19. Giraffa, G., Pepe, G., Locci, F., Neviani, E., and Carminati, D. (1995) Hemolytic activity, production of thermonuclease and biogenic amines by dairy enterococci. *Ital. J. Food Sci.* **7**, 341–349.

20. Bover Cid, S., Hugas, M., Izquierdo-Pulido, M., and Vidal-Carou, M. C. (2001) Amino acid-decarboxylase activity of bacteria isolated from fermented pork sausages. *Int. J. Food Microbiol.* **66**, 185–189.

21. Boulanger, J. M., Ford-Jones, E. L., and Matlow, A. G. (1991) Enterococcal bacteremia in a pediatric institution: a four-year review. *Rev. Infect Dis.* **13**, 847–856.

22. Bryan, C. S., Reynolds, K. L., and Brown, J. J. (1985) Mortality associated with enterococcal bacteremia. *Surg. Gynecol. Obstet.* **160**, 557–561.

23. Graninger, W. and Ragette, R. (1992) Nosocmial bacteremia due to *Enterococcus faecalis* without endocarditis. *Clin. Infect. Dis.* **15**, 49–57.

24. Gullberg, R. M., Homann, S. R., and Phair, J. P. (1989) Enterococcal bacteremia: analysis of 75 episodes. *Rev. Infect. Dis.* **11**, 74–85.

25. Jones, W. G., Barie, P. S., Yurt, R. W., and Goodwin, C. W. (1986) Enterococcal burn sepsis: a highly lethal complication in severely burned patients. *Arch. Surg.* **121**, 649–652.

26. Landry, S. L., Kaiser, D. L., and Wenzel, R. P. (1989) Hospital stay and mortality attributed to nosocomial enterococcal bacteremia: a controlled study. *Am. J. Infect. Control* **17**, 323–329.

27. Malone, D. A., Wagner, R. A., Myers, J. P., and Watanakunakorn, C. (1986) Enterococcal bacteremia in two large community teaching hospitals. *Am. J. Med.* **81**, 601–606.

28. Rimailho, A., Lampl, E., Riou, B., Richard, C. Rottman, E., and Auzepy, P. (1988) Enterococcal bacteremia in an intensive care unit. *Crit. Care Med.* **16**, 126–129.

29. Schlaes, D. M., Levy, J., and Wolinsky, E. (1981) Enterococcal bacteremia without endocarditis. *Arch. Intern. Med.* **141**, 578–581.

30. Zervos, M. J., Kauffman, C. A., Tharasse, P. M., Bergman, A. G., Mikesell, T. S., and Schaberg, D. R. (1987) Nosocomial infection by gentamicin-resistant *Streptococcus faecalis:* an epidemiologic study. *Ann. Intern. Med.* **106**, 687–691.

31. Vergis, E. N., Hayden, M. K., Chow, J. W., et al. (2001) Determinants of vancomycin resistance and mortality rates in enterococcal bacteremia: a prospective multicenter study. *Ann. Intern. Med.* **135**, 484–492.

32. Weinstein, M. P., Murphy, J. R., Reller, L. B., and Lichtenstein, K. A. (1983) The clinical significance of positive blood cultures: a comprehensive analysis of 500 episodes of bacteremia and fungemia in adults. II. Clinical observations, with reference to factors influencing prognosis. *Rev. Infect. Dis.* **5**, 54–69.

33. Jett, B. D., Jensen, H. G., Nordquist, R. E., and Gilmore, M. S. (1992) Contribution of the pAD1-encoded cytollysin to the severity of experimental *Enterococcus faecalis* endophtalmitis. *Infect. Immun.* **60**, 2445–2452.

34. Chow, J. W., Thal, L. A., Perri, M. B., et al. (1993) Plasmid-associated hemolysin and aggregation substance production contributes to virulence in experimental enterococcal endocarditis. *Antimicrob. Agents Chemother.* **37**, 2474–2477.

35. Galli, D. and Wirth, R. (1991) Comparative analysis of *Enterococcus faecalis* sex pheromone plasmids identifies a single homologous DNA region which codes for aggregation substance. *J. Bacteriol.* **173**, 3029–3033.

36. Kreft, B., Marre, R., Schramm, U., and Wirth, R. (1992) Aggregation substance of *Enterococcus faecalis* mediates adhesion to cultured renal tubular cells. *Infect. Immun.* **60**, 25–30.

37. Ember, J. A. and Hugli, T. E. (1989) Characterization of the human neutrophil response to sex pheromones from *Streptococcus faecalis*. *Am. J. Pathol.* **134**, 797–805.

38. Bhakdi, S., Klonisch, T., Nuber, P., and Fischer, W. (1991) Stimulation of monokine production by lipoteichoic acids. *Infect. Immun.* **59**, 4614–4620.

39. Ehrenfeld, E. E., Kessler, R. E., and Clewell, D. B. (1986) Identification of pheromone-induced surface proteins in *Streptococcus faecalis* and evidence of a role for lipoteichoic acid in formation of mating aggregates. *J. Bacteriol.* **168**, 6–12.

40. Mäkinen, P., Clewell, D. B., An, F., and Mäkinen, K. K. (1989) Purification and substrate specificity of a strongly hydrophobic extracellular metalloendopeptidase ("gelatinase") from *Streptococcus faecalis* (strain OG1-10). *J. Biol. Chem.* **264**, 3325–3334.

41. Su, Y. A., Sulavik, M. C., He, P., et al. (1991) Nucleotide sequence of the gelatinase gene (*gelE*) from *Enterococcus faecalis* aubsp. *Liquefaciens*. *Infect Immun.* **59**, 415–420.

42. Rosan, B. and Williams, N. B. (1964) Hyaluronidase production by oral enterococci. *Arch. Oral Biol.* **9**, 291–298.

43. Gálvez, A., Maqueda, M., Martinez-Bueno, M., and Valdivia, E. (1991) Permeation of bacterial cells, permeation of cytoplasmic and artificial membrane vesicles, and channel formation on lipid bilayers of peptide antibiotic AS-48. *J. Bacteriol.* **173**, 886–892.

44. Franz, C. M. A. P., Muscholl-Silberhorn, A. B., Yousif, N. M. K., Vancanneyt, M., Swings, J., and Holzapfel, W. H. (2001) Incidence of virulence factors and antibiotic resistance among *Enterococci* isolated from food. *Appl. Environ. Microbiol.* **67(9)**, 4385–4389.

45. Mundy, L. M., Sahm, D. F., and Gilmore, M. (2000) Relationship between enterococcal virulence and antimicrobial resistance. *Clin. Microbiol. Rev.* **13**, 513–522.

46. Clewell, D. B. (1990) Movable genetic elements and antibiotic resistance in enterococci. *Eur. J. Clin. Microbiol. Infect.* **9**, 90–102.

47. Noble, W., Virani, Z., and Crec, R. (1992) Co-transfer of vancomycin and other resistance genes from *Enterococcus faecalis* NCTC 12201 to *Staphylococcus aureus*. *FEMS Microbiol. Lett.* **93**, 195–198.

48. Chang, S., Sievert, D. M., Hageman, J. C., et al. (2003) Infection with vancomycin-resistant *Staphylococcus aureus* containing the *vanA* resistance gene. *N. Engl. J. Med.* **348**, 1342–1347.

49. Tenover, F. C., Weigel, L. M., Appelbaum, P. C., et al. (2004) Vancomycin-resistant *Staphylococcus aureus* isolate from a patient in Pennsylvania. *Antimicrob. Agents Chemother.* **48**, 275–280.

50. Franz, C. M., Holzapfel, W. H., and Stiles, M. E. (1999) Enterococci at the crossroads of food safety? *Int. J. Food Microbiol.* **47**, 1–24.

51. MMWR. (2004) Vancomycin-resistant *Staphylococcus aureus* New York—2004. *Morb. Mortal Wkly Rep.* **53**, 322.

52. Flannagan, S. E., Chow, J. W., Donabedian, S. M., et al. (2003) *Antimicrob. Agents Chemother.* **47**, 3954–3959.

53. Malani, P. N., Kauffman, C. A., and Zervos, M. J. (2002) Enterococcal disease, epidemiology, and treatment. In: *The Enterococci. Pathogenesis, Molecular Biology, and Antibiotic Resistance* (Gilmore, M. S., ed.), ASM, Washington, DC, pp. 385–408.

54. Wells, C. L., Maddaus, M. A., and Simmons, R. L. (1988) Proposed mechanisms for the translocation of intestinal bacteria. *Rev. Infect. Dis.* **10,** 958–978.

55. Kak, V. and Chow, J. W. (2002) Acquired antibiotic resistances in *Enterococci*. In: *The Enterococci. Pathogenesis, Molecular Biology, and Antibiotic Resistance* (Gilmore, M. S., ed.), ASM, Washington, DC, pp. 355–383.

56. Leclercq, R., Derlot, E., Duval, J., and Courvalin, P. (1988) Plasmid-mediated resistance to vancomycin and teicoplanin in *Enterococcus faecium*. *N. Engl. J. Med.* **319,** 157–161.

57. Hospital Infections Program. (1999) National Nosocomial Infections Surveillance (NNIS) System report, data summary from January 1990–May 1999, issued June 1999. *Am. J. Infect. Control.* **27,** 520–532.

58. Wells, C. L., Juni, B. A., Cameron, S. B., et al. (1994) Stool carriage, clinical isolation, and mortality during an outbreak of vancomycin-resistant enterococci in hospitalized medical and/or surgical patients. *Clin. Infect. Dis.* **21,** 45–50.

59. Beezhold, D. W., Slaughter, S., Hayden, M. K., et al. (1997) Skin colonization with vancomycin-resistant enterococci among hospitalized patients with bacteremia. *Clin. Infect. Dis.* **24,** 704–706.

60. Zervos, M. J., Terpenning, M. S., Schaberg, D. R., Therasse, P. M., Medendorp, S. V., and Kauffman, C. A. (1987) High-level aminoglycoside-resistant enterococci: colonization of nursing home and acute care hospital patients. *Arch. Intern. Med.* **147,** 1591–1594.

61. Chenoweth, C. E., Bradley, S. F., Trepenning, M. S., et al. (1994) Colonization and transmission of high-level gentamicin-resistant enterococci in a long-term care facility. *Infect. Control Hosp. Epidemiol.* **15,** 703–709.

62. Bonten, M. J., Slaughter, S., Ambergen, A. W., et al. (1998) The role of "colonization pressure" in the spread of vancomycin-resistant enterococci: an important infection control variable. *Arch. Intern. Med.* **158,** 1127–1132.

63. Carmeli, Y., Samore, M. H., and Huskins, C. (1999) The association between antecedent vancomycin treatment and hospital-acquired vancomycin-resistant enterococci. *Arch. Intern. Med.* **159,** 2461–2468.

64. Karanfil, L. V., Murphy, M., Josephson, A., et al. (1992) A cluster of vancomycin-resistant *Enterococcus faecium* in an intensive care unit. *Infect. Control Hosp. Epidemiol.* **13,** 195–200.

65. Morris, J. G., Jr., Shay, D. K., Hebden, J. N., et al. (1995) Enterococci resistant to multiple antimicrobial agents, including vancomycin. *Arch. Intern. Med.* **123,** 250–259.

66. Ostrowsky, B. E., Venkataraman, L., D'Agata, E. M., Gold, H. S., DeGirolami, P. C., and Samore, M. H. (1999) Vancomycin-resistant enterococci in intensive care units. *Arch. Intern. Med.* **159,** 1467–1472.

67. Tornieporth, N. G., Roberts, R. B., John, J., Hafner, A., and Riley, L. W. (1996) Risk factors associated with vancomycin-resistant *Enterococcus faecium* infection or colonization in 145 matched case-patients and control patients. *Clin. Infect. Dis.* **23,** 767–772.

68. Oprea, S. F., Zaidi, N., Donabedian, S. M., Balasubramaniam, M., Hershberger, E., and Zervos, M. J. (2004) Molecular and clinical epidemiology of vancomycin-resistant *Enterococcus faecalis*. *J. Antimicrob. Chemother.* **53,** 626–630.

69. Donskey, C. J., Chowdhry, T. K., Hecker, M. T., et al. (2000) Effect of antibiotic therapy on the density of vancomycin-resistant enterococci in the stool of colonized patients. *N. Engl. J. Med.* **343,** 1925–1932.

70. Loeb, M., Salama, S., Armstrong-Evans, M., Capretta, G., and Olde, J. (1999) A case–control study to detect modifiable risk factors for colonization with vancomycin-resistant enterococci. *Infect. Control Hosp. Epidemiol.* **20,** 760–763.

71. Boyce, J. M., Opal, S. M., Chow, J. W., et al. (1994) Outbreak of multi-drug resistant *Enterococcus faecium* with transferable VanB class vancomycin resistance. *J. Clin. Microbiol.* **32,** 1148–1153.

72. Mayhall, C. G. (1999) The epidemiology and control of VRE: still struggling to come of age. *Infect. Control Hosp. Epidemiol.* **20,** 650–652.

73. D'Agata, E. M. C., Green, W. K., Schulman, G., Li, H., Tang, Y.-W., and Schaffner, W. (2001) Vancomycin-resistant enterococci among chronic hemodialysis patients: a prospective study of acquisition. *Clin. Infect. Dis.* **32,** 23–29.

74. Montecalvo, M. A., Shay, D. K., Patel, P., Tacsa, L., Maloney, S. A., and Jarvis, W. R. (1996) Bloodstream infections with vancomycin-resistant enterococci. *Arch. Intern. Med.* **156,** 1458.

75. Lautenbach, E., Bilker, W. B., and Brennan, P. J. (1999) Enterococcal bacteremia: risk factors for vancomycin resistance and predictors of mortality. *Infect. Control Hosp. Epidemiol.* **20,** 318–323.

76. Linden, P. K., Pasculle, A. W., Manez, R., et al. (1996) Differences in outcomes for patients with bacteremia due to vancomycin-resistant *Enterococcus feacium* or vancomycin-sensitive *E. faecium. Clin. Infect. Dis.* **22,** 663–670.

77. Montecalvo, M. A., Horowitz, H., Gedris, C., Carbonaro, C., Teneover, F. C., and Isaah, A. (1994) Outbreak of vancomycin-, ampicillin-, and aminoglycoside-resistant *Enterococcus feacium* bacteremia in an adult oncology unit. *Antimicrob. Agents Chemother.* **38,** 1363.

78. Gonzales, R.D., Schreckenberger, P. C., Graham, M. B., Kelkar, S., DenBeste, K., and Quinn, J. P. (2001) Infections due to vancomycin-resistant *Enterococcus feacium* resistant to linezolid. *Lancet.* **357,** 1179.

79. Chow, J. W., Donabedian, S. M., and Zervos, M. J. (1997) Emergence of increased resistance to quinupristin/dalfopristin during therapy for *Enterococcus feacium* bacteremia. *Clin. Infect. Dis.* **24,** 90–91.

80. Dowzicky, M., Talbot, G. H., Feger, C., Prokocimer, P., Etienne, J., and Leclercq, R. (2000) Characterization of isolates associated with emerging resistance to quinupristin/dalfopristin (Synercid) during a worldwide clinical program. *Diagn. Microbiol. Infect. Dis.* **37,** 57–62.

81. Hershberger, E., Donabedian, S., Konstantinou, K., and Zervos, M. J., (2004) Quinupristin–dalfopristin resistance in Gram-positive bacteria: mechanism of resistance and epidemiology. *Clin. Infect. Dis.* **38,** 92–98.

82. McDonald, L. C., Rossiter, S., Mackinson, C., et al. (2001) Quinupristin–dalfopristin-resistant *Enterococcus feacium* on chicken and in human stool specimens. *N. Engl. J. Med.* **345,** 1155–1160.

83. Simjee, S., White, D. G., Meng, J., et al. (2002) Prevalence of streptogramin resistance genes among *Enterococcus* isolates from retail meats in the greater Washington DC area. *J. Antimicrob. Chemother.* **50,** 877–882.

84. Carpenter, C. F. and Chambers, H. F. (2004) Daptomycin: another novel agent for treating infections due to drug-resistant Gram-positive pathogens. *Clin. Infect. Dis.* **38,** 994–1000.

85. Moellering, R. C., Linden, P. K., Reinhardt, J., Blumberg, E. A., Bompart, F., and Talbot, G. H. (1999) The efficacy and safety of quinupristin–dalfopristin for the treatment of infections caused by vancomycin-resistant *Enterococcus faecium.* Synercid Emergency Use Study Group. *J. Antimicrob. Chemother.* **44,** 251–261.

86. Hershberger, E., Donabedian, S., Konstantinou, K., and Zervos, M. J. (2003) Quinupristin–dalfopristin resistance in Gram-positive bacteria: mechanism of resistance and epidemiology. *Clin. Infect. Dis.* **38,** 92–98.

87. Drexler, M. (2002) *Secret Agents: The Menace of Emerging Infections.* Joseph Henry, Washington, DC.

88. McDonald, L. C., Rossiter, S., Mackinson, C., et al. (2001) Quinupristin–dalfopristin-resistant *Enterococcus faecium* on chicken and in human stool specimens. *N. Engl. J. Med.* **345,** 1155–1160.

89. Werner, G., Klare, I., Heier, H., et al. (2000) Quinupristin/dalfopristin-resistant enterococci of the satA (vatD) and satG (vatE) genotypes from different ecological origins in Germany. *Microb. Drug Resist.* **6,** 37–47.

90. Ronconi, M. C., Merino, L. A., and Fernandez, G. (2002) Detection of *Enterococcus* with high-level aminoglycoside and glycopeptide resistance in *Lactuca sativa* (lettuce). *Enfermed. Infec. Microbiol. Clin.* **20,** 380–383 (in Spanish).
91. Curtis, G. D. and Bowler, I. C. (2001) Prevalence of glycopeptide and aminoglycoside resistance in *Enterococcus* and *Listeri* spp. in low microbial load diets of neutropenic hospital patients. *Int. J. Food Microbiol.* **64,** 41–49.
92. Aarestrup, F. M. (1995) Occurrence of glycopeptide resistance among *Enterococcus faecium* isolates from conventional and ecological poultry farms. *Microb. Drug Res.* **1,** 255–257.
93. Bates, J., Jordens, J., and Griffiths, D. (1994) Farm animals as a putative reservoir for vancomycin-resistant enterococcal infections in man. *J. Antimicrob. Chemother.* **34,** 507–516.
94. Klare, I., Heier, H, Claus, H., and Witte, W. (1993) Environmental strains of *Enterococcus faecium* with inducible high-level resistance to glycopeptides. *FEMS Microbiol. Lett.* **80,** 23–29.
95. Klare, I., Heier, H, Claus, H., Reissbrodt, R., and Witte, W. (1995) *vanA*-mediated high-level glycopeptide resistance in *Enterococcus faecium* from animal husbandry. *FEMS Microbiol. Lett.* **125,** 165–171.
96. Bates, J. (1997) Epidemiology of vancomycin-resistant enterococcus in the community and the relevance of farm animals to human infection. *J. Antimicrob. Chemother.* **37,** 89–101.
97. Aarestrup, F. M., Seyfarth, A. M., Emborg, H. D., Pedersen, K., Hendriksen, R. S., and Bager, F. (2001) Effect of abolishment of the use of antimicrobial agents for growth promotion on occurrence of antimicrobial resistance in fecal enterococci from food animals in Denmark. *Antimicrob. Agents Chemother.* **45,** 2054–2059.
98. Bengmark, S. (2000) Colonic food: Pre- and Probiotics. *Am. J. Gastroenter.* **95,** S5–S7.
99. WHO. (1994) WHO Scientific Working Group on monitoring and management of bacterial resistance to antimicrobial agents, Geneva.

7

Foodborne Viral Hepatitis

Hazel Appleton, Malcolm Banks, Catherine M. Dentinger, and Chong Gee Teo

1. INTRODUCTION

The infectious nature of hepatitis has been recognized since the eighth century. Epidemiological observations and human volunteer studies conducted during the twentieth century initially revealed two types of hepatitis. One type followed a long incubation period and was parentally transmitted; at first called serum hepatitis, it is now referred to as hepatitis B. The other type, with a short incubation period, was enterically transmitted and very contagious; it was first called infectious hepatitis, but subsequently hepatitis A *(1)*. We now know that there are two rather than one major enterically transmitted hepatitides: hepatitis A and hepatitis E. This chapter reviews the progress made in the understanding of the epidemiology of these two enterically transmitted diseases and of their causative pathogens: hepatitis A virus (HAV) and hepatitis E virus (HEV). Emphasis is on foodborne aspects of the diseases.

2. HEPATITIS A

2.1. Characteristics and Classification of HAV

The virus causing hepatitis A was first identified in 1973 by immune electron microscopy of fecal samples and sera derived from volunteers who had been experimentally infected with the MS-1 strain of hepatitis A *(2)*. HAV is an isometric particle with a diameter of 25–30 nm. It has no envelope and appears as a featureless sphere under the electron microscope. The virus is a member of the *Picornaviridae* family. Within this family it was originally classified as an *Enterovirus*. However, because of its unique genome organization and the very small size of the internal VP4 protein, it is now classified as a separate genus, *Hepatovirus (3–5)*. The core of the virion contains a 7.5-kb single-strand RNA with a small protein (VPg) linked to the 5′-end and a poly(A) tail at the 3′-end. The capsid consists of 60 identical protomers each comprising three surface proteins: VP1, VP2, and VP3. VP4 is a small internal protein.

There are two strains (or biotypes) of HAV: human HAV which infects humans, chimpanzees, tamarins and owl monkeys; and simian HAV which infects Old World monkeys such as the African green monkey and the cynomolgus monkey. Although these two strains are phylogenetically distinct and have biotype-specific epitopes that can be distinguished by monoclonal antibodies, they share antigenic properties and form

From: *Infectious Disease: Foodborne Diseases*
Edited by: S. Simjee © Humana Press Inc., Totowa, NJ

only one serotype *(6–8)*. There appears to be a single conformational immunogenic site, composed of amino-acid residues of VP1 and VP3, on the surface of the virion. Four genotypes based on nucleotide sequence differences have been identified from human isolates. The only other virus that is tentatively classified as a member of the *Hepatovirus* genus is the avian encephalomyelitis-like virus *(5,9)*.

2.2. HAV Stability and Resistance

2.2.1. Survival of HAV in the Environment

HAV is a stable virus that can survive for long periods in the environment, in seawater, fresh waters, groundwaters, and soils *(10–15)*. It has been detected in groundwater supplies implicated in hepatitis A outbreaks *(16–18)*. In one outbreak, there was molecular confirmation that the HAV from the patients was the same strain as that found in the water supply *(19)*. Viable HAV has been detected in groundwater as long as 17 mo after the original contamination of wells associated with outbreaks *(20)*. It has also been shown that HAV can survive in mineral water for more than a year at 4°C *(21)*.

HAV survives on inanimate surfaces as would be found in food preparation areas, on hands, and in dried fecal suspensions *(22–26)*. Experimental studies have demonstrated that HAV can be transferred from contaminated hands to foods and surfaces *(23,25)*. HAV has survived for longer than 30 d at 20°C on both porous and nonporous surfaces *(24)*. Relative humidity has little effect on survival at 5°C, but at 20°C survival is the longest if the relative humidity is low *(27)*.

2.2.2. Acid Resistance

In common with all enteric viruses, hepatitis A is acid-stable and able to retain infectivity below pH 3 *(28,29)*. At pH 1, it survives for 8 h at room temperature and for up to 90 min at 38°C *(30)*. HAV is therefore likely to survive food processing and preservation conditions used to inhibit bacterial spoilage of food, such as pickling in vinegar, and fermentation to produce foods such as yoghurt. In experimentally infected mussels, HAV has survived exposure to an acid marinade at pH 3.75 for over 4 wk *(31)*.

2.2.3. Temperature Resistance

HAV remains infectious after refrigeration and freezing; consequently, frozen fruits can and have been implicated in hepatitis A outbreaks. HAV is destroyed by conventional cooking processes, but retains infectivity after heating to 60°C for 30 min *(32–35)*. It is therefore uncertain whether it would be inactivated completely in some pasteurization processes, particularly as proteins in foods such as milk and shellfish may protect it. In one study, when HAV was suspended in milk at 62.5°C for 30 min, 0.1% of virus infectivity remained *(33)*. As ingestion of partially cooked HAV-contaminated shellfish has led to outbreaks, there is the likelihood that the temperatures achieved during cooking were inadequate to inactivate the virus *(35–38)*. Studies on the heat inactivation of HAV in cockles have shown that the internal temperature of the meat must reach 85 to 90°C and be retained for at least 1 min to completely inactivate the virus *(35)*.

Modified atmosphere packaging (MAP) is used increasingly by the retail food industry for salad items and vegetables. The survival of HAV on lettuce in MAP stored at room temperature and at 4°C has been investigated *(25)*. Survival at 4°C was the same in MAP as under normal packaging conditions, but at room temperature viral survival was slightly better in the higher carbon dioxide levels of MAP.

2.2.4. Chlorine Resistance

Chlorine-based compounds are commonly used for disinfection but there are conflicting reports on the efficacy for inactivation of HAV, which probably relate to the different experimental conditions used. With environmental contamination HAV is likely to be associated with organic matter, which rapidly inactivates disinfectants such as chlorine and ozone, and hence protects the virus from inactivation. In the United States, a level of 5 mg/L free chlorine with a contact time of 1 min is recommended for general disinfection *(20)*. In a comparative study of a range of disinfectants, sodium hypochlorite containing >5000 ppm free chlorine reduced the HAV titer by >99.9% *(39)*. Of 20 formulations tested, only 2% glutaraldehyde and a quaternary ammonium cleaning agent containing 23% HCl were equally effective. In another study, fecal suspensions containing HAV were dried on polystyrene discs: the HAV titer was reduced 1000-fold after contact with sodium hypochlorite containing 1250 ppm free chlorine at pH 9.56 for 1 min *(24)*. The persistence of HAV on surfaces and the potential problem that it poses for food preparation and health care settings prompted a study on the efficacy of a range of hand washing agents *(40)*. The results were disappointing, in that HAV remained on the hands after using most products.

Chlorine is used in commercial washing processes for fruit and vegetables, but a review by Seymour and Appleton *(41)* concluded that there is little information either on the efficacy of this process in removing viruses or the survival of viruses on these foods. Subsequent studies showed that the concentrations of chlorine used in the food industry are unlikely to remove significant amounts of virus and, furthermore, virus survives on fresh fruits and vegetables for longer than the shelf-life of the product *([42–44]*; H. Appleton, unpublished). Hutin and colleagues *(42)* reported that strawberries responsible for one outbreak had been treated in a wash containing 12 ppm chlorine. It has been suggested that viruses may be protected from direct contact with disinfectants by lodging in the crevices formed by the rough or irregular surfaces of fruits and vegetables *(45,46)*.

Although chlorine is the most widely used disinfectant for washing fresh produce, there are concerns about the health effects of chlorine products. In the United States, the Food and Drug Administration currently permits the use of chlorine in the food and vegetable industry, but its use is prohibited in some European countries. Alternatives such as ozone are being considered *(41,47)*.

2.3. General Epidemiology

Hepatitis A occurs worldwide. Most transmission is from person to person, via the fecal–oral route. Infection is prevalent in conditions of overcrowding and poor sanitation. Fecal contamination of food or water can result in common-source outbreaks, but such outbreaks are reported infrequently.

Hepatitis A was once a common childhood infection. It is still common in less-developed countries where most children have acquired antibody by the age of 5 yr *(48,49)*. Infection in young children is frequently asymptomatic and the severity of infection increases with age. In developed countries, rates of infection have fallen progressively, with the average age of infection rising as living standards improve. Less than 20% of younger adults in countries including Japan, Australia, New Zealand, Canada, United States, and most of Europe have serologic evidence of immunity to HAV *(49–53)*; thus, large susceptible populations now inhabit these areas.

In countries where infection has become relatively uncommon, there is often a cyclic pattern of transmission *(53–55)*. In the United Kingdom, the last peak in the incidence of hepatitis A occurred between 1990 and 1992 when there were over 7000 reported cases in each of these years. The incidence has now fallen to about 1000 cases per year *(53)*. Among the native American populations, which once experienced the highest rate of HAV infection in the United States, epidemics now occur every 5–7 yr *(55)*. In epidemic periods, infection can be widespread in the community, particularly among children and young adults of low socio-economic status *(56,57)*. Cases can cluster in families and in settings where the potential for fecal–oral spread is high, such as daycare centers, nurseries and primary schools *(53)*. Patterns of infection are changing with increasing use of hepatitis A vaccine *(57a)*. Following the introduction of routine childhood hepatitis A vaccination to American Indians and Alaska Natives, the predicted cyclic increase of cases has not occurred and rates of disease fell 20-fold from 1990 to 2001 *(58)*.

Travel to countries where infection is common also increases the risk of becoming infected. Between 1990 and 2000, 25% of laboratory-confirmed cases of hepatitis A in the United Kingdom had traveled abroad, mainly to the Indian subcontinent *(53)*. Sexual transmission of HAV can occur via fecal–oral transmission, accounting for infection among men who have sex with men *(59)*. Among illicit drug users, transmission may occur from injecting with contaminated works *(60,61)*. On rare occasions, infection occurs in persons receiving blood products that have been collected during the brief viremic phase of infection *(62–64)*.

There is some evidence that humans have acquired hepatitis A from handling infected nonhuman primates *(65,66)*. However, it is not clear whether infections in these animals were natural, or whether the primates had become infected with human strains from close contact with humans or by being treated with therapeutic products of human origin.

2.4. Food and Waterborne Infection

Food and waterborne outbreaks of hepatitis A have been recognized for over 40 yr, but are infrequently reported *(67)*. Unless there is a clearly defined point-source outbreak, it is difficult to link infections to a food or water source. There are several reasons these infections can be difficult to detect. First, in communities with a high incidence of ongoing HAV transmission, persons who acquire infection from contaminated food or water may be difficult to detect. Reporting of hepatitis A cases may be incomplete, or cases may occur over diverse geographic areas and be difficult to link. Persons with asymptomatic or mild infections may never come to medical attention and, thus, may not be reported. Once cases are identified, recalling food exposures during the 2–6 wk prior to infection can be difficult. Moreover, where initial cases are infected through contaminated food or water, secondary person-to-person spread to susceptible individuals may occur, particularly in countries with low levels of immunity, thereby compounding the difficulty of identifying a food or waterborne outbreak. Despite the relatively few documented reports, outbreaks have occurred that involved large numbers of people, as is discussed below.

In the United States, approx 10% of reported hepatitis A cases are related to suspected food or waterborne transmission *(68)*. Approx 55% of reported hepatitis A cases in have no identifiable source *(69,70)*, but some of these could be owing to unrecognized foodborne transmission. Current estimates of food-related illness in the United States suggest that hepatitis A is the fourth leading cause of viral foodborne illness *(71)*.

Table 1
Risk Factors for Food and Waterborne Hepatitis A

Source of virus	Foods implicated	Preventative measures
Sewage discharge and leaks	Water	Treatment of sewage to remove viruses Treatment of water (e.g., chlorination)
Sewage discharge into rivers/sea	Bivalve molluscs	Treat sewage to remove viruses prior to discharge Monitor microbiological quality of shellfish Relaying, depuration and heat treatment of shellfish
Contaminated irrigation water	Fruits, salads, and vegetables	Adequate sewage treatment to prevent water contamination
Use of sewage sludge on agricultural land		Only treated sewage to be applied to agricultural land
Food handlers	Mainly cold foods or foods that do not receive further cooking after handling, e.g., salads, sandwiches, fruits, cakes, and cream	Good basic hygiene (hand washing, keeping food preparation areas clean, etc.) Education
Imported foods from endemic areas	Fruits, salads, and shellfish	Selection of safe food sources by food industry Education of consumer on washing fruits and salads Development of effective washing and sanitization processes by food industry

The majority of hepatitis A outbreaks in England and Wales are small and occur in families. The last recorded foodborne outbreak was in 1992 and was associated with shellfish ingestion (53,72). From 1992 to 1999, the Public Health Laboratory Service (now the Health Protection Agency) received 19,747 laboratory-confirmed reports of hepatitis A. The source of most infections was unknown, and only 155 cases were recorded as being foodborne (41). A summary of risk factors for food and waterborne hepatitis A is shown in Table 1.

2.4.1. Waterborne Outbreaks

HAV is excreted in the stools, and hence it enters the sewage system. While outbreaks have been associated with contamination of private wells (18,73), transmission from contaminated water is very rare in areas with adequate public water treatment systems. Waterborne outbreaks of hepatitis occur more frequently in regions with poor sanitation, but laboratory facilities may not always be available to distinguish between hepatitis A and E. In one outbreak in East Africa, evidence of infection with both viruses was reported (74).

Recreational waters have been a source of infection for a number of viruses (75,76). Although outbreaks of hepatitis A have been linked to swimming in chlorinated pools and fresh water lakes, such incidents are identified infrequently (77,78).

2.4.2. Foodborne Outbreaks

Food may be contaminated with HAV at any point from cultivation to service. Crops may be contaminated, if irrigated with sewage-polluted water or if fertilizer carrying HAV is used in the cultivation process. Shellfish harvested from sewage-polluted water have been the source of many outbreaks of both viral gastroenteritis and hepatitis A. Foods may also be contaminated during harvesting, processing, preparation, and serving by infectious persons.

2.4.2.1. From Shellfish. Common-source outbreaks of shellfish-associated viral hepatitis have been recognized for many years and continue to be problematic *(79)*. Some outbreaks have been very large. The largest foodborne outbreak of hepatitis A ever recorded occurred in Shanghai in 1988 where almost 300,000 people were infected after consuming clams harvested from polluted waters *(37,80)*. Outbreaks have occurred in developed countries, such as Italy, Spain, Australia, France, Japan, the United Kingdom and the United States, both from locally grown and imported shellfish, as well as in countries with lower standards of hygiene *(81–92)*.

Although the number of reported shellfish-associated outbreaks of hepatitis A have been few in recent years, it is important to remain vigilant. An outbreak of oyster-associated hepatitis A in Australia in 1997 was the first such outbreak in that country for a number of years and affected over 400 people *(90)*.

Shellfish-associated cases of hepatitis A are believed to be underreported *(86)*. A number of retrospective seroepidemiological studies have been undertaken to determine risk factors for hepatitis A *(93–97)*. Several, but not all, identified shellfish consumptions as significant. The variable findings could be related to the prevalence of hepatitis A in the local community at the time of the study. Shellfish have been consistently identified as a risk in Italy; recent estimates have suggested that shellfish are responsible for 70% of hepatitis A cases *(98)*. One outbreak in Italy was estimated to have cost the health service 0.4% of its total expenditure for the region for that year and 0.04% of the gross domestic product of the entire region *(99)*. In a survey of shellfish in Puglia in southern Italy *(100)*, HAV RNA could be detected in 20% of nondepurated and 11% of depurated mussels, and in 23% of other samples collected from the shellfish markets. HAV could be cultured from 34% of the samples that were polymerase chain reaction (PCR)-positive. Bacteriological testing is used routinely for monitoring the safety of shellfish, but in that study there was no correlation between bacterial and viral contamination.

The main shellfish implicated in the outbreaks of viral illness are the bivalve molluscs (oysters, mussels, cockles, and clams). Bivalve molluscs grow in shallow coastal and estuarine waters that are polluted with sewage in many parts of the world. They feed by filtering seawater to obtain nutrients, and, where water is contaminated, can accumulate microorganisms, including HAV, within their tissues. Viruses do not replicate in the shellfish but may be concentrated at levels greater than those in the surrounding water *(101,102)*. Thorough cooking will inactivate virus, but this renders the shellfish meat tough and unpalatable. These shellfish are frequently consumed raw or are only very lightly cooked; such cooking may not inactivate microbial contaminants *(35,36,38,43)*. Scallops are rarely implicated: they are free-living organisms that inhabit deeper offshore waters which are less polluted.

2.4.2.2. From Fruits and Vegetables. Several hepatitis A outbreaks associated with eating raw or partially cooked fresh produce have been recognized, but to-date no investigation has definitively confirmed the point of produce contamination *(41)*. Fruit and vegetable crops can be contaminated with viruses in their growing or harvesting areas from contact with polluted water and from inadequately or untreated sewage sludge used for irrigation and fertilization. Contamination could also result from handling by people during processing, packaging, or distribution steps, or if HAV-contaminated water is used for rinsing and icing of produce.

Outbreaks of hepatitis A associated with fresh produce, particularly soft fruits and salads, have been reported from several countries. Iceberg lettuce *(103)*, strawberries *(42, 104)*, tomatoes *(105)*, salad items *(106,107)* and green onions *(108,109)* have all been implicated. In a multifocal outbreak associated with commercially distributed lettuce, it was suggested that contamination might have arisen from dirty water used for growing or irrigation, or from the use of night soil, although this remained unproven *(103)*. In a study in Costa Rica, HAV and rotavirus were detected in lettuces acquired from local markets; it was suggested that the source of contamination was the discharge of sewage into river water used for irrigating crops *(110)*. Investigations into two separate US multistate outbreaks associated with fresh frozen strawberries revealed that the straw-berries had been extensively handled before processing. In both instances, stems from strawberries were removed by hand or by fingernail before being sent to processing plants *(42,104)*. In the United Kingdom, a number of outbreaks have been traced to frozen raspberries. There was the suggestion that the raspberries implicated in these outbreaks were contaminated by infected fruit pickers *(111–113)*. The investigation of an outbreak of hepatitis A in New Zealand associated with blueberries revealed multiple opportunities for contamination of the blueberries by pickers *(114)*.

Outbreaks of hepatitis A have been associated with green onions in the United States *(108,109)*. In one outbreak, it was observed that the harvesting process required extensive field handling. Onions were cut by machine and gathered into piles by hand, the outer skin of each onion was removed by hand, onions were bundled and bound with rubber bands and, finally, the stems and roots of onion bundles were trimmed by hand. Several workers were required to complete these steps *(108)*. No ill-workers were reported during the exposure period.

In an increasingly global market, where fruits and vegetables are exported from endemic countries to developed countries that often have low levels of immunity to hepatitis A, it is likely that further outbreaks related to the consumption of these foods will result. A survey of the microbiological status of ready-to-eat fruit and vegetables undertaken in England and Wales *(115)* showed that between 1992 and 1999, fruit and vegetables accounted for 5.5% (83 out of 1518) of the total foodborne outbreaks reported. Of the 83 outbreaks, 16% were caused by norovirus. The causative agent for 28% of the outbreaks was unknown, but clinical and epidemiological features suggest that the majority were probably viral. While HAV was not identified in this particular survey, the study exemplifies the importance of fresh fruit and vegetables as vehicles in the transmission of enteric viruses.

2.4.2.3. From Food Handlers. Food handlers are not at higher risk for hepatitis A than other persons because of their occupation. However, HAV-infected foodhandlers can potentially be a vehicle for viral transmission to consumers. From 2001 to 2003, 5.8% of the 342 investigated hepatitis A cases reported to the CDC's Sentinel Counties Viral Hepatitis Surveillance System worked as food handlers during the 2 wk before to 1 wk after their illness onset (CDC, unpublished data). Despite the relatively large number of hepatitis A cases who worked as food handlers during their incubation period, few food handler-associated hepatitis A outbreaks have been reported. Thus, while HAV-infected food handlers can contaminate foods, thereby potentially exposing large numbers of people to HAV, transmission from such contamination does not seem to occur readily. Nonetheless, when outbreaks do occur, case identification and implementation of control measures can result in considerable expense to public health departments *(116)*.

Cold food items that require much handling during preparation together with poor personal hygiene by infectious food handlers favor food handler-associated outbreaks. Sandwiches, salad items and fresh fruit prepared by infectious food handlers have been implicated in hepatitis A outbreaks *(117–121)*. If food is prepared centrally by commercial caterers and then distributed widely to different venues, a large number of people may be at risk. One outbreak in the United States involved a catering company which employed a food handler who was infected by HAV; 91 hepatitis A cases were subsequently identified from 21 out of the 41 events catered for by that company *(122)*. In an outbreak in Milwaukee, Wisconsin, an employee working in two sandwich shops infected 230 people *(121)*. Two similar community-wide outbreaks occurred among persons who ate bakery products contaminated by the handler who applied sugar glaze to the baked goods *(123,124)*. In the United Kingdom, more than 50 residents in a group of villages became infected after eating unwrapped bread, rolls or sandwiches supplied from one shop, where the owner had been ill. It was concluded that the bread was contaminated by the owner whose hands were inadequately washed because of the presence of painful skin lesions *(125)*. Drinking glasses have also been implicated in transmitting infection to consumers. An outbreak occurred in Thailand among college students who filled their glasses by dipping them into an overflow water reservoir in the college cafeteria: it was postulated that someone with fecally contaminated fingers had contaminated the water with HAV *(126)*. In another outbreak, associated with a public house in the United Kingdom, infection was attributed to an infectious barman who probably contaminated the glasses *(127)*.

2.5. Clinical Characteristics

The clinical spectrum of HAV infection ranges from nil or few symptoms to fulminant hepatitis *(128)*. In most persons, hepatitis A is a self-limited illness. The characteristic symptom is jaundice, which typically develops 1–2 wk after the onset of prodromal symptoms. Pruritus, tenderness, and mild enlargement of the liver may accompany jaundice. Milder symptoms of nausea and general malaise with or without jaundice are common. Other clinical features may include fever, abdominal pain, anorexia, arthralgia, and diarrhea. Recovery is usually complete, although patients may feel unwell for several weeks. Most children less than 6 yr of age have asymptomatic infections or mild, non-specific symptoms including nausea and vomiting, malaise and diarrhea, but do not develop jaundice *(129,130)*.

Complications of acute hepatitis A include coagulopathy, encephalopathy, and liver failure *(70)*. Fulminant hepatitis is the most severe form of infection, but is rare *(131)*. The overall rate in the United States is <1% of all reported cases, increasing with age and with the coexistence of other underlying liver diseases, such as hepatitis B or C *(132,133)*. Children very occasionally develop liver failure requiring liver transplantation *(134)*.

Persons with prolonged or relapsing hepatitis A have been recognised *(135,136)*, but chronic hepatitis A does not occur. There is, therefore, no carrier state as such.

In developed countries, deaths from hepatitis A are rare. In the United States, 0.6% of hepatitis A reported in 2004 resulted in death *(69)*. In the United Kingdom, the mortality data from the Office for National Statistics record 39 deaths in the 5-yr period 1993–1997 *(53)*, of which only one death in a patient <40 yr of age. The case fatality rate increases progressively with age, with a fatality rate of 2% in those aged between 50 and 59 yr rising to 12.8% in those >70 yr.

The most common biochemical feature of hepatitis A is elevation in the concentrations of the serum aminotransferases. During the prodromal phase, the serum alanine and aspartate aminotransferases typically rise to a peak of 500–5000 IU/L, returning to normal within a mean of 7 wk (range 1–29 wk) *(137)*. Serum bilirubin levels are not markedly elevated (usually <10 mg/dL) except in cases complicated by cholestasis or hepatic failure *(138)*.

2.6. Pathogenesis

HAV is transmitted primarily via the fecal–oral route, although the infective oral dose is not well defined. HAV is not neutralized by gastric acid and is thought to be transported across the intestinal epithelium to be taken up by hepatocytes *(139,140)*. The main target organ is the liver. Viral replication occurs in the cytoplasm of the infected hepatocytes *(141)*. It is not clear whether there is also some limited replication in intestinal cells. Injury to the hepatocytes is thought to be related to host immune response rather than to cytopathic effects of virus, although the precise mechanism of cell injury is not known *(142)*. From the liver, HAV is transported through the biliary tree to the intestine, and from there is shed in the stools *(143)*. Infection with HAV induces lifelong immunity.

The incubation period of hepatitis A is 15–50 d with a mean of about 30 d *(144)*. Viremia occurs soon after infection, persisting throughout the period of liver enzyme elevation *(145)*. Viral shedding in stool, and thus peak infectivity, occurs during the 2-wk period before the onset of jaundice or elevation of liver enzymes, when the concentration of fecal virus is the highest, declining by the time jaundice appears *(146,147)*. Children and infants may shed HAV for longer periods than adults, up to several months after the onset of clinical illness *(147,148)*. Chronic shedding of virus does not occur; however, persons with relapsing disease may shed virus for up to 6 mo *(149)*.

2.7. Diagnosis and Detection of HAV

Hepatitis A cannot be differentiated from other forms of acute viral hepatitis on the basis of clinical symptoms, clinical pathological tests, or epidemiologic features. Diagnosis of hepatitis A infection is made by detection of specific IgM antibody rather than the isolation of virus, because the virus excretion has largely ceased by the time illness becomes apparent. IgM anti-HAV becomes detectable 5–10 d before the onset of symptoms in most persons with acute hepatitis A and can persist for up to 6 mo after infection *(150)*. There are several commercial immunoassays available for detecting serum anti-HAV IgM. Total anti-HAV (IgG + IgM) appears early in the course of infection and remains detectable indefinitely. Since IgM disappears in most persons after 6 mo, it is not possible to distinguish between current or past infection based on the presence of total anti-HAV alone. Measuring total anti-HAV is useful to determine immunity.

The virus can be cultured in the laboratory, but, for primary isolation, monolayer cell culture is a long and unreliable procedure *(151–155)*. Reverse-transcriptase (RT)-PCR assays have been developed and can be used for detecting viral antigen in fecal specimens should early specimens be available. PCR assays have mainly been used for detecting virus in water, shellfish, soft fruits and environmental samples *(12,17–19,73,85,91,92, 100,114,156–162)*. These samples are often inhibitory in PCR reactions, so complex procedures for the extraction of viral RNA are required; consequently, these tests are not used routinely. It is not clear whether viral nucleic acid detected by PCR necessarily indicates the presence of infectious viruses, as under experimental conditions some workers have been unable to recover viruses from samples positive by PCR *(163–165)*. This consideration will have implications should PCR for enteric viruses be introduced into routine monitoring protocols. Protocols have been developed that integrate cell culture with RT-PCR which may prove to be useful *(166–167)*.

Because of the long incubation period of hepatitis A, implicated food items are rarely available for testing. Fresh fruits, salads, and shellfish have a short shelf-life. Although virus may survive in frozen produce, as has been demonstrated in a number of outbreaks associated with frozen raspberries and strawberries, testing of implicated frozen items has not resulted in the detection of HAV to-date *(42,104,111–113)*.

2.8. Prevention and Control

2.8.1. General Measures

Prevention of hepatitis A requires effective public and personal hygiene practices. Public sanitation, including adequate human-waste disposal and water-treatment systems, is important in the maintenance of a low hepatitis A incident rate in industrialized countries *(168)*. Attention to personal hygiene, particularly strict adherence to hand washing in settings where contact with human waste may be common, including daycare centers and residential homes, is critical to the prevention of HAV transmission. Careful attention to personal and public sanitation among all persons handling food and fresh produce, including agriculture workers, food processors, and food handlers, should decrease the risk of HAV transmission. Table 1 also summarizes preventative steps that may be implemented to reduce the risks of food and waterborne infection.

2.8.2. Active Immunoprophylaxis

Two hepatitis A vaccines, which are formalin-inactivated preparations of cell-culture adapted virus, have been licensed since the mid-1990s by the US Food and Drug Administration for use in persons ≥2 yr of age. They are highly immunogenic. Protective concentrations of anti-HAV are measurable in 54–62% of persons by 2 wk and in 90% of vaccinees by 4 wk after one dose of vaccine *(169,170)*. A second dose of vaccine, given 6 mo after the first dose, confers long-term protection. Vaccine efficacy is 94–100% *(68)*. Follow-up studies of vaccine recipients indicate that the duration of anti-HAV and protection from clinical illness lasts for at least 12 yr *(171)*. Kinetic models of anti-HAV persistence suggest that anti-HAV may persist in the circulation for ≥20 yr *(172)*. Booster doses are not recommended. Studies are in progress to evaluate the long-term efficacy of vaccination. Residual, passively acquired maternal anti-HAV interferes with the anti-HAV response in children born to previously infected mothers. Therefore, although the vaccine is safe for infants, most countries have restricted its use to children above

1 or 2 yr of age *(173)*. In 2005, vaccines were licensed in the United States for use in children ≥1 yr *(68)*.

Hepatitis A vaccine is currently targeted at persons with an increased risk for HAV infection. These populations include travelers to endemic countries, illicit drug users, men who have sex with men, and persons who regularly receive blood products *(53,68)*. In the United States, routine hepatitis A vaccination is also recommended for all children *(68)*, in order to reduce the rates of HAV infection among children and curtail HAV transmission to older children and adults *(174)*. In addition, hepatitis A vaccine is recommended for persons who are at increased risk of serious outcomes from HAV infection including those with chronic liver disease *(68)*.

The question of whether vaccination is efficacious for postexposure prophylaxis remains unresolved. One study *(175)* suggested that it could be, but no comparison with immune globulin was done.

2.8.3. Passive Immunoprophylaxis

Immune globulin preparations, available in the United States as "immune serum immunoglobulin G," and in the United Kingdom as "human normal immunoglobulin," when given intramuscularly provide short-term (1–2 mo) protection from hepatitis A *(176,177)*. These preparations are antibodies concentrated from pooled human plasma. Postexposure immunoglobulin prophylaxis is >85% effective in preventing hepatitis A if administered within 2 wk after exposure to HAV, but the efficacy is highest when administered early in the incubation period *(177,178)*.

Immune globulin is recommended for prevention of hepatitis A in susceptible persons who have had specific exposures to HAV. These exposures include household or sexual contact with a hepatitis A case during a time when that case is likely to be infectious (i.e., 2 wk before through 1 wk after the onset of illness) and whose last contact is within the previous 2 wk. In addition, susceptible food service workers in the same establishment as an infected food-service worker may receive immune globulin if they have been in contact with the infected worker during the infectious period of latter. Persons who have received at least one dose of hepatitis A vaccine ≥1 mo previously or who have a history of laboratory-confirmed HAV infection should be considered immune and do not require passive immunization. Passive immunoprophylaxis is not recommended for persons whose only exposure to a person with hepatitis A occurred >1 wk after the onset of jaundice, since the patient is unlikely to be infectious by this stage. Immune globulin can be administered as pre-exposure prophylaxis for suitably aged children and adults who are planning travel to endemic areas within 2–4 wk and who require immediate protection *(179)*, and as pre-exposure prophylaxis for children <1 yr for whom the vaccine is not licensed.

2.8.4. Prevention and Control Measures of Foodborne Transmission

Prevention of foodborne hepatitis A is dependent on access to safe food and water, and good personal hygiene. Sewage pollution is a major factor in the contamination of filter feeding shellfish and of fruits and vegetables that are not cooked before consumption.

2.8.4.1. Reducing Risk From Shellfish. Filter-feeding shellfish are often harvested from coastal waters that receive sewage discharges. Developed countries require treatment of such shellfish by relaying, depuration or heating before marketing *(180,181)*. Shellfish such as oysters that are eaten raw are either relayed to unpolluted coastal or estuarine

Table 2
Classification of Harvesting Areas for Bivalve Molluscs in the European Community

Water quality category	*E. coli* /100 g shellfish meat	Fecal coliforms/ 100 g shellfish meat	Shellfish treatment
A	<230	<300	May go for direct consumption
B	<4600 in 90% samples	<6000 in 90% of samples	Must be depurated, heat treated, or relayed to meet Category A
C		<60,000	Must be relayed for long period (2 mo) to meet categories A or B; and/or may be heat treated by an approved method
D		>60,000	Harvesting prohibited

Table 3
Classification of Harvesting Areas for Bivalve Molluscs in the United States

Classification	Fecal coliforms/100 mL water	Shellfish treatment
Approved	GM <14 and 43 in 90% of samples	May go for direct consumption
Restricted	GM <88 and <260 in 90% of samples	Must be depurated or relayed
	Above levels exceeded	Harvesting prohibited

GM, geometric mean.

waters until bacterial counts have fallen to acceptable levels, or they may be placed in land-based depuration tanks where the water is treated with UV irradiation or ozone. Bacterial contaminants are rapidly eliminated during the natural feeding process but viruses may remain. Alternatively, shellfish may be heat-treated, but care must be taken to ensure that the internal temperature of the shellfish meat reaches the required level to inactivate viruses *(35,163,182)*. In Europe, the level of fecal contamination of shellfish is monitored by measuring bacterial counts in the shellfish meat (Table 2), but absence of bacterial indicators does not guarantee the absence of viruses *(100,162,183)*. Growing waters are classified on the basis of contamination levels in the shellfish and harvesting from heavily polluted sites is prohibited. In the United States, the harvesting waters, rather than the shellfish, are tested for bacterial contamination (Table 3).

2.8.4.2. Reducing Risk From Fruits and Vegetables. Untreated or inadequately treated sewage discharged into natural waters can cause contamination of crops. Sewage sludge is sometimes applied to agricultural land, with the benefit that useful plant nutrients are recycled to the soil. Guidelines issued by the World Health Organization state that fruits and vegetables to be eaten raw should not be fertilized with sewage or irrigated with contaminated water *(184)*. Application of untreated sewage sludge to agricultural land is prohibited in the United Kingdom. There is guidance on the minimum acceptable level of treatment for any sewage sludge applied to agricultural land and regulations on the time interval from applying sludge to harvesting food crops *(185,186)*. Controls such as these may reduce the risk of microbial contamination in developed countries, but do not address the potential problem of imported produce from countries with different standards for organic fertilizers or irrigation water.

2.8.4.3. Indicator Organisms. The difficulty of detecting and working with pathogenic viruses such as HAV in food, water, and environmental samples, the lack of standardized techniques, and the failure of conventional bacterial indicators to predict viral contamination have led to alternative indicator organisms being sought. Enteroviruses, rotavirus, and adenoviruses have been considered *(83,187)*. However, most interest has been in the possible use of bacteriophages (reviewed by Lees *[79]*). Male-specific RNA (F+) bacteriophage and the bacteriophage of *Bacteroides fragilis* have been considered as candidate viral indicators of water pollution *(188,189)*. These bacteriophages appear to be promising indicators of enteric virus contamination of shellfish *(190–192)*. Studies on the elimination kinetics of F+ bacteriophage during depuration of shellfish have shown that it behaves similarly to enteric viruses *(193,194)*. Recent studies on the survival and removal of MS2 bacteriophage from fruit and vegetables suggest that the use of bacteriophages may be suitable for the investigation and design of washing and sanitization processes *(195)*.

2.8.4.4. Reducing Risk in Food Handlers. Workers with symptoms suggestive of hepatitis A should be excluded from food handling. However, as the most infectious period for hepatitis A is in the 2 wk before illness becomes apparent, exclusion may be too late to control transmission. Meticulous attention to good food-handling practices and education is, therefore, essential. There should be the provision of adequate toilet and hand-washing facilities, not only in the catering and retail industry, but also in the wholesale and agricultural industry. The wearing of disposable gloves is recommended if foods are to be manipulated by hand, but this does not prevent the transfer of viruses to gloves by touching contaminated surfaces.

Cliver *(196)* has suggested that the food handlers in the United States should be vaccinated, but this would have considerable economic implications *(197,198)*. Current epidemiological data indicate that food-handlers do not pose a significantly greater risk of transmitting HAV than other people, and use of vaccine in this group may not be cost-effective. Vaccine may be used for preventing secondary cases and outbreaks, but the latest time beyond which the vaccine is likely to be effective in preventing disease in contacts is probably no more than 7 d from the onset of illness in the primary case. If the time from exposure is longer than 7 d, immune globulin may be given. Immune globulin is effective at preventing disease when given within 2 wk of exposure; and even if the time from exposure is longer, it may modify the severity of illness *(53)*. The burden of post-exposure prophylactic interventions can be considerable *(199)*.

3. HEPATITIS E

3.1. Characteristics and Classification of HEV

A form of enterically transmitted hepatitis distinct from hepatitis A has been recognized for many decades *(200,201)*. In developing parts of Africa, the Middle East, Asia including the Indian subcontinent, and Central America, the disease periodically appears as epidemics each of which may involve thousands of cases. Early electron microscopic studies reported virus-like structures in stool samples collected from patients with acute non-A, non-B (NANB) hepatitis; furthermore, stool suspensions could lead to liver disease when inoculated to nonhuman primates *(202,203)*. In 1987, Bradley et al. *(204)* reported that the disease could be serially transmitted to cynomolgus monkeys. Recombinant cDNA libraries subsequently constructed from infectious bile of the monkeys and from stools

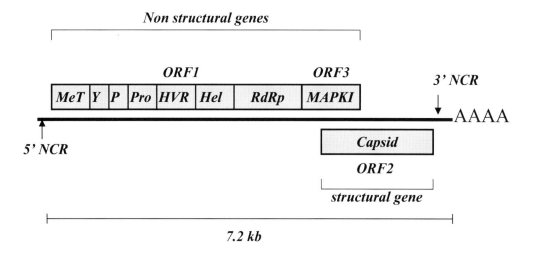

Fig. 1. Schematic of HEV genome organization showing coding regions. *Abbreviations*: Hel, helicase; HVR, hypervariable region; MAPKI, mitogen-activated protein kinase inhibitor; MeT, methyl transferase; NCR, noncoding region; P, protease; Pro, proline hinge; RdRp, RNA-dependent RNA polymerase; Y, Y-domain.

of infected humans permitted the identification of an RNA-directed RNA polymerase-like amino-acid motif *(205)*. The putative virus from which the genome was partially characterized was conferred the name hepatitis E virus. The full genomic sequence of the virus was reported soon thereafter *(206)*.

HEV is a positive-sense, single-stranded, 7.2-kb RNA virus with a spherical, non-enveloped capsid. The HEV genome is polyadenylated and has a 7-methyl guanosine cap, but the replication strategy of HEV is unknown. The genome encodes three open-reading frames (ORFs) (Fig. 1). ORF-1 codes for a polyprotein comprising nonstructural genes and ORF-2 for the capsid protein which contains the major antigenic determinants. ORF-3 encodes a protein that is involved in the regulation of cellular protein kinase activity *(207)*. Electron micrographs reveal symmetrical, 27–34-nm diameter viral particles with cup-shaped depressions on the surface. HEV was originally classified as a calicivirus. More recent genetic characterization studies indicate that the genome is sufficiently different from that of caliciviruses for HEV to be classified as a separate genus, *Hepevirus (208)*.

The native virus grows poorly, if at all, in vitro *(209,210)*. Most preparations for inoculations or other experimental manipulations are produced from fecal material. The recent construction of infectious clones of HEV *(211,212)* leads to the possibility of generating these recombinant viruses to high titers in vitro, thereby facilitating studies to reproduce disease in nonhuman primates and providing material for vaccine development and other investigative manipulations.

The presence of hepatitis E antibodies and RNA in swine in Nepal *(213)* was reported in 1995; and in 1997, sequence data were published indicating that the swine virus was closely related to the human HEV *(214)*. Shortly thereafter, two human HEV strains in the United States were shown to be almost identical in sequence to the pig strains *(215)*. These data suggested the very close relationship human HEV has with swine HEV.

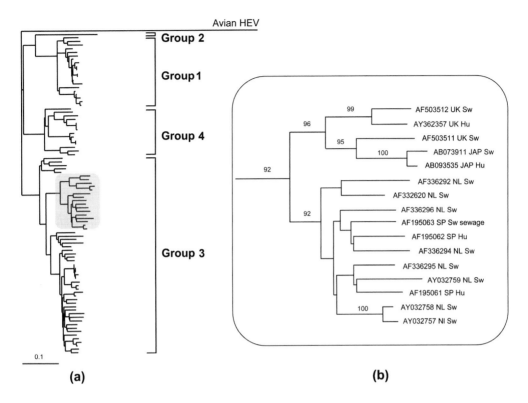

Fig. 2. HEV genomic phylogenetic relationships as determined by partial sequencing. (**a**) Outline genotypic groupings of human and swine HEV deduced from partial sequencing of the ORF-2 gene arranged according to the scheme of Schlauder et al. *(215)*. Avian HEV is included as an outgroup. (**b**) Selected detail of genotype-3 region showing the close genetic relationship between human and swine strains of HEV. GenBank accession numbers, country of origin, and host are indicated. Significant bootstrap values of >70% are indicated. *Abbreviations*: SW, swine; Hu, human; JAP, Japan; NL, Netherlands; SP, Spain.

Four genotype groupings of human and swine HEV are recognized (Fig. 2). Type 1 is the prototype strain found in human cases of hepatitis E in hyperendemic areas. Type 2 is found almost exclusively in human cases in Central America. Types 3 and 4 are found in humans and pigs in developed regions, and in pigs in some hyperendemic regions.

More recently, an HEV-like virus associated with big liver and spleen disease has been identified in chickens in Australia *(216)*, and a related virus causing hepato-splenomegaly in chickens has been discovered in the United States *(217)*. Avian HEV show clear genetic and biologic differences from the human and swine viruses *(218,219)*.

3.2. General Epidemiology

In the hyperendemic regions, HEV is often the principal cause of epidemic and sporadic acute hepatitis. Apart from the waterborne hepatitis E outbreaks in India reported earlier *(200,201)*, outbreaks have been reported from other parts of India *(220–222)*, and from Pakistan *(223)*, Burma/Myanmar *(224,225)*, and Nepal *(203,226)*. Labrique et al. *(227)* have reviewed the epidemiology of HEV epidemics in the Indian subcontinent.

Table 4
List of Developed Countries with Reported Autochthonous Hepatitis E

UK	ITALY
USA	ISRAEL
SPAIN	JAPAN
FRANCE	KOREA
HOLLAND	TAIWAN
AUSTRIA	AUSTRALIA
GERMANY	NEW ZEALAND

Epidemics were also reported in Borneo and Java in Indonesia *(228,229)*, and in Vietnam *(230)*, Ethiopia *(231)*, Somalia *(232)*, Morocco *(233)*, Namibia *(234)*, Botswana *(235)*, and Mexico *(236)*. The largest outbreak recorded occurred in the Xinjiang region of China in 1986–1988 involving >100,000 cases *(237)*. Sporadic infections occur in hyperendemic regions at rates much higher than in nonhyperendemic regions: it is associated with, e.g., 40–50% of acute hepatitis in Sudanese children *(238)*, and >60% of hospitalized adult cases of acute hepatitis in certain parts of India *(239–241)*.

In both epidemic and sporadic hepatitis E in hyperendemic regions, clinical attack rates are highest among young adults (i.e., in the 15–25 yr age range). In developed regions, there is collected evidence from Japan *(242–244)* and Europe *(245,245a)* that clinical attack rates are highest among older adults (>50 yr). Whether the high rates in older patients reflect particular susceptibility to infection or disposition to develop symptoms on infection is unknown.

Secondary and intra-familial spread may occur but are uncommon (<5%) *(246,247)*. Such low figures contrast with the >50% attack rates among susceptible household contacts of hepatitis A cases *(248)*. If, however, the primary case is a pregnant woman, the risk of vertical transmission is significant, ranging from one-third to 100% *(249–252)*. Post-transfusion hepatitis E has been reported in both developed and developing regions *(253–256)*.

Anti-HEV seroprevalences in the human population are high in most hyperendemic areas, e.g., between 15 and 30% in Nepal *(257)*, 40% in Vietnam *(258)*, and >60% in Egypt *(259)*. In India, the population distribution of anti-HEV contrasts markedly with that of anti-HAV: one study found that while anti-HAV could be detected in almost all children <5 yr, anti-HEV was detected infrequently in young children, increasing to ~40% during young adulthood without significant rises in later life *(260)*. In Egypt, by contrast, the population distribution pattern of anti-HEV has been observed to match that of anti-HAV, the prevalences being high by early childhood, plateauing from young adulthood *(259)*. Early childhood exposure to HEV, with the consequent attainment of high levels of herd immunity, probably explains why epidemic hepatitis E has not been reported from Egypt even though sporadic infection does occur *(261)*.

Infection may be imported to developed countries by people who have traveled to hyperendemic areas (reviewed by Schwartz et al. *[262]*). There is increasing awareness that sporadic hepatitis E in nonhyperendemic countries is not always travel-associated. The number of developed countries reporting the cases of autochthonous hepatitis E *(215,245,245a,262,263–274)* is steadily increasing (Table 4). The substantially high anti-HEV prevalences in these countries cannot be attributed only to imported infection. In a serosurvey conducted in eight states in the United States, 20–25% of swine veterinarians were found to be anti-HEV seropositive, but this was against a background of 18% positives amongst blood donors *(275)*. Among Korean blood donors, the seroprevalence is similar in magnitude to that of American blood donors *(276)*. These figures contrast with lower figures found in comparatively less-developed countries, e.g., Mexico (10%) *(277)*, Brazil (3–4%) *(278)*, Bangladesh (7%) *(279)*, and Vietnam (9%) *(280)*. While the sensitivity and specificity of the HEV antibody assays may contribute to the disparate high seroprevalence rates observed *(281,282)*, the possibility that people living in regions with good sanitation truly possess HEV antibodies deserves closer scrutiny, as nonwaterborne transmission is then implicated.

3.2.1. HEV in the Environment

The fecal contamination of drinking water is the predominant means by which HEV is transmitted in hyperendemic regions. This is exemplified by the observation of significantly higher attack rates of hepatitis E, during an epidemic in Somalia, among the residents of villages supplied by riverine water compared to those of villages that depended on ponds or wells *(232)*. Moreover, an ecological analysis of hepatitis E transmission patterns in Vietnam, Laos, and Indonesia revealed that epidemics tended to be centered in riverine environments rather than urbanized districts, and that higher inter-epidemic anti-HEV seroprevalences were significantly associated with the usage of river water for drinking, cooking, washing or human excreta disposal *(283)*.

HEV RNA has been detected by RT-PCR in human sewage from India *(284,285)*, slaughter-house sewage from Spain *(286,287)*, and urban sewage from Spain, France, and the United States *(287)*. While such detection need not necessarily correlate with infectivity, HEV rescued from sewage has been shown to be capable of infecting rhesus monkeys *(288)*.

Data on the resistance of HEV to temperature are now available *(288a)*. The virus is more labile than HAV, accounting for the low attack rates of secondary hepatitis E. Boiling of water prior to drinking has consistently been associated with reduction in the risk of developing hepatitis E *(280,283)*.

3.2.2. Animal Reservoirs of HEV

Worldwide, antibodies to HEV have been detected in a wide range of mammals including pigs, boars, deer, monkeys, cattle, sheep, goats, dogs, cats, rats, and mice *(289–296)*. These potentially act as reservoirs of HEV whether in hyperendemic or in the less endemic regions. The precise relationship of the viruses infecting these animals to human and swine HEV and whether cross-species transmission occurs are unknown. Under experimental conditions, human HEV has been transmitted to sheep *(297)*. One strain identified in rats was characterized to have a 95–96% homology with genotype 1 human strains circulating in Nepal from where it was identified *(292)*. A study of homeless people attending a free clinic in downtown Los Angeles revealed 13% to be seropositive for anti-HEV, this high figure being attributed to exposure to rats *(294)*.

Because of the close homology between human and swine HEV strains *(214,215)*, with up to 97% amino-acid identity in the proteins encoded, pigs are considered the principal natural reservoir of infection. That HEV might be zoonotically transmitted is supported by the finding of an anti-HEV seroprevalence rate of 20%-25% among American swine veterinarians *(275)*, and 10% in American swine workers *(298)*. In a study conducted in Moldavia, where no human hepatitis E has been reported, the anti-HEV seroprevalence was 51% among swine farmers compared to 25% among those not occupationally exposed to swine *(299)*.

Little is known about the epidemiology of HEV in pigs, although the raised profile of pigs in association with human hepatitis E has led to an increase in strain typing and seroprevalence studies. There appears to be high anti-HEV seroprevalences in developing *(300–302)* as well as in some developed countries, e.g., 50% in Korean swine and 60% in Canadian swine *(301)*. In the United Kingdom, the current seroprevalence of about 80% of pigs has been maintained at that level since the early 1990s *(303)*. Endemicity in pigs is likely to be sustained by fecal–oral transmission *(304)*.

Infectivity studies in pigs and rhesus macaques have confirmed cross-infectivity between swine and human HEV *(305)*, and, certainly in developed countries, human and swine strains cluster together phylogenetically (Fig. 2). Further evidence for possible interspecies HEV transmission between pigs and humans comes from studies in Taiwan which show that HEV subgenomic sequences obtained from pigs and humans form a monophyletic group *(265,304)*. In China, Japan, the UK and the Netherlands, genomic sequences of HEV derived from autochthonous cases of human hepatitis E show the greatest similarity to those of swine HEV *(245a 263,266,307)*. Such findings contradict the findings of a molecular epidemiological study conducted in India which showed that human HEV segregated to genotype 1, but swine HEV to genotype 4 *(308)*.

3.3. Foodborne Infection

Hepatitis E associated with foodborne transmission has been reported from Japan, Italy, China, Israel, Germany, and Vietnam *(309–314)*. Some cases of infection acquired in Europe are attributed to the ingestion of uncooked or inadequately cooked shellfish, but the association between shellfish consumption and hepatitis E is not firmly established.

Evidence has recently emerged from Japan that hepatitis E may be transmitted via the consumption of raw or undercooked meat or viscera from pigs and other mammals. A cluster of cases of acute hepatitis E centered around a Japanese family that had consumed raw Sika deer meat some 3 wk earlier. Sequence data from unconsumed meat kept in the freezer showed 100% identity with the sequence obtained from stool samples of the affected family members *(314)*. Moreover, the seroprevalence of anti-HEV IgG was found to be significantly higher in Japanese who had consumed raw deer meat compared to those who had not *(315)*. In another investigation *(316)*, a man presented at hospital with acute hepatitis, 5 wk after consuming uncooked liver from a wild boar. His companion at the same meal presented with fulminant hepatitis in a different hospital at the same city, where he subsequently died. Hepatitis E was diagnosed in each case. In one other investigation, two Japanese men belonging to a seniors association were diagnosed with acute hepatitis E, 39 d after consuming undercooked wild boar meat at a barbecue. After serological testing, 11 of the 12 men present at the barbecue were shown to possess IgG antibodies against HEV, with 8 men still possessing anti-HEV IgM 6 mo after consuming that meal. None had recently traveled to an HEV-hyperendemic area. Sequencing of the HEV genome amplified from the blood of the two original patients showed a 99.4% homology to each other but 92% homology to other genotype 3 HEV sequences originating from Japan *(317)*. A more recent study reported that, of 10 patients presenting with acute or fulminant hepatitis E, nine of them had consumed grilled or undercooked pig liver 2–8 wk prior to the onset of the disease symptoms. Follow-up survey of retail meat outlets revealed that 1.9% of raw pig livers carried detectable HEV *(318)*. Further evidence of the potential significance of the feral mammal reservoir of HEV for interspecies transmission was demonstrated by a survey in Japan that showed antibody to HEV in populations of deer and wild boar and the presence of HEV genotype-3 sequences in the boar samples *(296)*. Moreover, a HEV strain isolated from a wild boar in Japan had a 99.7% sequence homology to HEV isolated from a wild sika deer in the same forest and also to the strain isolated from humans who had consumed the sika deer meat and developed symptoms of hepatitis E *(319)*.

The eating of raw or undercooked meat and viscera may therefore be a major risk factor for nontravel associated hepatitis E in Japan, and it is this activity, rather than contact with pigs *per se*, which may be the route of interspecies HEV transmission. Evidence implicating the consumption of pork as a risk factor is also emerging from countries other than Japan. The islands of Lombok and Surabaya are, like most of Indonesia, predominantly Muslim- populated. An exception is Bali, where the population is approx 90% of Hindus. Hindus eat pork whereas Muslims do not. The HEV seroprevalence amongst apparently healthy subjects was 4.0% in Lombok and 0.5% in Surabaya but 20% in Bali; in contrast, the infection rates of hepatitis A did not differ significantly among these three islands *(300)*. Whether for nonhyperendemic regions the ingestion of inadequately cooked animal products is also a risk factor for hepatitis E merits further investigation.

3.4. Clinical Characteristics and Pathogenicity

3.4.1. In Humans

In humans the disease is that of an acute, icteric hepatitis. The incubation period ranges from 2 to 9 wk (average: 6 wk). Following exposure, viral RNA may be detected in blood and stools about a week before symptoms. Viremia seldom lasts for >2 wk and is often absent by the time illness becomes manifest. Virus is shed into the feces longer but generally stops by the time jaundice resolves. Serum liver enzyme levels are elevated from 4 wk postexposure, usually peaking by about 6 wk and returning to normal by 10 wk. There is a tendency toward cholestasis during the acute phase of infection. In patients with severe disease, hepatocytic necrosis can be extensive *(320)*. HEV infection carries a high risk of mortality when superimposed on existing chronic liver disease, whether of viral or nonviral origin *(321,322)*.

The mortality rate ranges from 0.5 to 2%. Women infected in the third trimester of pregnancy are at particular high risk of progressing to fulminant hepatic failure and death, the mortality rate being as high as 25% *(224,231,250,323,324)*. The mechanisms behind the high mortality rate in pregnant women are unclear, but it is of note that this outcome is peculiar to humans and cannot be reproduced in pregnant nonhuman primates *(325)* or pregnant sows *(326)*.

Evidence is emerging from Japan of increased morbidity and mortality associated with autochthonous hepatitis E *(242,327,328)*, but conflicting findings have been reported *(244)*. The apparent susceptibility of older people to infection may be related to the transmission route and the high inoculation load introduced in the process of ingesting organ(s) that carry replicating virus, such load being much higher compared to virus suspended in

Table 5
HEV Extra-Hepatic Sites of Infection in Pigs,
After Banks et al. *(269)*

Tissue sampled	Pig 1	Pig 2
Bone marrow	+	+
Salivary Gland	–	+
Mesenteric LN	+	–
Inguinal LN	+	NT
Tonsil	–	–
Kidney	+	–
Testes	NT	–
Ileum	–	NT
Serum	+	+
Plasma	+	+
Feces	+	+
Urine	+	–

NT, not tested; LN, lymph node.

contaminated water. It is also possible that there are significant strain differences to pathogenicity, though this has yet to be demonstrated.

3.4.2. In Nonhuman Primates

Many experimental studies on the enteric NANB agent and on HEV have been conducted in primates, including rhesus and cynomolgus macaques, tamarins, African green monkeys, and chimpanzees (*see* reviews by Bradley *[329]*, and Purcell and Emerson *[330]*). Intravenous inoculations with stool suspensions from infected patients (oral inoculations were seldom successful) were shown to lead to hepatic damage, the extent of which varies according to the type of animals used, and also within species *(331)*. Typically, viremia and fecal shedding appear before liver abnormalities. Whether fecal shedding originates from virus accumulating in the bile (which can do so to high concentrations in rhesus monkeys) or whether intestinal viral replication contributes as well is unresolved. Large challenge doses were required to produce disease.

3.4.3. In Pigs

Although antibodies and virus have been detected in pigs worldwide, no specific disease has yet been observed in these animals, despite the very high seroprevalences observed in many developed and developing countries. Most seroconversions occur between 8 and 12 wk of age and may be related to the decline of maternal immunity. Following experimental infection, seroconversion occurs about 18–20 d postinfection, virus is excreted in feces for 3–4 wk, and the pigs are viremic for 1–3 wk. In addition to the liver, HEV may be detected in many other tissues (Table 5); there is evidence that replication is occurring in these tissues *(332,333)*.

Cross-infection studies have revealed the ability of swine strains to infect nonhuman primates; there is also little difference in the limited multifocal lymphohistiocytic hepatitis observed when pigs are infected either with swine strains or with human strains *(305,334)*. One study of experimentally infected pregnant-gilts *(324)* showed that vertical transmission may be prevented, and HEV-free pigs may be derived by early weaning and segregation of the piglets and dams. A minority of the infected dams exhibited mild hepatitis and interstitial nephritis and no effect on the growth rate of the piglets was observed.

3.5. Diagnosis and Detection of HEV

Infection with HEV results in the development of specific antibodies that appear within 2–4 wk after infection. Diagnosis of hepatitis E in humans and pigs may be made by demonstrating circulating HEV antibodies by enzyme immunoassay, for which there are several commercial kits available, using as antigens recombinant capsid protein expressed in insect cells or *Escherichia coli*. Acute infection is confirmed by an IgM assay. This diagnosis may be backed up by RT-PCR testing for viral RNA and subsequent sequencing for epidemiological purposes. Whole blood, serum, or plasma is a suitable source for RT-PCR studies, but fecal samples may be preferable because the period of fecal excretion exceeds the period of viremia in both humans and pigs. HEV has just a single serotype, so HEV capsid antigens derived from viruses of human and swine origin are expected to be equally efficient for detecting anti-HEV by enzyme immunoassay *(301)*. Anti-HEV IgG and protection against reinfection may persist for many years following infection *(335)*, although some reports describe the disappearance of IgG antibodies within 6 mo following primary infection. In a case report from France of an immunologically silent, autochthonous hepatitis E *(271)*, the HEV RNA was detected by PCR in serum and stools from a patient with fever and jaundice, and confirmed to be closely related to other genotype 3 strains identified from the region. Whether this phenomenon reflects the inability of the patient to mount an antibody response or of the insensitivity of the assays used to detect antibodies warrants further investigation.

HEV antigens in liver histological sections may be detected by immunochemistry and in feces by immune electron microscopy, both using convalescent serum as detecting reagent. Such procedures are not routinely used.

RT-PCR approaches have been developed to detect the presence of the HEV genome in water samples with *(336)* and without *(337,338)* prior concentration, and in sewage samples *(284)*. HEV RNA is amplifiable from meat from various animals and from pig liver, blood, and feces *(314,317,318)*. The RT-PCR has also been used in demonstrating transfusion transmission of HEV both in endemic developing and nonendemic developed regions *(253–256)*.

3.6. Treatment and Prevention

There are no reports of specific antiviral therapies for the treatment of hepatitis E. However, the demonstration of the in vitro efficacy of hammerhead ribozymes appears promising as a therapeutic option. Catalytic RNA motifs present in these ribozymes are found in a relatively small number of plant and animal genomes and some viruses (including HEV), and are associated with cellular or viral replication. Inhibition of the 3′-end cis-acting replication regions of the HEV genome by complementary hammerhead motif RNA can inhibit expression from the region, and hence viral replication, by up to 60% *(339)*.

Studies conducted in India and China suggest that immunoglobulin intended for pre-or postexposure prophylaxis against hepatitis E is not efficacious *(310,335,340)*. Plasma collected from donors in less endemic countries will be even less likely to contain

Table 6
Candidate Hepatitis E Vaccines

HEV vaccines
(1) Subunit
Recombinant ORF2 prokaryotic/*E. coli*
Recombinant ORF2 eukaryotic/baculovirus
Recombinant ORF2 eukaryotic/genetically modified tomato
(2) DNA
Chimeric vaccine of ORF-2 & CTLA4
ORF-2 DNA with IL2 and GM-CSF expression plasmids + recombinant
ORF-2 protein booster
(3) Ribozymes

VLP, virus-like particles; CTLA4, cytotoxic T-lymphocyte antigen-4; GM-CSF, granulocyte macrophage colony-stimulating factor.

sufficient titers of protective antibody. Nonetheless, in an epidemic, immunizing pregnant women with high-titered anti-HEV immunoglobulin has been advocated *(340)*.

The development of a vaccine against HEV has been hampered by the difficulty to propagate the virus in vitro, but recent advances in recombinant protein production have led to the development of subunit vaccines, mostly based on the immunodominant ORF-2-encoded capsid protein. At least one of these vaccines has proven to be safe and efficacious in mice and nonhuman primates, and has completed stage III clinical trials in Nepal *(341)*. DNA vaccines based on ORF-2 delivered by gene gun have also been shown to elicit protective immunity in macaques *(342)* and HEV ORF-2-expressing transgenic tomatoes are being evaluated as candidate HEV vaccines in China *(343)*. Most of the HEV DNA vaccines under development are delivered either as a chimera with immuno-stimulating protein genes or as a cocktail of ORF-2 expression plasmids plus immuno-stimulating gene expression plasmids *(344,345)*. Table 6 lists some of the HEV vaccines currently under development.

4. SUMMARY

The incidence of hepatitis A is declining in many countries as standards of living rise. Most hepatitis A is attributable to contact between HAV-infected and susceptible persons. Foodborne transmission of hepatitis A is reported infrequently, although some foodborne transmission probably goes unrecognized. Recognized foodborne hepatitis A has been associated with shellfish, soft fruits, and salad items. Prevention and control are dependent on good personal hygiene and production of safe food and drinking water. Ready access to hand-washing facilities and education in basic hygiene are required for all persons working with food from the harvesting and packing sites to commercial caterers and to the home kitchen. Inadequately treated sewage and contaminated water must be prevented from coming into contact with food crops. Where there is a danger that foods such as shellfish may be contaminated, regulations for treatment to ensure a safe product for the consumer would need to be strictly enforced.

Waterborne hepatitis E continues to be a major cause of epidemic and sporadic hepatitis in sanitation-poor countries. In developed countries, hepatitis E tends to be imported, but shellfish consumption has been implicated. Recent reports from Japan suggest that the disease can be acquired from eating raw or inadequately cooked pig liver, pork, venison and boar meat. A more proactive approach to diagnosis involving routine HEV testing of nontravel-associated non-A, non-B and non-C hepatitis cases is recommended to determine the true extent of autochthonous hepatitis E. In particular, its relationship with the consumption of inadequately cooked animal produce needs to be ascertained. Comprehensive preventative measures against foodborne hepatitis E can only be implemented when the routes of HEV transmission from swine (and other animals) to humans have been precisely defined.

REFERENCES

1. MacCallum, F. O. (1947) Homologous serum jaundice. *Lancet* **ii,** 691–692.
2. Feinstone, S. M., Kapikian, A. Z., and Purcell, R. H. (1973) Hepatitis A: detection by immune electron microscopy of a virus-like antigen associated with acute illness. *Science* **182,** 1026–1028.
3. Ticehurst, J. R. (1986) Hepatitis A virus: clones, cultures and vaccines. *Sem. Liver Dis.* **6,** 46–55.
4. Minor, P. D. (1991) *Picornaviridae.* In: *Classification and Nomenclature of Viruses: Fifth Report of the International Committee on the Taxonomy of Viruses* (Franki, R. I. B., Fauquet, C. M., Knudson, D. L., and Brown, F., eds.), Springer, Wien, pp. 320–326.
5. King, A. M. Q., Brown, F., Christian, P., et al. (2000) *Picornaviridae.* In: *Seventh Report of the International Committee on the Taxonomy of Viruses* (Regenmortel, M. H. V., Fauquet, C. M., Bishop, D. H. L., et al., eds.), Academic, San Diego, CA, pp. 657–683.
6. Lemon, S. M. and Binn, L. N. (1983) Antigenic relatedness of two strains of hepatitis A virus determined by cross-neutralization. *Infect. Immun.* **42,** 418–420.
7. Stapleton, J. T. and Lemon, S. M. (1987) Neutralization escape mutants define a dominant immunogenic neutralization site on hepatitis A virus. *J. Virol.* **61,** 491–498.
8. Lemon, S. M., Chao, S. F., Jansen, R. W., Binn, L. N., and LeDuc, J. W. (1987) Genomic heterogeneity among human and nonhuman strains of hepatitis A virus. *J. Virol.* **61,** 735–742.
9. Marvil, P., Knowles, N. J., Mockett, A. P., Britton, P., Brown, T. D., and Cavanagh, D. (1999) Avian encephalomyelitis virus is a picornavirus and is most closely related to hepatitis A virus. *J. Gen. Virol.* **80,** 653–662.
10. Sobsey, M. D., Shields, P. A., Hauchman, F. H., Hazard, R. L., and Caton, L. W. (1986) Survival and transport of hepatitis A in soils, groundwater and wastewater. *Water Sci. Technol.* **18,** 97–106.
11. Sobsey, M. D., Shields, P. A., Hauchman, F. H., Davis, A. L., Rullman, V. A., and Bosch, A. (1988) Survival and persistence of hepatitis A in environmental samples. In: *Viral Hepatitis and Liver Disease* (Zuckerman, A. J., ed.), Alan Liss, New York, pp. 121–124.
12. Deng, M. Y., Day, S. P., and Oliver, D. O. (1994) Detection of hepatitis A virus in environmental samples by antigen-capture PCR. *Appl. Environ. Microbiol.* **60,** 1927–1933.
13. Nasser, A. M. (1994) Prevalence and fate of hepatitis A in water. *Crit. Rev. Environ. Sci. Technol.* **24,** 281–323.
14. Tsai, Y. L., Sobsey, M. D., Sangermano, L. R., and Palmer. C. J. (1993) Simple method of concentrating enteroviruses and hepatitis A virus from sewage and ocean water for rapid detection by reverse transcriptase-polymerase chain reaction. *Appl. Environ. Microbiol.* **59,** 3488–3491.
15. Callahan, K. M., Taylor, D. J., and Sobsey, M. D. (1995) Comparative survival of hepatitis A, poliovirus and indicator viruses in geographically diverse seawaters. *Water Sci. Technol.* **31,** 189–193.

16. Hejkal, T. W., Keswick, B., LaBelle, R. L., et al. (1982) Viruses in a community water supply associated with an outbreak of gastroenteritis and infectious hepatitis A. *J. Amer. Waterworks Assoc.* **74,** 318–321.

17. Sobsey, M. D., Oglesbee, S. E., and Wait, D. A. (1985) Evaluation of methods for concentrating hepatitis A virus from drinking water. *Appl. Environ. Microbiol.* **50,** 1457–1463.

18. Bloch, A. B., Stramer, S. L., Smith, J. D., et al. (1990) Recovery of hepatitis A from a water supply responsible for a common source outbreak of hepatitis A. *Am. J. Public Health* **80,** 428–430.

19. de Serres, G., Cromeans, T. L., Levesque, B., et al. (1999) Molecular confirmation of hepatitis virus from well water: epidemiology and public health implications. *J. Infect. Dis.* **179,** 37–43.

20. Cromeans, T. L., Favorov, M. O., Nainian, O. V., and Margolis, H. S. (2001) Hepatitis A and E viruses. In: *Foodborne Disease Handbook* (Hui, Y. H., Sattar, S. A., Murrell, K. D., Nip, W-K., and Stanfield, P. S., eds.), 2nd edn, Vol. 2, Marcel Dekker, New York, pp. 23–76.

21. Biziagos, E., Passagot, J., Crance, J. M., and Deloince, R. (1988) Long-term survival of hepatitis A virus and poliovirus type 1 in mineral water. *Appl. Environ. Microbiol.* **54,** 2705–2710.

22. McCaustland, K. A., Bond, W. W., Bradley, D. W., Ebert, J. W., and Maynard, J. E. (1982) Survival of hepatitis A in feces after drying and storage for one month. *J. Clin. Microbiol.* **16,** 957–958.

23. Mbithi, J. N., Springthorpe, V. S., Boulet, J. R., and Sattar, S. A. (1992) Survival of hepatitis A on human hands and its transfer on contact with inanimate surfaces. *J. Clin. Microbiol.* **30,** 757–763.

24. Abad, F. X., Pinto, R. M., and Bosch, A. (1994) Survival of enteric viruses on environmental fomites. *Appl. Environ. Microbiol.* **60,** 3704–3710.

25. Bidawid, S., Farber, J. M., and Sattar, S. A. (2000) Contamination of foods by food handlers: experiments on hepatitis A transfer to food and its interruption. *Appl. Environ. Microbiol.* **66,** 2759–2763.

26. Sattar, S. A., Tetro, J., Bidawid, S., and Farber, J. (2000) Foodborne spread of hepatitis A: recent studies on virus survival, transfer and inactivation. *Can. J. Infect. Dis.* **11,** 159–163.

27. Mbithi, J. N., Springthorpe, V. S., and Sattar, S. A. (1991) Effect of relative humidity and air temperature on survival of hepatitis A on environmental surfaces. *Appl. Environ. Microbiol.* **57,** 1394–1399.

28. Provost, P. J., Oswald, L., Ittensohn, V. M., Villarejos, J. A., Argueda, G., and Hilleman, R. (1973) Etiologic relationship of marmoset propagated CR326 hepatitis A virus to hepatitis in man. *Proc. Soc. Exp. Med. Biol.* **148,** 532–539.

29. Siegl, G., Weitz, M., and Kronauer, G. (1984) Stability of hepatitis A virus. *Intervirology* **22,** 218–226.

30. Scholtz, E., Heinricy, C., and Flemhig, B. (1989) Acid stability of hepatitis A virus. *J. Gen. Virol.* **70,** 2481–2485.

31. Hewitt, J. and Greening, G. E. (2004) Survival and persistence of norovirus, hepatitis A virus, and feline calicivirus in marinated mussels. *J. Food Protect.* **67,** 1743–1750.

32. Provost, P. J., Wolanski, B. S., Miller, W. J., Ittensohn, O. L., McAleer, W. J., and Hilleman, M. R. (1975) Biophysical and biochemical properties of CR326 human hepatitis A virus. *Am. J. Med. Sci.* **270,** 87–92.

33. Parry, J. V. and Mortimer, P. P. (1984) The heat sensitivity of hepatitis A virus determined by a simple tissue culture method. *J. Med. Virol.* **14,** 277–283.

34. Fleming, B., Billing, A., Vallbracht, A., and Botzenhart, K. (1985) Inactivation of hepatitis A by heat and formaldehyde. *Water Sci. Technol.* **17,** 43–45.

35. Millard, J., Appleton, H., and Parry, J. V. (1987) Studies on heat inactivation of hepatitis A virus with special reference to shellfish. *Epid. Infect.* **98,** 397–414.

36. Koff, R. S. and Sears, H. S. (1967) Internal temperature of steamed clams. *N. Engl. J. Med.* **276,** 737–739.
37. Tang, Y. W., Wang, J. X., Zu, Z. Y., Guo, Y. F., Qian, W. H., and Zu, J. X. (1991) A serologically confirmed case–control study of a large outbreak of hepatitis A in China associated with consumption of clams. *Epidemiol. Infect.* **107,** 651–658.
38. Appleton, H. and Pereira, M. S. (1977) A possible virus aetiology in outbreaks of food poisoning from cockles. *Lancet* **i,** 780–781.
39. Mbithi, J. N., Springthorne, V. S., and Sattar, S. A. (1990) Chemical disinfection of hepatitis A on environmental surfaces. *Appl. Environ. Microbiol.* **56,** 3601–3604.
40. Mbithi, J. N., Springthorpe, V. S., and Sattar, SA. (1993) Comparative in vivo efficiencies of hand-washing agents against hepatitis A virus (HM-175) and poliovirus type 1 (Sabin). *Appl. Environ. Microbiol.* **59,** 3463–3469.
41. Seymour, I. J. and Appleton, H. (2001) Foodborne viruses and fresh produce. *J. Appl. Microbiol.* **91,** 759–773.
42. Hutin, Y. J., Pool, V., Cramer, E. H., et al. (1999) A multistate, foodborne outbreak of hepatitis A. National Hepatitis A Investigation Team. *N. Engl. J. Med.* **340,** 595–602.
43. Croci, L., De Medici, D., Scalfaro, C., Fiore, A., and Toti, L. (2002) The survival of hepatitis A in fresh produce. *Int. J. Food Microbiol.* **73,** 29–34.
44. Dawson, D. (2003) *The survival and decontamination of viruses on fresh produce.* Report MB\REP\50260\1. UK Food Standards Agency, London.
45. Badaway, A. S., Gerba, C. P., and Kelley, L. M. (1985) Survival of rotavirus SA-11 on vegetables. *Food Microbiol.* **2,** 199–205.
46. Beuchat, L. R. (1992) Surface disinfection of raw produce. *Dairy Food Environ. Sanitat.* **12,** 6–9.
47. Graham, D. M. (1997) Use of ozone for food processing. *Food Technol.* **51,** 72–75.
48. Hadler, S. C. (1991) Global impact of hepatitis A virus infection changing patterns. In: *Viral Hepatitis and Liver Disease* (Hollinger, F. B., Lemon, S. M., and Margolis, H. S., eds.) Williams and Wilkins, Baltimore, pp. 14–20.
49. Jacobsen, K. H. and Koopman, J. S. (2004) Declining hepatitis A seroprevalence: a global review and analysis. *Epidemiol. Infect.* **132(6),** 1005–1022.
50. Gay, N. J., Morgan-Capner, P., Wright, J., Farrington, C. P., and Miller, E. (1994) Age-specific antibody to hepatitis A in England: implications for disease control. *Epidemiol. Infect.* **113,** 113–120.
51. Harrison, T. J., Dusheiko, G. M., and Zuckerman, A. J. (2000) Hepatitis viruses. In: *Principles and Practice of Clinical Virology* (Zuckerman, A. J., Banatvala, J. E., and Patterson, J. R., eds.), 4th edn, John Wiley, Chichester, UK, pp. 187–233.
52. Termorshuizen, F., Dorigo–Zetsma, J. W., de Melker, H. E., van den Hof, S., and Conyn-Van Spaendonck, S. M. (2000) The prevalence of antibodies to hepatitis A and its determination in the Netherlands: a population based survey. *Epidemiol. Infect.* **124,** 459–466.
53. Crowcroft, N. S., Walsh, B., Davison, K. L., and Gungabissoon, U., on behalf of the PHLS Advisory Committee on Vaccination and Immunisation. (2001) Guidelines for the control of hepatitis A virus infection. *Comm. Dis. Public Health* **4,** 213–227.
54. Shaw, F. E., Hadler, S. C., Maynard, J. E., et al. (1989) Hepatitis A in Mauritius: an apparent transition from endemic to epidemic transmission patterns. *Ann. Trop. Med. Parisitol.* **83,** 179–185.
55. Shaw, F. E., Shapiro, C. N., Welty, T. K., Dill, W., Reddington, J., and Hadler, S. C. (1990) Hepatitis transmission among the Sioux Indians of South Dakota. *Am. J. Public Health* **80,** 1091–1094.
56. Shaw, F. E., Sunman, J. H., Smith, S. M., et al. (1986) A community-wide outbreak of hepatitis A in Ohio. *Am. J. Epidemiol.* **123,** 1057–1065.

57. Crusberg, T. C., Burke, W. M., Reynolds, J. T., Morse, L. E., Reilly, J., and Hoffman, A. H. (1978) The reappearance of classical epidemic infectious hepatitis in Worcester, Massachusetts. *Am. J. Epidemiol.* **107,** 545–551

57a. Wasley, A., Samandari, T., and Bell, B. P. (2005) Incidence of hepatitis A in the United States in the era of vaccination. *JAMA* **294,** 194–201.

58. Bialek, S. R., Thoroughman, D. A., Hu, D., et al. (2004) Hepatitis A incidence and hepatitis A vaccination among American Indians and Alaska Natives, 1990–2001. *Am. J. Public Health* **94,** 996–1001.

59. Corey, L. and Holmes, K. K. (1980) Sexual transmission of hepatitis A in homosexual men: Incidence and mechanism. *N. Engl. J. Med.* **302,** 435–438.

60. Shaw, D. D., Whiteman, D. C., Merritt, A. D., et al. (1999) Hepatitis A outbreaks among illicit drug users and their contacts in Queensland, 1997. *Med. J. Aust.* **170,** 584–587.

61. Hutin, Y. J., Sabin, K. M., Hutwagner, L. C., et al. (2000) Multiple modes of hepatitis A virus transmission among methamphetamine users. *Am. J. Epidemiol.* **152,** 186–192.

62. Mannucci, P. M., Gdovin, S., Gringeri, A., et al. (1994) Transmission of hepatitis A to patients with haemophilia by factor VIII concentrates treated with organic solvent and detergent to inactivate viruses. *Ann. Intern. Med.* **120,** 1–7.

63. Soucie, J. M., Robertson, B. H., Bell, B. P., McCaustland, K. A., and Evatt, B. L. (1998) Hepatitis A virus infection associated with clotting factor concentrate in the United States. *Transfusion* **38,** 573–579.

64. Chudy, M., Budek, I., Keller–Staisalawski, B., et al. (1999) A new cluster of hepatitis A infection in haemophiliacs traced to a contaminated plasma pool. *J. Med. Virol.* **57,** 91–99.

65. Hillis, W. D. (1961) An outbreak of hepatitis among chimpanzee handlers at a United States Airforce Base. *Am. J. Hyg.* **73,** 316–328.

66. Krushale, D. H. (1970) Application of preventative health measures to curtail chimpanzee-associated infectious hepatitis in handlers. *Lab. Animal Care* **20,** 52–56.

67. Mosley, J. W. (1959) Water-borne infectious hepatitis. *N. Engl. J. Med.* **261,** 703–708.

68. Centers for Disease Control and Prevention. (2006) Prevention of Hepatitis A Through Active or Passive Immunization. Recommendations of the Advisory Committee on Immunization Practices. *MMWR* **55 RR07,** 1–23.

69. Centers for Disease Control and Prevention. (2006) Hepatitis surveillance report No. 61. US Department of Health and Human Services, Public Health Service, Atlanta.

70. Bell, B. P., Shapiro, C. N., Alter, M. J., et al. (1998) The diverse patterns of hepatitis A epidemiology in the United States—implications for vaccination strategies. *J. Infect. Dis.* **178,** 1579–1584.

71. Mead, P. S., Slutsker, L., Dietz, V., et al. (1999) Food-related illness and death in the United States. *Emerg. Infect. Dis.* **5,** 607–625.

72. Sockett, P. N., Cowden, J. M., LeBaigue, S., Ross, D., Adak, G., and Evans, H. (1993) Foodborne disease surveillance in England and Wales: 1989–1992. *Comm. Dis. Report* **3,** 159–174.

73. Bowen, G. S. and McCarthy, M. A. (1983) Hepatitis A associated with a hardware store water fountain and contaminated well in Lancaster County, Pennsylvania, 1980. *Amer. J. Epidemiol.* **117,** 695–705.

74. Coursaget, P., Buisson, Y., N'Gawara, M. N., Van Cuyck-Gandre, H., and Roue, R. (1998) Role of hepatitis E virus in sporadic cases of acute and fulminant hepatitis in an endemic area (Chad). *Am. J. Trop. Med. Hyg.* **58,** 330–334.

75. Gray, J., Green, J., Gallimore, C., Lee, J. V., Neal, K., and Brown, D. W. (1997) Mixed genotype SRSV infections among a party of canoeists exposed to contaminated recreational water. *J. Med. Virol.* **52,** 425–429.

76. Sellwood, J., Appleton, H., and Wyn-Jones, P. (1999) Review of sources of enteroviruses in the environment. *EC Bathing Water Directive Enterovirus Research Report.* Ref. No. 99/WW/11/1. UK Water Industry Research, London.

77. Bryan, J. A., Lehmann, J. D., Setiady, I. F., and Hatch, M. H. (1974) An outbreak of hepatitis A associated with recreational lake water. *Am. J. Epidemiol.* **99**, 145–154.

78. Mahoney, F. J., Farley, T. A., Kelso, K. Y., Wilson, S. A., Horan, J. M., and McFarland, L. M. (1992) An outbreak of hepatitis A associated with swimming in a public pool. *J. Infect. Dis.* **165**, 613–618.

79. Lees, D. (2000) Viruses and bivalve shellfish. *Int. J. Food Microbiol.* **59**, 81–116.

80. Halliday, M. L., Kang, Y. L., Zhou, T. E., et al. (1991) An epidemic of hepatitis A attributable to ingestion of raw clams in Shanghai, China. *J. Infect. Dis.* **164**, 852–859.

81. Bostock, A. D. (1979) Hepatitis A infection associated with the consumption of mussels. *J. Infect.* **1**, 171–177.

82. Fujiyama, S., Akahoshi, M., Sagara, K., Sato, T., and Tsurusaki, R. (1985) An epidemic of hepatitis A related to ingestion of raw oysters. *Gastroenterol. Jpn.* **20**, 6–13.

83. Richards, G. P. (1985) Outbreaks of shellfish-associated enteric virus illness in the United States: requisite for development of viral guidelines. *J. Food Protect.* **48**, 815–823.

84. Sockett, P. N., West, P. A., and Jacobs, M. (1985) Shellfish and Public Health. *PHLS Microbiol. Digest* **2**, 29–35.

85. Desenclos, J. C., Klontz, K. C., Wilder, M. H., Nainan, O. V., Margolis, H. S., and Gunn, R. A. (1991) A multistate outbreak of hepatitis A caused by the consumption of raw oysters. *Am. J. Public Health* **81**, 1268–1272.

86. Rippey, S. R. (1994) Infectious diseases associated with molluscan shellfish consumption. *Clin. Microbiol. Rev.* **7**, 419–425.

87. Apairemarchais, V., Robertson, B. H., Aubineauferre, V., et al. (1995) Direct sequencing of hepatitis A virus strains isolated during an epidemic in France. *Appl. Environ. Microbiol.* **61**, 3977–3980.

88. Malfait, P., Lopalco, P. L., Salmaso, S., et al. (1996) An outbreak of hepatitis A in Puglia, Italy. *Euro Surveill.* **1**, 33–35.

89. Leoni, E., Bevini, C., Degli–Esposti, S., and Graziano, A. (1998) An outbreak of intra-familiar hepatitis A associated with clam consumption: epidemic transmission to a school community. *Eur. J. Epidemiol* **14**, 187–192.

90. Conaty, S., Bird, P., Bell, G., Kraa, E., Grohmann, G., and McAnulty, J. M. (2000) Hepatitis A in New South Wales, Australia from consumption of oysters: the first reported outbreak. *Epidemiol. Infect.* **124**, 121–130.

91. Bosch, A., Sanchez, G., Le-Guyader, F., Vanaclocha, H., Haugarreau, L., and Pinto, R. M. (2001) Human enteric viruses in Coquina clams associated with a large hepatitis A outbreak. *Water Sci. Technol.* **43**, 61–65.

92. Sanchez, G., Pinto, R. M., Vanaclocha, H., and Bosch, A. (2002) Molecular characterization of hepatitis A virus isolates from a transcontinental shell-fish-borne outbreak. *J. Clin. Microbiol.* **40**, 4148–4155.

93. Koff, R. S., Grady, G. F., Chalmers, T. C., Mosley, J. W., and Swartz, B. L. (1967) Viral hepatitis in a group of Boston hospitals. *N. Engl. J. Med.* **276**, 703–710.

94. Stille, W. B., Kunkel, B., and Nerger, K. (1972) Oyster-transmitted hepatitis. *Dtsch. Med. Wochenschi.* **97**, 145–147.

95. O'Mahoney, M. C., Gooch, C. D., Smythe, D. A., Thrussell, A. J., Bartlett, C. L. R., and Noah, N. D. (1983) Epidemic hepatitis A from cockles. *Lancet* **i**, 518–520.

96. Kiyosawa, K., Gibo, Y., Sodeyama, T., et al. (1987) Possible infectious causes in 651 patients with acute viral hepatitis during a ten-year period 1976–1985. *Liver* **7**, 163–168.

97. Maguire, H. C., Handford, S., Perry, K. R., et al. (1995) A collaborative case control study of sporadic hepatitis A in England. *Commun. Dis. Report* **5**, 33–40.

98. Salamina, G. and D'Argenio, P. (1998) Shellfish consumption and awareness of risk of acquiring hepatitis A among Neopolitan families. *Euro Surveill.* **3**, 97–98.

99. Lucioni, C., Cipriani, V., Mazzi, S., and Panunzio, M. (1998). Cost of an outbreak of hepatitis A in Puglia, Italy. *Pharmacoeconomics* **13**, 257–266.

100. Chironna, M., Germinario, C., De-Medici, D., Di-Pasquale, S., Quarto, M., and Barbuti, S. (2002) Detection of hepatitis A virus in mussels from different sources marketed in Puglia (Southern Italy). *Int. J. Food Microbiol.* **75,** 12–18.

101. Mitchell, J., Presnell, N., Akin, E., Cummins, J., and Liu, O. (1966). Accumulation and elimination of poliovirus by the eastern oyster. *Am. J. Epidemiol.* **84,** 40–50.

102. Enriquez, R., Frosner, G. G., Hochstein-Mintzel, V., Riedermann, S., and Reinhardt, G. (1992) Accumulation and persistence of hepatitis A in mussels. *J. Med. Virol.* **37,** 174–179.

103. Rosenblum, L. S., Mirkin, I. R., Allen, D. T., Safford, S., and Hadler, S. C. (1990) A multifocal outbreak of hepatitis A traced to commercially distributed lettuce. *Am. J. Public Health* **80,** 1075–1079.

104. Niu, M. T., Polish, L. B., Robertson, B. H., et al. (1992) Multistate outbreak of hepatitis A associated with frozen strawberries. *J. Infect. Dis.* **166,** 518–524.

105. Williams, I. T., Bell, B., Berry, D., and Shapiro, C. N. (1995) *Foodborne outbreak of hepatitis A, Arkansas.* Epidemic Intelligence Service 44th Annual Conference, 27–31 March 1994. Centers for Disease Control and Prevention, Atlanta, p. 19.

106. Pebody, R. G., Leino, T., Ruuttu, P., et al. (1998) Foodborne outbreaks of hepatitis A in a low endemic country: an emerging problem? *Epidemiol. Infect.* **120,** 55–59.

107. Nygard, K., Andersson, Y., Lindkvist, P., et al. (2001) Imported rocket salad partly responsible for increased incidence of hepatitis A cases in Sweden, 2000–2001. *Euro Surveil.* **6,** 151–153.

108. Dentinger, C. M., Bower, W. A., Nainan, O. V., et al. (2001) An outbreak of hepatitis A associated with green onions. *J. Infect. Dis.* **183,** 1273–1276.

109. Wheeler, C.,Vogt, T. M., Armstrong, G. L., et al. (2005) An outbreak of hepatitis A associated with green onions. *N. Engl. J. Med.* **353,** 890–897.

110. Hernandez, F., Monge, R., Jimenez, C., and Taylor, L. (1997) Rotavirus and hepatitis A virus in market lettuce (*Lactuca sativa*) in Costa Rica. *Int. J. Food Microbiol.* **37,** 221–223.

111. Noah, N. D. (1981) Foodborne outbreaks of hepatitis A. *Med. Lab. Sci.* **38,** 428–434.

112. Reid, T. M. S. and Robinson, H. G. (1987) Frozen raspberries and hepatitis A. *Epidemiol. Infect.* **98,** 109–112.

113. Ramsay, C. N. and Upton, P. A. (1989) Hepatitis A and frozen raspberries. *Lancet* **i,** 43–44.

114. Calder, L., Simmons, G., Thornley, C., et al. (2003) An outbreak of hepatitis A associated with the consumption of raw blueberries. *Epidem. Infect.* **131,** 745–751.

115. Long, S. M., Adak, G. K., O'Brien, S. J., and Gillespie, I. A. (2002) General outbreaks of infectious intestinal disease linked with salad vegetables and fruit, England and Wales, 1992–2000. *Comm. Dis. Public Health* **5,** 101–105.

116. Dalton, C. B., Haddix, A., Hoffman, R. E., and Mast, E. E. (1996) The cost of a food-borne outbreak of hepatitis A in Denver, Colorado. *Arch. Int. Med.* **156,** 1013–1016.

117. Levy, B. S., Fontaine, R. E., Smith, C. A., et al. (1975) A large food-borne outbreak of hepatitis A. *JAMA* **234,** 289–294.

118. Meyers, J. D., Romm, F. J., Tihen, W. S., and Bryan, J. (1975) Food-borne hepatitis A in a general hospital: epidemiologic study of an outbreak attributed to sandwiches. *JAMA* **231,** 1049–1053.

119. Hooper, R. R., Juels, C. W., Routenberg, J. A., et al. (1977) An outbreak of type A viral hepatitis at the Naval Training Center, San Diego: epidemiologic evaluation. *Am. J. Epidemiol.* **105,** 148–155.

120. Lowry, P. W., Levine, R., Stroup, D. F., Gunn, R. A., Wilder, M. H., and Konisberg, C. (1989) Hepatitis A outbreak on a floating restaurant in Florida, 1986. *Am. J. Epidemiol.* **129,** 155–164.

121. Centers for Disease Control and Prevention. (1993) Foodborne hepatitis A—Missouri, Wisconsin, and Alaska, 1990–1992. *MMWR* **42,** 526–529.

122. Massoudi, M. S., Bell, B. P., Paredes, V., Insko, J., Evans, K., and Shapiro, C. N. (1999) An outbreak of hepatitis associated with an infected foodhandler. *Public Health Rep.* **114,** 157–164.

123. Schoenbaum, S. C., Baker, O., and Jezek, Z. (1976) Common-source epidemic of hepatitis due to glazed and iced pastries. *Am. J. Epidemiol.* **104**, 74–80.

124. Weltman, A. C., Bennett, N. M., Ackman, D. A., et al. (1996) An outbreak of hepatitis A associated with a bakery, New York 1994: the 1968 "West Branch, Michigan" outbreak repeated. *Epidemiol. Infect.* **117**, 333–341.

125. Warburton, A. R. E., Wreghitt, T. G., Rampling, A., et al. (1991) Hepatitis A outbreak involving bread. *Epidemiol. Inf.* **106**, 199–202.

126. Poonawagul, U., Warintrawat, S., Snitbhan, R., Kitisriwarapoj, S., Chaiyakunt, V., and Foy, H. M. (1995) An outbreak of hepatitis A traced to contaminated water reservoir in cafeteria. *Southeast Asian J. Trop. Med. Public Health* **26**, 705–708.

127. Sundkvist, T., Hamilton, G. R., and Hourihan, M. (2000) Outbreak of hepatitis A spread by contaminated drinking glasses in a public house. *Commun. Dis. Public Health* **3**, 60–62.

128. Krugman, S. and Giles, J. P. (1970) Viral hepatitis: new light on an old disease. *JAMA* **212**, 1019–1029.

129. Smith, P. F., Grabau, J. C., Werzberger, A., et al. (1997) The role of young children in a community-wide outbreak of hepatitis A. *Epidemiol. Infect.* **118**, 243–252.

130. Staes, C. J., Schlenker, T. L., Risk, I., et al. (2000) Sources of infection among persons with acute hepatitis A and no identified risk factors during a sustained community-wide outbreak. *Pediatrics* **106**, E54.

131. Taylor, R. M., Davern, T., Munoz, S., et al. US Acute Liver Failure Study Group. (2006) Fulminant hepatitis A virus infection in the United States: incidence, prognosis, and outcomes. *Hepatology* **44**, 1589–1597.

132. Vento, S., Garofano, T., Renzini, C., et al. (1998) Fulminant hepatitis associated with hepatitis A superinfection in patients with chronic hepatitis C. *N. Engl. J. Med.* **338**, 286–290.

133. Reiss, G. and Keeffe, E. B. (2004) Review article: hepatitis vaccination in patients with chronic liver disease. *Aliment. Pharmacol. Ther.* **19**, 715–727.

134. Debray, D., Cullufi, P., Devictor, D., Fabre, M., and Bernard, O. (1997) Liver failure in children with hepatitis A. *Hepatology* **26**, 1018–1022.

135. Glikson, M., Galun, E., Oren, R., Tur-Kaspa, R., and Shouval, D. (1992) Relapsing hepatitis A. Review of 14 cases and literature survey. *Medicine* **71**, 14–23.

136. Ciocca, M. (2000) Clinical course and consequences of hepatitis A infection. *Vaccine* **18**, S71–S74.

137. Tong, M. J., el-Farra, N. S., and Grew, M. I. (1995) Clinical manifestations of hepatitis A: recent experience in a community teaching hospital. *J. Infect. Dis.* **171**, S15–S18.

138. Gordon, S. C., Reddy, K. R., Schiff, L., and Schiff, E. R. (1984) Prolonged intrahepatic cholestasis secondary to acute hepatitis A. *Ann. Intern. Med.* **101**, 635.

139. Krawczynski, K. Z., Bradley, D. W., Murphy, B. L., et al. (1981) Pathogenic aspects of hepatitis A virus infection in enterally inoculated marmosets. *Am. J. Pathol.* **76**, 698–706.

140. Taylor, G. M., Goldin, R. D., Karayiannis, P., and Thomas, H. C. (1992) In situ hybridization studies in hepatitis A infection. *Hepatology* **16**, 642–648.

141. Karayiannis, P., Jowett, T., Enticott, M., et al. (1986) Hepatitis A virus replication in tamarins and host immune response in relation to pathogenesis of liver cell damage. *J. Med. Virol.* **18**, 261–276.

142. Vallbracht, A. and Fleischer, B. (1992) Immune pathogenesis of hepatitis A. *Archiv. Virol.* **4**, S3–S4.

143. Stapleton, J. C. (1995) Host immune response to hepatitis A virus. *J. Infect. Dis.* **171**, S9–S14.

144. Skinhøj, P., Mathiesen, L. R., Kryger, P., and Møller, A. M. (1981) Faecal excretion of hepatitis A virus in patients with symptomatic hepatitis A infection. *Scand. J. Gastroenterol.* **16**, 1057–1059.

145. Deinstag, J. L., Feinstone, S. M., Kapikian, A. Z., and Purcell, R. H. (1975) Faecal shedding of hepatitis A antigen. *Lancet* **i**, 765–767.

146. Tassopoulos, N. C., Papaevangelou, G. J., Ticehurst, J. R., and Purcell, R. H. (1986) Fecal excretion of Greek strains of hepatitis A virus in patients with hepatitis A and in experimentally infected chimpanzees. *J. Infect. Dis.* **154,** 231–237.

147. Rosenblum, L. S., Villarino, M. E., Nainan, O. V., et al. (1991) Hepatitis A in a neonatal intensive care unit: risk factors for transmission and evidence of prolonged viral excretion among preterm infants. *J. Infect. Dis.* **164,** 476–482.

148. Robertson, B. H., Averhoff, F., Cromeans, T. L., et al. (2000) Genetic relatedness of hepatitis A virus isolates during a community-wide outbreak. *J. Med. Virol.* **62,** 144–150.

149. Yotsuyanangi, H., Koike, K., Yasuda, K., et al. (1996) Prolonged fecal excretion of hepatitis A virus in adults with hepatitis A as determined by polymerase chain reaction. *Hepatology* **24,** 10–13.

150. Liaw, Y. F., Yang, C. Y., Chu, C. M., and Huang, M. J. (1986) Appearance and persistence of hepatitis A IgM antibody in acute clinical hepatitis A observed in an outbreak. *Infection* **14,** 156–158.

151. Provost, P. J. and Hilleman, M. R. (1979) Propagation of human hepatitis A virus in cell culture in vitro. *Proc. Soc. Exp. Biol. Med.* **160,** 213–221.

152. Frosner, G. G., Deinhart, F., Scheid, R., et al. (1979) Propagation of human hepatitis A virus in a hepatoma cell line. *Infection* **7,** 303–305.

153. Daemer, R. J., Feinstone, S. M., Gust, I. D., and Purcell, R. J. (1981) Propagation of human hepatitis A virus in African green monkey kidney cell culture: primary isolation and serial passage. *Infect. Immun.* **32,** 388–393.

154. Flehmig, B. (1980) Hepatitis A-virus in cell culture: I. Propagation of different hepatitis A-virus isolates in a fetal rhesus monkey kidney cell line (Frhk-4). *Med. Microbiol. Immunol.* **168,** 239–248.

155. Gauss-Muller, V., Frosner, G. G., and Deinhardt, F. (1981) Propagation of hepatitis A virus in human embryo fibroblasts. *J. Med. Virol.* **7,** 233–239.

156. Jaykus, L. A., DeLeon, R., and Sobsey, M. D. (1993) Application of RT-PCR for the detection of enteric viruses in oysters. *Water Sci. Technol.* **27,** 49–53.

157. Tsai, Y. L., Tran, B., Sangermano, L. R., and Palmer, C. J. (1994) Detection of poliovirus, hepatits A virus, and rotavirus from sewage and ocean water by triplex reverse transcriptase PCR. *Appl. Environ. Microbiol.* **60,** 2400–2407.

158. Romalde, J. L., Torrado, I., Ribao, C., and Barja, J. L. (2001) Global market: shellfish imports as a source of re-emerging hepatitis A virus infection in Spain. *Int. Microbiol.* **4,** 223–226.

159. Dubois, E., Agier, C., Traore, O., et al. (2002) Modified concentration method for the detection of enteric viruses on fruits and vegetables by reverse transcriptase-polymerase chain reaction or cell culture. *J. Food Protect.* **65,** 1962–1969.

160. Kingsley, D. H., Meade, G. K., and Richards, G. P. (2002) Detection of both hepatitis A virus and Norwalk-like virus in imported clams associated with food-borne illness. *Appl. Environ. Microbiol.* **68,** 3914–3918.

161. Morace, G., Aulicino, F. A., Costanzo, L., Donadio, F., and Rapicetta, M. (2002) Microbial quality of wastewater: detection of hepatitis A by reverse transcriptase-polymerase chain reaction. *J. Appl. Microbiol.* **92,** 828–836.

162. Romalde, J. L., Area, E., Sanchez, G., et al. (2002) Prevalence of enterovirus and hepatitis A virus in bivalve mollusks from Galicia (NW Spain): inadequacy of EU standards of microbiology quality. *Int. J. Food Microbiol.* **74,** 119–130.

163. Slomka, M. J. and Appleton, H. (1998) Feline calicivirus as a model system for heat inactivation studies of small round structured viruses in shellfish. *Epidemiol. Infect.* **121,** 401–407.

164. Sobsey, M. D., Battigelli, D. A., Shin, D. A., and Newland, S. (1999) RT-PCR amplification detects inactivated viruses in water and wastewater. *Water Sci. Technol.* **38,** 91–94.

165. Nuanualsuwan, S. and Cliver, D. O. (2003) Capsid function of inactivated human picorna-viruses and feline calicivirus. *Appl. Environ. Microbiol.* **69**, 350–357.

166. Reynolds, K. A., Gerba, C. P., Abbaszadegan, M., and Pepper, L. L. (2001) ICC/PCR detection of enteroviruses and hepatitis A virus in environmental samples. *Can. J. Microbiol.* **47**, 153–157.

167. Jiang, Y. J., Liao, G. Y., Zhao, W., et al. (2004) Detection of infectious hepatitis A virus by integrated cell culture/strand-specific reverse transcriptase-polymerase chain reaction. *J. Appl. Microbiol.* **97**, 1105–1112.

168. Shapiro, C. N. and Margolis, H. S. (1993) Worldwide epidemiology of hepatitis A virus infection. *J. Hepatol.* **18**, S11–S14.

169. Shouval, D., Ashur, Y., Adler, R., et al. (1993) Safety, tolerability and immunogenicity of an inactivated hepatitis A vaccine: effects of single and booster injections, and comparison to administration of immune globulin. *J. Hepatol.* **18**, S32–S37.

170. Van Damme, P., Mathei, C., Thoelen, S., Meheus, A., Safary, A., and Andre, F. E. (1994) Single dose inactivated hepatitis A vaccine: rationale and clinical assessment of the safety and immunogenicity. *J. Med. Virol.* **44**, 435–441.

171. Van Herck, K., Van Damme, P., Lievens, M., and Stoffel, M. (2004) Hepatitis A vaccine: indirect evidence of immune memory 12 years after the primary course. *J. Med. Virol.* **72**, 194–196.

172. Bovier, P. A., Bock, J., Loutan, L., Farinelli, T., Glueck, R., and Herzog, C. (2002) Long-term immunogenicity of an inactivated virosome hepatitis A vaccine. *J. Med. Virol.* **68**, 489–493.

173. Fiore, A. E., Shapiro, C. N., Sabin, K., et al. (2003) Hepatitis A vaccination of infants: effect of maternal antibody status on antibody persistence and response to a booster dose. *Pediatr. Infect. Dis.* **22**, 354–359.

174. Armstrong, G. L. and Bell, B. P. (2002) Hepatitis A virus infections in the United States: model-based estimates and implications for childhood immunizations. *Pediatrics* **109**, 839–845.

175. Sagliocca, l., Amoroso, P., Stroffolini, T., et al. (1999) Efficacy of hepatitis A vaccine in prevention of secondary hepatitis A infection: a randomised trial. *Lancet* **353**, 1136–1139.

176. Stokes, J. Jr. and Neefe, J. R. (1945) The prevention and attenuation of infectious hepatitis by gamma globulin: preliminary note. *JAMA* **127**, 144–145.

177. Winokur, P. L. and Stapleton, J. T. (1992) Immunoglobulin prophylaxis for hepatitis A. *Clin. Infect. Dis.* **14**, 580–586.

178. Sonder, G. J., Steenbergen, J. E., Bovee, L. P. M. J., Peerbooms, P. G. H., Coutinho, R. A., and van den Hoek, A. (2004) Hepatitis A immunity and seroconversion among contacts of acute hepatitis A patients in Amsterdam, 1996–2000: an evaluation of current prevention policy. *Am. J. Public Health* **94**, 1620–1626.

179. Centers for Disease Control and Prevention. (2003) *Health information for international travel 2003–2004.* Centers for Disease Control and Prevention, US Department of Health and Human Services, Public Health Service, Atlanta.

180. National Shellfish Sanitation Program. (1989) *Manual of Operations. Part 1. Sanitation of shellfish growing areas.* Food and Drug Administration, US Department of Health and Human Services, Public Health Service, Washington, DC.

181. Council of the European Communities. (1991) Council Directive of 15 July 1991 laying down the health conditions for the production and placing on the market of live bivalve molluscs (91/492/EEC). *Official J. Euro. Comm.* **L268**, pp. 1–14.

182. Croci, I., Ciccozzi, M., DeMedici, D., et al. (1999) Inactivation of hepatitis A in heat-treated mussels. *J. Appl. Microbiol.* **87**, 884–888.

183. Croci, L., De Medici, D., Scalfaro, C., et al. (2000) Determination of enteroviruses, hepatitis A virus, bacteriophages and *Escherichia coli* in Adriatic Sea mussels. *J. Appl. Microbiol.* **88**, 293–298.

184. Beuchat, L. R. (1998) Surface decontamination of fruits and vegetables eaten raw: a review. WHO/FSF/FOS/98.2. Food Safety Unit, World Health Organisation, Geneva.

185. Ministry of Agriculture, Fisheries and Food. (1998) *The soil code: Code of good agricultural practice for the protection of soil.* Welsh Office Agricultural Department, Ministry of Agriculture, Fisheries and Food, London.

186. Agricultural Development and Advisory Service. (2001) The safe sludge matrix. Ref. no. AMPU 81234/C/2. ADAS, Wolverhampton.

187. Pina, S., Puig, M., Lucena, F., Jofre, J., and Girones, R. (1998) Viral pollution in the environment and in shellfish—human adenovirus detection by PCR as an index of human viruses. *Appl. Environ. Microbiol.* **64,** 3376–3382.

188. Havelaar, A. H. (1987) Bacteriophages as model organisms in water treatment. *Microbiol. Sci.* **4,** 362–364.

189. Tartera, C. and Joffre, J. (1987) Bacteriophages active against bacteriooides fragilis in sewage-polluted water. *Appl. Environ. Microbiol.* **53,** 1632–1637.

190. Lucena, F., Lasobras, J., Mcintosh, D., Forcadell, M., and Joffre, J. (1994) Effect of distance from the polluting focus on relative concentrations of *Bacteroides fragilis* and coliphages in mussels. *Appl. Environ. Microbiol.* **60,** 2272–2277.

191. Chung, H., Jaykus, L. A., Lovelace, G., and Sobsey, M. D. (1998) Bacteriophages and bacteria as indicators of enteric viruses in oysters and their harvest waters. *Water Sci. Technol.* **38,** 37–44.

192. Dore, W. J., Henshilwod, K., and Lees, D. N. (2000) Evaluation of F-specific bacteriophage as a candidiate human enteric virus indicator for bivalve molluscan shellfish. *Appl. Environ. Microbiol.* **66,** 1280–1285.

193. Power, U. F. and Collins, J. K. (1989) Differential depuration of poliovirus, *Escherichia coli* and a coliphage by the common mussel *Mytilus edulis. Appl. Environ. Microbiol.* **55,** 1386–1390.

194. Dore, W. J. and Lees, D. N. (1995) Behaviours of *Escherichia coli* and male specific bacteriophage in environmentally contaminated bivalve mollusks before and after depuration. *Appl. Environ. Microbiol.* **61,** 2830–2834.

195. Dawson, D. J., Paish, A., Staffell, L. M., Seymour, I. J., and Appleton, H. (2005) Survival of viruses on fresh produce, using MS2 as a surrogate for norovirus. *J. Appl. Microbiol.* **98,** 203–209.

196. Cliver, D. O. (1997) Virus transmission via food. *Food Technol.* **51,** 71–78.

197. Meltzer, M. I., Shapiro, C. N., Mast, E. E., and Arcari, C. (2001) The economics of vaccinating restaurant workers against hepatitis A. *Vaccine* **19,** 2138–2145.

198. Jacobs, R. J., Grover, S. F., Meyerhoff, A. S., and Paivana, T. A. (2000) Cost effectiveness of vaccinating food service workers against hepatitis A infection. *J. Food Prot.* **63,** 768–774.

199. Tricco, A. C., Pham, B., Duval, B., et al. (2006) A review of interventions triggered by hepatitis A infected food-handlers in Canada. *BMC Health Serv. Res.* **6,** 157.

200. Vishwanathan, R. (1957) Infectious hepatitis in Delhi (1955–56). A critical study. Epidemiology. *Indian J. Med. Res.* **45,** S1–S29.

201. Khuroo, M. S. (1980) Study of an epidemic of non-A, non-B hepatitis. Possibility of another human hepatitis virus distinct from post-transfusion non-A, non-B type. *Am. J. Med.* **68,** 818–824.

202. Balayan, M. S., Andjaparidze, A. G., Savinskaya, S. S., et al. (1983) Evidence for a virus in non-A, non-B hepatitis transmitted via the fecal-oral route. *Intervirology* **20,** 23–31.

203. Kane, M. A., Bradley, D. W., Shrestha, S. M., et al. (1984) Epidemic non-A, non-B hepatitis in Nepal. Recovery of a possible etiologic agent and transmission studies in marmosets. *JAMA* **252,** 3140–3145.

204. Bradley, D. W., Krawczynski, K., Cook, E. H., et al. (1987) Enterically transmitted non-A, non-B hepatitis: serial passage of disease in cynomolgus macaques and tamarins and recovery of disease-associated 27- to 34-nm virus like particles. *Proc. Natl Acad. Sci. USA* **84,** 6277–6281.

205. Reyes, G. R., Purdy, M. A., Kim, J. P., et al. (1990) Isolation of a cDNA from the virus responsible for enterically transmitted non-A, non-B hepatitis. *Science* **247**, 1335–1339.
206. Tam, A. W., Smith, M. M., Guerra, M. E., et al. (1991) Hepatitis E virus (HEV): molecular cloning and sequencing of the full-length viral genome. *Virology* **185**, 120–131.
207. Kar-Roy, A., Korkaya, H., Oberoi, R., Lal, S. K., and Jameel, S. (2004) The hepatitis E virus ORF3 protein activates ERK through binding and inhibition of the MAPK phosphatase. *J. Biol. Chem.* **279**, 28,345–28,357.
208. Emerson, S. U., Anderson D., Arankalle, A., et al. (2004) *Hepevirus.* In: *Virus taxonomy: eighth report of the International Committee on Taxonomy of Viruses* (Fauquet, C. M., Mayo, M. A., Maniloff, J., Desselberger, U., and Ball, L. A., ed.), Elsevier/Academic Press, London. pp. 851–855.
209. Kazachkov, Y. A., Balayan, M. S., Ivannikova, T. A., et al. (1992) Hepatitis E virus in cultivated cells. *Arch. Virol.* **127**, 399–402.
210. Tam, A. W., White, R., Reed, E., et al. (1996) In vitro propagation and production of hepatitis E virus from in vivo-infected primary macaque hepatocytes. *Virology* **215**, 1–9.
211. Panda, S. K., Ansari, I. H., Durgapal, H., Agrawal, S., and Jameel, S. (2000) The in vitro-synthesized RNA from a cDNA clone of hepatitis E virus is infectious. *Virology* **74**, 2430–2437.
212. Emerson, S. U., Zhang, M., Meng, X. J., et al. (2001) Recombinant hepatitis E virus genomes infectious for primates: importance of capping and discovery of a cis-reactive element. *Proc. Natl Acad. Sci. USA* **98**, 15,270–15,275.
213. Clayson, E. T., Innis, B. L., Myint, K. S., et al. (1995) Detection of hepatitis E virus infections among domestic swine in the Kathmandu Valley of Nepal. *Am. J. Trop. Med. Hyg.* **53**, 228–232.
214. Meng, X. J., Purcell, R. H., Halbur, P. G., et al. (1997) A novel virus in swine is closely related to the human hepatitis E virus. *Proc. Natl Acad. Sci. USA* **94**, 9860–9865.
215. Schlauder, G. G., Dawson, G. J., Erker, J. C., et al. (1998) The sequence and phylogenetic analysis of a novel hepatitis E virus isolated from a patient with acute hepatitis reported in the United States. *J. Gen. Virol.* **79**, 447–456.
216. Payne, C. J., Ellis, T. M., Plant, S. L., Gregory, A. R., and Wilcox, G. E. (1999) Sequence data suggests big liver and spleen disease virus (BLSV) is genetically related to hepatitis E virus. *Vet. Microbiol.* **68**, 119–125.
217. Haqshenas, G., Shivaprasad, H. L., Woolcock, P. R., Read, D. H., and Meng, X. J. (2001) Genetic identification and characterization of a novel virus related to human hepatitis E virus from chickens with hepatitis-splenomegaly syndrome in the United States. *J. Gen. Virol.* **82**, 2449–2462.
218. Sun, Z. F., Larsen, C. T., Dunlop, A., et al. (2004) Genetic identification of avian hepatitis E virus (HEV) from healthy chicken flocks and characterization of the capsid gene of 14 avian HEV isolates from chickens with hepatitis-splenomegaly syndrome in different geographical regions of the United States. *J. Gen. Virol.* **85**, 693–700.
219. Sun, Z. F., Larsen, C. T., Huang, F. F., et al. (2004) Generation and infectivity titration of an infectious stock of avian hepatitis E virus (HEV) in chickens and cross-species infection of turkeys with avian HEV. *J. Clin. Microbiol.* **42**, 2658–2662.
220. Skidmore, S. J., Yarbough, P. O., Gabor, K. A., and Reyes, G. R. (1992) Hepatitis E virus: the cause of a waterbourne hepatitis outbreak. *J. Med. Virol.* **37**, 58–60.
221. Naik, S. R., Aggarwal, R., Salunke, P. N., and Mehrotra, N. N. (1992) A large waterborne viral hepatitis E epidemic in Kanpur, India. *Bull. World Health Organ.* **70**, 597–604.
222. Dilawari, J. B., Singh, K., Chawla, Y. K., et al. (1994) Hepatitis E virus: epidemiological, clinical and serological studies of north Indian epidemic. *Indian J. Gastroenterol.* **13**, 44–48.
223. Rab, M. A., Bile, M. K., Mubarik, M. M., et al. (1997) Water-borne hepatitis E virus epidemic in Islamabad, Pakistan: a common source outbreak traced to the malfunction of a modern water treatment plant. *Am. J. Trop. Med. Hyg.* **57**, 151–157.

224. Myint, H., Soe, M. M., Khin, T., Myint, T. M., and Tin, K. M. (1985) A clinical and epidemiological study of an epidemic of non-A non-B hepatitis in Rangoon. *Am. J. Trop. Med. Hyg.* **34**, 1183–1189.

225. Uchida, T., Aye, T. T., Ma, X., et al. (1993) An epidemic outbreak of hepatitis E in Yangon of Myanmar: antibody assay and animal transmission of the virus. *Acta Pathol. Jpn.* **43**, 94–98.

226. Clayson, E. T., Vaughn, D. W., Innis, B. L., Shrestha, M. P., Pandey, R., and Malla, D. B. (1998) Association of hepatitis E virus with an outbreak of hepatitis at a military training camp in Nepal. *J. Med. Virol.* **54**, 178–182.

227. Labrique, A. B., Thomas, D. L., Stoszek, S. K., and Nelson, K. E. (1999) Hepatitis E: an emerging infectious disease. *Epidemiol. Rev.* **21**,162–179.

228. Corwin, A., Putri, M. P., Winarno, J., et al. (1997) Epidemic and sporadic hepatitis E virus transmission in West Kalimantan (Borneo), Indonesia. *Am. J. Trop. Med. Hyg.* **57**, 62–65.

229. Sedyaningsih-Mamahit, E. R., Larasati, R. P., Laras, K., et al. (2002) First documented outbreak of hepatitis E virus transmission in Java, Indonesia. *Trans. R. Soc. Trop. Med. Hyg.* **96**, 398–404.

230. Corwin, A. L., Khiem, H. B., Clayson, E. T., et al. (1996) A waterborne outbreak of hepatitis E virus transmission in southwestern Vietnam. *Am. J. Trop. Med. Hyg.* **54**, 559–562.

231. Tsega, E., Hansson, B. G., Krawczynski, K., and Nordenfelt, E. (1992) Acute sporadic viral hepatitis in Ethiopia: causes, risk factors, and effects on pregnancy. *Clin. Infect. Dis.* **14**, 961–965.

232. Bile, K., Isse, A., Mohamud, O., et al. (1994) Contrasting roles of rivers and wells as sources of drinking water on attack and fatality rates in a hepatitis E epidemic in Somalia. *Am. J. Trop. Med. Hyg.* **51**, 466–474.

233. Benjelloun, S., Bahbouhi, B., Bouchrit, N., et al. (1997) Seroepidemiological study of an acute hepatitis E outbreak in Morocco. *Res. Virol.* **148**, 279–287.

234. Isaacson, M., Frean, J., He, J., Seriwatana, J., and Innis, B. L. (2000) An outbreak of hepatitis E in Northern Namibia, 1983. *Am. J. Trop. Med. Hyg.* **62**, 619–625.

235. Byskov, J., Wouters, J. S., Sathekge, T. J., and Swanepoel, R. (1989) An outbreak of suspected water-borne epidemic non-A non-B hepatitis in northern Botswana with a high prevalence of hepatitis B carriers and hepatitis delta markers among patients. *Trans. R. Soc. Trop. Med. Hyg.* **83**, 110–116.

236. Velazquez, O., Stetler, H. C., Avila, C., et al. (1990) Epidemic transmission of enterically transmitted non-A, non-B hepatitis in Mexico, 1986–1987. *JAMA* **263**, 3281–3285.

237. Zhuang, H. (1992) Hepatitis E and strategies for its control. In: *Viral Hepatitis in China: Problems and Control strategies, Monogr. Virol. Vol. 19* (Wen, Y-M., Xu, Z-Y., and Melnick, J. L., eds.), Karger, Basel, pp. 126–139.

238. Hyams, K. C., Purdy, M. A., Kaur, M., et al. (1992) Acute sporadic hepatitis E in Sudanese children: analysis based on a new western blot assay. *J. Infect. Dis.* **165**, 1001–1005.

239. Bansal, J., He, J., Yarbough, P. O., Sen, S., Constantine, N. T., and Sen, D. (1998) Hepatitis E virus infection in eastern India. *Am. J. Trop. Med. Hyg.* **59**, 258–260.

240. Das, K., Agarwal, A., Andrew, R., Frosner, G. G., and Kar, P. (2000) Role of hepatitis E and other hepatotropic virus in aetiology of sporadic acute viral hepatitis: a hospital based study from urban Delhi. *Eur. J. Epidemiol.* **16**, 937–940.

241. Chadha, M. S., Walimbe, A. M., Chobe, L. P., and Arankalle, V. A. (2003) Comparison of etiology of sporadic acute and fulminant viral hepatitis in hospitalized patients in Pune, India, during 1978–1981 and 1994–1997. *Indian J Gastroenterol.* **22**, 11–15.

242. Okamoto, H., Takahashi, M., and Nishizawa, T. (2003) Features of hepatitis E virus infection in Japan. *Intern. Med.* **42**, 1065–1071.

243. Yamamoto, T., Suzuki, H., Toyota, T., Takahashi, M., and Okamoto, H. (2004) Three male patients with sporadic acute hepatitis E in Sendai, Japan, who were domestically infected with hepatitis E virus of genotype III or IV. *J. Gastroenterol.* **39**, 292–298.

244. Sainokami, S., Abe, K., Kumagai, I., et al. (2004) Epidemiological and clinical study of sporadic acute hepatitis E caused by indigenous strains of hepatitis E virus in Japan compared with acute hepatitis A. *J. Gastroenterol.* **39,** 640–648.

245. Widdowson, M. A., Jaspers, W. J., van der Poel, W. H., et al. (2003) Cluster of cases of acute hepatitis associated with hepatitis E virus infection acquired in the Netherlands. *Clin. Infect. Dis.* **36,** 29–33.

245a. Ijaz, S., Arnold, E., Banks, M., et al. (2005) Non-travel-associated hepatitis E in England and Wales: demographic, clinical, and molecular epidemiological characteristics. *J. Infect. Dis.* **192,** 1166–1172.

246. Arankalle, V. A., Chadha, M. S., Mehendale, S. M., and Tungatkar, S. P. (2000) Epidemic hepatitis E: serological evidence for lack of intrafamilial spread. *Indian J. Gastroenterol.* **19,** 24–28.

247. Somani, S. K., Aggarwal, R., Naik, S. R., Srivastava, S., and Naik, S. (2003) A serological study of intrafamilial spread from patients with sporadic hepatitis E virus infection. *J. Viral Hepat.* **10,** 446–449.

248. Greco, D., De Giacomi, G., Piersante, G. P., Bibby, L., Nicastro, M., and Cavalcanti, P. (1986) A person to person hepatitis A outbreak. *Int. J. Epidemiol.* **15,** 108–111.

249. Khuroo, M. S., Kamili, S., and Jameel, S. (1995) Vertical transmission of hepatitis E virus. *Lancet* **345,** 1025–1026.

250. Kumar, A., Beniwal, M., Kar, P., Sharma, J. B., and Murthy, N. S. (2004) Hepatitis E in pregnancy. *Int. J. Gynaecol. Obstet.* **85,** 240–244.

251. Kumar, R. M., Uduman, S., Rana, S., Kochiyil, J. K., Usmani, A., and Thomas, L. (2001) Sero-prevalence and mother-to-infant transmission of hepatitis E virus among pregnant women in the United Arab Emirates. *Eur. J. Obstet. Gynecol. Reprod. Biol.* **100,** 9–15.

252. Singh, S., Mohanty, A., Joshi, Y. K., Deka, D., Mohanty, S., and Panda, S. K. (2003) Mother-to-child transmission of hepatitis E virus infection. *Indian J. Pediatr.* **70,** 37–39.

253. Khuroo, M. S., Kamili, S., and Yattoo, G. N. (2004) Hepatitis E virus infection may be transmitted through blood transfusions in an endemic area. *J. Gastroenterol. Hepatol.* **19,** 778–784.

254. Fukuda, S., Sunaga, J., Saito, N., et al. (2004) Prevalence of antibodies to hepatitis E virus among Japanese blood donors: identification of three blood donors infected with a genotype 3 hepatitis E virus. *J. Med. Virol.* **73,** 554–561.

255. Mitsui, T., Tsukamoto, Y., Yamazaki, C., et al. (2004) Prevalence of hepatitis E virus infection among hemodialysis patients in Japan: evidence for infection with a genotype 3 HEV by blood transfusion. *J. Med. Virol.* **74,** 563–572.

256. Matsubayashi, K., Nagaoka, Y., Sakata, H., et al. (2004) Transfusion-transmitted hepatitis E caused by apparently indigenous hepatitis E virus strain in Hokkaido, Japan. *Transfusion* **44,** 934–940.

257. Clayson, E. T., Shrestha, M. P., Vaughn, D. W., et al. (1997) Rates of hepatitis E virus infection and disease among adolescents and adults in Kathmandu, Nepal. *J. Infect. Dis.* **176,** 763–766.

258. Tran, H. T., Ushijima, H., Quang, V. X., et al. (2003) Prevalence of hepatitis virus types B through E and genotypic distribution of HBV and HCV in Ho Chi Minh City, Vietnam. *Hepatol. Res.* **26,** 275–280.

259. Fix, A. D., Abdel-Hamid, M., Purcell, R. H., et al. (2000) Prevalence of antibodies to hepatitis E in two rural Egyptian communities. *Am. J. Trop. Med. Hyg.* **62,** 519–523.

260. Arankalle, V. A., Tsarev, S. A., Chadha, M. S., et al. (1995) Age-specific prevalence of antibodies to hepatitis A and E viruses in Pune, India, 1982 and 1992. *J. Infect. Dis.* **171,** 447–450.

261. Hyams, K. C., McCarthy, M. C., Kaur, M., et al. (1992) Acute sporadic hepatitis E in children living in Cairo, Egypt. *J. Med. Virol.* **37,** 274–277.

262. Schwartz, E., Jenks, N. P., Van Damme, P., and Galun, E. (1999) Hepatitis E virus infection in travelers. *Clin. Infect. Dis.* **29,** 1312–1314.

263. Takahashi, K., Kang, J. H., Ohnishi, S., Hino, K., and Mishiro, S. (2002) Genetic heterogeneity of hepatitis E virus recovered from Japanese patients with acute sporadic hepatitis. *J. Infect. Dis.* **185,** 1342–1345.

264. Mizuo, H., Suzuki, K., Takikawa, Y., et al. (2002) Polyphyletic strains of hepatitis E virus are responsible for sporadic cases of acute hepatitis in Japan. *J. Clin. Microbiol.* **40,** 3209–3218.

265. Hsieh, S. Y., Meng, X. J., Wu, Y. H., et al. (1999) Identity of a novel swine hepatitis E virus in Taiwan forming a monophyletic group with Taiwan isolates of human hepatitis E virus. *J. Clin. Microbiol.* **37,** 3828–3834.

266. Van der Poel, W. H., Verschoor, F., van der Heide, R., et al. (2001) Hepatitis E virus sequences in swine related to sequences in humans, The Netherlands. *Emerg. Infect. Dis.* **7,** 970–976.

267. McCrudden, R., O'Connell, S., Farrant, T., Beaton, S., Iredale, J. P., and Fine, D. (2000) Sporadic acute hepatitis E in the United Kingdom: an underdiagnosed phenomenon? *Gut* **46,** 732–733.

268. Levine, D. F., Bendall, R. P., and Teo, C. G. (2000) Hepatitis E acquired in the UK. *Gut* **47,** 740.

269. Banks, M., Bendall, R., Grierson, S., Heath, G., Mitchell, J., and Dalton, H. (2004) Human and porcine hepatitis E virus strains, United Kingdom. *Emerg. Infect. Dis.* **10,** 953–955.

270. Wang, Y., Levine, D. F., Bendall, R. P., Teo, C. G., and Harrison, T. J. (2001) Partial sequence analysis of indigenous hepatitis E virus isolated in the United Kingdom. *J. Med. Virol.* **65,** 706–709.

271. Mansuy, J. M., Peron, J. M., Abravanel, F., et al. (2004) Hepatitis E in the south west of France in individuals who have never visited an endemic area. *J. Med. Virol.* **74,** 419–424.

272. Mansuy, J. M., Peron, J. M., Bureau, C., Alric, L., Vinel, J. P., and Izopet, J. (2004) Immunologically silent autochthonous acute hepatitis E virus infection in France. *J. Clin. Microbiol.* **42,** 912–913.

273. Schlauder, G. G., Desai, S. M., Zanetti, A. R., Tassopoulos, N. C., and Mushahwar, I. K. (1999) Novel hepatitis E virus (HEV) isolates from Europe: evidence for additional genotypes of HEV. *J. Med. Virol.* **57,** 243–251.

274. Buti, M., Clemente-Casares, P., Jardi, R., et al. (2004) Sporadic cases of acute autochthonous hepatitis E in Spain. *J. Hepatol.* **411,** 26–31.

275. Meng, X. J., Wiseman, B., Elvinger, F., et al. (2002) Prevalence of antibodies to hepatitis E virus in veterinarians working with swine and in normal blood donors in the United States and other countries. *J. Clin. Microbiol.* **40,** 117–122.

276. Choi, I. S., Kwon, H. J., Shin, N. R., and Yoo, H. S. (2003) Identification of swine hepatitis E virus (HEV) and prevalence of anti-HEV antibodies in swine and human populations in Korea. *J. Clin. Microbiol.* **41,** 3602–3608.

277. Alvarez-Munoz, M. T., Torres, J., Damasio, L., Gomez, A., Tapia-Conyer, R., and Munoz, O. (1999) Seroepidemiology of hepatitis E virus infection in Mexican subjects 1 to 29 years of age. *Arch. Med. Res.* **30,** 251–254.

278. Goncales, N. S., Pinho, J. R., Moreira, R. C., et al. (2000) Hepatitis E virus immunoglobulin G antibodies in different populations in Campinas, Brazil. *Clin. Diagn. Lab. Immunol.* **7,** 813–816.

279. Sheikh, A., Sugitani, M., Kinukawa, N., et al. (2002) Hepatitis E virus infection in fulminant hepatitis patients and an apparently healthy population in Bangladesh. *Am. J. Trop. Med. Hyg.* **66,** 721–724.

280. Hau, C. H., Hien, T. T., Tien, N. T., et al. (1999) Prevalence of enteric hepatitis A and E viruses in the Mekong River delta region of Vietnam. *Am. J. Trop. Med. Hyg.* **60,** 277–280.

281. Mast, E. E., Alter, M. J., Holland, P. V., and Purcell, R. H. (1998) Evaluation of assays for antibody to hepatitis E virus by a serum panel. Hepatitis E Virus Antibody Serum Panel Evaluation Group. *Hepatology* **27,** 857–861.

282. Wang, Y., Zhang, H., Li, Z., et al. (2001) Detection of sporadic cases of hepatitis E virus (HEV) infection in China using immunoassays based on recombinant open reading frame 2 and 3 polypeptides from HEV genotype 4. *J. Clin. Microbiol.* **39,** 4370–4379.
283. Corwin, A. L., Tien, N. T., Bounlu, K., et al. (1999) The unique riverine ecology of hepatitis E virus transmission in South-East Asia. *Trans. R. Soc. Trop. Med. Hyg.* **93,** 255–260.
284. Jothikumar, N., Aparna, K., Kamatchiammal, S., Paulmurugan, R., Saravanadevi, S., and Khanna, P. (1993) Detection of hepatitis E virus in raw and treated wastewater with the polymerase chain reaction. *Appl. Environ. Microbiol.* **59,** 2558–2562.
285. Vaidya, S. R., Chitambar, S. D., and Arankalle, V. A. (2002) Polymerase chain reaction–based prevalence of hepatitis A, hepatitis E and TT viruses in sewage from an endemic area. *J. Hepatol.* **37,** 131–136.
286. Pina, S., Buti, M., Cotrina, M., Piella, J., and Girones, R. (2000) HEV identified in serum from humans with acute hepatitis and in sewage of animal origin in Spain. *J. Hepatol.* **33,** 826–833.
287. Clemente-Casares, P., Pina, S., Buti, M., et al. (2003) Hepatitis E virus epidemiology in industrialized countries. *Emerg. Infect. Dis.* **9,** 448–454.
288. Pina, S., Jofre, J., Emerson, S. U., Purcell, R. H., and Girones, R. (1998) Characterization of a strain of infectious hepatitis E virus isolated from sewage in an area where hepatitis E is not endemic. *Appl. Environ. Microbiol.* **64,** 4485–4488.
288a. Emerson, S. U., Arankalle, V. A., and Purcell, R. H. (2005) Thermal stability of hepatitis E virus. *J. Infect. Dis.* **192,** 930–933.
289. Arankalle, V. A., Joshi, M. V., Kulkarni, A. M., et al. (2001). Prevalence of anti-hepatitis E virus antibodies in different Indian animal species. *J. Viral Hepat.* **8,** 223–227.
290. Hirano, M., Ding, X., Li, T. C., et al. (2003) Evidence for widespread infection of hepatitis E virus among wild rats in Japan. *Hepatol. Res.* **27,** 1–5.
291. Hirano, M., Ding, X., Tran, H. T., et al. (2003) Prevalence of antibody against hepatitis E virus in various species of non-human primates: evidence of widespread infection in Japanese monkeys (*Macaca fuscata*). *Jpn. J. Infect. Dis.* **56,** 8–11.
292. He, J., Innis, B. L., Shrestha, M. P., et al. (2002) Evidence that rodents are a reservoir of hepatitis E virus for humans in Nepal. *J. Clin. Microbiol.* **40,** 4493–4498.
293. Mushahwar, I. K., Dawson, G. J., Bile, K. M., and Magnius, L. O. (1993) Serological studies of an enterically transmitted non-A, non-B hepatitis in Somalia. *J. Med. Virol.* **40,** 218–221.
294. Smith, H. M., Reporter, R., Rood, M. P., et al. (2002) Prevalence study of antibody to ratborne pathogens and other agents among patients using a free clinic in downtown Los Angeles. *J. Infect. Dis.* **186,** 1673–1676.
295. Wang, Y. C., Zhang, H. Y., Xia, N. S., et al. (2002) Prevalence, isolation, and partial sequence analysis of hepatitis E virus from domestic animals in China. *J. Med. Virol.* **67,** 516–521.
296. Sonoda, H., Abe, M., Sugimoto, T., et al. (2004) Prevalence of Hepatitis E Virus (HEV) infection in wild boars and deer and genetic identification of a genotype 3 HEV from a boar in Japan. *J. Clin. Microbiol.* **42,** 5371–5374.
297. Usmanov, R. K., Balaian, M. S., Dvoinikova, O. V., et al. (1994) An experimental infection in lambs by the hepatitis E virus. *Vopr. Virusol.* **39,** 165–168.
298. Withers, M. R., Correa, M. T., Morrow, M., et al. (2002) Antibody levels to hepatitis E virus in North Carolina swine workers, non-swine workers, swine, and murids. *Am. J. Trop. Med. Hyg.* **66,** 384–388.
299. Drobeniuc, J., Favorov, M. O., Shapiro, C. N., et al. (2001) Hepatitis E virus antibody prevalence among persons who work with swine. *J. Infect. Dis.* **184,** 1594–1597.
300. Garkavenko, O., Obriadina, A., Meng, J., et al. (2001) Detection and characterisation of swine hepatitis E virus in New Zealand. *J. Med. Virol.* **65,** 525–529.

301. Engle, R. E., Yu, C., Emerson, S. U., Meng, X. J., and Purcell, R. H. (2002) Hepatitis E virus (HEV) capsid antigens derived from viruses of human and swine origin are equally efficient for detecting anti-HEV by enzyme immunoassay. *J. Clin. Microbiol.* **40,** 4576–4580.

302. Wibawa, I. D., Muljono, D. H., Mulyanto, et al. (2004) Prevalence of antibodies to hepatitis E virus among apparently healthy humans and pigs in Bali, Indonesia: identification of a pig infected with a genotype 4 hepatitis E virus. *J. Med. Virol.* **73,** 38–44.

303. Banks, M., Heath, G. S., Grierson, S. S., et al. (2004) Evidence for the presence of hepatitis E virus in pigs in the United Kingdom. *Vet. Rec.* **154,** 223–227.

304. Kasorndorkbua, C., Guenette, D. K., Huang, F. F., Thomas, P. J., Meng, X. J., and Halbur, P. G. (2004) Routes of transmission of swine hepatitis E virus in pigs. *J. Clin. Microbiol.* **42,** 5047–5052.

305. Meng, X. J., Halbur, P. G., Shapiro, M. S., et al. (1998) Genetic and experimental evidence for cross-species infection by swine hepatitis E virus. *J. Virol.* **72,** 9714–9721.

306. Wu, J. C., Chen, C. M., Chiang, T. Y., et al. (2000) Spread of hepatitis E virus among different-aged pigs: two-year survey in Taiwan. *J. Med. Virol.* **66,** 488–492.

307. Nishizawa, T., Takahashi, M., Mizuo, H., Miyajima, H., Gotanda, Y., and Okamoto, H. (2003) Characterization of Japanese swine and human hepatitis E virus isolates of genotype IV with 99% identity over the entire genome. *J. Gen. Virol.* **84,** 1245–1251.

308. Arankalle, V. A., Chobe, L. P., Walimbe, A. M., Yergolkar, P. N., and Jacob, G. P. (2003) Swine HEV infection in south India and phylogenetic analysis (1985–1999). *J. Med. Virol.* **69,** 391–396.

309. Caredda, F., Antinori, S., Re, T., Pastecchia, C., Zavaglia, C., and Moroni, M. (1985) Clinical features of sporadic non-A, non-B hepatitis possibly associated with faecal-oral spread. Lancet **ii,** 444–445.

310. Zhuang, H., Cao, X. Y., Liu, C. B., and Wang, G. M. (1991) Epidemiology of hepatitis E in China. *Gastroenterol. Jpn.* **26,** S35–S38.

311. Cacopardo, B., Russo, R., Preiser, W., Benanti, F., Brancati, G., and Nunnari, A. (1997) Acute hepatitis E in Catania (eastern Sicily) 1980–1994. The role of hepatitis E virus. *Infection* **25,** 313–316.

312. Hartmann, W. J., Frosner, G. G., and Eichenlaub, D. (1998) Transmission of hepatitis E in Germany. *Infection* **26,** 409.

313. Mechnik, L., Bergman, N., Attali, M., et al. (2001). Acute hepatitis E virus infection presenting as a prolonged cholestatic jaundice. *J. Clin. Gastroenterol.* **33,** 421–422.

314. Tei, S., Kitajima, N., Takahashi, K., and Mishiro, S. (2003) Zoonotic transmission of hepatitis E virus from deer to human beings. *Lancet* **362,** 371–373.

315. Tei, S., Kitajima, N., Ohara, S., et al. (2004) Consumption of uncooked deer meat as a risk factor for hepatitis E virus infection: an age- and sex-matched case–control study. *J. Med. Virol.* **74,** 67–70.

316. Matsuda, H., Okada, K., Takahashi, K., and Mishiro, S. (2003) Severe hepatitis E virus infection after ingestion of uncooked liver from a wild boar. *J. Infect. Dis.* **188,** 944.

317. Tamada, Y., Yano, K., Yatsuhashi, H., Inoue, O., Mawatari, F., and Ishibashi, H. (2004) Consumption of wild boar linked to cases of hepatitis E. *J. Hepatol.* **40,** 869–870.

318. Yazaki, Y., Mizuo, H., Takahashi, M., Nishizawa, T., Sasaki, N., and Gotanda, Y. (2003) Sporadic acute or fulminant hepatitis E in Hokkaido, Japan, may be foodborne, as suggested by the presence of hepatitis E virus in pig liver as food. *J. Gen. Virol.* **84,** 2351–2357.

319. Takahashi, K., Kitajima, N., Abe, N., and Mishiro, S. (2004) Complete or near-complete nucleotide sequences of hepatitis E virus genome recovered from a wild boar, a deer, and four patients who ate the deer. *Virology* **330,** 501–515.

320. Gupta, D. N., and Smetana, H. F. (1957) The histopathology of viral hepatitis as seen in the Delhi epidemic (1956–56). *Indian J. Med. Res.* **45,** S101–S113.

321. Coursaget, P., Buisson, Y., Enogat, N., et al. (1998) Outbreak of enterically transmitted hepatitis due to hepatitis A and hepatitis E viruses. *J. Hepatol.* **28,** 745–750.

322. Ramachandran, J., Eapen, C. E., Kang, G., et al. (2004) Hepatitis E superinfection produces severe decompensation in patients with chronic liver disease. *J. Gastroenterol. Hepatol.* **19,**134–138.

323. Jaiswal, S. P., Jain, A. K., Naik, G., Soni, N., and Chitnis, D. S. (2001) Viral hepatitis during pregnancy. *Int. J. Gynaecol. Obstet.* **72,** 103–108.

324. Khuroo, M. S. and Kamili, S. (2003) Aetiology, clinical course and outcome of sporadic acute viral hepatitis in pregnancy. *J. Viral Hepat.* **10,** 61–69.

325. Tsarev, S. A., Tsareva, T. S., Emerson, S. U., et al. (1995) Experimental hepatitis E in pregnant rhesus monkeys: failure to transmit hepatitis E virus (HEV) to offspring and evidence of naturally acquired antibodies to HEV. *J. Infect. Dis.* **172,** 31–37.

326. Kasorndorkbua, C., Thacker, B. J., Halbur, P. G., et al. (2003) Experimental infection of pregnant gilts with swine hepatitis E virus. *Can. J. Vet. Res.* **67,** 303–306.

327. Suzuki, K., Aikawa, T., and Okamoto, H. (2002) Fulminant hepatitis E in Japan. *N. Engl. J. Med.* **347,** 1456.

328. Ohnishi, S., Kang, J. H., Maekubo, H., Takahashi, K., and Mishiro, S. (2003) A case report: two patients with fulminant hepatitis E in Hokkaido, Japan. *Hepatol Res.* **25,** 213–218.

329. Bradley, D. W. (1990) Enterically-transmitted non-A, non-B hepatitis. *Br. Med. Bull.* **46,** 442–461.

330. Purcell, R. H. and Emerson, S. U. (2001) Animal models of hepatitis A and E. *ILAR J.* **42,** 161–177.

331. McCaustland, K. A., Krawczynski, K., Ebert, J. W., et al. (2000) Hepatitis E virus infection in chimpanzees: a retrospective analysis. *Arch. Virol.* **145,** 1909–1918.

332. Williams, T. P. E., Kasorndorkbua, C., Halbur, P. G., et al. (2001) Evidence of extrahepatic sites of replication of the hepatitis E virus in a swine model. *J. Clin. Microbiol.* **39,** 3040–3046.

333. Ha, S. K. and Chae, C. (2004) Immunohistochemistry for the detection of swine hepatitis E virus in the liver. *J. Viral Hepat.* **11,** 263–267.

334. Halbur, P. G., Kasorndorkbua, C., Gilbert, C., et al. (2001) Comparative pathogenesis of infection of pigs with hepatitis E viruses recovered from a pig and a human. *J. Clin. Microbiol.* **39,** 918–923.

335. Khuroo, M. S., Kamili, S., Dar, M. Y., Moecklii, R., and Jameel, S. (1993) Hepatitis E and long-term antibody status. *Lancet* **341,** 1341–1355.

336. Jothikumar, N., Khanna, P., Paulmurugan, R., Kamatchiammal, S., and Padmanabhan, P. (1995) A simple device for the concentration and detection of enterovirus, hepatitis E virus and rotavirus from water samples by reverse transcription–polymerase chain reaction. *J. Virol. Methods* **55,** 401–415.

337. Jothikumar, N., Paulmurugan, R., Padmanabhan, P., Sundari, R. B., Kamatchiammal, S., and Rao, K. S. (2000) Duplex RT-PCR for simultaneous detection of hepatitis A and hepatitis E virus isolated from drinking water samples. *J. Environ. Monit.* **2,** 587–590.

338. Grimm, A. C. and Fout, G. S. (2002) Development of a molecular method to identify hepatitis E virus in water. *J. Virol. Methods* **101,** 175–188.

339. Sriram, B., Thakral, D., and Panda, S. K. (2003) Targeted cleavage of hepatitis E virus 3′ end RNA mediated by hammerhead ribozymes inhibits viral RNA replication. *Virology* **312,** 350–358.

340. Arankalle, V. A., Chadha, M. S., Dama, B. M., Tsarev, S. A., Purcell, R. H., and Banerjee, K. (1998) Role of immune serum globulins in pregnant women during an epidemic of hepatitis E. *J. Viral Hepat.* **5,** 199–204.

341. Purcell, R. H., Nguyen, H., Shapiro, M., et al. (2003) Pre-clinical immunogenicity and efficacy trial of a recombinant hepatitis E vaccine. *Vaccine* **21,** 2607–2615.

342. Kamili, S., Spelbring, J., Carson, D., and Krawczynski, K. (2004) Protective efficacy of hepatitis E virus DNA vaccine administered by gene gun in the cynomolgus macaque model of infection. *J. Infect. Dis.* **189,** 258–264.
343. Ma, Y., Lin, S. Q., Gao, Y., et al. (2003) Expression of ORF2 partial gene of hepatitis E virus in tomatoes and immunoactivity of expression products. *World J. Gastroenterol.* **9,** 2211–2115.
344. Emerson, S. U. and Purcell, R. H. (2001) Recombinant vaccines for hepatitis E. *Trends Mol. Med.* **7,** 462–426.
345. Worm, H. C. and Wirnsberger, G. (2004) Hepatitis E vaccines: progress and prospects. *Drugs* **64,** 1517–1531.

8

Gastroenteric Viruses

Miren Iturriza-Gómara, Chris I. Gallimore, and Jim Gray

1. INTRODUCTION

In recent years, viruses have been recognized increasingly as an important cause of foodborne infections. More than 160 enteric viruses are excreted in the feces of infected individuals, and some may also be present in the vomitus. Food and water are directly contaminated with fecal material, through the use of sewage sludge in agriculture, sewage pollution of shellfish culture beds, or may be contaminated by infected food-handlers.

Several groups of viruses cause gastroenteritis. The most common etiological agents are rotaviruses (RVs), human caliciviruses, which include noroviruses (NVs) and sapoviruses (SVs), astroviruses (ASVs), and enteric adenoviruses (ADVs, types 40 and 41).

Among the human caliciviruses, NVs are a leading cause of acute viral gastroenteritis worldwide and are responsible for sporadic cases and outbreaks of gastroenteritis affecting all age groups. Outbreaks in semiclosed environments such as hospitals, cruise ships, and homes of elderly persons [1,2] are frequent. As with all enteric viruses, transmission is predominantly person-to-person, but transmission via contaminated food, water, or the environment has often been demonstrated.

RVs are the most common cause of endemic acute infantile gastroenteritis. Mostly in the developing world, they are responsible for approx 600,000–800,000 deaths every year in children aged below 5 yr [3,4], and in developed countries they remain the most common cause of pediatric hospitalization in children aged below 2 yr [4]. Outbreaks involving other age groups, in particular the elderly, are frequent in semiclose environments such as hospitals and nursing homes. Sporadic cases in young adults are usually resulting from the contact with infected children. Foodborne transmission has been implicated in rotavirus outbreaks [5].

SVs, ASVs, and ADVs are mostly associated with sporadic cases of gastroenteritis in children aged below 5 yr. Outbreaks of gastroenteritis associated with these viruses can also occur in nurseries, schools, and pediatric hospital wards, and occasionally may involve adults in residential and nursing homes. Foodborne transmission has not been well documented for these viruses.

Viruses such as coronaviruses (CoVs) and toroviruses (ToVs) have been described but their role in acute gastroenteritis is not fully understood [6–9]. The severe acute respiratory syndrome (SARS)-CoV was also associated with enteric symptoms and is excreted in the feces of infected patients [10]. Picobirnaviruses and Aichi viruses have been found in the feces of individuals with gastroenteritis [11–16], but their significance

From: *Infectious Disease: Foodborne Diseases*
Edited by: S. Simjee © Humana Press Inc., Totowa, NJ

as causative agents of gastroenteritis in humans and their role in foodborne diseases remain unclear.

Other enteric viruses, not associated with gastroenteritis, such as hepatitis A virus, hepatitis E virus, and enteroviruses, including polioviruses are excreted in the feces of infected individuals and are also transmitted via contaminated food and water.

2. CLASSIFICATION AND IDENTIFICATION

2.1. Human Caliciviruses

Human caliciviruses, noroviruses (NV; formerly known as Norwalk-like or small round-structured viruses), and sapoviruses (SV; formerly known as Sapporo-like viruses) are members of the Caliciviridae family, which are nonenveloped viruses with a genome of positive-sense ssRNA. NVs and SVs can be distinguished morphologically and this allowed the first classification scheme for these viruses *(17)*. Both are approx 30–35 nm in diameter, but NVs have an amorphous structure with a ragged outer edge, and the SVs or "classical" caliciviruses display the characteristic cup-shaped structures from which the Caliciviridae family derives their name (*calix* = cup in latin).

NVs are currently classified into two and possibly three genogroups: GI, GII and GIII, based on the sequence diversity within the capsid *(18)*. Within GI, seven genotypes have been identified to date, including GI-1 (Norwalk/1968/US; accession number M87661), which is the prototype strain for the NV genus. Eight different genotypes have been identified to date within GII, and a single genotype constitutes GIII (Table 1) *(19,20–21)*.

Among SVs, three genogroups have been proposed *(22–24)*; genogroup 1 is represented by the Sapporo/1982/JP strain (accession number U65427), genogroup 2 by the London/1992/UK (accession number U95645) strain, and genogroup 3 by the Houston/1990/US strain (accession number U95644).

2.2. Rotaviruses

RVs are members of the Reoviridae family, and are nonenveloped triple-layered viruses, which possess a segmented genome consisting of 11 dsRNA segments. By EM, particles are approx 75 nm in diameter with a wheel-like structure from which they derive their name.

RVs are classified into groups A–E based on the antigenic differences of the viral middle layer *(25)*. Group A RVs are the most common cause of human gastroenteritis, but groups B and C RVs also infect humans. Group A RVs are further classified into subgroups (SG) based on the immunological reactivities of the middle layer protein VP6, and into G and P types according to the diversity of the outer layer proteins VP7 (*Glycoprotein*) and VP4 (*Protease-sensitive protein*), respectively *(25)*. Four different SGs (I, II, I+II and nonI/nonII), 14 or 15 G types (G1–G15) and 20 P types (P[1]–P[20]) have been identified to date among Group A RVs *(25)*.

2.3. Astroviruses

ASVs are members of the Astroviridae family. They are nonenveloped viruses with a genome of positive-sense ssRNA. By EM, they appear as spherical particles of 35–40 nm in diameter with the characteristic 5–6 pointed Star of David configuration which gives these viruses their name.

Table 1
Classification of Noroviruses

Genogroup	Genotype	Reference virus	Examples
I	1	Norwalk/1968/US	KY/89/JP
	2	Southampton/1991/UK	White Rose, Crawley
	3	Desert Shield 395/1990/SA	Birmingham 291, Potsdam
	4	Chiba 407/1987/JP	Thistle Hall, Valetta, Malta
	5	Musgrove/1989/UK	Butlins
	6	Hesse 3/1997/GE	Sindlesham, Mikkeli, Lord Harris
	7	Winchester/1994/UK	Lwymontley
II	1	Hawaii/1971/US	Wortley, Girlington, Port Canaveral 1994, Richmond 1994, Westover 1994, Honolulu 1994, Miami 1986, Stepping Hill, Pfaffenhofen
	2	Melksham/1994/UK	Snow Mountain, Melksham
	3	Toronto 24/1991/CA	Mexico, Auckland, Rotterdam, New Orleans 279
	4	Bristol/1993/UK	Lordsdale, Camberwell, Pilgrim, SymGreen, Grimsby, Burwash Landing 1995, Miami Beach 1995, Withybush
	5	Hillingdon/1990/UK	White River, Welterhof, New Orleans 1994
	6	Seacroft/1990/UK	Florida 1993, Baltimore 1993
	7	Leeds/1990/UK	Gwynedd, Venlo, Creche, Bridlington, HCALV
	8	Amsterdam/1998/NL	
III	1	Alphatron/1998/NL	Fort Lauderdale, Saint Cloud

Modified from ref. *21*

ASVs are classified into eight serotypes *(26)* based on the serological tests using type-specific antibodies. Phylogenetic analysis of sequences from a region of the ORF2 has shown that genotypes correlate serotypes *(27)*.

2.4. Adenoviruses

Human ADVs are members of the Adenoviridae family. They are nonenveloped icosahedral particles of 80 nm in diameter and possess a genome of dsDNA. ADVs are classified into six different subgroups or species (A–F) and within these subdivided into 51 distinct serotypes according to immunological, biochemical, and biological differences *(28)*. Among these, ADVs of subgroup F, serotypes 40 and 41 have been associated with gastroenteritis, and these are termed enteric or fastidious ADVs *(29,30)*.

2.5. Toroviruses, Coronaviruses, Picobirnaviruses, Aichi Viruses, and Small Round Viruses

CoVs and ToVs are two genus within the Coronaviridae family. CoVs are enveloped particles of 120–160 nm in diameter with an internal icosahedral core of approx 65 nm in diameter and a helical nucleocapsid. They have large surface projections with stem and globular portions which give them their characteristic appearance from which their name derives (*corona* is the latin word for crown) *(28)*.

Table 2
Quantity of Virus Excreted in Feces at the Peak of Infection
and the Probability of Detection by EM

Viruses	Quantity in feces (per g)	Probability of detection
Rotaviruses Group A	10^{8-12}	++++
Rotaviruses Group C	10^{5-7}	+++
Noroviruses	$\leq 10^{7-8}$	+/−
Sapoviruses	$\leq 10^{7-8}$	+/−
Astroviruses	10^{7-8}	+
Adenoviruses	10^{7-10}	++
Enteroviruses	$<10^6$	−

CoVs are classified into three genogroups, of which genogroups 1 and 2 have human CoVs representatives (229E in genogroup 1 and OC43 in genogroup 2), and the SARS CoVs which initially appeared not to belong to any of the three known genotypes may possibly be classified within genogroup 2 *(31)*.

ToVs differ morphologically from CoVs in that their nucleocapsid has a tubular appearance and the particles may be disk-, kidney-, or rod-like shaped. ToVs are grouped in a single genogroup, which contains bovine, equine, porcine, and human viruses *(27)*.

Picobirnaviruses are a new genus of the Birnaviridae family *(28)*. These are nonenveloped round viruses of 24–41 nm diameter with a bi-segmented dsRNA genome. Picobirnaviruses have been found in human and animal feces *(12–15)*.

Aichi virus is a member of the Picornaviridae family, recently included in a separate genus, kobuvirus, which also includes a bovine virus (32,33). They are small nonenveloped viruses of 22–30 nm diameter with a genome of positive-sense ssRNA.

Small round viruses or parvo-like virus particles found in human feces *(34,35)* are DNA viruses of approx 22 nm diameter.

3. DIAGNOSIS

Although EM has traditionally been used for the detection of enteric viruses in the feces of infected individuals, this is a labor-intensive and relatively insensitive method, as detection requires approx 10^6/g virus particles in feces. This is a problem particularly for the detection of caliciviruses (Table 2). Immune EM, which increases sensitivity and allows virus characterization when type-specific antibodies are used, has also been used for the detection of enteric viruses. Serological methods have been developed for the detection of some of these viruses. Enzyme immunosorbent assays (EIA) and passive particle agglutination tests (PPAT), some of which are available commercially, provide sensitivity comparable to, or better than, EM for the detection of RVs, NVs, ASVs, and ADVs. More recently, molecular methods, reverse-transcription polymerase chain reaction (RT-PCR), PCR, or nucleic acid-based sequence amplification (NASBA) assays have been developed for the detection of enteric viruses. These methods provide improved sensitivity for the detection of all enteric viruses, but have had a major impact on the detection of human caliciviruses (Table 3). Viruses do not replicate in food or water, and

Table 3
Detection of Enteric Viruses By EM Compared With PCR/RT-PCR

Virus	Number detected by EM	Percent	Number detected by PCR/RT-PCR[a]	Percent change (PCR-EM/EM) × 100
Rotavirus	70	25.8	86	+22.9
Norovirus	6	2.2	46	+666.7
Adenovirus	12	4.4	40	+233.4
Sapovirus	1[b]	0.4	8	+700
Astrovirus	3	1.1	7	+133.3
Virus detected	92	33.9	187	+103.3
No virus detected	179	66.1	111	−38.0
Total	**271**	**100.0**	**298**	

[a]Includes detection of dual and triple infections. EM detected one dual infection (rotavirus and adenovirus).
[b]Appearance of "classical calicivirus."
From ref. *64*

the concentration of virus particles in contaminated products is likely to be very small and not distributed homogeneously throughout the foodstuff. Testing for the presence of viruses in food, water, or environmental samples has only been possible since the development of very sensitive molecular methods, which include virus elution from the foodstuff, followed by concentration *(36)* efficient nucleic acid extraction methods for the removal of inhibitors of amplification.

One frequent source of foodborne enteric virus infections is shellfish. The development of a method for dissecting the stomach and digestive diverticula of shellfish *(37)* followed by nucleic acid extraction and DNA amplification-based methods *(37–41)* allows reliable and sensitive detection of enteric viruses in contaminated shellfish.

3.1. Human Caliciviruses

EM has been used for first-line diagnosis of NVs and SVs in clinical samples *(17,42, 43)*, however in recent years, EM has been replaced in many laboratories with in-house or commercial EIAs for the detection of NVs *(44–46)*. Molecular diagnosis using nucleic-acid extraction and RT-PCR assays has been introduced into many laboratories for the detection of NVs and SVs *(23,47–51)*. With the development of sensitive nested PCR assays, detection of NVs contamination of foodstuffs has recently been feasible. In particular the detection of NVs in oysters and other shellfish has been widely reported *(40,52–57)*. Some foods, including raspberries, are contaminated with NVs *(58)*. The detection of NVs in other foodstuffs is still in its developmental stage although a few studies have been undertaken *(59–61)*.

3.2. Rotaviruses

Because the number of RVs particles that are excreted at the peak of infection may be as high as 10^{12}/g in feces, diagnosis can be made using EM. EM will not however distinguish between RVs of different groups, and for this immune EM can be used. Most laboratories use EIA or PPAT, which use broadly reacting capture antibodies directed against epitopes of Group A RV VP6, for the routine diagnosis of RV infections. Commercially available assays have a sensitivity for detection comparable to EM, but are more prone to nonspecific reactions (reviewed in ref. *[62]*). The use of RT-PCR for the detection and characterization of RVs provides increased sensitivity and specificity *(62–64)*. Molecular methods for the detection of RVs are not routinely used in diagnostic laboratories, but their increased sensitivity make them useful for detecting low viral loads in asymptomatic infections, virus in samples that have been collected late after the onset of symptoms, or virus in environmental or food and water samples *(40,65–68)*.

3.3. Astroviruses

ASVs were first detected by EM *(69,70)*. Initially, characterization of ASVs was carried out by immune EM, but the development of EIA and RT-PCR assays for the detection and typing of ASV has made the detection of ASV available to diagnostic laboratories *(27, 71–73)*. In addition to increased sensitivity, the RT-PCR, used in conjunction with DNA sequencing, provides genotyping data which are vital for molecular epidemiological studies, and could be used for outbreak tracking, whether foodborne or otherwise.

3.4. Enteric Adenoviruses

ADVs were first identified by EM in the feces of children with gastroenteritis *(74)*. These viruses induced typical cytopathic effects in cell culture, but could not be passaged or typed *(75)*, for this reason, these viruses were designated fastidious ADVs. Later, the cell line 293 was shown to support the propagation of enteric ADVs, and permitted the development of neutralization assays *(76)*. This method is, however, time consuming and relatively insensitive, and most diagnostic laboratories have a number of rapid serological tests available (immunofluorescence, EIA, and PPAT assays) many with a sensitivity of detection comparable to EM (reviewed in *[77]*). In recent years, broadly reactive PCR methods have been developed for detecting ADVs, and in conjunction with restriction endonuclease analysis provide a sensitive tool for ADV characterization *(78)*. PCRs which use primers specific to a region of the genome (the long-fiber gene), highly conserved between ADV 40 and 41 but significantly divergent between these and other human ADVs have also been developed for the specific detection of enteric ADVs *(79)*.

3.5. Aichi Virus

EM is inappropriate for the detection and identification of these viruses in clinical samples. Aichi viruses cannot be differentiated from other enteroviruses excreted in feces. An enzyme-linked immunosorbent assay was developed for the detection of antibody responses to Aichi virus infection and RT-PCR assays have also been developed to detect the RNA genome of the virus in fecal and oyster samples *(80,81)*.

3.6. Toroviruses, Coronaviruses, Picobirnaviruses, and Small Round Viruses

Many novel gastroenteric viruses in humans were first discovered by EM including ToVs *(82,83)*, CoVs *(6)*, and small round viruses *(34)*. The picobirnaviruses were first detected using polyacrylamide gel electrophoresis (PAGE) of RNA derived from rat and human feces *(15,84)*. Molecular methods for the detection of these viruses have been developed *(85,86)*, although further studies are required.

4. RESERVOIRS

Humans are the principal reservoir of many of the enteric viruses, and person-to-person spread is the major route of transmission. The members of many virus families

Table 4
Enteric Virus Families and Genera That Infect Humans and Also Other Animal Species

Virus family	Genus	Animal host species	Evidence for zoonotic transmission
Caliciviridae	Norovirus	Bovine, porcine	No
	Sapovirus	None	
Reoviridae	Rotavirus A	Avian, bovine, canine, equine, feline, lapine, murine, ovine, porcine, simian	Yes
	Rotavirus B	Bovine, porcine	No
	Rotavirus C	Bovine, porcine	No
Adenoviridae	Adenovirus	Bovine, canine, equine, murine, ovine, porcine	No
Astroviridae	Astrovirus	Bovine, duck, feline, ovine, porcine, turkey	No
Coronaviridae	Coronavirus	Bovine, canine, feline, murine, porcine, turkey	Yes
	Torovirus	Bovine, equine, porcine	No
Picornaviridae	Kobuvirus	Bovine	No
Birnaviridae		Bovine, lapine, rat	No

infect animal species (Table 4), although zoonotic transmission is rare *(87)* with the exception of RVs *(88)*. Evidence of interspecies transmission of RVs has been obtained through comparative analysis of the genes derived from RV isolates from humans or animals either by whole-genome hybridization methods *(89)* or by gene sequencing and phylogenetic analysis (reviewed in ref. *[90]*), and many RV genotypes are shared among different species (Table 5).

The recent SARS epidemic is thought to have originated through transmission of the SARS CoV from an animal reservoir *(91)*, highlighting the potential for CoVs to cross the species barrier.

The lack of evidence for other viruses crossing the species barrier and of recombination/reassortment between animal and human pathogens needs to be addressed. Concomitant studies of disease in humans and animals in the same geographical locations are required.

5. FOODBORNE OUTBREAKS

Foods that are consumed raw or minimally processed, such as fruit, vegetables, and shellfish, are typically implicated as vehicles for the transmission of enteric viruses. However, a wide variety of foods have been implicated in foodborne viral gastroenteritis outbreaks (Table 6).

Enteric viruses can be present in foodstuffs through direct contamination with untreated sewage-sludge used in agriculture or sewage polluting shellfish culture beds. Food can also become contaminated during processing either by the use of polluted water in the preparation process or by infected food-handlers. Food-handlers have been shown to contaminate food during presymptomatic, symptomatic, and postsymptomatic infections *(92–96)*.

Table 5
**Rotavirus Genotypes Found in Human and Commonly Found
in Other Animal Species**

Genotype found in humans	Other host animal species
G3	Cats, dogs, monkeys, goats
G5	Pigs, horses
G6	Calves
G8	Calves
G9	Lams
G10	Calves
P[6]	Pigs
P[9]	Cats
P[11]	Calves, horses
P[14]	Pigs
P[19]	Pigs

Modified from ref. *97*

The majority of food- or waterborne outbreaks in which a virus is identified are caused by NVs *(58,97–107)*. RVs, and possibly ASVs, have also been implicated in food- or waterborne outbreaks *(5,105,106)*, but much less frequently. Aichi virus was first isolated in 1989 in BS-C1 cells from patients in outbreaks of oyster-associated gastroenteritis, in Japan *(16)*. Later studies have also showed a link between oyster-associated gastroenteritis and the acquisition of Aichi virus-specific antibodies in some patients *(110)*.

SVs and ADVs have yet to be confirmed as the cause of any food- or waterborne gastroenteritis outbreaks. Recently, outbreaks suspected of being foodborne have been detected among passengers on cruise ships. Multiple enteric viruses, SVs, ADVs, NVs, and RVs were detected in symptomatic patients suggesting the ingestion of fecally contaminated food or water (unpublished data).

6. PATHOGENICITY

Gastroenteritis viruses infect mainly the epithelial cells of the proximal part of the small intestine. The intestinal lumen is lined with a layer of polarized epithelial cells (enterocytes), which cover the villi and crypts. The enterocytes lining the villi are non-dividing absorptive cells, and those lining the crypts are undifferentiated proliferative cells that differentiate in order to renew the absorptive enterocytes of the villi. Some enteric viruses, RVs, ADVs, and ASVs infect the mature enterocytes exclusively, CoVs and ToVs infect the crypt and basal villus enterocytes, and enteric parvoviruses infect the crypt cells in the animal models (reviewed in *[109]*).

Viral diarrhea is caused by several factors:

- Primary malabsorption that originates from decreased absorption due to mature enterocyte cell death, which results in shortening of the villi, and also induces loss of enzymes leading to the accumulation of undigested carbohydrates and proteins.
- Reactive crypt hyperplasia which leads to increased secretion into the intestinal lumen.
- Decreased intestinal motility induced by the autonomous central nervous system.

In RV diarrhea, the symptoms precede the appearance of any histological changes *(112)*. This suggested that, in RV infection at least, other mechanisms in addition to the

Table 6
Foods Implicated in the Transmission of Gastroentetritis (Additional Data Obtained From http://www.cdc.gov/foodborneoutbreaks/)

Food category	Product	Virus identified
Vegetables	Green salad	NV, RV
	Green beans	NV
	Brocoli	NV
Dairy	Ice cream	NV
	Cheese	NV
	Cream	NV
Meat	Chicken	NV
	Beef	NV
	Hamburger	NV
Fish	Oyster[a]	NV, RV
	Shrimp	NV
Confectionary	Cheesecake	NV
	Lemonade	NV
	Slush drink	NV
Bakery products	Sandwich	NV, RV
	Donought	NV
Fruit	Berries	NV
	Melon	NV
	Pinaple	NV
	Grapes	NV
	Grapefruit	NV
Other	Eggs	NV
	Pasta	NV
	Rice	NV
	Pizza	NV

[a]Multiple NV genotypes have often been associated with outbreaks following the consumption of oysters.

ones described above must exist. One of the RV nonstructural proteins (NSP4), and a short peptide derived from it (aa 114–135) were shown to induce diarrhea in a dose-dependent manner in the neonatal mouse model *(113)*. NSP4 is the first identified viral enterotoxin, which has the capacity to mobilize intracellular calcium and increase the cellular membrane chloride permeability *(114,115)*, and NSP4 is also secreted from the infected enterocytes in early infection *(116)*. Recently, it has been observed that RV evokes intestinal fluid and electrolyte secretion by activation of the enteric nervous system *(117)*.

7. CLINICAL CHARACTERISTICS

7.1. Caliciviruses

The incubation period for NVs is 24–48 h, and the mean duration of illness is 12–60 h. The clinical manifestations of NV are characterized by nausea, projectile vomiting, diarrhea, and abdominal cramps. Fever, chills, and lethargy can also occur *(29)*. Vomiting is usually more common in children, and diarrhea is the main symptom in adults *(118,119)*.

The incubation period for SVs is 24–36 h, with illness lasting for 1–4 d. Symptoms include diarrhea (95% cases) and vomiting (60%), as well as fever and abdominal pain *(120)*.

7.2. Rotaviruses

The incubation period is usually 2 d, with the illness lasting for an average of 3–8 d *(29)*. Vomiting and watery diarrhea are the predominant symptoms, and fever and abdominal pain are also frequent. Extraintestinal spread of RVs has also been reported on numerous occasions, and may be associated to neurological disease *(121,122)*.

7.3. Astroviruses

The incubation period is between 24 and 36 h, with illness lasting for 1–4 d. Symptoms include vomiting, diarrhea, fever, and abdominal pain *(29)*.

7.4. Enteric Adenoviruses

The incubation period can vary between 3 and 10 d, with illness often lasting for more than 1 wk. Diarhea is more prominent than vomiting or fever *(29)*.

7.5. Aichi Virus

With Aichi virus gastroenteritis, diarrhea has been demonstrated in 58% of patients, and others include abdominal pain (92%), vomiting (71%), and fever (58%) *(81)*. However, these viruses have not been identified as a cause of gastroenteritis outside Japan, and their importance and spread remain unclear.

8. CHOICE OF TREATMENT

Viral gastroenteritis is usually self-limiting and its symptoms resolve without significant sequelae. In the cases of prolonged diarrhea, especially in infantile RV gastroenteritis, rehydration therapy with oral rehydration salt (ORS) solution (WHO formula; http://www.who.int/medicines/organization/par/edl/expcom13/ors.doc) or in severe cases, intravenous rehydration is indicated.

9. SUMMARY AND CONCLUSIONS

Enteric viruses are transmitted mainly from person-to-person. However, as these viruses are excreted in the feces of infected individuals, food and water can become contaminated with fecal material directly (sewage pollution) or indirectly by infected food-handlers.

The transmission of foodborne enteric virus infection by food-handlers can be prevented through the instigation of good hygiene practices. Symptomatic food-handlers should remain away from work for 48 h after the last episode of vomiting or diarrhea and should not prepare food for others during this period.

Procedures should be in place to address an incident of vomiting in the workplace. Exposed food and food that has been handled by an infected person should be destroyed. All contaminated areas, including vertical surfaces, must be thoroughly cleaned and attention to hand-washing procedures should be emphasized.

The extent of foodborne infection is not fully known *(123,124)*, a study conducted in Sweden estimated the annual incidence of foodborne illness at 38–79 per 1000 inhabitants

(125). More alarming estimates from the United States attribute 76 million illnesses, 325,000 hospitalizations, and 5000 deaths to foodborne infections *(126).* Mead et al. estimate that 34% of the hospitalizations attributable to foodborne transmission have a viral etiology.

To date, most foodborne viral gastroenteritis outbreaks have been associated with NV infections. However, it is likely that the other enteric viruses will also be transmitted via contaminated food and water with similar frequencies. A study assessing viral contamination of several shellfish beds in France that lasted more than 3 yr detected ASV in 17% of the samples, NV in 23%, enterovirus in 19%, and RV in 27% *(40).* Similarly, in another study ADV contaminated shellfish was found in 47% of the samples tested and included sampling in areas considered unpolluted according to current methods for the determination of microbiological quality based on coliform counts *(127).*

There may be several confounding factors that affect the detection of foodborne viral gastroenteritis. They include:

- Many laboratories will only investigate for the presence of NVs in suspected foodborne outbreaks.
- NVs are an extremely diverse group of viruses *(128),* and although several broadly reactive NV-specific assays are available, there is no single test that will detect all NVs with the same efficiency. Also, geographical differences detected among NV genotypes have been observed (unpublished data). Contaminated foodstuffs can be sourced from all over the world, and it is possible that the methods available in any one country, although being suitable for the locally endemic strains, may not be efficient for the detection of variants from other geographical regions.
- Immunity to NV is short-lived (~6 mo) and the number of individuals susceptible to symptomatic illness is constantly high. However, other enteric viruses (RV, ADV, and ASV) induce long-lasting immunity, which does not prevent reinfections but protects most adults from illness. Therefore, identifying a foodborne outbreaks caused by RV or ADV may be difficult as most people will not show symptoms or these will be very mild, and may not even give rise to the suspicion of a foodborne outbreak.

It is clear that further work is required in order to properly estimate the burden of food- and waterborne viral diseases. The advent of molecular methods of exquisite sensitivity provides the tools for the examination of food and water and may provide data on the zoonotic transmission of animal viruses into the human population via the food chain.

REFERENCES

1. Gallimore, C. I., Cubitt, D., du Plessis, N., and Gray, J. J. (2004) Asymptomatic and symptomatic excretion of noroviruses during a hospital outbreak of gastroenteritis. *J. Clin. Microbiol.* **42,** 2271–2274.
2. Gallimore, C. I., Richards, A., and Gray, J. J. (2003) Molecular diversity of noroviruses associated with outbreaks on cruise ships: comparisons with strains circulating in the UK. *Comm. Dis. Pub. Health* **6,** 285–293.
3. Parashar, U. D., Bresee, J. S., Gentsch, J. R., and Glass, R. I. (1998) Rotavirus. *Emerg. Infect. Dis.* **4,** 561–570.
4. Parashar, U. D., Hummelman, E. G., Bresee, J. S., Miller, M. A., and Glass, R. I. (2003) Global illness and deaths caused by rotavirus disease in children. *Emerg. Infect. Dis.* **9,** 565–572.
5. MMWR. (2000) Foodborne outbreak of Group A rotavirus gastroenteritis among college students–District of Columbia, March–April. *Morb. Mort. Wkly Rep.* **49,** 1131–1133.

6. Caul, E. O., Paver, W. K., and Clarke, S. K. R. (1975) Coronavirus particles in faeces from patients with gastroenteritis. *Lancet* **1**, 1192.

7. Clarke, S. K. R., Caul, E. O., and Egglestone, S. I. (1979) The human enteric coronaviruses. *Postgrad. Med. J.* **55**, 135–142.

8. Koopmans, M. and Horzinek, M. C. (1994) Toroviruses of animals and humans: a review. *Adv. Virus. Res.* **43**, 233–273.

9. Koopmans, M. P., Goosen, E. S., Lima, A. A., et al. (1997) Association of torovirus with acute and persistent diarrhea in children. *Pediatr. Infect. Dis. J.* **16**, 504–507.

10. Leung, W. K., To, K. F., Chan, P. K., et al. (2003) Enteric involvement of severe acute respiratory syndrome-associated coronavirus infection. *Gastroenterology* **125**, 1011–1017.

11. Ludert, J. E. and Liprandi, F. (1993) Identification of viruses with bi- and trisegmented double-stranded RNA genome in faeces of children with gastroenteritis. *Res. Virol.* **144**, 219–224.

12. Gallimore, C. I., Appleton, H., Lewis, D., Green, J., and Brown, D. W. G. (1995) Detection and characterisation of bisegmented dsRNA viruses (picobirnaviruses) in human faeces. *J. Med. Virol.* **45**, 135–140.

13. Gallimore, C. I., Lewis, D., and Brown, D. W. G. (1993) Detection and characterization of a novel bisegmented double-stranded RNA virus (picobirnavirus) from rabbit faeces. *Arch. Virol.* **133**, 63–73.

14. Grohmann, G. S., Glass, R. I., Pereira, H. G., et al. (1993) Enteric viruses and diarrhea in HIV-infected patients. *N. Engl. J. Med.* **329**, 14–20.

15. Pereira, H. G., Fialho, A. M., Flewett, T. H., Teixeira, J. M. S., and Andrade, Z. P. (1988) Novel viruses in human faeces. *Lancet* **2**, 103–104.

16. Yamashita, T., Kobayashi, S., Sakae, K., et al. (1991) Isolation of cytopathic small round viruses with BC-C-1 cells from patients with gastroenteritis. *J. Infect. Dis.* **164**, 954–957.

17. Caul, E. O. and Appleton, H. (1982) The electron microscopical and physical characteristics of small round structured fecal viruses: an interim scheme for classification. *J. Med. Virol.* **9**, 257–265.

18. Koopmans, M., van Strien, E., and Vennema, H. (2003). Molecular epidemiology of human caliciviruses. In: *Viral Gastroenteritis. Prespectives in Medical Virology* (Desselberger, U. and Gray, J. J., eds.), Elsevier, Amsterdam, pp. 523–554.

19. Green, K. Y., Ando, T., Balayan, M. S., et al. (2000) Taxonomy of the caliciviruses. *J. Infect. Dis.* **181**, S322–S330.

20. Mayo, M. A. (2002) A summary of taxonomic changes recently approved by ICTV. *Arch. Virol.* **147**, 1655–1663.

21. Green, K., Chanock, R., and Kapilian, A. (2001) Human caliciviruses. In: *Fields Virology* (Knipe, D. M., Howeley, M. M., et al., eds.), 4th edn. Lippincott Williams and Wilkins, Philadelphia:841–874.

22. Berke, T., Golding, B., Jiang, X., et al. (1997) Phylogenetic analysis of the caliciviruses. *J. Med. Virol.* **52**, 419–424.

23. Jiang, X., Cubitt, W. D., Berke, T., et al. (1997) Sapporo-like human caliciviruses are genetically and antigenically diverse. *Arch. Virol.* **142**, 1813–1827.

24. Noel, J. S., Liu, B. L., Humphrey, C. D., et al. (1997) Parkville virus: a novel genetic variant of human calicivirus in the Sapporo virus clade, associated with an outbreak of gastroenteritis in adults. *J. Med. Virol.* **52**, 173–178.

25. Estes, M. (2001) Rotaviruses and their replication. In: *Fields Virology* (Knipe, D. M., Howley, P. M., et al., eds.), 4th edn, Lippincott Williams & Wilkins, Philadelphia, pp. 1747–1785.

26. Lee, T. W. and Kurtz, J. B. (1994) Prevalence of human astrovirus serotypes in the Oxford region 1976–92, with evidence for two new serotypes. *Epidemiol. Infect.* **112**, 187–193.

27. Noel, J. S., Lee, T. W., Kurtz, J. B., Glass, R. I., and Monroe, S. S. (1995) Typing of human astroviruses from clinical isolates by enzyme immunoassay and nucleotide sequencing. *J. Clin. Microbiol.* **33**, 797–801.

28. van Regenmortel, M. H. V., Fauquet, C. M., Bishop, D. H. L., et al. (2000) *Virus Taxonomy, Seventh Report of the International Committee on Taxonomy of Viruses*, Academic, San Diego, CA.

29. Desselberger, U. and Gray, J. (2004) Viruses associated with acute diarrhoeal disease. In: *Principles and Practice of Clinical Virology* (Zuckerman, A. J., Banatvala J. E., Pattison, J. R., Griffiths, P., and Schoub B., eds.), 5th edn, Wiley, Chichester, UK, pp. 249–270.

30. Echevarria, M. (2004) Adenovirus. In: *Principles and Practice of Clinical Virology* (Zuckerman, A. J., Banatvala J. E., Pattison, J. R., Griffiths, P., and Schoub B., eds.), 5th edn, Wiley, Chichester, UK, pp. 249–270.

31. Zhu, G. and Chen, H. W. (2004) Monophyletic relationship between severe acute respiratory syndrome coronavirus and group 2 coronaviruses. *J. Infect. Dis.* **189,** 1676–1678.

32. Yamashita, T., Sakae, K., Tsuzuki, H., et al. (1998) Complete nucleotide sequence and genetic organization of Aichi virus, a distinct member of the Picornaviridae associated with acute gastroenteritis in humans. *J. Virol.* **72,** 8408–8412.

33. Yamashita, T., Ito, M., Kabashima, Y., Tsuzuki, H., Fujiura, A., and Sakae, K. (2003) Isolation and characterization of a new species of kobuvirus associated with cattle. *J. Gen. Virol.* **84,** 3069–3077.

34. Paver, W. K., Ashley, C. R., Caul, E. O., and Clarke, S. K. R. (1973) A small virus in human faeces. *Lancet* **1,** 237–239.

35. Paver, W. K. and Clarke, S. K. R. (1976) Comparison of human fecal and serum parvo-like viruses. *J. Clin. Microbiol.* **4,** 67–70.

36. Dubois, E., Agier, C., Traore, O., et al. (2002) Modified concentration method for the detection of enteric viruses on fruits and vegetables by reverse transcriptase-polymerase chain reaction or cell culture. *J. Food. Prot.* **65,** 1962–1969.

37. Atmar, R. L., Neill, F. H., Romalde, J. L., et al. (1995) Detection of Norwalk virus and hepatitis A virus in shellfish tissues with the PCR. *Appl. Environ. Microbiol.* **61,** 3014–3018.

38. Atmar, R. L., Metcalf, T. G., Neill, F. H., and Estes, M. K. (1993) Detection of enteric viruses in oysters by using the polymerase chain reaction. *Appl. Environ. Microbiol.* **59,** 631–635.

39. Atmar, R. L., Neill, F. H., Woodley, C. M., et al. (1996) Collaborative evaluation of a method for the detection of Norwalk virus in shellfish tissues by PCR. *Appl. Environ. Microbiol.* **62,** 254–258.

40. Le Guyader, F., Haugarreau, L., Miossec, L., Dubois, E., and Pommepuy, M. (2000) Three-year study to assess human enteric viruses in shellfish. *Appl. Environ. Microbiol.* **66,** 3241–3248.

41. Schwab, K. J., Neill, F. H., Fankhauser, R. L., et al. (2000) Development of methods to detect "Norwalk-like viruses" (NLVs) and hepatitis A virus in delicatessen foods: application to a food-borne NLV outbreak. *Appl. Environ. Microbiol.* **66,** 213–218.

42. Curry, A., Bryden, A., Morgan-Capner, P., et al. (1999) A rationalised virological electron microscope specimen testing policy. PHLS North West Viral Gastroenteritis and Electron Microscopy Subcommittee. *J. Clin. Path.* **52,** 471–474.

43. Lewis, D., Ando, T., Humphrey, C. D., Monroe, S. S., and Glass, R. I. (1995) Use of solid-phase immune electron microscopy for classification of Norwalk-like viruses into six antigenic groups from 10 outbreaks of gastroenteritis in the United States. *J. Clin. Microbiol.* **33,** 501–504.

44. Hale, A. D., Crawford, S. E., Ciarlet, M., et al. (1999) Expression and self-assembly of Grimsby virus: antigenic relationship to Norwalk and Mexico virus. *Clin. Diag. Lab. Immunol.* **6,** 142–145.

45. Richards, A. F., Lopman, B. A., Gunn, A., et al. (2003) Evaluation of a commercial ELISA for detecting Norwalk-like virus antigen in faeces. *J. Clin. Virol.* **26,** 109–115.

46. Vipond, I. B., Pelosi, E., Williams, J., et al. (2000) A diagnostic EIA for detection of the prevalent SRSV strain in United Kingdom outbreaks of gastroenteritis. *J. Med. Virol.* **61,** 132–137.

47. Ando, T., Monroe, S. S., Gentsch, J. R., et al. (1995) Detection and differentiation of antigenically distinct small round structured viruses (Norwalk-like viruses) by reverse transcription PCR and Southern hybridization. *J. Clin. Microbiol.* **33,** 64–71.

48. Green, J., Gallimore, C. I., Norcott, J. P., Lewis, D., and Brown, D. W. G. (1995) Broadly reactive reverse transcriptase polymerase chain reaction (RT-PCR) for the diagnosis of SRSV-associated gastroenteritis. *J. Med. Virol.* **47,** 392–398.

49. Green, S. M., Lambden, P. R., Deng, Y., et al. (1995) Polymerase chain reaction detection of small round-structured viruses from two related hospital outbreaks of gastroenteritis using inosine-containing primers. *J. Med. Virol.* **45,** 197–202.

50. Matson, D. O., Zhong, W., Nakata, S., et al. (1995) Molecular characterisation of a human calicivirus with sequence relationships closer to animal caliciviruses than other known human caliciviruses. *J. Med. Virol.* **45,** 215–222.

51. Vinje, J., Deijl, H., van der Heide, R., et al. (2000) Molecular detection and epidemiology of Sapporo-like viruses. *J. Clin. Microbiol.* **44,** 113–118.

52. Green, J., Henshilwood, K., Gallimore, C. I., Brown, D. W. G., and Lees, D. N. (1998) A nested reverse transcriptase PCR assay for detection of small round-structured viruses in environmentally contaminated molluscan shellfish. *Appl. Environ. Microbiol.* **64,** 858–863.

53. Henshilwood, K., Green, J., Gallimore, C. I., Brown, D. W. G., and Lees, D. N. (1998) The development of polymerase chain reaction assays for detection of small round structured and other human enteric viruses in molluscan shellfish. *J. Shellfish Res.* **17,** 1675–1678.

54. Le Guyader, F. S., Neill, F. H., Dubois, E., et al. (2003) A semiquantitative approach to estimate Norwalk-like virus contamination of oysters implicated in an outbreak. *Int. J. Food Microbiol.* **87,** 107–112.

55. Lees, D. (2000) Viruses and bivalve shellfish. *Int. J. Food Microbiol.* **59,** 81–116.

56. Lees, D. N., Henshilwood, K., Gallimore, C. I., Green, J., and Brown, D. W. G. (1995) Detection of small round structured viruses in shellfish by RT-PCR. *Appl. Environ. Microbiol.* **61,** 4418–4424.

57. Nishida, T., Kimura, H., Saitoh, M., et al. (2003) Detection, quantitation, and phylogenetic analysis of noroviruses in Japanese oysters. *Appl. Environ. Microbiol.* **69,** 5782–5786.

58. Ponka, A., Maunula, L., von Bonsdorff, C. H., and Lyytikainen, O. (1999) An outbreak of calicivirus associated with consumption of frozen raspberries. *Epidemiol. Infect.* **123,** 469–474.

59. Gouvea, V., Santos, N., do Carmo Timenetsky, M., and Estes, M. K. (1994) Identification of Norwalk virus in artificially seeded shellfish and selected foods. *J. Virol. Meth.* **48,** 177–187.

60. Sair, A. I., D'Souza, D. H., Moe, C. L., and Jaykus, L. A. (2002) Improved detection of human enteric viruses in foods by RT-PCR. *J. Virol. Meth.* **100,** 57–69.

61. Schwab, K., J., Neill, F. H., Le Guyader, F., Estes, M. K., and Atmar, R. L. (2001) Development of a reverse transcription-PCR-DNA enzyme immunoassay for detection of "Norwalk-like" viruses and hepatitis A virus in stool and shellfish. *Appl. Environ. Microbiol.* **67,** 742–749.

62. Iturriza-Gómara, M., Green, J., and Gray, J. J. (2000) Methods of rotavirus detection, sero-and genotyping, sequencing and phylogenetic analysis. In: *Rotaviruses: Methods and Protocols. Methods in Molecular Medicine* (Gray, J. J. and Desselberger, U., eds.), Humana, Totowa, NJ, pp. 189–217.

63. Iturriza-Gómara, M., Wong, C., Blome, S., Desselberger, U., and Gray, J. (2002) Molecular characterisation of VP6 genes of human rotavirus isolates: correlation of genogroups with subgroups and evidence of independent segregation. *J. Virol.* **76,** 6596–6601.

64. Simpson, R., Aliyu, S., Iturriza-Gómara, M., Desselberger, U., and Gray, J. (2003) Infantile viral gastroenteritis: on the way to closing the diagnostic gap. *J. Med. Virol.* **70,** 258–262.

65. Kang, G., Iturriza-Gómara, M., Wheeler, J. G., et al. (2004) Quantitation of Group A rotavirus RNA by real time reverse-transcription polymerase chain reaction: correlation with clinical severity in children in South India. *J. Med. Virol.* **73,** 118–122.

66. Muniain-Mujika, I., Girones, R., and Lucena, F. (2000) Viral contamination of shellfish: evaluation of methods and analysis of bacteriophages and human viruses. *J. Virol. Meth.* **89,** 109–118.

67. Villena, C., El-Senousy, W. M., Abad, F. X., Pinto, R. M., and Bosch, A. (2003) Group A rotavirus in sewage samples from Barcelona and Cairo: emergence of unusual genotypes. *Appl. Environ. Microbiol.* **69,** 3919–3923.

68. Jean, J., Blais, B., Darveau, A., and Fliss, I. (2002). Rapid detection of human rotavirus using colorimetric nucleic acid sequence-based amplification (NASBA)-enzyme-linked immunosorbent assay in sewage treatment effluent. *FEMS Microbiol. Lett.* **210,** 143–147.

69. Appleton, H., Buckley, M., Thom, B. T., Cotton, J. L., and Henderson, S. (1977) Virus-like particles in winter vomiting disease. *Lancet* **19,** 409–411.

70. Madeley, C. R. and Cosgrove, B. P. (1975) 28 nm particles in faeces in infantile gastroenteritis. *Lancet* **6,** 451–452.

71. Herrmann, J. E., Nowak, N. A., Perron-Henry, D. M., Hudson, R. W., Cubitt, W. D., and Blacklow, N. R. (1990) Diagnosis of astrovirus gastroenteritis by antigen detection with monoclonal antibodies. *J. Infect. Dis.* **161,** 226–229.

72. Jonassen, T. O., Monceyron, C., Lee, T. W., Kurtz, J. B., and Grinde, B. (1995) Detection of all serotypes of human astrovirus by the polymerase chain reaction. *J. Virol. Methods.* **52,** 327–334.

73. Saito, K., Ushijima, H., Nishio, O., et al. (1995) Detection of astroviruses from stool samples in Japan using reverse transcription and polymerase chain reaction amplification. *Microbiol. Immunol.* **39,** 825–828.

74. Flewett, T. H., Bryden, A. S., Davies, H. A., and Morris, C. A. (1973) Epidemic viral enteritis in a long-stay children's ward. *Lancet* **1,** 4–5.

75. Madeley, C. R., Cosgrove, B. P., Bell, E. J., and Fallon, R. J. (1977) Stool viruses in babies in Glasgow. I. Hospital admissions with diarrhoea. *J. Hyg.* **78,** 261–273.

76. Takiff, H. E., Straus, S. E., and Garon, C. F. (1981) Propagation and in vitro studies of previously non-cultivable enteral adenoviruses in 293 cells. *Lancet* **17,** 832–834.

77. de Jong, J. C. (2003) Epidemiology of enteric adenoviruses 40 and 41 and other adenoviruses in immunocompetent and immunodeficient individuals. In: *Viral Gastroenteritis. Prespectives in Medical Virology* (Desselberger, U. and Gray, J. J., eds.), Elsevier, Amsterdam, pp. 407–446.

78. Allard, A., Albinsson, B., and Wadell, G. (2001) Rapid typing of human adenoviruses by a general PCR combined with restriction endonuclease analysis. *J. Clin. Microbiol.* **39,** 498–505.

79. Tiemessen, C. T. and Nel, M. J. (1996) Detection and typing of subgroup F adenoviruses using the polymerase chain reaction. *J. Virol. Meth.* **59,** 73–82.

80. Yamashita, T., Sugiyama, M., Tsuzuki, H., Sakae, K., Suzuki, Y., and Miyazaki, Y. (2000) Application of a reverse transcription-PCR for identification and differentiation of Aichi virus, a new member of the Picornavirus family associated with gastroenteritis in humans. *J. Clin. Microbiol.* **38,** 2955–2961.

81. Yamashita, T., Ito, M., Tsuzuki, H., and Sakae, K. (2001) Identification of Aichi virus infection by measurement of immunoglobulin responses in an enzyme-linked immunosorbent assay. *J. Clin. Microbiol.* **39,** 4178–4180.

82. Beards, G. M., Green, J., Hall, C., and Flewett, T. H. (1984) An enveloped virus in stools of children and adults with gastroenteritis that resembles the Breda virus of calves. *Lancet* **1,** 1050–1052.

83. Beards, G. M., Brown, D. W. G., Green, J., and Flewett, T. H. (1986) Preliminary characterisation of torovirus-like particles of humans: comparison with Berne virus of horses and Breda virus of calves. *J. Med. Virol.* **20,** 67–78.

84. Pereira, H. G., Flewett, T. H., Candeias, J. A. N., and Barth, O. M. (1988) A virus with a bisegmented double-stranded RNA genome in rat (*Oryzomys nigripes*) intestines. *J. Gen. Virol.* **69,** 2749–2754.

85. Duckmanton, L., Luan, B., Devenish, J., Tellier, R., and Petric, M. (1997) Characterization of torovirus from human fecal specimens. *Virology* **239,** 158–168.

86. Rosen, B. I., Fang, Z. Y., Glass, R. I., and Monroe, S. S. (2000) Cloning of human picobirnavirus genomic segments and development of an RT-PCR detection assay. *Virology* **277,** 316–329.

87. Oliver, S. L., Dastjerdi, A. M., Wong, S., et al. (2003) Molecular characterization of bovine enteric caliciviruses: a distinct third genogroup of noroviruses (Norwalk-like viruses) unlikely to be of risk to humans. *J. Virol.* **77,** 2789–2798.

88. Cook, N., Bridger, J., Kendall, K., Iturriza-Gómara, M. I., El-Attar, L., and Gray, J. (2004) The zoonotic potential of rotavirus. *J. Infect.* **48,** 289–302.

89. Nakagomi, T. and Nakagomi, O. (2000) Human rotavirus HCR3 possesses a genomic RNA constellation indistinguishable from that of feline and canine rotaviruses. *Arch. Virol.* **145,** 2403–2409.

90. Iturriza-Gómara, M., Desselberger, U., and Gray, J. J. (2003) Molecular epidemiology of rotaviruses: genetic mechanisms associated with diversity. In: *Viral Gastroenteritis. Prespectives in Medical Virology* (Desselberger, U. and Gray, J. J., eds.), Elsevier, Amsterdam, pp. 317–344.

91. Guan, Y., Zheng, B. J., He, Y. Q., et al. (2003) Isolation and characterization of viruses related to the SARS coronavirus from animals in southern China. *Science* **302,** 276–278.

92. Gaulin, C., Frigon, M., Poirier, D., and Fournier, C. (1999) Transmission of calicivirus by a foodhandler in the pre-symptomatic phase of illness. *Epidemiol. Infect.* **123,** 475–478.

93. Lo, S. V., Connolly, A. M., Palmer, S. R., Wright, D., Thomas, P. D., and Joynson, D. (1994) The role of pre-symptomatic food handler in a common source of food-borne SRSV gastroenteritis in a group of hospitals. *Epidemiol. Infect.* **113,** 513–521.

94. Parashar, U. D., Dow, L., Fankhauser, R. L., et al. (1998). An outbreak of viral gastroenteritis associated with consumption of sandwiches: implications for the control of transmission by food handlers. *Epidemiol. Infect.* **121,** 615–621.

95. Patterson, T., Hutchings, P., and Palmer, S. (1993) Outbreak of SRSV gastroenteritis at an international conference traced to food handled by a post-symptomatic caterer. *Epidemiol. Infect.* **111,** 157–162.

96. Patterson, W., Haswell, P., Fryers, P. T., and Green, J. (1997) Outbreak of small round structured virus gastroenteritis arose after kitchen assistant vomited. *Comm. Dis. Rep.* **7,** R101–R103.

97. Desselberger, U., Iturriza-Gómara, M., and Gray, J. J. (2001) Rotavirus epidemiology and surveillance. In: Chadwick, D., Goode, J., eds. Novartis Foundation Symposium No. 238. In: Viral Gastroenteritis. Chichester: John Wiley & Sons,: 125–147.

98. Kuritsky, J. N., Osterholm, M. T., Greenberg, H. B., et al. (1984) Norwalk gastroenteritis: a community outbreak associated with bakery product consumption. *Ann. Int. Med.* **100,** 519–521.

99. Fleissner, M. L., Herrman, J. E., Booth, J. W., Blacklow, N. R., and Nowak, N. A. (1989) Role of Norwalk virus in two foodborne outbreaks of gastroenteritis: definitive virus association. *Am. J. Epidemiol.* **129,** 165–172.

100. Herwaldt, B. L., Lew, J. F., Moe, C. L., et al. (1994) Characterization of a variant strain of Norwalk virus from a food-borne outbreak of gastroenteritis on a cruise ship in Hawaii. *J. Clin. Microbiol.* **32,** 861–866.

101. Kilgore, P. E., Belay, E. D., Hamlin, D. M., et al. (1996) A university outbreak of gastroenteritis due to a small round structured virus: application of molecular diagnosis to identify the etiologic agent and patterns of transmission. *J. Infect. Dis.* **173,** 787–793.

102. Christensen, B. F., Lees, D. N., Henshilwood, K., Bjergskov, T., and Green, J. (1998) Human enteric viruses in oysters causing a large outbreak of human foodborne infection in 1996/97. *J. Shellfish Res* **17,** 1633–1635.

103. Dowell, S. F., Groves, C., Kirkland, K. B., et al. (1995). A multistate outbreak of oyster-associated gastroenteritis: implications for interstate tracing of contaminated shellfish. *J. Infec. Dis.* **171,** 1497–1503.

104. McDonnell, S., Kirkland, K. B., Hlady, W. G., et al. (1997) Failure of cooking to prevent shellfish-associated viral gastroenteritis. *Arch. Int. Med.* **157,** 111–116.

105. Otsu, R. (1999) Outbreaks of gastroenteritis caused by SRSVs from 1987 to 1992 in Kyushu, Japan: four outbreaks associated with oyster consumption. *Eur. J. Epidemiol.* **15,** 175–180.

106. Simmons, G., Greening, G., Gao, W., and Campbell, D. (2001) Raw oyster consumption and outbreaks of viral gastroenteritis in New Zealand: evidence for risk to the public's health. *Aust. N Z J Public Health* **25,** 234–240.

107. Sugieda, M., Nakajima, K., and Nakajima, S. (1996) Outbreaks of Norwalk-like virus associated gastroenteritis traced to shellfish: coexistence of two genotypes in one specimen. *Epidemiol. Infect.* **116,** 339–346.

108. Oishi, I., Yamazaki, K., Kimoto, T., et al. (1994) A large outbreak of acute gastroenteritis associated with astrovirus among students and teachers in Osaka, Japan. *J. Infect. Dis.* **170,** 439–443.

109. Villena, C., Gabrieli, R., Pinto, R. M., et al. (2003) A large infantile gastroenteritis outbreak in Albania caused by multiple emerging rotavirus genotypes. *Epidemiol. Infect.* **131,** 1105–1110.

110. Yamashita, T., Sakae, K., Ishihara, Y., Isomura, S., and Utagawa, E. (1993) Prevalence of newly isolated, cytopathic small round virus (Aichi strain) in Japan. *J. Clin. Microbiol.* **31,** 2938–2943.

111. Michelangeli, F. and Ruiz, M. C. (2003) Physiology and pathophysiology of the gut in relation to viral diarrhoea. In: *Viral Gastroenteritis. Prespectives in Medical Virology* (Desselberger, U. and Gray, J. J., eds.), Elsevier, Ámsterdam, pp. 23–50.

112. Estes, M. K. (2003) The rotavirus NSP4 enterotoxin: current status and challenges. In: *Viral Gastroenteritis. Prespectives in Medical Virology*, (Desselberger, U. and Gray, J. J., eds.), Elsevier, Amsterdam, pp. 207–224.

113. Ball, J. M., Tian, P., Zeng, C. Q., Morris, A. P., and Estes, M. K. (1996) Age-dependent diarrhea induced by a rotaviral nonstructural glycoprotein. *Science* **272,** 101–104.

114. Tian, P., Hu, Y., Schilling, W. P., Lindsay, D. A, Eiden, J., and Estes, M. K. (1994) The nonstructural glycoprotein of rotavirus affects intracellular calcium levels. *J. Virol.* **68,** 251–257.

115. Tian, P., Estes, M. K., Hu, Y., Ball, J. M., Zeng, C. Q., and Schilling, W. P. (1995) The rotavirus nonstructural glycoprotein NSP4 mobilizes Ca^{2+} from the endoplasmic reticulum. *J. Virol.* **69,** 5763–5772.

116. Zhang, M., Zeng, C. Q., Morris, A. P, and Estes, M. K. (2000) A functional NSP4 enterotoxin peptide secreted from rotavirus-infected cells. *J. Virol.* **74,** 11,663–11,670.

117. Lundgren, O., Peregrin, A. T., Persson, K., Kordasti, S., Uhnoo, I., and Svensson, L. (2000) Role of the enteric nervous system in the fluid and electrolyte secretion of rotavirus diarrhea. *Science* **287,** 491–495.

118. Kaplan, J. E., Gary, G. W., Baron, R. C., et al. (1982) Epidemiology of Norwalk gastroenteritis and the role of Norwalk virus in outbreaks of acute nonbacterial gastroenteritis. *Ann. Int. Med.* **96,** 756–761.

119. Kaplan, J. E., Schonberger, L. B., Varano, G., Jackman, N., Bied, J., and Gary, G. W. (1982) An outbreak of acute nonbacterial gastroenteritis in a nursing home. Demonstration of person-to-person transmission by temporal clustering of cases. *Am. J. Epidemiol.* **116,** 940–948.

120. Rockx, B., De Wit, M., Vennema, H., et al. (2002) Natural history of human calicivirus infection: a prospective cohort study. *Clin. Infect. Dis.* **35,** 246–253.

121. Lynch, M., Lee, B., Azimi, P., et al. (2001) Rotavirus and central nervous system symptoms: cause or contaminant? Case reports and review. *Clin. Infect. Dis.* **33,** 932–938.

122. Iturriza-Gómara, M., Auchterlonie, I. A., Zaw, W., Molyneaux, P., Desselberger, U., and Gray, J. (2002) Rotavirus gastroenteritis and CNS infection: detection and characterisation of the VP7 and VP4 genes of rotavirus strains isolated from paired faecal and CSF samples from a child with CNS symptoms. *J. Clin. Microbiol.* **40,** 4797–4799.

123. Koopmans, M., Vennema, H., Heersma, H., et al. (2003). Early identification of common-source foodborne virus outbreaks in Europe. *Emerg. Infect. Dis.* **9,** 1136–1142.

124. Lopman, B., Van Duynhoven, Y., Hanon, F. X., Reacher, M., Koopmans, M., and Brown, D. W. (2002) Laboratory capability in Europe for foodborne viruses. *Euro. Surv.* **7,** 61–65.

125. Lindqvist, R., Andersson, Y., Lindback, J., et al. (2001) A one-year study of foodborne illnesses in the municipality of Uppsala, Sweden. *Emerg. Infect. Dis.* **7,** 588–592.

126. Mead, P. S, Slutsker, L., Dietz, V., et al. (1999) Food-related illness and death in the United States. *Emerg. Infect. Dis.* **5,** 607–625.

127. Muniain-Mujika, I., Calvo, M., Lucena, F., and Girones, R. (2003) Comparative analysis of viral pathogens and potential indicators in shellfish. Int J Food Microbiol. **83,**(1):75–85.

128. Gallimore, C. I., Green, J., Lewis, D., et al. (2004) Diversity of noroviruses cocirculating in the North of England from 1998 to 2001. *J. Clin. Microbiol.* **42,** 1396–1401.

9

Cryptosporidium

Huw V. Smith and Rosely A. B. Nichols

Abstract

Seven species of the intracellular, protozoan parasite, *Cryptosporidium*, cause diarrhea in humans. *Cryptosporidium* completes its life cycle in individual hosts, and its infectious dose is small (ID_{50} = 9–1042 oocysts). Its virulence and pathogenicity are poorly understood. The drug of choice for immunocompetent individuals is nitazoxanide and modifications of treatment regimens may also increase its usefulness for immunocompromised individuals. Immune reconstitution using highly active antiretroviral therapy and secondary prophylaxis should be considered in HIV-infected individuals. Transmission to humans can occur via any mechanism by which material contaminated with feces containing infectious oocysts from infected human beings or non-human hosts can be swallowed by a susceptible host. Biotic reservoirs include all potential hosts of human-infectious *Cryptosporidium* species, while abiotic reservoirs include all vehicles that contain sufficient infectious oocysts to cause human infection, the most commonly recognized being food and water.

Both foodborne and waterborne outbreaks have been documented. In three out of the six foodborne outbreaks documented, contaminated foodstuffs were implicated as the vehicles of transmission, but in another two, foodhandlers, rather than indigenous contamination of foodstuff, were implicated in disease transmission. Increased global sourcing and rapid transport of soft fruit, salad vegetables, and seafood can enhance both the likelihood of oocyst contamination and oocyst survival. Standardized methods for detecting oocysts on foods must be maximized as there is no method to augment parasite numbers prior to detection. Oocyst contamination of food can be on the surface of, or in, the food matrix and products at greatest risk of transmitting infection include those that receive no, or minimal, heat treatment after they become contaminated. Heating at ≥64.2°C for 2 min and exposure to UV light ablates *Cryptosporidium parvum* infectivity for neonatal mice, while drying/desiccation for 4 h or exposure to 0.03% H_2O_2 for ≥2 h ablates oocyst viability. Disinfectants and other treatment processes used in the food industry may be detrimental to oocyst survival or lethal, but further research in this important area is required.

Key Words: *Cryptosporidium*; oocysts; occurrence; detection; outbreaks; foodborne; environment.

1. CLASSIFICATION AND IDENTIFICATION

1.1. Classification

1.1.1. Historical

In 1907, Tyzzer described organisms found in the gastric crypts of an experimental laboratory mouse, which he named *Cryptosporidium muris (1)*. In 1910, he described many stages of the *C. muris* life cycle and in 1912, he described much of the morphology and life cycle of a second species of *Cryptosporidium* found in the small intestine of

From: *Infectious Disease: Foodborne Diseases*
Edited by: S. Simjee © Humana Press Inc., Totowa, NJ

laboratory mice, which had smaller sized oocysts, and which he named *Cryptosporidium parvum (2)*. In 1955, *Cryptosporidium* was recognized as a cause of morbidity and mortality in young turkeys (*C. meleagridis*) and as a cause of scouring in calves in 1971, but it was not until 1976 that the first two human cases of cryptosporidiosis were described.

Cryptosporidium is a member of the phylum Apicomplexa, class Sporozoasida, subclass Coccidiasina, order Eucoccidiorida, suborder Eimeriorina, and family Cryptosporidiidae. The genus name describes the transmissive stage (the oocyst), which contains four sporozoites that are not contained within sporocysts (naked), hence the name *Cryptosporidium*, derived from the Greek and meaning hidden spores (sporocysts). The genus *Cryptosporidium* belongs to the family Cryptosporidiidae *(3)* from which it adopts its characters. The characters of the family Cryptosporidiidae are as follows: "development just under the surface membrane of the host cell or within its striated border and not in the cell proper. Oocysts and meronts with a knob-like attachment organelle at some point on their surface. Oocysts without sporocysts, with four naked sporozoites. Monoxenous. Microgametes without flagella" *(4)*.

Cryptosporidium is often described as a "coccidian" parasite. Coccidia are protozoan parasites that develop in the gastrointestinal tract of vertebrates through all or part of their life cycle. All coccidia produce meronts, gamonts, and sporonts in their life cycles and some coccidia have developmental stages in tissues other than the intestinal tract (these are known as the tissue cyst-forming coccidia and include the genus *Toxoplasma*). Organisms of the genus *Cryptosporidium* make up the order Eucoccidiorida with four other genera: *Cyclospora*, *Isospora*, *Sarcocystis*, and *Toxoplasma*, all of which cause disease in humans. The genera *Cryptosporidium* and *Sarcocystis* differ from other coccidia whose oocysts require a period of maturation (sporulation) outside the host to become infectious for the next host. *Cryptosporidium* oocysts are fully sporulated and infectious for other susceptible hosts when excreted.

Three sets of data suggest that *Cryptosporidium* differs significantly from other coccidia and probably constitutes an early emerging branch of the Apicomplexa.

(a) Recent phylogenetic analysis based on sequences of the 18S ribosomal RNA (rRNA) gene casts doubt on the affinity of the genus *Cryptosporidium* to other taxa within the coccidia *(5,6)*. Carreno et al. *(5)* constructed a phylogenetic tree with *C. parvum* and three genera of Gregarina and showed the gregarines forming a monophyletic clade, that is, a sister group of *Cryptosporidium*. To reevaluate the phylogenetic position of *Cryptosporidium*, Zhu et al. *(6)* constructed phylogenetic trees using distance-based neighbor-joining, maximum-parsimony, and maximum-likelihood as well as Slow—Fast analyses of both the small subunit (SSU) and the fused large subunit (LSU)/SSU RNA genes and six other protein genes (α-tubulin, β-tubulin, elongation factor 1-α, heat shock protein (*hsp*) 70, cyclin-dependent protein kinase, and dihydrofolate reductase (*dhfr*)-thymidylate synthase) avoiding the possible bias of single gene analysis. Using SSU and fused LSU/ SSU rRNA data, the authors *(6)* concluded that *Cryptosporidium* belongs to the phylum Apicomplexa, but does not form a monophyletic clade with any of the Eucoccidia, supporting a hypothesis that the genus *Cryptosporidium* constitutes an early emerging branch of the Apicomplexa. These analyses are in agreement with new observations on the *Cryptosporidium* life cycle *(7)*.

(b) Hijawi et al. *(7)* reported the presence of gregarine-like life forms in two *Cryptosporidium* species (*C. parvum* and *C. andersoni*). The gregarine-like life forms were observed both in vivo and in vitro (*see* Section 1.1.2), which points towards the need for reevaluating the phylogenetic position of *Cryptosporidium* and elucidating its evolutionary path.

Table 1
Differences Between Some Species Within the Genus *Cryptosporidium*

Species	Dimensions of oocysts (μm)	Site of infection	Major hosts
C. hominis	4.5 × 5.5	Small intestine	Humans
C. parvum	4.5 × 5.5	Small intestine	Cattle, livestock, humans
C. muris	5.6 × 7.4	Stomach	Rodents
C. suis	(4.9–4.4 × 4.0–4.3)	Small and large intestine	Pigs
C. felis	4.5 × 5.0	Small intestine	Cats
C. canis	4.71 × 4.95	Small intestine	Dogs
C. meleagridis	4.5–4.0 × 4.6–5.2	Small intestine	Turkey, humans
C. wrairi	4.9–5.0 × 4.8–5.6	Small intestine	Guinea pigs
C. bovis	4.7–5.3 × 4.2–4.8	Small intestine	Cattle
C. andersoni	5.5 × 7.4	Abomasum	Cattle, bactrian camel
C. baileyi	4.6 × 6.2	Bursa	Poultry
C. galli	8.25 × 6.3	Proventriculus	Finches, chicken
C. serpentis	5.6–6.6 × 4.8–5.6	Stomach	Lizards, snakes
C. saurophilum	4.2–5.2 × 4.4–5.6	Stomach and small intestine	Lizards
C. scophthalmi	3.7–5.0 × 3.0–4.7	Intestine (and stomach)	Fish
C. molnari	4.7 × 4.5	Stomach (and intestine)	Fish

(c) Database mining of both the *Cryptosporidium* genome sequence database (CryptoDB) *(8,9)* and a comparative apicomplexan database *(10)* reveals the differences between *Cryptosporidium* and other apicomplexans, including the lack of an apicoplast, reliance on adenosine salvage through the adenosine kinase pathway, the lack of identifiable genes-encoding enzymes for pyrimidine *de novo* synthesis, the possible lateral transfer of some genes from a proteobacterium and the presence of unique *Cryptosporidium* oocyst wall protein (COWP) and thrombospondin-related anonymous protein genes encoding for oocyst wall constituents and locomotory, and invasion proteins, respectively *(10)*.

Host species, site of development, and oocyst morphometry (size and shape) have been used to determine *Cryptosporidium* species (Table 1). Of late, molecular methods for species identity have also been described and should be used in conjunction with more conventional biological parameters. Oocysts of various *Cryptosporidium* species vary in shape and size (Table 1) although overlap in size can occur, and for the majority of known species, oocyst morphometry cannot be used to ascribe *Cryptosporidium* species.

On the basis of the animal hosts from which they were isolated, more than 20 "species" of *Cryptosporidium* have been described; however, host specificity as a criterion for speciation appears to be ill-founded as some species lack such specificity. Species definition and identification of this genus are constantly changing, with the addition of "new" species based primarily on molecular criteria and host species.

1.1.2. Current Classification

Current classification is based on a variety of parameters, including host preference and cross-transmissibility, morphological differences, sites of infection, etc., and molecular taxonomic methods. There are 16 'valid' *Cryptosporidium* species, namely: *C. hominis* found primarily in humans (previously known as *C. parvum* Type 1), *C. parvum*, found in humans and other mammals (previously known as *C. parvum* Type 2), *C. andersoni and C. bovis* in cattle, *C. canis* in dogs, *C. muris* in mice, *C. felis* in cats, *C. suis* in pigs, *C. wrairi* in guinea-pigs, *C. meleagridis* in turkeys and human beings, *C. baileyi* in chickens, *C. galli* in finches and chicken, *C. saurophilum* in lizards, *C. serpentis* in snakes and lizards, and *C. molnari* and *C. scophthalmi* in fish.

Caution should be noted given the debate surrounding the choice of both "host preference" and "host specificity" as criteria for speciation. Some hosts may be more permissive than others based on the expression of immune response and other background genes, age, immunocompetence, concomitant infection/disease, etc. For example, *C. parvum* (*senso stricto*) readily infects a variety of neonatal mouse genotypes, which have been used to good effect as *in vivo* surrogates, yet, at best, generates low-grade infections on their adult counterparts. The low abundance of a surface expressed, ileal enterocyte antigen (p57) has been suggested for the inability of *Cryptosporidium* to establish infection in adult mice *(13)*.

Prior to the restructuring of the genus *Cryptosporidium*, isolates from humans were normally referred to as *C. parvum* or *Cryptosporidium* sp. Recent genetic analyses reveal that seven species of *Cryptosporidium* can infect susceptible, immunocompetent, and immunocompromised human hosts. *C. parvum* and *C. hominis* remain the most common species infecting man, but *C. meleagridis*, *C. felis*, *C. canis*, *C. suis*, *C. muris* and the *Cryptosporidium* and monkey genotypes also infect humans *(11,12)*. As *C. parvum* can infect humans and other mammalian hosts this species is normally, but not exclusively, associated with zoonotic transmission of disease, and although *C. hominis* infects human hosts, primarily, there are sporadic reports of *C. hominis* infections in cattle and sheep (Table 2). Additionally, within *Cryptosporidium* up to over 40 genotypes have been described, some of which may be cryptic species, and may cause human disease. In livestock, *C. parvum*, *C. andersoni*, *C. suis*, *C. baileyi*, *C. meleagridis*, and *C. galli* have been reported to cause morbidity and/or outbreaks of disease.

1.1.3. Life Cycle

Of the *Cryptosporidium* species investigated, the C. *parvum* life cycle has been studied most intensively. *C. parvum* is an obligate intracellular parasite and monoxenous, completing its life cycle in a single host. All stages of the life cycle occur either in epithelial cells lining the intestine or, less frequently, in extraintestinal epithelia, particularly the biliary tree and the respiratory tract, within parasitophorous vacuoles situated in the brush border between the plasma membrane and the cytoplasm, or in the lumen.

The infectious stage is the sporulated oocyst. The *C. parvum* oocyst is spherical or sub-spherical, smooth walled, measures 4.5–5.5 µm in diameter and contains four naked motile sporozoites (Fig. 1A,B).

The banana-shaped sporozoites are released through the suture in the oocyst wall following exposure to body temperature, acid, trypsin and bile salts (Fig. 1A), and attach themselves intimately to the surface of adjacent enterocytes (the epithelial cells which

Table 2
Some Features that Distinguish *C. parvum* from *C. hominis*

	C. parvum	*C. hominis*
Oocyst size	3.7–5.0 μm × 5.0–5.5 μm (mean 4.3 × 5.2 μm)	4.4–5.4 μm × 4.4–5.9 μm (mean 4.86 × 5.2 μm)
Major hosts	Cattle, sheep, goats, and humans	Humans
Minor hosts	Deer, mice, and pigs	Dugong, lambs, and cattle (experimental infections in gnotobiotic pigs)
Transmission studies	Infectious to neonatal rats, neonatal mice and nude mice, cats, and dogs	Not infectious to mice, nude mice and rats, cats, dogs, and cattle
Pathogenicity in gnotobiotic pig model	• Parasites observed throughout the length of the small and large intestine • Moderate lymphoid hyperplasia with scattered neutrophils in the duodenum, jejunum, ileum, and colon • Mild to moderate mucosal attenuation throughout intestine and colon	• Parasites observed in the ileum and colon • Mild lymphoid hyperplasia in the ileum and colon • Only mild or no mucosal attenuation restricted to ileum and colon
Prepatent period in gnotobiotic pigs	5.4 d	8.8 d
Oocyst shedding pattern in humans	• Significant greater proportion of stool samples with small number of oocysts (UK study) • Low duration of oocyst shedding in stool and low intensity of infections (Peru study)	• Significant greater proportion of stool samples with large number of oocysts (UK study) • High duration of oocyst shedding in stool and high intensity of infections (Peru study)

Adapted from refs. *(11,14)*.

line the gastro-intestinal tract) (Fig. 2). Sporozoites invade enterocytes to initiate the asexual cycle of development. Sporozoites and all subsequent endogenous asexual and sexual stages develop within a parasitophorous vacuole which is intracellular, but extracytoplasmic (Fig. 2). Excysted *C. parvum* sporozoites range from 3.8 to 5.2 × 0.6–1.2 μm. Sporozoites have a clearly defined narrow apical end and a broader, rounded posterior end (Fig. 1A,B). Microtubules, which are important for gliding and invasion, are situated laterally under the plasma membrane. These are linked to the circular, apical polar ring, and run horizontally from the apex to the middle body of zoites conferring their shape. Within the sporozoite are spherical to ovoid-shaped dense granules, micronemes, nucleus and nucleolus, as well as a Golgi complex, conoidal complex, a single rhoptry, double unit membrane, and endoplasmic reticulum. The oval nucleus is located at the mid-terminal posterior end of the organism. A mitochondrion-like organelle has been observed juxtaposed to the nucleus (Fig. 1B).

Sporozoites differentiate into spherical trophozoites and nuclear division results in the production of the multinucleated schizont stage (schizogony; Fig. 2). Type I schizonts contain six to eight nuclei which mature into 6–8 merozoites. Merozoites from type I

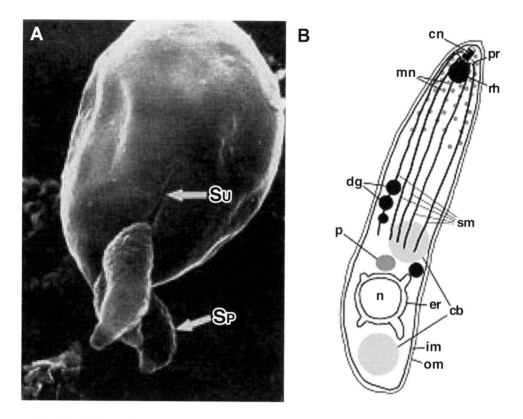

Fig. 1. (**A**) Scanning electron micrograph of an excysting *C. parvum* oocyst. Note two sporozoites (Sp) emerging by their anterior ends through the suture (Su) of the oocyst. From reference *(15)*. With permission, Blackwell Publishing. (**B**) Stylized drawing of excysted *C. parvum* sporozoite. Key: cb, crystalloid bodies; cn, conoid; dg, dense granules; er, endoplasmic reticulum; im, inner membrane; mn, micronemes; n, nucleus; om, outer membrane; p, putative plastid-like organelle; pr, polar ring; rh, rhoptry; sm, sub-pellicular microtubules. From reference *(16)*. With permission, Elsevier Publishing.

schizonts can either infect neighboring cells, where they recycle and undergo an asexual multiplication cycle, similar to that described for the trophozoite stage, and produce further type I merozoite progeny, or develop into a type II schizont (Fig. 2).

Each maturing type II schizont develops into four type II merozoites which are thought to initiate the sexual cycle. In sexual multiplication (gametogony), individual merozoites produce either microgamonts or macrogamonts. Nuclear division in the microgamont leads to the production of numerous microgametes which are released from the parasitophorous vacuole and each macrogamont is fertilized by a microgamete. The product of fertilization, the zygote, develops into an oocyst (Fig. 2). The zygote differentiates into four sporozoites (sporogony) within the oocyst and fully sporulated oocysts (each containing four sporozoites) are released into the lumen of the intestine and pass out of the body in feces where they are infectious for other susceptible hosts. Some of the oocysts released in the lumen of the gut have been reported to cause autoinfection in the same parasitized host by liberating their sporozoites in the gut lumen (Fig. 2). The released sporozoites undergo the developmental processes of schizogony, gametogony, and sporogony in enterocytes of the same

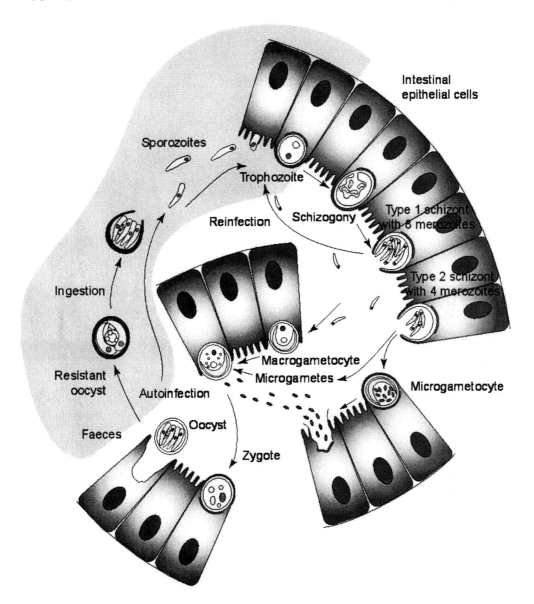

Fig. 2. *Cryptosporidium parvum* life cycle. From reference *(16)*. With permission, Elsevier Publishing. (Please see color insert.)

infected host. In this monoxenous life cycle, both the recycling of merozoites to produce further type I generations of schizonts and the endogenous reinfection from thin-walled oocysts ensure that a large number of infective (thick-walled) oocysts are excreted in feces.

Recent findings challenge this accepted *Cryptosporidium* life cycle *(7,17)*. Two previously unknown extracellular forms (stages 1 and 2) of the parasite have been described both in vitro and in infected hosts in *C. andersoni* and *C. parvum*. The morphology of the "stage one" form is elongated, and it appears to arise from sporozoites, possessing a posterior nucleus, and an anterior round structure, which may play a role in attachment. This "resistant" form can survive in vitro for 7 d at 5°C following isolation from the murine gut. The "stage two" form may represent a gregarine-like gamont and was observed to encyst in vitro to produce a form resembling an unsporulated oocyst *(7)*. The complete development of *Cryptosporidium* in cell-free culture has been reported recently *(17)*, which further complicates our understanding of this intracellular proto-zoan parasite. Culture in vitro in cell-free media should assist in the development of novel drugs and vaccines to control cryptosporidiosis.

1.2. Identification

There are no pathognomonic signs, and laboratory identification is required to confirm diagnosis.

1.2.1. Detection in Feces

Oocysts are excreted in the feces of human and animal hosts and can be preserved in 10% formalin, 2.5% potassium dichromate, or sodium acetate–acetic acid–formalin fixatives. Oocyst viability is retained following storage in 2.5% potassium dichromate. Formalin is a known inhibitor of the polymerase chain reaction (PCR). Fresh or preserved stools can be concentrated to increase the yield of oocysts by sedimentation, using the formol–ether or formol–ethyl acetate techniques or by any conventional fecal parasite flotation method, including zinc sulfate, saturated sodium chloride, and sucrose flota-tion *(18,19)*. Because oocysts are small and lack distinctive visible internal or external structures, staining techniques are required to differentiate them from similarly sized objects present in feces. The modified Ziehl–Neelsen (mZN) stain or the auramine phenol fluorescent stain is commonly used *(20)*, although safranin–methylene blue, Wright–Giemsa, Kinyoun modification of mZN (KmZN), and quinacrine stains have also been recommended. Epifluorescence microscopy detection, using commercially available fluorescein isothiocyanate-conjugated anti-*Cryptosporidium* monoclonal anti-bodies (FITC-*C*-mAb), has the advantage of both genus specificity and requiring lower total magnifications for oocyst identification, thus enabling preparations to be scanned more rapidly. Although oocysts can be excreted in very large numbers from both human and nonhuman hosts (up to 10^7/g) during the acute phase of infection, it should be appreciated that the threshold for 100% detection efficiency using conventional methods can be as high as 5×10^4/g oocysts.

Cryptosporidium spp. antigens can be sought in fecal samples (coproantigens) with or without oocyst concentration and a variety of commercial kits based on enzyme-linked immunoassay (ELISA) and lateral flow immunochromographic (IC) formats are available. Identified benefits of coproantigen detection immunoassays over microscopy are that they are less time-consuming and easier to perform, and do not require experienced microscopists and the observation of intact organisms. IC kits greatly increase the speed of antigen-antibody interaction (seconds to minutes) as all fluids, including soluble *Cryptosporidium* antigens in the test sample, are drawn through a membrane enclosed in a cassette, by a wicking action. Positive reactions are qualitative and are seen as a band of colour at a specific location on the membrane, normally identified on the cassette. IC kits provide diagnostic laboratories with a convenient alternative method for performing antigen detection assays for *Cryptosporidium* on stool samples.

Some commercial ELISA kits are superior at detecting infection than some commercial FITC-*C*-mAb kits, although reports from different research groups vary (reviewed in ref. *[160]*). The consensus opinion is that coproantigen detection immunoassays appear

Table 3
Some Features of the Biology of *C. parvum* Which Can Facilitate Waterborne Transmission

Feature	*Cryptosporidium*
Large numbers of oocysts excreted by infected hosts	Up to 10^{10} oocysts excreted during symptomatic infection. Up to 10^7/g oocysts of feces in infected calves
Lack of, or reduced, host specificity increases the potential for environmental spread and contamination	*C. parvum* infections reported from a variety of mammals, including human beings, domestic livestock, pets and feral animals
Robust nature of oocysts enhances their survival for long periods of time in favorable environments before ingestion by potential hosts	Oocyst survival is enhanced in moist cold environments (such as those found in temperate regions),e.g., a small proportion of viable oocysts can survive for 12 mo suspended in water
Environmental robustness of oocysts enables them to survive some water treatment processes	Waterborne outbreaks indicate that oocysts can survive physical treatment and disinfection. Oocysts are insensitive to disinfectants commonly used in water treatment
Small size of oocysts aid their penetration through sand filters	4.5–6 μm
Low infectious dose means that few viable oocysts need to be ingested for infection to establish in susceptible hosts	Small numbers of *C. parvum* oocysts can cause infection in humans. Infectious dose is dependent upon oocyst isolate [ID_{50} = 9–1042 oocysts]; 10 oocysts can cause infection in juvenile nonhuman primates; 5 oocysts can cause infection in gnotobiotic lambs
Excretion of oocysts into water courses facilitates spread to, and entrapment in, freshwater and marine shellfish	Viable oocysts can accumulate within shellfish. Up to 5×10^3 *C. parvum* oocysts have been reported per shellfish. *C. parvum*, *C. hominis*, and *C. meleagridis* oocysts have been detected in various shellfish
Excretion of oocysts in feces facilitates spread to water by water-roosting refuse feeders	Viable oocysts excreted by transport hosts such as seagulls and waterfowl

Adapted from ref. *(21)*. With permission.

to offer no increase in sensitivity over microscopy. One advantage that coproantigen detection immunoassays have over microscopy is that they can be invaluable in cases of infection in the absence of detectable oocysts (reviewed in ref. *[160]*).

The high threshold for 100% detection efficiency results in the under diagnosis of infection and consequently fails to determine sources contributing to environmental and thereby water and food contamination (*see* Table 3). Although complete development of the life cycle has been described in vitro, in vitro culture is not routinely used for clinical diagnostic purposes.

PCR is more sensitive than microscopy, but its reliance on enzymatic amplification of nucleic acids makes it susceptible to inhibitors present in feces, waters and foods (*see* below). Its increased discriminatory powers will prove beneficial for epidemiological and disease tracking proposes, but, currently, there are few published instances of its usefulness in the diagnostic setting. PCR methods which capitalize on our current knowledge of the genetic variants that constitute the genus *Cryptosporidum* can be designed to identify specific human-infective genotypes and sub-genotypes in stool and environmental samples.

When *Cryptosporidium* oocysts/antigens/DNA cannot be found in stools, diagnosis can be augmented by light microscopy of hematoxylin- and eosin-stained intestinal biopsies. Transmission electron microscopy (TEM) of biopsies reveals the presence of parasitophorous vacuoles containing the replicating stages. Neither light microscopy nor TEM examination of tissue is routinely performed for diagnostic purposes, because of the invasive nature of the biopsy procedure. The demonstration of specific circulating antibodies is only of diagnostic benefit if seroconversion from negative to positive serology or elevation in titer can be ascribed to a current or recent infection. In the absence of demonstrating seroconversion, titer elevation, or antibody isotype switching, positive reactions can indicate either current infection, past exposure, or both. Accordingly, serological methods tend to be used for epidemiological purposes.

1.2.2. Sensitivity of Detection in Feces

In unconcentrated fecal smears, a detection limit of 10^6 oocysts/mL of feces was reported using KmZN. Following stool concentration, between 1×10^4 and 5×10^4 oocysts/g of unconcentrated stool are necessary to obtain a 100% detection efficiency using either KmZN or a commercially available FITC-*C*-mAb method *(22)*. For bovine stools, the threshold for 100% detection efficiency using auramine phenol or FITC-*C*-mAb is 1×10^3 oocysts/g *(23)*. Variations in fecal consistency influence the ease of detection, with oocysts being more easily detected in concentrates made from watery, diarrheal specimens than from formed stool specimens, because oocysts excreted in diarrheal stools are less likely to become attached to particulates than oocysts excreted in partially formed or formed stools. Various immunological methods have also been developed for detecting *Cryptosporidium* oocysts. The use of an FITC-*C*-mAb offers no significant increase in sensitivity over conventional stains. In addition to microscopical techniques, antigen capture enzyme linked immunosorbent assays have been reported with the detection limits in the region of 3×10^5–10^6 oocysts/mL, which is similar in sensitivity to microscopy *(23)*. Antigenic variability between *Cryptosporidium* isolates could compromise immunodiagnostic tests. Although effective for diagnosing symptomatic infection, these methods remain insensitive and unable to detect the smaller number of oocysts present in asymptomatic infections. In addition, anecdotal evidence suggests that smaller or minimal number of oocysts may be excreted in the chronic disease stages in immunocompromised individuals indicating that oocyst detection methods may not be the ideal option for all situations.

1.2.3. Alternative Detection Procedures

Flow cytometry coupled with cell sorting (FCCS) has been used to detect oocysts in human stools with a fourfold increase in sensitivity over direct fluorescence antibody tests, and offering, in some instances, a detection limit of 1×10^3 oocysts/mL of sample *(24)*. FCCS is a costly procedure that is not available in most diagnostic parasitology laboratories, and being based on the use of FITC-*C*-mAb is subject to antigenic variability in oocyst epitopes.

PCR is more sensitive than conventional and immunological assays for the coprological detection of infection (25–27). PCR has also the potential for identifying *Cryptosporidium* DNA in the stools of asymptomatic individuals and in fomites, food and water contaminated with small numbers of oocysts. Rapid detection of asymptomatic individuals could limit clinical sequelae in "at-risk" groups and reduce environmental spread. Further analysis of specifically targeted PCR products, by restriction fragment length polymorphism (RFLP) or sequencing, can provide invaluable data for source and disease tracking, particularly for water and food contamination.

1.2.3.1. Potential PCR Inhibitors. Prior to adoption in diagnostic laboratories, both the variability between methods and the recognized difficulties in amplifying nucleic acids by PCR from feces, water and food must be overcome. Fecal samples can contain many PCR inhibitors, including bilirubin, bile salts, and complex polysaccharides, which are significant inhibitors of *Taq* polymerase and adversely affect PCR sensitivity (28). In environmental (e.g., water) samples, humic or fulvic acids are the most commonly encountered inhibitory substances (29,30), and plant-derived polyphenolics and polysaccharides occur in fruit extracts (31).

Immunomagnetizable separation (IMS) of oocysts, in which magnetizable beads are coated with an anti-*Cryptosporidium* monoclonal antibody which binds to, and concentrates, suspended oocysts from contaminated matrices is an useful approach for concentrating small number of where oocysts (32). This method is the most effective for both the concentration of oocysts and the elimination of inhibitors for PCR-based detection of oocysts in water (33,34) and food (35,36) and has also been used in the authors' laboratory for concentrating small number of oocysts from minimal amount of stool sample in order to identify the species of *Cryptosporidium* present in index cases during outbreak investigations. IMS-purified oocysts are separated from magnetizable beads by acid dissociation prior to DNA extraction (33,34). Alternatively, dissociated oocysts can be placed on a membrane filter or microscope slide, and their morphology and morphometry determined by microscopy (following staining with FITC-*C*-mAb and the fluorogenic DNA intercalator, 4′,6-diamidino-2-phenylindole (DAPI)) prior to oocyst disruption and DNA extraction directly from the filter or from slide scrapings (37,38). IMS-captured oocysts can be left attached onto beads during DNA extraction (39) which can increase the sensitivity of detection by reducing oocyst losses during acid dissociation; however, oocyst enumeration, oocyst morphometry, and assessment of oocyst integrity by microscopy are not possible with this method. Purified oocysts can be disrupted by either mechanical (glass-bead beating) or physical (exposure to freeze—thawing cycles) injury to the oocyst wall to release sporozoites from which DNA is extracted (Table 4).

Extraction of oocyst DNA from whole feces, water or food, without IMS purification may be necessary in some situations, particularly when more than one parasite is sought, parasites have been stored frozen on/in their original matrices for retrospective or epidemiologic analysis, or where oocysts cannot be dissociated from their matrix. Commercial methods, which contain substances that relieve the effects of inhibitory substances on *Taq* polymerase, can be used to prepare (extraction and purification) oocyst DNA, either alone or in conjunction with a purification method, so that nucleic acids suitable for PCR amplification are obtained (Table 4).

Table 4
**Some Methods for Extracting *Cryptosporidium* Oocyst DNA
from Feces, Water, and Food**

Matrix	DNA extraction and purification methods from oocysts	Facilitators	PCR Target (detection limit)	Sensitivity (% positives); nos of positives/ total	Ref.
Unpreserved human stool (180–200 mg)	QIAamp® DNA stool mini kit[a], boil 5 min, freeze–thaw 3 × (liquid nitrogen 1 min, boil 2 min)	Inhibit-EX-tablet	nested COWP (5×10^2 oocysts)	(97%); 86/89	(40)
Frozen pig and calf stools (diluted 1:4 w/v in PBS)	Freeze–thaw 3 × (liquid nitrogen for 2 min, 75°C for 2 min), centrifuge, and add lysis buffer (Tris-EDTA–SDS-proteinase K) to pellet, glassmilk[b]	Bovine serum albumin (BSA)	Nested; Laxer et al. (1991); (100 seeded oocysts)	NA	(41)
Human stool (0.3–0.5 g) preserved in 2.5% potassium dichromate	Modified FastDNA Prep kit[c] and QIAquick spin column	PVP in extraction method. 11.3% (24 out of 213 samples) were inhibitory	Direct; 18S rRNA	(100%); 53/53 plus 31 negatives by micro-scopy	(26)
Unpreserved solid (0.4–0.5 g) or liquid (200 µL) human stool	Water–ether extraction and washes in lysis buffer (Tris–EDTA–SDS)[d], freeze–thawing ×15 (liquid nitrogen for 1 min, 65°C for 1 min)	BSA, Tween 20 and PVP in-corporated in PCR	Single tube nested COWP & nested 18S rRNA	(97.8%); 90/92 (98.9%); 91/92	(161)
Storm waters (filtered and purified by sucrose–Percoll flotation)	Oocysts concentrated from a 0.5 mL pellet of storm water concentrated by IMS. Five cycles of freeze–thawing without dissociating oocysts from mag-netizable beads,	None	Nested 18S rRNA	27/29 include 12 samples negative by micro-scopy	(42)

Table 4 (*Continued*)

Matrix	DNA extraction and purification methods from oocysts	Facilitators	Target (detection limit)	Sensitivity (% positives); nos of positives/ total	Ref.
			PCR		
Raw milk (50 mL samples)	proteinase K digestion, and diluted with ethanol, QIAamp® DNA mini kit Centrifugation, IMS, 5 × freeze–thawing (−80°C, 5 min, 95°C, 5 min), QIAGEN LB + proteinase K digestion, QIAmp spin column[e]	None	18S rRNA & COWP (<10 seeded oocysts)	NA	*(36)*
Apple juice (50 mL samples)	Centrifugation, sucrose gradient, IMS, proteinase digestion, QIA mp blood kit	None	COWP (30–100 seeded oocysts)	NA	*(43)*

[a]QIAamp® DNA stool mini kit is based on the use of proprietary buffers and DNA purification through a silica column.

[b]Glassmilk (GENECLEAN® Bio 101) is based on DNA adsorption to silica suspensions that allows subsequent washing of bound DNA to remove inhibitors.

[c]Modified FastDNA kit uses FP120 FastPrep Cell Disruptor to disrupt oocysts in a proprietary buffer that minimizes adsorption of DNA to fecal particles (Cell lysis/DNA Solubilizing Solution). Polyvinylpyrrolidone (final concentration of 0.5% w/v) used in this step precipitates polyphenolic compounds and the solubilized DNA is bound to the binding matrix in the presence of chaotropic salt which is washed and then eluted. The final purification of DNA is performed in a QIAquick spin column.

[d]This method uses semi-purified oocysts and no DNA purification. The presence of 0.5% SDS in the oocyst lysate is inactivated by the addition of 2% Tween 20 to the PCR mixture.

[e]DNA bound to spin columns is washed twice (instead of the recommended single wash) with two different wash buffers to improve the purity of the DNA.

1.2.3.2. PCR Targets. Many PCR protocols target the *Cryptosporidium* 18S rRNA gene *(33,34,37,44,45)*. This multicopy gene offers high-detection sensitivity, because each oocyst contains 20 copies. Because it contains conserved and variable regions, it is suitable for PCR-RFLP assays for species identification. The single-copy COWP gene is heterogeneous throughout its length and is also used for species identification; however, problems remain with designing suitable primers capable of amplifying all described species within the genus *(46,47)*. Other targets for detection, species discrimination, and sequence analysis for phylogenetic studies include the *dhfr* gene *(48)*, the *hsp* analog gene *(49)*, the actin gene *(50)*, the gp60 highly heterogenous protein-coding gene *(51)*, and unidentified, sequenced DNA targets, reported to offer high-detection sensitivity *(52,53)*.

1.2.3.3. Sensitivity of Detection by PCR. The sensitivity of detection depends on maximizing DNA release, identifying ideal PCR targets, while optimizing primer choice and minimizing inhibitory effects. Comparisons between published methods are difficult, because no standardized, optimized procedures for DNA extraction or relieving inhibitory effects on *Taq* exist (Table 4). At present, in the diagnostic laboratory, PCR is a more of a tool used to assist epidemiological investigations than a routine method.

Quantitative real-time PCR methods use patented technology. Both TaqMan™ (Applied Biosystems) and LightCycler™ (Roche Molecular Biochemicals) systems have been used to detect *Cryptosporidium* DNA. For both methods, fluorogenic dyes are incorporated into the amplicon during PCR; therefore, amplicon fluorescence increases as more PCR product is generated. The TaqMan™ system uses a fluorescent hydrolysis probe system with a fluorescent dye at the 5′-end and a quencher at the 3′-end of a probe that binds to the PCR product and is cleaved during PCR by the action of the *Taq* polymerase. The LightCycler™ can be used with a hydrolysis probe, but it can also detect the binding of a fluorescent hybridization probe or the incorporation of a fluorogenic dye (usually SYBR® Green I) into the amplicon. Both systems can detect PCR products during the initial cycles of the PCR when amplification is exponential, thus enabling quantitative analysis of fluorescent product.

One application uses the TaqMan™ approach for the quantitative detection of *Cryptosporidium* oocysts in environmental water samples. Using primers targeting, the "Laxer amplicon" *(52)*, a probe labeled at the 5′-end with the 6-carboxyfluorescein reporter dye and at the 3′-end with the 6-carboxy-tetramethylrhodamine quencher dye, five purified oocysts and eight oocysts spiked into water samples were detected by this IMS real-time PCR method when DNA purification (Nanosep® centrifugal device) was incorporated following the IMS oocyst concentration step *(54)*. Guy et al. *(55)* used more difficult matrices, such as sewage (1 L samples) and environmental water samples (2 L samples) to detect the *Cryptosporidium* COWP gene by TaqMan™ assay. DNA extraction using the DNeasy tissue kit (Qiagen, Germany) was optimized by modifying the commercial protocol and the introduction of a three cycle freeze–thaw step and sonication (3×20 s bursts). Inhibitory effects were counteracted using 20% Chelex 100 (Bio-Rad) and 2% PVP 360 (ICN, Ohio): in some instances, bovine serum albumin proved useful in eliminating inhibition present in lake water. Tenfold dilutions of purified DNA detected the equivalent of one *C. parvum* oocyst (40 fg); however, in experimentally seeded sewage, a minimum of 100 oocysts were required to produce a positive signal. No *Cryptosporidium* DNA was detected by either a multiplex real-time PCR or immunofluorescence microscopy in sewage or water.

The LightCycler™ system was used to identify different *Cryptosporidium* species by exploiting the genetic polymorphism of the 18S rRNA gene for devising probes with differing melting temperatures *(56)*. After completion of the PCR, the melting curve analysis is determined following rapid denaturation, and annealing using stepped increases in temperature. This technique can detect a minimum of five oocysts and differentiate between *C. parvum*, *C. hominis*, *C. felis*, *C. canis*, and *C. meleagridis* by their melting curves. Tanriverdi *et al.* *(57)* discriminated between *C. hominis* and *C. parvum* by SYBR® Green I incorporation and fluorescent probe melting curve analysis, on the basis of single nucleotide polymorphism of both the β-tubulin and the gp900/poly(T) genes,

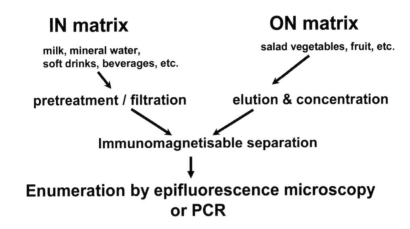

Protozoan parasite contamination of foods

Fig. 3. Methods for detecting protozoan parasite contamination of foods.

using the LightCycler™. This method detected a single oocyst introduced into the PCR by micromanipulation.

The increased sensitivity of real-time PCR guarantees increased speed of detection and qualitative diagnosis, whereas the quantitative nature of the assay will be invaluable in estimating levels of contamination. In addition, the "closed-tube" assay format eliminates the danger of contamination from "carry over". These characteristics identify real-time PCR as a useful tool for the future, particularly for the detection of parasites in food and water matrices, once the issues surrounding matrix inhibition are overcome.

Fluorescence *in situ* hybridization (FISH) was used to identify *C. parvum* oocysts in a laboratory-based approach. Probes to unique regions of *C. parvum* rRNA were synthesized and used to detect oocysts on glass microscope slides. Two probes, with similar hybridization characteristics, were labeled with the fluorescent reporter, 6-carboxy-fluorescein phosphoramadite (excitation 488 nm and emission 522 nm) and viewed by epifluorescence microscopy *(58)*.

1.2.4. Detection in/on Foods

Infective (sporulated) *Cryptosporidium* spp. oocysts cannot multiply in/on foodstuffs. Food becomes a potential vehicle for human infection by contamination during production, collection, transport, and preparation (e.g., fruit, vegetables, etc.) or during processing. The sources of such contamination include feces, fecally contaminated soil, irrigation water, water for direct incorporation into product (ingredient water), and/or for washing produce (process water) and infected food handlers. *Cryptosporidium* spp. oocysts normally occur in low numbers in/on food matrices and in vitro culture techniques that increase parasite numbers prior to identification are not available for protozoan parasites in food and water matrices. Concentration by size alone, through membrane or depth filters, results in accumulation in the retentate, not only of oocysts, but also of similar- and larger-sized particles, and as *C. parvum* oocysts are 4–6 μm in diameter, filters with

Table 5
Some Reported Recovery Rates of *Cryptosporidium* Oocysts from Foods

Food matrices	Extraction and concentration methods	Detection methods	Recovery rates	Ref.
Cabbage and lettuce leaves	200 g seeded with *Cryptosporidium* ssp. oocysts (1/g). FDA method: sonication in 1% SDS and 0.1% Tween 80, centrifugation	Immuno-fluorescence	1%	*(61)*
• Milk • Orange juice • White wine	Seed 10–1000; *C. parvum* oocysts in 70–200 mL; filtration	Immuno-fluorescence on filters	• 4–9.5% • <10% • >40%	*(62)*
• Cilantro leaves • Cilantro roots • Lettuce	Rinse; centrifugation	Koster stain	• 5.0% (four samples) • 8.7% (seven samples) • 2.5% (two samples)	*(63)*
Raw milk	Bacto-Trypsin & Triton X-100 treatment of seeded samples (20 mL); centrifugation	PCR and probe hybridization	10 seeded *C. parvum* oocysts	*(64)*
Lettuce leaves	Seeded with 100 *C. parvum* oocysts. • Rinse in tap water; centrifugation	Acid-fast stain, Immuno-fluorescence and direct wet direct mount	25–36%	*(65)*
110 vegetables and herbs from 13 markets; • 62 vegetables and herbs from 15 markets	Rinse in tap water; centrifugation	Acid-fast stain, Immunofluorescence and direct wet mount	14.5%; • 19.35%	*(65)*
Homogenized milk	100 mL samples seeded with 10^4–1 *C. parvum* oocysts. Centrifugation & IMS	Direct PCR	10 oocysts	*(66)*
Apple juice	Seeded with 10^5–1 *C. parvum* oocysts • Flotation and IMS • Flotation • Flotation	 • Direct PCR • Immuno-fluorescence • Acid-fast stain	 • 10–30 oocysts • 30–100 oocysts • 3000 to 10,000 oocysts	*(67)*

Table 5 (*Continued*)

Food matrices	Extraction and concentration methods	Detection methods	Recovery rates	Ref.
• Four leafy vegetables and strawberries • Bean sprouts	Seeded with 46–165 *C. parvum* oocysts. Rotation and sonication in elution buffer; centrifugation & IMS	Immuno-fluorescence	• 42% • 22–35%	*(68)*
Lettuce leaves	Seeded with 100 *C. parvum* oocysts. Elution buffer (pH 5.5) • Pulsification centrifugation and IMS • Stomaching, centrifugation, and IMS	Immuno-fluorescence	• 40% • 59%	*(69)*

a smaller pore size (1–2 µm) are employed to entrap them, resulting in the accumulation of large volumes of extraneous particulate material. Such particles interfere with oocyst detection and identification; therefore, a clarification step which separates oocysts from other contaminating particles is employed. The present methods can be subdivided into the following component parts: (a) sampling, (b) desorption of oocysts from the matrix and their concentration, and (c) identification (Fig. 3), and are those, or modifications of those, which apply to water *(59,60)* (Table 5).

Methods for detecting oocyst contamination in liquid foods (e.g., drinks and beverages) and on solid foods (e.g. fruit and vegetables) are similar, but can differ in the techniques used to separate oocysts from the food matrix (Table 5). Differences are primarily dependent on the turbidity of the matrix and for non-turbid liquid samples, such as bottled waters, either small or large volume filtration (1–20,000 L samples have been tested in the authors' laboratory) can be used. For turbid samples (naturally turbid, e.g., milk, fruit juices, or following extraction from surfaces of vegetables and fruit), further oocyst concentration using IMS is recommended.

Sampling methods, volumes/weights of product depend on the reasons for sampling (quality control/assurance, ascertainment, compliance, incident/outbreak investigation, etc.). Desorption can be accomplished by mechanical agitation, stomaching, pulsifying, and sonication of leafy vegetables or fruit suspended in a liquid that encourages desorption of oocysts from the food matrix. Detergents including Tween 20, Tween 80, sodium lauryl sulfate, and Laureth 12 have been used to encourage desorption (Table 5). Lowering or elevating pH affects surface charge which can also increase desorption. Depending on the turbidity and pH of the eluate, oocysts eluted into non-turbid, neutral pH eluates can be concentrated by filtration through a 1-µm flat-bed cellulose acetate membrane *(70)*, whereas oocysts eluted into turbid eluates can be concentrated by IMS. Oocysts can be identified and enumerated following staining with FITC-*C*-mAb and their morphology can be assessed by Nomarski differential interference contrast (DIC) microscopy (Table 5). There is only one published, optimized *(162)* and validated *(163)* method for detecting *C. parvum* on lettuce and raspberries. It is based on four stages: (i) extraction of oocysts from the foodstuffs, (ii) concentration of the extract and separation of the oocysts from food materials, (iii) staining of the oocysts to allow their visualization, and (iv) identification of oocysts by microscopy. The method to detect *C. parvum* on lettuce recovered 59.0 ± 12.0% (n = 30) of artificially contaminated oocysts. The method to detect *C. parvum* on raspberries recovered 41.0 ± 13.0% (n = 30) of artificially contaminated oocysts *(162)*. Recently, sensitive PCR-based methods have been used to determine the presence of *Cryptosporidium* DNA in eluates, although there is little standardization of methodology available at present *(71)* and PCR is sensitive to inhibitory factors present in foodstuffs and waters (*see* earlier discussion and Table 4). Neither the DNA-based nor the microscopy-based detection methods identified previously can be used to determine oocyst viability.

1.2.5. Determination of Oocyst Viability

Little can be inferred about the likely impact of oocysts detected in water or food concentrates on public health without knowing whether they are viable or not. The conventional techniques of animal infectivity and excystation in vitro are not applicable to the small number of organisms found in food and water concentrates, and much effort has been expended on the development of surrogate techniques, which can address, accurately, the viability of individual oocysts (reviewed in ref. *[72]*).

1.2.5.1. Determination by morphology. Rapid, objective estimates of oocyst viability revolve around the microscopical observation of fluorescence inclusion or exclusion of specific fluorogens into individual oocysts *(72)*. Campbell et al. *(73)* developed a fluorogenic vital dye assay to determine *C. parvum* oocyst viability, based upon the inclusion/exclusion of two fluorogenic vital dyes, DAPI and propidium iodide (PI). Four classes of oocysts were identified using the assay: (i) viable oocysts that include DAPI but exclude PI, (ii) nonviable oocysts that include DAPI and PI, (iii) potentially infectious oocysts that do not include either DAPI or PI, but have "viable type" contents by DIC microscopy (and that take up DAPI, but not PI, following a further trigger), and (iv) empty oocysts which have no recognizable contents by DIC microscopy and which include neither DAPI nor PI *(73,74)*. The addition of FITC-*C*-mAb, which binds to the outer oocyst walls, 30 min before the completion of the assay allows rapid identification of oocysts. A UV filter block is used to visualize DAPI (excitation 350 nm and emission 450 nm) and a green filter block for PI (excitation 535 nm and emission >610 nm). Results correlated closely with optimized in vitro excystation *(75)*.

SYTO®9 and SYTO®59 are both nucleic-acid intercalators which permeate damaged membranes, staining damaged or dead oocysts. SYTO®9 fluoresces light green-yellow under the blue filter (–excitation 480 nm and –emission 520 nm) and SYTO®59 fluoresces red under the green filter block (–excitation 535 nm and emission >610 nm) of an epifluorescence microscope *(76)*. Belosovic et al. *(76)* and Neumann et al. *(77)* showed that *C. parvum* oocyst viability following heat treatment (70°C, 30 min) and chemical disinfection correlated with infectivity in CD-1 neonate mice, but not with in vitro excystation. However, similar experiments employing untreated, ozonized, and high-intensity UV-treated *C. parvum* oocysts showed that viability determined uisng SYTO®9 staining does not correlate with CD-1 neonatal mouse infectivity *(72)*, SYTO®9 overestimating the number of nonviable oocysts. Bukhari et al. *(78)* compared four viability assays with CD-1 neonatal mouse infectivity to assess the effect of low levels of ozonation on fresh and aged oocysts and reported that SYTO®59 assay showed minimal change on oocyst viability, whereas DAPI-PI, in vitro excystation and SYTO®9 assays, performed on sub-samples of the same ozonated oocysts, showed marginal reduction on viability. All in vitro tests developed underestimate reduction in viability when compared with in vivo infectivity.

Biophysical methods have also been used to determine oocyst viability. Both dielectro-phoresis and electrorotation have been used to demonstrate differences between viable and nonviable oocysts *(79)* and are reviewed in O'Grady and Smith *(72)*.

1.2.5.2. Determination by Molecular Methods. The detection of excysted sporozoites following in vitro excystation (refer to in vitro excystation in ref. *[72]*) has been used as the first stage in developing PCR-based viability assays and two approaches have been described for *C. parvum* oocysts. In the first approach, sporozoites are excysted in vitro according to a standardized protocol, lysed, and the DNA is amplified using the "Laxer" primers *(80)*. In the second approach, IMS is used to concentrate oocysts from the inhibitory fecal matrix, the sporozoites excysted, their DNA released, and ampli-fied using a nested PCR (IC-PCR) *(81)*. Although the sensitivity of the first approach was approx 25 oocysts (100 sporozoites) in an experimental system, no information was available on its sensitivity with environmental samples. The sensitivity of the IC-PCR was between 1 and 10 oocysts in purified samples and between 30 and 100 oocysts inoculated into feces.

The detection of labile messenger RNA (mRNA) species that are denatured or destroyed soon after the death of the organism can be used to infer viability. A reverse-transcription PCR (RT-PCR), which amplifies *C. parvum hsp* 70 mRNA was used to determine the oocyst viability in four different water types *(82)*. Synthesis of *hsp* 70 mRNA was induced following 20-min incubation at 45°C, mRNA was released by rupturing oocysts using five freeze–thaw cycles (1 min liquid N_2 and 1 min 65°C), which was then hybridized to oligo $(dT)_{25}$-linked magnetizable beads. Amplicons from the RT reaction were detected visually on gels and by chemiluminescent detection of Southern blot hybridizations. Although no differences in detection levels were observed between induced and non-induced viable organisms (either because the oocysts were already stressed or the *hsp* 70 mRNA is constitutively expressed in *Cryptosporidium*), only viable oocysts were detected. Importantly, all genomic DNA should be removed prior to RT-PCR, since very few *Cryptosporidium* genes possess introns. Failure to remove genomic DNA can lead to the genomic DNA instead of cDNA being detected which leads to the detection of false positives when undigested DNA remains, eroding assay confidence, and imposing limits to assay applicability.

The *C. parvum* amyloglucosidase gene has also been targeted in an RT-PCR assay. Oocysts were exposed to 1.7% sodium hypochlorite for 10 min at room temperature and disrupted by three freeze–thaw cycles (dry ice/ethanol, 37°C) for RNA extraction and purification using Trizol reagent (Gibco, BRL). Purified RNA was treated with DNase I and subsequently recovered by phenol–chloroform–ethanol precipitation. The maxi-mum detection sensitivity was 10^3 oocysts and detection of the $[^{32}P]dCTP$ incorporated into the RT-PCR, was following electrophoresis in 5% nondenaturing polyacrylamide gels. Both in vitro and animal infectivity assays correlated with the inability to detect amyloglucosidase mRNA following 9 mo of oocyst storage at 4°C, whereas a gradual decline in mRNA detection occurred between 5 and 7 mo of storage *(83)*.

More recently, a *C. parvum* nucleic acid-based amplification assay (NASBA) using the *hsp* 70 mRNA target has been described *(84)*. The proprietary NASBA assay (Biomérieux, UK, Ltd.) utilizes a cocktail of three enzymes for the transcription and amplification of specific mRNAs in vitro in the presence of DNA during an isothermal reaction (usually at 41°C) followed by probe hybridization and detection by electro-chemiluminescence. Baeumner et al. *(84)* reported detecting five viable *C. parvum* oocysts after optimum heat shock induction achieved by pre-incubating oocysts for 20 min at 42–43°C.

FISH has also been used to determine *C. parvum* oocyst viability *(58)*. The fluorescently labeled oligonucleotide probe targets a specific sequence in *C. parvum* 18S rRNA and causes viable sporozoites (capable of in vitro excystation) to fluoresce. Neither dead oocysts nor organisms other than *C. parvum* fluoresced following *in situ* hybridization. FISH-stained oocysts did not fluoresce sufficiently to allow their detection in environmental water samples; however, simultaneous detection and viability determination could be performed when FISH was used in combination with a FITC-*C*-mAb.

1.2.6. Determination of Infectivity

1.2.6.1. In vitro Infectivity. While incomplete development in vitro (primarily asexual developmental stages) has been documented in a variety of human- and animal-derived cell lines, reports of viable oocyst production in vitro are scarce *(70,85)*. However, Hijawi et al. *(86)* described the complete culture of *C. parvum* over 25 d in HCT-8 cells, and more recently the morphology of extracellular life forms in both *C. parvum* and *C. andersoni* during in vitro culture *(7)*. Further observations included the complete development of *C. parvum* in host cell-free cultures, with the development of asexual and sexual stages and the in vitro production of oocysts that were subsequently infectious to neonatal ARC/Swiss mice *(17)*.

1.2.6.2. In vivo Infectivity. A limitation of the neonatal mouse model is that it can only readily be used to determine the infectivity of *C. parvum* isolates, as it does not support the growth of other human-infective species effectively. In vivo infectivity of *C. parvum* isolates can be determined in neonatal or infant mice: 4–7-d-old CD-1 strain mice are commonly used, because the strain is outbred and has large litter sizes. Either conventional hematoxylin and eosin histology of the small intestine, to demonstrate endogenous stages and oocysts; or homogenization of the small intestine, followed by concentration and clarification of oocysts and visualization by epifluorescence microscopy using FITC-*C*-mAb can be used to demonstrate infection. The ID_{100} for CD-1 strain mice is approx 300 oocysts and the ID_{50} ranges between 60 and 87 oocysts of either an Iowa (bovine) or Moredun (cervine–ovine) strain of *C. parvum* *(87)*. The usefulness of in vivo infectivity is reviewed in ref. *(72)*.

2. TRANSMISSION AND RESERVOIRS OF INFECTION

2.1. Routes of Exposure

The transmission of *Cryptosporidium* to humans can occur via any mechanism by which material contaminated with feces containing infectious oocysts from infected humans or nonhuman hosts can be swallowed by a susceptible host. Both biotic and abiotic reservoirs of infection exist. Biotic reservoirs include all potential hosts of human-infectious *Cryptosporidium* species. Transmission of *C. hominis* is primarily, but probably not exclusively, person-to-person, while transmission of *C. parvum*, *C. meleagridis*, *C. felis*, *C. canis*, *C. suis* and *C. muris*, and the *C. parvum* cervine and monkey genotypes can include both person-to-person and animal-to-person transmission. Abiotic reservoirs include all vehicles that contain sufficient infectious oocysts to cause human infection, and the most commonly recognized are food and water.

2.1.1. Person-to-Person

The initial cases of human disease were believed to have been acquired from animals (zooanthroponosis), but person-to-person transmission of *Cryptosporidium* spp. is the major route. Transmission via the venereal route has also been reported. Secondary cases and asymptomatic excretors are a source of infection for other susceptible persons. Person-to-person transmission has been documented between family/household members, sexual partners, health workers and their patients, and children in daycare centers and other institutions. Transmission is particularly common among children in daycare centers probably because of the lower standards of personal hygiene exhibited by preschool children and their tendency to put many objects that they handle in their mouths. The robustness of oocysts, the low-infectious dose (the median infectious dose is between 9 and 1042 oocysts in human volunteer studies), and the variable state of immunity elicited by infection contribute to *Cryptosporidium* being a widespread intestinal protozoan parasite of humans occurring commonly in industrialized nations and more frequently in developing countries, particularly where infection can be endemic and sanitation frequently minimal.

2.1.2. Animal-to-Person (Zoonotic)

C. parvum or *C. parvum*-like parasites have been reported in over 150 mammalian species *(11)*, including domestic animals, livestock and wildlife, and companion animals which can be reservoirs of human infection. Domestic and feral animals can be the important reservoirs of human infection and infected calves and lambs can excrete up to 10^{10} oocysts during the course of infection. The broad host-range exemplified by *C. parvum* and the high output of infective oocysts from numerous mammalian hosts ensure a high level of environmental contamination. In addition, redistribution of oocysts by transport hosts, including coprophagous animals and bivalves, further enhance environmental contamination and zoonotic transmission.

Contributions from agricultural practices include storage and spread of muck and slurry, discharge of oocyst contaminated dirty water to land or to water courses, pasturing of livestock in land-adjoining water sources, and disposal of fecally contaminated wastes from livestock markets and abattoirs. Furthermore, practices such as hosing down rearing pens and sluices release recently excreted oocysts into an aquatic environment where the survival is prolonged. Such oocysts are likely to have a higher viability than those excreted onto grazing land which take time to percolate through substrata into water courses *(88)*. Development and regular review of catchment control policies can reduce the potential for the contamination of water courses and foods by these routes. Waterborne, airborne, and foodborne (*see* below and Section 3) transmission have been documented, and outbreaks following visits to city farms together with some waterborne outbreaks of disease further testify to the importance of the zoonotic route of transmission. Animals can be infected, experimentally, with *C. parvum* oocysts of human origin, and may acquire infection naturally from man (anthropozoonosis).

2.1.3. Drinking Water

Compared with other waterborne protozoan parasites, *Cryptosporidium* spp. oocysts have the greatest potential for transmission through drinking water because (a) human infective oocysts are widely distributed in the environment, (b) oocysts can penetrate physical barriers in water treatment processes and are chlorine resistant, and (c) *Cryptosporidium* has a low infectious dose for humans.

As both infected humans and other infected mammals are reservoirs of human infection, the contamination of potable water supplies with either human or animal sewage can lead to the transmission of this organism through drinking water. Numerous outbreaks traced to contamination of drinking water by both human and animal sewage have been reported *(88–92)*. Oocysts can survive several months in water at 4°C and are one of the most chlorine-resistant pathogens known *(88)*. Waterborne outbreaks of cryptosporidiosis have been reported from numerous countries *(91,93)* as has the occurrence of oocysts in drinking water in the absence of disease in the community *(91,93)*. Outbreaks have been associated with untreated drinking water, water receiving only chlorination, and water receiving conventional treatment (e.g., coagulation, sedimentation, sand filtration, and chlorination) *(88)*. Because of their small size (4.5–5.5 μm in diameter), the extent to which oocysts present in the raw water are removed during water treatment is dependent on the water-treatment processes utilized. The efficacy of various water-treatment processes in removing *Cryptosporidium* oocysts is reviewed in refs. *(21,88)*.

2.1.4. Airborne

The small size of *Cryptosporidium* oocysts render them readily aerosolisable. Zoonotic airborne transmission, owing to the inhalation of droplets containing oocysts following the insertion of a gastric tube into an experimentally infected calf, has been documented *(94)*.

2.1.5. Food

Both source contamination of produce and contamination from water used in food preparation are transmission routes that are significant to the food industry. Surface contamination can be direct, following contamination by the infected host, or indirect, following contamination by transport (birds, flies, etc.) hosts, the use of manure and contaminated water for irrigation, fumigation, and pesticide application, etc. Oocyst-contaminated irrigation water *(95)* as well as oocyst-contaminated water used for spraying or misting produce present specific problems to food producers and increase risk to the customer, particularly when the produce is eaten raw.

Oocyst-contaminated river waters can be a source of contamination of the marine environment, and rivers polluted by anthropogenic and livestock fecal discharges *(21)* (Tables 3 and 6) can play a major role in oocyst contamination of shellfish in estuaries and coastal environments (Tables 3 and 6). Experimentally, *C. parvum* oocysts can survive in seawater at salinities of up to 30 ppt for up to 1 mo *(75,96)* and in artificial seawater for at least 1 yr under moderate oxygenation *(97)*. *C. parvum* oocysts can survive in marine and freshwater shellfish. Oocysts with a *C. parvum*

Table 6
Possible Sources and Routes of Food Contamination
with *Cryptosporidium* spp. Oocysts

- Use of oocyst contaminated feces (night soil), farmyard manure and slurry
  as fertilizer for crop cultivation
- Pasturing infected livestock near crops
- defecation of infected feral hosts onto crops
- direct contamination of foods following contact with oocyst-contaminated
  feces transmitted by coprophagous transport hosts (e.g., birds and insects)
- Use of contaminated wastewater for irrigation
- Aerosolization of contaminated water used for insecticide and fungicide
  sprays and mists
- Aerosols from slurry spraying and muck spreading
- Poor personal hygiene of food handlers
- Washing "salad" vegetables, or those consumed raw, in contaminated water
- Use of contaminated water for making ice and frozen/chilled foods
- Use of contaminated water for making products which receive minimum
  heat or preservative treatment
- Ingestion of viable oocysts from raw or undercooked shellfish which accumulate
  human infectious oocysts from their contaminated aquatic environment

morphology occur in mussels (*Mytilus edulis*) from Ireland and in bent mussels (*Ischadium recurvum*) from Chesapeake Bay in the United States. The eastern oyster (*Crassostrea virginica*), the freshwater benthic clam (*Carbicula fluminea*), and different species of mussel (*M. edulis* and *Mytilus galloprovincialis*) can filter *C. parvum* oocysts from water and accumulate them on the gills and inside hemocytes in the hemolymph under laboratory conditions *(96,98)*. In Spain, mussels (*M. galloprovincialis*) and cockles (*Cerastoderma edule*) are reservoirs of *C. parvum* oocysts *(99)*. Viable *C. parvum* oocysts have been recovered from experimentally infected freshwater clams (*C. fluminea*), marine eastern oysters (*C. virginica*), and mussels (*M. edulis* and *M. galloprovincialis*).

Oocyst contamination of shellfish is also dependent on the hydrography of river mouths, estuaries, and adjacent shores can greatly affect the distribution of human-infectious oocysts at the seashore. Mussels, cockles and clams represent an important reservoir of *C. parvum* infection for humans, as in addition to being detected on the gills, oocysts can also accumulate in the hemolymph, gills, and intestinal tract of mussels, eastern oysters, and freshwater clams. Contamination can be as high as 5×10^3 oocysts per shellfish, which enhances the potential risk of infection of humans. Shellfish are usually depurated prior to consumption, and standards regulating the quality of the waters used in the culture and depuration process are usually legislated in most developed countries, but depuration rules are normally regulated only with respect to coliform contamination. Gomez-Bautistam et al. *(99)* recommend that a depuration process of 72 h could be adopted to remove *C. parvum* oocysts from mussels.

Whether seasonal variation in surface contamination of foods occurs, requires further investigation; however, seasonal peaks in parasitism will influence when water and foods become surface contaminated. The rapid transportation of foods acquired from global markets and their chilling and wetting can enhance parasite survival. Water and food enhance survival of environmental stages by preventing their desiccation. The widespread distribution of oocysts in the environment enhances the possibility of food-borne transmission of cryptosporidiosis (Table 6).

2.1.6. Other Routes of Exposure

Sexual transmission of *Cryptosporidium* has been reported among homosexual males *(100,101)*. The significance of the recreational water route as a vehicle for transmission has gained prominence over the last few years. Recreational water venues including swimming pools, water parks, interactive, and sprinkler fountains that use recirculating water *(102,103)*, lakes, rivers, and oceans can be contaminated with infectious oocysts and have been implicated as vehicles of transmission. Swimming pools, particularly children's paddling pools, have been incriminated in the transmission of cryptosporidiosis, and the ingestion of swimming pool water following either accidental defaecation events in swimming and paddling pools or cross-contamination of swimming pool water with sewage effluent have resulted in outbreaks *(104–106)*. The increase in the use of recreational waters for immersion sports, in oocyst contaminated recreational fresh and marine waters, also increases the potential for this transmission route.

Oocysts can be redistributed to other uncontaminated sites associated with human activities by transport hosts, such as coprophagous animals (e.g., pigs, dogs, chicken, seagulls, and flies; *[18, 107–111]*). Filth flies can play a role in the transmission of *Cryptosporidium* spp. oocysts because of their association with feces and their ability to travel over large distances. Filth flies ingest 1–3 mg feces over 2–3 h *(111)*, and can transmit *Cryptosporidium* oocysts either attached to their exoskeleton or following their ingestion, in excrement *(109,110)*. Experimentally, recently emerged adult flies that visited a slurry of oocyst contaminated bovine feces deposited up to 108/cm^2 oocysts on surfaces they visited and each adult fly could harbor up to 267 oocysts on its exoskeleton *(109)*. In another study, flies from 2 out of 62 households (3%) were positive for oocysts in a Pueblo Joven community of Lima, Peru *(107)*, whereas wild-caught synanthropic flies harbored up to a mean of 13.4 oocysts in their guts and 7.4 on their exoskeleton *(110)*. Furthermore, parasites with both human and nonhuman reservoir hosts (e.g. *C. parvum*) can augment environmental contamination following infection of coprophagous hosts. Therefore, the sources of contamination can be point sources, such as infected hosts, wastewater effluents or non-point sources, such as muck spreading, slurry spraying, and run off from contaminated land. Rapid climate changes due to global warming, including temperature fluctuations, rainfall, and changes in water table levels are also likely to influence the distribution of oocysts in the environment.

2.2. Survival in Food and Related Environments

C. parvum oocysts can survive for more than 12 mo in water at 4°C *(112)*; however, air drying (18–20°C), for 4 h kills oocysts *(75)*. A proportion of *C. parvum* oocysts can survive at −20°C for 12 h *(75)*. Nichols and Smith *(113)* cite data on the resistance of *C. parvum* oocysts to physical and chemical treatments. Some experimental survival and inactivation data for *Cryptosporidium* oocysts on/in foods and related environments are presented in Table 7.

Table 7
Experimental Survival and Inactivation of *Cryptosporidium* Oocysts in Foods and Related Environments

Matrix	Treatment	Survival/inactivation	Method of assessing viability/infectivity	Ref.
Whole milk and water	Pasteurization[1]; 71°C for 5, 10, 15 s	Non-infectious to suckling BALB/*c* mice	In vivo infectivity	(114)
Yoghurt from pasteurized milk (pH 4.8)	48 h at 37°C; 8 d storage at 4°C	Viability declined from 83 to 60% after 48 h; and to 58% after 8 d	In vitro; PI exclusion	(35)
Cider (seeded with a bovine isolate of *C. parvum* oocysts)	Flash pasteurization • 70°C/71.7°C for 10 or 20 s • 70°C for 5 s • 71.7°C for 5 s	• 4.9 log reduction (99.999%) • 3.0 log reduction (99.9%) • 4.8 log reduction (99.998%)	In vitro infectivity in MDCK cells[2]	(115)
Fresh apple cider (2 L seeded with Iowa isolate *C. parvum* oocysts)	UV irradiation; 14.32 mJ/cm² for 1.2–1.9 s (CiderSure 3500A apparatus equipped with 8 LPH lamps and UV sensor)	Treated oocysts non-infectious to both interferon γ gene knockout and infant BALB/cByJ mice	In vivo infectivity; oocysts in feces and intestinal tissue detected by PCR	(116)
Apple, orange, purple and white grape juices (seeded with sodium hypochlorite treated *C. parvum* oocysts)	0.03% H₂O₂ for ≥2 h	≥5.9 log reduction in infectivity and log reduction in percentage excystation ranging from 0.54 to 0.60	In vitro infectivity in HCT-8 cells and in vitro excystation	(117)
Food and beverage related media (seeded with *C. parvum* and *C. hominis*)	• Buffered media pH 7.0, 3.6 and 2.6; 4 and 20°C • High salt (4.5% w/v), glycerol (20% v/v), sucrose (50% w/v) and ethanol (9 and 40% v/v)	• Viable sporozoites detected after 14 d by assessing sporozoite ratio • Lower sporozoite ratio; reduction in in vitro infectivity ranging from 53 to 100%	In vitro infectivity in MRC-5 cells[3] and sporozoite ratio during in vitro excystation	(118)

(Continued)

257

Table 7 (Continued)

Matrix	Treatment	Survival/inactivation	Method of assessing viability/infectivity	Ref.
Four mineral waters of TDS values ranging from 91 to 430 mg/L (seeded with *C. parvum*; human and cervine–ovine isolates)	Storage at: • 20°C • 4°C	• 30% viable after 12 week • Viability unaltered up to 16 weeks	In vitro; DAPI/PI fluorogenic viability assay	(119)
Pesticides; Fungicides (captan 50% W.P., benomyl 50% W.P.) and insecticide (diazinon 4E 47.5%) (seeded with Iowa isolate *C. parvum* oocysts)	Storage at 23°C in dilutions of pesticides above and below the recommended dose: • 1–24 h • 1 week	• No effect on infectivity • Highest concentration decreased infectivity from 30 to 50%	In vivo infectivity in CD-1/ICR mice	(120)

[1]The commercial pasteurization treatment of milk by high-temperature-short-time is 15 s at 71.7°C.
[2]Madin-Darby bovine kidney cell line.
[3]MRC–5 fibroblast cell line.

3. FOODBORNE OUTBREAKS

Early reports highlighted a role for milk-borne transmission. Foodborne transmission following the ingestion of uncooked bovine offal, which had been stored frozen prior to it being thawed out for use was also reported, casting doubt on the efficacy of freezing for the inactivation of oocysts.

Six outbreaks of foodborne disease have been documented, five of which occurred in the USA and one in Australia. Two occurred following the consumption of nonalcoholic, pressed, apple cider, in 1993 and 1996 affecting a total of 185 individuals. In the 1993 outbreak, apples were collected from an orchard in which an infected calf grazed. Some apples had fallen onto the ground (windfalls) and had probably been contaminated with infectious oocysts then *(121)*. The source of oocysts in the second outbreak is less clear as windfalls were not used and waterborne, as well as, other routes of contamination were suggested *(122)*. A foodborne outbreak, which affected 15 individuals, occurred in 1995 with chicken salad, contamination by a food handler, being the probable vehicle of transmission *(123)*. In 1997, an outbreak was documented in Spokane, Washington. Amongst 62 attendees of a banquet dinner, 54 (87%) became ill. Eight out of 10 stool specimens obtained from ill banquet attendees were positive for *Cryptosporidium*. Epidemiological investigation suggested that foodborne transmission occurred through a contaminated ingredient in multiple menu items *(124)*.

During September and October 1998, a cryptosporidiosis outbreak, affecting 152 individuals, occurred at a university campus in Washington, DC, USA. A case–control study with 88 case patients and 67 control subjects revealed that eating in one out of the two cafeterias was associated with illness. One food-handler, positive for *Cryptosporidium*, had prepared raw produce between September 20–22, 1998. All *Cryptosporidium* fecal samples from the 1997 banquet dinner outbreak and the 1998 university campus (25 cases, including the food-handler) analyzed by molecular methods contained *C. hominis* oocysts *(125)*.

In three out of the six outbreaks, contaminated foodstuffs were implicated as the vehicles of transmission, but in the 1995 chicken salad outbreak (1995) and the Washington, DC university campus outbreak (1998), foodhandlers were implicated in the transmission of disease, implying that contamination occurred probably during food preparation and that, until then, the foodstuffs were free of contaminating infectious oocysts. That transmission was not due to indigenous contamination of the foodstuffs, but to mishandling by foodhandlers manipulating the foodstuffs during food preparation, is an important distinction that is not always recognized. Immaterial of the difference, such outbreaks are categorized as foodborne, because the vehicle of transmission was food, but such scenarios place an unfair public perception on specific food production practices that are effectively quality assured.

In the Sunshine Coast, Queensland outbreak (August and September 2001), the only outbreak-associated exposure for all eight laboratory-confirmed cryptosporidiosis cases was children who had drank commercially obtained unpasteurized milk in the 2 wk prior to the onset of disease. From a further 10 samples of the commercially obtained unpasteurized milk analyzed, five contained *Cryptosporidium* sp. antigen determined using the ELISA SYSTEMS™ kit *(126)*.

4. PATHOGENICITY (VIRULENCE FACTORS)

4.1. Infectious Dose

Most human volunteer infectivity studies have used *C. parvum* oocyst isolates. The infectious dose for humans is small, but previous exposure confers some protection against reinfection. Of 29 healthy human volunteers, with no evidence of previous serological *Cryptosporidium* infection, 20% became infected following an oral dose of 30 *C. parvum* (Iowa isolate, bovine) oocysts *(127)*. A dose of 300 oocysts caused infection in 88%, and 1000 oocysts produced infection in 100% of volunteers tested and the median infective dose was 132 oocysts. Of the volunteers who excreted oocysts, 39% developed diarrhea and one other enteric symptom. Those with diarrhea excreted more oocysts than those without diarrhea, and were more likely to excrete oocysts on consecutive days *(128)*. One year later, the same volunteers were each given a challenge dose of 500 oocysts. Reduced clinical severity as well as a decline in the number of subjects shedding oocysts occurred, although the rate of diarrhea was comparable. The serum antibody response did not correlate with the presence or absence of infection. A 14-fold increase in ID_{50} occurred in volunteers with pre-existing anti-*C. parvum* serum IgG *(129)*. The infectivity of different *C. parvum* isolates can vary in healthy human adult volunteers. Isolates differ in their ID_{50}, attack rate, and duration of diarrhea they induce *(130)*. The median infectious dose is nine oocysts for the "Texas A&M University" (TAMU, equine) isolate, 132 oocysts for the Iowa (bovine) isolate, and 1042 oocysts for the "Ungar *Cryptosporidium parvum*" (UCP, bovine) isolate of *C. parvum (130)*.

C. hominis infectivity was studied in 21 healthy adult volunteers experimentally infected with 10–500 oocysts (strain TU502). Sixteen individuals (76.2%) had evidence of infection; the ID_{50} was estimated to be 10 to 83 oocysts using clinical and microbiological definitions of infection, respectively. Diarrhoea oocurred in 40% of individuals challenged with 10 oocysts and this percentage increased stepwise to 75% in those receiving 500 oocysts. An immune response, characterised by IgG secretion occured in those receiving more than 30 oocysts and, contrary to *C. parvum* responses, *C. hominis* elicited a serum IgG response in most infected individuals *(164)*.

One hundred oocysts produced infection in 22% of mice exposed *(131)* and the ID_{50} in CD1 neonatal mice is between 87 and 60 oocysts of the Iowa isolate of *C. parvum (87)*. Ten oocysts produced infection in two out of two infant nonhuman primates tested *(132)* and five *C. parvum* oocysts (cervine/ovine, MD isolate) produced disease in gnotobiotic lambs *(133)*.

4.2. Pathogenicity

Histopathology of intestinal tissue reveals loss of villus height, villus oedema, and an inflammatory reaction. Both the loss of microvilli and the decrease in the levels of microvillar disaccharidases interfere with absorption and contribute to malnutrition. Local cellular infiltrates are usually of plasma cells and neutrophils but also contain subepithelial macrophages and lymphocytes. In moderate to severe infections, intraepithelial neutrophils occur. An increase in lamina propria macrophages-producing tumor necrosis factor occurs in *Cryptosporidium*-infected piglets. Peyer's patches appear reactive. The mechanism by which *Cryptosporidium* infections cause severe diarrhea remains unclear and an intensive search for an enterotoxin has been inconclusive. The secretory diarrhea has also been attributed to a prostaglandin-dependent effect; however, experiments in infected piglets and calves, designed to measure the electrical resistance of the intestinal epithelium in vivo, demonstrate little change in permeability. This is supported by the observation that cell disruption is not a normal finding and is believed that the epithelium is repaired. The loss of epithelial barrier integrity has been observed in cell lines infected with sporozoites in vitro, yet pathological studies on the intestines of infected animals and humans indicate that diarrhea is the consequence of malabsorption, possibly due to a reduction of lactase activity.

4.3. Immune Response

Both immunocompetent- and immunocompromised-infected individuals can mount antibody (humoral) responses to *Cryptosporidium* antigens. Infected individuals respond to infection by producing IgM, IgG, and IgA antibody isotypes which persist after infection thus, seropositive-uninfected individuals reflect the fact that they have been exposed/infected, or both, previously. Our understanding of the significance of the humoral immune response of an individual to eradicate *Cryptosporidium* infections is still unclear. Chronic cryptosporidiosis can occur in hypogammaglobulinemic individuals in the absence of detectable antibodies. Chronically infected AIDS patients produce higher serum and salivary IgA antibody titers than non-chronically infected AIDS patients or uninfected immunocompetent controls, though they are unable to eradicate infections, indicating that other factors are involved in an effective protective immune response.

Several episodes of cryptosporidiosis can occur in the same individual, and one report of a longitudinal cohort study in HIV-negative children in Peru illustrates this *(134)*. Out of 119 children, 13 (11%) had more than one episode of infection with the median time of 10 mo (range 2.1–26 mo) between episodes. Four children who suffered nine sequential episodes of infection were studied and the species of *Cryptosporidium* determined by molecular techniques revealed that these episodes were, alternatively, caused by *C. hominis* (six episodes), *C. meleagridis* (two episodes), or *C. canis* (one episode). One child had three consecutive episodes, each caused by one of these three species *(134)*. These three immunocompetent children were considered neither stunted, wasted nor malnourished, when compared with the other children in the study and lived under the same sanitary conditions. Human infections with species other than *C. parvum* or *C. hominis* may be more common in such settings where the presence of sources of mixed *Cryptosporidium* species is part of the endemic environment that these individuals are exposed to.

A complex array of innate, humoral, and cell-mediated immune responses is required within the intestinal mucosa to remove *Cryptosporidium* and the interactions between the cytokines and the chemokines that regulate these responses and preserve the integrity of the intestinal mucosa are important. Activation of lymphocytes by *Cryptosporidium* antigens probably occurs in mucosal lymphoid follicles, particularly Peyer's patches, and CD4 T cells within the intraepithelial lymphocyte population appear to be important effector cells. In AIDS patients with cryptosporidiosis, the rapid repopulation of the intestinal mucosa with CD4 T cells, which occurs following treatment with highly active antiretroviral therapy (HAART), appears to correlate with parasite clearance.

Clearance of infection requires cell-mediated immunity, and T-helper 1 (Th1) responses are important. In in vitro cell cultures, increased innate expression of enterocyte-derived chemokine responses interleukin (IL)-8, growth-regulated oncogene-α and regulated on activation: normal T cell expressed/secreted (RANTES) occurs within 24–48-h infection. These are probably important initiators of the intestinal lamina propria inflammatory response. The neutrophil attracting chemokines IL-8 and growth-related oncogene (GRO)-α are upregulated in *C. parvum*-infected monolayers of intestinal epithelial cells (HCT8, CaCO$_2$) and basolateral secretion of IL-8 into the underlying mucosa provides a mechanism for neutrophil accumulation at the site of infection.

C. parvum infection leads to the triggering of apoptotic mechanisms within infected enterocytes that may represent an attempt by the host to limit the spread of infection. Innate production of the proinflammatory cytokine interferon-γ not only controls the early stages of infection, but also, together with IL-12, polarizes T-helper cells towards the Th1 phenotype. Humans recovering from *C. parvum* infection have increased intestinal mucosal expression of interferon-γ. The Th2 cytokines, IL-4 and IL-10, are also upregulated in the infected human intestinal mucosa *(135)*. Therefore, the protective T-helper response may not be as polarized as observed for other intracellular parasites. The inflammatory response induced by cytokines also contributes to infection-associated pathology, and regulatory cytokines such as transforming growth factor (TGF)-β1, which helps limit parasite-induced enterocyte damage, may help limit mucosal damage by downregulating pro-inflammatory cytokine activity *(135)*.

4.4. Virulence

Differences in the virulence of *C. parvum* and *C. hominis* infections, characterized by duration of illness, intensity of gut colonization, or variations in prepatent period, have not been thoroughly studied, although reports indicate that patients infected with *C. hominis* shed higher numbers of oocysts than patients infected with *C. parvum* exist *(134,136)* (Table 2). Risk factors for sporadic cryptosporidiosis include age (children under 5 and, to a lesser extent young adults who presumably have greater likelihood of contact with these patients) travelling abroad, contact with a diarrhoeic individual, and contact with farm animals. Swimming in fresh water or public swimming pools are positively associated in Australian and US studies *(165)*. Separate risks have been identified for *C. hominis* (travel abroad and contact with diarrhoeic individual) and *C. parvum* (contact with cattle). Joint and eye pain, recurrent headaches, dizzy spells and fatigue occurred significantly more often in *C. hominis* cases than in *C. parvum* cases *(165)*. The evidence that exists for greater numbers of *C. hominis*, than *C. parvum*, oocysts being excreted could prove important when estimating the impact of human sewage as a source of oocysts into the environment. Eating tomatoes and carrots was strongly negatively associated with infection in these studies, reinforcing similar data from outbreaks *(165)*. This negative association has not been explained adequately, although consumption of fruit and vegetables contaminated with low numbers of parasites may cause subclinical infection and, in turn, augment protective immunity. One study, conducted among Peruvian HIV-negative children with cryptosporidiosis, showed that the duration of oocyst shedding was longer for *C. hominis* than for *C. parvum* (mean 13.9 and 6.4 d, respectively; $p=.004$); however, no differences between *C. hominis* and *C. parvum* infections were observed with respect to duration of, or percentage with diarrhea, age or antecedent stunting. In the endemic setting of this study, the largest percentage of infection and illness was found in children infected with *C. hominis* although *C. meleagridis* infections were as common as *C. parvum* infections and *C. felis* and *C. canis* infections also occurred *(134)*.

5. CLINICAL CHARACTERISTICS

There are no pathognomonic signs. Some features that distinguish *C. parvum* cryptosporidiosis from *C. hominis* cryptosporidiosis are presented in Table 2. Cryptosporidiosis is associated with profuse watery diarrhea, rapid weight loss, dehydration, and abdominal cramps. Diarrhea is the predominant symptom (80–90% of cases) and less frequent symptoms include low grade fever, nausea, vomiting, anorexia, and general fatigue. Gastrointestinal symptoms, which may be accompanied by a "flu-like" illness (20–40% of cases) include vomiting, anorexia, and flatulence. The incubation period, the time from ingestion of organisms to the manifestation of symptoms, ranges from 5 to 10 d but, can be as much as 28 d. Prolonged excretion of oocysts is unusual. In immunocompetent individuals, both duration and severity of disease can vary but the diarrhea is self-limiting and the infection is limited to the small and large intestine. Infection can be asymptomatic, but the ratio of symptomatic to asymptomatic cases is not known.

In immunocompromised individuals, especially those with acquired immunodeficiency syndrome (AIDS), infection can lead to dehydration, electrolyte imbalance, and eventually death. Infection in AIDS patients can spread to the esophagus, stomach, gall bladder, common bile duct, rectum, appendix, and into the respiratory tract. Immunocompromised individuals include those with AIDS, primary immune deficiencies, other acquired abnormalities of T-lymphocytes, hypo- and agammaglobulinemia, X-linked hyperimmunoglobulin M syndrome, severe combined immunodeficiency syndrome, leukemia (especially during aplastic crises), those receiving immunosuppressive drugs for transplantation and chemotherapy and those with severe malnutrition where infection may be associated with measles. Except in those individuals for whom immunosuppressive therapy is reversed, symptoms can persist unabated. HIV-infected individuals with CD4 counts >200/mm^3 experience self-limiting disease, whereas those with counts <100/mm^3 frequently experience chronic illness or extraintestinal, most frequently biliary, disease. Fulminant cryptosporidiosis occurs when the CD4 count is <50/mm^3 *(137)*.

Young children are more susceptible to infections and have more severe clinical signs due to an immature immune system and poorer hygiene habits. Breast-fed babies are protected from a variety of intestinal infections, including *Cryptosporidium*. This protection arises from various specific and nonspecific mechanisms, including antibodies, colostrum and milk, and the production of organic acids from the metabolism of intestinal anaerobes (especially bifidobacteria) stimulated by human milk substances. After weaning, the acquisition of infection is frequently due to exposure to contaminated individuals, fluids, and food.

Asymptomatic infections appear to be more prevalent in children, young adults, and AIDS patients in endemic areas of disease. Children infected with *Cryptosporidium* suffer retarded growth. In a study that measured the weight of 1064 children with cryptosporidiosis from Guinea-Bissau, a loss of 392 g in boys and 294 g in girls at the age of 2 yr was observed compared with uninfected controls. Weight loss was not compensated by time, and the study suggested that cryptosporidiosis in infancy has a permanent deleterious effect on growth *(138)*. Asymptomatic infections, which cause weight loss in young Peruvian children, are reported to be more common than symptomatic infections. Out of 207 children (aged 0–3 mo) studied, 45% became infected over the following 2 yr of life. Fifty-seven infected children were assessed, of which 63% were asymptomatic. Symptomatic cases lost an average of 342 g during the first month of infection, whereas asymptomatic children lost an average of 162 g, compared with uninfected controls. Since asymptomatic infections were more common in the population studied, they may have accounted for retarded growth *(139)*. Asymptomatic infections appeared to be common (39.7%) in a rural population of southern Indians with *Cryptosporidium* being one of the commonest parasites found in this highly parasitized population (of 97.4% individuals infected, 74.3% had multiple intestinal parasites). In 377 individuals studied, asymptomatic *Cryptosporidium* infections in 5–19-yr olds of the northern Bolivian Altiplano were frequent (31.6%) *(140)*.

6. CHOICE OF TREATMENT

There is a history of inadequate treatments for cryptosporidiosis and outcomes from in vitro and in vivo models have not always paralleled studies in humans. Current options include public health education for those at greatest risk, supportive therapy, such as rehydration and treatment of symptoms and, where possible, reversal of any underlying immunosuppression. Anti-apicomplexan and coccidial drugs, effective for other members of the phylum, have proved to be of little value in treating cryptosporidiosis. The unique location of its endogenous stages, residing within a parasitophorous vacuole that is intracellular yet extracytoplasmic, has been postulated for the failure of some candidate drugs *(141)*.

Drug treatment of cryptosporidiosis falls broadly into three categories: antimicrobial therapy, immunotherapy, symptomatic anti-diarrheal treatment, and drugs from more than one of these groups have been used in combination *(142)*. HAART has reduced the incidence of disease in HIV-positive patients, probably affecting both viral and parasite loads. In immunocompromised patients with HIV-related disease, immune reconstitution using HAART, which acts prophylactically is the treatment of choice. Protease inhibitors used in HAART reduce *C. parvum* sporozoite host cell invasion and parasite development in vitro and inhibition is enhanced in combination with paromomycin (PRM) *(143)*. HAART in combination with antiparasitic therapy helps parasite clearance, irrespective of CD4 count *(144)*. In non-HIV immunosuppressed patients with cryptosporidiosis, reducing immunosuppression can lead to clinical and parasitological improvements *(145)*.

Drugs reported to be beneficial in the treatment of human cryptosporidiosis include spiramycin *(146,147)*, PRM *(148)*, nitazoxanide (NTZ; 2-acetyloxy-*N*-(5-nitro-2-thiazolyl)benzamide) *(149,150)* and azithromycin *(151)*. Recently, the most promising agent is NTZ, a synthetic antiparasitic drug for oral administration. NTZ is rapidly hydrolyzed to its active metabolite tizoxanide and its antiprotozoal activity is probably due to the inhibition of pyruvate/ferredoxin oxidoreductase *(152)*, which is a vital enzyme in parasite metabolism. NTZ is a broad range anti-parasitic drug that has been used since 1984.

In a randomized, double-blind, placebo-controlled study, NTZ was shown to be effective in the treatment of mild cryptosporidiosis of children and adults. NTZ was administered in doses of 500, 200, or 100 mg twice daily for 3 d in adults and adolescents: children aged 4–11 yr or aged 1–3 yr, respectively. Seven days after initiation of therapy, diarrhea had resolved in 80% of patients in the NTZ group, compared with 41% in the placebo group ($p<0.0001$). Diarrhea resolved in most patients receiving NTZ within 3 or 4 d of the initiation of treatment. NTZ-treatment reduced the duration of both diarrhea ($p<0.0001$) and oocyst shedding ($p<0.0001$) (153).

The effect of NTZ versus placebo was tested in a randomized-controlled trial on 100 children with cryptosporidiosis (50 HIV-seronegative and 50 HIV-seropositive). The parameters examined were: clinical response on d 7 after the start of treatment, parasitological response by d 10, and mortality at d 8. A 3-d course of NTZ (100 mg twice daily orally for 3 d) given to HIV-seronegative children significantly improved the resolution of diarrhea (difference 33%, 95%CI 7–59; $p=0.037$), the disappearance of oocysts from stools (difference 38%, 95%CI 14–63; $p=0.007$) and mortality (-18%, 95%CI -34 to 2; $p=0.041$) when compared with placebo. Although HIV-seropositive children had no significant improvement during primary treatment, further treatment of nonresponders with an additional 3-d course of open-label NTZ showed 92% clinical or parasitological response, suggesting that a higher dose may be necessary for the treatment of HIV-positive malnourished children (154).

In a Medline search on studies (emphasis on randomized, double-blind, placebo-controlled trials from 1996 to February 2004) performed with NTZ for giardiasis and cryptosporidiosis treatment, most studies in immunocompetent patients reported clinical and parasitological response rates close to 80 and 70%, respectively, for both parasites, yet response rates were always lower in immunocompromised patients (155).

NTZ is safely administered to humans in doses of up to 4 g (156). Further investigations into the use of a single dose medication (for practical and economic reasons) would greatly facilitate the use of the drug in endemic regions where cryptosporidiosis causes stunting and malnutrition in young children (152).

NTZ has US Food and Drug Administration approval for the treatment of both cryptosporidiosis (the first drug approved for the treatment of cryptosporidiosis) and giardiasis in children aged from 1 to 11 yr; however, no studies have been performed with children under 1 yr, patients with hepatic or renal impairment or on the effect the drug may have on pregnancy or on breast-feeding individuals. Both the doses required and the duration of treatment of immunocompromised patients still have to be established (157).

7. RESISTANCE EPIDEMIOLOGY

Despite the fact that many hundred of candidate drugs and combinations thereof have been tested in vitro or in vivo for activity against *Cryptosporidium*, no treatment has been effective at ameliorating both clinical and parasitological responses, particularly in the immunocompromised, in large cohort studies. Given that there is no effective chemo- or immunotherapeutic intervention available, there is no emerging resistance epidemiology. Oocysts are resistant to many disinfectants used in the water industry (88) – a property attributed to the biochemistry, biophysics, and physiology of the oocyst wall (158,159).

8. SUMMARY AND CONCLUSIONS

Cryptosporidium is an obligate intracellular parasite that completes its life cycle in a single host and whose taxonomy is under review. Commonly referred to as a coccidian parasite, it is now thought to have diverged from the remainder of the Apicomplexa early on in evolutionary terms. Both SSU rRNA sequence analysis and the recognition of additional life cycle stages appear to support this concept. Of 16 "valid" species of *Cryptosporidium*, genetic analyses reveal that seven species can infect susceptible, immunocompetent, and immunocompromised human hosts. *C. parvum*, found in humans and other mammals, and *C. hominis*, found predominantly in humans, remain the most common species infecting man, but *C. meleagridis*, *C. felis*, *C. canis*, *C. suis*, *C. muris*, and the *Cryptosporidium* cervine and monkey genotypes also infect humans. Humans do not become solidly immune to reinfection following their first exposure, but do develop an anamnestic response. Different *C. parvum* isolates evoke different clinical outcomes in human volunteers, indicating that parasite virulence and/or pathogenicity factors affect clinical outcome. Parasite factors and the inflammatory immune response drive both parasite clearance and immunopathology. Presently, there are no recognized, specific, chemotherapeutic drugs for controlling cryptosporidiosis although NTZ shows promise, particularly for immunocompetent individuals. Modifications of treatment regimens may also increase the usefulness of NTZ for immunocompromised individuals.

Transmission to humans can occur via any mechanism by which material contaminated with feces containing infectious oocysts from infected human beings or other non-human hosts can be swallowed by a susceptible host. Both biotic and abiotic reservoirs of infection exist. Biotic reservoirs include all potential hosts of human-infectious *Cryptosporidium* species, whereas abiotic reservoirs include all vehicles that contain sufficient infectious oocysts to cause human infection, the most commonly recognized being food and water. Transport hosts, including gulls, wildfowl, filth flies, and bivalves should also be borne in mind when developing risk assessment questionnaires and performing risk assessments.

While conventional detection methods are useful for detecting symptomatic human cases, they are not sufficiently sensitive for detecting asymptomatic carriers. Furthermore, they cannot discriminate between the species that commonly infect human or nonhuman hosts. Here, molecular methods that can determine species, genotypes, and subtypes of *Cryptosporidium* are required and a variety of PCR-based methods are available. PCR-RFLP and DNA sequence-based methods have been used to identify species/genotypes in clinical and environmental samples, but there is increasing interest in real-time PCR methods as they can be more sensitive and rapid.

A variety of oocyst viability/infectivity methods have been described which indicates that, currently, no method has received universal approval. Both in vitro and in vivo methods have been advocated and, depending on the requirement, each has its advantages and limitations. Standardized methods for detecting low densities of oocysts in water and food are available as are molecular methods for determining their species and genotypes. Unidentified interferents in water and food matrices reduce the efficiency of detection by PCR, and methods to relieve such interferents should be assessed in order to maximize detections from low densities of oocysts. Immaterial of the method used, effective quality assurance, quality control, and validation are necessary prior to adoption.

Both foodborne and waterborne outbreaks of cryptosporidiosis have been documented. In three out of the six outbreaks documented, contaminated foodstuffs were implicated as the vehicles of transmission, but in another two, foodhandlers, rather than indigenous contamination of foodstuff, were implicated in the transmission of disease. Whether seasonal variation in surface contamination of foods occurs, requires further investigation; however, seasonal peaks in parasitism will influence when water and foods become surface contaminated. Water and food enhance survival of environmental stages by preventing their desiccation and the widespread distribution of oocysts in the environment enhances the possibility of foodborne transmission.

Increased demand, global sourcing, and rapid transport of soft fruit, salad vegetables, and seafood can enhance both the likelihood of oocyst contamination and oocyst survival. A risk-based assessment based upon standard, validated methods is required for detecting *Cryptosporidium* oocysts on/in food. In addition to determining genus and species of protozoan parasites present on/in foods and whether they are infectious to humans, subtyping methods are required to track outbreaks of disease and incidents and to determine the risk associated with specific genotypes. A clearer understanding of the population biology of the parasite will assist in unravelling occurrence and prevalence of genotypes, whereas a multidisciplinary approach will assist in unravelling the impact of foodborne protozoan parasites on human health.

Methods for detecting oocysts on foods are modifications of those used for detecting oocysts in water, and, as such, have recovery efficiencies ranging from 1 to 59%. Specificity, sensitivity, and reproducibility are paramount as, unlike pre-enrichment methods that increase organism numbers for prokaryotic pathogens and indicators, there is no method to augment parasite numbers prior to detection. Oocyst contamination of food can be on the surface of, or in, the food matrix and products at greatest risk of transmitting infection to man include those that receive no, or minimal, heat treatment after they become contaminated. Examples of surface contamination include salad vegetables and fruit, whereas examples of contamination within the food matrix include drinks, beverages, milk, and other foodstuffs containing naturally contaminated produce such as bivalves. Bivalves can remove *Cryptosporidium* oocysts from water and accumulate them on the gills and inside hemocytes in the hemolymph for protracted periods of time. Heating at $\geq$64.2°C for 2 min will ablate the infectivity of *C. parvum* oocysts to neonatal mice, and air drying/desiccation for 4 h or exposure to 0.03% H_2O_2 for $\geq$2 h will ablate oocyst viability. In addition to temperature elevation ($\geq$64.2°C for 2 min), disinfectants and other treatment processes used in the food industry may be detrimental to oocyst survival or lethal, but further research in this important area is required. In the absence of specific data, approaches such as hazard analysis and critical control points should be used to identify and control risk.

REFERENCES

1. Tyzzer, E. E. (1907) A sporozoan found in the peptic glands of the common mouse. *Proc. Soc. Exp. Biol. Med.* **5,** 12–13.
2. Tyzzer, E. E. (1912) *Cryptosporidium parvum* (sp. nov.) a coccidian found in the small intestine of the common mouse. *Arch. Protisenkd.* **26,** 394–412.
3. Todd, K. S. and Ernst, J. V. (1977) Coccidia of mammals except man. In: *Parasitic protozoa, Vol. III, Gregarines, Haemogregarines, Coccidia, Plasmodia and Haemoproteids* (Kreier, J. P., ed.)., Academic Press, New York and London, pp. 7–99.

4. Levine, N. D. (1973) Introduction, history, and taxonomy. In: *The Coccidia* (Hammond, D. M. and Long, P. L., eds.), University Park Press, Baltimore, Maryland, pp. 1–22.

5. Carreno, R. A., Martin, D. S., and Barta, J. R. (1999). *Cryptosporidium* is more closely related to the gregarines than to coccidia as shown by phylogenetic analysis of apicomplexan parasites inferred using small-subunit ribosomal RNA gene sequences. *Parasitol. Res.* **85,** 899–904.

6. Zhu, G., Keithly, J. S., and Philippe, H. (2000) What is the phylogenetic position of *Cryptosporidium*? *Int. J. Syst. Evol. Microbiol.* **50,** 1673–1681.

7. Hijjawi, N. S*.,* Meloni, B. P., Ryan, U. M., Olson, M. E., and Thompson, R. C. A. (2002) Successful in vitro cultivation of *Cryptosporidium andersoni*: evidence for the existence of novel extracellular stages in the life cycle and implications for the classification of *Cryptosporidium*. *Int. J. Parasitol.* **32,** 1719–1726.

8. Abrahamsen, M. S., Templeton, T. J., Enomoto, S., et al. (2004) Complete genome sequence of the Apicomplexan, *Cryptosporidium parvum*. *Science* **304,** 441–445.

9. Puiu, D., Enomoto, S., Buck, G. A., Abrahamsen, M. S., and Kissinger, J. C. (2004) CryptoDB: the *Cryptosporidium* genome resource. *Nucleic Acids Res.* **32**(Database issue)**,** D329–D331.

10. Streipen, B. and Kissinger, J. C. (2004). Genomics meets transgenics in search of the elusive *Cryptosporidium* drug target. *Trends Parasitol.* **20,** 355–3258.

11. Xiao, L., Fayer, R., Ryan, U., and Upton, S. J. (2004) *Cryptosporidium* taxonomy: recent advances and implications for public health. *Clin. Microbiol. Rev.* **17,** 72–97.

12. Ryan, U. M., Monis, P., Enemark, H. L., et al. (2004) *Cryptosporidium suis.* n. spp. (Apicomplexa: Cryptosporidiidae) in pigs (*Sus scrofa*). *J. Parasitol.* **90,** 769–773.

13. Nesterenko, M. V., Woods, K., and Upton S. J. (1999) Receptor/ligand interactions between *Cryptosporidium parvum* and the surface of the host cell. *Biochem. Biophys. Acta* **1454,** 165–173.

14. Morgan-Ryan, U. M., Fall, A., Ward, L.A., et al. (2002) *Cryptosporidium hominis* n. sp. (Apicomplexa: Cryptosporidiidae) from *Homo sapiens. J. Euk. Microbiol.* 433–440.

15. Fayer, R., Speer, C.A., and Dubey, J. P. (1990) General biology of *Cryptosporidium.* In: *Cryptosporidiosis of Man and Animals* (Dubey, J. P., Speer, C. A., and Fayer, R., eds.), CRC Press, Boca Raton, Florida, pp. 1–29.

16. Smith, H. V., Nichols, R. A. B., and Grimason, A. M. (2005) *Cryptosporidium* excystation and invasion mechanisms: getting to the guts of the matter. *Trends Parasitol.* **21,** 133–142.

17. Hijawi, N. S*.,* Meloni, B. P., Ng'anzo, M., et al. (2004) Complete development of *Cryptosporidium parvum* in host cell-free culture. *Int. J. Parasitol.* **34,** 769–777.

18. Smith, H. V. (1992) Intestinal protozoa. In *Medical Parasitology: A Practical Approach* (Hawkey, P. M. and Gillespie, S. H., eds.), IRL Press, pp. 79–118.

19. Smith, H. V. (1999) Detection of parasites in the environment. In *Infectious Diseases Diagnosis: Current Status and Future Trends* (Smith, H. V. and Stimson, W. H. eds., Chappel, L. H. co-ordinating ed.), *Parasitology* **117,** S113–S141.

20. Casemore, D. P. (1991) Laboratory methods for diagnosing cryptosporidiosis (ACP broadsheet 128). *J. Clin. Pathol.* **44,** 445–451.

21. Smith, H. V., Robertson, L. J., and Ongerth, J. E. (1995) Cryptosporidiosis and giardiasis, the impact of waterborne transmission. *J.W.S.R.T. – Aqua* **44,** 258–274.

22. Weber, R., Bryan, R. T., Bishop, H. S., Wahiquist, S. P., Sullivan, J. J., and Juranek, D. D. (1991) Threshold of detection of *Cryptosporidium* oocysts in human stool specimens: evidence for low sensitivity of current diagnostic methods. *J. Clin. Microbiol.* **29,** 1323–1327.

23. Webster, K. A., Smith, H. V., Giles, M., Dawson, L., and Robertson, L. J. (1996) Detection of *Cryptosporidium* parvum oocysts in faeces: comparison of conventional coproscopical methods and the polymerase chain reaction. *Vet. Parasitol.* **61,** 5–13.

24. Valdez, L. M., Dang, H., Okhuysen, P. C., and Chappell, C. L. (1997) Flow cytometric detection of *Cryptosporidium* oocysts in human stool samples. *J. Clin. Microbiol.* **35,** 2013–2017.

25. Morgan, U. M., Pallant, L., Dwyer, B. W., Forbes, D. A., Rich, G., and Thompson, R. C. A. (1998) Comparison of PCR and microscopy for detection of *Cryptosporidium parvum* in human fecal specimens: clinical trial. *J. Clin. Microbiol.* **36,** 995–998.

26. da Silva, A. J., Bornay-Llinares, F. J., Moura, I. N. S., Slemenda, S. B., Tuttle, J. L., and Pieniazek, N. J. (1999) Fast and reliable extraction of protozoan parasite DNA from fecal specimens. *Mol. Diagn.* **4,** 57–64.

27. Gibbons, C. L., Ong, C. S. L., Miao, Y., Casemore, D. P., Gazzard, B. G., and Awad-El-Kariem, F. M. (2001) PCR-ELISA: a new simplified tool for tracing the source of cryptosporidiosis in HIV-positive patients. *Parasitol. Res.* **87,** 1031–1034.

28. Monteiro, L., Bonnemaison, D., Vekris, A., et al. (1997) Complex polysaccharides as PCR inhibitors in feces: *Helicobacter pylori* model. *J. Clin. Microbiol.* **35,** 995–998.

29. Tebbe, C. C. and Vahjen, W. (1993) Interference of humic acids and DNA extracted directly from soil in detection and transformation of recombinant DNA from bacteria and a yeast. *Appl. Environ. Microbiol.* **59,** 2657–2665.

30. Sluter, S. D., Tzipori, S., and Widmer, G. (1997) Parameters affecting polymerase chain reaction detection of waterborne *Cryptosporidium parvum* oocysts. *Appl. Microbiol. Biotechnol.* **48,** 325–330.

31. Jones, C. S., Iannetta, P. P. M., Woodhead, M., Davies, H. V., McNicol, R. J., and Taylor M. A. (1997) The isolation of RNA from raspberry (*Rubus idacus*) fruit. *Mol. Biotechnol.* **8,** 219–221.

32. Campbell, A. T. and Smith, H V. (1997) Immunomagnetic separation of *Cryptosporidium* oocysts from water samples: round robin comparison of techniques. *Water Sci. Technol.* **35,** 397–401.

33. Johnson, D. W., Pieniazek, N. J., Griffin, D. W., Misener, L., and Rose, J. B. (1995) Development of a PCR protocol for sensitive detection of *Cryptosporidium* in water samples. *Appl. Environ. Microbiol.* **61,** 3849–3855.

34. Lowery, C. J., Moore, J. E., Millar, B. C., et al. (2000) Detection and speciation of *Cryptosporidium* spp. in environmental water samples by immunomagnetic separation, PCR and endonuclease restriction. *J. Med. Microbiol.* **49,** 779–785.

35. Deng, M. Q. and Cliver, D. O. (1999) *Cryptosporidium parvum* studies with dairy products. *Int. J. Food Microbiol.* **46,** 113–121.

36. Di Pinto, A. and Tantillo, M. G. (2002) Direct detection of *Cryptosporidium parvum* oocysts by immunomagnetic separation – polymerase chain reaction in raw milk. *J. Food Prot.* **65,** 1345–1348.

37. Nichols, R. A. B., Campbell, B. M., and Smith, H. V. (2003) Identification of *Cryptosporidium* spp. oocysts in United Kingdom noncarbonated natural mineral waters and drinking waters by using a modified nested PCR-restriction fragment length polymorphism assay. *Appl. Environ. Microbiol.* **69,** 4183–4189.

38. Smith, H. V., Nichols, R. A. B., and Campbell, B. M. (2003) Molecular fingerprinting of *Cryptosporidium* oocysts isolated during regulatory monitoring (DWI0832). Foundation for Water Research, Allen House, The Listons, Liston Road, Marlow, Bucks, SL7 1FD, UK. 53 pp.

39. Zhou, L., Singh, A., Jiang, J., and Xiao, L. (2003) Molecular surveillance of *Cryptosporidium* spp. in raw wastewater in Milwaukee: implications for understanding outbreak occurrence and transmission dynamics. *J. Clin. Microbiol.* **41,** 5254–5257.

40. Bialek, R., Binder, N., Dietz, K., Joachim, A., Knobloch, J., and Zelck, U. E. (2002) Comparison of fluorescence, antigen and PCR assays to detect *Cryptosporidium parvum* in faecal specimens. *Diagn. Microbiol. Infect. Dis.* **43,** 283–288.

41. Ward, L.A. and Wang, Y. (2001) Rapid methods to isolate *Cryptosporidium* DNA from frozen feces for PCR. *Diagn. Microbiol. Infect. Dis.* **41,** 37–42.

42. Xiao, L., Alderisio, K., Limor, J., Royer, M., and Lal, A. A. (2000) Identification of species and sources of *Cryptosporidium* oocysts in storm waters with a small-subunit rRNA-based diagnostic and genotyping tool. *Appl. Environ. Microbiol.* **66,** 5492–5498.

43. Deng, M. Q. and Cliver, D. O. (2000) Comparative detection of *Cryptosporidium parvum* oocysts from apple juice. *Int. J. Food Microbiol.* **54,** 155–162.
44. Leng, X., Mosier, D. A., and Oberst, R. D. (1996) Differentiation of *Cryptosporidium parvum, C. muris, and C. baileyi* by PCR-RFLP analysis of the 18S rRNA gene. *Vet. Parasitol.* **62,** 1–7.
45. Xiao, L., Escalante, L., Yang, C., et al. (1999) Phylogenetic analysis of *Cryptosporidium* parasites based on the small-subunit rRNA gene locus. *Appl. Environ. Microbiol.* **65,** 1578–1583.
46. Spano, F., Putignani, L., McLauchlin, J., Casemore, D. P., and Crisanti, A. (1997) PCR-RFLP analysis of the *Cryptosporidium* oocyst wall protein (COWP) gene discriminates between *C. wrairi* and *C. parvum*, and between *C. parvum* isolates of human and animal origin. *FEMS Microbiol. Lett.* **150,** 209–217.
47. Homan, W., van Gorkom, T., Kan, Y. Y., and Hepener, J. (1999) Characterization of *Cryptosporidium parvum* in human and animal feces by single-tube nested polymerase chain reaction and restriction analysis. *Parasitol. Res.* **85,** 707–712.
48. Gibbons, C. L., Gazzard, B. G., Ibrahim M. A. A., Morris-Jones, S., Ong, C. S. L., and Awad-El-Kariem, F. M. (1998) Correlation between markers of strain variation in *Cryptosporidium parvum*: Evidence of clonality. *Parasitol. Int.* **47,** 139–147.
49. Sulaiman, I. M., Morgan, U. M., Thompson, R. C. A., Lal, A. A., and Xiao, L. (2000) Phylogenetic relationship of *Cryptosporidium* parasites based on the 70-kilodalton heat shock protein (HSP70) gene. *Appl. Environ. Microbiol.* **66,** 2385–2391.
50. Sulaiman, I. M., Lal, A. A., and Xiao, L. (2002) Molecular phylogeny and evolutionary relationships of *Cryptosporidium* parasites at the actin locus. *J. Parasitol.* **88,** 388–394.
51. Glaberman, S., Moore, J. E., Lowery, C. J., et al. (2002) Three drinking water-associated cryptosporidiosis outbreaks, Northern Ireland. *Emerg. Infect. Dis.* **8,** 631–633.
52. Laxer, M. A., Timblin, B. K., and Patel, R. J. (1991) DNA sequences for the specific detection of *Cryptosporidium parvum* by the polymerase chain reaction. *Am. J. Trop. Med. Hyg.* **45,** 688–694.
53. Morgan, U. M., O'Brien, P. A., and Thompson, R. C. A. (1996) The development of diagnostic PCR primers for *Cryptosporidium* using RAPD-PCR. *Mol. Biochem. Parasitol.* 103–108.
54. Fontaine, M. and Guillot, E. (2003) An immunomagnetic separation-real-time PCR method for quantification of *Cryptosporidium parvum* in water samples *J. Microbiol. Methods* **54,** 29–36.
55. Guy, R. A., Payment, P., Krull, U. J., and Horgen, P. A. (2003) Real-time PCR for quantification of *Giardia* and *Cryptosporidium* in environmental water samples and sewage. *Appl. Environ. Microbiol.* **69,** 5178–5185.
56. Limor, J. R., Lal, A. A., and Xiao, L. (2002) Detection and differentiation of *Cryptosporidium* parasites that are pathogenic for humans by real-time PCR. *J. Clin. Microbiol.* **40,** 2335–2338.
57. Tanriverdi, S., Tanyeli, A., Baslamish F., et al. (2002) Detection and genotyping of oocysts of *Cryptosporidium parvum* by real-time PCR and melting curve analysis. *J. Clin. Microbiol.* **40,** 3237–3244.
58. Vesey, G., Ashbolt, N., Fricker, E. J., et al. (1998) The use of a ribosomal RNA targeted oligonucleotide probe for fluorescent labelling of viable *Cryptosporidium parvum* oocysts. *J. Appl. Microbiol.* **85,** 429–440.
59. Anonymous (1999) *Isolation and identification of Cryptosporidium oocysts and Giardia cysts in waters 1999. Methods for the examination of waters and associated materials.* HMSO, London, 44pp.
60. Anonymous (1999). *UK Statutory Instruments 1999 No. 1524. The Water Supply (Water Quality) (Amendment) Regulations 1999.* The Stationery Office, Ltd, 5pp.
61. Bier, J. W. (1990) Isolation of parasites on fruits and vegetables. *S.E. Asian J. Trop. Med. Pub. Health* **22**(Supplement)**,** 144–145.

62. Bankes, P. (1995) The detection of *Cryptosporidium* oocysts in milk and beverages. In *Protozoan Parasites in Water* (Betts, W. B., Casemore, D., Fricker, C. R., Smith, H. V., and Watkins, J., eds.), The Royal Society of Chemistry, Cambridge, UK, CB4 4WF, pp. 152–153.

63. Monge, R. and Chinchilla, M. (1996) Presence of *Cryptosporidium* oocysts in fresh vegetables. *J. Food Prot.* **59**, 202–203.

64. Laberge, I., Ibrahim, A., Barta, J. R., and Griffiths, M. W. (1996) Detection of *Cryptosporidium parvum* in raw milk by PCR and oligonucleotide probe hybridisation. *Appl. Environ. Microbiol.* **62**, 3259–3264.

65. Ortega, Y. R., Roxas, C. R., Gilman, R. H., et al. (1997) Isolation of *Cryptosporidium parvum* and *Cyclospora cayetanensis* from vegetables collected in markets of an endemic region in Peru. *Am. J. Trop. Med. Hyg.* **57**, 683–686.

66. Deng, M. Q., Lam, K. M., and Cliver, D. O. (2000) Immunomagnetic separation of *Cryptosporidium parvum* oocysts using MACS MicroBeads and high gradient separation columns. *J. Microbiol. Methods* **40**, 11–17.

67. Deng, M. Q. and Cliver, D. O. (2000) Comparative detection of *Cryptosporidium parvum* oocysts from apple juice. *Int. J. Food Microbiol.* **54**, 155–162.

68. Robertson, L. J. and Gjerde, B. (2000) Isolation and enumeration of *Giardia* cysts, *Cryptosporidium* oocysts, and *Ascaris* eggs from fruits and vegetables. *J. Food Prot.* **63**, 775–778.

69. Wilkinson, N., Paton, C. A., Nichols, R. A. B., Cook, N., and Smith, H. V. (2000) Development of a standard method to detect parasitic protozoa on fresh vegetables. In: *87th Annual Meeting of the International Association of Food Protection (formerly IAMFES)*, Atlanta, GA, USA, August 6–9.

70. Girdwood, R. W. A. and Smith, H. V. (1999) *Cryptosporidium*. In: *Encyclopaedia of Food Microbiology* (Robinson, R., Batt, C., and Patel, P. eds.), Academic Press, London and New York, pp. 487–497.

71. Nichols, R. A. B. and Smith, H. V. (2004) Optimisation of DNA extraction and molecular detection of *Cryptosporidium parvum* oocysts in natural mineral water sources. *J. Food Prot.* **67**, 524–532.

72. O'Grady, J. E. and Smith, H. V. (2002) Methods for determining the viability and infectivity of *Cryptosporidium* oocysts and *Giardia* cysts. In: *Detection methods for algae, protozoa and helminths* (Ziglio, G. and Palumbo, F., eds.), John Wiley and Sons, Chichester, UK, pp. 193–220.

73. Campbell, A. T., Robertson, L. J., and Smith, H. V. (1992) Viability of *Cryptosporidium parvum* oocysts: correlation of *in vitro* excystation with inclusion or exclusion of fluorogenic vital dyes. *Appl. Environ. Microbiol.* **58**, 3488–3493.

74. Campbell, A. T., Robertson, L. J., and Smith, H. V. (1993) Effects of preservatives on viability of *Cryptosporidium parvum* oocysts. *Appl. Environ. Microbiol.* **59**, 4361–4362.

75. Robertson, L. J., Campbell A. T., and Smith, H. V. (1992) Survival of *Cryptosporidium parvum* oocysts under various environmental pressures. *Appl. Environ. Microbiol.* **58**, 3494–3500.

76. Belosevic, M., Guy, R. A., Taghi-Kilani, R., et al. (1997) Nucleic acid stains as indicators of *Cryptosporidium parvum* oocyst viability. *Int. J. Parasitol.* **27**, 787–798.

77. Neumann, N. F., Gyurek, L. L., Finch, G. R., and Belosevic, M. (2000) Intact *Cryptosporidium parvum* oocysts isolated after *in vitro* excystation are infectious to neonatal mice. *FEMS Microbiol. Lett.* **183**, 331–336.

78. Bukhari, Z., Marshall, M. M., Korich, D. G., et al. (2000) Comparison of *Cryptosporidium parvum* viability and infectivity following ozone treatment of oocysts. *Appl. Environ. Microbiol.* **66**, 2972–2980.

79. Goater, A. D. and Pethig, R. (1999) Electrorotation and dielectrophoresis. *Parasitology* **117**, S177–S189.

80. Wagner-Wiening, C. and Kimmig, P. (1995) Detection of viable *Cryptosporidium parvum* oocysts by PCR. *Appl. Environ. Microbiol.* **61**, 4514–4516.

81. Deng, M. Q., Cliver, D. O., and Mariam, T. W. (1997) Immunomagnetic capture PCR to detect viable *Cryptosporidium parvum* oocysts from environmental samples. *Appl. Environ. Microbiol.* **63,** 3134–3138.

82. Stinear, T., Matusan, A., Hines, K., and Sandery, M. (1996) Detection of a single viable *Cryptosporidium parvum* oocyst in environmental water concentrates by reverse transcription-PCR. *Appl. Environ. Microbiol.* **62,** 3385–3390.

83. Jenkins, M., Trout, J., Abrahamsen, M. S., Lancto, C. A., Higgins, J., and Fayer, R. (2000) Estimating viability of *Cryptosporidium parvum* oocysts using reverse transcriptase-polymerase chain reaction (RT-PCR) directed at mRNA encoding amyloglucosidase. *J. Microbiol. Methods* **43,** 97–106.

84. Baeumner, A. J., Humiston, M. C., Montagna, R. A., and Durst, R. A. (2001) Detection of viable oocysts of *Cryptosporidium parvum* following nucleic acid sequence based amplification. *Anal. Chem.* **73,** 1176–1180.

85. Upton, S. J., Tilley, M., and Brillhart, D. B. (1994) Comparative development of *Cryptosporidium parvum* (Apicomplexa) in 11 continuous host cell lines. *FEMS Microbiol. Lett.* **118,** 233–236.

86. Hijjawi, N. S., Meloni, B. P., Morgan U. M., and Thompson, R. C. A. (2001) Complete development and long-term maintenance of *Cryptosporidium parvum* human and cattle genotypes in cell culture. *Int. J. Parasitol.* **31,** 1048–1055.

87. Korich, D. G., Marshall, M. M., Smith, H. V., et al. (2000) Inter-laboratory comparison of the CD-1 neonatal mouse logistic dose–response model for *Cryptosporidium parvum* oocysts. *J. Eukaryot. Microbiol.* **47,** 294–298.

88. Smith, H. V. and Grimason, A. M. (2003) *Giardia* and *Cryptosporidium* in water and wastewater. In: *The Handbook of Water and Wastewater Microbiology* (Mara, D. and Horan, N., eds.), Elsevier Science Limited, Oxford, UK, pp. 619–781.

89. Fayer, R., Morgan, U. M., and Upton, S. J. (2000) Epidemiology of *Cryptosporidium*: transmission, detection and identification. *Int. J. Parasitol.* **30,** 1305–1322.

90. Dillingham, R. A., Lima, A. A., and Guerrant, R. L. (2002). Cryptosporidiosis: epidemiology and impact. *Microbes Infec.* **4,** 1059–1066.

91. Smith, H. V. and Rose, J. B. (1998) Waterborne cryptosporidiosis: current status. *Parasitol. Today* **14,** 14–22.

92. Meinhardt, P. L., Casemore, D. P., and Miller, K. B. (1996) Epidemiologic aspects of human cryptosporidiosis and the role of waterborne transmission. *Epidemiol. Rev.* **18,** 5940–5942.

93. Gold, D. and Smith, H. V. (2002) Pathogenic protozoa in fresh and drinking water. In: *Detection methods for algae, protozoa and helminths* (Ziglio, G. and Palumbo, F., eds.), John Wiley and Sons, Chichester, UK, pp. 143–166.

94. Hojlyng, N., Holten-Andersen, W., and Jepsen, S. (1987) Cryptosporidiosis: a case of airborne transmission. *Lancet* **ii,** 271–272.

95. Thurston-Enriquez, J. A., Watt, P., Dowd, S. E., Enriquez, R., Pepper, I. L., and Gerba, C. P. (2002). Detection of protozoan parasites and Microsporidia in irrigation waters used for crop production. *J. Food Prot.* **65,** 378–382.

96. Fayer, R., Graczyk, T. K., Lewis, E. J., Trout, J. M., and Farley, C. A. (1998) Survival of infectious *Cryptosporidium parvum* oocysts in seawater and eastern oyster (*Cassostrea virginica*) in the Chesapeake Bay. *Appl. Environ. Microbiol.* **64,** 1070–1074.

97. Tamburrini, A. and Pozio, E. (1999) Long term survival of *Cryptosporidium parvum* oocysts in seawater and in experimentally infected mussels (*Mytilus galloprovincialis*). *Int. J. Parasitol.* **29,** 711–715.

98. Graczyk, T. K., Fayer, R., Cranfield, M. R., and Conn, D. B. (1998) Recovery of waterborne *Cryptosporidium parvum* oocysts by freshwater benthic clams (*Corbicula fluminea*). *Appl. Environ. Microbiol.* **64,** 427–430.

99. Gomez-Bautistam, M. Ortega-Mora, L. M., Tabares, E., Lopez-Rodas, V., and Costas, E. (2000) Detection of infectious *Cryptosporidium parvum* oocysts in mussels (*Mytilus galloprovincialis*) and cockles (*Cerastoderma edule*). *Appl Environ Microbiol.* **66**,1866–1870.

100. Navin, T. R. and Hardy, A. M. (1987) Cryptosporidiosis in patients with AIDS. *J. Infect. Dis.* **155**, 150.

101. Crawford, F. G. and Vermund, S. H. (1998) Human cryptosporidiosis. *CRC Crit. Rev. Microbiol.* **16**, 113–159.

102. Anonymous (2000) Surveillance for waterborne-disease outbreaks – United States, 1997–1998. In: *CDC Surveillance Summaries (May), MMWR* **49** (no. SS-4).

103. Anonymous (2000) Outbreak of gastroenteritis associated with an interactive water fountain at a beachside park – Florida, 1999. *MMWR*, **49(25)**, 565–568.

104. Joce, R. E., Bruce, J., Kiely, D., et al. (1991) An outbreak of cryptosporidiosis associated with a swimming pool. *Epidemiol. Infec.* **107**, 497–508.

105. Sorvillo, F. J., Fujioka, K., Nahlen, B., Tormey, M. P., Kebabjian, R., and Mascola, L. (1992) Swimming-associated cryptosporidiosis. *Am. J. Public Health* **82**, 742–744.

106. Rose, J. B., Lisle, J. T., and LeChevallier, M. (1997) Waterborne cryptosporidiosis: incidence, outbreaks and treatment strategies. In: *Cryptosporidium and Cryptosporidiosis* (Fayer, R. ed.), CRC Press, Boca Raton, FL, pp. 93–110.

107. Sterling, C. R., Miranda, E., and Gilman, R. H. (1987) The potential role of flies (*Musca domestica*) in the mechanical transmission of *Giardia* and *Cryptosporidium* in a Pueblo Joven community of Lima, Peru. *Am. Soc. Trop. Med. Hyg.* **349**, 233.

108. Smith, H. V., Brown, J., Coulson, J. C., Morris, G. P., and Girdwood, R. W. A. (1993) Occurrence of *Cryptosporidium* sp. oocysts in *Larus* spp. gulls. *Epidemiol. Infec.* **110**, 135–143.

109. Graczyk, T. K., Cranfield, M. R., Fayer, R., and Bixler, H. (1999) House flies (*Musca domestica*) as transport hosts of *Cryptosporidium parvum*. *Am. J. Trop. Med. Hyg.* **61**, 500–504.

110. Graczyk, T. K., Grimes, B. H., Knight, R., Da Silva, A. J., Pieniazek, N. J., and Veal, D. A. (2003) Detection of *Cryptosporidium parvum* and *Giardia lamblia* carried by synanthropic flies by combined fluorescent in situ hybridization and a monoclonal antibody. *Am. J. Trop. Med. Hyg.* **68**, 228–232.

111. Greenberg, B. (1973) Flies and disease. In: *Biology and Disease Transmission, Volume II*, Princeton University Press, Princeton, 447pp.

112. Smith, H. V. (1992) *Cryptosporidium* and water – a review. *J.I.W.E.M.* **6**, 443–451.

113. Nichols, R. A. B. and Smith, H. V. (2002) *Cryptosporidium, Giardia* and *Cyclospora* as foodborne pathogens. In *Foodborne Pathogens: Hazards, Risk and Control* (Blackburn, C. and McClure, P., eds.), Woodhead Publishing Limited, Cambridge, UK, pp. 453–478.

114. Harp, J. A., Fayer, R., Pesch, B. A., and Jackson, G. J. (1996) Effect of pasteurization on infectivity of *Cryptosporidium parvum* oocyst in water and milk. *Appl. Environ. Microbiol.* **62**, 2866–2868.

115. Deng, M. Q. and Cliver, D. O. (2001) Inactivation of *Cryptosporidium parvum* oocysts in cider by flash pasteurization. *J. Food Prot.* **64**, 523–527.

116. Hanes, D. E., Worobo, R. W., Orlandi, P. A., et al. (2002) Inactivation of *Cryptosporidium parvum* oocysts in fresh apple cider by UV irradiation. *Appl. Environ. Microbiol.* **68**, 4168–4172.

117. Kniel, K. E., Sumner, S. S., Lindsay, D. S., et al. (2003) Effect of organic acids and hydrogen peroxide on *Cryptosporidium parvum* viability in fruit juices. *J. Food Prot.* **66**, 1650–1657.

118. Dawson, D. J., Samuel, C. M., Scrannage, V., and Atherton, C. J. (2004) Survival of *Cryptosporidium* species in environments relevant to foods and beverages. *J. Appl. Microbiol.* **96**, 1222–1229.

119. Nichols, R. A. B., Paton, C. A., and Smith, H. V. (2004) Survival of *Cryptosporidium parvum* oocysts after prolonged exposure to still natural mineral waters. *J. Food Prot.* **67**, 517–523.

120. Sathyanarayanan, L. and Ortega, Y. (2004) Effects of pesticides on sporulation of *Cyclospora cayetanensis* and viability of *Cryptosporidium parvum*. *J. Food Prot.* **67,** 1044–1049.

121. Millard, P. S., Gensheimer, K. F., Addiss, D. G., et al. (1994) An outbreak of cryptosporidiosis from fresh-pressed apple cider. *J. Am. Med. Assoc.* **272,** 1592–1596.

122. Anonymous (1997). Outbreaks of *Escherichia coli* 0157:H7 infection and cryptosporidiosis associated with drinking unpasteurized apple cider – Connecticut and New York, October 1996. *MMWR* **46(1),** 4–8.

123. Anonymous (1996). Foodborne outbreak of diarrhoea illness associated with *Cryptosporidium parvum* – Minnesota, 1995. *MMWR* **45(36),** 783–784.

124. Anonymous (1998b). 'Foodborne outbreak of cryptosporidiosis – Spokane, Washington, 1997'. *MMWR* **47(27),** 565–567.

125. Quiroz, E. S., Bern, C., MacArthur, J. R., et al. (2000) An outbreak of cryptosporidiosis linked to a foodhandler. *J. Infect. Dis.* **181,** 695–700.

126. Harper, C. M., Cowell, N. A., Adams, B. C., Langley, A. J., and Wohlsen, T. D. (2002) Outbreak of *Cryptosporidium* linked to drinking unpasteurised milk. *Commun. Dis. Intell.* **26,** 449–500.

127. DuPont, H. L., Chappell, C. L., Sterling, C. R., Okhuysen, P. C., Rose, J. B., and Jakubowski, W. (1995) The infectivity of *Cryptosporidium parvum* in health volunteers. *N. Engl. J. Med,* **332,** 855–859.

128. Chappell, C. L., Okhuysen, P. C., Sterling, C. R., and DuPont, H. L. (1996) *Cryptosporidium parvum*: intensity of infection and oocyst excretion patterns in healthy volunteers. *J. Infect. Dis.* **173,** 232–236.

129. Chappell, C. L., Okhuysen, P. C., Sterling, C. R., Wang, C., Jakubowski, W., and Dupont, H. L. (1999) Infectivity of *Cryptosporidium parvum* in healthy adults with pre-existing anti-*C. parvum* serum immunoglobulin G. *Am. J. Trop. Med. Hyg.* **60,** 157–164.

130. Okhuysen, P. C., Chappell, C. L., Crabb, J. H., Sterling, C. R., and DuPont, H. L. (1999) Virulence of three distinct *Cryptosporidium parvum* isolates for healthy adults. *J. Infect. Dis.* **180,** 1275–1281.

131. Ernest, J. A., Blagburn, B. L., Lindsay, D. S., and Current, W. L. (1987) Dynamics of *Cryptosporidium parvum* (Apicomplexa: Cryptosporidiidae) in neonatal mice (*Mus musculus*). *J. Parasitol.* **75,** 796–798.

132. Miller, R. A., Brondson, M. A., and Morton, W. R. (1990) Experimental cryptosporidiosis in a primate model. *J. Infect. Dis.* **161,** 312–315.

133. Blewett, D. A., Wright, S. E., Casemore, D. P., Booth, N. E., and Jones, C. E. (1993) Infective dose size studies on *Cryptosporidium parvum* using gnotobiotic lambs. *Water Sci. Technol.* **27,** 61–64.

134. Xiao, L., Bern, C., Limor, J., et al. (2001) Identification of 5 types of *Cryptosporidium* parasites in children in Lima, Peru. *J. Infect. Dis.* **183,** 492–497.

135. Lean, I.-S., McDonald, V., and Pollock, R. C. G. (2002) The role of cytokines in the pathogenesis of *Cryptosporidium* infection. *Curr. Opin. Infect. Dis.* **15,** 229–234.

136. McLauchlin, J., Pedraza-Diaz, S., Amar-Hoetzeneder, C., and Nichols G. L. (1999) Genetic characterization of *Cryptosporidium* strains from 218 patients with diarrhea diagnosed as having sporadic cryptosporidiosis. *J. Clin. Microbiol.* **37,** 3153–3158.

137. Chen, X. M., Keithly, J. S., Paya, C. V., and LaRusso, N. F. (2002) Cryptosporidiosis. *N. Engl. J. Med.* **346,** 1723–1731.

138. Molbak, K., Andersen, M., Aaby, P., et al. (1997) *Cryptosporidium* infection in infancy as a cause of malnutrition: a community study from Guinea-Bissau, West Africa. *Am. J. Clin. Nutr.* **65,** 149–152.

139. Checkley, W., Gilman, R. H., Epstein, L. D., et al. (1997) Asymptomatic and symptomatic cryptosporidiosis: their acute effect on weight gain in Peruvian children. *Am. J. Epidemiol.* **145,** 156–163.

140. Esteban, J. G., Aguirre, C., Flores, A., Strauss, W., Angles, R., MasComa, S. (1998) High *Cryptosporidium* prevalences in healthy Aymara children from the northern Bolivian Altiplano. *Am. J. Trop. Med. Hyg.* **58,** 50–55.

141. Mead, J. (2002). Cryptosporidiosis and the challenges of chemotherapy. *Drug Resist. Update* **5,** 47–57.

142. Smith, H. V. and Corcoran, G. D. (2004) New drugs and treatment for cryptosporidiosis. *Curr. Opin. Infect. Dis.* **17,** 557–564.

143. Hommer, V., Eicholz, J., and Petry, F. (2003) Effect of antiretroviral protease inhibitors alone, and in combination with paromomycin, on the excystation, invasion and in vivo development of *Cryptosporidium parvum*. *J. Antimicrob. Chemother.* **52,** 359–364.

144. Maggi, P., Larocca, A. M., Ladisa, N., et al. (2001) Opportunistic parasitic infections of the intestinal tract in the era of highly active antiretroviral therapy: is the CD4(+) count so important? *Clin. Infect. Dis.* **33,** 1609–1611.

145. Abdo, A., Klassen, J., Urbanski, S., Raber, E., and Swain, M. G. (2003) Reversible sclerosing cholangitis secondary to cryptosporidiosis in a renal transplant patient. *J. Hepatol.* **38,** 688–691.

146. Portnoy, D., Whiteside, M. E., Buckley, III E., and MacLeod, C. L. (1984) Treatment of intestinal cryptosporidiosis with spiramycin. *Ann. Int. Med.* **101,** 202–204.

147. Collier, A. C., Miller, R. A., and Meyers, J. D. (1984) Cryptosporidiosis after marrow transplantation: person-to-person transmission and treatment with spiramycin. *Ann. Int. Med.* **101,** 205–206.

148. White, A. C., Chappell, C. S., Hayat, C. S., Kimball, K. T., Flanigan, T. P., and Goodgame, R. W. (1994) Paramomycin for cryptosporidiosis in AIDS: a prospective, double-blind trial. *J. Infect. Dis.* **170,** 419–424.

149. Doumbo, O., Rossignol, J. F., and Pichard, E. (1997) Nitazoxanide in treatment of cryptosporidial diarrhea and other intestinal parasitic infections associated with acquired immunodeficiency syndrome in tropical Africa. *Am. J. Trop. Med. Hyg.* **56,** 637–639.

150. Rossignol, J. F., Hidalgo, H., Feregrino, M., et al. (1998) A double-blind placebo-controlled study of nitazoxanide in the treatment of cryptosporidial diarrhoea in AIDS patients in Mexico. *Trans. R. Soc. Trop. Med. Hyg.* **92,** 663–666.

151. Russell, T. S., Lynch, J., and Ottolini, M. G. (1998) Eradication of *Cryptosporidium* in a child undergoing maintenance chemotherapy for leukemia using high dose azithromycin therapy. *J. Ped. Hematol. Oncol.* **20,** 83–85.

152. Gilles, H. M. and Hoffman, P. S. (2002) Treatment of intestinal parasitic infections: a review on nitazoxanide. *Trends Parasitol.* **18,** 95–97.

153. Rossignol, J. F., Ayoub, A., and Ayers, M. S. (2001) Treatment of diarrhea caused by *Cryptosporidium parvum*: a prospective randomized, double-blind, placebo-controlled study of nitazoxanide. *J. Infect. Dis.* **184,** 103–106.

154. Amadi, B., Mwiya, M., Musuku, J., et al. (2002) Effect of nitazoxanide on morbidity and mortality in Zambian children with cryptosporidiosis: a randomised controlled trial. *Lancet* **360,** 1375–1380.

155. Bailey, J. M. and Erramouspe, J. (2004) Nitazoxanide treatment for giardiasis and cryptosporidiosis in children. *Ann. Pharmacother.* **38,** 634–640.

156. Stockis, A., Allemon, A. M., De Bruyn, S., and Gengler, C. (2002) Nitazoxanide pharmacokinetics and tolerability in man using single ascending oral doses. *Int. J. Clin. Pharmacol. Therapeut.* **40,** 213–220.

157. Anonymous (2003). Treatment of cryptosporidiosis. *CRD Weekly* **13**(46).

158. Smith, V. H. and Ronald, A. (2002) *Cryptosporidium*: the analytical challenge. In: *Cryptosporidium: The Analytical Challenge* (Smith, M. and Thompson, K., eds.), The Royal Society of Chemistry, Cambridge, UK, pp. 1–43.

159. Ward, H., Bhat, N., O'Connor, R., et al. (2004) Structural physiology of the *Cryptosporidium* oocyst wall. *Final Report to the American Water Works Association Research Foundation, Denver, CO. 80235-3098, USA*, 88pp.

160. Smith, H. V. (2007) Diagnosis of human and livestock cryptosporidiosis. In: *Cryptosporidium* and *cryptosporidiosis. 2nd Edition* (eds. Fayer, R. and Xiao, L.) Taylor and Francis, Boca Raton, Florida, USA.

161. Nichols, R. A. B., Moore, J. E., and Smith, H. V. (2006) A rapid method for extracting oocyst DNA from *Cryptosporidium* positive human faeces for outbreak investigations. *J. Microbiol. Methods.* **65,** 512–524.

162. Cook, N., Paton, C. A., Wilkinson, N., Nichols, R. A. B., Barker, K., and Smith, H. V. (2006a) Towards standard methods for the detection of *Cryptosporidium parvum* on lettuce and raspberries. Part 1: Development and optimization of methods. *Int. J. Food Microbiol.* **109,** 215–221.

163. Cook, N., Paton, C. A., Wilkinson, N., Nichols, R. A. B., Barker, K., and Smith, H. V. (2006b) Towards standard methods for the detection of *Cryptosporidium parvum* on lettuce and raspberries. Part 2: Validation. *Int. J. Food Microbiol.* **109,** 222–228.

164. Chappell, C. L., Okhuysen, P. C., Langer-Curry, R., et al. (2006) *Cryptosporidium hominis*: experimental challenge of healthy adults. *Am. J. Trop. Med. Hyg.* **75,** 851–857.

165. Cacciò, S. M., Thompson, R. C. A., McLauchlin, J., and Smith, H. V. (2005) Unravelling *Cryptosporidium* and *Giardia* epidemiology. *Trends Parasitol.* **21,** 430–437.

10

Cyclospora

Huw V. Smith

Abstract

Cyclospora cayetanensis is a diarrhea-causing, intracellular, intestinal parasite of humans. It completes its life cycle in individual hosts, but many aspects of its biology are not well understood, including its infectious dose and its virulence and/or pathogenicity. The drug of choice is trimethoprim-sulfamethoxazole and immune reconstitution using HAART and secondary trimethoprim-sulfamethoxazole prophylaxis should be considered in HIV-infected individuals.

Both foodborne and waterborne outbreaks have been documented. Foodborne clusters have affected large numbers of individuals in north America, with cases occurring most commonly in springtime and early summer. Increased global sourcing and rapid transport of soft fruit, salad vegetables, and seafood can enhance both the likelihood of oocyst contamination and the oocyst survival. Standardized methods for detecting oocysts on foods must be maximized as, unlike pre-enrichment methods that increase organism numbers for prokaryotic pathogens and indicators, there is no method to augment parasite numbers prior to detection. Oocyst contamination of food can be on the surface of, or in, the food matrix and products at greatest risk of transmitting infection to man include those that receive no, or minimal, heat treatment after they become contaminated. Temperature elevation kills oocysts and disinfectants and other treatment processes used in the food industry may be detrimental to oocyst survival or lethal, but further research in this important area is required.

Key Words: *Cyclospora*; oocysts; occurrence; detection; outbreaks; foodborne; environment.

1. CLASSIFICATION AND IDENTIFICATION

1.1. Classification

1.1.1. Historical

In 1979, Ashford *(1)* reported the finding of unsporulated, *Isospora*-like oocysts in routine stool samples from three individuals from Papua New Guinea and described them as undistinctive, uniformly sized, and easily confused with fungal spores. From 1985 onwards, 8–10 µm oocysts of an unknown *Cryptosporidium muris*-like parasite were reported from expatriates and native Peruvians suffering from extended bouts of chronic diarrhea, weight loss, and fatigue *(2,3)*. Further reports from the USA *(4)* and Kathmandu *(5)* identified the local occurrence of this parasite and confirmed its broad geographical distribution. Since that time, these "*Isospora*-like" oocysts, also known as "Big Crypto" or "Crypto Grande", have been described as a flagellate *(2)*, a blue-green alga *(6)*, cyanobacterium-like body *(7)*, coccidian-like body, coccidian-like organism or cyclospora-like body *(8)*, and identified as a cause of prolonged diarrhea in both the immunocompetent and the immunocompromised. Finally classified in 1994 *(9)*,

From: *Infectious Disease: Foodborne Diseases*
Edited by: S. Simjee © Humana Press Inc., Totowa, NJ

C. cayetanensis oocysts have been described in the stools of residents in, and travelers returning from, developing nations, and in association with diarrheal illness in individuals from north, central, and south America, the Caribbean, Africa, the Indian sub-continent, southeast Asia, Australia, and Europe. *C. cayetanensis* was named after the location of the authors' principal studies, Cayetano Heredia University in Lima, Peru.

1.1.2. Current Classification

Originally described in the intestines of moles by Eimer in 1870, *Cyclospora* is related taxonomically to other coccidian parasites, including *Eimeria* and the human pathogens *Cryptosporidium* and *Toxoplasma*. *Cyclospora* is a member of the subphylum Apicomplexa, class Sporozoasida, subclass Coccidiasina, and family Eimeriidae. Organisms of the genus *Cyclospora* have an oocyst with two sporocysts, each of which contains two sporozoites *(10)*. Seventeen species have been described: *C. cayetanensis* (from humans) *(9)*, *C. cercopitheci* (from *Cerocopithecus aethopis*, African green or vervet monkey), *C. colobi* (from *Colobus guereza*, colobus monkey), *C. papionis* (from *Papio anubis*, olive baboon), *C. angimurinensis* (from *Chaetodipus hispidus hispudus*), *C. ashtabulensis* (from *Parascalops breweri*), *C. babaulti* (from *Vipera berus*), *C. caryolytica* (from *Talpea europea, Talpea micrura coreana* and possibly *P. breweri*), *C. glomericola,* (from *Glomeris* species), *C. megacephali* (from *Scalopus aquaticus*), *C. ninae* (from *Ninia s. sebac*), *C. parascalopi* (from *P. breweri*), *C. scinci* (from *Scincus officinalis*), *C. talpea* (from *Talpea europaea*), *C. mopidonori* (from *Natrix natrix* and *Natrix stolata*), *C. viperae* (from *Viperia aspis* and possibly *Coluber scalaris, Coronella austraca* and *Natrix viperinus*) and *C. zamenis* (from *Coluber v. viridiflavus*) *(11)*. Four species infect primates, namely *C. cayetanensis (9)*, and based on morphological and molecular analyses, three futher species from non-human primates (*Cyclospora cercopitheci sp.n., C. colobi* sp.n., and *C. papionis* sp.n. *(12)*.

Oocyst size and shape, sporulation characteristics, life cycle, host range, and anatomical site(s) of infection are the important criteria for classifying coccidia; however, reliance on phenotypic characteristics for classifying *Cyclospora* has identified certain limitations resulting in misclassification *(2–8,13)*. Molecular methods have advanced our understanding of the phylogenetic relationships among closely related organisms and have helped us to resolve some previous limitations. Comparative 18S rRNA phylogenetic analysis reveals that *C. cayetanensis* is most closely related to the genus *Eimeria (14)*. The 18S rRNA sequence data suggest that the relationship between *C. cayetanensis* and *Eimeria* is as close as that between some *Eimeria* species and have prompted speculation that nonhuman reservoirs of *C. cayetanensis* may exist *(11)* and that *Cyclospora* might be an *Eimeria* species *(15)*. Taxonomic revisions such as these are difficult to reconcile completely because the sporulation characteristics for each genus are quite distinct: *Eimeria* oocysts have four sporocysts, each containing two sporozoites; while *Cyclospora* oocysts have two sporocysts, each containing two sporozoites. *Eimeria* are host-species specific, nonhuman pathogens whose oocysts sporulate outside the host and complete their asexual and sexual developmental life cycles within one host. The close molecular phylogenetic relationship between *Eimeria* and *Cyclospora* predicts that they may share similar phenotypic characteristics beyond sporogony outside the host *(16,17)* and a monoxenous life cycle *(18)*. Variability in nucleotide sequences in the first internal-transcribed spacer (ITS1) region in isolates from different geographical origins has been reported, and is suggestive of the existence of multiple strains *(19,20)*.

Sequencing and alignment of *C. cercopitheci*, *C. colobi*, and *C. papionis* 18S rRNA demonstrate the high homology among *Cyclospora* species and for further species discrimination, other hyper-variable regions must be analyzed. The ITS1 region *(19,20)* can be used to distinguish between *C. cayetanensis* and *C. papionis (20)*. Out of the 17 known species of *Cyclospora*, sequence data exist for only four species (*C. cayetanensis, C. cercopitheci, C. colobi*, and *C. papionis*) and for these species, only the 18S rRNA, ITS, and 5.8S regions have been sequenced. Phylogenetic trees using the 18S rRNA region of sequenced *Cyclospora* and *Eimeria* species indicate that, while there is a great deal of relatedness among the genera, *Cyclospora* species appear to be more related to each other than to *Eimeria (11)*.

1.1.3. Life Cycle

Cyclospora completes its life cycle within one host (monoxenous), a phenomenon also characteristic of *Cryptosporidium* and *Eimeria* species, yet, many details of the *C. cayetanensis* life cycle remain unknown. Endogenous stages are intracytoplasmic and contained within a vacuole *(9)*, and the transmissive stage, the oocyst, is excreted in the stool. Sparsely distributed intracellular stages occur both at the luminal surface and in glandular clefts of the small intestine *(18)*. Infective oocysts contain two sporocysts and each sporocyst contains two sporozoites. Excysted sporozoites infect the epithelial cells (enterocytes) lining the small intestine, the preferred site of infection being the jejunum. Intracellular sporozoites divide by multiple fission to form meronts, which contain varying numbers of merozoites. Two different morphological types of fully developed (asexual) meronts have been described: type I, with 8–12 fully mature merozoites (0.5 × 3–4 μm) and type II, with four merozoites (0.7–0.8 × 12–15 μm) *(18)*. The final generation of merozoites infect further jejunal enterocytes and initiate the sexual component of the life cycle. In sexual multiplication (gametogony), individual merozoites produce either microgamonts or macrogamonts. Nuclear division in the microgamont leads to the production of numerous microgametes which are released from host cells and fertilize macrogametocytes. Macrogametocytes contain types I and II wall-forming bodies *(18)*. The product of fertilization, the zygote, develops into an unsporulated oocyst (Fig. 1A) which is released into the lumen of the intestine. Unsporulated oocysts pass out of the body in feces where further development including the development of sporocysts and sporozoites (sporogony/sporulation) occurs in the presence of the higher concentrations of atmospheric oxygen (Fig. 1B). Sporulation is dependent on ambient temperature *(16,17)*. While these descriptions of intestinal asexual and sexual stages comply with accepted coccidian development, description of the complete life cycle is awaited (Fig. 2).

C. cayetanensis oocysts are spherical, measuring 8–10 μm in diameter. Oocysts are unsporulated when excreted in the stool and sporulate to infectivity in the environment. Unsporulated oocysts contain a central morula-like structure consisting of a variable number of inclusions (Fig. 1A), whilst sporulated oocysts contain two ovoid sporocysts (Fig. 1B). Within each sporocyst reside two crescentic sporozoites, each sporozoite measuring 1.2 × 9 μm. Sporulated oocysts excyst following incubation in an excystation medium at 37°C for up to 40 min *(16)*.

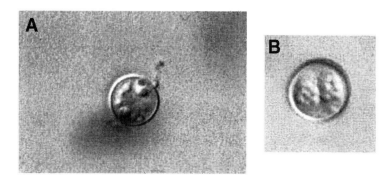

Fig. 1. (**A**) Nomarski differential interference contract photomicrograph of an unsporulated *C. cayetanensis* oocyst. (**B**) Nomarski differential interference contract photomicrograph of a sporulated *C. cayetanensis* oocyst.

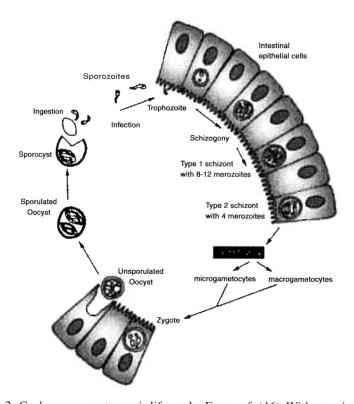

Fig. 2. *Cyclospora cayetanensis* life cycle. From ref. *(16)*. With permission.

1.2. Identification

There are no pathogenomonic signs and laboratory identification is required to confirm diagnosis.

1.2.1. Detection in Feces

Oocysts are excreted in relatively low numbers in human feces and can be preserved in 10% formalin, 2.5% potassium dichromate, or sodium acetate–acetic acid–formalin

(SAF) fixatives. Oocyst viability is retained following storage in 2.5% potassium dichromate. Formalin is a known inhibitor of the polymerase chain reaction (PCR). Concentration methods are recommended to increase *C. cayetanensis* oocyst yield and fresh or preserved stools can be concentrated by centrifugation, formol ether sedimentation or Sheather's sucrose flotation *(21)*. One-step discontinuous Percoll gradient concentration can be more effective for concentrating oocysts than Sheather's sucrose flotation or formol–ether sedimentation *(22)*. The microscopical detection of oocysts in fecal samples is the mainstay of diagnosis, but oocysts have also been reported from jejunal aspirates *(8)*. Microscopy of unstained wet mounts or stained smears can be performed on fresh or fixed (10% formalin, polyvinyl alcohol; SAF solution; 2.5% potassium dichromate) samples.

Oocysts seen in stool samples are normally unsporulated. In wet mounts, oocyst walls appear as well-defined non-refractile spheres measuring 8–10 μm in diameter by bright-field microscopy, and within an oocyst is a central morula-like structure (*see* Fig. 1A) which appears refractile, exhibiting a greenish tinge at higher (×400) magnification. Oocysts are remarkably uniform in size *(1,7)* and occasionally, oocysts which have collapsed into crescents are encountered. Epifluorescence microscopy enhances detection as the oocyst wall autofluoresces blue when visualized with a 330–380 nm ultraviolet excitation filter *(23)* or green when using a 450–490 nm dichroic mirror excitation filter *(24)*. Sporulation of unfixed oocysts confirms diagnosis. Nomarski differential interference contrast (DIC) microscopy can improve detection of unstained *Cyclospora* oocysts *(25)* (Fig. 1A,B) and sensitivities of 75, 30, and 23% have been reported using microscopy of saline wet mounts, safranin, and auramine rhodamine staining, respectively *(26)*. Oocysts do not stain with Gram, Giemsa, hematoxylin and eosin, Lugol's iodine, methylene blue, and Grocott–Gomori silver stains *(6,26)*. Staining air-dried fecal smears can aid identification, and the rapid dimethyl sulfoxide-modified acid-fast-staining method is more effective than either the Kinyoun or the modified Ziehl–Neelsen method *(27)*. Oocysts stain variably with acid-fast stains ranging from deep red to unstained. A modified safranin method (microwaving followed by safranin staining and malachite green or methylene blue counterstaining) stains oocysts a brilliant reddish orange *(28)*.

C. cayetanensis oocysts are approximately twice the diameter of *Cryptosporidium* oocysts (8–10 μm vs. 4.5–5.5 μm); however, laboratory misdiagnosis (i.e., reporting of false positives) has resulted in recent reports of outbreaks of pseudoinfection with *Cyclospora (29)*. There is a need for effective training of personnel and independent confirmation of positive results from laboratories that are beginning or increasing surveillance for this organism *(29)*.

When stored in 2.5% potassium dichromate or deionized water, up to 24% of purified *Cyclospora* oocysts will sporulate by 14 d at temperatures ranging from 22 to 37°C, but not at 4°C *(16,17)*. Sporulation is reduced at the higher temperatures and occurs most rapidly at 22–30°C. The observation that a small percentage (9–12%) of unsporulated oocysts held at 4°C for 1–2 mo are able to sporulate by 7 d when stored at 30°C may prove to be epidemiologically significant when evaluating outbreaks associated with imported produce contaminated with *Cyclospora (17)*. Under- and misdiagnosis of infection occurs and consequently environmental contributions leading to water and food contamination are probably underestimated (*see* Table 1).

Table 1
Some Features of the Biology of *C. cayetanensis* Which Can Facilitate
Environmental Transmission

Feature	*Cyclospora cayetanensis*
Robust nature of oocysts enhances their survival for long periods of time in favorable environments before ingestion by potential hosts	Oocyst survival is enhanced in moist cold environments. Sporulation is retarded at 4°C and survival is enhanced by storage in dark, moist, cool microclimates
Environmental robustness of oocysts enables them to survive some water treatment processes	Oocysts can survive physical treatment and pass through treatment barriers. Oocysts are insensitive to disinfectants commonly used in water treatment
Small size of oocysts aid their penetration through sand filters	8–10 µm
Few viable oocysts need to be ingested for infection to establish in susceptible hosts	No human infectivity data are available, but the infectious dose is thought to be low (10–100 oocysts). Foodborne transmission is suggestive of low infectious doses due to low level contamination of foodstuffs
Excretion of oocysts into water courses facilitate spread to, and entrapment in shellfish	Experimentally, viable *C. cayetanensis* oocysts can accumulate within freshwater clams
Excretion of oocysts in feces facilitates spread to water by water-roosting refuse feeders, and environmental spread by filth flies	Currently, no evidence is available for mechanical transmission of *Cyclospora*, but evidence exists for *Cryptosporidium* and *Giardia*

Where oocysts cannot be found in stools, diagnosis can be augmented by light microscopy of hematoxylin- and eosin-stained intestinal biopsies. Transmission electron microscopy (TEM) of biopsies reveals the presence of parasitophorous vacuoles containing the replicating stages. Neither light microscopy nor TEM examination of tissue is performed routinely for diagnostic purposes, because of the invasive nature of the biopsy procedure. Whether examination of jejunal juice is more or less predictive than microscopy of fecal concentrates, is not known.

1.2.2. Sensitivity of Detection in Feces

No specific data are available for the sensitivity of detection in feces. Being intermediate in size between *Cryptosporidium* spp. oocysts (4–6 µm and 6–8 µm) and *Giardia* spp. cysts (8–16 × 7–10 µm), the sensitivity of detection is expected to be >1 × 10⁴/g oocysts of unconcentrated stool following stool concentration. *C. cayetanensis* oocysts are excreted in lower numbers than those of *Cryptosporidium* and the concentration methods should be used to increase the likelihood of detection (*see* above). The estimated sensitivity of acid-fast staining when compared with autofluorescence microscopy is approx 78% *(23)*. Whether variations in fecal consistency influence the ease of detection is not known, but, as for *Cryptosporidium*, oocysts are probably more readily detected in concentrates made from watery, diarrheal specimens than from formed stool specimens. There are no standardized immunological methods for detecting *Cyclospora* oocysts. Whilst current detection methods are effective for diagnosing symptomatic infection, they remain insensitive and unable to detect the smaller number of oocysts present in asymptomatic infections. As for *Cryptosporidium*, oocyst detection methods may not be the ideal option for all situations.

1.2.3. Alternative Detection Procedures

PCR is more sensitive than conventional and immunological assays for the coprological detection of infection. The reliance of PCR on enzymatic amplification of nucleic acids makes it susceptible to inhibitors present in feces, waters, and foods. Its increased discriminatory powers will prove beneficial for epidemiological, source, and disease tracking proposes; but, currently, there are few published instances of its usefulness. Issues pertaining to inhibitors in feces, water, and foods together with methods for relieving their inhibitory effects, PCR targets, the sensitivity of detection, and methods for extracting DNA are presented in Section 1.2.3 (alternative detection procedures) of the *Cryptosporidium* chapter. (Please *see* page 242.)

A two-step nested PCR assay targeting a small 18S ribosomal RNA (18S rRNA) gene locus and restriction fragment length polymorphism (RFLP) analysis detects *Cyclospora* DNA *(30)*. The nested PCR primers also amplify other coccidian DNA, particularly those belonging to the genus *Eimeria (30,31)*, but this is not problematic for coprological diagnosis, as the genus *Eimeria* is not known to infect humans. Oocyst purification by sucrose flotation or cesium chloride gradient centrifugation reduces PCR inhibitors present in feces, while glass-bead disruption of the oocyst wall releases DNA for PCR amplification *(32)*. The nested PCR product is approx 300 bp and sensitivity can be 10 oocysts *(30)*. Based on a small number of samples, a specificity of 100% and a sensitivity of 62% were reported *(31)*. The 10 specimens which were negative by PCR, but positive by microscopy contained either few or moderate number of *Cyclospora* oocysts.

These primers distinguish *Cyclospora* 18S rRNA from that of human, *C. parvum*, *Toxoplasma gondii*, and *Babesia microti*, but do not differentiate *Cyclospora* from the closely related *Eimeria* species: coccidians that are known to infect domestic fowl, cattle, and rodents *(30)*. PCR products must be further characterized by RFLP to confirm the nature of the amplified DNA. RFLP analysis of second round amplicons with the restriction endonuclease *Mnl*I can distinguish *Cyclospora* from *Eimeria (33)*, and can be used for sequencing. The conserved 18S rRNA gene sequence is not sufficiently variable to distinguish reliably between *C. cayetanensis* genotypes *(34)* and less conserved loci, including the ITS region, may prove more valuable in genotype/subtype typing *(30,35)*. The development of sequence-based, strain-specific genotyping techniques, and databanks should prove to be a valuable tool in the study of the epidemiology of *Cyclospora* outbreaks. The sensitivity of this PCR is 62% and the specificity is 100% *(30)*. A limitation of this method for analyzing food or environmental specimens is its cross-reactivity with *Eimeria* DNA which can complicate specific detection of *Cyclospora* DNA *(30,31)*.

A PCR oligonucleotide-ligation assay (PCR/OLA) distinguishes between three *Cyclospora*, three *E. tenella*, and one *E. mitis* strains, and the ratio of positive:negative spectrophotometric absorbance (A_{490}) values for each strain ranged from 4.086 to 15.280

(median 9.5) indicating that PCR/OLA offers a rapid, reliable, spectrophotometric alternative to the above PCR-RFLP *(36)*. Further analysis of nucleotide sequences revealed that a different primer set coupled with RLFP using the restriction enzyme *Alu*I could better differentiate the amplified target sequence from *C. cayetanensis* from others that cross-react *(37)*. As few as one oocyst seeded into an autoclaved pellet flocculated from 10 L of surface water was detected and the authors suggested that their method obviated the need for microscopic examination for detecting *C. cayetanensis* in environmental samples *(37)*. When attempting to determine and verify the presence of protozoan pathogens, such as *C. cayetanensis* in environmental samples, for public health purposes, corroboration using as many different methods as possible increases the validity of the result. Obviating the need for a confirmatory method (microscopy) reduces available information (e.g., sporulation state) and fails to identify whether PCR positivity might be the result of the presence of naked DNA in the sample.

Real-time PCR and fluorescence microscopy were used to detect *C. cayetanensis* amplicons in extracts from stools from diarrheic travelers. *C. cayetanensis* was detected in four cases both by real-time PCR and by fluorescence microscopy, but one additional sample was positive only by real-time PCR. Examination of several additional slides by fluorescence microscopy confirmed the presence of *C. cayetanensis* oocysts *(39)*.

There is no commercially available, validated immunomagnetizable (IMS) method available for concentrating and isolating *Cyclospora* oocysts, and oocyst concentration is based upon their physicochemical properties (e.g., buoyant density: sucrose flotation or cesium chloride-gradient centrifugation). Magnetizable bead-based DNA hybridization assays, to concentrate *Cyclospora* DNA released from oocysts in lysis buffer, whereby bead-bound probes hybridize to complimentary single-stranded DNA, have been developed; however, they lack sensitivity and some lack specificity.

A multi-locus approach characterizing *Cyclospora* isolates at species and subspecies levels is essential. Further developments, involving multiplexing, real-time PCR, and melting curve analysis offer prospects for multiple species and genotype detection in automated procedures (reviewed in *Cryptosporidium* chapter). (Please *see* page 233) However, the public health value of these molecular tools depends on being able to correlate results in the context of mapping transmission patterns in defined endemic foci, and, as such, require further research and consolidation.

1.2.4. Detection in/on Foods

Non-infective (unsporulated) and infective (sporulated) *Cyclospora* oocysts cannot multiply on/in foodstuffs. Food becomes a potential vehicle for human infection by contamination during production, collection, transport, and preparation (e.g., fruit, vegetables, etc.) or during processing. The sources of such contamination include feces, fecally contaminated soil, irrigation water, water for direct incorporation into product (ingredient water), and/or for washing produce (process water) and infected food handlers. *Cyclospora* oocysts normally occur in low numbers in/on food matrices and in vitro culture techniques that increase parasite numbers prior to identification are not available for protozoan parasites in food and water matrices. Concentration by size alone, through membrane or depth filters, results in accumulation in the retentate, not only of oocysts, but also of similar- and larger-sized particles. *C. cayetanensis* oocysts are spherical, measuring 8–10 μm in diameter, and are frequently sought for in environmental

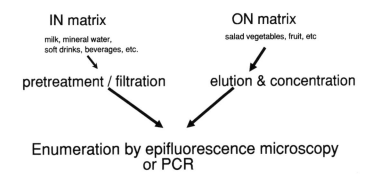

Fig. 3. Methods for detecting *Cyclospora* spp. oocyst contamination of foods.

samples in conjunction with *Giardia* spp. cysts ($8–16 \times 7–10$ μm) and *Cryptosporidium* spp. oocysts (4–6 μm and 6–8 μm in diameter). Filters with a smaller pore size (1–2 μm, particularly when *Cryptosporidium* is also sought) are employed to entrap *Cyclospora* oocysts, resulting in the accumulation of large volumes of extraneous particulate material. Such particles interfere with oocyst detection and identification, therefore, a clarification step which separates oocysts from other contaminating particles is employed. Current methods can be subdivided into the following component parts: (a) sampling, (b) desorption of oocysts from the matrix and their concentration, and (c) identification (Fig. 3) and are those, or modifications of those, which apply to water *(40,41)* (Table 2).

There is no IMS method available for concentrating and isolating *Cyclospora* oocysts, and oocyst concentration is based on their physicochemical properties. The use of lectin-coated paramagnetic beads did not significantly increase the recovery efficiency of *Cyclospora* oocysts from mushrooms, lettuce, raspberries, or bean sprouts (*see* Table 2), they did generate a smaller, cleaner final volume which reduces analysis time and may increase the sensitivity of detection *(42)*. Magnetizable bead-based DNA hybridization assays are an option for concentrating and isolating *Cyclospora* oocyst DNA from food matrices, but are not available commercially.

Methods for detecting oocyst contamination in liquid foods (e.g., drinks and beverages) and on solid foods (e.g., fruits and vegetables) are similar, but can differ in the techniques used to separate oocysts from the food matrix (Table 2). Differences are primarily dependent on the turbidity of the matrix and for nonturbid liquid samples either small or large volume filtration can be used.

Sampling methods, volumes/weights of product can depend on the reasons for sampling (quality control/assurance, ascertainment, compliance, incident/outbreak investigation, etc.). Desorption can be accomplished by mechanical agitation, stomaching, pulsifying, and sonication of leafy vegetables or fruit suspended in a liquid that encourages desorption of oocysts from the food matrix. Lowering or elevating pH affects surface charge and may increase desorption. Depending on the turbidity and pH of the eluate, oocysts eluted into nonturbid, neutral pH eluates can be concentrated by filtration through a 1-μm flat bed cellulose acetate membrane, whereas oocysts eluted into turbid eluates can be

Table 2
Some Reported Recovery Rates of *Cyclospora* Oocysts from Foods

Food matrices	Extraction and concentration methods	Detection methods	Recovery rates	Ref.
Lettuce leaves seeded with 50 *C. cayetanensis* oocysts	Rinse in tap water, centrifugation	Acid-fast stain, autofluorescence and direct wet mount	13–15%	*(45)*
(i) 110 vegetables and herbs from 13 markets; (ii) 62 vegetables and herbs from 15 markets	Rinse in tap water, centrifugation	Acid-fast stain, autofluorescence and direct wet mount	(i) 1.8%; (ii) 1.6%	*(45)*
Individual strawberries seeded with 170 *C. cayetanensis* oocysts	Vortex gently in reverse osmosis water, centrifugation	Autofluorescence and DIC microscopy	50–66%	*(17)*
• Mushrooms, lettuce, and raspberries, • Bean sprouts	Washing, sonication, and separation using lectin-coated para-magnetic bean	Autofluorescence and DIC microscopy	• ~12% • ~ 4%	*(42)*
100 g samples of raspberries	Washing, centrifugation, glass wool column, and DNA sequestration on FTA filters	PCR	30 oocysts 100/g detection limit	*(44)*

concentrated by flotation methods. Oocysts can be identified by their UV autofluorescence (UV filter block: excitation, 350 nm and –emission, 450 nm) and enumerated, and their morphology can be assessed by Nomarski DIC microscopy (Table 2).

Since the foodborne outbreaks of cyclosporiasis in 1990s, much effort has been directed at developing sensitive PCR-based methods that can detect small number of oocysts in food and other environmental samples *(33,36,37,43)* and standardized methods are available (http://vm.cfsan.fda.gov/~mow/kjcs19c.html; http://vm.cfsan.fda.gov/~ebam/bam-19a.html).

Flinders Technology Associates (FTA®) filter paper, a cotton-based cellulose membrane impregnated with denaturants, chelating agents, and free-radical traps which causes most cell types to lyse on contact sequestering their DNA within the matrix, has been used as to sequester *C. cayetanensis* DNA *(44)*. Cell remnants, sample debris, and other factors that may interfere with PCR are removed by briefly washing the filters in purification buffer. Eluates from berries seeded with *C. cayetanensis* oocysts were concentrated and applied onto FTA filters, washed in FTA purification buffer, dried at 56°C, and 6-mm discs were used directly as template. FTA-PCR detected a DNA equivalent of 30 *C. cayetanensis* oocysts in 100 g sample of raspberries *(44)*. The authors recommended FTA-PCR for detecting various pathogens in food, environmental samples, and clinical specimens.

Cyclospora oocysts have been isolated from vegetables and herbs *(45)*, lettuce *(46)* and green leafy vegetables *(47)*, and detected by microscopy. Consensus opinion indicates that PCR-based methods provide better insights and outcomes than microscopy-based methods, including increased sensitivity of detection and the potential for source- and disease-tracking using validated molecular subtyping and fingerprinting tools. There remains a need to develop more effective extraction methods (*see* Table 3) and to determine the sensitivity of PCR-based detection methods using a variety of produce, particularly those from which oocysts are difficult to extract and/or those that enhance PCR inhibition. Neither the DNA-based or the microscopy-based detection methods identified previously can be used to determine oocyst viability.

1.2.5. Determination of Oocyst Viability

Little can be inferred about the likely impact of oocysts detected in water or food concentrates on public health without knowing whether they are viable or not. There is no animal (human surrogate) model available and excystation in vitro is not applicable to the small number of organisms found in food and water concentrates.

Sporulated oocysts can be mechanically ruptured and the liberated sporocysts incubated in a "coccidian" excystation medium containing 0.5% trypsin and 1.5% sodium taurocholate in phosphate-buffered saline *(16)* to determine the ratio of excysted:unexcysted sporozoites. Conversely, oocysts can be incubated in excystation medium (see *Cryptosporidium* chapter) (Please *see* page 233), placed on a microscope slide and mechanical pressure placed on the coverslip to accomplish rupture of the oocyst wall and sporozoite excystation from sporocysts.

1.2.5.1. Determination by morphology. No rapid, objective estimate of oocyst viability revolving around the microscopical observation of fluorescence inclusion or exclusion of specific fluorogens into individual oocysts, and which can be used for determining contamination in/on food and in water, is available. Biophysical methods have also been used to determine *Cyclospora* sporulation state. Both dielectrophoresis and electrorotation have been used to demonstrate the differences between unsporulated and sporulated oocysts *(48,49)*.

1.2.5.2. Determination by molecular methods. Currently, there is no validated molecular method to determine *Cyclospora* oocyst viability. Candidate methods include those described in the chapters on *Cryptosporidium* and *Giardia*. (Please see page 233 for Chapter 9 and page 303 for Chapter 11.)

1.2.6. Determination of Infectivity

1.2.6.1. *In vitro* infectivity. Currently, there is no method available to determine in vitro infectivity. The ability to culture *C. cayetanensis* life cycle stages in vitro would provide us with an in vitro infectivity system.

1.2.6.2. *In vivo* infectivity. No in vivo infectivity model is available, because of the inability to identify a suitable animal surrogate for *C. cayetanensis* human infectivity *(50)*. A human volunteer trial failed to demonstrate infection in seven *Cyclospora*-seronegative, immunocompetent, adults (22–53 yr of age) following infection with between 200 and 49,000 oocysts, of which between 67 and 94% were sporulated, over a period of 16 weeks. No evidence of gastroenteritis was documented and no oocysts were detected in stool samples. Oocyst sources were from infected individuals from Haiti and the USA *(51)*. Further work is required to identify suitable human surrogates and to determine specific parasite- and human-related criteria that cause human infection, prior to instigating further human volunteer trials.

2. TRANSMISSION AND RESERVOIRS OF INFECTION

Routes of transmission for *Cyclospora* are poorly understood, although transmission via water and food are probably major routes. Direct (person-to-person) and indirect (abiotic) contacts with recently voided feces are unlikely to be a significant risk because of the requirement for external sporulation. Adults from non-endemic areas may be more likely to develop symptomatic infections. Discharges from sewage treatment works (STWs) and waste stabilization ponds (WSPs), into water used for drinking or irrigation can pose a risk *(47)* as can the use of contaminated, untreated feces as fertilizer. Analysis of 11 samples from a primary oxidation lagoon serving a Peruvian shantytown revealed unsporulated oocysts in seven samples and sporulated oocysts in two samples *(52)*. Oocysts in one sample sporulated after 2 wk at ambient temperature.

2.1. Routes of Exposure

A transmission of *Cyclospora* to human can occur via any mechanism by which material contaminated with feces containing infectious oocysts from infected humans can be swallowed by a susceptible host. Abiotic reservoirs include all vehicles that contain sufficient infectious oocysts to cause human infection, and the most commonly recognized are food and water. Abiotic reservoirs are important because of the requirement for oocysts to sporulate to infectivity in the environment.

2.1.1. Person-to-Person

Direct person-to-person transmission is not deemed significant (see earlier), whereas indirect transmission is the major route. Secondary cases and asymptomatic excretors are a source of infection for other susceptible persons. The robustness of oocysts and the variable state of immunity elicited by infection contribute to *C. cayetanensis* being widespread, particularly in developing countries where infection can be endemic and sanitation frequently minimal.

2.1.2. Animal-to-Person (Zoonotic)

No evidence exists for zoonotic transmission of *C. cayetanensis*, where the nonhumans host is capable of maintaining the parasite's life cycle. Redistribution of human-derived oocysts by transport hosts including coprophagous animals and bivalves enhance environmental contamination and identify further foodborne transmission routes.

2.1.3. Drinking Water

Waterborne transmission of *Cyclospora* has been documented *(53–56)*. Currently, neither the number of species of *Cyclospora* infective to human beings is known nor is it known whether human-derived oocysts are infectious to nonhuman hosts; however, the primary sources of pollution will be oocyst-contaminated human feces. As *Cyclospora* sp. oocysts are larger than *C. parvum* oocysts, but smaller than *G. duodenalis* cysts; it is likely that they will be discharged with final effluents from WSPs and

STWs. Oocysts take up to 14 d to sporulate in the laboratory, sporulating more rapidly at higher temperatures. Sporulation time in the environment will depend upon ambient temperature and sporulated oocysts may be found distant from the pollution source in the aquatic environment. Sources of pollution with unsporulated oocysts are likely to be effluent discharges from STWs and WSPs with detention times of <1 wk.

As for *C. parvum* oocysts and *G. duodenalis* cysts, oocysts of *Cyclospora* sp. are likely to survive longer at lower temperatures when suspended in water. No data are available regarding survival and transport in soil, however, *Cyclospora* sp. oocysts stored 4°C do not sporulate *(17)*, but a proportion of oocysts stored at 4°C for up to 2 mo will sporulate when incubated at temperatures between 22 and 30°C.

An outbreak amongst house staff and employees in hospital dormitory in Chicago occurred following the failure of dormitory's water pump. Illness was associated with the ingestion of water in the 24 h after the pump failure and *Cyclospora* sp. oocysts were detected in the stools of 11 out of 21 persons who developed diarrhea *(27,55)*. An outbreak occurred amongst British soldiers and dependants stationed in a small military detachment in Nepal and 12 out of 14 persons developed diarrhea. *Cyclospora* sp. oocysts were detected in stool samples from six out of eight patients. Oocysts were also detected microscopically in a concentrate from a 2 L water sample. Drinking water for the camp consisted of a mixture of river water and chlorinated municipal water. Chlorine residuals of 0.3–0.8 ppm were measured before and during the outbreak. No coliforms were detected in the drinking water *(54)*. Despite the role of contaminated potable water in disease transmission, oocysts confirmed as *Cyclospora* have seldom been identified from water concentrates. Methods developed for isolating *Cryptosporidium* and *Giardia* including filtration and flocculation (Please *see* pages 254, 316) have also been used to isolate *Cyclospora* oocysts from water. Out of 93 water samples collected from Nepal, which were filtered through PTFE membrane (pore size 5 μm) and examined microscopically, two were positive for *Cyclospora* oocysts *(57)*.

2.1.4. Airborne

The small size of *Cyclospora* spp. oocysts renders them aerosolizable in large particle droplets (>5 μm diameter).

2.1.5. Food

Both source contamination of produce and contamination from water used in food preparation are transmission routes that are significant to the food industry. Surface contamination can be direct, following contamination by the infected host, or indirect, following contamination by transport (birds, filth flies, etc.) hosts, the use of manure and contaminated water for irrigation, fumigation and pesticide application, etc. Oocyst-contaminated irrigation water as well as oocyst-contaminated water used for spraying or misting produce present specific problems to food producers and increases risk to the customer, particularly when the produce is eaten raw.

Oocyst-contaminated river waters can be a source of contamination of the marine environment, and, as for *Cryptosporidium* and *Giardia*, rivers polluted by human fecal discharges can play a major role in oocyst contamination of shellfish in estuaries and coastal environments. Experimentally infected Asian freshwater clams (*Corbicula fluminea*) concentrate *C. cayetanensis* oocysts onto their gills and into their hemolymph. Oocysts were detected at decreasing densities for up to 13 d following an initial 24 h exposure to $2.6 \times 10^3/l$ oocysts *(58)*. In the presence of sufficient aquatic oocysts, natural contamination of edible bivalves is possible.

Whether seasonal variation in surface contamination of foods occurs, requires further investigation; however, seasonal peaks in parasitism will influence when water and foods become surface contaminated. The rapid transportation of foods acquired from global markets and their chilling and wetting can enhance parasite survival. Water and food enhance survival of environmental stages by preventing their desiccation. The environmental occurrence of oocysts enhances the possibility of foodborne transmission of cyclosporiasis (Table 3).

2.1.6. Other Routes of Exposure

The recreational water route may be a vehicle for transmission. Recreational water venues, particularly those receiving discharges from STWs and WSPs in endemic areas of infection and/or where water retention time is of sufficient duration to allow oocysts to sporulate (including water parks, sprinkler fountains that use recirculating water, lakes, rivers, and oceans) can be contaminated with infectious oocysts and be vehicles of transmission. Oocysts may be redistributed to other uncontaminated sites associated with human activities by transport hosts such as coprophagous animals, filth flies, which play a role in the transmission of fecal parasites, such as *Cryptosporidium* and *Giardia*, may also transmit *C. cayetanensis* oocysts (reviewed in *Cryptosporidium* and *Giardia* chapters). (Please *see* pages 256, 318)

2.2. Survival in Food and Related Environments

Few survival data are available for *C. cayetanensis* oocysts. Chilling extends oocyst survival. *C. cayetanensis* oocysts, stored between 22 and 37°C, sporulated within 14 d, whilst oocysts stored at 4°C did not. Up to 30°C, neither storage in deionized water nor potassium dichromate affected the rate of sporulation. Storage at 4 and 37°C for 14 d retarded sporulation. The maximum percentage sporulation occurred at 22 and 30°C, irrespective of the suspending medium. Up to 12% of oocysts previously stored at 4°C for 1–2 mo sporulated when stored for 6–7 d at 30°C *(17)*. The risks of contracting infection in endemic areas can be minimized by boiling, filtering (pore size ≤5 µm) or avoiding the consumption of cold, chlorinated potable water, and avoiding eating uncooked vegetables and unpeeled fruits.

3. FOODBORNE OUTBREAKS

Documented foodborne outbreaks are more common than documented waterborne outbreaks and are associated with soft skinned fruit and leafy vegetables, such as berries and lettuce, that receive no, or minimal heat treatment. To date, no frozen, processed, or peeled fruits and vegetables have been implicated in disease transmission, although oocysts and/or *Cyclospora* DNA have been detected in produce that was later frozen (*see* below). The majority of documented outbreaks occurred in north America *(59–63)*, but one occurred in Germany *(64)*. In 1996, numerous outbreaks, totalling 1465 cases, were reported in the USA and Canada, with approx 50% being clustered cases and the remainder sporadic cases. Most infections were reported from immunocompetent adults, with only 40 childhood cases. The 55 clusters occurred between early May

Table 3
Possible Sources and Routes of Food Contamination with Sporulated,
Human-Infectious *Cyclospora* Oocysts

- Use of oocyst-contaminated feces (night soil) as fertilizer for crop cultivation.
- Direct contamination of foods following contact with oocyst contaminated feces transmitted by coprophagous transport hosts (e.g., birds and insects).
- Use of contaminated wastewater for irrigation.
- Aerosolization of contaminated water used for insecticide and fungicide sprays and mists.
- Washing "salad" vegetables, or those consumed raw, in contaminated water.
- Use of contaminated water for making ice and frozen/chilled foods.
- Use of contaminated water for making products which receive minimum heat or preservative treatment.
- Ingestion of viable oocysts from raw or undercooked shellfish which accumulate human infectious oocysts from their contaminated aquatic environment.

and mid-June following events at which Guatemalan berries, primarily raspberries, were served *(61)*. In 1997, outbreaks in the USA were also associated with imported raspberries, and later that year, with contaminated basil and lettuce *(61–63)*. In 1998, clusters of cases, again associated with fresh berries from Guatemala, were recorded in Ontario, Canada *(65)*. The review by Herwaldt *(63)* identifies likely vehicles of transmission as well as the difficulty in tracing outbreak sources in the absence of supportive parasitological tools.

Other fresh fruits have also been implicated in north American outbreaks, including non-Guatemalan raspberries, or fruits other than raspberries or blackberries; multiple combinations of many fruits, including raspberries and blackberries of undetermined sources; fresh Guatemalan blackberries, frozen Chilean raspberries, and fresh U.S. strawberries; non-Guatemalan raspberries (both foreign and domestic sources); imported blackberries (Guatemalan?), strawberries, and blueberries *(63)*. Both fresh mesclun lettuce (also known as spring mix or baby greens) and basil have also been implicated in outbreaks *(63)* and the Missouri (1999) outbreak was the first U.S. outbreak where *Cyclospora* was found, by both microscopy and PCR, in frozen leftovers of an epidemiologically implicated food item *(66)*. A wedding cake, containing a cream filling that included raspberries, was the food item most strongly associated with illness in 68.4% of 79 interviewed wedding party members and guests *(66)*. Leftover frozen cake was positive for *Cyclospora* DNA and sequencing of the amplicons confirmed that the DNA was *Cyclospora cayetanensis*. The millenium year (2000) was the fifth year that spring outbreaks cyclosporiasis, definitely or probably associated with Guatemalan raspberries, had occurred in North America. Thai basil, imported via the United States, was implicated in the May 2001 outbreak in British Colombia, Canada. Of a total of 17 reported cases, 12 were interviewed and 11 (92%) reported consuming Thai basil *(83*, Huang et al., 2005).

The first documented European foodborne outbreak of cyclosporiasis occurred in December 2000 *(64)*. Illness was reported in 34 persons who attended luncheons at a German restaurant and the overall attack rate was 85% (34 out of 40). The only foods associated with significant disease risk were two salad side dishes prepared from lettuce imported from southern Europe and spiced with fresh green leafy herbs (dill, chives, parsley, and green onions). All 25 persons who ate a salad consisting of butterhead lettuce, mixed lettuce (lollo rosso, lollo bianco, oak leaf, and romaine lettuce), red cabbage, white cabbage, carrots, cucumbers, and celery became ill, and the vehicle of transmission was deduced to be one or more of the lettuce varieties. The butterhead lettuce was grown in southern France and the mixed lettuce, dill, parsley, and green onions in southern Italy. The chives were grown in Germany. The most probable routes of contamination included the use of human waste as fertilizer; fecally contaminated water used to irrigate crops, prepare pesticides, to refresh/clean produce at their origin; and inappropriate sanitary facilities for seasonal field workers (64).

In February 2004, clusters of cases of cyclosporiasis associated with events held in Texas and Illinois and affecting approx 95 cases were identified. Fresh produce, including basil and mesclun lettuce/spring mix salad products, were served at these events in Texas and Illinois. During May and June, outbreaks and sporadic cases of cyclosporiasis occurring in four U.S. states and one Canadian province were also reported (http://www.dsf.health.state.pa.us/health/cwp/view.asp?A=171&Q=239018). An outbreak which occurred in Pennsylvania, USA between June and July 2004 was reported from a single residential facility. Attendance at five special events was linked to the onset of symptoms and 96 cases attended at least one of the events. The onset of illness was 1-14 days after consuming food or beverages at one or more of these events. Laboratory diagnosis confirmed 40 cases (Pennsylvanian health officials and CDC) and 56 cases were assessed as being probable for cyclosporiasis. The source of the infection was traced to a single container of Guatemalan snow peas used in the preparation of a pasta salad and was the first documented outbreak linked to the consumption of raw snow peas (84, Anon., 2004).

4. PATHOGENICITY (VIRULENCE FACTORS)

Virulence and characteristics of *Cyclospora* necessary to infect human hosts are unknown. The sequence variability present in the ITS1 region of the 18S rRNA gene of isolates from different geographical origins suggests the existence of multiple strains (19,20).

4.1. Infectious Dose

The infectious dose has not been established but is thought to be low. A human volunteer trial failed to demonstrate enteritis or oocyst shedding in seven *Cyclospora*-seronegative adults infected with between 200 and 49,000 oocysts over a 16-wk period (51). Adams et al. (68) suggest that between 100 oocysts and possibly as few as 10 sporulated oocysts can cause infection, although, as for *Cryptosporidium*, different isolates may have different infectious doses (69).

4.2. Pathogenicity

The clinical presentation associated with *Cyclospora* infection reflects small bowel involvement. Duodenal and jejunal biopsies reveal a moderate to marked erythema of the distal duodenum with acute and chronic inflammation of the lamina propria and evidence of surface epithelial injury (70). The mixed inflammatory infiltrates are primarily lymphocytic and, to a lesser extent, eosinophilic and there is an extensive lymphocytic infiltration into the epithelium, particularly at the tip of shortened villi (71). Plasma cell numbers also increase (71). Crypt hyperplasia and villous blunting with villus:crypt ratios of <50% of normal have also been reported *(69)*. Intracellular stages can be visualized with hematoxylin-based stains *(71,72)*, but little is known of pathogenic mechanisms and as yet, no virulence factors have been described.

4.3. Immune Response

Both immunocompetent and immunocompromised individuals are susceptible to infection, but the immunology of cyclosporiasis is poorly understood. Currently, very little information exists regarding the presence of *Cyclospora* antibody levels and their clinical significance.

4.4. Virulence

No virulence factors have been identified to date, but different genotypes (strains) may possess different virulence determinants.

5. CLINICAL CHARACTERISTICS

There are no pathognomonic signs. The incubation period is between 2 and 11 d *(73)* with moderate numbers of unsporulated oocysts being excreted for up to 60 d or more. In immunocompetent individuals, the symptoms are self-limiting and oocyst excretion is associated with clinical illness; whereas in immunocompromised individuals, diarrhea may be prolonged. Symptoms include a flu-like illness, diarrhea with weight loss, low-grade fever, fatigue, anorexia, nausea, vomiting, dyspepsia, abdominal pain, and bloating *(16,55)*. The self-limiting watery diarrhea can be explosive, but leukocytes or red blood cells are usually absent. Often, diarrhea can last longer than 6 weeks in immuno-competent individuals. The diarrheal syndrome may be characterized by remittent periods of constipation or normal bowel movements *(16)*. Malabsorption with abnormal D-xylose levels has also been reported *(71)*. *Cyclospora*, like *Campylobacter jejuni*, should be added to the list of infectious agents that elicit an immune response resulting in Guillain–Barre syndrome *(75)*.

Cyclospora infections are well described in patients with HIV/AIDS. In Haiti, a cohort study of HIV-positive patients with diarrhea of at least 3 wk duration described *Cyclospora* oocysts in 11% of stool samples *(26)* with clinical manifestations similar to those described for immunocompetent patients, including diarrhea, fever, abdominal pain, and weight loss. Although not a component of the Centers for Disease Control criteria for AIDS diagnosis, many of these *Cyclospora*-infected patients met or soon developed clinical manifestations that met the case definition of AIDS *(26)*. The differences cited in the prevalence of *Cyclospora* infection in the HIV-infected population may be a reflection of several factors, including local prescribing practices, i.e., sulfa-based prophylactic regimens for *Pneumocystis carinii* pneumonia, level of expertise in identifying *Cyclospora* oocysts and endemnicity of infection, and exposure to sources of contamination *(26)*.

In another study, 12 out of 235 patients with diarrhea who presented to a tertiary care hospital in Mexico City had *Cyclospora* infection *(76)*. Three out of the 12 patients were diabetic and seven were infected with HIV. In the latter group, lower CD4 lymphocyte counts seemed to be associated with more severe infection. Weight loss and duration of illness were greater in the HIV-infected patients than in HIV-negative patients, although the latter experienced more bowel movements daily. Interestingly, two out of the seven patients with HIV co-infection and diarrhea had also evidence of gall bladder infection that responded to trimethoprim-sulfamethoxazole therapy, suggesting that, as with human cryptosporidiosis, the biliary tract may also be involved in human-associated *Cyclospora* infections *(76)*.

In one case-controlled investigation of diarrhea conducted at two outpatient clinics in Nepal, a significant number of expatriates infected with *Cyclospora* experienced prolonged diarrhea often associated with nausea, increased gas, and loss of appetite *(53)*. Criteria for selecting controls included: no history of a diarrheal illness in the preceding 2 wk, no previous evidence of *Cyclospora* infection and ability to provide a stool specimen *(53)*. Only 1 out of 96 control patients was infected with *Cyclospora* and that individual was later excluded from the study due to the development of diarrhea. Conversely, in endemic areas, infected members of an indigenous population may be asymptomatic. In two temporally distinct prospective cohort studies, totaling 377 children in Lima, Peru, only 11 out of 41 *Cyclospora*-infected individuals had diarrhea *(16)*. In a cross-sectional epidemiologic study of children <18 yr of age living in a Peruvian shantytown, 1.1% of almost 6000 fecal samples collected contained *Cyclospora* oocysts *(77)*. Approximately two-thirds of these infected children did not have symptoms of diarrhea, abdominal pain, or anorexia. In addition, the prevalence of *Cyclospora* infection in this endemic area appeared to be inversely related to age *(77)*. Though these findings suggest that protective immunity may be responsible for fewer infections in previously exposed individuals later in life, more than one episode of infection has been reported in some individuals *(16)*.

C. cayetanensis infections show a marked seasonality, and a relationship with warm and rainy seasons has been noted *(16,53)*. In north America, cyclosporiasis has a marked seasonality, being a disease of springtime and early summer. Sporadic cases of cyclosporiasis have been reported from many countries and oocysts have been identified in stools from immunocompetent individuals from non-endemic areas *(27)*.

6. CHOICE OF TREATMENT

Trimethoprim-sulfamethoxazole is useful both for treating *Cyclospora* infection and for decreasing the duration of oocyst shedding *(77–79)*. One double-blind placebo-controlled study of 40 *Cyclospora*-infected expatriates in Nepal revealed that double strength co-trimoxazole taken daily for 7 d was effective *(79)*. An additional 7 d was necessary to eradicate oocysts in the one patient who remained infected after 7 d of therapy. A more recent double-blind, randomized, placebo-controlled trial evaluated a 3 d course of twice daily trimethoprim-sulfamethoxazole therapy for *Cyclospora*-infected Peruvian children <18 yr of age *(77)*. The period of oocyst shedding was significantly reduced in treated children, but the sample size was very small to draw definitive conclusions about treatment effects on the length of symptomatic illness.

Pape et al. *(26)* studied the effects of oral double strength trimethoprim-sulfamethoxazole administered four times daily for 10 d in 43 patients with *Cyclospora* and HIV co-infection. After 5-d therapy, none of the patients with chronic diarrhea who were evaluated were noted to have abdominal pain or diarrhea, and no microbiological evidence of *Cyclospora* infection existed in any of the patients by the completion of therapy. Of 28 successfully treated patients who were followed beyond the therapeutic trial, 12 developed diarrhea again secondary to *Cyclospora* within 3-mo treatment. A further course of treatment with trimethoprim-sulfamethoxazole followed by thrice weekly "maintenance" therapy was instituted and resulted in only one more episode of recurrent infection, suggesting that secondary prophylaxis should be considered in HIV-infected individuals *(26)*. Other investigators have reported similar results with trimethoprim-sulfamethoxazole in HIV-infected patients with cyclosporiasis, including infections that involve the biliary tract *(76)*.

To date, no reliably effective therapy has been determined for sulfa-allergic patients *(13,27)*. Heightened clinical awareness of *Cyclospora* infection in patients returning from endemic areas who have failed quinolone therapy, the empirical drug(s) of choice for many with travel-associated diarrhea, is necessary. Hydration is a major component of supportive care for patients with cyclosporiasis, and, in the absence of data, no recommendations can be made regarding diarrheal antimotility agents *(80)*.

In immunocompromised patients with HIV-related disease, immune reconstitution using HAART, which acts prophylactically is the treatment of choice. HAART reduces viral load and may also reduce parasite load. Protease inhibitors used in HAART reduce *C. parvum* sporozoite host-cell invasion and parasite development in vitro, and inhibition is enhanced in combination with paromomycin *(81)*. HAART, in combination with antiparasitic therapy, may enhance *Cyclospora* clearance. In non-HIV immuno-suppressed patients with cyclosporiasis, reducing immunosuppression prior to specific chemotherapy is an useful option.

7. RESISTANCE EPIDEMIOLOGY

No emerging resistance epidemiology has been described.

8. SUMMARY AND CONCLUSIONS

Cyclospora is an obligate intracellular parasite that completes its life cycle in a single host and whose taxonomy is under review. *Cyclospora* is closely related to the genus *Eimeria* and of 17 known species, four infect primates and one, *C. cayetanensis*, infects humans. There is very little known about *C. cayetanensis* infectious doses or whether different isolates possess different virulence and/or pathogenicity factors and how they might affect clinical outcome. A preliminary human volunteer trial failed to demonstrate infection, clinical signs or symptoms. As for other coccidia, parasite factors and the inflammatory immune response probably drive both parasite clearance and immunopatho-logy. Trimethoprim-sulfamethoxazole is useful both for treating *Cyclospora* infection and for decreasing the duration of oocyst shedding. Immune reconstitution using HAART and secondary Trimethoprim-sulfamethoxazole prophylaxis should be considered in HIV-infected individuals.

Transmission to man can occur *via* any mechanism by which material contaminated with faeces containing infectious oocysts from infected human beings or other non-human hosts can be swallowed by a susceptible host. No convincing evidence of non-human reservoir hosts or zoonotic transmission of *C. cayetanensis* has been described, to date, indicating that *C. cayetanensis* may be human-specific. Recently, *C. cayetanensis* oocysts and DNA were detected in the stools of two dogs and one monkey in Nepal, but as no histology was performed on these animals, their roles as a natural reservoir host for *C. cayetanensis* remains to be determined *(82)*.

Abiotic reservoirs include all vehicles that contain sufficient infectious oocysts to cause human infection, the most commonly recognised being food and water. Transport hosts including refuse-feeding birds, wildfowl and filth flies should also be borne in mind when developing risk assessment questionnaires and performing risk assessments.

While conventional detection methods are useful for detecting symptomatic human cases, they are less effective for detecting asymptomatic carriers. Here, molecular methods that can determine species, genotypes and subtypes of *Cyclospora* are required and a variety of PCR-based methods are available. PCR-RFLP and DNA sequence based methods have been used to identify species/genotypes in clinical and environmental samples, but there is increasing interest in real time PCR methods as they can be more sensitive and rapid.

No validated oocyst viability/infectivity method has been described and further focus is required in this area of research. Standardised methods for detecting low densities of oocysts in water and food are available as are molecular methods for determining their species. Unidentified interferents in water and food matrices reduce the efficiency of detection by PCR, and further research into methods that relieve such interferents should be assessed in order to maximise detections from low densities of oocysts. Immaterial of the method used, effective quality assurance, quality control and validation are necessary prior to adoption.

Both foodborne and waterborne outbreaks of cyclosporiasis have been documented. Foodborne clusters have affected large numbers of individuals in north America, and cases occur more commonly in springtime and early summer. Whether seasonal variation in surface contamination of foods occurs, requires further investigation however, seasonal peaks in parasitism will influence when water and foods become surface contaminated. Water and food enhance survival of environmental stages by preventing their desiccation and the widespread distribution of oocysts in the environment enhances the possibility of foodborne transmission.

Increased demand, global sourcing, and rapid transport of soft fruit, salad vegetables and seafood can enhance both the likelihood of oocyst contamination and oocyst survival. A risk based assessment based upon standard, validated methods is required for detecting *Cyclospora* oocysts on/in food. In addition to determining genus and species of protozoan parasites present on/in foods and whether they are infectious to humans, subtyping methods are required to track outbreaks of disease and incidents and to determine the risk associated with specific genotypes. A clearer understanding of the population biology of the parasite will assist in unravelling occurrence and prevalence of genotypes, while a multidisciplinary approach will assist in unravelling the impact of foodborne protozoan on human health.

Methods for detecting oocysts on foods are modifications of those used for detecting oocysts in water, and, as such, have recovery efficiencies ranging from 1–59%. Specificity, sensitivity and reproducibility are paramount as, unlike pre-enrichment methods that increase organism numbers for prokaryotic pathogens and indicators, there is no method to augment parasite numbers prior to detection. Oocyst contamination of food can be on the surface of, or in, the food matrix and products at greatest risk of transmitting infection to man include those that receive no, or minimal, heat treatment after they become contaminated. Bivalves can remove *Cyclospora* oocysts from water and accumulate them on the gills and in the haemolymph. Temperature elevation kills oocysts and disinfectants and other treatment processes used in the food industry may be detrimental to oocyst survival or lethal, but further research in this important area is required.

In the absence of specific data, approaches such as hazard assessment and critical control point (HACCP) should be used to identify and control risk.

ACKNOWLEDGMENT

I thank Drs. David A. Relman, Departments of Medicine, and Microbiology & Immunology, Stanford University School of Medicine, Palo Alto VA Health Care System, PAVAHCS 154T, 3801 Miranda Avenue, Palo Alto, CA 94304, USA and Fred A. Lopez, Infectious Diseases Section, Louisiana State University School of Medicine, 1542 Tulane Avenue, New Orleans, LA 70112-2822, USA for their assistance with this chapter.

REFERENCES

1. Ashford, R. W. (1979) Occurrence of an undescribed coccidian in man in Papua New Guinea. *Ann. Trop. Med. Parasitol.* **73,** 497–500.
2. Narango, J., Sterling, C. R., and Gilman, R. H. (1989) *Cryptosporidium* muris-like objects from fecal samples of Peruvians. *Proc. 38th meeting Am. Soc. Trop. Med. Hyg.* Honolulu. Allen Press, Lawrence, KS. Abstract 324, p.243.
3. Knight, P. (1995) Once misidentified human parasite is a cyclopsoran. *ASM News* **61,** 520–522.
4. Soave, R., Dubey, J. P., Ramos, L. J., and Tummings, M. (1986) A new intestinal pathogen? *Clin. Res.* **34,** Abstract 533.
5. Shlim, D. R., Cohen, M. T., Eaton, M., Rajah, R., Long, E. G., and Ungar, B. L. P. (1991) An alga-like organism associated with an outbreak of prolonged diarrhea among foreigners in Nepal. *Am. J. Trop. Med. & Hyg.* **45,** 383–389.
6. Long, E. G., Ebrahimzadeh, A., White, E. H., Swisher, B., and Callaway, C. S. (1990) Alga associated with diarrhea in patients with acquired immunodeficiency syndrome and in travellers. *J. Clin. Microbiol.* **28,** 1101–1104.
7. Centers for Disease Control and Prevention. (1991) Outbreaks of diarrhoeal illness associated with cyanobacteria (blue-green algae)-like bodies - Chicago and Nepal, 1989 and 1990. *MMWR* **40,** 325–327.
8. Bendall, R. P., Lucas, S., Moody, A., Tovey, G., and Chiodini, P. L. (1993) Diarrhoea associated with cyanobacterium-like bodies: a new coccidian enteritis of man. *Lancet* **341,** 590–592.
9. Ortega, Y. R., Gilman, R. H., and Sterling, C. R. (1994) A new coccidian parasite (Apicomplexa: Eimeriidae) from humans. *J. Parasitol.* **80,** 625–629.
10. Levine, N. D. (1973) Introduction, history and taxonomy, in *The coccidia*: Eimeria, Isospora, Toxoplasma *and related genera* (Hammond, D. M. and Long, P. L., eds.), University Park Press, Baltimore, Maryland, USA. pp. 1–22.
11. Shields, J. M. and Olson, B. H. (2003) *Cyclospora cayetanensis:* a review of an emerging parasitic coccidian. *Int. J. Parasitol.* **33,** 371–391.
12. Eberhard, M. L., da Silva, A. J., Lilley, B. G., and Pieniazek, N. J. (1999) Morphologic and molecular characterization of new *Cyclospora* species from Ethiopian monkeys: *C. cercopitheci* sp.n., *C. colobi* sp.n., and *C papionis* sp.n. *Emerg. Infect. Dis.* **5,** 651–658.
13. Wurtz, R. M., Kocka, F. E., Peters, C. S., Weldon-Linne, C. M., Kuritza, A., and Yungbluth, P. (1993) Clinical characteristics of seven cases of diarrhea associated with a novel acid-fast organism in the stool. *Clin. Infect. Dis.* **16,** 136–138.
14. Relman, D. A., Schmidt, T. A., Gajadhar, A., et al. (1996) Molecular phylogenetic analysis of *Cyclospora*, the human intestinal pathogen, suggests that it is closely related to *Eimeria* species. *J. Infect. Dis.* **173,** 440–445.
15. Pieniazek, N. J. and Herwaldt, B. L. (1997) Reevaluating the molecular taxonomy: Is human-associated *Cyclospora* a mammalian *Eimerian* species? *Emerg. Infect. Dis.* **3,** 381–383.

16. Ortega, Y. R., Sterling, C. R., Gilman, R. H., Cama, V. A., and Diaz, F. (1993) *Cyclospora* species: a new protozoan pathogen of humans. *N. Engl. J. Med.* **328,** 1308–1312.

17. Smith, H. V., Paton, C. A., Mtambo, M. M. A., and Girdwood, R. W. A. (1997) Sporulation of *Cyclospora* sp. oocysts. *Appl. Environ. Microbiol.* **63,** 1631–1632.

18. Ortega, Y. R., Nagle, R., Gilman, R. H., et al. (1997) Pathologic and clinical findings in patients with cyclosporiasis and a description of intracellular parasite life-cycle stages. *J. Infect. Dis.* **176,** 1584–1589.

19. Adam, R. D., Ortega, Y. R., Gilman, R. H., and Sterling, C. R. (2000) Intervening transcribed spacer region 1 variability in *Cyclospora cayetanensis. J. Clin. Microbiol.* **38,** 2339–2343.

20. Olivier, C., van de Pas, S., Lepp, P. W., Yoder, K., and Relman, D. A. (2001) Sequence variability in the first internal transcribed spacer region within and among *Cyclospora* species is consistent with polyparasitism. *Int. J. Parasitol.* **31,** 1475–1487.

21. Garcia, L. S. and Bruckner, D. A. (1997) *Diagnostic Medical Parasitology. 3rd ed.* Washington, D.C. American Society for Microbiology.

22. Medina-De La Garza, C. E., Garcia-Lopez, H. L., Salinas-Carmona, M. C., and Gonzalez-Spencer, D. J. (1997) Use of discontinuous Percoll gradients to isolate *Cyclospora* oocysts. *Ann. Trop. Med. Parasitol.* **91,** 319–321.

23. Berlin, O. G. W., Peter, J. B., Gagne, C., Conteas, C. N., and Ash, L. R. (1998) Autofluorescence and the detection of *Cyclospora* oocysts. *Emerg. Infect. Dis.* **4,** 127–128.

24. Ortega, Y. R., Sterling, C. R., and Gilman, R. H. (1998) *Cyclospora* cayetanensis. *Adv. Parasitol.* **40,** 399–418.

25. Eberhard, M. L., Pieniazek, N. J., and Arrowood, M. J. (1997) Laboratory diagnosis of *Cyclospora* infections. *Arch. Pathol. Lab. Med.* **121,** 792–797.

26. Pape, J. W., Verdier, R.-I., Boncy, M., Boncy, J., and Johnson, W. D., Jr. (1994) *Cyclospora* infection in adults infected with HIV. Clinical manifestations, treatment, and prophylaxis. *Ann. Intern. Med.* **121,** 654–657.

27. Wurtz, R. (1994) *Cyclospora*: A newly identified intestinal pathogen of humans. *Clin. Infect. Dis.* **18,** 620–623.

28. Visvesvara, G. S., Moura, H., Kovacs-Nace, E., Wallace, S., and Eberhard, M. L. (1997) Uniform staining of *Cyclospora* oocysts in fecal smears by a modified safranin technique with microwave heating. *J. Clin. Microbiol.* **35,** 730–733.

29. Centers for Disease Control and Prevention. (1997) Outbreaks of pseudo-infection with *Cyclospora* and Cryptosporidium-Florida and New York City, 1995. *MMWR* **46,** 354–358.

30. Relman, D. A., Schmidt, T. A., Gajadhar, A., et al. (1996) Molecular phylogenetic analysis of *Cyclospora*, the human intestinal pathogen, suggests that it is closely related to *Eimeria* species. *J. Infect. Dis.* **173,** 440–445.

31. Pieniazek, N. J., Slemenda, S. B., Da Silva, A. J., Alfano, E. M., and Arrowood, M. J. (1996) PCR confirmation of infection with *Cyclospora cayetanensis. Emerg. Infect. Dis.* **2,** 342–343.

32. Yoder, K. E., Sethabutr, O., and Relman, D. A. (1996) PCR-based detection of the intestinal pathogen *Cyclospora*. in *PCR Protocols for Emerging Infectious Diseases* (Persing D. H., ed), Washington, D.C.: American Society for Microbiology, pp. 169–175.

33. Jinneman, K. C., Wetherington, J. W., Adams, A. M., et al. (1996) Differentiation of *Cyclospora* sp. and Eimeria spp. by using the polymerase chain reaction amplification products and restrictionfragment length polymorphisms. FDA *Laboratory Information Bulletin,* 4044. http://vm.cfsan.fda.gov/~mow/kjcs l9c.html.

34. Lopez, F. A., Manglicmot, J., Schmidt, T. M., Yeh, C., Smith, H. V., and Relman, D. A. (1999) Molecular characterization of *Cyclospora*-like organisms from baboons. *J. Infect. Dis.* **179,** 670–676.

35. Tsolaki, A. G., Miller, R. F., Underwood, A. P., Banerji, S., and Wakefield, A. E. (1996) Genetic diversity at the internal transcribed spacer regions of the rRNA operon among isolates of Pneumocystis carinii from AIDS patients with recurrent pneumonia. *J. Infect. Dis.* **174,** 141–156.

36. Jinneman, K. C., Wetherington, J. H., Hill, W. E., et al. (1999). An oligonucleotide-ligation assay for the differentiation between *Cyclospora* and *Eimeria* spp. polymerase chain reaction amplification products. *J. Food Prot.* **62**, 682–685.

37. Shields, J. M. and Olson, B. H. (2003) PCR-restriction fragment length polymorphism method for detection of *Cyclospora cayetanensis* in environmental waters without microscopic confirmation. *Appl. Environ. Microbiol.* **69**, 4662–4669.

38. Orlandi, P. A., Carter, L., Brinker, A. M., et al. (2003) Targeting single-nucleotide polymorphisms in the 18S rRNA gene to differentiate *Cyclospora* species from Eimeria species by multiplex PCR. *Appl. Environ. Microbiol.* **69**, 4806–4813.

39. Verweij, J. J., Laeijendecker, D., Brienen, E. A., van Lieshout, L., and Polderman, A. M. (2003) Detection of *Cyclospora cayetanensis* in travellers returning from the tropics and subtropics using microscopy and real-time PCR. *Int. J. Med. Microbiol.* **293**, 199–202.

40. Anonymous. (1999) *Isolation and identification of* Cryptosporidium *oocysts and Giardia cysts in waters 1999. Methods for the examination of waters and associated materials.* London: HMSO. 44pp.

41. Anonymous. (1999) *UK Statutory Instruments 1999 No. 1524. The Water Supply (Water Quality) (Amendment) Regulations 1999.* The Stationery Office, Ltd. 5pp.

42. Robertson, L. J., Gjerde, B., and Campbell, A. T. (2000) Isolation of *Cyclospora* oocysts from fruits and vegetables using lectin-coated paramagnetic beads. *J. Food Prot.* **63**, 1410–1414.

43. Jinneman, K. C., Wetherington, J. H., Hill, W. E., et al. (1998) Template preparation for PCR and RFLP of amplification products for the detection and identification of *Cyclospora* sp. and *Eimeria* spp. oocysts directly from raspberries. *J. Food. Prot.* **61**, 1497–1503.

44. Orlandi, P. A. and Lampel, K. A. (2000) Extraction-free, filter-based template preparation for rapid and sensitive PCR detection of pathogenic parasitic protozoa. *J. Clin. Microbiol.* **38**, 2271–2277.

45. Ortega, Y. R., Roxas, C. R., Gilman, R. H., et al. (1997) Isolation of *Cryptosporidium parvum* and *Cyclospora cayetanensis* from vegetables collected in markets of an endemic region in Peru. *Am. J. Trop. Med. Hyg.* **57**, 683–686.

46. Abou el Naga, I. (1999) Studies on a newly emerging protozoal pathogen: *Cyclospora cayetanensis.* *J. Egypt. Soc. Parasitol.* **29**, 575–586.

47. Sherchand, J. B., Cross, J. H., Jimba, M., Sherchand, S., and Shrestha, M. P. (1999) Study of *Cyclospora cayetanensis* in health care facilities, sewage water and green leafy vegetables in Nepal. *Southeast Asian J. Trop. Med. Public Health* **30**, 58–63.

48. Dalton, C. A., Goater, A. D., Pethig, R., and Smith, H. V. (2001) Viability of *Giardia intestinalis* cysts and viability and sporulation state of *Cyclospora cayetanensis* oocysts determined by electrorotation. *Appl. Environ. Microbiol.* **67**, 586–590.

49. Dalton, C., Goater, A. G., Burt, J. P. H., and Smith, H. V. (2004) Analysis of parasites by electrorotation. *J. Appl. Microbiol.* **96**, 24–32.

50. Eberhard, M. L., Ortega, Y. R., Haynes, D. E., et al. (2000) Attempts to establish experimental *Cyclospora cayetanensis* infection in laboratory animals. *J. Parasitol.* **86**, 577–582.

51. Alfano-Sobsey, E. M., Eberhard, M. L., Seed, J. R., et al. (2004) Human challenge pilot study with *Cyclospora cayetanensis.* *Emerg. Infect. Dis.* **10**, 726–728.

52. Sturbaum, G. D., Ortega, Y. R., Gilman, R. H., Sterling, C. R., Cabrera, L., and Klein, D. (1998) Detection of *Cyclospora* cayetanensis in wastewater. *Appl. Environ. Microbiol.* **64**, 2284–2286.

53. Hoge, C. W., Shlim, D. R., Rajah, R., et al. (1993) Epidemiology of diarrhoeal illness associated with coccidian-like organism among travellers and foreign residents in Nepal. *Lancet* **341**, 1175–1179.

54. Rabold, J. G., Hoge, C. W., Shlim, D. R., Kefford, C., Rajah, R., and Echeverria, P. (1994) *Cyclospora* outbreak associated with chlorinated drinking water, [letter] *Lancet* **344**, 1360–1361.

55. Huang, P., Weber, J. T., Sosin, D.M., et al. (1995) The first reported outbreak of diarrheal illness associated with *Cyclospora* in the United States. *Ann. Intern. Med.* **123,** 409–414.

56. Centers for Disease Control and Prevention. (1996) Update: outbreaks of *Cyclospora cayetanensis* infection - United States and Canada, 1996. *MMWR* **45,** 611–612.

57. Kimura, K., Rai, S. K., Rai, G., Insisiengmay, S., Kawabata, M., and Uga, S. (2004) Detection of *Cyclospora* and *Isospora* from diarrheal feces and tap water collected from Nepal and Lao PDR, in *4th International* Giardia *Conference and First Combined Giardia and* Cryptosporidium *conference,* Amsterdam, the Netherlands, 20–24 September 2004, abstract, p. 42.

58. Graczyk, T. K., Ortega, Y. R., and Conn, D. B. (1998) Recovery of waterborne oocysts of *Cyclospora cayetanensis* by Asian freshwater clams *(Corbicula flumined). Am. J. Trop. MedHyg.* **59,** 928–932.

59. Centers for Disease Control and Prevention. (1997) Update, outbreaks of cyclosporiasis-United States and Canada, 1997. *MMWR* **46,** 521–523.

60. Centers for Disease Control and Prevention. (1997) Outbreak of cyclosporiasis -Northern Virginia-Washington, D.C.-Baltimore, Maryland, Metropolitan Area, 1997. *MMWR* **46,** 689–691.

61. Herwaldt, B. L., Ackers, M., and the *Cyclospora* working group. (1997) An outbreak of cyclosporiasis associated with imported raspberries. *N. Engl. J. Med.* **336,** 1548–1556.

62. Herwaldt, B. L., Beach, M. J., and the *Cyclospora* Working Group. (1999) The return of *Cyclospora* in 1997: another outbreak of cyclosporiasis in North America associated with imported raspberries. *Ann. Intern. Med.* **130,** 210–220.

63. Herwaldt, B. L. (2000) *Cyclospora* cayetanensis: a review focusing on the outbreaks of cyclosporiasis in the 1990s. *Clin. Infect. Dis.* **31,** 1040–1057.

64. Doller, P. C., Dietrich, K., Filipp, N., et al. (2002) Cyclosporiasis outbreak in Germany associated with the consumption of salad. *Emerg. Infect. Dis.* **8,** 992–994.

65. Centers for Disease Control and Prevention. (1998) Outbreak of cyclosporiasis -Ontario, Canada, May 1998. *MMWR* **47,** 806–809.

66. Lopez, A. S., Dodson, D. R., Arrowood, M. J., et al. (2001) Outbreak of cyclosporiasis associated with basil in Missouri in 1999. *Clin. Infect. Dis.* **32,** 1010–1017.

67. Ho, A. Y., Lopez, A. S., Eberhart, M. G., et al. (2002) Outbreak of cyclosporiasis associated with imported raspberries, Philadelphia, Pennsylvania, 2000. *Emerg. Infect. Dis.* **8,** 783–788.

68. Adams, A. M., Jinneman, K. C., and Ortega, Y. R. (1999) Cyclospora, in *Encyclopaedia of Food Microbiology* (Robinson R., Batt C., and Patel P., eds.), Academic Press, London and New York. Volume **1,** pp. 502–513.

69. Okhuysen, P. C., Chappell, C. L., Crabb, J. H., Sterling, C R., and DuPont, H. L. (1999) Virulence of three distinct *Cryptosporidium parvum* isolates for healthy adults. *J. Infect. Dis.* **180,** 1275–1281.

70. Connor, B. A., Shlim, D. R., Scholes, J. V., Rayburn, J. L., Reidy, J., and Rajah, R. (1993) Pathologic changes in the small bowel in nine patients with diarrhea associated with a coccidia-like body. *Ann. Intern Med.* **119,** 377–382.

71. Ortega, Y. R., Nagle, R., Gilman, R. H., et al. (1997) Pathologic and clinical findings in patients with cyclosporiasis and a description of intracellular parasite life-cycle stages. *J. Infect. Dis.* **176,** 1584–1589.

72. VanNhieu, J. T., Nin, F., Fleury-Feith, J., Chaumette, M.-T., Schaeffer, A., and Bretagne, S. (1996) Identification of intracellular stages of *Cyclospora* species by light microscopy of thick sections using haematoxylin. *Hum. Pathol.* **27,** 1107–1109.

73. Soave, R. (1996) *Cyclospora*: an overview. *Clin. Infec. Dis.* **23,** 429–437.

74. Fleming, C. A., Carron, D., Gunn, J. E., and Barry, M. A. (1998) A foodborne outbreak of *Cyclospora cayetanensis* at a wedding. *Arch. Intern. Med.* **158,** 1121–1125.

75. Richardson, R. F., Remler, B. F., Katirji, B., and Murad, M. H. (1998) Guillain-Barre Syndrome after *Cyclospora* infection. *Muscle Nerve* **21,** 669–671.

76. Sifuentes-Osornio, J., Porras-Cortes, G., Bendall, R. P., Morales-Villareal, F., Reyes-Teran, G., and Ruiz-Palacios, G. M. (1995) *Cyclospora cayetanensis* infection in patients with and without AIDS: Biliary disease as another clinical manifestation. *Clin. Infect. Dis.* **21,** 1092–1097.

77. Madico, G., McDonald, J., Gilman, R. H., Cabrera, L., and Sterling, C. R. (1997) Epidemiology and treatment of *Cyclospora cayetanensis* infection in Peruvian children. *Clin. Infect. Dis.* **24,** 977–981.

78. Madico, G., Gilman, R. H., Miranda, E., Cabrera, L., and Sterling, C. R. (1993) Treatment of *cyclospora* infections with co-trimoxazole. [letter] Lancet **342,** 122–123.

79. Hoge, C. W., Shlim, D. R., Ghimire, M., et al. (1995) Placebo-controlled trial of co-trimoxazole for cyclospora infections among travellers and foreign residents in Nepal. *Lancet* **345,** 691–693.

80. Taylor, A. P., Davis, L. J., and Soave, R. (1997) Cyclospora. *Curr. Clin. Top. Infect. Dis.* 17,256–17,268.

81. Hommer, V., Eicholz, J., and Petry, F. (2003) Effect of antiretroviral protease inhibitors alone, and in combination with paromomycin, on the excystation, invasion and in vivo development of *Cryptosporidium parvum. J. Antimicrob. Chemother.* **52,** 359–364.

82. Chu, D-M. T., Sherchand, J. B., Cross, J. H., and Orlandi, P. A. (2004) Detection of *Cyclospora cayetanensis* in animal fecal isolates from Nepal using an FTA filter-base polymerase chain reaction method. *Am. J. Trop. Med. Hyg.* **71,** 373–379.

Huw V. Smith and Tim Paget

Abstract

Two assemblages of the extracellular, intestinal parasite, *Giardia*, cause diarrhea in humans. *Giardia* completes its life cycle in individual hosts and its infectious dose can be small (25–100 cysts). Many aspects of its virulence and pathogenicity are poorly understood. The drug of choice is metronidazole. Transmission to humans can occur via any mechanism in which material contaminated with feces containing infectious cysts from infected human beings or nonhuman hosts can be swallowed by a susceptible host. Biotic reservoirs include all potential hosts of human-infectious *Giardia* species, while abiotic reservoirs include all vehicles that contain sufficient infectious cysts to cause human infection, the most commonly recognized being food and water.

Both foodborne and waterborne outbreaks have been documented. In two out of eight outbreaks, contaminated foodstuffs were implicated as the vehicles of transmission but, in six, foodhandlers were implicated, implying that contamination occurred probably during food preparation and that, until then, the foodstuffs were free of infectious cysts. Increased global sourcing and rapid transport of soft fruit, salad vegetables, and seafood can enhance both the likelihood of cyst contamination and cyst survival. Standardized methods for detecting cysts on foods must be maximized as, unlike pre-enrichment methods that increase organism numbers for prokaryotic pathogens and indicators, there is no method to augment parasite numbers prior to detection. Cyst contamination of food can be on the surface of, or in, the food matrix and products at greatest risk of transmitting infection to man include those that receive no, or minimal, heat treatment after they become contaminated. Temperature elevation (≥64.2°C for 2 min), chlorine dioxide, ozone, UV light, and other treatment processes used in the food industry may be detrimental to cyst survival or lethal, but further research in this important area is required.

Key Words: *Giardia*; cysts; occurrence; detection; outbreaks; foodborne; environment.

1. CLASSIFICATION AND IDENTIFICATION

1.1. Classification

1.1.1. Historical

Giardia was first described by Anthony van Leeuwenhoek (1632–1723) in 1681 when he discovered in his watery excrement "... small animalcules a-moving very prettily; some of 'em a bit bigger, others a bit less, than a blood globule, but all of one and the same make, ..." which Dobell in 1932 thought were the vegetative (trophozoite) stage of the infection. In 1859, the Czech physician Vilem Lambl (1824–1895) rediscovered *Giardia* and described some of the morphological details of the trophozoite in its intestinal environment, which was subsequently named *Lamblia*. In 1952, Filice published his morphometric findings of the genus *Giardia*. Even at this time in *Giardia* taxonomy, there was controversy

From: *Infectious Disease: Foodborne Diseases*
Edited by: S. Simjee © Humana Press Inc., Totowa, NJ

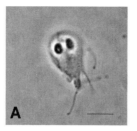

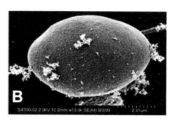

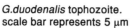

G.duodenalis cyst.
G.duodenalis tophozoite.   Scale bar represents 2.3 μm
scale bar represents 5 μm   SEM courtesy of S.L. Erlandsen
                                  and E.L. Jarroll

Fig. 1. *Giardia duodenalis* trophozoite (**A**) and cyst (**B**).

over its naming. *Cercomonas intestinalis* was the first name of the organism we now know as *Giardia*, but isolates of similar morphology from different animal hosts frequently adopted the species name from which host they were described (e.g., *Dimorphus muris* from mice). The description of *Hexamita duodenalis* from rabbits introduced the species name *duodenalis*, and the genus name *Giardia* was first used by Kunstler in 1882 *(1)* for a protozoan isolated from tadpoles which he named *Giardia agilis*.

Filice *(2)* first characterized three major type species, based primarily on the morphology and morphometry of the vegetative (trophozoite) form (Fig. 1A). He named these *G. agilis*, *G. muris*, and *G. duodenalis*, but *G. duodenalis* proved to be the most problematic. The genus *Giardia* was created by Kunstler in 1882 and ascribed to the family Hexamitidae. The members of the Hexamitidae are characterized by having bilateral symmetry, two nuclei lying side-by-side and six or eight flagella. Although a few species of the five genera within the Hexamitidae are free living, the majority are parasites of animals. However, it should be noted that more modern classifications have created a new kingdom, the Archezoa which embraces those eukaryotic organisms which lack mitochondria, plastids, hydrogenosomes, peroxisomes, and Golgi bodies. Thus, as members of this kingdom, *Giardia* are not protozoans *sensu stricto*. In these newer classifications, *Giardia* are in the order Diplomonadida of the class Trepomonadea and phylum Metamonada.

1.1.2. Current Classification

Currently, six species of *Giardia* are recognized on the basis of morphological characteristics and host preference (Tables 1 and 2). The taxonomy of *G. duodenalis*, based on Filice's type species descriptions of trophozoites, is unsatisfactory, as it relates solely to morphology and morphometry (Tables 1 and 2) and lacks the discriminatory power required for epidemiology. The "*duodenalis*" type organisms have also been named "*lamblia*" (for organisms isolated from humans) and "*intestinalis*" (for those isolated from mammals including humans) which, rather than clarifying the issue, adds further confusion to it. The numerous genetic variants within this one "*duodenalis*-type species" make this a very diverse group.

The lack of morphological differences between genetic variants found in mammals (Table 1) has resulted in an informal categorization of these genotypes based on genetic differences (Table 2). Polymerase chain reaction (PCR)-based techniques targeting

Table 1
Differences Between Some Species Within the Genus *Giardia*

Species	Host(s)
G. duodenalis	A wide range of mammals, including livestock, pets, and humans
G. muris	Mice
G. microti	Voles and muskrats
G. psittaci	Budgerigars
G. ardeae	Heron
G. agilis	Frogs

Data from ref. *(3)*.

Table 2
Molecular Characterization of *Giardia* Found in Mammals

Genotype	Host range
Assemblage A	Humans, slow loris, livestock, deer, cats, dogs, beavers, muskrats, voles, guinea pigs, ferret
Assemblage B	Humans, siamang, slow loris, livestock, chinchillas, dogs, beavers, muskrats, voles, rats, marmoset
Assemblage C/D	Dogs, coyotes
Assemblage E	Alpaca, cattle, goats, pigs, sheep
Assemblage F	Cats
Assemblage G	Rats

Data from ref. *(3)*.

specific genes, such as the small subunit (SSU) rRNA, surface protein genes, glutamate dehydrogenase, triosephosphate isomerase, and other catabolic enzyme genes as well as restriction fragment length polymorphism (RFLP) *(4–10)*, sequence polymorphism and karyotype analysis have been used to subdivide this species into six "assemblages" (Table 2). Some of these assemblages are genetically very distinct and/or have a limited or very specific host range, and these may be distinct (cryptic) species (Table 2). This variation may also be due to genetic plasticity within this species and it is tempting to presume that genetic flexibility may, in part, be due to transmission between a variety of hosts *(3)*.

Nearly 300 yr elapsed between the discovery of *Giardia (11)* and the recognition that *Giardia* is an etiological agent of disease *(12)*. Pathogenicity in humans was formally established and Koch's postulates fulfilled <20 yr ago *(13)*. Koch's postulates identify the rules for proving that an organism causes disease. There are several reasons for this prolonged time period between identifying *Giardia* and realizing its potential for pathogenicity. Significant amongst these is the problem of assigning specific symptoms/pathology to one organism in an individual with multiple infections. Another confounding factor may be the long association between the organism and its host; *Giardia* probably represents one of the most ancient lineages amongst (but the recent finding of a mitochondrial relic, the mitosome questions this assumption) eukaryotes *(14,15)*, which probably allowed a near commensal relationship to be established. The presence or absence of clinical signs

ORGANISMS IN MAN

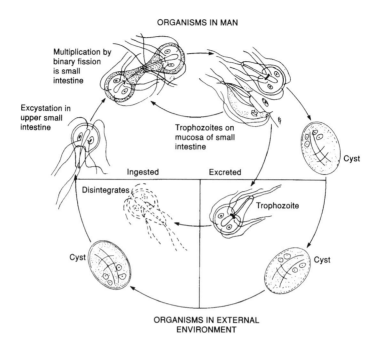

Fig. 2. *Giardia duodenalis* life cycle. (From ref. *[20].*)

and symptoms will be due, in part, to the plethora of different *Giardia* isolates which have evolved, and together with differences in the genetic backgrounds of the hosts it parasitises, differences within and between isolates can account for the variations documented in host range, immune responses, and virulence.

1.1.3. Life Cycle

Giardia exists in two forms: the parasitic, feeding, multiplying trophozoite and the environmentally resistant, infective cyst (Figures 1A,B and 2). Nuclear division, but not cytoplasmic division, occurs to produce the mature, quadrinucleate cyst. Either binucleate or quadrinucleate cysts are excreted in the feces depending on transit time through the intestinal tract. Infection is initiated following the ingestion of mature, infective cysts *(16)*. As cysts pass through the acidic stomach, the low pH and elevated CO_2 followed by slightly alkaline environment of the proximal small intestine induce excystation (cyst-trophozoite transformation) *(17)*. One trophozoite emerges from each quadrinucleate cyst, which rapidly undergoes cytoplasmic, but not nuclear division to form two binucleate trophozoites.

Trophozoites attach themselves to the luminal surface of duodenal and jejunal enterocytes (the epithelial cells that line the intestine), and, once attached, undergo further division by asexual binary fission. Motile trophozoites rotate around their longitudinal axis displaying both a tumbling movement resembling that of a falling leaf and an up- and down movement referred to as "skipping". After a period of about 4–7 d, trophozoites detach from enterocytes to form cysts in a process known as encystation. Exposure to bile salts and alkaline pH, as trophozoites pass down the small intestine are thought to be the major triggers for encystation *(18)*. During encystation, trophozoites become rounded, and the external architecture (flagella and ventral disc) is resorbed into the trophozoite body, forming the binucleate immature cyst. In mature cysts, the four nuclei are located at one pole and other trophozoite structures (axostyles, remnants of the ventral disc) are also present. Environmentally resistant cysts *(18,19)* are voided in the feces and are the infective and disseminating stage. Cysts are infectious to other susceptible hosts soon after excretion (cysts may require a period of maturation before becoming infectious).

1.2. Identification

There are no pathogenomonic signs and laboratory identification is required to confirm diagnosis.

1.2.1. Detection in Feces

The fact that a variety of laboratory tests are required to diagnose human giardiasis indicates that no one test alone has a consistently high diagnostic index. Fortunately, many laboratory diagnosticians recognize that these tests are complimentary. Currently, diagnosis of giardiasis is dependent upon the demonstration of intact parasites (cysts and/or less frequently trophozoites) by microscopy, parasite products by immunoassay and/or parasite DNA in feces or duodenal/jejunal aspirates, and serology.

Fresh or preserved stools can be concentrated to increase the yield of cysts by sedimentation, using the formol–ether or formol–ethyl acetate techniques or by any conventional fecal parasite flotation method *(18,21)*. Both trophozoites and cysts can occur in feces and can be sought for by examination of a wet mount or a permanently stained smear. Wet mounts can be fecal suspensions or concentrates and may be unstained (saline or formalin–ether/formalin–ethyl acetate) or stained (iodine, which is primarily a stain for cysts, or other temporary stains). Cyst measurements range from 8 to 16×7–10 μm (length × width).

Concentration techniques are unsatisfactory for trophozoites, and the direct examination of recently excreted, emulsified (hot) stools, or the examination of permanently stained smears is necessary for their demonstration *(21)*. Stools are concentrated by formalin–ether or formalin–ethyl acetate and concentrates are analyzed by brightfield microscopy *(21)*. Epifluorescence microscopy, using commercially available fluorescein isothiocyanate labeled monoclonal antibodies that react with surface-exposed, genus-specific cyst wall epitopes (FITC-*G*-mAb) is reported to be more effective for detecting cysts than light microscopy. The absence of cysts in a single stool sample cannot exclude infection, therefore at least three stool specimens should be examined by a competent microscopist prior to suggesting other diagnostic procedures. The erratic nature of cyst excretion, the requirement for experienced staff for microscopic identification, and the relatively low detection rate have precipitated the development of alternative methods for the diagnosis of giardiasis.

When trophozoites or cysts cannot be found following the examination of a reasonable number of stool samples, and giardiasis is suspected clinically, the examination of duodenal or jejunal fluid may be indicated. Trophozoites may be demonstrated in duodenal/jejunal aspirates or biopsies, when stool microscopy is negative, but it should be noted that the converse is not always true.

1.2.1.1. Antigen Detection in Feces. The enzyme-linked immunosorbent assay (ELISA) is used to detect *Giardia* antigen in feces. Numerous methods have been described, and target antigens range from the multitude present in aqueous trophozoite extracts to individual *Giardia*-specific molecules, such as *Giardia* stool antigen, present in trophozoites and cysts, and which has a relative molecular mass of 65 kDa (GSA 65; *[22]*). The methods described in the literature are antigen capture ELISAs which utilize antibodies to capture *Giardia* stool antigen on the solid phase and to detect its presence and develop the colored reaction product.

Detection of antigen in feces can overcome the diagnostic issues associated with erratic cyst excretion, trophozoite, and cyst disintegration in vivo, the low detection rate associated with microscopy, and the inability to detect trophozoites in jejunal juice or biopsy, particularly if the ELISA detects antigens that are common to both trophozoites and cysts. The reported sensitivity of ELISA is similar to *(23)* or better than *(22,24–26)* microscopy, and antigen detection ELISAs have been used to diagnose infection when standard microscopic techniques have failed. Furthermore, ELISAs offer much needed quality assurance to diagnosis, simplify screening procedures in suspected outbreak settings (e.g., waterborne *[27]*) and are also helpful for treatment follow-up.

1.2.2. Sensitivity of Detection in Feces

Diagnosis, based on the detection of cysts in fecal concentrates, can be inefficient *(28)*. For example, the likelihood of detecting cysts, by brightfield microscopy, in a single stool sample from an infected human is between 35 and 50%, whereas between 6 and 10 stool specimens from an infected human need to be examined to achieve a detection rate of 70–90% *(29)*, reflecting the erratic nature of cyst excretion in symptomatic patients.

The sensitivity of ELISA for detecting *Giardia* antigens in stools has been evaluated on numerous occasions *(26)*. The sensitivity and specificity of the commercially available ELISA-GSA 65 kit are comparable with those of microscopic examination for cysts in stool. All studies using ELISA-GSA 65 reported greater sensitivity over the microscopic examination of a single specimen *(26)*, and sensitivity varies between 95 and 100%, with 100% specificity reported when used with stools from patients infected with other intestinal parasites *(24,25)*. ELISA-GSA 65 can detect *Giardia* infection in at least 30% more cases than microscopy *(22)*.

1.2.3. Alternative Detection Procedures

Brightfield microscopic examination of either unconcentrated or concentrated stools remains the first line approach in the diagnosis of human giardiasis. Commercially available immunofluorescence assays, using FITC-*G*-MAbs, are used to detect cysts in either unconcentrated or concentrated stools, and for maximum sensitivity, sample concentration is recommended as formalin does not interfere with assay sensitivity. Although this method is more time-consuming than brightfield microscopy, translucent cysts and cysts with poor morphology can be detected. Cyst organelles are distorted on air-dried smears, reducing the potential for definitive identification by recognizing organelles using Nomarski differential interference contrast (DIC) microscopy. If a permanent record is required, a separate sample should be placed in the chemical preservative of choice. If the test result is negative, giardiasis cannot be excluded, and further stool samples should be obtained and tested (*see* above). FITC-*G*-MAbs are especially useful for environmental samples.

1.2.3.1. Molecular Detection Methods. PCR is more sensitive than conventional and immunological assays for the coprological detection of infection. Various PCR-based methods, with sensitivities from 1 to 10 cysts have been cited *(30,31)*, as have their abilities to discriminate between pathogenic and non-pathogenic organisms *(20,32,33)*. The reliance of PCR on enzymatic amplification of nucleic acids makes it susceptible to inhibitors present in feces, waters, and foods. Its increased discriminatory powers will prove beneficial for epidemiological, source, and disease-tracking proposes, but currently, there are few published instances of its usefulness. Issues pertaining to inhibitors in feces, water, and foods together with methods for relieving their inhibitory effects, PCR targets, the sensitivity of detection and methods for extracting DNA are presented in Section 1.2.3. (alternative detection procedures) of the *Cryptosporidium* chapter. (Please *see* page 242.) Currently, PCR methods which capitalize on our current knowledge of the genetic variants that constitute the *G. duodenalis* group (Table 2), have been designed to identify specific human-infective assemblages in stool samples *(34–36)* and sewage *(37)*.

Immunomagnetizable separation (IMS) is useful for capturing protozoan parasites from inhibitory matrices and to concentrate and process them in a buffer free of PCR inhibitors, thus increasing the sensitivity of detection *(38)*. Mahbubani et al. *(39)* demonstrated that *Giardia muris* cyst DNA extracted directly from environmental surface waters with moderate or high turbidity failed to amplify following PCR of a 0.171 kbp segment the giardin gene. However, cysts separated from the same samples by IMS, whose DNA was extracted with a freeze-boil Chelex®100 treatment, could be detected (3 and 30 cysts/ml^{-1}, respectively).

The simultaneous detection of multiple parasites by TaqMan (Applied Biosystems) real-time PCR assay has been reported. *Entamoeba histolytica*, *Giardia lamblia*, and *Cryptosporidium parvum* were detected using specific primers and probes by real-time PCR with 100% specificity and sensitivity ($n=60$; 20 for each group of individuals with confirmed microscopic identification of one out of the three parasites) *(40)*. The correlation between PCR positivity and cyst numbers in PCR-positive, cyst-negative, stools is not known.

PCR has also the potential for identifying *Giardia* DNA in the stools of asymptomatic individuals and in fomites, food, and water contaminated with small numbers of cysts. Rapid detection of asymptomatic individuals could limit clinical sequelae in "at risk" groups and reduce environmental spread. At present, in the clinical diagnostic laboratory, PCR is more likely to become a tool to assist epidemiological investigations than a routine method, but given a better understanding of the relationship between specific *Giardia* genotypes and host specificity, improved molecular tools for epidemiology, source and disease tracking will be developed.

Guy et al. *(41)* used sewage and environmental water matrices to detect the *G. duodenalis* β-giardin gene by TaqMan™ assay. DNA extraction using a DNeasy tissue kit (Qiagen, Germany) was optimized by modifying the commercial protocol and by the introduction of a three cycle freeze–thaw step and sonication (three 20-s bursts). The inhibitory effects of these matrices were counteracted with 20% (v/v) Chelex 100 (Bio-Rad) and 2% (v/v) PVP 360 (ICN, Ohio); in some cases, bovine-serum-albumin proved useful in eliminating inhibition present in lake water. DNA equivalent to that present in a single cyst was detected. Cyst DNA was detected and quantified in 1 L sewage samples using a multiplex real-time PCR (qPCR) and the numbers obtained using qPCR assay were comparable with those obtained using an immunofluorescence microscopy. Both assemblage A and B genotypes were detected in sewage by qPCR analysis.

1.2.3.2. Alternative Clinical Approaches. In the absence of laboratory evidence confirming *Giardia* infection (no trophozoites, cysts parasite antigen or DNA detected following the examination of a minimum of three stool samples), and giardiasis remains a clinical suspicion, the examination of duodenal, or jejunal fluid may be indicated. Endoscopy samples are normally liquid, whereas Enterotest[R] samples are mucus-saturated nylon string. The Enterotest[R] is a rubber-lined, weighted, gelatin capsule containing a nylon string which is swallowed, whilst the free end of the string is taped to the outside of the cheek. The Enterotest[R] is left in place for 4–8 h or overnight by which time the string extends to its full length, and its distal half becomes saturated with bile-stained mucus which can contain trophozoites. The mucus is scraped off the string, mixed with saline, and any trophozoites present in the mucus are detected microscopically. Endoscopy and Enterotest[R] specimens should be examined immediately as the chances of detecting motile trophozoites decrease with the passage of time. Storage of the specimen at low temperature (e.g., 2–10°C) or delay in getting the sample to the laboratory will reduce trophozoite motility, and can make identification more difficult. There is no agreement as to whether centrifugation, to concentrate trophozoites, is advantageous; however, centrifuging at speeds that do not rupture trophozoites produces a pellet of smaller volume which can be analyzed more rapidly. For definitive diagnosis, the examination of intestinal fluid for the presence of trophozoites has been reported as both more and less reliable than the examination of stools. Thus, the examination of intestinal fluid for the presence of trophozoites should not replace, but supplement the examination of stools for the presence of trophozoites and cysts.

Histopathological detection of morphological forms confirms diagnosis, but is often a last resort because of the invasive nature of the biopsy procedure, histopathology can reveal both the presence of parasites and the underlying pathology.

1.2.4. Detection in/on Foods

Infective *G. duodenalis* cysts cannot multiply in/on foodstuffs. Food becomes a potential vehicle for human infection by contamination during production, collection, transport, and preparation (e.g., fruits, vegetables, etc.) or during processing. The sources of contamination are usually feces, fecally contaminated soil, irrigation water, water for direct incorporation into products (ingredient water), and/or for washing produce (process water) or infected foodhandlers. *G. duodenalis* cysts normally occur in low numbers in/on food matrices and in vitro culture techniques that increase parasite numbers prior to identification are not available for protozoan parasites in food and water matrices. Concentration by size alone, through membrane or depth filters, results in accumulation in the retentate, not only of cysts, but also of similar and larger sized particles, and as *G. duodenalis* cysts are (8–16 × 7–10 μm in diameter) filters with a smaller pore size are employed to entrap them. As *Giardia* and *Cryptosporidium* are often sought for at the same time, filters with pore size of 1–2 μm are employed to entrap both *Giardia* cysts and *Cryptosporidium* oocysts (*Cryptosporidium* oocysts are 4–6 μm in diameter), which results in the accumulation of large volumes of extraneous particulate material. Such particles interfere with cyst detection and identification; therefore, a clarification step which separates cysts from other contaminating particles is employed. Current methods can be subdivided into the following component parts: (a) sampling, (b) desorption of cysts from the matrix and their concentration, (c) identification (Fig. 3) and are those, or modifications of those, which apply to water *(42,43)* (Table 3).

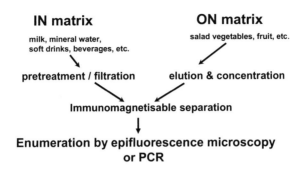

Fig. 3. Methods for detecting protozoan parasite contamination of foods.

Methods for detecting cyst contamination in liquid foods (e.g., drinks and beverages) and on solid foods (e.g., fruits and vegetables) are similar, but differ in the techniques used to separate cysts from the food matrix (Fig. 3). Differences are primarily dependent on the turbidity of the matrix and for non-turbid liquid samples, such as bottled waters, either small or large volume filtration (1–20,000 L samples have been tested in the authors' laboratory) can be used. For turbid samples (naturally turbid, e.g., milk, fruit juices, or following extraction from surfaces of vegetables and fruits), further cyst concentration using IMS is recommended.

Sampling methods, volumes/weights of product can depend on the reasons for sampling (quality control/assurance, ascertainment, compliance, incident/outbreak investigation, etc.). Desorption can be accomplished by mechanical agitation, stomaching, pulsifying, and sonication of leafy vegetables or fruit suspended in a liquid that encourages desorption of cysts from the food matrix. Detergents including Tween 20, Tween 80, sodium lauryl sulphate, and Laureth 12 have been used for desorption. Lowering or elevating pH affects surface charge and can also increase desorption. Depending on the turbidity and pH of the eluate, cysts eluted into non-turbid, neutral pH eluates can be concentrated by filtration through a 1 μm flat bed cellulose acetate membrane *(44)*, whereas cysts eluted into turbid eluates can be concentrated by IMS. Cysts can be identified and enumerated by immunofluorescence microscopy and their morphology can be assessed by Nomarski differential interference contrast microscopy.

PCR-based methods developed for detecting *Giardia* DNA in feces have been used to determine the presence of *Giardia* DNA in food eluates, but currently, there is little standardization available. Furthermore, the presence of inhibitors in foodstuffs and waters limits the application of these methods. Inhibitors can be reduced using kit-based technologies (Cartagen products, USA *Plant DNA Isolation Kit and Food DNA Isolation Kit*; Qiagen UK, automated DNA extraction), but the cost of these methods may influence their usefulness for routine analysis. Heller et al. *(46)* compared four commercial kits for extracting DNA from seeded *Escherichia coli* O157:H7 from four foodstuffs and found that they all worked similarly. These commercial kits may prove useful for extracting DNA from foodborne *Giardia* cysts. Neither the DNA-based nor microscopy-based detection methods identified earlier can be used for determining cyst viability.

Table 3
Reported Experimental Recovery Rates of *Giardia* Cysts from Foods

Food matrices	Extraction and concentration methods	Detection method	Recovery rates	Reference
• Four leafy vegetables and strawberries	(Experimental recoveries) Rotation and	Immuno-fluorescence	• 67%	*(45)*
• Bean sprouts	sonication in elution buffer; centrifugation and IMS		• 4–42%	

1.2.5. Determination of Cyst Viability

Little can be inferred about the likely impact of cysts detected in water or food concentrates on public health without knowing whether they are viable or not. *Giardia* cyst infectivity assessments can be undertaken in vivo in neonatal mice or adult gerbils *(47)*, whereas *Giardia* cyst viability (reviewed in ref. *[48]*) can be assessed in vitro by (a) excystation *(49,50)*, (b) fluorogenic vital dyes *(20,50–55)*, (c) propidium iodide vital dye staining and morphological assessment of cysts observed under DIC optics *(50)*, or (d) reverse transcriptase (RT)-PCR, amplifying *Giardia* heat shock protein 70 (*hsp 70*) messenger RNA (mRNA) *(31,56)*.

1.2.5.1. Determination by Morphology. Rapid, objective estimates of parasite viability revolve around the microscopical observation of the inclusion or exclusion of specific fluorogens into individual cysts (reviewed in ref. *[48]*). The fluorogens, fluorescein diacetate (FDA), and propidium iodide (PI) have been used to determine *G. muris* cyst viability. Cysts which included FDA and hydrolyzed it to free fluorescein caused infection in neonatal mice, whereas cysts which included PI had damaged membranes and were incapable of causing infection in neonatal mice *(51)*. *G. muris* cysts are often used as surrogates for *G. duodenalis* cysts, because they are readily available. There is good correlation between *G. muris* cyst morphology, animal infectivity, and inclusion/exclusion of the fluorogenic vital dyes *(51–53)*, but not between FDA inclusion and the viability of purified, human-derived *G. duodenalis* cysts as determined by in vitro excystation *(50)*. Compared with in vitro excystation, Smith and Smith *(50)* found that PI inclusion consistently underestimated dead human-derived *G. duodenalis* cysts, and suggested that a combination of PI inclusion/exclusion and assessment of morphology by DIC, according to the criteria of Schupp and Erlandsen *(52)* was the most suitable method to determine whether cysts were dead. *Giardia* cysts exposed to lethal doses of chlorine fail to take up PI *(54)*, and the exclusion of PI immediately following chlorine disinfection is probably due to the gradual loss of trophozoite membrane integrity following chlorine killing *(21)*. Therefore, a combination of PI inclusion/exclusion and assessment of morphology by DIC, according to the criteria of Schupp and Erlandsen *(52)* is the most satisfactory way to proceed *(59)*.

Taghi-Kilani et al. *(55)* sought alternatives to PI staining, and assessed the viability of untreated, heat killed, and chemically inactivated *G. muris* cysts by in vitro excystation and animal infectivity in CD-1 mice. The nucleic-acid stain SYTO®9 (Molecular

Probes, Eugene, Oregon, USA) stained dead cysts brightly, and had a relatively slow rate of decay in its visible light emission following DNA binding. Furthermore, staining correlated with animal infectivity. The Molecular Probes Live/Dead BacLight kit also showed correlation with animal infectivity and both staining regimes showed better correlation with infectivity than with in vitro excystation. Care should be taken when yeasts are also present in the same sample, as the ability to differentiate these from cysts requires careful morphological and morphometric assessment.

Metabolic activity can be measured with tetrazolium salts, and the commercially available fluorescent formazan 5-cyano-2,3-ditolyl tetrazolium chloride (CTC) has been used to determine the presence of viable bacteria and their metabolic activity in various aquatic and non-aquatic environments (58). CTC reduction by viable *Giardia* cysts (58) could become a rapid cyst viability assay and, when combined with FITC-*G*-mAbs, would enable the simultaneous detection of viable (redox activity) *Giardia* cysts.

Biophysical methods have also been used to determine *G. duodenalis* cyst viability: both dielectrophoresis and electrorotation have been used to demonstrate differences between viable and non-viable cysts (59). Currently, there is no consensus as to which method to adopt as each has its limitations (reviewed in ref. [48]).

1.2.5.2. Determination by Molecular Methods. The lack of consistency with fluorogenic vital dyes has led to the investigation of other approaches. mRNA is reported to have a short half-life within viable cells, is rapidly degraded by specific enzymes (RNAses) and appears to be a highly suitable target for detecting viable cells. mRNA increases following excystation, and PCR amplification of giardin mRNA before and after excystation was used to discriminate between live and dead *Giardia* cysts (60). External insults lead to the synthesis of *hsp 70* mRNA in viable organisms and RT-PCR of *Giardia hsp 70* (GHSP) mRNA has been used to determine cyst viability (31,56). While Abbasedegan et al. (58) could detect one *Giardia lamblia* cyst, the limit of sensitivity for cyst viability was 10 cysts, but Kaucner & Stinear (31) reported that this RT-PCR using the GHSP primers was inconsistent. *Giardia* cysts and *C. parvum* oocysts were detected simultaneously in water and wastewater by RT-PCR (31), with a sensitivity of a single viable organism in raw water concentrates and a comparison of methods showed that *Giardia* spp. cyst detection by immunofluorescence was 24% when compared with 69% by RT-PCR.

1.2.6. Determination of Infectivity

1.2.6.1. In Vitro Culture. Cysts which are relatively free from bacterial contaminants are required to initiate new isolates of axenically cultured trophozoites, but in vitro culture is not an option for the small number of cysts recovered from water and food. *G. duodenalis* trophozoites are aerotolerant anaerobes that can be cultured, axenically in a medium-containing cysteine and mammalian bile. Keister's modification of TYI-S-33 (61) supplemented with either 10% (v/v) sterile bovine or equine serum supports the growth of *G. duodenalis* isolates from human and nonhuman hosts. Cysteine protects trophozoites from the lethal effects of oxygen, and in combination with ascorbic acid, this effect is enhanced. However, only cysteine supports the growth of trophozoites. Cysteine initiates trophozoite attachment to glass or plastic in vitro, and the subsequent adherence (62). Attached trophozoites are detached by chilling the culture vessel for 10–15 min in ice-water, then the vessel is inverted gently several times. Suspended trophozoites are transferred aseptically to sterile tubes containing complete medium pre-warmed to room temperature or 35°C. Inocula of 4×10^3 trophozoites/ml yield 1×10^6 trophozoites/ml in 72 h, and subculturing can be performed twice weekly.

1.2.6.2. In Vivo Infectivity. A variety of laboratory animals have been used to determine *Giardia* infectivity. Both *G. muris* cysts from infected laboratory hosts and human-infective *G. duodenalis* cysts have been used to determine the disinfection sensitivity of *Giardia* cysts, although both the infection models and disinfectant sensitivity can differ.

Giardia lamblia infectivity can be determined in the Mongolian gerbil (*Meriones unguiculatus*) *(63–68)*, whereas *G. muris* infectivity can be determined in the C3H/HeN mouse model *(68,69)*. Neonatal mice *(70)* and rats *(71)* can be infected with *G. duodenalis* cysts. The minimum number of infective cysts required to establish infection in 50% of gerbils tested (ID_{50}) is 100 *(66)*, whereas the ID_{50} in gerbils, calculated by probit analysis, is 2.45 *(67)*. The *G. muris* mouse model has a similar sensitivity *(69)*. O'Grady and Smith *(48)* review *Giardia* in vivo infectivity.

2. TRANSMISSION AND RESERVOIRS OF INFECTION

2.1. Routes of Exposure

The transmission of *Giardia* to humans can occur through any mechanism by which material contaminated with feces containing infectious cysts from infected human beings or animals can be swallowed by a susceptible host. Some risk factors associated with *Giardia* infection are presented in Table 4.

Giardia is the most commonly detected intestinal protozoan parasite in the world, indicating its adaptation to numerous situations. The prevalence of giardiasis in developing countries is approximately 20% and approx 5% in the developed world *(73)*. Between 100,000 and 2.5 million *Giardia* infections occur annually in the United States *(74)*. The variable course of the infection *(13,75,76)* and the difficulty in diagnosing infections because of intermittent cyst excretion patterns are two important factors that contribute to the complexity of *Giardia* epidemiology and the difficulties in tracing sources and routes of infection.

Chronic and/or repeated infections appear to be common *(77)* and the development of immunity is variable. In instances of partial immunity, prolonged carriage may occur. In addition, the frequency of asymptomatic *Giardia* infections in not well known. The importance of asymptomatic excretors and symptomatic, undiagnosed excretors in disseminating infection is not known. Giardiasis affects all age groups, but in non-endemic areas a bimodal age distribution is often seen, with the incidence being highest in children aged 0–5 yr, followed by adults aged 31–40 yr. This correlates with reports of disease prevalence being higher than average among children who attend daycare centers together with the family members and daycare workers who care for the infected children *(78)*. *Giardia* prevalence ranges from 2 to 5% in industrialized nations and from 20 to 30% in developing nations and up to 35% prevalence has been reported among children attending daycare centers in the United States in non-outbreak settings *(79,80)*.

Giardiasis can be seasonal, but its seasonality is not as pronounced as cryptosporidiosis. In the USA, the increased transmission occurs during the summer months, which coincides with the summer recreational water season possibly reflecting the increased use of community bathing venues by young children and the higher prevalence of

Table 4
Risk Factors Associated with *Giardia* Infection

Risk factor	Association	Postulated reasons
Age	Children are more likely to be infected than adults and excrete cysts in larger numbers	1. Behaviour: greater risk of exposure 2. Immunology: lack of immunity 3. Diagnosis: infections in adults may more frequently be asymptomatic and thus less likely to be diagnosed
Gender	Some studies indicate males are more likely to be infected than females	1. Behaviour: greater risk of exposure 2. False result due to biased sampling
Nutritional status	Malnourished individuals more likely to be infected than well-nourished individuals	1. Hypochlorhydria 2. Reduced intestinal immune functions 3. Low enzyme activity 4. Poor intestinal motility
Breast-feeding	Non-breast-fed infants more likely to be infected than breast-fed infants	1. Contamination: greater risk of exposure to cysts in bottled milk 2. Immunology: acquisition of antibodies in breast milk 3. Cysticidal properties of breast milk
Diet	Consumption of leafy vegetables may be a risk factor	1. Often consumed raw 2. May be washed in contaminated water
Urban/rural	Some studies suggest city dwellers more likely to be infected than those in rural areas	1. High population density and overcrowding 2. Poverty and poor sanitation
Seasonality	Incidence may increase during cooler and wetter seasons	Cysts more likely to survive under cool, damp conditions
Socio-economic status	Associated with lower socio-economic status	1. Poverty and poor sanitation 2. Overcrowding 3. Inadequate water supplies 4. Lack of health education 5. Use of night soil as fertilizer
Socio-economic status	Associated with higher socio-economic status	1. Underreporting in low socio-economic groups as reduced access to healthcare 2. Access to international travel

From ref. *(72)*.

diapered children in swimming venues *(74)*. Limiting factors for cyst survival include high temperatures and desiccation *(28)*.

Warhurst and Smith *(81)* commented upon the global cost of *Giardia* infection, both in human and financial terms (in the USA in 1991, giardiasis accounted for >4000 hospital admissions annually costing over US$5 million) and emphasized the need to acquire pertinent data to enable strategic targeting of research funds with the ultimate aims being prevention and control. The ability to genotype and subtype *Giardia* fulfils some of these requirements by providing molecular tools to identify sources of infection within populations and routes of transmission.

2.1.1. Person-to-Person

Person-to-person transmission of *G. duodenalis* is the major route of infection. Person-to-person transmission has been documented between family/household members, homosexual partners, health workers, and their patients, children in daycare centers and other institutions and travelers to areas where disease is endemic. Persons at greatest risk of exposure are children in daycare centers, their close contacts, homosexual partners, and travelers. Transmission in daycare centers is particularly common *(78)* probably due to the lower standards of personal hygiene exhibited by preschool children and their tendency to put numerous objects that they handle in their mouths.

2.1.2. Animal-to-Person (Zoonotic)

Whether zoonotic transmission occurs frequently remains a topic of controversy. The widespread distribution of *G. duodenalis* in a variety of domestic and wild animals is suggestive of the potential for this route of transmission and has aroused considerable debate, curiosity, and some measure of concern, yet definitive evidence that this route of transmission is a public health concern remains elusive. Transmission of nonhuman isolates to humans has been demonstrated experimentally *(82)*, but the regularity with which this occurs outside the laboratory environment remains unknown. Evidence for and against zoonotic transmission has been reviewed frequently *(83–90)*.

Molecular evidence incriminating domestic animals is available *(91,92)* and the focus of attention has largely been upon livestock and feral animals. Although a range of aquatic animals harbor *G. duodenalis* cysts, their role in the transmission of giardiasis to humans requires further elucidation.

In a waterborne outbreak of giardiasis in British Columbia, Canada, cysts were collected from outbreak cases, the contaminated water supply, and beaver epidemiologically linked to the outbreak or inhabiting lodges close to where the outbreak occurred and analyzed by isoenzyme analysis and pulsed-field gel electrophoresis (PFGE) *(93)*. Isoenzyme electrophoresis indicated that all samples were in the same zymodeme and PFGE revealed that they were of the same karyotype, although PFGE discriminated between isolates within zymodemes. These results provided evidence of the potential of mammals, such as beavers for transmitting *Giardia* to humans, and suggests that a beaver was probably the source of the drinking water contamination and hence of the waterborne outbreak of giardiasis.

Transmission among nonhuman hosts, which also augments environmental contamination, is most likely to be direct, between infected animals, as environmental levels of cysts on farms appear to be too low to account for the high levels of infection seen in cattle *(3,94)*. While assemblage A & B (III) genotypes (Table 2) pose the greatest risk for zoonotic transmission, it should be noted that the occurrence of these genotypes in human and nonhuman hosts is not conclusive evidence of zoonotic transmission. Interestingly, the nonhuman-specific genotypes appear to be more host limited, and, as yet, there is no epidemiological evidence to indicate that they occur (infrequently) in humans.

2.1.3. Drinking Water

Giardia is the most commonly identified agent of waterborne disease in the USA with over 120 waterborne outbreaks affecting more than 25,000 persons, since 1965 *(44)*. According to Bennett et al. *(95)*, 60% of all *Giardia* infections in the USA are acquired from contaminated water, and both human and nonhuman sources of contamination have been implicated in waterborne outbreaks of giardiasis *(96,97)*. In the USA, *Giardia* is a significant waterborne pathogen, and was the most commonly identified pathogen in waterborne outbreaks, accounting for 18% of outbreaks resulting in 24,124 cases during the period 1971–1985 *(98)*. From 1976 to 1980, waterborne outbreaks of giardiasis accounted for 25% of optionally reported cases *(98–100)*.

Giardia cysts occur commonly in surface waters *(101,102)* (>80% of surface water samples in North America can contain *Giardia* cysts) *(103)*. Inadequate water treatment, ineffective filtration or pre-treatment of surface waters, and inadequate disinfection when disinfection was the only treatment are given as the deficiencies responsible for the majority of these waterborne outbreaks *(97,98)*. Although filtration is effective in removing *Giardia* cysts, in the USA as many as 21 million individuals may be at risk of giardiasis since their potable water comes from unfiltered water supplies, and in developing countries relatively few individuals will have access to drinking water that has received any treatment.

2.1.4. Air

The small size of *G. duodenalis* cysts renders them aerosolizable in large particle droplets (>5 µm diameter).

2.1.5. Food

Foodborne transmission of giardiasis was suggested in the 1920s *(104,105)*, and anecdotal evidence from outbreaks frequently implicated foodhandlers and contaminated fruits and vegetables *(107)*. Foodborne outbreaks are reviewed in Section 3.

The widespread distribution of cysts in the environment, enhance the possibility of food-borne transmission of giardiasis. Whilst the waterborne route is clearly of major importance, the potential for foodborne transmission should not be underestimated and several cases of foodborne giardiasis have been documented (*see* Section 3.). It is frequently difficult to associate an outbreak with a particular food item and how the food implicated became contaminated. Because of these difficulties, acquisition of *Giardia* infection via the food-borne route is almost certainly underdetected, probably by a factor of 10 or more *(107)*.

Other sources of food contamination include washing salad vegetables in water-containing infectious cysts, the use of excrement (night soil) for fertilizer, contaminated irrigation water in the cultivation of food crops, and the dissemination of cysts from feces to food by filth flies *(108,109)*. Vegetables can be surface contaminated with *Giardia* cysts *(110–112)*, although it is often unclear whether cyst contamination was due to cultivation practices, contaminated water, insects or other animal vectors, or following handling by individuals with cyst-contaminated hands. Raw wastewater, used as fertilizer, but not treated wastewater or fresh water, was responsible for contaminating coriander, mint, carrots, and radish in a study conducted in Marrakech, Morocco *(113)*.

Cyst-contaminated river waters act as a source of marine environment contamination and rivers polluted by anthropogenic and livestock fecal discharges can play a major role in oocyst contamination of shellfish in estuaries and coastal environments (see also Tables 3 and 6 of the *Cryptosporidium* chapter). Cysts of the common human genotype (*G. duodenalis* Assemblage A) were identified in clam (*Macoma balthica* and *M. mitchelli*) tissues. *Macoma* spp. clams burrow in mud or sandy-mud substrata and preferentially feed on the surface sediment layer. Although *Macoma* spp. clams have no economic value, they can serve as biologic indicators of sediment contamination with *Giardia* sp. cysts of public health importance *(114)*.

Whether seasonal variation in surface contamination of foods occurs, requires further investigation; however, seasonal peaks in parasitism will influence when water and foods become surface contaminated. The widespread distribution of cysts in the environment, the rapid transportation of foods from global markets, and their chilling and wetting, which augment parasite survival, enhance the likelihood of foodborne transmission. Water and food enhance cyst survival by preventing their desiccation.

2.1.6. Other Routes of Exposure

Venereal transmission of *Giardia* has been reported among homosexual males *(80)*. Travel to endemic areas of disease, backpacking, and camping, drinking untreated water from shallow wells, lakes, rivers, springs, ponds, and streams, and transmission via the recreational water route (swimming pools, water parks, fountains that use recirculating water, lakes, rivers, and oceans) are significant routes of exposure. *Giardia* cysts are less resistant to disinfectants that *C. parvum* oocysts but swimming pools, particularly children's paddling pools and the ingestion of pool water following either accidental defecation events have resulted in outbreaks. The increased use of recreational waters for immersion sports, in cyst-contaminated fresh and marine waters, also increases the likelihood of occurrence of this transmission route.

Cysts are redistributed to uncontaminated sites associated with human activities by coprophagous transport hosts (e.g., pigs, dogs, chicken, refuse feeding and aquatic birds and filth flies) *(21,109)*. Filth fly and avian transport hosts can travel over large distances. Filth flies ingest 1–3 mg feces over 2–3 h *(115)*, and can transmit *G. duodenalis* cysts attached to their exoskeleton and in their excrement *(109)*. The sources of contamination of human-infectious *G. duodenalis* are primarily point sources, such as infected hosts and waste water effluents but muck spreading, slurry spraying, and run off from contaminated land, etc., containing cysts from nonhuman hosts, also contaminate food and water. Rapid climate changes due to global warming, including temperature fluctuations, rainfall, and changes in water table levels also influence the distribution of cysts in the environment.

2.2. Cyst Survival in Water

Nichols and Smith *(116)* cite data on the resistance of *Giardia* cysts to physical and chemical treatments. *G. muris* cysts can survive for more than 3 mo in cold surface water; however, freezing (−13°C) for 14 d and thawing reduces cyst viability to <1%, whereas storage of cysts at 8°C for 77 d reduces *G. intestinalis* excystation to <5%. Some experimental survival and inactivation data for *Giardia* cysts are presented in Table 5.

3. FOODBORNE OUTBREAKS

The most frequent mechanism by which food becomes contaminated with infective cysts is from foodhandlers who have either been in contact with infected feces from another individual or are themselves infected (Table 6). Foltz and Harding *(117)* suggest that foodhandlers frequently come from lower socio-economic groups where personal hygiene and public health knowledge are poor. Whether this is correct, foodhandlers with asymptomatic or minimally symptomatic infections may not recognize their risk in transmitting infection. Here, education is the obvious option. An interesting route is the ingestion of cysts in an undercooked animal product, where the animal itself was infected,

Table 5
Some Conditions for Inactivating *Giardia* cysts

Physical inactivation	Chemical inactivation
• Freezing (−13°C) for 14 d and thawing reduces cyst viability to <1%	• Free chlorine: up to 142 mg/l min^{-1} required for 99% *G. intestinalis* inactivation
• Storage of *G. intestinalis* cysts at 8°C for 77 d reduces excyation to <5%	• At 5°C (pH 8), 2 mg/ml for 30 min produces <30% *G. intestinalis* cyst inactivation. At 25°C (pH 8), 1.5 mg/ml for 10 min produces >99% cyst inactivation
• *G. muris* cysts survive for 3 mon in cold raw water sources	• Chlorine dioxide: 11.2 mg/l min^{-1} required for 99% cyst inactivation
	• Ozone: up to 2.57 mg/l min^{-1} required for up to 4 $\log_{10}$ *G. intestinalis* inactivation

Adapted from ref. *(116)*.

Table 6
Some Studies of Occurrence of Giardiasis Amongst Foodhandlers

Location of foodhandlers	Prevalence of giardiasis detected amongst foodhandlers	Reference
Collegiate institution, USA	11.7%	*(119)*
U.S. Army, Korea	4.4%	*(120)*
Santiago, Chile	15.9%	*(121)*
Hospitals in Santiago, Chile	8.7%	*(122)*
Concepcion, Chile	7.7%	*(123)*
UK food factory	3.5%	*(124)*

as suggested in a case in which members of two families developed giardiasis, apparently as a result of consuming undercooked sheep tripe in soup *(118)*.

Eight outbreaks of foodborne disease have been documented (Table 7) and the potential for foodborne transmission should not be underestimated. In two out of the eight outbreaks, contaminated foodstuffs were implicated as the vehicles of transmission; but, in the remaining six, foodhandlers were implicated, implying that contamination occurred probably during food preparation and that, until then, the foodstuffs were free of contaminating, infectious cysts. That transmission was not due to indigenous contamination of the foodstuffs, but to mishandling by foodhandlers manipulating the foodstuffs during food preparation, is an important distinction that is not always recognized. Immaterial of this difference, such outbreaks are categorized as foodborne, because the vehicle of transmission was food, but such scenarios place an unfair public perception on specific food production practices that are effectively quality assured.

Other sources of contamination include washing salad vegetables in cyst-contaminated water, use of excrement (night soil) for fertilizer in the cultivation of food crops and dissemination of cysts from feces to food by mechanical vectors *(108,109)*. Vegetables sold in markets have been contaminated with *Giardia* cysts *(110,111)*.

Table 7
Some Documented Foodborne Outbreaks of Giardiasis

No. of persons affected	Suspected food-stuff	Probable/possible source of infection	Reference
3	Christmas pudding	Rodent feces	*(125)*
29	Home-canned salmon	Foodhandler	*(126)*
13	Noodle salad	Foodhandler	*(127)*
88	Sandwiches	–	*(128)*
10	Fruit salad	Foodhandler	*(129)*
–	Tripe soup	Infected sheep	*(118)*
27	Ice	Foodhandler	*(130)*
26	Raw sliced vegetables	Foodhandler	*(131)*

4. PATHOGENICITY (VIRULENCE FACTORS)

4.1. Infectious Dose

In a human volunteer study, the median infectious dose for *Giardia* was between 25 and 100 cysts, although as few as 10 cysts initiated infection in two out of two volunteers *(75,76)*. One volunteer study demonstrated that a human-source isolate can vary in its ability to colonize other humans *(13)*, suggesting that certain isolates may be less infectious to humans, or cause fewer clinical signs and symptoms than others.

4.2. Pathogenicity

Much debate has focused on the pathogenicity of *Giardia*. Signs and symptoms range from self-limiting diarrhea to severe chronic disease. Immunocompetent individuals with giardiasis can exhibit some or all the signs and symptoms identified in Section 5, but infected individuals with immunodeficiency or underlying protein malnutrition can develop more severe disease, including interference with fat and fat-soluble vitamin absorption, retarded growth, weight loss, and a "coeliac-disease-like" syndrome *(28)*.

Trophozoites are not invasive and while they cause malabsorption by physically occluding the enterocyte brush border; some pathology is due to inflammation targeted at enterocytes *(132)*. Inflammation results in an increased turnover rate of enterocytes and immature, replacement enterocytes have less functional surface area and less digestive and absorptive ability than the mature ones they replace *(132)*. Toxin production *(133)* may also upregulate inflammation and increase intestinal barrier permeability. Colonization reduces the mucosal absorptive surface area in the small intestine, which leads to disaccharidase deficiencies and decreased absorption of electrolytes, nutrients, and water. Together, these abnormalities are responsible for malabsorption and diarrhea *(134,135)*. Iron deficiency is commonly associated with a high incidence of parasitism in children in developing countries, and iron malabsorption and anemia are commonly outcomes of giardiasis *(136,137)*.

4.2.1. Effects on Epithelia

Disaccharidase deficiencies have been reported in giardiasis patients with only mild abnormalities of mucosal architecture *(138–141)*. Brush border enzyme abnormalities result from a diffuse shortening of brush border microvilli along the entire villus axis, both at the sites of trophozoite attachment and in other uncolonized areas *(140,142)*. The diffuse loss of overall absorptive area also reduces sodium-coupled glucose absorption, active glucose uptake, and water absorption along the small intestine *(142)*. These factors combine to produce diarrhea. Infection can also result in reduced lipase activity *(143)* which explains the occasional steatorrhea associated with giardiasis. *Giardia*-induced alterations in epithelial permeability are associated with upregulated enterocyte apoptosis *(144)*, but the significance of this remains unclear.

4.2.2. Effects on the Epithelial Cytoskeleton

The loss of microvillus height rather than villus atrophy appears to be the rate-limiting injury responsible for impairing brush border digestive enzymatic activity and reducing active epithelial absorption in giardiasis. The enterocyte cytoskeleton supports the cell structure and helps maintaining epithelial polarity, but rearrangement of cytoskeletal F-actin, following *Giardia* exposure, leads to epithelial pathophysiology *(145)*. Infection induces focal loss of peripheral α-actinin, and trophozoites may cause rearrangement of other cytoskeletal components *(145,146)*. These observations indicate that *Giardia* excretory-secretory products significantly alter enterocyte cysoskeletal proteins, which contribute to changes seen in mucosal architecture. Fecal neopterin concentrations (a marker of gut inflammation) and lactulose mannitol (L:M) absorption ratio (a measure of intestinal permeability) are useful indicators of the long-term effects of *Giardia* and other intestinal parasites on infant development *(147)*.

4.3. Immune Response

Host–parasite interactions in human giardiasis are complex. Intestinal trophozoite populations can be controlled by both innate (natural) and acquired immunity. Normal human milk, conjugated bile salts, unsaturated fatty acids, and free fatty acids are giardicidal, however, intestinal mucus protects trophozoites from this giardicidal effect. Trophozoites are killed by products of lipolysis present in human duodenal and upper jejunal fluid, and human neutrophil defensins and indolicidin in vitro. In acquired immunity, both humoral and cellular arms of the immune system play roles in controlling infection. IgM-, IgA-, and IgG-specific antibodies play major roles, as do T-cell subsets, macrophages, and neutrophils. Secretory IgA antibody is present in human milk. The human humoral antibody response is isolate dependent *(13)*. Complement alone, or in the presence of IgM anti-trophozoite antibody kills >98% of trophozoites in vitro (reviewed in ref. *[26]*). The immune response is also involved in intestinal mucosal pathology, but few studies have assessed the role of cytokines.

Symptomatic individuals have elevated humoral antibody responses to trophozoite antigens. IgG antibodies to surface expressed trophozoite antigens occurred in 81% of symptomatic cases and 12% of controls *(148)*, and, following chemotherapy, IgG antibody remained detectable in most cases for up to 18 mon. In two out of three symptomatic individuals studied longitudinally, IgM values fell to control levels between 2 and 3 wk after chemotherapy, indicating that IgM antibody might be an useful indicator of current infection *(149)*. Both immunofluorescence and ELISA are comparable for detecting IgA and IgG anti-*Giardia* antibodies in the sera of individuals with proven infection. ELISA sensitivity varies with the antibody isotype detected, IgM being most sensitive *(26)*, and intact trophozoites provide greater sensitivity than trophozoite extracts when used as antigen in the IgG ELISA. The western blot assay cannot detect antibodies in all samples from patients with proven giardiasis, although higher circulating antibody titers occurred in sera from symptomatic patients when compared with sera from asymptomatic patients. Western blot assay sensitivity increases when purified *Giardia* proteins are used as antigens *(26)*.

The usefulness of serological assays for diagnosing human giardiasis is debatable because (a) different geographical isolates have different antigenic identity and (b) in chronic disease, parasites cause immunodepression, in the absence of the acute, immuno-responsive stage of the disease, and (c) antigenic variation downregulates antibody production *(26)*. Serology fails to reveal differences in serum antibody responses between symptomatic and asymptomatic patients *(26)*. Even though numerous immunoreactive *G. duodenalis* trophozoite antigens have been described, a common, non-variant, immuno-dominant antigen, useful for serodiagnosis, has not been identified. Palm et al. *(150)* suggest that the *Giardia*-specific α-giardins could prove useful for the serodiagnosis of acute giardiasis. While drug intervention in infected cases reduced antibody titers to recombinant α-giardins, treatment failure or chronic infection or reinfection resulted in higher antibody titers *(150)*.

The prevalence of giardiasis is increased in malnourished patients *(151)*, and while serum antibody responses in malnutrition are frequently normal, the level of secretory IgA antibody on mucosal surfaces is reduced *(151)*, which adversely affects parasite elimination.

Serological evidence of recent *Giardia* infection could be an useful epidemiological tool for water and foodborne outbreak investigations, where giardiasis is not endemic. Seroepidemiological surveys of trophozoite surface-specific serum IgG antibody prevalence indicate rates ranging from 18% in an inner city area in the USA to 48% in a rural area of Panama *(152)*. *Giardia*-specific serum IgA antibody responses have been used to determine exposure to *Giardia* contaminated water and illness from giardiasis during waterborne outbreaks of diarrheal disease *(153)*. The recombinant protein α-giardin *(150)* might prove a useful target for further studies, depending upon the antibody isotype, subclass and avidity responses developed against them during infection and disease.

4.4. Virulence

Variation in the infectivity and virulence of *G. duodenalis* isolates *(13,66)* and different clones from a single isolate *(154)* has been described. Some isolates instigate disaccharidase deficiencies in CaCo2 cell monolayers, whereas others, of similar challenge inocula, do not *(155)*.

4.4.1. Giardia *Toxin*

The possibility that some pathophysiological changes observed at the intestinal mucosa during giardiasis is caused by a *Giardia* toxin is an attractive hypothesis. The presence of a *Giardia* gene encoding a sarafotoxin-like protein, which shared 57% similarity with the gene encoding the precursor of sarafotoxins has been reported *(156)*, and a 58 kDa excretory–secretory product (ESP) has enterotoxic activity *(133)*. Epithelial cells incubated with purified ESP showed elongated morphology, then lysed, therefore, 58-kDa ESP is a candidate for the *Giardia* toxin. The 58-kDa ESP is immunogenic and anti-58kDa ESP antibodies, raised against the purified glycoprotein, cross-react with the binding subunit of commercially available cholera toxin *(157)*. Although the mechanism of ESP-induced cytopathic and enterotoxic effects is not yet understood, *G. duodenalis* virulence may be dependent on the parasite's environment. Interaction between parasite and intestinal microbiota may be important in augmenting pathogenicity *(157,158)*.

5. CLINICAL CHARACTERISTICS

Since its discovery of 300 yr ago, *G. duodenalis* is recognized as a cosmopolitan protozoan parasite of humans, and is ranked in the top 10 of human parasitic diseases *(159)*. Giardiasis is primarily a disease of the upper small intestine caused by noninvasive, reproductive trophozoites which attach onto enterocytes. In most instances, the disease is self-limiting. Clinically, there are two disease phases: the acute and the chronic. The acute phase is usually short-lived, characterized by flatulence with sometimes sulfurous belching, and abdominal distention with cramps. Diarrhea is initially frequent and watery but later becomes bulky, sometimes frothy, greasy, and offensive, and stools may float on water. Blood and mucus are usually absent and pus cells are not a feature on microscopy *(18,72,134,159)*.

In chronic giardiasis, malaise, weight loss, and other features of malabsorption become prominent. By this time, stools are usually pale or yellow and are frequent and of small volume. Occasionally episodes of constipation intervene with nausea and diarrhea precipitated by the ingestion of food. Malabsorption of vitamins A and B12 and D-xylose can occur. Disaccharidase deficiencies (most commonly lactase) are frequently detected in chronic cases. In young children, "failure-to-thrive" is frequently due to giardiasis, and all infants being investigated for causes of malabsorption should have a diagnosis of giardiasis excluded *(44,135)*.

The prepatent period (the time from infection to the initial detection of parasites in stools) is on average 9.1 days *(75,76)* and the incubation period is usually 1–2 wk. As the prepatent period can exceed the incubation period, initially a patient can have symptoms in the absence of cysts in the feces. Cyst excretion can approach 10^7/g feces *(160)*. In severe infections up to 14 billion parasites can occur in a diarrheal stool, whereas in a moderate infection up to 300 million parasites are present.

There are few reports on giardiasis in immunocompromised hosts. Infection is more severe in hypogammaglobulinemic individuals, but not normally in those with HIV/AIDS. Cyst prevalence is significantly higher in hypogammaglobulinemics than in immunocompetent hosts *(26)*, and approx 90% of hypogammaglobulinemics passing cysts are symptomatic with chronic diarrhea *(161)*. Giardiasis is always symptomatic in hypogammaglobulinemic children *(162)* and more prevalent in malnourished patients *(151)*.

6. CHOICE OF TREATMENT

6.1. Drugs

Giardia infections can be self-limiting, however, a number of factors predispose to treatment. Infection, if left untreated, can last several months to years, re-infection is not uncommon and symptoms associated with long-term infection can be debilitating. In the immunosuppressed, infection can become increasingly chronic. Thus, chemotherapy is of major importance. Several drugs, including nitroimidazole compounds, quinacrine and furazolidone are available for treatment, with reported cure rates between 75 and 95% *(80,163)*. Metronidazole has been the treatment of choice for over 40 yr *(164)*. Its mode of action is thought to involve drug activation by a one electron-reduction step, involving the enzyme pyruvate ferredoxin oxidreductase and ferredoxin *(165–167)* (Fig. 4).

A combination of metronidazole and quinacrine can be used to treat refractory cases *(168)*. Other nitroimidazoles, such as tinidazole, are also effective and widely used, globally. Tinidazole is not approved for use in the USA.

Nitazoxanide (NTZ) is as effective as metronidazole and NTZ has recently received FDA approval for the treatment of giardiasis in children *(169,170)*. Alternative anti-*Giardia* drugs include furazolidone, which can cause hemolysis in individuals with glucose-6-phosphate dehydrogenase deficiency, and quinacrine, which is poorly tolerated because of nausea, vomiting, and cramping side effects. Paromomycin has been used in a number of small-scale studies *(170)*.

Other drugs which are currently under investigation include ciprofloxacin *(171)* and the benzimidazole, albendazole. Albendazole, a broad spectrum anti-parasitic agent *(172)*, is as effective as metronidazole with fewer side effects among children aged 2–12 yr *(173)*. Paromomycin, a nonabsorbed aminoglycoside, is a less-effective treatment, but is used for treating pregnant women, where the potential teratogenicity of more effective drugs limits their usefulness *(174)*.

6.2. Vaccination

Mice immunized with *G. muris* trophozoite antigens and challenged with *G. muris* showed significant levels of protection, and suggested that *G. duodenalis* infections might be preventable using immunotherapy. Subunit vaccines can reduce *Giardia* infectivity and vaccination has the potential to protect animals from infection and reduce clinical signs *(175–177)*. A commercial *Giardia* vaccine (GiardiaVax™, Fort Dodge Animal Health, Overland Park, Kansas, USA) is licensed for use in dogs and cats in the USA. Efficacy studies indicated that GiardiaVax™ successfully reduced the duration of cyst shedding and the number of cysts shed in the feces of experimentally infected animals when compared with controls. At the end of the study, no trophozoites were found in vaccinated, infected animals but all unvaccinated, infected (control) animals had trophozoites throughout length of their small intestine.

Clearly, effective immunoprophylaxis should be beneficial to human and nonhuman hosts. Vaccines should eliminate the parasite or, at least, reduce the parasite load in the host, thereby preventing/reducing cyst shedding, which in turn reduces both transmission and environmental contamination of food and water. Vaccination of domestic animals should reduce zoonotic transmission.

7. RESISTANCE EPIDEMIOLOGY

Metronidazole failures have been reported in immunodeficient individuals, including those with AIDS *(178)*. In such situations, some experts repeat the treatment course with a higher metronidazole dose. In five out of six patients with giardiasis refractory to metronidazole treatment, a combination regimen of quinacrine and metronidazole resulted in cure *(170)*. NTZ could become the preferred drug as more experience is gained with its use, especially in light of the not infrequent failures of metronidazole therapy *(169)*.

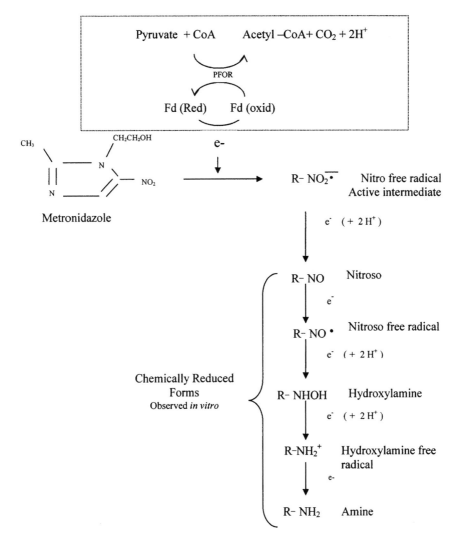

Fig. 4. Proposed pathway of metroniazole reduction in anaerobic protozoa. PFOR–Pyruvate ferredoxin oxidreductase; Fd (Red)–reduced ferredoxin; Fd (Oxid)–oxidised ferredoxin.

The in vivo mechanisms of metronidazole resistance in *Giardia* are unclear, however, they are likely to be multifactorial. The enzymatic activity of both pyruvate ferredoxin oxidreductase and the electron acceptor ferredoxin are reduced in resistant *G. duodenalis* *(166)*, however, other mechanisms including increased NAD(P)H oxidase activity (competition for reductive power) *(179)*, transport changes and gene rearrangements also appear to be involved in the development of resistance to metronidazole in *Giardia (180)*.

Effective diagnostic tools to determine resistance to metronidazole remain elusive and resistance does not correlate with specific *Giardia* genotypes *(181)*. Downregulation of ferredoxin oxidoreductase mRNA transcription is an indicator of drug resistance and induction, however, given the multifactorial nature of drug resistance *(182)*; it is unlikely that a single diagnostic test for metronidazole resistance in *Giardia* will detect all clinically resistant isolates. The fact that the putative assay would have to be performed on cysts further challenges current technology.

8. SUMMARY AND CONCLUSIONS

Members of the genus *Giardia* are all obligate parasites of animals. Morphological analysis identifies six species: *G. duodenalis*, *G. muris*, *G. microti*, *G. psittaci*, *G. ardeae*, and *G. agilis*. Of these species, the *duodenalis* type is associated with human infection. *Giardia* completes its life cycle in a single host and its taxonomy is under review. *Giardia* represents one of the most ancient lineages amongst eukaryotes and the long association between *Giardia* and its hosts has allowed, in some instances, an almost commensal relationship to be established. The lack of morphological differences between genetic variants found in mammals has resulted in an informal categorization of *G. duodenalis* based on genetic differences in specific genes. There are six assemblages (A–F) within *G. duodenalis* which suggests some host adaptation. Human infective isolates are contained within assemblage A (gp. I) & B (gps. III & IV).

Humans do not become solidly immune to reinfection following their first exposure, but do develop an anamnestic response. Different *G. duodenalis* assemblages isolates evoke different clinical outcomes in human volunteers, indicating that parasite virulence and/or pathogenicity factors affect clinical outcome. Parasite factors and the inflammatory immune response drive both parasite clearance and immunopathology. Metronidazole, tinidazole, albendazole, and NTZ appear effective in controlling giardiasis. Some evidence exists for metronidazole resistance, and NTZ, albendazole or tinidazole can be options in such situations.

Transmission to humans can occur via any mechanism by which material contaminated with feces containing infectious cysts from infected human beings or other nonhuman hosts can be swallowed by a susceptible host. Both biotic and abiotic reservoirs of infection exist. Biotic reservoirs include all potential hosts of human-infectious *G. duodenalis* assemblages, whereas abiotic reservoirs include all vehicles that contain sufficient infectious cysts to cause human infection, the most commonly recognized being food and water. Transport hosts, including gulls, wildfowl, and filth flies, should also be borne in mind when developing risk assessment questionnaires and performing risk assessments. No one diagnostic test alone has a consistently high diagnostic index, and microscopy and antigen detection tests are frequently used in conjunction to provide higher diagnostic indices. While conventional detection methods are useful for detecting symptomatic human cases, they are not sufficiently sensitive for detecting asymptomatic carriers. Furthermore, they cannot discriminate between the *G. duodenalis* assemblages that commonly infect human or nonhuman hosts. Here, molecular methods that can determine *Giardia* species, assemblages and subtypes are required and a variety of PCR-based methods are available. PCR-RFLP and DNA sequence-based methods have been used to identify species/ genotypes in clinical and environmental samples, but there is increasing interest in real-time PCR methods as they can be more sensitive and rapid.

A variety of cyst viability/infectivity methods have been described, but currently, no one method has received universal approval. Both in vitro and in vivo methods have been advocated and, depending on the requirement, each has its advantages and limitations. Standardized methods for detecting low densities of cysts in water and food are available, as are molecular methods for determining their species and assemblages. As for *Cryptosporidium* oocysts, unidentified interferents in water and food matrices reduce the efficiency of detection by PCR, and methods to relieve such interferents should be assessed in order to maximize detections from low densities of cysts. Immaterial of the method used, effective quality assurance, quality control, and validation are necessary prior to adoption. Both foodborne and waterborne outbreaks of giardiasis have been documented. In two out of eight outbreaks, contaminated foodstuffs were implicated as the vehicles of transmission but, in six, foodhandlers were implicated, implying that contamination occurred probably during food preparation and that, until then, the foodstuffs were free of infectious cysts. That transmission was not due to indigenous contamination of the foodstuffs, but to mishandling by foodhandlers manipulating the foodstuffs during food preparation, is an important distinction that is not always recognized. Whether seasonal variation in surface contamination of foods occurs, requires further investigation; however, seasonal peaks in parasitism will influence when water and foods become surface contaminated. Water and food enhance cyst survival by preventing their desiccation and the widespread distribution of cysts in the environment enhances the possibility of foodborne transmission.

Increased demand, global sourcing, and rapid transport of soft fruit, salad vegetables, and seafood can enhance both the likelihood of cyst contamination and cyst survival. A risk-based assessment based upon standard, validated methods are required for detecting *Giardia* cysts on/in food. In addition to determining genus and species of protozoan parasites present on/in foods and whether they are infectious to humans, subtyping methods are required to track outbreaks of disease and incidents and to determine the risk associated with specific assemblages/subtypes. A clearer understanding of the population biology of the parasite will assist in unraveling occurrence and prevalence of assemblages/subtypes, whereas a multidisciplinary approach will assist in unraveling the impact of foodborne protozoan on human health.

Methods for detecting cysts on foods are modifications of those used for detecting cysts in water, and, as such, have recovery efficiencies ranging from 4 to 67%. Specificity, sensitivity, and reproducibility are paramount as, unlike pre-enrichment methods that increase organism numbers for prokaryotic pathogens and indicators, there is no method to augment parasite numbers prior to detection. Cyst contamination of food can be on the surface of, or in, the food matrix and products at greatest risk of transmitting infection to man include those that receive no, or minimal, heat treatment after they become contaminated. Examples of surface contamination include salad vegetables and fruit, whereas examples of contamination within the food matrix include drinks, beverages, milk, and other foodstuffs containing naturally contaminated produce, such as bivalves. Bivalves can remove *Giardia* cysts from water and accumulate them on the gills and inside hemocytes in the hemolymph for contracted periods of time. In addition to temperature elevation ($\geq 64.2°C$ for 2 min), disinfectants (Table 5) and other treatment processes used in the food industry may be detrimental to cyst survival or lethal, but further research in this important area is required. In the absence of specific data, approaches such as hazard assessment and critical control point (HACCP) should be used to identify and control risk.

REFERENCES

1. Kunstler, J. (1882) Sur cinq protozoaires parasites nouveaux. *C. R. Seances Soc. Biol.* **95,** 347–349.

2. Filice, F. P. (1952) Studies on the cytology and life history of a *Giardia* from the laboratory rat. *Univ. Calif. Publ. Zool.* **57**, 53–146.

3. Olson, M. E., O'Handley, R. M., Ralston, B., and Thompson, R. C. A. (2004) Emerging issues of *Cryptosporidium* and *Giardia* infections in cattle. *Trends Parasitol.* **20**, 185–191.

4. Andrews, R. H., Adams, M., Boreham, P. F. L., Mayrhofer, G., and Meloni, B. P. (1989) *Giardia intestinalis*: electrophoretic evidence for a species complex. *Int. J. Parasitol.* **19**, 183–190.

5. Homan, W. L. and Mank, T. G. (2001) Human giardiasis: genotype linked differences in clinical symptomatology. *Int. J. Parasitol.* **31**, 822–826.

6. Homan, W. L., van Enckevort, F. H., Limper, L., et al. (1992) Comparison of *Giardia* isolates from different laboratories by isoenzyme analysis and recombinant DNA probes. *Parasitol. Res.* **78**, 316–323.

7. Hopkins, R. M., Meloni, B. P., Groth, D. M., Wetherall, J. D., Reynoldson, J. A., and Thompson, R. C. A. (1997) Ribosomal RNA sequencing reveals differences between the genotypes of *Giardia* isolates recovered from humans and dogs living in the same locality. *J. Parasitol.* **83**, 44–51.

8. Meloni, B. P., Lymbery, A. J., and Thompson, R. C. A. (1995) Genetic characterization of isolates of *Giardia duodenalis* by enzyme electrophoresis: implications for reproductive biology, population structure, taxonomy, and epidemiology. *J. Parasitol.* **81**, 368–383.

9. Monis, P. T., Andrews, R. H., Mayrhofer, G., and Ey, P. L. (1999) Molecular systematics of the parasitic protozoan *Giardia intestinalis*. *Mol. Biol. Evol.* **16**, 1135–1144.

10. Thompson, R. C. A. (2000) Giardiasis as a re-emerging infectious disease and its zoonotic potential. *Int. J. Parasitol.* **30**, 1259–1267.

11. van Leeuwenhoek, A. (1681) Cited in *The collected letters of Antoni van Leeuwenhoek. Edited, illustrated and annotated by a Committee of Dutch Scientists* **III,** Amsterdam Swets and Zeitlinger Ltd, 1948, pp. 345–371.

12. Kulda, J. and Nohnykova, E. (1978) Flagellates of the human intestine and of intestines of other species. In: *Parasitic Protozoa II* (Kreier, J. P., ed.), Academic Press, New York, pp. 1–138.

13. Nash, T. E., Herrington, D. A., Losonsky, G. A., and Levine, M. M. (1987) Experimental human infections with *Giardia lamblia*. *J. Infect. Dis.* **156**, 974–984.

14. Sogin, M. L., Gunderson, J. H., Elwood, H. J., Alonso, R. A., and Peattie, D. A. (1989) Phylogenetic meaning of the kingdom concept: an unusual ribosomal RNA from *Giardia lamblia*. *Science* **6**, 75–77.

15. Kabnick, K. S. and Peattie, D. A. (1991) *Giardia:* a missing link between prokaryotes and eukaryotes. *Am. Scientist* **79**, 34–43.

16. Feeley, D., Holbertson, D., and Erlandsen, S. (1990) The biology of *Giardia*. In: *Human Parasitic Diseases, Vol. 3, Giardiasis* (Meyer, E. A., ed.), Elsevier, Amsterdam, pp. 11–50.

17. Gillin, F. D., Reiner, D. S., and McCaffery, J. M. (1996) Cell biology of the primitive eukaryote *Giardia lamblia*. *Ann. Rev. Microbiol.* **50**, 679–705.

18. Smith, H. V. (1992) Intestinal protozoa. In: *Medical Parasitology: A Practical Approach* (Hawkey, P. M. and Gillespie, S. H., eds.), IRL Press, New York, pp. 79–118.

19. Paget, T. A., Jarroll, E. L., Manning, P., Lindmark, D. G., and Lloyd, D. (1989) Respiration in the cysts and trophozoites of *Giardia muris*. *J. Gen. Microbiol.* **135**, 145–154.

20. Meyer, E. A. and Jarroll, E. L. (1980) Giardiasis. *Am. J. Epidemiol.* **111**, 1–12.

21. Smith, H.V. (1998) Detection of parasites in the environment. In: *Infectious Diseases Diagnosis: Current Status and Future Trends* (Smith, H. V. and Stimson, W. H., eds., co-ordinating ed. Chappel, L. H.), *Parasitology* **117**, S113–S141.

22. Rosoff, J. D., Sanders, C. A., Seema, S. S., et al. (1989) Stool diagnosis of giardiasis using a commercially available enzyme immunoassay to detect *Giardia*-specific antigen 65 (GSA 65). *J. Clin. Microbiol.* **27**, 1997–2002.

23. Goldin, A. J., Apt, W., Aguilera, X., Zulantay, I., Warhurst, D. C., and Miles, M. A. (1990) Efficient diagnosis of giardiasis among nursery and primary school children in Santiago,

Chile by capture ELISA for the detection of faecal *Giardia* antigens. *Am. J. Trop. Med. Hyg.* **42**, 538–545.

24. Behr, M. A., Kokoskin, E., Gyorkos, T. W., Cédilotte, L., Faubert, G. M., and MacLean, J. D. (1997) Laboratory diagnosis for *Giardia lamblia* infection: a comparison of microscopy, coprodiagnosis and serology. *Can. J. Infect. Dis.* **8**, 33–38.

25. Rosoff, J. D. and Stibbs, H. H. (1986) Isolation and identification of a *Giardia lamblia*-specific stool antigen (GSA-65) useful in coprodiagnosis of giardiasis. *J. Clin. Microbiol.* **23**, 905–910.

26. Faubert, G. (2000) Immune response to *Giardia duodenalis. Clin. Microbiol. Rev.* **13**, 35–54.

27. Green, E., Warhurst, D., Williams, J., Dickens, T., and Miles, M. (1990) Application of a capture enzyme immunoassay in an outbreak of waterborne giardiasis in the United Kingdom. *Eur. J. Clin. Microbiol. Infect. Dis.* **9**, 424–428.

28. Adam, R. D. (1991) The biology of *Giardia* spp. *Microbiol. Rev.* **55**, 706–732.

29. Sawitz, W. G. and Faust, E. C. (1942) The probability of detecting intestinal protozoa by successive stool examinations. *Am. J. Trop. Med.* **22**, 130–136.

30. Rochelle, P. A., Ferguson, D. M., Handojo, T. J., De Leon, R., Stewart, M. H., and Wolfe, R. L. (1997) An assay combining cell culture with reverse transcriptase PCR to detect and determine the infectivity of waterborne *Cryptosporidium parvum. Appl. Environ. Microbiol.* **63**, 2029–2037.

31. Kaucner, C. and Stinear, T. (1998) Sensitive and rapid detection of viable *Giardia* cysts and *Cryptosporidium parvum* oocysts in large-volume water samples with wound fibreglass cartridge filters and reverse transcription-PCR. *Appl. Environ. Microbiol.* **64**, 1743–1749.

32. Mahbubani, M. H., Bej, A. K., Perlin, M., Schaefer III, F. W., Jakubowski, W., and Atlas, R. M. (1992) Differentiation of *Giardia duodenalis* from other *Giardia* spp. cysts by using polymerase chain reaction and gene probes. *J. Clin. Miocrobiol.* **30**, 74–78.

33. Slifco, T. R., Smith, H. V., and Rose, J. B. (2000) Emerging parasite zoonoses associated with food and water. *Int. J. Parasitol.* **30**, 1379–1393.

34. Caccio, S. M., De Giacomo, M., and Pozio, E. (2002) Sequence analysis of the β-giardin gene and development of a polymerase chain reaction-restriction fragment length polymorphism assay to genotype *Giardia duodenalis* cysts from human faecal samples. *Int. J. Parasitol.* **32**, 1023–1030.

35. Yong, T., Hank, K., and Park, S. (2002) PCR-RFLP analysis of *Giardia intestinalis* using a *Giardia*-specific gene, GLORF-C4. *Parasite* **9**, 65–70.

36. Amar, C. F. L., Dear P. H., and McLauchlin, J. (2003) Detection and genotyping by real-time PCR/RFLP analyses of *Giardia duodenalis* from human faeces. *J. Med. Microbiol.* **52**, 681–683.

37. Caccio, S. M., De Giacomo, M., Aulicino, F. A., and Pozio, E. (2003) *Giardia* cysts in wastewater treatment plants in Italy. *Appl. Environ. Microbiol.* **69**, 3393–3398.

38. Smith, H. V. (1996) Detection of *Cryptosporidium* and *Giardia* in water. In: *Molecular Approaches to Environmental Microbiology* (Pickup, R. W. and Saunders, J. R., eds.), Ellis-Horwood, Hemel Hempstead, HP2 7EZ, UK, pp. 195–225.

39. Mahabubani, M. H., Schafer III, F. W., Jones, D. D., and Bej, A. K. (1998) Detection of *Giardia* in environmental waters by immuno-PCR amplification methods. *Curr. Microbiol.* **36**, 107–113.

40. Verweij, J. J., Blange, R. A., Templeton, K., et al. (2004) Simultaneous detection of *Entamoeba histolytica*, *Giardia lamblia*, and *Cryptosporidium parvum* in fecal samples by using multiplex real-time PCR. *J. Clin. Microbiol.* **42**, 1220–1223.

41. Guy, R. A., Payment, P., Krull, U. J., and Horgen, P. A. (2003) Real-time PCR for quantification of *Giardia* and *Cryptosporidium* in environmental water samples and sewage. *Appl. Environ. Microbiol.* **69**, 5178–5185.

42. Anonymous. (1998) Method 1623, *Cryptosporidium* in water by filtration/IMS/FA. United States Environmental Protection Agency, Office of Water, Washington. Consumer confidence reports final rule. Federal Register 63, 160.

43. Anonymous. (1999) *UK Statutory Instruments No. 1524. The Water Supply (Water Quality) (Amendment) Regulations 1999. The Stationery Office, Ltd*, 5pp.
44. Girdwood, R. W. A. and Smith, H. V. (1999) Giardia. In: *Encyclopaedia of Food Microbiology,* Volume 1 (Robinson, R., Batt, C., and Patel, P., eds.), Academic Press, London and New York, pp. 946–954.
45. Robertson, L. J. and Gjerde, B. (2000) Isolation and enumeration of *Giardia* cysts, *Cryptosporidium* oocysts, and *Ascaris* eggs from fruits and vegetables. *J. Food Prot.* **63,** 775–778.
46. Heller, L. C., Davis, C. R., Peak, K. K., et al. (2003) Comparison of methods for DNA isolation from food samples for detection of Shiga toxin-producing *Escherichia coli* by real-time PCR. *Appl. Environ. Microbiol.* **69,** 1844–1846.
47. Faubert, G. M. and Belosevic, M. (1990) Animal models for *Giardia duodenalis* type organisms. In: *Human Parasitic Diseases, Vol. 3, Giardiasis* (Meyer, E. A., ed.), Elsevier, Amsterdam, pp. 77–85.
48. O'Grady, J. E. and Smith, H. V. (2002) Methods for determining the viability and infectivity of *Cryptosporidium* oocysts and *Giardia* cysts. In: *Detection Methods for Algae, Protozoa and Helminths* (Ziglio, G. and Palumbo, F., eds.), Wiley, Chichester, UK, pp. 193–220.
49. Bhatia ,V. N. and Warhurst, D. C. (1981) Hatching and subsequent cultivation of cysts of *Giardia intestinalis* in Diamond's medium. *J. Trop. Med. Hyg.* **84,** 45.
50. Smith, A. L. and Smith, H. V. (1989) A comparison of fluorescein diacetate and propidium iodide staining and *in vitro* excystation for determining *Giardia intestinalis* cyst viability. *Parasitology* **99,** 329–331.
51. Schupp, D. E. and Erlandsen, S. L. (1987) A new method to determine *Giardia* cyst viability, correlation between fluorescein diacetate/propidium iodide staining and animal infectivity. *Appl. Environ. Microbiol.* **55,** 704–707.
52. Schupp, D. E. and Erlandsen, S. L. (1987) Determination of *Giardia muris* cyst viability by differential interference contrast, phase or bright field microscopy. *J. Parasitol.* **73,** 723–729.
53. Schupp, D. E., Januschka, M. M., and Erlandsen, S. L. (1988) Assessing *Giardia* cyst viability with fluorogenic dyes, comparisons to animal infectivity and cyst morphology by light and electron microscopy. In: *Advances in Giardia Research* (Wallis, P. M. and Hammond, B. R., eds.), University of Calgary Press, Calgary, Canada, pp. 265–269.
54. Sauch, J. F., Flannigan, D., Galvin, M. L., Berman, D., and Jakubowski, W. (1991) Propidium iodide as an indicator of *Giardia* cyst viability. *Appl. Environ. Microbiol.* **57,** 3243–3247.
55. Taghi-Kilani, R., Gyürék, L. L., Millard, P. J., Finch, G. R., and Belosevic, M. (1996) Nucleic acid stains as indicators of *Giardia muris* viability following cyst inactivation'. *Int. J. Parasitol.* **26,** 637–646.
56. Abbaszadegan, M., Huber, M. S., Gerba, C. P., and Pepper, I. (1997) Detection of viable *Giardia* cysts by amplification of heat shock-induced mRNA. *Appl. Environ. Microbiol.* **63,** 324–328.
57. Thompson, R. C. A. and Boreham, P. F. L. (1994) Discussants Report: Biotic and abiotic transmission. In: *Giardia: from Molecules to Disease* (Thompson, R. C. A., Reynoldson, J. A., and Lymbery, A. J. eds.), CAB International, Oxon, UK, p. 135.
58. Iturriaga, R., Zhang, S., Sonek, G. J., and Stibbs, H. (2001) Detection of respiratory enzyme activity in *Giardia* cysts and *Cryptosporidium* oocysts using redox dyes and immunofluorescent techniques. *J. Microbiol. Methods* **46,** 19–28.
59. Dalton, C., Goater, A. D., Burt, J. P., and Smith, H. V. (2004) Analysis of parasites by electrorotation. *J. Appl. Microbiol.* **96,** 24–32.
60. Mahbubani, M. H., Bej, A. K., Perlin, M., Schaefer III, F. W., Jakubowski, W. and Atlas, R. M. (1991) Detection of *Giardia* cysts using the polymerase chain reaction and distinguishing live from dead cysts. *Appl. Environ. Microbiol.* **57,** 3456–3461.

61. Keister, D. B. (1983) Axenic culture of *Giardia lamblia* in TYI-S-33 medium supplemented with bile. *Trans. R. Soc. Trop. Med. Hyg.* **77**, 487–488.

62. Gillin, F. D. (1984) The role of reducing agents and the physiology of trophozoite attachment. In: *Giardia and Giardiasis. Biology, Pathogenesis and Epidemiology* (Erlandsen, S. L. and Meyer, E. A., eds.), Plenum Press, New York and London, pp. 111–130.

63. Belosevic, M., Faubert, G. M., MacLean, J. D., Law, C., and Croll, N. A. (1983) *Giardia lamblia* infection in Mongolian gerbils: an animal model. *J. Infect. Dis.* **147**, 222–226.

64. Faubert, G. M., Belosevic, M., Walker, T. S., MacLean, J. D., and Meerovitch, E. (1983) Comparative studies of the pattern of infection with *Giardia spp.* in Mongolian gerbils. *J. Parasitol.* **69**, 802–805.

65. Gasser, R. B., Eckert, J., and Rohrer, L. (1987) Infectivity of Swiss *Giardia* isolates to jirds and mice, and *in vitro* cultivation of trophozoites originating from sheep. *Parasitol. Res.* **74**, 103–111.

66. Visvesvara, G. S., Dickerson, J. W., and Healy, G. R. (1988) Variable infectivity of human-derived *Giardia lamblia* for Mongolian gerbils (*Meriones unguiculatus*) *Int. J. Microbiol.* **26**, 837–841.

67. Shaefer, F. W., III, Johnson, C. H., Hsu, C. H., and Rice, E. W. (1991) Determination of *Giardia lamblia* cyst infective dose for the Mongolian gerbil (*Meriones unguiculatus*). *Appl. Environ. Microbiol.* **57**, 2408–2409.

68. Finch, G. R., Black, E. K., Labatiuk, C. W., Gyürék, L., and Belosevic, M. (1993) Comparison of *Giardia lamblia* and *Giardia muris* cyst inactivation by ozone. *Appl. Environ. Microbiol.* **59**, 3674–3680.

69. Labatiuk, C. W., Schaefer, F. W., III, Finch, G. R., and Belosevic, M. (1991) Comparison of animal infectivity, excystation, fluorogenic dye as measures of *Giardia muris* cyst activation by ozone. *Appl. Environ. Microbiol.* **11**, 3187–3192.

70. Upcroft, J. A., McDonnell, P. A., Gallagher, A. N., and Upcroft, P. (1997) Lethal *Giardia* from a wild caught, sulphur-crested cockatoo (*Cacatua galerita*) established *in vitro* chronically infects mice. *Parasitology* **114**, 407–412.

71. Cevallos, A., Carnaby, S., James, M., and Farthing, M. G. (1995) Small intestinal injury in a neonatal rat model is strain dependent. *Gastroenterology* **109**, 766–773.

72. Smith, H. V., Robertson, L. J., Campbell, A. T., and Girdwood, R. W. A. (1993) *Giardia* and giardiasis – what's in a name? *Microbiol. Eur.* **3**, 22–29.

73. WHO (2002) *Protozoan Parasites (Cryptosporidium, Giardia, Cyclospora). Addendum: Microbiological Agents in Drinking Water*, 2nd edition.

74. Furness, B. W., Beach, M. J., and Roberts, J. M. (2000) Giardiasis Surveillance – United States, 1992–1997. *MMWR* **49** (SS07), 1–13.

75. Rendtorff, R. C. (1954) The experimental transmission of human intestinal protozoan parasites. II. *Giardia lamblia* cysts given in capsules. *Am. J. Hyg.* **59**, 209–220.

76. Rendtorff, R. C. (1979) The experimental transmission of *Giardia lamblia* among volunteer subjects. In: *Waterborne Transmission of Giardiasis* (Jakubowski, W. and Hoff, J. C., eds.), U.S. Environmental Protection Agency, Office of Research and Development, Environmental Research Centre, Cincinnati, Ohio 45268, USA, EPA-600/9-79-001, pp. 64–81.

77. Gilman, R. H., Marquis, G. S., Miranda, E., Vestegui, M., and Martinez, H. (1988) Rapid reinfection by *Giardia lamblia* after treatment in a hyperendemic third world community. *Lancet* (**i**), 343–345.

78. Pickering, L. K. and Engelkirk, P. G. (1990) *Giardia* among children in day care. In: *Human Parasitic Diseases, Vol. 3, Giardiasis* (Meyer, E. A., ed.), Elsevier, Amsterdam, pp. 235–266.

79. Marshall, M. M., Naumovitz, D., Ortega, Y., and Sterling, C. R. (1997) Waterborne protozoan pathogens. *Clin. Microbiol. Rev.* **10**, 67–85.

80. Ortega, Y. R. and Adam, R. D. (1997) *Giardia*: overview and update. *Clin. Infect. Dis.* **25**, 545–550.

81. Warhurst, D. C. and Smith, H. V. (1992) Getting to the guts of the problem. *Parasitol. Today* **8,** 292–293.

82. Davies, R. B. and Hibler, C. P. (1979) Animal reservoirs and cross-species transmission of *Giardia.* In: *Waterborne Transmission of Giardiasis* (Jakubowski, W. and Hoff, J. C., eds.), U.S. Environmental Protection Agency, Office of Research and Development, Environmental Research Centre, Cincinatti, Ohio 45268, USA, EPA-600/9-79-001, pp. 104–126.

83. Bemrick, W. J. (1984) Some perspectives on the transmission of giardiasis. In *Giardia and Giardiasis* (Erlandsen, S. L. and Meyer, E. A., eds.), Plenum Press, New York, pp. 379–400.

84. Faubert, G. M. (1988) Evidence that giardiasis is a zoonosis. *Parasitol. Today* **4,** 66–68.

85. Bemrick, W. J. and Erlandsen, S. L. (1988) Giardiasis–is it really a zoonosis. *Parasitol. Today* **4,** 69–71.

86. Kaprzak, W. and Pawlowski, Z. (1989) Zoonotic aspects of giardiasis: a review. *Vet. Parasitol.* **32,** 101–108.

87. Healey, G. R. (1990) Giardiasis in perspective: the evidence of animals as a source of human *Giardia* infections. In: *Human Parasitic Diseases, Vol. 3, Giardiasis* (Meyer, E. A., ed.), Elsevier, Amsterdam, pp. 305–313.

88. Thompson, R. C. A., Lymbery, A. J., Meloni, B. P., and Binz, N. (1990) The zoonotic transmission of *Giardia* species. *Vet. Record* **19,** 513–514.

89. Thompson, R. C. A. (2000) Giardiasis as a re-emerging infectious disease and its zoonotic potential. *Int. J. Parasitol.* **30,** 1259–1267.

90. Monis, P. T., Andrews, R. H., Mayrhofer, G., and Ey, P. L. (2003) Genetic diversity within the morphological species *Giardia intestinalis* and its relationship to host origin. *Infect. Genet. Evol.* **3,** 29–38.

91. Meloni, B. P., Thompson, R. C. A., Hopkins, R. M., Reynoldson, J. A., and Gracey, M. (1993) The prevalence of *Giardia* and other intestinal parasites in children, dogs and cats from Aboriginal communities in the West Kimberley region of Western Australia. *Med. J. Aust.* **158,** 157–159.

92. Traub, R. J., Monis, P., Robertson, I. D., Mencke, N., and Thompson, R. C. A. (2004) Epidemiological and molecular evidence supports the zoonotic transmission of *Giardia* among humans and dogs living in the same community. *Parasitology* **128,** 253–262.

93. Isaac-Renton, J. L., Cordeiro, C., Sarafis, K., and Shahriari, H. (1993) Characterisation of *Giardia duodenalis* isolates from a waterborne outbreak. *J. Infect. Dis.* **167,** 431–440.

94. Huetink, R. E. C., van der Giessen, J. W. B., Noordhuizen, J. P. T. M., and Ploeger, H. W. (2001) Epidemiology of *Cryptosporidium* spp. and *Giardia duodenalis* on a dairy farm. *Vet. Parasitol.* **102,** 53–67.

95. Bennett, J. V., Holmberg, S. D., Rogers, M. F., and Solomon, S. L. (1987) Infectious and parasitic diseases data selection. *Am. J. Prevent. Med.* **3,** Suppl. 102–114.

96. Craun, G. C. and Jakubowski, W. (1987) Status of waterborne giardiasis outbreaks and monitoring methods. In: *International Symposium on Water-related Health Issues Proceedings* (Tate, C. L., ed.), American Water Resources Association, TPS87-3, Bethesda, Maryland, USA.

97. Jakubowski, W. (1990) The control of *Giardia* in water supplies. In: *Human Parasitic Diseases, Vol. 3, Giardiasis* (Meyer, E. A., ed.), Elsevier, Amsterdam, pp. 335–354.

98. Craun, G. F. (1990) Waterborne giardiasis. In: *Human Parasitic Diseases, Vol. 3, Giardiasis* (Meyer, E. A., ed.), Elsevier, New York, pp. 267–293.

99. Craun, G. F. (1986) Water not sole source of disease transmission. *JAWWA* **78,** 4.

100. Levine, W. C., Stephenson, W. T., and Craun, G. F. (1991) Waterborne disease outbreaks, 1986–1988. *J. Food Prot.* **54,** 71–78.

101. Gold, D. and Smith, H. V. (2002) Pathogenic protozoa and drinking water. In: *Detection Methods for Algae, Protozoa and Helminths* (Ziglio, G. and Palumbo, F., eds.), Wiley, Chichester, UK, pp. 143–166.

102. Smith, H. V. and Grimason, A. M. (2003) *Giardia* and *Cryptosporidium* in water and wastewater. In: *The Handbook of Water and Wastewater Microbiology* (Mara, D. and Horan, N., eds.), Elsevier, Oxford, UK, pp. 619–781.

103. LeChevallier, M. W., Norton, W. D., and Lee, R. G. (1991) Occurrence of *Giardia* and *Cryptosporidium* in surface water supplies. *Appl. Environ. Microbiol.* **57**, 2610–2616.

104. Musgrave, W. E. (1922) Flagellate infestations and infections. *J. Am. Med. Assoc.* **79**, 2219–2220.

105. Lyon, B. B. V. and Swalm, W. A. (1925) Giardiasis, its frequency, recognition and certain clinical factors. *Am. J. Med. Sci.* **170**, 348–364.

106. Barnard, R. J. and Jackson, G. J. (1984) *Giardia lamblia*. The transfer of human infections by foods. In: Giardia *and Giardiasis* (Erlandsen, S. L. and Meyer, E. A., eds.), Plenum Press, New York and London, pp. 365–377.

107. Casemore, D. P. (1990) Foodborne protozoal infection. *Lancet* **336**, 1427–1432.

108. Gupta, S. R., Rao, C. K., Biswas, H., Krishnaswami, A. K., Wattal, B. L., and Raghavan, N. G. S. (1972) Role of the housefly in the transmission of intestinal parasitic cysts/ova. *Indian J. Med. Res.* **60**, 1120–1125.

109. Graczyk, T. K., Grimes, B. H., Knight, R., Da Silva, A. J., Pieniazek, N. J., and Veal, D. A. (2003) Detection of *Cryptosporidium parvum* and *Giardia lamblia* carried by synanthropic flies by combined fluorescent in situ hybridization and a monoclonal antibody. *Am. J. Trop. Med. Hyg.* **68**, 228–232.

110. Mastrandrea, G. and Micarelli, A. (1968) Search for parasites in vegetables from the local markets in the city of Rome. *Arch. Ital. Sci. Med. Trop. Parasitol.* **49**, 55–59.

111. De Oliveira, C. A. and Germano, P. M. (1992) Presence of intestinal parasites in vegetables sold in the metropolitan area of Sao Paulo, Brazil. II. Research on intestinal protozoans. *Rev. Saude Publica* **26**, 332–335.

112. Robertson, L. J. and Gjerde, B. (2001) Occurrence of parasites on fruits and vegetables in Norway. *J. Food Prot.* **64**, 1793–1798.

113. Amahmid, O., Asmama, S., and Bouhoum, K. (1999) The effect of waste water reuse in irrigation on the contamination level of food crops by *Giardia* cysts and *Ascaris* eggs. *Int. J. Food Microbiol.* **49**, 19–26.

114. Graczyk, T. K., Thompson, R. C. A., Fayer, R., Adams, P., Morgan, U. M., and Lewis, E. J. (1999) *Giardia duodenalis* cysts of genotype A recovered from clams in the Chesapeake Bay subestuary, Rhode River. *Am. J. Trop. Med. Hyg.* **61**, 526–529.

115. Greenberg, B. (1973) *Flies and Disease. Volume II. Biology and Disease Transmission.* Princeton University Press, Princeton, USA.

116. Nichols, R. A. B. and Smith, H. V. (2002) *Cryptosporidium, Giardia* and *Cyclospora* as foodborne pathogens. In: *Foodborne Pathogens: Hazards, Risk and Control* (Blackburn, C. and McClure, P., eds.), Woodhead Publishing Limited, Cambridge, UK, pp. 453–478.

117. Foltz, E. E. and Harding, H. B. (1968) A continuing survey of parasitism in a midwestern college community. *Clin. Avi. Aero. Med.* **39**, 74–81.

118. Karabiber, N. and Aktas, F. (1991) Foodborne giardiasis. *Lancet* **377**, 376–377.

119. Wenrich, D. H. and Arnett, J. H. (1938) The results of a six year protozoological survey of food-handlers in a collegiate institution. *J. Parasitol.* **24**, 8.

120. Wilks, N. E. and Sonnenberg, B. (1954) Intestinal parasites in food-handlers returned from Korea. *Am. J. Trop. Med. Hyg.* **3**, 131–135.

121. Reyes, H., Olea, M., and Hernandez, R. (1972) Enteroparasitoses in foodhandlers in the public health sector east of Santiago. *Bol. Chil. Parasitol.* **27**, 115–116.

122. Reyes, H. and Munoz, V. (1975) Intestinal parasitoses in hospital handlers. *Rev. Med. Chil.* **103**, 477–479.

123. Dall'Orso, L. M., Pinilla, N., Parra, G., and Bull, F. (1975) Intestinal parasites and commensal protozoa in food handlers from the central area of the city of Concepcion, Chile. *Bol. Chil. Parasitol.* **30**, 30–31.

124. Hall, A. P., Ridley, D. S., and Thomas, J. R. L. (1976) Pathogenic parasites in food handlers. *Br. Med. J.* **1,** 1542.

125. Conroy, D. A. (1960) A note on the occurrence of *Giardia* sp. in a Christmas pudding. *Rev. Iber. Parasitol.* **20,** 567–571.

126. Osterholm, M. T., Forfang, J. C., Ristinen, T. L, et al. (1981) An outbreak of foodborne giardiasis. *N. Engl. J. Med.* **304,** 24–28.

127. Petersen, L. R., Cartter, M. L., and Hadler, J. L. (1988) A food-borne outbreak of *Giardia lamblia. J. Infect. Dis.* **157,** 846–848.

128. White, K. E., Hedberg, C. W., Edmonson, L. M., Jones, D. B. W., Osterholme, M. T., and MacDonald, K. L. (1989) An outbreak of giardiasis in a nursing home with evidence for multiple modes of transmission. *J. Infect. Dis.* **160,** 298–304.

129. Porter, J. D. H., Gaffney, C., Heymann, D., and Parkin, W. (1990) Food-borne outbreak of *Giardia lamblia. Am. J. Public Health* **80,** 1259–1260.

130. Quick, R., Paugh, K., Addiss, D., Kobayashi, J., and Baron, R. (1992) Restaurant associated outbreak of giardiasis. *J. Infect. Dis.* **166,** 673–676.

131. Mintz, E. D., Hudson-Wragg, M., Mshar, P., Cartr, M. L., and Hadler, J. L. (1993) Foodborne giardiasis in a corporate office setting. *J. Infect. Dis.* **167,** 250–253.

132. Buret, A. G., Chin, A. C., and Scott, K. G. E. (2003) Infection of human and bovine epithelial cells with *Cryptosporidium andersoni* induces apoptosis and disrupts tight junction ZO-1: effects of epidermal growth factor. *Int. J. Parasitol.* **33,** 1363–1371.

133. Kaur, H., Ghosh, S., Samra, H., Vinayak, V. K., and Ganguly, N. K. (2001) Identification and characterization of an excretory–secretory product from *Giardia lamblia. Parasitology* **123,** 347–356.

134. Farthing, M. J. (1997) The molecular pathogenesis of giardiasis. *J. Paediatr. Gastro. Nutr.* **24,** 79–88.

135. Buret, A., Scott, K. G.-E., and Chin, A. C. (2004) Giardiasis: pathophysiology and pathogenesis. In: *Giardia* the Cosmopolitan Parasite (Olsen, B. E., Olsen, M. E., and Wallis, P. M., eds.), CAB International, Wallingford, UK, pp. 109–127.

136. De Vizia, B., Poggi, V., Vajro, P., Cuschiara, S., and Campora, S. (1985) Iron malabsorption in giardiasis. *J. Pediatr.* **107,** 75–78.

137. Olivares, J. L., Fernandez, R., Fleta, J., Ruiz, M. Y., Clavel, A., and Moreno, L. A. (2004) Iron deficiency in children with *Giardia lamblia* and *Enterobius vermicularis. Nutr. Res.* **24,** 1–5.

138. Gillon, J. and Ferguson, A. (1984) Changes in small intestinal mucosa in giardiasis. In: Giardia *and Giardiasis* (Erlandsen, S. L. and Meyer, E. A., eds.), Plenum Press, New York, pp. 63–185.

139. Ferguson, A., Gillon, J., and Munro, G. (1990) Pathology and pathogenesis of intestinal mucosal damage in giardiasis. In: *Human Parasitic Diseases, Vol. 3, Giardiasis* (Meyer, E. A., ed.), Elsevier, Amsterdam, pp. 153–174.

140. Buret, A., Gall, D. G., and Olson, M. E. (1991) Growth, activities of enzymes in the small intestine and ultrastructure of microvillus border in gerbils infected with *Giardia duodenalis. Parasitol. Res.* **77,** 109–114.

141. Buret, A. (1994) Pathogenesis–how does *Giardia* cause disease? In: *Giardia: from Molecules to Disease* (Thompson, R. C. A., Reynoldson, J. A., and Lymbery, A. L., eds.), CAB International, Wallingford, UK, pp. 293–315.

142. Buret, A., Hardin, J. A., Olson, M. E., and Gall, D. G. (1992) Pathophysiology of small intestinal malabsorption in gerbils infected with *Giardia lamblia. Gastroenterology* **103,** 506–513.

143. Gupta, R. K. and Mehta, S. (1973) Giardiasis in children: a study of pancreatic functions. *Indian J. Med. Res.* **61,** 743–748.

144. Chin, A. C., Mitchell, K. D., Teoh, D. A., Scott, K. G. E., Macnaughton, W., and Buret, A. (2000) Epithelial barrier dysfunction is associated with induction of enterocyte apoptosis

in human duodenal epithelial monolayers exposed to *Giardia lamblia*. *Gastroenterology* **118** (suppl. 2), A692, 3789.

145. Teoh, D. A., Kamieniecki, D., Pang, G., and Buret. A. G. (2000) *Giardia lamblia* rearranges F-actin and actinin in human colonic and duodenal monolayers and reduces transepithelia electrical resistance. *J. Parasitol.* **86**, 800–806.

146. Teoh, D. A. and Buret, A. (1999) Decreased electrical resistance in duodenal monolayers exposed to *Giardia lamblia* is associated with re-localisation of villin and ezrin *in vitro*. *Gastroneterology* **116**, A831.

147. Campbell, D. I., McPhail, G., Lunn, P. G., Elia, M., and Jeffries, D. J. (2004) Intestinal inflammation measured by fecal neopterin in Gambian children with enteropathy associated with growth failure, *Giardia lamblia* and intestinal permeability. *J. Pediatr. Gastr. Nutr.* **39**, 153–157.

148. Smith, P. D. (1984) Human immune responses to *Giardia lamblia*. In: *Giardia and Giardiasis. Biology, Pathogenesis and Epidemiology.* (Erlandsen, S. L. and Meyer, E. A., eds.), Plenum Press, New York and London, pp. 201–218.

149. Goka, A. K. J., Roltson, D. D. K., Mathan, V. I., and Farthing, M. J. G. (1986) Evaluation of specific serum anti-*Giardia* IgM antibody response in diagnosis of giardiasis in children. *Lancet* (**ii**), 184.

150. Palm, J. E. D., Weiland, M. E-L., Griffiths, W. J., Ljungström, I., and Svärd, S. G. (2003) Identification of immunoreactive proteins during acute human giardiasis. *J. Infect. Dis.* **187**, 1849–1859.

151. Chandra, R. K. (1984) Parasitic infection, nutrition, and immune response. *Fed. Proc.* **43**, 251–255.

152. Miotti, P. G., Gilman, R. H., Santosham, M., Ryder, R. W., and Yolken, R. H. (1986) Age-related rate of seropositivity of antibody to *Giardia lamblia* in four diverse populations. *J. Clin. Microbiol.* **24**, 972–975.

153. Birkhead, G., Janoff, E. N., Vogt, R. L., and Smith, P. D. (1989) Elevated levels of immuno-globulin A to *Giardia lamblia* during a waterborne outbreak of gastroenteritis. *J. Clin. Microbiol.* **27**, 1707–1710.

154. Udezulu, I. A., Visvesvara, G. A., Moss, D. M., and Leitch, G. J. (1992) Isolation of *Giardia lamblia* (WB strain) clones with distinct surface protein and antigenic profiles and differing infectivity and virulence. *Infect. Immun.* **60**, 2274–2280.

155. Favennec, L., Magne, D., and Gobert, J.-G. (1991) Cytopathogenic effect of *Giardia intestinalis in vitro*. *Parasitol. Today* **7**, 141.

156. Chen, N., Upcroft, J. A., and Upcroft, P. (1995) A *G. duodenalis* gene encoding a protein with multiple repeats of a toxin homolog. *Parasitology* **111**, 423–431.

157. Shant, J., Bhattacharyya, S., Ghosh Nirmal, S., Ganguly, S., and Majumdar, S. (2002) A potentially important excretory–secretory product of *Giardia lamblia*. *Exp. Parasitol.* **102**, 178–186.

158. Torres, M. F., Uetanabaro, A. P., Costa, A. F., et al. (2000) Influence of bacteria from the duodenal microbiota of patients with symptomatic giardiasis on the pathogenicity of *Giardia duodenalis* in gnotoxenic mice. *J. Med. Microbiol.* **49**, 209–215.

159. Farthing, M. J. G. (1994) Giardiasis as a disease. In: *Giardia: from Molecules to Disease* (Thompson, R. C. A., Reynoldson, J. A., and Lymbery, A. L., eds.), CAB International, Wallingford, UK, pp. 15–37.

160. Danciger, M. and Lopez, M. (1975) Numbers of *Giardia* in the feces of infected children. *Am. J. Trop. Med. Hyg.* **24**, 237–242.

161. Ament, M. E. and Rubin, C. E. (1972) Relation of giardiasis to abnormal intestinal structure and function in gastrointestinal syndrome. *Gastroenterology* **62**, 216–226.

162. Perlmutter, D. H., Leichtner, A. M., Goldman, H., and Winter, H. S. (1985) Chronic diarrhea associated with hypogammaglobulinemia and enteropathy in infants and children. *Dig. Dis. Sci.* **30**, 1149–1155.

163. Reynoldson, J. A. (2004) Therapeutics and new drug targets for giardiasis. In: *Giardia, the Cosmopolitan Parasite* (Olson, B. E., Olson, M. E., and Wallis, P. M., eds.), CAB International, Wallingford, UK, pp. 159–177.

164. Upcroft, P. and Upcroft, J. A. (2001) Drug targets and mechanisms of resistance in the anaerobic protozoa. *Clin. Microbiol. Rev.* **14,** 150–164.

165. Müller, M. (1988) Energy metabolism of protozoa without mitochondria. *Ann. Rev. Microbiol.* **42,** 465–488.

166. Townson, S. M., Boreham, P. F. L., Upcroft, P., and Upcroft, J. A. (1994) Resistance to nitroheterocylclic drugs. *Acta Trop.* **56,** 173–194.

167. Brown, D. M., Upcroft, J. A., and Upcroft, P. (1995) Free radical detoxification in *Giardia duodenalis. Mol. Biochem. Parasitol.* **72,** 47–56.

168. Wenman, W. M. and Tyrrell, D. L. J. (1987) Combined metronidazole and quinacrine hydrochloride therapy for chronic giardiasis. *Can. Med. Assoc. J.* **136,** 1179–1180.

169. Abboud, P., Lemee, V., Gargala, G., et al. (2001) Successful treatment of metronidazole- and albendazole-resistant giardiasis with nitazoxanide in a patient with acquired immuno-deficiency syndrome. *Clin. Infect. Dis.* **32,** 1792–1794.

170. Nash, T. E., Ohl, C. A., Thomas, E., Subramanian, G., Keiser, P., and Moore, T. A. (2001) Treatment of patients with refractory giardiasis. *Clin. Infect. Dis.* **33,** 22–28.

171. Sousa, M. C. and Poiares-da-Silva, J. (2001) The cytotoxic effects of ciprofloxacin in *Giardia lamblia* trophozoites. *Toxicol. in Vitro* **15,** 297–301.

172. Harris, J. C., Plummer, S., and Lloyd, D. (2001) Antigiardial drugs. *Appl. Microbiol. Biotech.* **57,** 614–619.

173. Dutta, A. K., Phadke, M. A., Bagade, A. C., et al. (1994) A randomized multicenter study to compare the safety and efficacy of albendazole and metronidazole in the treatment of giardiasis in children. *Indian J. Pediatr.* **61,** 689–693.

174. Kreutner, A. K., Del Bene, V. E., and Amstey, M. S. (1981) Giardiasis in pregnancy. *Am. J. Obstet. Gynecol.* **40,** 895–901.

175. Olson, M. E., Mock, D. W., and Ceri, H. (1996) The efficacy of a *Giardia lamblia* vaccine in kittens. *Can. J. Vet. Res.* **60,** 249–256.

176. Olson, M. E., Mock, D. W., and Ceri, H. (1998) Preliminary data on the efficacy of a *Giardia* vaccine in puppies. *Can. Vet. J.* **38,** 777–779.

177. Vinayak, V. K., Kum, K., Khanna, R., and Khuller, M. (1992) Systemic-oral immunization with 56 kDa molecule of *Giardia lamblia* affords protection in experimental mice. *Vaccine* **10,** 21–27.

178. Romero-Cabello, R., Guerrero, L. R., Munoz-Garcia, M. R., and Geyne-Cruz, A. (1997) Nitazoxanide for the treatment of intestinal protozoan and helminthic infections in Mexico. *Trans. R. Soc. Trop. Med. Hyg.* **91,** 701–703.

179. Ellis, J. E., Wingfield, J. M., Cole, D., Boreham, P. F. L., and Lloyd, D. (1992) Oxygen affinities of metronidazole-resistant and -sensitive stocks of *Giardia intestinalis. Int. J. Parasitol.* **23,** 35–39.

180. Upcroft, P. and Upcroft, J. A. (2001) Drug targets and mechanisms of resistance in the anaerobic protozoa. *Clin. Microbiol. Rev.* **14,** 150–164.

181. Upcroft, J. A., Boreham, P. F. L., Campbell, R. W., Shepherd, R. W., and Upcroft, P. (1995) Biological and genetic analysis of a longitudinal collection of *Giardia* samples derived from humans. *Acta Trop.* **60,** 35–46.

182. Upcroft, J. A. and Upcroft, P. (1999) Keto-acid oxidoreductases in the anaerobic protozoa. *J. Euk. Microbiol.* **46,** 447–449.

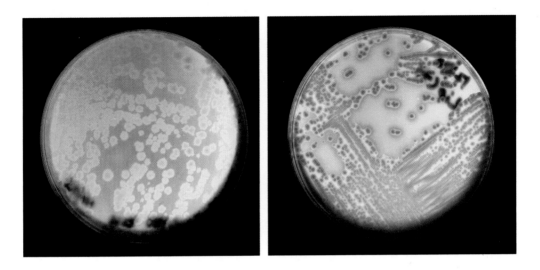

Color Plate 1. Fig. 3.1. *See* legend on p. 45.

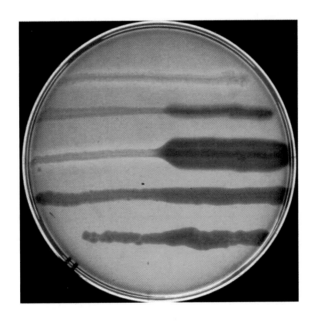

Color Plate 2. Fig. 3.3. *See* legend on p. 61.

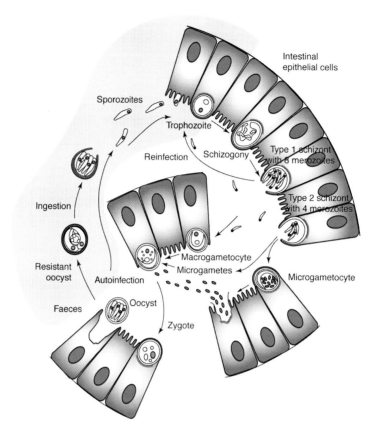

Color Plate 3. Fig. 9.2. *See* legend on p. 239.

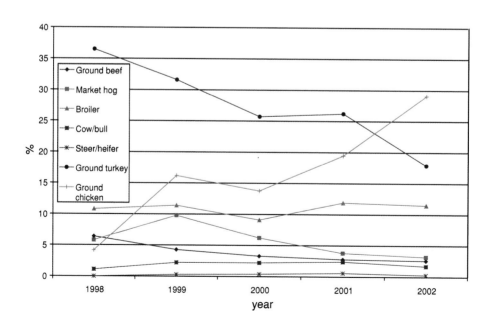

Color Plate 4. Fig. 15.1. *See* legend on p. 391.

12

Toxoplasma gondii

Dolores E. Hill, Chirukandoth Sreekumar, Jeffrey Jones, and J. P. Dubey

1. INTRODUCTION

Infection with the protozoan parasite *Toxoplasma gondii* is one of the most common parasitic infections of man and other warm-blooded animals *(1)*. It has been found worldwide from Alaska to Australia. Nearly one-third of humanity has been exposed to this parasite *(1)*. In most adults it does not cause serious illness, but it can cause blindness and mental retardation in congenitally infected children, blindness in persons infected after birth, and devastating disease in immunocompromised individuals. Consumption of raw or undercooked meat products and contamination of food or drink with oocysts are the major risk factors associated with *T. gondii* infection.

2. CLASSIFICATION AND IDENTIFICATION

T. gondii is a coccidian parasite with cats as the definitive host, and warm-blooded animals as intermediate hosts *(2)*. It is one of the most important parasites of animals. It belongs to:

Phylum: Apicomplexa; Levine, 1970
Class: Sporozoasida; Leukart, 1879
Subclass: Coccidiasina; Leukart, 1879
Order: Eimeriorina; Leger, 1911
Family: Toxoplasmatidae; Biocca, 1956

There is only one species of *Toxoplasma*, *T. gondii*.

Coccidia in general have complicated life cycles. Most coccidia are host-specific, and are transmitted via a fecal–oral route. Transmission of *T. gondii* occurs via the fecal–oral route, as well as through consumption of infected meat, and by transplacental transfer from mother to fetus *(1,2)*.

The name *Toxoplasma* (toxon = arc, plasma = form) is derived from the crescent shape of the tachyzoite stage (Fig. 1). There are three infectious stages of *T. gondii*: the tachyzoites (in groups), the bradyzoites (in tissue cysts), and the sporozoites (in oocysts).

The tachyzoite is often crescent-shaped and is approximately the size (2×6 μm) of a red blood cell (Fig. 1A). The anterior end of the tachyzoite is pointed, and the posterior end is round. It has a pellicle (outer covering), several organelles including subpellicular

From: *Infectious Disease: Foodborne Diseases*
Edited by: S. Simjee © Humana Press Inc., Totowa, NJ

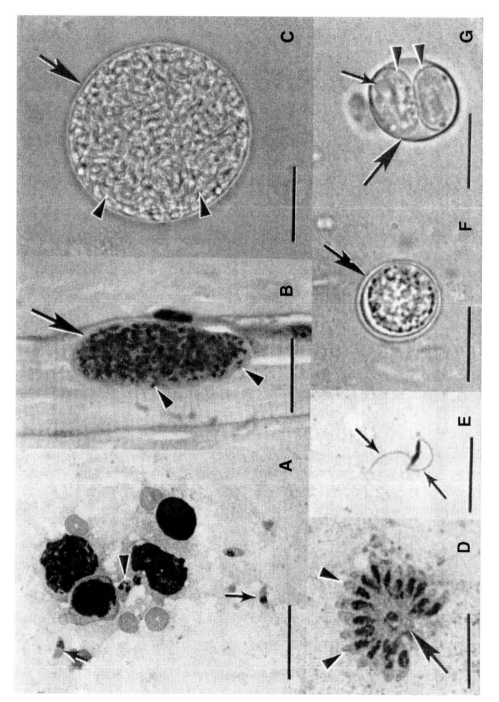

Fig. 1.

microtubules, mitochondrium, smooth and rough endoplasmic reticulum, a Golgi apparatus, apicoplast, ribosomes, a micropore, and a well-defined nucleus. The nucleus is usually situated toward the central area of the cell.

The tachyzoite enters the host cell by active penetration of the host cell membrane and can tilt, extend, and retract as it searches for a host cell. After entering the host cell, the tachyzoite becomes ovoid in shape and is surrounded by a parasitophorous vacuole. *T. gondii* in a parasitophorous vacuole is protected from host defense mechanisms. The tachyzoite multiplies asexually within the host cell by repeated divisions in which two progeny form within the parent parasite, consuming it (Fig. 1A). Tachyzoites continue to divide until the host cell is filled with parasites. Cells rupture, and free tachyzoites infect neighboring cells and the cycle is repeated. After an unknown number of cycles, *T. gondii* forms tissue cysts. Tissue cysts vary in size from 5 to 70 μm and remain intracellular (Fig. 1B,C). The tissue cyst wall is elastic, thin (< 0.5 μm), and may enclose hundreds of the crescent-shaped, slender *T. gondii* stage known as bradyzoites (*[3]*; Fig. 1C). The bradyzoites are approx 7 × 1.5 μm. Bradyzoites differ structurally only slightly from tachyzoites. They have a nucleus situated toward the posterior end whereas the nucleus in tachyzoites is more centrally located. Bradyzoites are more slender than are tachyzoites and they are less susceptible to destruction by proteolytic enzymes than are tachyzoites. Although tissue cysts containing bradyzoites may develop in visceral organs, including lungs, liver, and kidneys, they are more prevalent in muscular and neural tissues (Fig. 1B), including the brain (Fig. 1C), eye, skeletal, and cardiac muscle. Intact tissue cysts probably do not cause any harm and can persist for the life of the host.

All coccidian parasites have an environmentally resistant stage in their life cycle, called the oocyst. Oocysts of *T. gondii* are formed only in cats, probably in all members of the Felidae (Figs. 2 and 3). Cats shed oocysts after ingesting any of the three infectious stages of *T. gondii*, i.e., tachyzoites, bradyzoites, and sporozoites (*4–6*). Prepatent periods (time to the shedding of oocysts after initial infection) and frequency of oocyst shedding vary according to the stage of *T. gondii* ingested. Prepatent periods are 3–10 d after ingesting tissue cysts and 18 d or more after ingesting tachyzoites or oocysts (*4–7*). Less than 50% of cats shed oocysts after ingesting tachyzoites or oocysts, whereas nearly all cats shed oocysts after ingesting tissue cysts (*5*).

After the ingestion of tissue cysts by cats, the tissue cyst wall is dissolved by proteolytic enzymes in the stomach and small intestine. The released bradyzoites penetrate the

---

Fig. 1. Stages of *Toxoplasma gondii*. (**A**) Tachyzoites in impression smear of lung. Note crescent-shaped individual tachyzoites (arrows), dividing tachyzoites (arrowheads) compared with size of host red blood cells and leukocytes; Giema stain. (**B**) Tissue cysts in section of muscle. The tissue cyst wall is very thin (arrow) and encloses many tiny bradyzoites (arrowheads); H&E stain. (**C**) Tissue cyst separated from host tissue by homogenization of infected brain. Note tissue cyst wall (arrow) and hundreds of bradyzoites (arrowheads); Unstained. (**D**) Schizont (arrow) with several merozoites (arrowheads) separating from the main mass. Impression smear of infected cat intestine; Giemsa stain. (**E**) A male gamete with two flagella (arrows). Impression smear of infected cat intestine; Giemsa stain. (**F**) Unsporulated oocyst in fecal float of cat feces; Unstained. Note double-layered oocyst wall (arrow) enclosing a central undivided mass. (**G**) Sporulated oocyst with a thin oocyst wall (large arrow), two sporocysts (arrowheads). Each sporocyst has four sporozoites (small arrow) which are not in complete focus; Unstained. Scale bar: A–D = 20 μm; E–G = 10 μm.

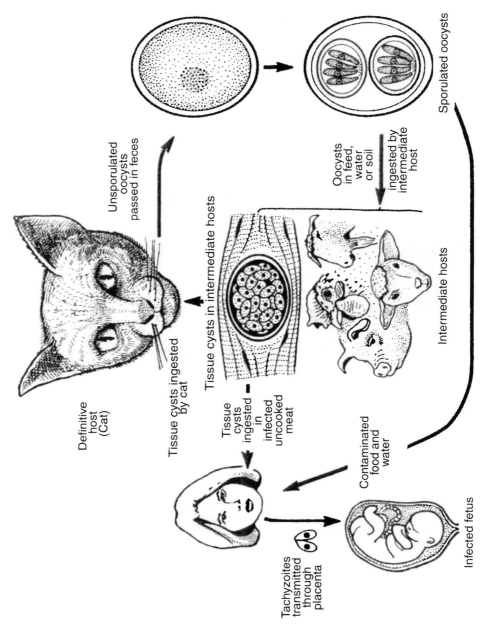

Unsporulated oocysts passed in feces

Sporulated oocysts

Definitive host (Cat)

Tissue cysts ingested by cat

Tissue cysts in intermediate hosts

Tissue cysts ingested in infected uncooked meat

Oocysts in feed, water or soil

ingested by intermediate host

Intermediate hosts

Contaminated food and water

Tachyzoites transmitted through placenta

Infected fetus

Fig. 2. Life cycle of *Toxoplasma gondii*.

340

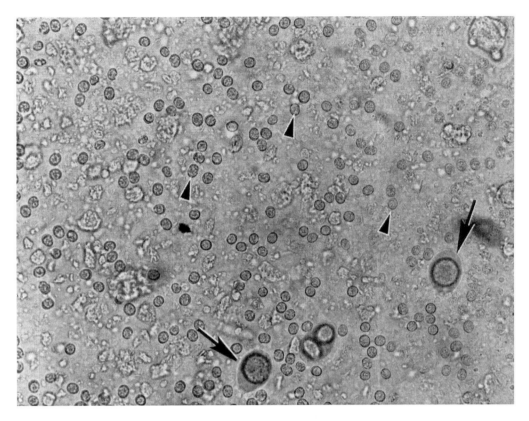

Fig. 3. *Toxoplasma gondii* oocysts in sugar fecal float of an infected cat. Note many spherical *T. gondii* oocysts (arrowheads). Also note oocysts of *Isospora felis* (arrows) which are often present in cat feces and are about four times the size of *T. gondii* oocysts.

epithelial cells of the small intestine and initiate development of numerous generations of asexual and sexual cycles of *T. gondii (4)*. *T. gondii* multiplies profusely in intestinal epithelial cells of cats (entero-epithelial cycle) and these stages are known as schizonts (Fig. 1D). Organisms (merozoites) released from schizonts form male and female gametes. The male gamete has two flagella (Fig. 1E), and it swims to and enters the female gamete. After the female gamete is fertilized by the male gamete (Fig. 1E), oocyst wall formation begins around the fertilized gamete. When oocysts are mature, they are discharged into the intestinal lumen by the rupture of intestinal epithelial cells.

In freshly passed feces, oocysts are unsporulated (noninfective). Unsporulated oocysts are subspherical to spherical and are 10×12 μm in diameter (Fig. 1F). They sporulate (become infectious) outside the cat within 1–5 d depending on aeration and temperature. Sporulated oocysts contain two ellipsoidal sporocysts (Fig. 1G). Each sporocyst contains four sporozoites. The sporozoites are 2×6 to 8 μm in size.

As the entero-epithelial cycle progresses, bradyzoites penetrate the lamina propria of the feline intestine and multiply as tachyzoites. Within a few hours after infection of cats, *T. gondii* may disseminate to extra-intestinal tissues via the lymphatics and the bloodstream. *T. gondii* persists in intestinal and extra-intestinal tissues of cats for at least several months, and possibly for the life of the cat.

Unlike many other microorganisms and in spite of a wide host range and worldwide distribution, *T. gondii* has a low genetic diversity. *T. gondii* strains have been classified into three genetic types (I, II, and III), based on antigens, isoenzymes, and restriction fragment length polymorphism *(8–11)*. Type I strains are considered highly virulent in outbred laboratory mice, whereas types II and III are considered less virulent for mice *(8,12–14)* but there is no correlation between virulence in mice to clinical disease in other animals or humans. Genetic typing of isolates has not provided clues to sources of infection for humans or animals.

3. TRANSMISSION OF *T. GONDII*

Toxoplasmosis may be acquired by ingestion of oocysts or by ingestion of tissue-inhabiting stages of the parasite. Contamination of the environment by oocysts is widespread because oocysts are shed by domestic cats and other felids *(1,2)*. Domestic cats are probably the major source of contamination because oocyst formation is greatest in domestic cats, which are extremely common. Widespread natural infection of the environment is possible because a cat may excrete millions of oocysts after ingesting as few as one bradyzoite or one tissue cyst, and many tissue cysts may be present in one infected mouse *(2,15)*. Sporulated oocysts survive for long periods under most ordinary environmental conditions and even in harsh environment for months. They can survive in moist soil, for example, for months and even years *(1,16)*. Oocysts in soil can be mechanically transmitted by invertebrates such as flies, cockroaches, dung beetles, and earthworms, which can spread oocysts onto human food and animal feeds.

Infection rates in cats are determined by the rate of infection in local avian and rodent populations because cats are thought to become infected by eating these animals. The more oocysts in the environment, the more likely it is that prey animals would be infected, and this in turn would increase the infection rate in cats.

In certain areas of Brazil, approx 60% of 6–8-yr-old children have antibodies to *T. gondii* linked to the ingestion of oocysts from the environment heavily contaminated with *T. gondii* oocysts *(17)*. Infection in aquatic mammals indicates contamination and survival of oocysts in sea water *(16,18–20)*. The largest recorded outbreak of clinical toxoplasmosis in humans was epidemiologically linked to drinking water from a municipal water reservoir in British Columbia, Canada *(21)*. This water reservoir was thought to be contaminated with *T. gondii* oocysts excreted by cougars (*Felis concolor*). Although attempts to recover *T. gondii* oocysts from water samples in the British Columbia outbreak were unsuccessful, methods to detect oocysts were reported *(22)*. At present there are no commercial reagents available to detect *T. gondii* oocysts in the environment.

Increased risk for *T. gondii* infection has been associated with many food-related factors, including: eating raw or undercooked pork, mutton, lamb, beef, or mincemeat products *(23–26)*, eating raw or unwashed vegetables, raw vegetables outside the home, or fruits *(23)*, washing kitchen knives infrequently *(25)*, and having poor hand-hygiene *(23)*. Decreased risk for *T. gondii* infection has been found to be associated with eating a meat-free diet *(27)*. Outbreaks of toxoplasmosis have been attributed to ingestion of raw or undercooked beef, lamb, pork, and venison *(28–34)*; and consumption of raw goat's milk *(35)*.

In the United States, infection in humans is probably most often the result of ingestion of tissue cysts contained in undercooked meat *(1,27,36)*, though the exact contribution of foodborne toxoplasmosis vs oocyst induced toxoplasmosis to human infection is currently unknown. *T. gondii* infection is common in many animals used for food, including sheep, pigs, goats, and rabbits. Birds and other domesticated and wild animals can also become infected *(1)*. Animals that survive infection harbor tissue cysts, and can therefore transmit *T. gondii* infection to human consumers *(37,38)*. In one study, viable *T. gondii* tissue cysts were isolated from 17% of 1000 adult pigs (sows) from a slaughter plant in Iowa *(39)*. Serological surveys of pigs from pig farms in Illinois indicate an infection rate of about 3% in market weight animals and 20% for breeding pigs, suggesting that age is a factor for pigs acquiring *Toxoplasma* infection *(40)*. Serological surveys of pigs on New England farms revealed an overall infection rate of 47% *(41)*, and from one farm *T. gondii* was isolated from 51 of 55 market age (feeder) pigs *(42)*. Infection in cattle is less prevalent than in sheep or pigs in the United States, however, recent surveys in several European countries using serology and PCR to detect parasite DNA have shown that the infection rates in pigs and horses are negligible, compared to sheep and cattle that ranges from 1 to 6% *(43,44)*. Serological surveys in eastern Poland revealed that 53% of cattle, 15% of pigs, and 0–6% of chickens, ducks, and turkeys were positive for *T. gondii* infection; nearly 50% of the people in the region were also serologically positive for *T. gondii* infection *(45)*. The prevalence of *T. gondii* infection in commercially raised chickens in the United States and elsewhere has not been investigated; however, most chicken meat in the United States is cooled to near freezing or is completely frozen at the packing plant *(46)*, which would kill organisms in tissue cysts *(47)*. The relative contributions of undercooked pork, beef, and chicken to *T. gondii* infection in humans was unknown; a nationwide retail meat survey was conducted to determine the risk to US consumers of purchasing pork, beef, and chicken containing viable *T. gondii* tissue cysts at the retail level *(95)*. The national retail meats survey for *T. gondii* collected 6,282 meat samples; 2094 each of beef, chicken, and pork, from 698 randomly selected retail outlets in 28 major geographic regions in the U.S. The survey determined that viable *T. gondii* was present in 0.4% of retail pork. There was little risk of acquiring *T. gondii* after ingestion of beef and chicken.

T. gondii infection is also prevalent in game animals. Among wild game, *T. gondii* infection is most prevalent in black bears and in white-tailed deer. Serological surveys of white-tailed deer in the United States have demonstrated seropositivity of 30–60% *(48–50)*, and viable *T. gondii* can be demonstrated in half of seropositive deer *(51)*. A recent study reported the occurrence of clinical toxoplasmosis and necrotizing retinitis in deer hunters with a history of consuming undercooked or raw venison *(34)*. Approx 80% of black bears are infected in the United States *(52)*, and about 60% of raccoons have antibodies to *T. gondii* *(53,54)*. Because raccoons and bears scavenge for their food, infection in these animals is a good indicator of the prevalence of *T. gondii* in the environment.

Virtually all edible portions of an animal can harbor viable *T. gondii* tissue cysts, and tissue cysts can survive in food animals for years. The number of *T. gondii* in meat from food animals is very low. It is estimated that as few as one tissue cyst may be present in 100 g of meat. Since it is not practical to detect this low level of *T. gondii* infection in meat samples, digestion of meat samples in trypsin or pepsin is used to concentrate *T. gondii* for detection *(55)*. Digestion in trypsin and pepsin ruptures the *T. gondii* tissue cyst wall, releasing hundreds of bradyzoites. The bradyzoites survive in the digests for several hours. Even in the digested samples, only a few *T. gondii* are present and their identification by direct microscopic examination is not practical. Therefore, the digested material is bioassayed in mice *(55)*. The mice inoculated with digested material have to be kept for 6–8 wk before *T. gondii* infection can be detected reliably—this procedure is not practical for mass scale samples. The detection of *T. gondii* DNA in meat samples by PCR has been reported *(56)*, but there are no data on specificity and sensitivity of this method to detect *T. gondii*. A highly sensitive method using a Real-Time PCR and fluorogenic probes was found to detect *T. gondii* DNA from as few as four bradyzoites in meat samples *(57)*. This method is now being tested to detect *T. gondii* in meat samples obtained from slaughtered animals. Cultural habits of people may affect the acquisition of *T. gondii* infection *(26)*. For example, in France the prevalence of antibody to *T. gondii* is very high in humans. Though 84% of pregnant women in Paris have antibodies to *T. gondii*, only 32% of pregnant women in New York City and 22% in London have such antibodies *(1)*. Jones et al. *(58)* have shown a seroprevalence of 23% in the United States in the National Health and Nutrition Examination Survey (NHANES), which is a representative sample of the US noninstitutionalized civilian population. The high incidence of *T. gondii* infection in humans in France appears to be related in part to the French habit of eating some of their meat products undercooked or uncooked. In contrast, the high prevalence of *T. gondii* infection in Central and South America is probably due to high levels of contamination of the environment with oocysts *(1,17,59,60)*. Having said this, it should be noted that the relative frequency of acquisition of toxoplasmosis from eating raw meat and that related to ingestion of food or water contaminated by oocysts from cat feces is very difficult to determine and as a result, statements on the subject are at best controversial. There are no tests at the present time to determine the source of infection in a given person. There is little, if any, danger of *T. gondii* infection by drinking cow's milk and, in any case, cow's milk is generally pasteurized or even boiled, but infection has followed drinking unboiled goat's milk *(1)*. Raw hens' eggs, although an important source of *Salmonella* infection, are extremely unlikely to transmit *T. gondii* infection.

4. PATHOGENICITY

T. gondii can multiply in virtually any cell in the body. How *T. gondii* is destroyed in immune cells is not completely known *(61)*. All extracellular forms of the parasite are directly affected by antibody but intracellular forms are not. It is believed that cellular factors, including lymphocytes and lymphokines, are more important than humoral factors in immune-mediated destruction of *T. gondii (61)*.

Immunity does not eradicate infection. *T. gondii* tissue cysts persist several years after acute infection. The fate of tissue cysts is not fully known. Whether bradyzoites can form new tissue cysts directly without transforming into tachyzoites is not known. It has been proposed that the tissue cysts may at times rupture during the life of the host. The released bradyzoites may be destroyed by the host's immune responses, or there may be formation of new tissue cysts.

In immunosuppressed patients, such as those given large doses of immunosuppressive agents in preparation for organ transplants and in those with AIDS, rupture of a tissue cyst may result in transformation of bradyzoites into tachyzoites and renewed multiplication. The immunosuppressed host may die from toxoplasmosis unless treated. It is not known how corticosteroids cause relapse, but it is unlikely that they directly cause rupture of the tissue cysts.

Pathogenicity of *T. gondii* is determined by the virulence of the strain and the susceptibility of the host species. *T. gondii* strains may vary in their pathogenicity in a given host. Certain strains of mice are more susceptible than others and the severity of infection in individual mice within the same strain may vary. Certain species are genetically resistant to clinical toxoplasmosis. For example, adult rats do not become ill, while young rats can die of toxoplasmosis. Mice of any age are susceptible to clinical *T. gondii* infection. Adult dogs, like adult rats, are resistant, whereas puppies are fully susceptible to clinical toxoplasmosis. Cattle and horses are among the hosts more resistant to clinical toxoplasmosis, whereas certain marsupials and New World monkeys are highly susceptible to *T. gondii* infection *(1)*. Nothing is known concerning genetically determined susceptibility to clinical toxoplasmosis in higher mammals, including humans.

5. CLINICAL CHARACTERISTICS

T. gondii infection is widespread in humans though its prevalence varies widely from place to place. In the United States and in the United Kingdom, it is estimated that 16–40% of people are infected *(58)*, whereas in Central and South America and continental Europe, estimates of infection range from 50 to 80% *(1,17)*. Most infections in humans are asymptomatic but at times the parasite can produce devastating disease. Infection may be congenitally or postnatally acquired. Congenital infection occurs only when a woman becomes infected during pregnancy. Congenital infections acquired during the first trimester are more severe than those acquired in the second and third trimester *(62,63)*. While the mother rarely has symptoms of infection, she does have a temporary parasitemia. Focal lesions develop in the placenta and the fetus may become infected. At first there is generalized infection in the fetus. Later, infection is cleared from the visceral tissues and may localize in the central nervous system. Although most children are asymptomatic at birth *(64)*, a wide spectrum of clinical diseases can occur in congenitally infected children *(62,65)* or develop later in life *(66)*. Mild disease may consist of slightly diminished vision, whereas severely diseased children may have the full tetrad of signs: retinochoroiditis (inflammation of the inner layers of the eye), hydrocephalus (big head), convulsions, and intracerebral calcification. Of these, hydrocephalus is the least common but most dramatic result of toxoplasmosis (Fig. 4). By far the most common sequela of congenital toxoplasmosis is ocular disease *(62,63)*. In addition to ocular infection that occurs with congenital disease, up to 2% of adults newly infected with *T. gondii* develop ocular lesions, and because most people are infected with *T. gondii* after birth, authorities now believe that the majority of Toxoplasma-induced ocular disease is a result of infection with *T. gondii* after birth *(67)*.

The socio-economic impact of toxoplasmosis in human suffering and the cost of care of sick children, especially those with mental retardation and blindness, are enormous *(68,69)*. The testing of all pregnant women for *T. gondii* infection is compulsory in some European countries including France and Austria. The cost benefits of such mass screening are being debated in many other countries *(63)*. A number of recent studies have raised questions about the effectiveness of treating acutely infected pregnant

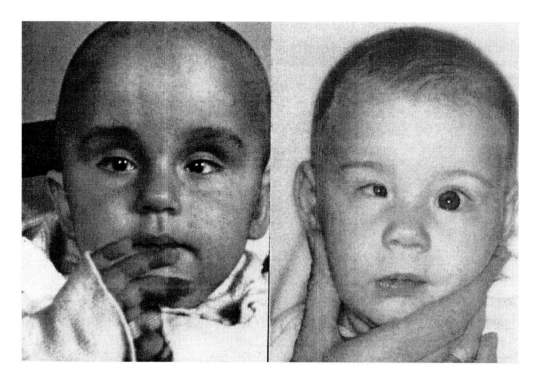

Fig. 4. Congenital toxoplasmosis in children. Hydrocephalus with bulging forehead (left) and microophthalmia of the left eye (right).

women to prevent transmission to the fetus and or prevent sequela in infants *(70–72)*. Newborn screening is another option for identifying infected infants and has been used in two states in the United States *(64)*, but infected newborns that are identified by screening require a year of follow-up and treatment with potentially toxic drugs and the efficacy of treating infants with congenital toxoplasmosis has not been documented in well-controlled studies *(73)*.

Postnatally acquired infection may be localized or generalized. Oocyst-transmitted infections may be more severe than tissue cyst-induced infections *(1,74–76)*. Enlarged lymph nodes are the most frequently observed clinical form of toxoplasmosis in humans (Table 1). Lymphadenopathy may be associated with fever, fatigue, muscle pain, sore throat, and headache. Although the condition may be benign, diagnosis of *T. gondii*-associated lymphadenopathy is important in pregnant women because of the risk to the fetus. In a British Columbia outbreak, of 100 people who were diagnosed with acute infection, 51 had lymphadenopathy and 20 had retinitis *(21,77)*. Encephalitis is an important manifestation of toxoplasmosis in immunosuppressed patients because it causes the most severe damage to the patient *(1,78)*. Infection may occur in any organ. Patients may have headache, disorientation, drowsiness, hemiparesis, reflex changes, and convulsions, and many become comatose.

Toxoplasmosis ranked high on the list of diseases which lead to death of patients with acquired immunodeficiency syndrome (AIDS) in the United States; approx 10% of AIDS patients in the United States and up to 30% in Europe were estimated to die from toxoplasmosis *(78)* before prophylactic medications such as trimethoprim-sulfamethoxazole,

Table 1
**Frequency of Symptoms in People With Postnatally Acquired
Toxoplasmosis From Oocyst Ingestion**[a]

	Patients with symptoms (%)	
Symptoms	Atlanta outbreak[b] (35 Patients)	Panama outbreak[c] (35 Patients)
Fever	94	90
Lymphadenopathy	88	77
Headache	88	77
Myalgia	63	68
Stiff neck	57	55
Anorexia	57	NR[d]
Sore throat	46	NR
Artharlgia	26	29
Rash	23	0
Confusion	20	NR
Earache	17	NR
Nausea	17	36
Eye pain	14	26
Abdominal pain	11	55

[a]Both the outbreaks thought to be caused by infection with oocysts.
[b]From ref. (74).
[c]From ref. (75).
[d]Not reported.

and the treatment for HIV infection with highly active antiretroviral therapy (HAART) were widely available. However, since use of prophylactic therapy and HAART became common in the mid-1990s, the number of persons with AIDS dying of toxoplasmosis has markedly declined (79). Although in AIDS patients any organ may be involved, including the testis, dermis, and the spinal cord, infection of the brain is most frequently reported. Most AIDS patients suffering from toxoplasmosis have bilateral, severe, and persistent headache which responds poorly to analgesics. As the disease progresses, the headache may give way to a condition characterized by confusion, lethargy, ataxia, and coma. The predominant lesion in the brain is necrosis, especially of the thalamus (61).

Diagnosis is made by biologic, serologic, or histologic methods or by a combination of the above. Clinical signs of toxoplasmosis are nonspecific and are not sufficiently characteristic for a definite diagnosis. Detection of *T. gondii* antibody in patients may aid diagnosis. There are numerous serologic procedures available for detection of humoral antibodies. These include the Sabin–Feldman dye test, the indirect hemagglutination assay, the indirect fluorescent antibody assay (IFA), the direct agglutination test, the latex agglutination test, the enzyme-linked immunoabsorbent assay (ELISA), and the immunoabsorbent agglutination assay test (IAAT). The IFA, IAAT and ELISA have been modified to detect IgM antibodies (63). The IgM antibodies appear sooner after infection than the IgG antibodies and the IgM antibodies disappear faster than IgG antibodies after recovery (63). Detection of IgM and IgG antibodies, along with a panel of other serologic tests including the avidity test, have been found to be very helpful in diagnosing acute infection in pregnant women when conducted at a reference laboratory *(80,81)*.

6. CHOICE OF TREATMENT

In general, physicians are most likely to consider treatment for *T. gondii* infection in four circumstances: (1) pregnant women with acute infection to prevent fetal infection, (2) congenitally infected infants, (3) immunosuppressed persons, usually with reactivated disease, and (4) acute and recurrent ocular disease *(82)*. Although well-designed clinical trials have demonstrated the effectiveness of treatment in immunosuppressed persons for reactivated disease *(83–86)*, there is less evidence for the effectiveness of treatment in the other circumstances listed above *(70–73,87,88)*.

Sulfadiazine and pyrimethamine (Daraprim) are two drugs widely used for treatment of toxoplasmosis *(64,89)*. While these drugs have a beneficial action when given in the acute stage of the disease process when there is active multiplication of the parasite, they will not usually eradicate infection. It is believed that these drugs have little effect on subclinical infections, but the growth of tissue cysts in mice has been restrained with sulfonamides. Certain other drugs, diaminodiphenylsulfone, atovaquone, azithromycin, clarithromycin, dapson, spiramycin, and clindamycin are also used to treat toxoplasmosis in difficult cases, often in combination with pyrimethamine. Medications are also prescribed for preventive or suppressive treatment in HIV-infected persons and have been quite effective when used for this purpose *(90)*.

7. PREVENTION AND CONTROL

To prevent infection of human beings by *T. gondii*, the hands of people handling meat should be washed thoroughly with soap and water before they go to other tasks *(1,36)*. All cutting boards, sink tops, knives, and other materials coming in contact with uncooked meat should be washed with soap and water also. Washing is effective because of the physical removal of material from the hands and because the stages of *T. gondii* in meat are killed by contact with soap and water *(1)*.

T. gondii organisms in meat can be killed by exposure to extreme cold or heat. Tissue cysts in meat are killed by heating the meat throughout to 67°C *(91)*. *T. gondii* in meat is killed by cooling to −13°C *(47)*. *Toxoplasma* in tissue cysts are also killed by exposure to 0.5 krad of γ irradiation *(92)*. Meat should be cooked to 145°F for beef, 160°F for pork, ground meat, and wild game, and 180°F for poultry (in thigh, to ensure doneness) before consumption, and tasting meat while cooking or while seasoning should be avoided. Pregnant women, especially, should avoid contact with cats, soil, and raw meat. Pet cats should be fed only dry, canned, or cooked food. The cat litter box should be emptied every day, preferably not by a pregnant woman or an immunosuppressed person. Pregnant women and immunosuppressed persons should wear gloves while gardening or changing cat litter (if no one else can change the litter) and wash their hands thoroughly afterwards. Fruits and vegetables should be washed thoroughly before eating because they may have been contaminated with cat feces or soil containing oocysts from cat feces. Untreated water should not be consumed, particularly in developing countries. Women of childbearing age and expectant mothers should be aware of the dangers of toxoplasmosis *(36,93,94)*. At present there is no vaccine to prevent toxoplasmosis in humans.

8. SUMMARY

Infection by the protozoan parasite *T. gondii* is widely prevalent in humans and animals. Although it usually causes asymptomatic infection in immune competent adults, *T. gondii* can cause devastating disease in congenitally infected children and those with depressed immunity. To prevent human infection, all meat should be cooked well and fruits and vegetables washed before consumption, precautions such as wearing gloves and washing hands should be taken during and after gardening to prevent exposure to soil contaminated with *T. gondii* oocysts excreted in cat feces, and ingestion of untreated water should be avoided, especially in developing countries.

REFERENCES

1. Dubey, J. P. and Beattie, C. P. (1988) *Toxoplasmosis of Animals and Man*, CRC, Boca Raton, FL.
2. Frenkel, J. K., Dubey, J. P., and Miller, N. L. (1970) *Toxoplasma gondii* in cats: fecal stages identified as coccidian oocysts. *Science* **167,** 893–896.
3. Dubey, J. P., Lindsay, D. S., and Speer, C. A. (1998) Structure of *Toxoplasma gondii* tachyzoites, bradyzoites, and sporozoites, and biology and development of tissue cysts. *Clin. Microbiol. Rev.* **11,** 267–299.
4. Dubey, J. P. and Frenkel, J. K. (1972) Cyst-induced toxoplasmosis in cats. *J. Protozool.* **19,** 155–177.
5. Dubey, J. P. and Frenkel, J. K. (1976) Feline toxoplasmosis from acutely infected mice and the development of *Toxoplasma* cysts. *J. Protozool.* **23,** 537–546.
6. Dubey, J. P. (1996) Infectivity and pathogenicity of *Toxoplasma gondii* oocysts for cats. *J. Parasitol.* **82,** 957–961.
7. Dubey, J. P. (2002) Tachyzoite-induced life cycle of *Toxoplasma gondii* in cats. *J. Parasitol.* **88,** 713–717.
8. Howe, D. K. and Sibley, L. D. (1995) *Toxoplasma gondii* comprises three clonal lineages: Correlation of parasite genotypes with human disease. *J. Infect. Dis.* **172,** 1561–1566.
9. Guo, Z. G. and Johnson, A. M. (1996) DNA polymorphisms associated with murine virulence of *Toxoplasma gondii* identified by RAPD-PCR. *Curr. Top. Microbiol. Immunol.* **219,** 17–26.
10. Darde, M. L., Bouteille, B., and Pestre-Alexandre, M. (1988) Isoenzyme characterization of seven strains of *Toxoplasma gondii* by isoelectrofocusing in polyacrylamide gels. *Am. J. Trop. Med. Hyg.* **39,** 551–558.
11. Terry, R. S., Smith, J. E., Duncanson, P., and Hide, G. (2001) MGE-PCR: a novel approach to the analysis of *Toxoplasma gondii* strain differentiation using mobile genetic elements. *Int. J. Parasitol.* **31,** 155–161.
12. Howe, D. K., Honore, S., Derouin, F., and Sibley, L. D. (1997) Determination of genotypes of *Toxoplasma gondii* strains isolated from patients with toxoplasmosis. *J. Clin. Microbiol.* **35,** 1411–1414.
13. Mondragon, R., Howe, D. K., Dubey, J. P., and Sibley, L. D. (1998) Genotypic analysis of *Toxoplasma gondii* isolates from pigs. *J. Parasitol.* **84,** 639–641.
14. Owen, M. R. and Trees, A. J. (1999) Genotyping of *Toxoplasma gondii* associated with abortion in sheep. *J. Parasitol.* **85,** 382–384.
15. Dubey, J. P. (2001) Oocyst shedding by cats fed isolated bradyzoites and comparison of infectivity of bradyzoites of the VEG strain *Toxoplasma gondii* to cats and mice. *J. Parasitol.* **87,** 215–219.
16. Dubey, J. P. (2004) Toxoplasmosis—a waterborne zoonosis. *Vet. Parasitol.* **126(1–2),** 57–72.
17. Bahia-Oliveira, L. M., Jones, J. L., Azevedo-Silva, J., Alves, C. C., Orefice, F., and Addiss, D. G. (2003) Highly endemic, waterborne toxoplasmosis in north Rio de Janeiro state, Brazil. *Emerg. Infect. Dis.* **9,** 55–62.

18. Cole, R. A., Lindsay, D. S., Howe, D. K., et al. (2000) Biological and molecular characterizations of *Toxoplasma gondii* strains obtained from southern sea otters (*Enhydra lutris nereis*). *J. Parasitol.* **86,** 526–530.

19. Miller, M. A., Gardner, I. A., Kreuder, C., et al. (2002) Coastal freshwater runoff is a risk factor for *Toxoplasma gondii* infection of southern sea otters (*Enhydra lutris nereis*). *Int. J. Parasitol.* **32,** 997–1006.

20. Dumetre, A. and Darde, M. L. (2003) How to detect *Toxoplasma gondii* oocysts in environmental samples? *FEMS Microbiol. Rev.* **27,** 651–661.

21. Bowie, W. R., King, A. S., Werker, D. H., et al. (1997) Outbreak of toxoplasmosis associated with municipal drinking water. *Lancet* **350,** 173–177.

22. Isaac-Renton, J., Bowie, W. R., King, A., et al. (1998) Detection of *Toxoplasma gondii* oocysts in drinking water. *Appl. Environ.* **64,** 2278–2280.

23. Baril, L., Ancelle, T., Goulet, V., Thulliez, P., Tirard-Fleury, V., and Carme, B. (1999) Risk factors for *Toxoplasma* infection in pregnancy: a case–control study in France. *Scand. J. Infect. Dis.* **31,** 305–309.

24. Weigel, R. M., Dubey, J. P., Dyer, D., and Siegel, A. M. (1999) Risk factors for infection with *Toxoplasma gondii* for residents and workers on swine farms in Illinois. *Am. J. Trop. Med. Hyg.* **60,** 793–798.

25. Kapperud, G., Jenum, P. A., Stray-Pedersen, B., Melby, K. K., Eskild, A., and Eng, J. (1996) Risk factors for *Toxoplasma gondii* infection in pregnancy, results of a prospective case–control study in Norway. *Am. J. Epidemiol.* **144,** 405–412.

26. Cook, A. J. C., Gilbert, R. E., Buffolano, W., et al. (2000) Sources of *Toxoplasma* infection in pregnant women: European multicentre case control study. *Br. Med. J.* **321,** 142–147.

27. Roghmann, M. C., Faulkner, C. T., Lefkowitz, A., Patton, S., Zimmerman, J., and Morris, J. G. Jr. (1999) Decreased seroprevalence for *Toxoplasma gondii* in Seventh Day Adventists in Maryland. *Am. J. Trop. Med. Hyg.* **60,** 790–792.

28. Kean, B. H., Kimball, A. C., and Christenson, W. N. (1969) An epidemic of acute toxoplasmosis. *J. Am. Med. Assoc.* **208,** 1002–1004.

29. Lord, W. G., Boni, F., Bodek, A., Hilberg, R. W., Rosini, R., and Clack, F. B. (1975) Toxoplasmosis–Pennsylvania. *Morb. Mort. Wkly Rep.* **24,** 285–286.

30. Masur, H., Jones, T. C., Lempert, J. A., and Cherubini, T. D. (1978) Outbreak of toxoplasmosis in a family and documentation of acquired retinochoroiditis. *Am. J. Med.* **64,** 396–402.

31. Fertig, A., Selwyn, S., and Tibble, M. J. (1977) Tetracycline treatment in a food-borne outbreak of toxoplasmosis. *Br. Med. J.* **1,** 1064.

32. Choi, W. Y., Nam, H. W., Kwak, N. H., et al. (1997) Foodborne outbreaks of human toxoplasmosis. *J. Infect. Dis.* **175,** 1280–1282.

33. Sacks, J. J., Delgato, D. G., Lobel, H. O., and Parker, R. L. (1983) Toxoplasmosis infection associated with eating undercooked venison. *Am. J. Epidemiol.* **118,** 832–838.

34. Ross, R. D., Stec, L. A., Werner, J. C., Blumenkranz, M. S., Glazer, L., and Williams, G. A. (2001) Presumed acquired ocular toxoplasmosis in deer hunters. *Retina.* **21,** 226–229.

35. Sacks, J. J., Roberto, R. R., and Brooks, N. F. (1982) Toxoplasmosis infection associated with raw goat's milk. *JAMA* **248,** 1728–1732.

36. Lopez, A., Dietz, V. J., Wilson, M., Navin, T. R., and Jones, J. L. (2000) Preventing congenital toxoplasmosis. *Morb. Mort. Wkly Rep.* **49,** 59–75.

37. Dubey, J. P. (1986) A review of toxoplasmosis in pigs. *Vet. Parasitol.* **19,** 181–223.

38. Nogami, S., Tabata, A., Moritomo, T., and Hayashi, Y. (1999) Prevalence of anti-*Toxoplasma gondii* antibody in wild boar, *Sus scrofa riukiuanus*, on Iriomote Island, Japan. *Vet. Res. Commun.* **23,** 211–214.

39. Dubey, J. P., Thulliez, P., and Powell, E. C. (1995) *Toxoplasma gondii* in Iowa sows: comparison of antibody titers to isolation of *T. gondii* by bioassays in mice and cats. *J. Parasitol.* **81,** 48–53.

40. Weigel, R. M., Dubey, J. P., Siegel, A. M., et al. (1995) Prevalence of antibodies to *Toxoplasma gondii* in swine in Illinois in 1992. *J. Am. Vet. Med. Assoc.* **206**, 1747–1751.

41. Gamble, H. R., Brady, R. C., and Dubey, J. P. (1999) Prevalence of *Toxoplasma gondii* infection in domestic pigs in the New England states. *Vet. Parasitol.* **82**, 129–136.

42. Dubey, J. P., Gamble, H. R., Hill, D., Sreekumar, C., Romand, S., and Thuilliez, P. (2002) High prevalence of viable *Toxoplasma gondii* infection in market weight pigs from a farm in Massachusetts. *J. Parasitol.* **88**, 1234–1238.

43. Wyss, R., Sager, H., Muller, N., et al. (2000) The occurrence of *Toxoplasma gondii* and *Neospora caninum* as regards meat hygiene. *Schweiz Arch Tierheilkd.* **142**, 95–108.

44. Tenter, A. M., Heckeroth, A. R., and Weiss, L. M. (2000) *Toxoplasma gondii*: from animals to humans. *Int. J. Parasitol.* **30**, 1217–1258.

45. Sroka, J. (2001) Seroepidemiology of toxoplasmosis in the Lublin region. *Ann. Agric. Environ. Med.* **8**, 25–31.

46. Chan, K. F., Le Tran, H., Kanenaka, R. Y., and Kathariou, S. (2001) Survival of clinical and poultry-derived isolates of *Campylobacter jejuni* at a low temperature (4°C). *Appl. Environ. Microbiol.* **67**, 4186–4191.

47. Kotula, A. W., Dubey, J. P., Sharar, A. K., et al. (1991) Effect of freezing on infectivity of *Toxoplasma gondii* tissue cysts in pork. *J. Food Protection* **54**, 687–690.

48. Lindsay, D. S., Blagburn, B. L., Dubey, J. P., and Mason, W. H. (1991) Prevalence and isolation of *Toxoplasma gondii* from white-tailed deer in Alabama. *J. Parasitol.* **77**, 62–64.

49. Humphreys, J. G., Stewart, R. L., and Dubey, J. P. (1995) Prevalence of *Toxoplasma gondii* antibodies in sera of hunter-killed white-tailed deer in Pennsylvania. *Am. J. Vet. Res.* **56**, 172–173.

50. Vanek, J. A., Dubey, J. P., Thulliez, P., Riggs, M. R., and Stromberg, B. E. (1996) Prevalence of *Toxoplasma gondii* antibodies in hunter-killed white-tailed deer (*Odocoileus virginianus*) in four regions of Minnesota. *J. Parasitol.* **82**, 41–44.

51. Dubey, J. P., Graham, D. H., De Young, R. W., et al. (2004) Molecular and biologic characteristics of *Toxoplasma gondii* isolates from wildlife in the United States. *J. Parasitol.* **90**, 67–71.

52. Dubey, J. P., Humphreys, J. G., and Thulliez, P. (1995) Prevalence of viable *Toxoplasma gondii* tissue cysts and antibodies to *T. gondii* by various serologic tests in black bears (*Ursus americanus*) from Pennsylvania. *J. Parasitol.* **81**, 109–112.

53. Dubey, J. P., Weigel, R. M., Siegel, A. M., et al. (1995) Sources and reservoirs of *Toxoplasma gondii* infection on 47 swine farms in Illinois. *J. Parasitol.* **81**, 723–729.

54. Dubey, J. P. and Odening, K. (2001) *Parasitic Diseases of Wild Mammals*, Iowa State University Press, Ames, IA.

55. Dubey, J. P. (1988) Refinement of pepsin digestion method for isolation of *Toxoplasma gondii* from infected tissues. *Vet. Parasitol.* **74**, 75–77.

56. Warnekulasuriya, M. R., Johnson, J. D., and Holliman, R. E. (1998) Detection of *Toxoplasma gondii* in cured meats. *Int. J. Food Microbiol.* **45**, 211–215.

57. Jauregui, L. H., Higgins, J. A., Zarlenga, D. S., Dubey, J. P., and Lunney, J. K. (2001) Development of a real-time PCR assay for the detection of *Toxoplasma gondii* in pig and mouse tissues. *J. Clin. Microbiol.* **39**, 2065–2071.

58. Jones, J. L., Kruszon-Moran, D., Wilson, M., McQuillan, G., Navin, T., and McAuley, J. B. (2001) *Toxoplasma gondii* infection in the United States: seroprevalence and risk factors. *Am. J. Epidemiol.* **154**, 357–365.

59. Glasner, P. D., Silveira, C., Kruszon-Moran, D., et al. (1992) An unusually high prevalence of ocular toxoplasmosis in southern Brazil. *Am. J. Ophthalmol.* **114**, 136–144.

60. Neto, E. C., Anele, E., Rubim, R., et al. (2000) High prevalence of congenital toxoplasmosis in Brazil estimated in a 3-year prospective neonatal screening study. *Int. J. Epidemiol.* **29**, 941–947.

61. Renold, C., Sugar, A., Chave, J. P., et al. (1992) *Toxoplasma* encephalitis in patients with the acquired immunodeficiency syndrome. *Medicine* **71**, 224–239.

62. Desmonts, G. and Couvreur, J. (1974) Congenital toxoplasmosis. A prospective study of 378 pregnancies. *N. Engl. J. Med.* **290,** 1110–1116.
63. Remington, J. S., McLeod, R., Thulliez, P., and Desmonts, G. (2001) Toxoplasmosis. In: *Infectious Disease of the Fetus and Newborn* (Remington, J. S. and Klein, J. O., eds.), 5th edn, WB Saunders, Philadelphia, pp. 205–346.
64. Guerina, N. G., Hsu, H. W., Meissner, H. C., et al. (1994) Neonatal serologic screening and early treatment for congenital *Toxoplasma gondii* infection. The New England Regional *Toxoplasma* Working Group. *N .Engl. J. Med.* **330,** 1858–1863.
65. Dubey, J. P. (1997) Toxoplasmosis. In: *Microbiology and Microbial Infections, Vol. V: Parasitology,* Arnold, London.
66. Wilson, C. B., Remington, J. S., Stagno, S., and Reynolds, D. W. (1980) Development of adverse sequelae in children born with subclinical congenital *Toxoplasma* infection. *Pediatrics* **66,** 767–774.
67. Holland, G. N. (2003) Ocular toxoplasmosis: a global reassessment. Part 1: epidemiology and course of disease. *Am. J. Ophthalmol.* **136,** 973–988.
68. Roberts, T. and Frenkel, J. K. (1990) Estimating income losses and other preventable costs caused by congenital toxoplasmosis in people in the United States. *J. Am. Vet. Med. Assoc.* **196,** 249–256.
69. Roberts, T., Murrell, K. D., and Marks, S. (1994) Economic losses caused by foodborne parasitic diseases. *Parasitol. Today* **10,** 419–423.
70. Gilbert, R. and Gras, L. (2003) Effect of timing and type of treatment on the risk of mother to child transmission of *Toxoplasma gondii.* European Multicentre Study on Congenital Toxoplasmosis. *Br. J. Obstet. Gyn.* **110,** 112–120.
71. Wallon, M., Liou, C., Garner, P., and Peyron, F. (1999) Congenital toxoplasmosis: systematic review of evidence of efficacy of treatment in pregnancy. *Br. Med. J.* **318,** 1511–1514.
72. Gilbert, R., Dunn, D., Wallon, M., et al. (2001) Ecological comparison of the risks of mother-to-child transmission and clinical manifestations of congenital toxoplasmosis according to prenatal treatment protocol. *Epidemiol. Infect.* **127,** 113–120.
73. Petersen, E. and Schmidt, D. R. (2003) Sulfadiazine and pyrimethamine in the postnatal treatment of congenital toxoplasmosis: what are the options. *Expert Rev. Anti-Infect. Ther.* **1,** 175–182.
74. Teutsch, S. M., Juranek, D. D., Sulzer, A., Dubey, J. P., and Sikes, R. K. (1979) Epidemic toxoplasmosis associated with infected cats. *N. Engl. J. Med.* **300,** 695–699.
75. Benenson, M. W., Takafuji, E. T., Lemon, S. M., Greenup, R. L., and Sulzer, A. J. (1982) Oocyst-transmitted toxoplasmosis associated with ingestion of contaminated water. *N. Engl. J. Med.* **307,** 666–669.
76. Smith J. L. (1993) Documented outbreaks of toxoplasmosis: transmission of *Toxoplasma gondii* to humans. *J. Food Prot.* **56,** 630–639.
77. Burnett, A. J., Shortt, S. G., Isaac-Renton, J., King, A., Werker, D., and Bowie, W. R. (1998) Multiple cases of acquired toxoplasmosis retinitis presenting in an outbreak. *Ophthalmology* **105,** 1032–1037.
78. Luft, B. J. and Remington, J. S. (1992) Toxoplasmic encephalitis in AIDS. *Clin. Infect. Dis.* **15,** 211–222.
79. Jones, J. L., Sehgal, M., and Maguire, J. H. (2002) Toxoplasmosis-associated deaths among human immunodeficiency virus-infected persons in the United States, 1992–1998. *Clin. Infect. Dis.* **34,** 1161.
80. Montoya, J. G. (2002) Laboratory diagnosis of *Toxoplasma gondii* infection and toxoplasmosis. *J. Infect. Dis.* **185(Suppl. 1),** S73–S82.
81. Montoya, J. G., Liesenfeld, O., Kinney, S., Press, C., and Remington, J. S. (2002) VIDAS test for avidity of *Toxoplasma*-specific immunoglobulin G for confirmatory testing of pregnant women. *J. Clin. Microbiol.* **40,** 2504–2508.

82. Wilson, M., Jones, J. L., and McAuley, J. B. (2003) Toxoplasma. In: *Manual of Clinical Microbiology* (Murray, P. R., Baron, E. J., Jorgensen, J. H., Pfaller, M. A., and Yolken, R. H., eds.), 8th edn, ASM, Washington DC, pp. 1970–1980.

83. Leport, C., Chene, G., Morlat, P., et al. (1996) Pyrimethamine for primary prophylaxis of toxoplasmic encephalitis in patients with human immunodeficiency virus infection: a double-blind, randomized trial. ANRS 005-ACTG 154 Group Members. Agence Nationale de Recherche sur le SIDA. AIDS Clinical Trial Group. *J. Infect. Dis.* **173,** 91–97.

84. Morlat, P. and Leport, C. (1997) Prevention of toxoplasmosis in immunocompromised patients. *Ann. Med. Int.* **148,** 235–239.

85. Derouin, F., Gerard, L., Farinotti, R., Maslo, C., Leport, C. (1998) Determination of the inhibitory effect on *Toxoplasma* growth in the serum of AIDS patients during acute therapy for toxoplasmic encephalitis. *J. AIDS Hum. Retrovirol.* **19,** 50–54.

86. Leport, C. and Duval, X. (1999) Cerebral toxoplasmosis in an HIV-infected patient. Diagnosis, development, treatment and prevention. *Rev. Prat.* **49,** 2271–2274.

87. Gilbert, R. E. and Stanford, M. R. (2000) Is ocular toxoplasmosis caused by prenatal or postnatal infection? *Br. J. Ophthalmol.* **84,** 224–226.

88. Stanford, M., See, S. E., Jones, L. V., and Gilbert, R. E. (2003) Antibiotics for toxoplasmic retinochoroiditis, an evidence-based systematic review. *Ophthalmology* **110,** 926–931.

89. Chirgwin, K., Hafner, R., Leport, C., et al. (2002) Randomized phase II trial of atovaquone with pyrimethamine or sulfadiazine for treatment of toxoplasmic encephalitis in patients with acquired immunodeficiency syndrome: ACTG 237/ANRS 039 Study. *Clin. Infect. Dis.* **34,** 1243–1250.

90. Kaplan, J. E., Masur, H., and Holmes, K. K. (2002) Guidelines for preventing opportunistic infections among HIV-infected persons–2002. Recommendations of the US Public Health Service and the Infectious Diseases Society of America. *Morb. Mort. Wkly Rep.* **51(RR-08),** 1–46.

91. Dubey, J. P., Kotula, A. W., Sharar, A. K., Andrews, C. D., and Lindsay, D. S. (1990) Effect of high temperature on infectivity of *Toxoplasma gondii* tissue cysts in pork. *J. Parasitol.* **76,** 201–204.

92. Dubey, J. P. and Thayer, D. W. (1994) Killing of different strains of *Toxoplasma gondii* tissue cysts by irradiation under defined conditions. *J. Parasitol.* **80,** 764–767.

93. Foulon, W., Naessens, A., and Derde, M. P. (1994) Evaluation of the possibilities for preventing congenital toxoplasmosis. *Am. J. Perinatol.* **11,** 57–62.

94. Foulon, W., Naessens, A., and Ho-Yen, D. (2000) Prevention of congenital toxoplasmosis. *J. Perinat. Med.* **28,** 337–345.

95. Dubey, J. P., Hill, P. E., Jones, J. L., et al. (2005) Prevalence of viable *Toxoplasma gondii* in beef, chicken, and pork from retail meat stores in the United States: Risk assessment to consumers. *J. Parasitol.* **91,** 1082–1093.

13

Aflatoxins: Background, Toxicology, and Molecular Biology

J. W. Bennett, S. Kale, and Juijiang Yu

Abstract

Mycotoxins are mold poisons; aflatoxins are the best known and most widely studied mycotoxins. The contamination of foods and feeds with aflatoxin can have serious consequences for human and animal health. In general, aflatoxin exposure is most likely to occur in the developing countries where food handling and storage processes are suboptimal, where malnutrition is widespread, and where few regulations exist to protect the exposed populations. Depending on dose and other variables, aflatoxins can be mutagenic, carcinogenic, teratogenic, and immunosuppressive. Fundamental studies on the genetics, biosynthesis and molecular biology of aflatoxin producing fungi may offer insights into controlling this serious agricultural problem.

1. INTRODUCTION

Mycotoxins are difficult to define in a few words. All mycotoxins are low molecular weight natural products produced by filamentous fungi that are toxic to vertebrates in low concentrations. Many mycotoxins display overlapping toxicities to invertebrates, plants, and microorganisms [1,2]. Mycotoxins initially gained prominence in the early 1960s after a mysterious "Turkey X disease" killed approximately 100,000 turkey poults in England [3–5]. Turkey X disease was linked to a peanut (groundnut) meal contaminated with *Aspergillus flavus* and the toxic principles were named aflatoxins (*A. flavus* toxins). For a while, the study of toxic mold metabolites became a "hot topic" in agriculture. In fact the 15 yr between 1960 and 1975 were labeled as a "mycotoxin gold rush" because so many chemical prospectors joined the search for mycotoxins [6]. Eventually several hundred fungal metabolites with toxic properties were isolated. The best single compendium for accessing the structures and chemical profiles of these toxic compounds is the three-volume *Handbook of Secondary Fungal Metabolites [7]*.

Mycotoxins are commonly found in foods and feeds all over the world. It has been estimated that a quarter of the world's crops are contaminated to some extent with mycotoxins [8,9]. Kuiper-Goodman, a leading figure in the risk assessment field, ranks mycotoxins as the most important noninfectious, chronic dietary risk factor, higher than synthetic contaminants, plant toxins, food additives, or pesticide residues [10].

The mycotoxin literature is enormous. Since aflatoxins are the most important mycotoxin, each major monograph devotes considerable attention to the aflatoxin problem [11–27]. Many reviews focus specifically on mycotoxins and risks related to human health [2,9,28–34].

From: *Infectious Disease: Foodborne Diseases*
Edited by: S. Simjee © Humana Press Inc., Totowa, NJ

2. HISTORY OF AFLATOXIN RESEARCH

The classic monograph, *Aflatoxin. Scientific Background, Control, and Implications (35)* is an excellent source for learning about the early history of mycotoxin research, and summarizes the early chemical and toxicological studies. To reiterate, the aflatoxins were isolated and characterized after the Turkey X disease was traced to a family of metabolites found in mold contaminated feed. The major aflatoxins were called B_1, B_2, G_1, and G_2 (Fig. 1) on the basis of their blue or green fluorescence under ultraviolet light, and relative chromatographic mobility during silica gel thin-layer chromatography. In addition to the four major aflatoxins produced by mold metabolism, about a dozen other aflatoxins (e.g., P_1, Q_1, B_{2a}, and G_{2a}) were described, especially as mammalian biotransformation products of the major metabolites *(7,18,36)*. For example, cows metabolize aflatoxin B_1 from cattle feed into a hydroxylated derivative called aflatoxin M_1 that is then secreted in milk *(37)*.

Early toxicological studies focused on the acute toxic effects of aflatoxins on animals, and demonstrated that ducklings, hamsters, rabbits, trout, rats, and a number of other vertebrates were all susceptible. Soon it was discovered that aflatoxins administered in lower doses over longer periods of time could induce tumors, particularly in the liver. Rats and trout were highly susceptible to the carcinogenic effect of aflatoxin B_1. Ten percent of trout fed a diet containing 20 ppb of aflatoxin B_1 for as little as 3 d had developed hepatomas a year later; and rats fed a single oral does of 5–7 mg of aflatoxin B_1 developed liver tumors. On the other hand, mice, hamsters, and, by extrapolation, many other untested mammalian species were relatively resistant *(35,38)*. In addition, the mechanisms of acute toxicity and aflatoxin-mediated carcinogenicity seemed to be quite different. Rats were more susceptible to aflatoxin-induced hepatocarcinoma than were hamsters yet the acute LD_{50}s for the species were similar *(11)*. Aflatoxin B_1 was recognized as the most powerful naturally occurring carcinogen ever discovered *(39)*.

It should be pointed out that not all the authors distinguished between the term "aflatoxin" (the generic family of toxins) and "aflatoxin B_1" (usually the major aflatoxin produced by toxigenic strains of *Aspergillus*). Most toxicological studies have been conducted using aflatoxin B_1.

3. FUNGI AND PHYSIOLOGY

Aflatoxins enter the food chains when toxigenic molds grow on foods and feeds. For several decades, *A. flavus* and *Aspergillus parasiticus* were thought to be the only species capable of producing aflatoxins. Then, in 1987, it was reported by Kurtzman et al. *(40)* that *Aspergillus nomius*, a species closely related to *A. flavus*, was also aflatoxigenic. More recently, a number of other aflatoxin-producing species have been described: *A. bombycis (41)*, *A. ochraceoroseus (42,43)*, *A. pseudotamari (44)*, and *A. tamarii (45)*, as well as *Emericella astellata (46)* and *Emericella venezuelensis* (Klich, unpublished data). Compared to *A. flavus* and *A. parasiticus*, these species are less abundant in nature and are rarely encountered in agriculture.

Within *A. flavus* and *A. parasiticus*, different strains display a great deal of qualitative and quantitative difference in their toxigenic abilities. For example, it has been estimated

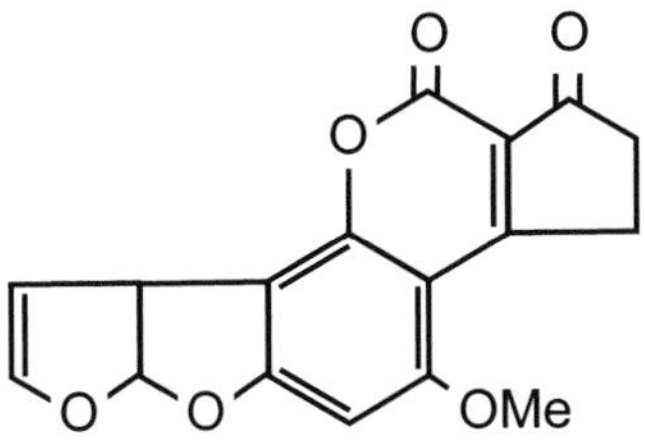

Fig. 1. Major aflatoxins.

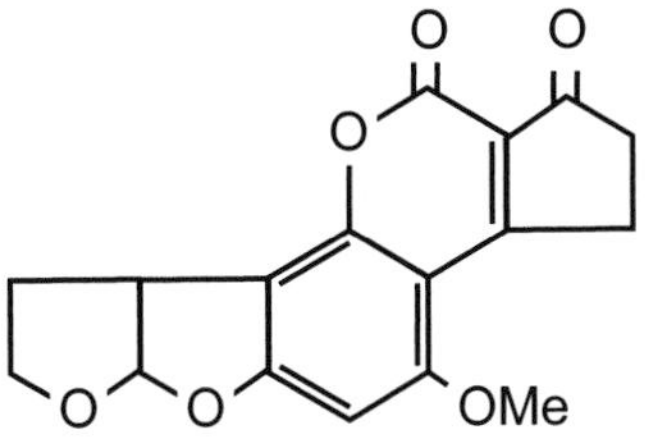

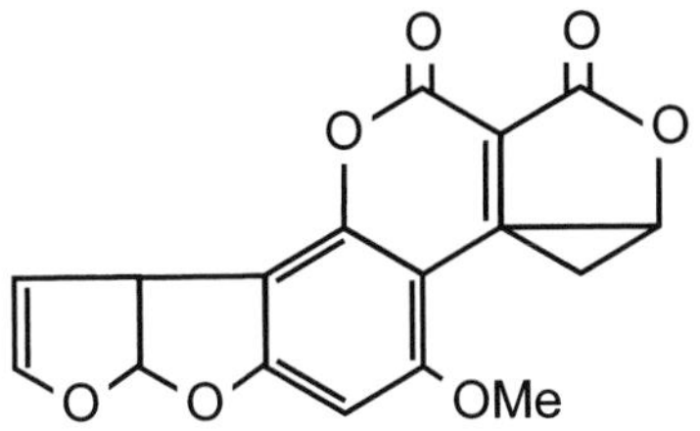

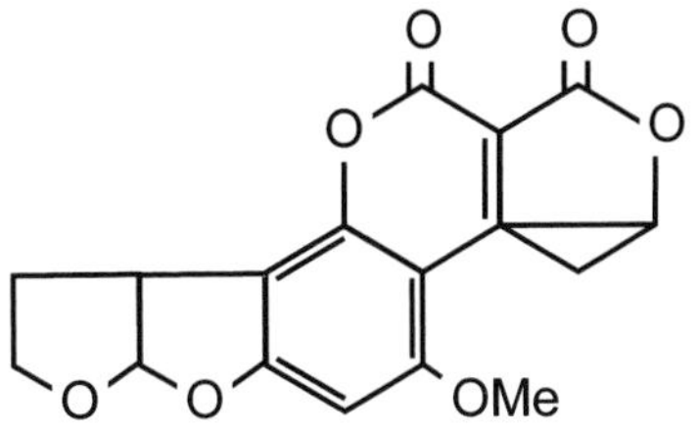

that only about the half of *A. flavus* isolates produce aflatoxins *(47)*. The total amount of toxin biosynthesized will vary with the strain of the fungus and with the growth conditions. Moisture, temperature, and insect damage are the most important environmental variables associated with aflatoxin contamination of agricultural commodities *(48)*. Corn, peanuts, cotton, tree nuts, rice, figs, tobacco, and spices are among the most frequently contaminated crops *(49,50)*.

Crops often become contaminated with aflatoxin in the field before harvest, and especially during drought years, the plants are weakened and become more susceptible to insect damage and other insults *(50–52)*. Once harvested, stored grains are also at a high risk of being contaminated by aflatoxins. In storage, the most important variables favoring mold growth are the moisture content of the substrate and the relative humidity of the surroundings *(49,53)*.

4. TOXICOLOGY

Cytochrome P450 enzymes convert aflatoxins to the reactive 8,9-epoxide form (originally called aflatoxin-2,3 epoxide), which in turn can bind to both DNA and proteins *(54)*. In DNA, the reactive aflatoxin epoxide binds to the N^7 position of guanines, and the resultant adducts can cause GC to TA transversions. A glutathione-*S*-transferase system catalyzes the conjugation of activated aflatoxins with reduced glutathione, leading to their excretion *(55)*. Variation in the level of the glutathione-transferase system, as well as variations in the cytochrome P450 system, is hypothesized to explain the differences observed in interspecific aflatoxin susceptibility *(38,54)*.

Like all toxicological syndromes, the diseases caused by aflatoxins are categorized as either acute or chronic. Acute toxicity has a rapid onset and clearly defined symptoms. Chronic toxicity is harder to diagnose and is characterized by low-dose exposure over a long time-period resulting in cancer and other generally irreversible effects (56). It is not always possible to distinguish between acute and chronic effects.

Aflatoxin is associated with both acute and chronic toxicities in human and animal populations (54,57–59). The disease syndromes caused by aflatoxin consumption are termed "aflatoxicoses." Acute aflatoxicosis results in death; chronic aflatoxicosis results in cancer, immune suppression, and other "slow" pathological conditions (60). In both cases, the liver is the primary target organ, and the sensitivity to aflatoxin differs from species to species. Within species, the extent of the response is influenced by age, diet, sex, exposure to pathogens, and the presence of other toxins. There is an abundance of literature on the effects of aflatoxin exposure on laboratory models and agriculturally important species (54,57,61).

The best-known aflatoxin episodes are manifestations of acute effects such as Turkey X syndrome, but the main human and veterinary health burden of aflatoxin exposure is related to chronic exposure (e.g., cancer induction, impairment of liver function, immune suppression). Establishing that a human disease is associated with aflatoxin poisoning is difficult. In general, epidemiologists look for a correlation between a suspected disease condition and the presence of aflatoxin in the diet. Then laboratory scientists attempt to reproduce the characteristic disease symptoms in animal models (60). Environmental and biological monitoring is important in assessing human aflatoxin exposure. In environmental monitoring, aflatoxins are assayed from foods and feeds; in biological monitoring, aflatoxin residues, adducts, and biotransformation products are assayed from blood, milk, tissues, feces, and urine samples (60,62). The aflatoxin B_1-N^7-guanine adduct is a reliable urinary biomarker for detecting recent exposure, and laboratory studies have shown that carcinogenic potency is highly correlated with the extent of total DNA adducts formed in vivo (54,63). Finally, it should be noted that aflatoxin effects are not limited to the liver. The nonhepatic effects of aflatoxin B_1 have been summarized by Coulombe in 1994 (64).

5. AFLATOXINS AND HUMAN DISEASE

Because of the differences in aflatoxin susceptibility in test animals, it has been difficult to extrapolate data from animal models to human disease. The general consensus is that humans are relatively resistant and that acute aflatoxicosis in *Homo sapiens* is rare (65). A 1974 outbreak of hepatitis in India, in which 100 people died, may have been caused by the consumption of maize that was heavily contaminated with aflatoxin (66). Subsequently, it was estimated that the acute lethal dose for adults is approximately 10–20 mg of aflatoxins (67). Nevertheless, at least one woman survived ingestion of over 40 mg purified aflatoxin in an unsuccessful suicide attempt (68).

Kwashiorkor, a severe malnutrition disease, has been called a form of pediatric aflatoxicosis (69), but animal data refute this conjecture (70). Similarly, early hypotheses that aflatoxin might be involved in Reye's syndrome (71) have not been substantiated.

Although the quantification of lifetime individual exposure to aflatoxin is extremely difficult, several epidemiological studies have linked liver cancer incidence to estimated aflatoxin consumption in the diet, particularly in individuals already exposed to hepatitis B infections *(59,72,73)*. Liver cancer incidence varies widely from country to country, but it is a common cancer in China, the Philippines, Thailand, and many African countries. The presence of hepatitis B virus infection complicates the epidemiological studies. In one case–control study involving more than 18,000 urine samples collected over 3.5 yr in Shanghai, it was estimated that the combination of aflatoxin and hepatitis B raised the cancer risk 30-fold over that for aflatoxin alone *(74)*. Because it is easier to vaccinate against hepatitis B virus than to remove aflatoxin from the diet, vaccination has been recommended as the most cost-effective strategy for lowering liver cancer in susceptible populations *(75,76)*.

A significant number of liver cancer patients in Africa and China have a mutation in the p53 tumor suppressor gene at codon 249 associated with a G to T transversion *(77,78)*. Because it is known that the reactive aflatoxin epoxide binds to the N^7 position of guanines and that aflatoxin B_1-DNA adducts can result in GC to TA transversions, these data add further support to the evidence that aflatoxin B_1 is a human carcinogen. Eaton and Gallagher *(63)* have written that this codon-specific change in the *p53* gene is the first example of a fixed "carcinogen-specific" biomarker.

In summary, there is no other natural product for which the evidence for human carcinogenicity is so compelling. Aflatoxin B_1 is classified as a Group I carcinogen by the International Agency for Research on Cancer *(79)*. This notoriety may explain why aflatoxins have been implicated as chemical warfare agents. In 1995, it was determined that Iraq had produced and deployed war instruments containing botulism toxin, anthrax spores and aflatoxins. International forensic teams showed that toxigenic strains of *A. flavus* and *A. parasiticus* were grown in Iraqi government sponsored facilities and aflatoxins were harvested to produce over 2300 L of concentrated toxin. The majority of this aflatoxin was used to fill warheads; the remainder was stockpiled *(80,81)*. Nonetheless, because of the large amounts of toxin necessary to cause disease and the relatively slow mode of action, aflatoxins were a strange choice for a bioterrorist *(82)*.

6. TREATMENT AND CONTROL

Most aflatoxicosis results from eating contaminated foods. Unfortunately, except for supportive therapy (e.g., diet and hydration) there are almost no treatments for aflatoxin exposure. Fink-Gremmels *(9)* has described a few methods for veterinary management of mycotoxicoses, and there is some evidence that some strains of *Lactobacillus* effectively bind dietary mycotoxins *(83–85)*. Similarly, clay-based enterosorbents have been used to bind aflatoxins in the gastrointestinal tract *(86,87)*. Selenium supplementation somewhat modified the negative effects of aflatoxin B_1 in Japanese quail *(88)*, and butylated hydroxytoluene gave some protection in turkeys *(89)*. Oltipraz, a drug originally used to treat schistosomiasis, has been tested in human populations in China with some apparent success *(76)*.

Methods for controlling aflatoxin exposure are largely prophylactic. Such methods include good agricultural practice, appropriate drying of crops after harvest and avoidance of moisture during storage *(90)*. Many agricultural scientists are trying to develop methods to minimize the preharvest contamination of crops. These approaches include developing host resistance through plant breeding and the use of biocontrol agents *(25)*.

Most efforts to address the mycotoxin problem involve analytic detection, government regulation, and diversion of mycotoxin-contaminated commodities from the food supply. Basic research on the biosynthesis and molecular biology of aflatoxins (*see* Section 8) has been a research priority because a full understanding of the fundamental biological processes may yield new control strategies for the abolition of aflatoxin contamination of food crops.

7. ECONOMICS, FOOD SAFETY, AND REGULATION

The economic consequences of aflatoxin contamination are extreme. In developed countries, crops with high amounts of aflatoxins are either destroyed or diverted into animal feeds; aflatoxins lower the value of grains as an animal feed and as an export commodity *(21)*. When susceptible animals are fed contaminated feeds it results in reduced growth rates, illness, and death; moreover, their meat and milk may contain toxic biotransformation products. Livestock owners often take farmers and feed companies to court; legal battles can involve considerable amounts of money *(91)*.

Numerous assay methods for detecting aflatoxins have been developed utilizing virtually all of the common tools of analytical chemistry including thin-layer chromatography, high-performance liquid chromatography, gas chromatography, mass spectrometry, immunoassays, capillary electrophoresis, and biosensors. Older methods usually require solvents for clean-up steps and chromatography for quantification; more recently, immunogenic assays that can be applied to samples with little or no clean up have been developed *(92)*. Aflatoxins are nonimmunogenic but they can be conjugated to a protein carrier; a number of inexpensive antibody-based kits are now commercially available. Methods for assaying aflatoxins and other mycotoxins have been reviewed *(26,93–95)*.

Because it is normally impossible to prevent entirely the formation of aflatoxins, complete elimination is an unattainable objective. Naturally occurring toxins such as aflatoxins are regulated quite differently than food additives *(96)*. In developed countries, human populations are protected because regular surveillance keeps contaminated foods out of the food supply. Unfortunately, in countries where populations are facing starvation, or where regulations are either nonexistent or unenforced, routine ingestion of aflatoxin is common *(97)*. A joint FAO/WHO/UNEP conference report pointed out that hungry people "cannot exercise the option of starving to death today in order to live a better life tomorrow" and statistics show that the incidence of liver cancer is 2–10 times higher in developing countries than in developed countries *(75)*.

Special committees and commissions have been established by many countries and international agencies to recommend guidelines, test standardized assay protocols, and maintain up-to-date information on regulatory statutes of aflatoxins and other mycotoxins. These guidelines are developed from epidemiological data and extrapolations from animal models, taking into account the inherent uncertainties associated with both types of analysis. Estimates of "safe doses" are usually stated as a "tolerable daily intake" *(10,31,98)*. For example, in the United States, the Food and Dug Administration guideline is 20 ppb total aflatoxin in food destined for human consumption and 100 ppb is the limit for breeding cattle and mature poultry *(99)*.

Different national guidelines for safe doses have been established, and hence, there is a need for worldwide harmonization of regulations *(100)*. A compendium summarizing worldwide regulations for mycotoxins has been published by the Food and Agriculture Organization of the United Nations *(96)*; an abbreviated version was given as an appendix by Weidenborner *(101)*.

The websites for the various commissions and organizations that study mycotoxins are excellent sources for the latest information: see, e.g., the Council for Agricultural Science and Technology (CAST) (www.cast-science.org); the American Oil Chemists Society Technical Committee on Mycotoxins (www.aocs.org); the Food and Agricultural Organization (FAO) of the United Nations (www.fao.org); the International Union for Pure and Applied Chemistry section on Mycotoxins and Phycotoxins (www.iupac.org); and the US Food and Drug Administration Committee on Additives and Contaminants (www.fda.gov).

8. BIOCHEMISTRY, MOLECULAR BIOLOGY, AND GENOMICS

The severity of the potential health effects, and the magnitude of the economic losses, have been the impetus for considerable research. Many scientists, including ourselves, believe that control can best be achieved by understanding the genetic basis of aflatoxin biosynthesis and the regulatory elements that control the biosynthetic pathway. To this end, the US Department of Agriculture and other US and international funding agencies have supported decades of basic research on the molecular biology of aflatoxin biosynthesis. More recently, the US Department of Agriculture has funded an *A. flavus* genome-sequencing project that is expected to be completed by early 2005.

Molecular research has targeted the genetics, biosynthesis, and regulation of aflatoxin formation in *A. flavus* and *A. parasiticus*. Aflatoxins are biosynthesized by a type II polyketide synthase; it has been known for a long time that the first stable step in the biosynthetic pathway is the norsolorinic acid, an anthraquinone *(102)*. A complex series of post-polyketide synthase steps follow, yielding a series of increasingly toxigenic anthraquinone and difurocoumarin metabolites *(103–113)*. Sterigmatocystin (ST) is a late metabolite in the aflatoxin pathway and is also produced as a final biosynthetic product by a number of species such as *Aspergillus versicolor* and *Aspergillus nidulans*. ST is less potent than aflatoxin but is nevertheless an important mycotoxin in its own right *(114)*. Perhaps more importantly, analysis of ST biosynthesis in the genetically traceable species *A. nidulans* was pivotal in accelerating research on the cognate pathway for aflatoxin. It is now known that ST and aflatoxins share almost identical biochemical pathways. The majority of the genes for both ST biosynthesis in *A. nidulans*, and aflatoxin pathway biosynthesis in *A. flavus* and *A. parasiticus* are homologous and clustered *(106,107,111,112,115–119)*. In *A. flavus* and *A. parasiticus*, a total of 25 genes involved in aflatoxin biosynthesis, along with four sugar utilization genes, are located together within a 70-kb region of DNA *(111,112,117,120)*. Recently, a new standardized system for naming the aflatoxin pathway genes has been introduced *(111)*. A diagram depicting the clustered genes of aflatoxin and ST biosynthesis and the verified post-polyketide biosynthetic steps in this pathway is shown in Fig. 2. The expression of the structural genes in both aflatoxin and ST biosynthesis is regulated by a regulatory gene, *aflR*, which encodes a GAL4-type C6 zinc binuclear DNA-binding protein. When *aflR* is disrupted, no structural gene transcript can be detected; introduction of an additional copy leads to overproduction of aflatoxin biosynthetic pathway intermediates *(121)*. The overall amino-acid identity is 31% between the *aflR* genes from *A. flavus* and *A. nidulans*, but the nuclear localization signal domain and the cys_6-Zn_2 domain are 71% identical. The immediate downstream linker region is also highly conserved; substitution of amino-acids in

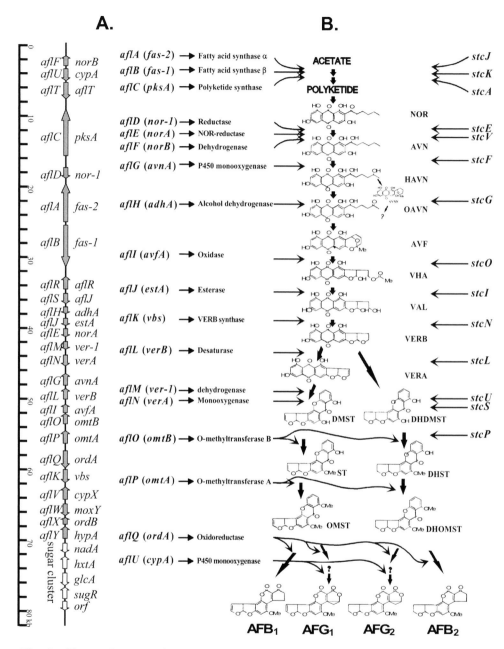

Fig. 2. Clustered genes (**A**) and the aflatoxin biosynthetic pathway (**B**). The generally accepted pathway for aflatoxin and sterigmatocystin (ST) biosynthesis is presented in Panel B. The corresponding genes and their enzymes involved in each bioconversion step are shown in Panel A. The vertical line represents the 82-kb aflatoxin biosynthetic pathway gene cluster and sugar utilization gene cluster in *A. parasiticus* and *A. flavus*. The new gene names are given on the left of the vertical line and the old names are given on the right. Arrows along the vertical line indicate the direction of gene transcription. The ruler at far left indicates the relative sizes of these genes in kilobase pairs. The ST biosynthetic pathway genes in *A. nidulans* are indicated at the right of Panel B. Arrows in Panel B indicate the connections from the genes to the enzymes they encode, from the enzymes to the bioconversion steps they are involved in, and

the linker region results in defective *aflR* expression *(122)*. Details of the promoter structure of *aflR* are reviewed by Yu et al. *(109)*.

A divergently transcribed gene, *aflJ*, is also involved in the regulation of the aflatoxin gene cluster; no aflatoxin pathway intermediates are produced when it is disrupted. The gene product of *aflJ* has no sequence homology to known proteins identified in databases *(123)*; it interacts with *aflR* but not the structural genes of the pathway *(124)*. It has been speculated that *aflJ* is an *aflR* coactivator *(109,111)*.

Aspergillus oryzae and *A. sojae* are nontoxigenic species that are widely used in Asian food fermentations such as soy sauce, miso, and sake. These food fungi are closely related to *A. flavus* and *A. parasiticus*. Although they never have been shown to produce aflatoxin, they do contain homologs of several aflatoxin biosynthetic pathway genes *(125,126)*. *A. sojae* contains a defective copy of *aflR* *(127,128)*. Other genetic defects have crippled the aflatoxin pathway in *A. oryzae* *(128,129)*.

The application of genomic DNA sequencing and functional genomics, powerful technologies that allow scientists to study a whole set of genes in an organism, is one of the most exciting developments in aflatoxin research *(130)*. The Food and Feed Safety Unit of the USDA–ARS, Southern Regional Research Center, New Orleans, LA, has sponsored an *A. flavus* expressed sequence tag (EST)/Microarray project. A normalized cDNA library was made. From over 26,000 clones sequenced, 7218 unique ESTs (genes) were identified after comparison and assembly *(131)*. Homology analysis by a BLAST search in the GenBank database indicated that 66% of these unique genes had identified homologs in the GenBank database and 34% unique ESTs had no identified homologs. Bioinformatics annotation identified many genes that are putatively involved in the aflatoxin process including signal transduction, global regulation, pathogenicity, virulence, stress response, and fungal development in addition to the genes of the biosynthetic pathway *(131)*.

These data will be useful in annotating the forthcoming genome sequence of *A. flavus* and in designing future microarray experiments to investigate the relationship between developmental and secondary metabolite genes (*see* Section 9). *A. flavus* microarrays have been constructed at The Institute for Genomic Research (TIGR), Rockville, MD. High density microarrays have been printed of 6684 short amplicons representing 5002 unique gene elements including 31 aflatoxin pathway genes. These microarrays are being used in time-course studies to detect sets of fungal genes transcribed under specific conditions at different developmental stages.

9. AFLATOXINS AND FUNGAL DEVELOPMENT

The association between fungal morphological development and secondary metabolism, including aflatoxin production, has been observed for many years *(132–135)*.

---

from the intermediates to the products in the aflatoxin bioconversion steps. *Abbreviations*: NOR, norsolorinic acid; AVN, averantin; HAVN, 5′-hydroxyaverantin; OAVN, oxoaverantin; AVNN, averufanin; AVF, averufin; VHA, versiconal hemiacetal acetate; VAL, versiconal; VERB, versicolorin B; VERA, versicolorin A; DMST, demethylsterigmatocystin; DHDMST, dihydrodemethylsterigmatocystin; ST, sterigmatocystin; DHST, dihydrosterigmatocystin; OMST, *O*-methylsterigmatocystin; DHOMST, dihydro-*O*-methylsterigmatocystin; AFB$_1$, aflatoxin B$_1$; AFB$_2$, aflatoxin B$_2$; AFG$_1$, aflatoxin G$_1$; and AFG$_2$, aflatoxin G$_2$.

The environmental conditions required for secondary metabolism and for sporulation are similar *(132,133)*, and both processes occur at about the same time *(105,136)*. Certain compounds in *A. parasiticus* that exhibit the ability to inhibit sporulation also inhibit aflatoxin formation *(137)*. Further, chemicals that inhibit polyamine biosynthesis in *A. parasiticus* and *A. nidulans* inhibit both sporulation and aflatoxin/ST biosynthesis *(138)*. Some sporulation deficient mutants of *A. parasiticus* are unable to produce aflatoxins *(139–141)*, and there is some evidence of an association between sclerotia and aflatoxin production in these species *(142)*. Similarly, a nonsporulating, "fluffy" mutant strain of *A. nidulans* is deficient in ST formation *(143,144)*. In fact, *A. nidulans*, long a well-known system for studying fungal development *(145)* is emerging as a model for studying the global regulation of both development and secondary metabolism.

Two distinct classes of *A. nidulans* mutants unable to make secondary metabolites were isolated in 1999 *(146)*. One group of mutants showed morphological defects, while the other had the wild-type parental morphology. Physiological and genetic complementation analyses of these mutants suggested that there were factors distinct from both the *aflR* gene and the developmental genes that controlled ST production. Almost simultaneously, these researchers reported that both asexual sporulation and ST production required the inactivation of proliferative growth through inhibition of the FadA (G-protein) signaling pathway *(136)* and identified a gene called *pkaA* (protein kinase A) as a component of this pathway *(147)*.

FadA is the α subunit of the *A. nidulans* heterotrimeric G-protein. When FadA was bound to GTP and in its active form, ST production and sporulation were repressed. However, in the presence of FlbA, the intrinsic GTPase activity of FadA was stimulated, thereby leading to GTP hydrolysis, inactivation of FadA-dependent signaling, and stimulation of ST production. In brief, the G-protein signal transduction pathway mediated by protein kinase A (PKA) regulated both aflatoxin/ST synthesis and sporulation *(136)*.

In the process of characterizing *A. nidulans* fluffy mutants, six loci were identified to be the results of recessive mutations in the fluffy genes *fluG, flbA, flbB, flbC, flbD,* and *flbE*. Two of these genes, *fluG* and *flbA* encoding protein factors FluG and FlbA, were involved in the regulation of both asexual development (conidiation) and ST biosynthesis in *A. nidulans (136,148)*. The *fluG* gene is involved in the synthesis of an extracellular diffusible factor that acts upstream of *flbA*. The *pkaA* gene encodes the catalytic subunit of a cyclic AMP (cAMP)-dependent protein kinase A, PkaA *(136,147)*. Over-expression of *pkaA* (PkaA) inhibits *brlA* and *aflR* expression *(147)*. The gene, *brlA*, in *A. nidulans* encodes a transcriptional regulator (BrlA) long believed to activate developmental genes *(149)*. A domain of the FlbA protein, the regulator of the G-protein signaling RGS, presumably is able to inhibit FadA *(148)*. In the overall scheme of the proposed G-protein signaling pathway, FadA and PkaA favor vegetative growth and inhibit conidiation and aflatoxin/ST production; while FluG and FlbA inhibit FadA and PkaA function and promote conidiation and aflatoxin/ST biosynthesis *(135,136,148)*. This G-protein signaling pathway involving FadA in the regulation of secondary metabolism may also exist in other aspergilli such as *A. parasiticus*.

The manner in which complex interactions among the components of this pathway (FlbA, FadA, and Pka proteins) and RasA (a member of the family of small GTP-binding proteins) influence *aflR* at both transcriptional and post-transcriptional levels is under investigation *(150)*.

Another link between sporulation and/or other aspects of development and aflatoxin biosynthesis has been studied using an unusual class of aflatoxin-negative mutants called *sec-* (for secondary metabolism minus). These strains were isolated after serial transfer of nonsporulating mycelial macerates. They exhibit reduced sporulation and no detectable aflatoxin production *(151)*. They are unable to bioconvert aflatoxin precursors to aflatoxins although Southern blot and polymerase chain reaction analysis demonstrated that the structural genes for pathway enzymes are present *(152)*. In the *sec-* strains, *aflR* expression is 5–10-fold lower than in the toxigenic strains from which they are derived, adding more evidence to the theory that *aflR* is necessary but not sufficient for aflatoxin production *(153)*. A different and possibly related morphological mutant in *A. parasiticus* called *fluP* causes a fluffy hyphal morphology, reduction of asexual spores, and a lowering of aflatoxin production *(154)*.

Finally, the *laeA* gene (for loss of *aflR* expression) is yet another intriguing discovery *(146)*. Originally isolated from *A. nidulans,* it encodes a putative nuclear methyltransferase and transcriptionally regulates several secondary metabolic pathways. Disruption of *laeA* (accession no. AY394722) in *A. nidulans* eliminates ST and penicillin biosynthesis due to the loss of gene expression *(aflR, stcU)* required for ST biosynthesis and a gene *(ipnA)* involved in penicillin biosynthesis. Disruption of the *laeA* in *A. fumigatus* (accession no. AY422723) and in *A. terreus* *(lovE)* eliminated gliotoxin and lovastatin biosynthesis, respectively *(155)*. The ST pathway regulator, AflR, the PKA, *(147)* and RasA, known to be involved in signal transduction and which negatively regulate sterimgatocystin biosynthesis and asexual sporulation in *A. nidulans (135,136)*, negatively regulate *laeA* expression *(155)*. It is possible that the LaeA protein is one of the global regulatory components in the signal transduction pathway that controls secondary metabolism pathways. The corresponding *laeA* gene in *A. flavus* and in *A. parasiticus* has been cloned. Although very low homology exists between *A. flavus*/*A. parasiticus* and *A. nidulans* at the nucleotide level, significant homology was observed at the amino-acid level (Yu, unpublished data). The *laeA* gene may have similar effects in regulating secondary metabolism pathway in *A. flavus*, e.g., aflatoxin biosynthesis.

The proliferation of new approaches to the study of secondary metabolism and morphological development, combined with the power of functional genomics, give us reason to hope that we are on the brink of a new era of molecular understanding of aflatoxin gene regulation.

10. SUMMARY

Aflatoxins are toxic and carcinogenic natural products biosynthesized by a polyketide pathway by certain members of the genus *Aspergillus*. When people and domestic animals eat aflatoxin contaminated foods, they could suffer both acute and chronic diseases. Of aflatoxigenic mold species, *A. flavus* and *A. parasiticus* are considered to be the most economically important. Aflatoxins can enter the food chain when these toxigenic species grow on commodities in the field, during storage, or at later points. Aflatoxin contamination is exacerbated whenever agricultural, storage, shipping, and food handling practices are conducive to mold growth. The acute and chronic effects of aflatoxin are largely avoided by preventative strategies good agricultural practice, government monitoring and regulation, and diversion of contaminated crops from the food supply. Unfortunately, strict limitation of aflatoxin contaminated food and feed is not always an option and there are almost no treatments for aflatoxin poisoning. It is hoped that the research on the molecular biology of aflatoxigenic fungi, including a forthcoming genome sequence and microarray analysis of the *A. flavus* genome, will lead to better methods for blocking the production of this family of terrible food borne toxins.

ACKNOWLEDGMENTS

At the Southern Regional Research Center we thank Deepak Bhatnagar, Jeff Cary, Perng-Kuang Chang, Ed Cleveland, Ken Ehrlich, Maren Klich and other members of the mycotoxin group for their many collaborations over the years. Gary Payne (North Carolina State) and Nancy Keller (University of Wisconsin) have also been instrumental contributors to our group efforts. Research in J. W. Bennett's lab has been supported by a cooperative agreement from the US Department of Agriculture; research in S. P. Kale's lab has been supported by the NIH (MBRS) grant no. 5SO6GM08008-31.

REFERENCES

1. Bennett, J. W. (1987) Mycotoxins, mycotoxicoses, mycotoxicology and Mycopathologia. *Mycopathlogia* **100**, 3–5.
2. Bennett, J. W. and Klich, M. (2003) Mycotoxins. *Clin. Microbiol Rev.* **16**, 497–516.
3. Blout, W. P. (1961) Turkey "X" disease. *Turkeys* **9**, 52, 55–58, 61, 77.
4. Forgacs, J. (1962) Mycotoxicoses—the neglected diseases. *Feedstuffs* **34**, 124–134.
5. Forgacs, J. and Carll, W. T. (1962) Mycotoxicoses. *Adv. Vet. Sci.* **7**, 273–382.
6. Maggon, K. K., Gupta, S. K., and Venkitasubramanian, T. A. (1977) Biosynthesis of aflatoxins. *Bacteriol. Rev.* **41**, 822–855.
7. Cole, R. J. and Schweikert, M. A. (2003) *Handbook of Secondary Fungal Metaabolites*, Vols. 1–3. Academic/Elsevier, Amsterdam.
8. Mannon, J. and Johnson, E. (1985) Fungi down on the farm. *New Sci.* **195**, 12–16.
9. Fink-Gremmels, J. (1999) Mycotoxins: their implications for human and animal health. *Vet. Quart.* **21**, 115–120.
10. Kuiper-Goodman, T. (1998) Food safety: mycotoxins and phycotoxins in perspective. In: *Mycotoxins and Phycotoxins—Developments in Chemistry, Toxicology and Food Safety* (Miraglia, M., van Edmond, H., Brera, C., and Gilbert, J., eds.), Alaken, Fort Collins, CO, pp. 25–48.
11. Wogan, G. N. (ed.) (1965) *Mycotoxins in Foodstuffs*, MIT, Cambridge, MA.
12. Ciegler, A., Kadis, S., and Ajl, S. J. (eds.) (1971) *Microbial Toxins, Vol. VI: Fungal Toxins*, Academic Press, New York.
13. Kadis, S., Ciegler, A., and Ajl, S. J. (eds.) (1972) *Microbial Toxins, Vol. VIII: Fungal Toxins*, Academic, New York.
14. Rodricks, J. V., Hesseltine, C. W., and Mehlman, M. A. (eds.) (1977) *Mycotoxins in Human and Animal Health*, Pathotox Publishers, Park Forest South, IL.
15. Wyllie, T. D. and Morehouse, L. G. (eds.) (1977) *Mycotoxic Fungi, Mycotoxins, Mycotoxicoses: An Encyclopedic Handbook*, Vol. I, Marcel Dekker, New York.
16. Wyllie, T. D. and Morehouse, L. G. (eds.) (1978) *Mycotoxic Fungi, Mycotoxins, Mycotoxicoses: An Encyclopedic Handbook*, Vol. II. Marcel Dekker, New York.
17. Steyn, P. S. (ed.) (1980) *The Biosynthesis of Mycotoxins: A Study in Secondary Metabolism*, Academic, New York.
18. Cole, R. J. and Cox, R. H. (1981) *Handbook of Toxic Fungal Metabolites*, Academic, New York.
19. Betina, V. (ed.) (1984) *Mycotoxins: Production, Isolation, Separation, and Purification*, Elsevier, Amsterdam.

20. Lacey, J. (1985) Factors affecting mycotoxin production. Teoksessa: Steyn, P. S. ja Vleggaar, R. (toim.). *Mycotoxins and Phytotoxins*. A Collection of Invited Papers Presented at the Sixth International IUPAC Symposium on Mycotoxins, Pretoria, RSA, 22–25 July 1985. Elsevier, Amsterdam, Sivut 65–76.

21. Smith, J. E. and Moss, M. O. (eds.) (1985) *Mycotoxins. Formation, Analyses and Significance*. John Wiley & Sons, Chichester, UK.

22. Betina, V. (1989) *Mycotoxins: Chemical, Biological and Environmental Aspects*, Elsevier, Amsterdam.

23. Smith, J. E. and Anderson, R. A. (eds.) (1991) *Mycotoxins and Animal Foods*, CRC, Boca Raton, FL.

24. Bhatnagar, D. and Cleveland, T. E. (1992) Molecular strategies for reducing aflatoxin levels in crops before harvest. In: *Molecular Approaches to Improving Food Quality and Safety* (Bhatnagar, D. and Cleveland, T. E., eds.), Chap. 9, Van Nostrand Reinhold, New York, pp. 205–228.

25. Sinha, K. K. and Bhatnagar, D. (eds.) (1998) *Mycotoxins in Agriculture and Food Safety*, Marcel Dekker, New York.

26. Trucksess, M. W. and Pohland, A. E. (eds.) (2001) *Mycotoxin Protocols*, Humana, Totowa, NJ.

27. De Vries, J. W., Trucksess, M. W., and Jackson, L. S. (eds.) (2002) *Mycotoxins and Food Safety*. Kluwer Academic/Plenum, New York.

28. CAST. (1989) Mycotoxins: *Economic and Health Risks*. Council for Agricultural Science and Technology Task Force Report No. 116, Ames, IA.

29. Robens, J. F. and Richard, J. L. (1992) Aflatoxins in animal and human health. *Rev. Environ. Contam. Toxicol.* **127,** 69–94.

30. Beardall, J. M. and Miller, J. D. (1994) Diseases in humans with mycotoxins as possible causes. In: *Mycotoxins in Grain: Compounds Other Than Aflatoxin* (Miller, J. D. and Trenholm, H. L., eds.), Eagan, St. Paul, MN, pp. 487–539.

31. Kuiper-Goodman, T. (1994) Prevention of human mycotoxicoses through risk assessment and risk management. In: *Mycotoxins in Grain: Compounds Other Than Aflatoxin* (Miller, J. D. and Trenholm, H. L., eds.), Eagan, St. Paul, MN, pp. 439–469.

32. Peraica, M., Radic, B., Lucic, A., and Pavlovic, M. (1999) Toxic effects of mycotoxins in humans. *Bull. World Health Organ.* **77,** 754–766.

33. Hussein, H. S. and Brasel, J. M. (2001) Toxicity, metabolism, and impact of mycotoxins on humans and animals. *Toxicology* **167,** 101–134.

34. Etzel, R. A. (2002) Mycotoxins. *J. Am. Med. Assoc.* **287,** 425–427.

35. Goldblatt, L. (ed.) (1969) *Aflatoxin. Scientific Background, Control, and Implications*, Academic, New York.

36. Heathcote, J. G. and Hibbert, J. R. (eds.) (1978) *Aflatoxins: Chemical and Biological Aspects*, Elsevier, Amsterdam.

37. Van Egmond, H. P. (1989) Aflatoxin M_1: occurrence, toxicity, regulation. In: *Mycotoxins in Dairy Products* (Van Egmond, H. P., ed.), Elsevier, London, pp. 11–55.

38. Eaton, D. L. and Ramsdel, H. S. (1992) Species and diet related differences in aflatoxin biotransformation. In: *Handbook of Applied Mycology* (Bhatnagar, D., ed.), Vol. 5, Marcel Dekker, New York, pp. 157–182.

39. Squire, R. A. (1981) Ranking animal carcinogens. A proposed regulatory approach. *Science* **214,** 877–880.

40. Kurtzman, C. P., Horn, B. W., and Hesseltine, C. W. (1987) *Aspergillus nomius*, a new aflatoxin-producing species related to *Aspergillus flavus* and *Aspergillus tamarii*. *Antonie Van Leeuwenhoek* **53,** 147–158.

41. Peterson, S. W., Ito, Y., Horn, B. W., and Goto, T. (2001) *Aspergillus bombycis*, a new aflatoxigenic species and genetic variation in its sibling species, *A. nomius*. *Mycologia* **93,** 689–703.

42. Klich, M., Mullaney, E., Daly, C. B., and Cary, J. W. (2000) Molecular and physiological aspects of aflatoxin and sterigmatocystin biosynthesis by *A. tamarii* and *A. ochraceoroseus. Appl Microbiol. Biotechnol.* **53,** 605–609.

43. Klich, M. A., Cary, J. W., Beltz, S. B., and Bennett, C. A. (2003) Phylogenetic and morphological analysis of *Aspergillus ochraceoroseus. Mycologia* **95,** 1252–1260.

44. Ito, Y., Peterson, S. W., Wicklow, D. T., and Goto, T. (2001) *Aspergillus pseudotamarii* a new aflatoxin producing species in *Aspergillus* section *Flavi. Mycol. Res.* **15,** 233–239.

45. Goto, T., Wicklow, D. T., and Ito, Y. (1996) Aflatoxin and cyclopiazonic acid production by a sclerotium-producing *Aspergillus tamarii* strain. *Appl. Environ. Microbiol.* **62,** 4036–4038.

46. Frisvad, J. C., Samson, R. A., and Smedsgaard, J. (2004) *Emericella astellata*, a new producer of aflatoxin B$_1$, B$_2$ and sterigmatocystin. *Lett. Microbiol.* **38,** 440–445.

47. Klich, M. A. and Pitt, J. I. (1988) Differentiation of *Aspergillus flavus* from *A. parasiticus* and other closely related species. *Trans. Br. Mycol. Soc.* **91,** 99–108.

48. CAST. (2003) *Mycotoxins: Risks in Plant, Animal and Human Systems.* Council for Agricultural Science and Technology Task Force Report No. 139, Ames, IA.

49. Detroy, R. W., Lillehoj, E. B., and Ciegler, A. (1971) Aflatoxin and related compounds. In: *Microbial Toxins, Vol. VI: Fungal Toxins* (Ciegler, A., Kadis, S., and Ajl, S. J., eds), Academic, New York, pp. 3–178.

50. Diener, U. L., Cole, R. J., Sanders, T. H., Payne, G. A., Lee, L. S., and Klich, M. A. (1987) Epidemiology of aflatoxin formation by *Aspergillus flavus. Annu. Rev. Phytopathol.* **25,** 249–270.

51. Klich, M. A. (1987) Relation of plant water potential at flowering to subsequent cottonseed infection by *Aspergillus flavus. Phytopathology* **77,** 739–741.

52. Dowd, D. (1998) Involvement of arthropods in the establishment of mycotoxigenic fungi under field conditions. In: *Mycotoxins in Agriculture and Food Safety* (Sinha, K. K. and Bhatnagar, D., eds.), Marcel Dekker, New York, pp. 307–350.

53. Wilson, D. M. and Payne, G. A. (1994) Factors affecting *Aspergillus flavus* group infection and aflatoxin contamination of crops. In: *The Toxicology of Aflatoxins. Human Health, Veterinary and Agricultural Significance* (Eaton, D. L and Groopman, J. D., eds.), Academic, San Diego, pp. 309–325.

54. Eaton, D. L. and Groopman, J. D. (eds.) (1994) *The Toxicology of Aflatoxins: Human Health, Veterinary, and Agricultural Significance*, Academic, San Diego.

55. Raj, H. G., Prasanna, H. R., Magee, P. N., and Lotlikar, P. D. (1986) Effect of purified rat and hamster hepatic glutathione S-transferases on the microsome mediated binding of aflatoxin B$_1$ to DNA. *Cancer Lett.* **33,** 1–9.

56. James, R. C. (1985) General principles of toxicology. In: *Industrial Toxicology* (Williams, P. L. and Burson, J. L., eds.), Van Nostrand Reinhold, New York, pp. 7–26.

57. Newberne, P. M. and Butler, W. H. (1969) Acute and chronic effect of aflatoxin B$_1$ on the liver of domestic and laboratory animals: a review. *Cancer Res.* **29,** 236–250.

58. Shank, R. C., Bhamarapravati, N., Gordon, J. E., and Wogan, G. N. (1972) Dietary aflatoxins and human liver cancer. IV. Incidence of primary liver cancer in two municipal populations in Thailand. *Food Cosmet. Toxicol.* **10,** 171–179.

59. Peers, F. G. and Linsell, C. A. (1973) Dietary aflatoxins and human liver cancer—a population study based in Kenya. *Br. J. Cancer* **27,** 473–484.

60. Hsieh, D. (1988) Potential human health hazards of mycotoxins. In: *Mycotoxins and Phytotoxins Third Joint FAO/WHO/UNEP International Conference of Mycotoxins* (Natori, S., Hashimoto, K., and Ueno, Y., eds.), Elsevier, Amsterdam, pp. 69–80.

61. Cullen, J. M. and Newberne, P. N. (1994) Acute heptoxotoxicity of aflatoxins. In: *The Toxicity of Aflatoxins. Human Health, Veterinary, and Agricultural Significance* (Eaton D. L. and Groopman, J. J., eds.), Academic, San Diego, pp. 3–26.

62. Sabbioni, G. and Sepai, O. (1994) Determination of human exposure to aflatoxins. In: *Mycotoxins in Agriculture and Food Safety* (Sinha, K. K. and Bhatnagar, D., eds.), Marcel Dekker, New York, pp. 183–226.

63. Eaton, D. L. and Gallagher, E. P. (1994) Mechanisms of aflatoxin carcinogenesis. *Ann. Rev. Pharmacol. Toxicol.* **34,** 135–172.

64. Coulombe, R. A. Jr. (1994) Nonhepatic disposition and effects of aflatoxin B1. In: *The Toxicology of Aflatoxins: Human Health, Veterinary, and Agricultural Significance* (Eaton, D. L. and Groopman, J. D., eds.), Academic, San Diego, pp. 89–101.

65. Harrison, J. C., Carvajal, M., and Garner, R. C. (1993) Does aflatoxin exposure in the United Kingdom constitute a cancer risk? *Environ. Helath Perspect.* **99,** 99–105.

66. Krishnamachari, K. A. V. R., Bhat, R. V., Nagarajan, V., and Tilnak, T. M. G. (1975) Hepatitis due to aflatoxicosis. An outbreak in Western India. *Lancet* **1,** 1061–1063.

67. Pitt, J. I. (2000) Toxigenic fungi: which are important? *Med. Mycol.* **38(Suppl. 1),** 17–22.

68. Willis, R. M., Mulvihill, J. J., and Hoofnagle, J. H. (1980) Attempted suicide with purified aflatoxin. *Lancet* **1,** 1198–1199.

69. Henrickse, R. G. (1997) Of sick turkeys, kwashiorkor, malaria, perinatal mortality, heroin addicts and food poisoning: research on the influence of aflatoxins on child health in the tropics. *Ann. Trop. Med. Parasitol.* **91,** 787–793.

70. Kocabuas, C. N., Coskun, T., Yudakok, M., and Hazlroglu, R. (2003) The effects of aflatoxin B_1 on the development of kwashiorkor in mice. *Human Exp. Toxicol.* **22,** 155–158.

71. Hayes, A. W. (1980) Mycotoxins: a review of biological effects and their role in human diseases. *Clin. Toxicol.* **17,** 45–83.

72. Van Rensburg, S. J., Cook-Mazaffari, P., van Schalkwyk, D. J., van der Watt, J. J., Vincent, T. J., and Purchase, I. F. (1985) Heptatocellular carcinoma and dietary aflatoxin in Mozambique and Transkei. *Br. J. Cancer* **51,** 713–720.

73. Li, F.-Q., Yoshizawa, T., Kawamura, S., Luo, S.-Y., and Li, Y.-W. (2001) Aflatoxins and fumonisins in corn from the high-incidence area for human hepatocellular carcinoma in Guangxi, China. *J. Agric. Food Chem.* **49,** 4122–4126.

74. Ross, R. K., Yuan, J. M., Yu, M. C., et al. (1992) Urinary aflatoxin biomarkers and risk of hepatocellular carcinoma. *Lancet.* **339,** 1413–1414.

75. Henry, S. H., Bosch, F. X., Troxell, T. C., and Bolger, P. M. (1999) Reducing liver cancer— global control of aflatoxin. *Science* **286,** 2453–2454.

76. Henry, S. H., Bosch, F. X., and Bowers, J. C. (2002) Aflatoxin, hepatitis and worldwide liver cancer risks. In: *Mycotoxins and Food Safety* (DeVries, J. W., Trucksess, M. W., and Jackson, L. S., eds.), Kluwer Academic/Plenum, New York, pp. 229–320.

77. Bressac, B., Kew, M., Wands, J., and Ozturk, M. (1991) Selective G to T mutations of p53 gene in hepatocellular carcinoma from southern Africa. *Nature* **350,** 429–431.

78. Hsu, I. C., Metcalf, R. A., Sun, T., Welsh, J. A., Wang, N. J., and Harris, C. C. (1991) Mutational hotspot in the p53 gene in human hepatocellular carcinomas. *Nature* **350,** 377–378.

79. IARC. (1982) The evaluation of the carcinogenic risk of chemicals to humans. *IARC Monograph Supplement 4*, International Agency for Research on Cancer, Lyon, France.

80. Zilinskas, R. A. (1997) Iraq's biological weapons. The past as future? *J. Am. Med. Assoc.* **276,** 418–424.

81. Stone, R. (2001) Down to the wire on bioweapons talks. *Science* **293,** 414–416.

82. Stone, R. (2002) Peering into the shadows: Iraq's bioweapons program. *Science* **297,** 1110–1112.

83. El-Nezami, H., Kankaanpaa, P. E., Salminen, S., and Ahokas, J. T. (1998) Physico-chemical alterations enhance the ability of dairy strains of lactic acid bacteria to remove aflatoxins from contaminated media. *J. Food Prot.* **61,** 466–468.

84. El-Nezami, H., Polychronaki, N., Salminen, S., and Mykkanen, H. (2002) Binding rather than metabolism may explain the interaction of two food-grade *Lactobacillus* strains with zearalenone and its derivative α–zearalenol. *Appl. Environ. Microbiol.* **68,** 3545–3549.

85. Kankaanpaa, P., Tuomola, E., El-Nezami, H., Ahokas, J., and Salminen, S. J. (2000) Binding of aflatoxin B_1 alters the adhesion properties of *Lactobacillus rhamnosus* strain GG in a caco-2 model. *J. Food Prot.* **63,** 412–414.

86. Phillips, T. D. (1999) Dietary clay in the chemoprevention of aflatoxin-induced disease. *Toxicol. Sci.* **52(2 Suppl.),** 118–126.

87. Phillips, T. D., Lemke, S. L., and Grant, P. G. (2002) Characterization of clay-based enterosorbents for the prevention of aflatoxicosis. *Adv. Exp. Med. Biol.* **54,** 157–171.

88. Jakhar, K. K. and Sadana, J. R. (2004) Sequential pathology of experimental aflatoxicosis in quail and the effect of selenium supplementation in modifying the disease process. *Mycopathologia* **157,** 99–109.

89. Klein, P. J., Van Vleet, T. R., Hall, J. O., and Coulombe, R. A. Jr. (2002) Dietary butylated hydroxytoluene protects against aflatoxicosis in turkeys. *Toxicol. Appl. Pharmacol.* **182,** 11–19.

90. Lisker, N. and Lillehoj, E. B. (1991) Prevention of mycotoxin contamination (principally aflatoxins and *Fusarium* toxins) at the preharvest stage. In: *Mycotoxins and Animals Foods* (Smith, J. E. and Henderson, R. S., eds.), CRC, Boca Raton, FL, pp. 689–719.

91. Pier, A. C., Richard, J. L., and Cysewski, S. J. (1980) Implications of mycotoxins in animal disease. *J. Am. Vet. Med. Assoc.* **176,** 719–724.

92. Chu, F. S. (1998) Mycotoxins—occurrence and toxic effect. *Encyclopedia of Food and Nutrition*, pp. 858–869.

93. Gilbert, J. (1993) Recent advances in analytical methods for mycotoxins. *Food Add. Contam.* **10,** 37–48.

94. Pohland, A. E. (1993) Mycotoxins in review. *Food Add. Contam.* **10,** 17–28.

95. Scott, P. M. (1995) Mycotoxin methodology. *Food Add. Contam.* **12,** 395–403.

96. FAO. (1997) Food and Agricultural Organization of the United Nations Food and Nutrition Paper 64. Worldwide Regulations for Mycotoxins. A Compendium, Rome.

97. Cotty, P. J., Bayman, P., Egel, D. S., and Elias, K. S. (1994) Agriculture, aflatoxins and *Aspergillus*. In: *The Genus Aspergillus* (Powell, K. A., Renwick, A., and Peberdy, J. F., eds.), Plenum, New York and London, pp. 1–27.

98. Smith, J. E., Solomons, G., Lewis, C., and Anderson, J. C. (1995) Role of mycotoxins in human and animal nutrition and health. *Natural Toxins* **3,** 187–192.

99. Food and Drug Administration, USA. (1988) Action levels for added poisonous or deleterious substances in food. *Notice Fed. Register* **53,** 5043–5044.

100. Wilson, D. M., Mubatanhema, W., and Jurjevic, Z. (2002) Biology and ecology of mycotoxigenic *Aspergillus* species as related to economic and health concerns. In: *Mycotoxins and Food Safety* (DeVries, J. W., Trucksess, M. W., and Jackson, L. S., eds.), Kluwer Academic/Plenum, New York and Dordrecht, pp. 3–17.

101. Weidenborner, W. (2001) *Encyclopedia of Food Mycotoxins*, Springer, Berlin.

102. Bennett, J. W., Chang, P.-K., and Bhatnagar, D. (1997) One gene to whole pathway: the role of norsolorinic acid in aflatoxin research. *Adv. Appl. Microbiol.* **45,** 1–15.

103. Cleveland, T. E. and Bhatnagar, D. (1992) Molecular strategies for reducing aflatoxin levels in crops before harvest. In: *Molecular Approaches to Improving Food Quality and Safety* (Bhatnagar, D. and Cleveland, T. E., eds.), Van Nostrand Reinhold, New York, pp. 205–228.

104. Townsend, C. A. (1997) Progress towards a biosynthetic rationale of the aflatoxin pathway. *Pure Appl. Chem.* **58,** 227–238.

105. Trail, F., Mahanti, N., and Linz, J. E. (1995) Molecular biology of aflatoxin biosynthesis. *Microbiology* **141,** 755–765.

106. Payne, G. and Brown, M. P. (1998) Genetics and physiology of aflatoxin biosynthesis. *Ann. Rev. Plant Path.* **36,** 329–362.

107. Hicks, J. K., Shimizu, K., and Keller, N. P. (2002) Genetics and biosynthesis of aflatoxins and sterigmatocystin. In: *The Mycota: Agricultural Applications* (Kempken, F., ed.), Vol. XI, Springer, Berlin, pp. 55–69.

108. Yabe, K. (2002) Pathway and genes of aflatoxin biosynthesis. In: *Microbial Secondary Metabolites: Biosynthesis, Genetics and Regulation, Research Signpost* (Fierro, F. and Martin, J. F., eds.), Kerala, India, pp. 227–251.

109. Yu, J., Bhatnagar, D., and Ehrlich, K. C. (2002) Aflatoxin biosynthesis. *Rev. Iberoam. Microbiol.* **19**, 191–200.

110. Yu, J. (2004) Genetics and biochemistry of mycotoxin Synthesis. In: *Handbook of Fungal Biotechnology in Agricultural, Food, and Environmental Applications* (Arora, D. K., ed.), Vol. 21, Marcel Dekker, New York, pp. 343–361.

111. Yu, J., Chang, P.-K., Ehrlich, K. C., et al. (2004) Clustered pathway genes in aflatoxin biosynthesis. *Appl. Environ. Microbiol.* **70**, 1253–1262.

112. Yu, J., Bhatnagar, D., and Cleveland, T. E. (2004) Completed sequence of aflatoxin pathway gene cluster in *Aspergillus parasiticus*. *FEBS Lett.* **564**, 126–130.

113. Yu, J., Bhatnagar, D., and Cleveland, T. E. (2004) Genetics and biochemistry of aflatoxin formation and genomics approach for eliminating aflatoxin contamination. In: *Recent Advances in Phytochemistry: Secondary Metabolism in Model Systems* (Romeo, J. T., ed.), Vol. 38, Elsevier, Amsterdam, pp. 224–242.

114. Berry, C. L. (1988) The pathology of mycotoxins. *J. Pathol.* **154**, 301–311.

115. Brown, D. H., Yu, J. H., Kelkar, H. S., et al. (1996) Twenty-five co-regulated transcripts define a sterigmatocystin gene cluster in *Aspergillus nidulans*. *Proc. Natl Acad. Sci. USA* **93**, 1418–1422.

116. Cary, J. W., Chang, P.-K., and Bhatnagar, D. (2001) Clustered metabolic pathway genes in filamentous fungi. In: *Applied Mycology and Biotechnology, Vol. I: Agriculture and Food Production* (Khachatourians, G. G. and Arora, D. K., eds.), Elsevier, Amsterdam, pp. 165–198.

117. Yu, J., Chang, P.-K., Cary, J. W., et al. (1995) Comparative mapping of aflatoxin pathway gene clusters in *Aspergillus parasiticus* and *Aspergillus flavus*. *Appl. Environ. Microbiol.* **61**, 2365–2371.

118. Yu, J., Chang, P.-K., Bhatnagar, D., and Cleveland, T. E. (2000) Genes encoding cytochrome P450 and monoooxygenase enzymes define one end of the aflatoxin pathway gene cluster in *Aspergillus parasiticus*. *Appl. Microbiol. Biotechnol.* **53**, 583–590.

119. Yu, J., Woloshuk, C. P., Bhatnagar, D., and Cleveland, T. E. (2000) Cloning and characterization of *avfA* and *omtB* genes involved in aflatoxin biosynthesis in three *Aspergillus* species. *Gene* **248**, 157–167.

120. Yu, J., Chang, P.-K., Bhatnagar, D., and Cleveland, T. E. (2000) Cloning of sugar utilization gene cluster in *Aspergillus parasiticus*. *Biochim. Biophys. Acta* **1493**, 211–214.

121. Chang, P.-K., Ehrlich, K. C., Yu, J., Bhatnagar, D., and Cleveland, T. E. (1995) Increased expression of *Aspergillus parasiticus* aflR ending a sequence-specific DNA binding protein relieves nitrate inhibition of aflatoxin biosynthesis. *Appl. Environ. Microbiol.* **61**, 2372–2377.

122. Ehrlich, K. E., Montalbano, B. G., Bhatnagar, D., and Cleveland, T. E. (1998) Alternation of different domains in *aflR* affects aflatoxin pathway metabolism in *Aspergillus parasiticus* transformants. *Fung. Genet. Biol.* **23**, 279–287.

123. Meyers, D. M., O'Brian, G., Du, W. L., Bhatnagar, D., and Payne, G. A. (1998) Characterization of *aflJ*, a gene required for conversion of pathway intermediates to aflatoxin. *Appl. Environ. Microbiol.* **64**, 3713–3717.

124. Chang, P.-K. (2003) The *Aspergillus parasiticus* protein AFLJ interacts with the aflatoxin pathway-specific regulator AFLR. *Mol. Genet. Genom.* **268**, 711–719.

125. Wei, D. L. and Jong, S. C. (1986) Production of aflatoxins by strains of the *Aspergillus flavus* group maintained in ATCC. *Mycopathologia* **93**, 19–24.

126. Klich, M. A., Yu, J., Chang, P.-K., Mullaney, E. J., Bhatnagar, D., and Cleveland, T. E. (1995) Hybridization of genes involved in aflatoxin biosynthesis to DNA of aflatoxigenic and non-aflatoxigenic aspergilli. *Appl. Microbiol. Biotechnol.* **44**, 439–443.

127. Matsushima, K., Chang, P.-K., Yu, J., Abe, K., Bhatnagar, D., and Cleveland, T. E. (2001) Pre-termination in aflR of *Aspergillus sojae* inhibits aflatoxin biosynthesis. *Appl. Microbiol. Biotechnol.* **55**, 585–589.

128. Takahashi, T., Chang, P.-K., Matsushima, K., et al. (2002) Nonfunctionality of *Aspergillus sojae aflR* in a strain of *Aspergillus parasiticus* with a disrupted *aflR* gene. *Appl. Environ. Microbiol.* **68**, 3737–3743.

129. Watson, A. J., Fuller, L. J., Jeens, D. J., and Archer, D. B. (1999) Homologues of aflatoxin biosynthesis genes and sequence of *aflR* in *Aspergillus oryzae* and *Aspergillus sojae*. *Appl. Environ. Microbiol.* **65**, 307–310.

130. Yu, J., Proctor, R. H., Brown, D. W., et al. (2004) Genomics of economically significant *Aspergillus* and *Fusarium* species. In: *Applied Mycology and Biotechnology* (Arora, K. D. and Khachatourians, G. G., eds.), Vol. 3, Elsevier, Amsterdam, pp. 249–283.

131. Yu, J., Whitelaw, C. A., Nierman, W. C., Bhatnagar, D., and Cleveland, T. E. (2004) *Aspergillus flavus* expressed sequence tags for identification of genes with putative roles in aflatoxin contamination of crops. *FEMS Lett.* (in press).

132. Bu'Lock, J. D. (1961) Intermediary metabolism and antibiotic synthesis. *Adv. Appl. Microbiol.* **3**, 293–342.

133. Sekiguchi, J. and Gaucher, G. M. (1977) Conidiogenesis and secondary metabolism in *Penicillium urticae*. *App. Environ. Microbiol.* **133**, 236–238.

134. Bennett, J. W. and Ciegler, A. (eds.) (1983) *Secondary Metabolism and Differentiation in Fungi*, Marcel Dekker, New York and Basel.

135. Calvo, A. M., Wilson, R. A., Bok, J. W., and Keller, N. P. (2002) Relationship between secondary metabolism and fungal development. *Microb. Mol. Biol. Rev.* **66**, 447–459.

136. Hicks, J. K., Yu, J. H., Keller, N. P., and Adams, T. H. (1997) *Aspergillus* sporulation and mycotoxin production both require inactivation of the FaddA G alpha protein-dependent signaling pathway. *EMBO J.* **16**, 4916–4923.

137. Reiss, J. (1982) Development of *Aspergillus parasiticus* and formation of aflatoxin B1 under the influence of conidiogenesis affecting compounds. *Arch. Microbiol.* **133**, 236–238.

138. Guzman-de-Pena, D., Aguirre, J., and Ruiz-Herrera, J. (1998) Correlation between the regulation of sterigmatocystin biosynthesis and asexual and sexual sporulation in *Emericella nidulans*. *Antonie Van Leeuwenhock* **73**, 199–205.

139. Bennett, J. W. (1981) Loss of norsolorinic acid and aflatoxin production by a mutant of *Aspergillus parasiticus*. *J. Gen. Microbiol.* **124**, 429–432.

140. Bennett, J. W., Leong, P. M., Kruger, S., and Keyes, D. (1986) Sclerotial and low aflatoxigenic morphological variants from haploid and diploid *Aspergillus parasiticus*. *Experientia* **42**, 841–851.

141. Bennett, J. W. and Papa, K. E. (1988) The aflatoxigenic *Aspergillus* species. In: *Genetics of Plant Pathogenic Fungi* (Ingram, D. S. and Williams, P. A., eds.), Academic, London, 264–280.

142. Chang P.-K., Bennett, J. W., and Cotty, P. J. (2001) Association of aflatoxin biosynthesis and sclerotial development in *Aspergillus parasiticus*. *Mycopathologia* **153**, 41–48.

143. Tamame, M., Antequera, F., Villanueva, J. R., and Santos, T. (1983) High frequency conversion of a "fluffy" developmental phenotype in *Aspergillus* spp. By 5-azacytidine treatment: evidence for involvement of a single nuclear gene. *Mol. Cell Biol.* **3**, 2287–2297.

144. Wieser, J., Lee, B. N., Fondon, J. W., and Adams, T. H. (1994) Genetic requirement for initiating asexual development in *Aspergillus nidulans*. *Curr. Genet.* **27**, 62–69.

145. Timberlake, W. E. (1991) Temporal and spatial controls of *Aspergillus* development. *Curr. Opin. Genet. Dev.* **1**, 351–357.

146. Butchko, R. A. E., Adams, T. H., and Keller, N. P. (1999) *Aspergillus nidulans* mutants defective in *stc* gene cluster regulation. *Genetics* **153**, 715–720.

147. Shimizu, K. and Keller, N. P. (2001) Genetic involvement of a cAMP-dependent protein kinase in a G protein signaling pathway regulating morphological and chemical transitions in *Aspergillus nidulans*. *Genetics* **157**, 591–600.

148. Yu, J. H., Weiser, J., and Adams, T. H. (1996) The *Aspergillus* FlbA RGS domain protein antagonizes G-protein signaling to block proliferation and allow development. *EMBO J.* **15**, 5184–5190.

149. Clutterbuck, A. J. (1969) A mutational analysis of conidial development in *Aspergillus nidulans. Genetics* **63,** 317–327.
150. Shimizu, K., Hicks, J. K., Huang, T.-P., and Keller, N. P. (2003) Pka, Ras and RGS protein interactions regulate activity of AflR, a Zn(II)2Cys6 transcription factor in *Aspergillus nidulans. Genetics* **165,** 1095–1104.
151. Kale, S. P., Bhatnagar, D., and Bennett, J. W. (1994) Isolation and characterization of morphological variants of *Aspergillus parasiticus* deficient in secondary metabolite production. *Mycol. Res.* **98,** 645–652.
152. Kale, S. P., Cary, J. W., Bhatnagar, D., and Bennett, J. W. (1996) Characterization of an experimentally induced, nonaflatoxigenic variant strains of *Aspergillus parasiticus. Appl. Environ. Microbiol.* **62,** 399–3404.
153. Kale, S. P., Cary, J. W., Baker, C., Walker, D., Bhatnagar, D., and Bennett, J. W. (2003) Genetic analysis of morphological variants of *Aspergillus parasiticus* deficient in secondary metabolite production. *Mycol. Res.* **107,** 831–840.
154. Zhou, R., Rasooly, R., and Linz, J. E. (2000) Isolation and analysis of *fluP*, a gene associated with hyphal grown and sporulation in *Aspergillus parasiticus. Mol. Gen. Genet.* **264,** 514–520.
155. Bok, J.-W. and Keller, N. P. (2004) *LaeA*, a regulator of secondary metabolism in *Aspergillus. Eukaryot. Cell* **3,** 527–535.

14

Scombroid Fish Poisoning

Elijah W. Stommel

1. CLASSIFICATION AND IDENTIFICATION

Scombroid fish poisoning, otherwise known as histamine fish poisoning (HFP), is the most common seafood poisoning in the United States related to the improper storage of fish *(1)*. Because of misdiagnosis and other factors, HFP is undoubtedly underreported *(2)*. Enteric bacteria (*Escherichia coli, Proteus morganii, Morganella morganii,* and *Proteus vulgaris*) act on the flesh of the poorly maintained fish to produce elevated histamine levels (as well as other bioactive amines) through the breakdown of the amino acid histadine *(3,4)*. These organisms are part of the normal flora of certain fish and not thought to be the contaminants *(5)*. Histamine production is correlated to the histadine content of fish, bacterial histadine decarboxylase, and environmental factors *(6)*. Histadine levels vary from 1 g/kg in herring to as much as 15 g/kg in tuna *(7)*. Unspoiled, fresh fish do contain small amounts of histamine (<0.1 mg/100 g *[8]*), but do not cause HFP.

2. RESERVOIRS

The vast majority of fish causing HFP are members of the families Scombroidae and Scomberesocidae (hence the name scombroid fish poisoning), which include tuna, mackerel, skipjack, bonita, suary, and seerfish. Several nonscombroid fish are also linked to HFP, including bluefish, mahi-mahi, herring, sardines, and anchovies *(9)*. These fish are nontoxic when afresh, can still appear very normal with the development of toxicity, and have no putrid odor *(10)*. Other species that have also been affected are of little concern because they are not commonly eaten. The fish that are affected are thought to be related closely to specific endogenous bacteria. Interestingly, dark meat, a major component of the muscle of these fish, seems to be more affected. Improper storage of the fish at insufficient low temperatures appears to initiate the bacterial growth. Freezing, cooking, smoking, curing, or canning does not eliminate the potential toxins but destroys the bacteria that produce the toxins *(9)*.

3. FOODBORNE OUTBREAKS

Outbreaks are worldwide and are found in places where potentially spoiled, improperly handled scombroid-like fish species are eaten. Of note is that HFP occurs with the ingestion of canned tuna, a mainstay of the American diet. Hence, wherever a can of tuna are sent creates the potential for HFP. Recreational fishermen tend to have less

From: *Infectious Disease: Foodborne Diseases*
Edited by: S. Simjee © Humana Press Inc., Totowa, NJ

stringent habits and regulations for storing their catch and are hence more likely to end up with fish capable of producing HFP. The Board of Health in New York City reports approx 70 cases of HFP per year, seen in emergency rooms or clinics, but clearly there are many more cases which are attested to a simple virus or food allergy (Dr. Herbert Schaumberg, personal communications). Last spring, there were 20 unreported HFP cases at Albert Einstein Medical Center alone after the Pediatric Department's farewell to the residents' banquet (Dr. Herbert Schaumberg, personal communications).

4. PATHOGENICITY (VIRULENCE FACTORS)

HFP is generally associated with high histamine levels (>50 mg/100 g) in spoiled fish that have high histidine content *(9)*. Histamine is endogenous to mast cells and basophils usually only having an effect when released in large amounts in response to an allergic reaction. Histamine has cell membrane receptors principally in skin, hematological, gastrointestinal, and respiratory systems *(11)*. Interestingly, consuming spoiled fish with histamine is more poisonous than taking an equivalent amount of pure histamine orally *(12)*, suggesting that there are other "scombroid toxins" acting with histamine *(13,14)*. The histamine-potentiator hypothesis is based on the observation that absorption, metabolism, and/or potency of one biogenic amine might be altered in the presence of another amine *(12,15)*. The biogenic amines putrescine and cadaverine are found in large quantities in toxic fish *(12)* and in significantly lower levels in nontoxic fish *(16)*. These amines, when given in higher ratios relative to histamine in toxic fish, have been shown to potentiate the effect of histamine in laboratory animals *(17)*. Some amines, such as cadaverine and putrescine along with a number of other tested chemicals found in spoiled tuna, may competitively inhibit diamine oxidase (DAO), the main histamine catabolizing enzyme in the intestinal tract, and histamine methyl transferase (HMT) *(18,19)*. The toxicity of histamine can be potentiated by cadaverine *(15)* and putrescine *(20)* in guinea pigs. A number of drugs inhibit DAO including iproniazid, isoniazid, pargyline, aminoguanidine, phenelzine, and tranylcypromine, which could predispose one to HFP *(21)*. Serotonin, tryptamine, and phenformin are good competitive inhibitors, whereas cimetidine and pheniprazine are noncompetitive inhibitors. Other antihistaminic drugs such as promethazine are less powerful inhibitors. Certain foods can inhibit DAO (alcohol and tyramine-containing foods such as strong cheeses) likely predisposing one to HFP as well *(12,22)*. HMT is inhibited by analogs of methylmethionine such as adenosyl-homocysteine, antimalarial drugs, and numerous agonists and antagonists of histamine receptors *(12)*. It should be noted that histamine has been implicated in the pathogenesis of migraine *(23,24)*, the intolerance of which may be based on a deficiency in the DAO enzyme *(25)*. A diet excluding alcohol and tyramine-containing foods is helpful for many patients with histamine-induced headaches. Regarding this fact, a worse reaction to HFP is expected if someone ingests wine with scombroid-laden fish, although no published data confirm this idea.

The "barrier disruption hypothesis," first proposed by Parrot and Nicot *(20)*, states that potentiators might interfere with mucin, which is known to bind histamine, thought to be an essential event in preventing the intestinal absorption of histamine and hence increases histamine absorption *(26,27)*. It has also been postulated that histamine may be released endogenously by an unknown scombroid toxin *(28)*. Urocanic acid, an imidazole compound and a histidine metabolite of spoiling fish, has been shown to induce histamine in vivo in mice *(29)* and to degranulate mast cells in human skin organ cultures *(30)*. The role of endogenous histamine release in HFP has not been established *(31)*.

A recent publication describes the sensitive polymerase chain reaction technique for detecting *M. morganii*, which was found to be present in fresh fish including mackerel, sardine, and albacore. The gill and skin were the main harbor sites of the bacterium in this study *(32)*. Conventional culturing techniques did not detect the bacterium. This study demonstrates that histamine-forming bacteria are endogenous to these fish and are likely to proliferate under the suitable conditions. These findings also emphasize the importance of proper handling of fish, including refrigeration, to prevent HFP.

5. CLINICAL CHARACTERISTICS

Fish affected by scombroid tastes "peppery," which is thought to be related to elevated histamine levels. Headache, a burning sensation in mouth and oropharynx, and nausea are common neurological symptoms of HFP *(4,33)*. Other symptoms include diffuse erythema (especially flushing of the head, neck, and upper torso), headache, vomiting, diarrhea, and abdominal cramping *(33–36)*. Occasional arrhythmias, hypotension, bronchospasm, and cardiovascular collapse also occur *(37,38)*. Anxiety has also been reported as a prominent symptom with HFP *(39,40)*. Generally, HFP is not a life-threatening condition, but in those patients with serious preexisting conditions, notable respiratory and cardiac complications do occur *(38,40)*. The symptoms of HFP usually last for 8–12 h *(33)* and can come on within minutes of ingestion of a toxic meal *(33,41)*.

With HFP, there is a remarkably wide spectrum of clinical symptoms (mild to severe). Symptoms can be confused with *Salmonella* infection *(40)* as well as food allergy *(4)*. The usual HFP-associated fish species cause HFP much more frequently than they cause true allergic reaction *(12)*. The clinical setting of the poisoning shows evidence to the etiology as if only one person is affected by a meal when several individuals have eaten the same fish, sheds doubt on the diagnosis of HFP. Nevertheless, the concentration of the histamine may not be evenly distributed through the fish and, hence, not everyone who eats the fish will necessarily get sick. In fact, food allergies to most of the common fish involved with HFP are rare.

Although the diagnosis can be confirmed by detecting histamine in contaminated fish, remnants of the meal, or similar fish from the same source *(12,17)* as well as other biogenic amines *(16)* and urocanic acid *(42,43)*, the diagnosis is normally made by the history and clinical presentation. There are numerous methods for the detection of histadine-decarboxylating bacteria that can be performed with specific media that are selective for this enzyme *(4,44,45)* although some of these media have given false positive results *(45)*. The biochemical methods for detecting histamine and other biogenic amines are extensive as well as controversial and are well outlined by Lehane and Olley *(31)*. Measurement of plasma histamine levels is not routinely available in most community-based hospitals. Measuring human serum levels requires that histamine is measured within 4 h of the ingestion of the fish *(46)*.

Other diagnoses to be considered with HFP are severe migraine, intracranial hemorrhage, and pheochromocytoma *(36)*. HFP may resemble an immunoglobulin E (IgE)-mediated allergic reaction; however, HFP is a foodborne intoxication related to elevated histamine levels, and therefore the patient can safely eat the same type of fish again with impunity.

6. CHOICE OF TREATMENT

HFP is self limited, and the need for long-term therapy is usually unnecessary *(3,12)*. HFP exposure responds to corticosteroids, charcoal, and histamine blockers *(4,41,47)*. The H2 blocker cimetidine at 300 mg iv has been effective in HFP *(41,47)*. There is a rational basis for blocking H1 and H2 receptors (H2 receptors are found in human blood vessels) *(3)* to minimize the vascular effects of histamine. Therefore, the addition of the H1 blocker diphenhydramine at 50 mg im, in addition to a H2 blocker, is recommended *(3)*. At least some case reports have not shown diphenhydramine hydrochloride 25–50 mg po or iv by itself to be effective as a treatment option *(41)*. In severe cases, aggressive symptomatic care is required including intravenous fluids and possibly steroid therapy *(41)*. Adrenaline, which is often helpful in allergic reactions, is of little benefit in HFP *(48)*.

7. RESISTANCE EPIDEMIOLOGY

Because HFP is a consequence of improper handling/storage of fish and there are effective testing methods to identify toxic fish, prevention and control of outbreaks are possible. Contamination with histidine-decarboxylating bacteria can occur immediately after the fish are caught on the fishing vessel, in the processing plant, in the distribution of the fish, and also with the consumer, such as at home or in a restaurant *(12)*. Key to the prevention of HFP is proper cooling of fish immediately after they have been caught *(31,49)*. Interestingly, most HFP in the United States results from the improper handling of fish by recreational fishermen who often neither have the knowledge nor the proper equipment to cool the fish properly *(50)*.

Many countries have set limits for the maximum-permitted levels of histamine in fish. The amount of histamine produced is a function of the fish type, the part of the fish sampled, temperature, and the types of bacteria found on the fish *(51)*. Normal fish has less than 100 ppm of histamine (1 mg/100 g of flesh). Although the toxic dose and symptoms of HFP are variable *(4,12,52)*, illness usually occurs at levels of 1000 ppm (100 mg/100 g of flesh), but lower levels (20 mg/100 g of flesh) can also cause illness in some individuals *(3)*.

8. SUMMARY AND CONCLUSIONS

HFP is a common form of fish poisoning that is often mistaken for allergic reactions. The affected Scombroidae and Scomberesocidae family fish are commonly eaten and poisonings can occur most anywhere in the civilized world. The mechanism of HFP is primarily related to the ingestion of high levels of histamine, but other related compounds and drug interactions are involved to produce the syndrome. Although HFP is not a fatal intoxication, it can be debilitating. HFP is easily recognized with a proper clinical history and foresight to its existence, and can be managed successfully with histamine blockers and supportive care. It is also an easily preventable poisoning through the proper regulation and handling of fish.

REFERENCES

1. Lipp, E. K. and Rose, J. B. (1997) The role of seafood infoodborne diseases in the United States of America. *Revue Scientifique et Technique*. **16**(2)**,** 620–640.
2. Wu, M. L., et al. (1997) Scombroid fish poisoning: an overlooked marine food poisoning. *Veterinary & Human Toxicology*. **39**(4)**,** 236–241.

3. Morrow, J. D., et al. (1991) Evidence that histamine is the causative toxin of scombroid-fish poisoning.[see comment]. *New England Journal of Medicine.* **324**(11), 716–720.

4. Taylor, S. L., Stratton, J. E., and Nordlee, J. A. (1989) Histamine poisoning (scombroid fish poisoning): an allergy-like intoxication. *Journal of Toxicology - Clinical Toxicology.* **27**(4–5), 225–240.

5. Lerke, P. A., et al. (1978) Scombroid poisoning. Report of an outbreak. *Western Journal of Medicine.* **129**(5), 381–386.

6. Ienistea, C. (1971) Bacterial production and destruction of histamine in foods, and food poisoning caused by histamine. *Nahrung.* **15**(1), 109–113.

7. Ijomah, P., et al. (1992) Further volunteer studies on scombrotoxicosis, in Pelagic Fish: The Resource and its Exploitation. (Burt, J. R., Hardy R. and Whittle K. J., eds.) Fishing News Books: Oxford. 194–199.

8. Yoshinaga, D. H. and Frank, H. A. (1982) Histamine-producing bacteria in decomposing skipjack tuna (Katsuwonus pelamis). *Applied & Environmental Microbiology.* **44**(2), 447–452.

9. Etkind, P., et al. (1987) Bluefish-associated scombroid poisoning. An example of the expanding spectrum of food poisoning from seafood. *Jama.* **258**(23), 3409–3410.

10. Sapin-Jaloustre, H. and Sapin-Jaloustre, J. (1957) A little known food poisoning: histamine poisoning from tuna. *Concours Med Paris.* **79**, 2705–2708.

11. Cavanah, D. K. and Casale, T. B. (1993) Histamine, in The mast cell in health and disease, (Kaliner M. A. and Metcalfe, D. D. eds.) Marcel Dekker: New York, Basel, Hong Kong. 321–342.

12. Taylor, S. L. (1986) Histamine food poisoning: toxicology and clinical aspects. *Critical Reviews in Toxicology.* **17**(2), 91–128.

13. Clifford, M. N., et al. (1991) Is there a role for amines other than histamines in the aetiology of scombrotoxicosis? *Food Additives & Contaminants.* **8**(5), 641–651.

14. Ijomah, P., et al. (1991) The importance of endogenous histamine relative to dietary histamine in the aetiology of scombrotoxicosis. *Food Additives & Contaminants.* **8**(4), 531–542.

15. Bjeldanes, L. F., Schutz, D. E., and Morris, M. M. (1978) On the aetiology of scombroid poisoning: cadaverine potentiation of histamine toxicity in the guinea-pig. *Food & Cosmetics Toxicology.* **16**(2), 157–159.

16. Mietz, J. L. and Karmas, E. (1977) Chemical quality index of canned tuna as determined by HPLC. *Journal of Food Science.* **42**, 155–158.

17. Lehane, L. (2000) Update on histamine fish poisoning. *Medical Journal of Australia.* **173**(3), 149–152.

18. Taylor, S. L., et al. (1979) Histamine production by Klebsiella pneumoniae and an incident of scombroid fish poisoning. *Applied & Environmental Microbiology.* **37**(2), 274–278.

19. Lyons, D. E., et al. (1983) Cadaverine and aminoguanidine potentiate the uptake of histamine in vitro in perfused intestinal segments of rats. *Toxicol Appl Pharmacol.* **70**(3), 445–458.

20. Parrot, J. and Nicot, G. (1966) Pharmacology of histamine, in Handbook of Experimental Pharmacology, (Eichler O. and Farah, S. eds.) Springer-Verlag: New York. 148–161.

21. Sattler, J., et al. (1989) Food-induced histaminosis under diamine oxidase (DAO) blockade in pigs: further evidence of the key role of elevated plasma histamine levels as demonstrated by successful prophylaxis with antihistamines. *Agents & Actions.* **27**(1–2), 212–214.

22. Stratton, J. E., Hutkins, R. W., and Taylor, D. L. (1991) Biogenic amines in cheese and other fermented foods. A review. *J Food Prot.* **54**, 460–470.

23. Millichap, J. G. and Yee, M. M. (2003) The diet factor in pediatric and adolescent migraine. *Pediatr Neurol.* **28**(1), 9–15.

24. Peatfield, R. C. (1995) Relationships between food, wine, and beer-precipitated migrainous headaches. *Headache.* **35**(6), 355–357.

25. Wantke, F., Gotz, M., and Jarisch, R. (1993) Histamine-free diet: treatment of choice for histamine-induced food intolerance and supporting treatment for chronic headaches. *Clin Exp Allergy.* **23**(12), 982–985.

26. Chu, C. H. (1981) Effects of diamines, polyamines and tuna fish extracts on the binding of histamine to mucin in vitro. *Journal of Food Science.* **47,** 79–88.

27. Jung, H. P. K. and Bjeldanes, L. F. (1978) Effects of cadaverine on histamine transport and metabolism in isolated gut sections of the guinea pig. *Food Cosmetic Toxicol.* **17,** 629–632.

28. Olley, J. (1972) Unconventional sources of fish protein. *CSIRO Food Research Quarterly.* **32**(2), 27–32.

29. Hart, P. H., et al. (1997) Histamine involvement in UVB- and cis-urocanic acid-induced systemic suppression of contact hypersensitivity responses. *Immunology.* **91**(4), 601–608.

30. Wille, J. J., Kydonieus, A. F., and Murphy, G. F. (1999) Cis-urocanic acid induces mast cell degranulation and release of preformed TNF-alpha: A possible mechanism linking UVB and cis-urocanic acid to immunosuppression of contact hypersensitivity. *Skin Pharmacology & Applied Skin Physiology.* **12**(1–2), 18–27.

31. Lehane, L. and Olley, J. (2000) Histamine fish poisoning revisited. *International Journal of Food Microbiology.* **58**(1–2), 1–37.

32. Kim, S. H., et al. (2003) Identification of the main bacteria contributing to histamine formation in seafood to ensure product safety. *Food Science and Biotechnology.* **12**(4), 451–460.

33. Gilbert, R. J., et al. (1980) Scombrotoxic fish poisoning: features of the first 50 incidents to be reported in Britain (1976–1979). *BMJ.* **281**(6232), 71–72.

34. Arnold, S. H. and Brown, W. D. (1978) Histamine toxicity from fish products. *Advances in Food Research.* **24,** 113–154.

35. Muller, G. J., et al. (1992) Scombroid poisoning. Case series of 10 incidents involving 22 patients. *South African Medical Journal.* **81**(8), 427–430.

36. Predy, G., et al. (2003) Was it something she ate? Case report and discussion of scombroid poisoning. *CMAJ Canadian Medical Association Journal.* **168**(5), 587–588.

37. Ascione, A., et al. (1997) [Two cases of "scombroid syndrome" with severe cardiovascular compromise]. *Cardiologia.* **42**(12), 1285–1288.

38. Shalaby, A. R. (1996) Significance of biogenic amines to food safety and human health. *Food Res Int.* **29**(7), 675–690.

39. Sabroe, R. A. and Kobza Black, A. (1998) Scombrotoxic fish poisoning. *Clinical & Experimental Dermatology.* **23**(6), 258–259.

40. Russell, F. E. and Maretic, Z. (1986) Scombroid poisoning: mini-review with case histories. *Toxicon.* **24**(10), 967–973.

41. Blakesley, M. L. (1983) Scombroid poisoning: prompt resolution of symptoms with cimetidine. *Annals of Emergency Medicine.* **12**(2), 104–106.

42. Caron, J. C., Martin, B., and Shroot, B. (1982) High-performance liquid chromatographic determination of urocanic acid isomers in biological samples. *J Chromatogr.* **230**(1), 125–130.

43. Baranowski, J. D. (1985) Low-Temperature Production of Urocanic Acid by Spoilage Bacteria Isolated from Mahimahi (Coryphaena hippurus). *Appl Environ Microbiol.* **50**(2), 546–547.

44. Omura, Y., Price, R. J., and Olcott, H. S. (1978) Histamine-forming bacteria isolated from spoiled skipjack tuna and jack mackerel. *Journal of Food Science.* **43,** 1779–1781.

45. Niven, C. F., Jr., Jeffrey, M. B., and Corlett, D. A. Jr. (1981) Differential plating medium for quantitative detection of histamine-producing bacteria. *Applied & Environmental Microbiology.* **41**(1), 321–322.

46. Bedry, R., Gabinski, C., and Paty, M. C. (2000) Diagnosis of scombroid poisoning by measurement of plasma histamine. *New England Journal of Medicine.* **342**(7), 520–521.

47. Guss, D. A. (1998) Scombroid fish poisoning: successful treatment with cimetidine. *Undersea & Hyperbaric Medicine.* **25**(2), 123–125.

48. Kerr, G. W. and Parke, T. R. (1998) Scombroid poisoning–a pseudoallergic syndrome. *Journal of the Royal Society of Medicine.* **91**(2), 83–84.

49. Ritchie, A. H. and Mackie, I. M. (1979) The formation of diamines and polyamines during storage of mackerel (Scomber scombrus), in Advances in Fish Science and Technology, (Connell, J. J. eds.) Fishing News Books: Surrey, UK. 489–494.

50. Gellert, G. A., et al. (1992) Scombroid fish poisoning. Underreporting and prevention among noncommercial recreational fishers. *Western Journal of Medicine.* **157**(6), 645–647.

51. Rawles, D. D., Flick, G. J., and Martin, R. E. (1996) Biogenic amines in fish and shellfish. *Adv Food Nutr Res.* **39,** 329–365.

52. Motil, K. J. and Scrimshaw, N. S. (1979) The role of exogenous histamine in scombroid poisoning. *Toxicol Lett.* **3,** 219–223.

15

Pathogen Control in Meat and Poultry Production: Implementing the USDA Food Safety and Inspection Service's Hazard Analysis and Critical Control Point System

Moshe S. Dreyfuss, Gerri M. Ransom, Mindi D. Russell, Kristina E. Barlow, Katrine M. Pritchard, Denise R. Eblen, Celine A. Nadon, Parmesh K. Saini, Nisha D. O. Antoine, Bonnie E. Rose, and Gerald W. Zirnstein

Abstract

Foodborne illness is the major public health concern for both the meat and the poultry industries in the United States and the U.S. Department of Agriculture' Food Safety and Inspection Service (FSIS), the agency that regulates the industry. FSIS introduced the Hazard Analysis and Critical Control Point (HACCP) Program as a means to allow flexibility in process design and control and to reduce foodborne pathogens in the food chain. This chapter will examine the historical changes brought by HACCP to evaluate the effectiveness of HACCP in controlling or reducing the presence of *E. coli* O157:H7, *Salmonella*, and *Listeria monocytogenes* on meat and poultry products, and explore the future of pathogen reduction in the meat and the poultry industries.

1. INTRODUCTION

Human foodborne illness is a complex public health challenge. The Centers for Disease Control and Prevention (CDC) estimates that all foodborne diseases are responsible for 76 million illnesses, 325,000 hospitalizations, and more than 5000 deaths in the US annually. These same national 1999 estimates show that the known pathogens, such as *Campylobacter, Salmonella, Escherichia coli* O157:H7, non-O157:H7 Shiga-toxin producing *E. coli*, and *Listeria monocytogenes* (*Lm*), are responsible for approx 14 million illnesses, over 60,000 hospitalizations, and approx 1800 deaths annually *(1)*. These illnesses are associated with water, seafood, vegetables, fruits, dairy products, meat, poultry, and egg products.

Various US government agencies are charged with protecting the public from illness caused by contaminated or adulterated food products. The role of regulating and inspecting meat, poultry, and egg products falls under the U.S. Department of Agriculture's (USDA), Food Safety and Inspection Service (FSIS).

The FSIS' mission is to "ensure that the nation's commercial supply of meat, poultry, and egg products is safe, wholesome, and correctly labeled and packaged" *(2)*. FSIS

From: *Infectious Disease: Foodborne Diseases*
Edited by: S. Simjee © Humana Press Inc., Totowa, NJ

inspects animals at slaughter and processed meat products at various stages of production, employing approx 7600 inspectors for approx 5500 slaughtering and/or processing establishments[1] nationwide. More than 850 red meat and 350 poultry slaughter plants were under federal inspection as of January 2004 *(3,4)*. To augment the inspection activities, the agency performs verification testing for microbiological and chemical agents in its three field laboratories.

The total US agricultural production in 2003 yielded 46.8 billion pounds of red meat *(3)*, 42.8 billion pounds of poultry, and 87.2 billion eggs for consumption *(5)*. It was estimated that meat- and poultry-related foodborne illnesses accounted for 27% of total food-related cases and outbreaks between 1990 and 2003 *(6)*.

In response to apparent increases in foodborne illnesses and outbreaks in 1990s, FSIS modified its inspection program to give it a more preventative public health focus and advanced the concept of reducing illnesses by targeting foodborne pathogens, such as *Salmonella*, *E. coli* O157:H7, and *Lm*. In 1996, the agency adopted the Hazard Analysis and Critical Control Point (HACCP) Program, an innovative approach to protect public health from hazards associated with meat and poultry products *(7)*. The goal was to reduce or eliminate foodborne illness that may be attributable to the FSIS-regulated products by mandating industry monitoring process controls of indicator pathogens and other hazards. HACCP shifted the focus of food safety control from detecting hazards that endanger human health in the end product to preventing the hazards from occurring; thereby, reducing the amount of contaminated food reaching the consumer, and reducing the incidence of foodborne illnesses.

This chapter will first review the FSIS-mandated historical changes in both the regulations associated with meat and poultry production and the microbiological assessment programs put into place to audit and monitor the presence of foodborne pathogens. All plants had to be operating under HACCP by January 2000 and the initial data from implementation through 2003 will be presented in the first part of this chapter. The trends of contamination by certain foodborne pathogens will be examined through assessment of microbiological data. This chapter will also examine the results of industry's changeover to a HACCP system, evaluate the success of using critical control points (CCPs) and the achievement of HACCP as a successful pathogen reduction tool and finally, comment on programmatic updates and new initiatives at FSIS since 2003.

2. FSIS ADOPTS HACCP

The National Academy of Sciences (NAS) first examined the issue of the role of microbiological criteria for foods in 1985 and concluded that a preventive system, such as HACCP, was essential for controlling foodborne pathogens and hazards *(8)*. NAS also recommended that FSIS create the National Advisory Committee on Microbiological Criteria for Foods (NACMCF) to develop and advise FSIS on criteria for food safety. In 1989, NACMCF produced its first recommendations for the implementation of food safety practices *(9)*. FSIS adopted the recommendations and accepted the principle that HACCP would provide the best approach for a new food safety control system.

---

[1] The words "establishments" and "plants" are used interchangeably and refer to all slaughter, processing, storage, and product finishing facilities under FSIS inspection authority.

NACMCF concluded that reliance on end-product testing only was not an effective means of monitoring the safety of food, because there were multiple processing steps at which contamination could occur. FSIS recognized that the prevention of contamination at these multiple steps was the most effective way to protect public health from foodborne pathogens *(10)*. Both NAS and NACMCF also recommended using microbiological testing to enumerate indicator organisms rather than detection of pathogens to determine prevalence *(9,11,12)*.

2.1. The PR/HACCP Final Rule

On July 25, 1996, FSIS promulgated the Pathogen Reduction/Hazard Analysis and Critical Control Points (PR/HACCP) System Final Rule *(7)*, revamping Sections 304, 308, 310, 320, 327, 381, 416, and 417 of Title 9, the U.S. Code of Federal Regulations. There are four main components to the FSIS PR/HACCP regulation:

1. Meat and poultry establishments are required to develop and implement Sanitation Standard Operating Procedures (SSOPs).
2. Slaughter establishments are required to conduct microbial testing of carcasses for generic *E. coli* to ensure process control.
3. Establishments producing certain raw meat and poultry products are required to meet performance standards for *Salmonella*.
4. All meat and poultry establishments are required to design and implement a HACCP system *(7)*.

Implementation of HACCP was phased in incrementally based on establishment size. Large plants, defined as those with 500 or more employees, were required to be operating their HACCP systems by January 1998; small plants, with 10–499 employees and greater than $2.5 million in annual sales, were to function under HACCP by January 1999, and very small plants, with less than 10 employees or less than $2.5 million in annual sales, by January 2000 *(7)*.

2.1.1. HACCP at FSIS

FSIS' HACCP program for meat and poultry establishments is based on the seven principles detailed in the 1989 NACMCF report, and has been tailored specifically to meet the needs of meat and poultry producers *(12)*. The following summarizes the general principles:

Principle 1: Conduct a hazard analysis.

Hazard analysis is the process of collecting and evaluating data on hazards associated with food and processing to determine those hazards that might be significant and must be further addressed in the HACCP plan *(see* Code of Federal Regulations Title 9 Section 417 [9 CFR 417]). This is accomplished by developing a list of hazards (biological, chemical, or physical factors) deemed reasonably likely to occur.

Principle 2: Identify CCPs.

A CCP is a point, step, or procedure in a food process at which control must be applied to prevent, eliminate, or reduce a food safety hazard to an acceptable level. All hazards identified in Principle 1 must be addressed and the establishment must carefully identify, develop, and document CCPs *(see* 9 CFR 417.2c). After analysis, these hazards are evaluated based on severity and potential to occur *(see* 9 CFR 417.2a). Identification of

CCPs for controlling biological, especially bacterial, hazards throughout production is essential as these are the primary cause of foodborne illness.

Principle 3: Establish critical limits for each CCP.

A critical limit is the maximum or minimum value to which a biological, chemical, or physical hazard must be controlled with a CCP to prevent or eliminate the identified food safety hazard or reduce it to an acceptable level. Critical limits were to be framed on the basis of FSIS regulations or guidelines. Interventions were to be put into place based on FDA tolerances and action levels; scientific data; or recommendations of recognized food safety process authority experts in the industry, academia, or trade associations (*see* 9 CFR 417) and on the process parameters, such as temperature, time, viscosity, and survival of target pathogens.

Principle 4: Establish CCP monitoring requirements.

FSIS requires that each monitoring step and its frequency be detailed in the HACCP plan. Monitoring the process step or intervention criteria and effectiveness ensure that the process is under control at each CCP. These activities consist of observations or measurements taken to assess whether a CCP is within the established critical limit parameters. Monitoring, continuously or frequently, must be sufficient to ensure the CCP is within the targeted range (*see* 9 CFR 417.7).

Principle 5: Establish corrective actions.

Corrective actions are mandatory when monitoring indicates that a deviation from an established critical limit has occurred. Such a deviation would indicate that the process step was inadequate and out of control and the intervention was not effective or properly applied. HACCP plans must identify pertinent corrective actions when a critical limit is not achieved (*see* 9 CFR 417.3 and 417.2(c) *[5]*).

Principle 6: Establish record keeping and documentation procedures.

The PR/HACCP regulation requires that all establishments maintain documentation of all HACCP data, including the hazard analysis and the written HACCP plan, records of monitoring CCPs, critical limits, verification activities, and the resolution of process deviations (*see* 9 CFR 417.5 and 417.2(c) *[6]*).

Principle 7: Establish verification procedures.

This principle incorporates two essential HACCP steps:

1. *Validation* ensures that HACCP plans are given sufficient forethought so that the plan does what it is designed to do: provide safe food products. Before implementation, scientific references and support, e.g., federal guidelines, scientific studies, and expert determinations, are required as supporting justification for CCPs.
2. *Verification* makes certain the HACCP plan is working as intended by ensuring continuous success in meeting all critical limits. This is accomplished by reviewing HACCP plans, monitoring CCP records and critical limits, and by performing microbial sampling and analysis.

2.2. Shifting FSIS and Industry Roles under HACCP

The introduction of PR/HACCP changed the regulatory environment, creating new roles and relationships for FSIS, and for the meat and poultry industries. PR/HACCP is overall a less prescriptive system of inspection than what existed previously. Under PR/HACCP, the focus of FSIS inspection is broadened to look at the production system in its entirety, with each critical aspect of production assessed for potential hazards.

FSIS assumes an oversight role under PR/HACCP and is responsible for establishing food safety and sanitation standards, conducting verification to ensure that these standards are met by the establishment, and taking enforcement actions when necessary.

Industry is expected to tailor food safety systems to meet standards set by FSIS *(7)*. Under HACCP, each establishment is given the responsibility for the assessment and ongoing control of CCPs. The design of their food safety system and its documentation based on each processing category, whether slaughter, raw/not ground, raw/ground, ready-to-eat (RTE), etc., is to be contained in a detailed written HACCP plan (*see* 9 CFR 417.2 (b)*(1)*).

By clarifying the respective roles of industry and government, the PR/HACCP Rule enabled FSIS to better target inspection resources and provide oversight to industry food safety systems and testing programs.

3. MICROBIOLOGICAL TESTING PROGRAMS

With the adoption of the PR/HACCP Final Rule, FSIS resolved that microbiological testing will be an essential component of HACCP-based safety systems and instituted microbiology-based requirements for *Salmonella* and generic *E. coli* in certain raw meat and poultry products. Microbial testing is an essential part of an effective program of monitoring pathogen presence in foods, and can be used to:

- Detect pathogens in high-risk foods.
- Verify effectiveness of process control measures.
- Collect baseline information for evaluation of sanitation programs and trend analysis in raw materials.
- Validate the effectiveness of pathogen interventions.
- Identify where and when modifications to control measures are needed.

In addition to the PR/HACCP *Salmonella* and generic *E. coli* programs, FSIS conducts a variety of other microbiological testing programs to either verify the overall effectiveness of food safety systems in federally inspected meat and poultry establishments or test for hazards in retail and imported products. These monitoring/verification testing programs (conducted by the FSIS laboratories using FSIS methods) include:

- Testing of both domestic and imported raw beef products for *E. coli* O157:H7;
- Testing of both domestic and RTE meat and poultry products for *Salmonella*, *Lm,* and, for certain products, *E. coli* O157:H7 (fermented sausages and cooked meat patties);
- Intensified risk-based sampling in RTE establishments that covers testing of product, product contact surfaces, and environmental surfaces for *Lm*;
- Follow-up testing of products as necessary to verify preventive and corrective actions following HACCP deviations;
- Testing of products as part of investigations into causes of outbreaks of foodborne illness.

3.1. **Salmonella** *and Generic* **E. coli** *Programs under HACCP*

3.1.1. Salmonella *Performance Standards*

The PR/HACCP Final Rule set *Salmonella* performance standards for processors to meet for certain carcasses and raw ground products in slaughter and grinding establishments.

Table 1
USDA/FSIS *Salmonella* Testing Compliance Standards
for Raw Meat and Poultry Commodities

Commodity	Maximum no. of *Salmonella*-positive samples permissible per sample set
Ground beef	5/53
Market hog	6/55
Broilers	12/51
Cow/bull	2/58
Steer/heifer	1/82
Ground turkey	29/53
Ground chicken	26/53

From ref. *(15)*.

The implementation of the PR/HACCP *Salmonella* performance standards was the first time that microbiological performance standards were incorporated into the meat and poultry regulatory systems. This was an initial step towards defining levels of food safety performance that meat and poultry establishments would be required to achieve consistently over time. These standards *(13)* are based on the prevalence of *Salmonella* as determined from the agency's nationwide microbiological baseline studies *(14)*.

Salmonella was targeted since it is an etiological agent of salmonellosis, one of the most common foodborne illnesses associated with meat and poultry products. It is also easily tested for by established laboratory methods. Under the PR/HACCP Final Rule, FSIS verifies that food safety systems effectively control *Salmonella* contamination, which may also have the added advantage of reducing other enteric pathogens.

Raw products covered by PR/HACCP performance standards include carcasses of cows/bulls, steers/heifers, market hogs, and broilers. Standards were also put in place for ground beef, ground chicken, and ground turkey. Industry guidance has more recently been established for turkey and goose carcasses. FSIS verifies that performance standards are met through sampling and analysis. PR/HACCP compliance testing is expressed in terms of the maximum number of *Salmonella*-positive samples that are allowed per sample set (Table 1) *(15)*. The number of samples in a sample set varies by product, and the performance standard provided an 80% probability of an establishment achieving acceptable levels when it operated at the standard. An initial sample set or a set that followed a passed set is designated an "A" set. "A" sample sets were collected at randomly selected establishments. If an establishment failed a set, corrective actions are implemented, and then the agency collected additional test sets, labeled "B," "C," and "D" sets as appropriate if compliance was not met. Once an establishment passes the set indicating the corrective actions were successful, the subsequent set will be labeled "A."

3.1.2. Using Generic E. coli *to Verify Process Control*

Generic *E. coli* (*E. coli* Biotype I), commonly found in an animal's intestinal tract, is an accepted indicator of fecal contamination *(13)*. The intestinal tract is also the primary source of contamination from additional pathogens, such as *E. coli* O157:H7 and *Salmonella* in FSIS-regulated products.

The generic *E. coli* testing component of the PR/HACCP Rule requires that slaughter establishments perform their own regular quantitative testing for this organism as an indicator of process control, i.e., control over sanitary carcass dressing procedures. FSIS developed *E. coli* performance criteria using baseline data, which provide the highest allowable microbial loads on carcasses when the slaughter process is in control *(16)*. A performance criterion based on a reference baseline provides the slaughter establishment with guidance on the effectiveness of its system in preventing contamination *(17)*.

Using generic *E. coli* as a direct measure of fecal contamination, FSIS has been able to assess how well industry's slaughter and dressing procedures controlled contamination by comparing *E. coli* levels from industry testing data against the agency-established criteria. Failing to meet these criteria serves as a catalyst for the establishment to review its processes, record its results, and initiate corrective actions. Failing the criteria may also trigger an additional FSIS inspection verification activity that may include review of SSOPs or HACCP records.

Evaluation of *E. coli* test results is done through a "moving window" approach *(18, 19)*. Sample results are accumulated until 13 have been accrued. As a new test result is added to the data, the oldest result is dropped and the new test is added to the most recent of 13 results *(16,20)*. Those test results in the "moving window" are considered in the evaluation of the process controls at a given time. FSIS requires industry to establish its own plant data and to use statistical process control (SPC) techniques to determine whether processing is under control when evaluate their results *(17)*. During testing, either one unacceptable result or more than three marginal results in the last 13 consecutive results should trigger action to review process controls, discover the cause, and prevent recurrence *(20)*.

SPC techniques are useful as they enable producers to: (1) understand variation in the process, (2) use that information to maintain process control, and (3) make improvements in performance on a continual basis *(21–23)*, which is integral to the HACCP concept *(17,24)*. SPC involves an initial evaluation of a process's capability, followed by determining the "normal" range of the process, setting control limits or thresholds (i.e., defining "in control"), monitoring regular production on a continual basis, and addressing any trends or signals that the process may not be in control. Generic *E. coli* data monitoring over time has proven extremely useful to ongoing process control, and evaluation of accumulated data is useful for identifying trends and occurrences, such as equipment failures.

3.1.3. NACMCF Performance Standards Evaluation

In an effort to continue food safety improvement, in 2001 FSIS sought guidance from NACMCF on the assessment of and on sound strategies for updating and improving performance standards. In response, four reports were produced focusing on the development and application of performance standards specific to ground beef, broiler chickens, ground chicken, and ground turkey *(16,25–27)*. From these reports, NACMCF affirmed that a microbiological approach to pathogen control was appropriate, stating "regardless of the approach taken to control the level of pathogenic microorganisms in raw meat and poultry, there should be an either explicit or implicit microbiological criterion underlying the approach taken *(16,25,26)*." The Committee noted that based on FoodNet data, CDC determined that overall human salmonellosis decreased 15% between 1996 and 2001. The report also underscored that performance standards stimulated the development and implementation of intervention technologies for reducing levels of pathogens on raw meat and poultry products.

3.2. FSIS Microbiological Sampling Program Results

Whereas it is important to consider all sampling program results to provide an overall view of establishment conformity with HACCP, government regulations, and food safety, this review cannot examine each and every program. Within any particular time period, programs will vary in their effectiveness and need to be modified or changed to react to microbial trends. Therefore, this section will look briefly at only the outcomes of two FSIS microbiological testing programs of raw products for food safety compliance since HACCP was implemented, namely, the *Salmonella* and *E. coli* O157:H7 programs.

3.2.1. Salmonella *PR/HACCP Testing Program*

From 1998 to 2002, the *Salmonella* PR/HACCP testing program, the percentage of *Salmonella*-positive samples in six out of the seven product categories either demonstrated a downward trend or remained about the same (Fig. 1)[1] *(28)*. Overall, for all sizes of establishments combined, the 2002 number of *Salmonella*-positive samples for broilers, market hogs, cows/bulls, steers/heifers, ground beef, and ground turkey decreased. Differences in pre- vs post-HACCP *Salmonella* prevalence reflect changes in industry practices in response to HACCP implementation.

Within the first 5 yr of PR/HACCP implementation, 90.1% of *Salmonella* random "A" sets tested passed. Examination of the data on a per establishment basis showed that 84.6% of establishments never failed an "A" set, whereas 15.4% failed at least one set *(28)*. Establishments that failed an "A" set were required to implement corrective action and improvements in pathogen-reduction programs and were then targeted for a "B" set. The percentage of all establishments that failed "B" sets averaged 4.4% across commodity products, ranging from 0.8% in steer/heifers to 9.2% in broilers.

3.2.2. E. coli O157:H7

As noted previously, FSIS requires assessments of various pathogens' presence in meat and poultry. Due to a number of outbreaks in the 1990s from *E. coli* O157:H7, the contamination of ground beef resulting in severe illness in children, FSIS declared it an adulterant in October 1994 and implemented a testing program for the presence of this pathogen in raw ground beef *(29)*. An overall decrease in *E. coli* O157: H7-positive raw ground beef samples were noted between October 1999 and September 2003 (Fig. 2)

---

[1]The data presented represent all samples collected as part of an "A" set during the indicated calendar year, with no consideration given as to whether a sample is part of a complete or an incomplete set, or a passed or failed "A" set.

It is necessary to note that these data have certain limitations that restrict the range of statistical inferences. The PR/HACCP verification testing program is strictly regulatory in nature and was designed to track establishment performance rather than to estimate nationwide prevalence of *Salmonella* in products. Because the program is not statistically designed, different establishments may be sampled from year to year, confounding rigorous trend analyses.

Furthermore, it is important to note that the prevalence estimates computed from the FSIS' pre-HACCP baseline studies and surveys were nationally representative, because they were weighted on the basis of the production volume of the sampled establishments. In contrast, the PR/HACCP *Salmonella* prevalence presented here represent unweighted test results from sampled establishments.

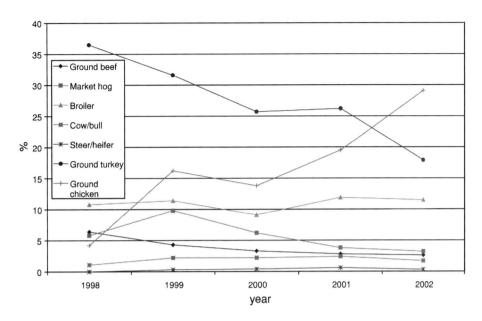

Fig. 1. Prevalence of *Salmonella* based on the percentage of PR/HACCP "A" *Salmonella* sets, 1998–2002 *(28)*. (Please see color insert.)

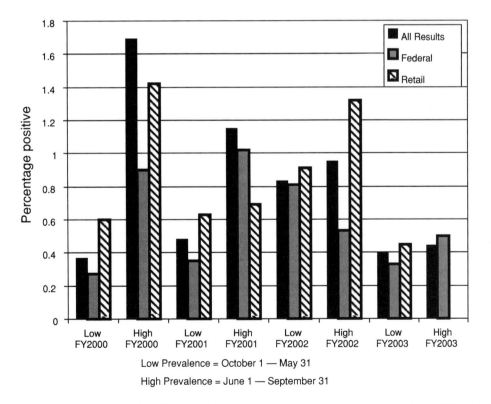

Fig. 2. Seasonal variation in FSIS microbiological regulatory testing for *Escherichia coli* O157:H7 in raw ground beef samples for all testing programs, verification testing in federally inspected establishments, and verification testing in retail outlets, FY2000–FY2003 *(30)*. Reprinted with permission.

(30). The data generated from more than 9 yr of FSIS testing showed that the overall percentage of positive remained below 1% for the samples analyzed. Out of the 26,521 raw ground beef samples tested from FY2000 to FY2003, 189 (0.71%) tested positive for *E. coli* O157:H7.

Additional decreases in *E. coli* O157:H7 were noted from FY2002 to FY2003, where a 50% reduction in the rate of positive ground beef samples was observed when controlling for season (95% CI=10–72% decrease; *p*=0.02). FSIS analysis demonstrated that these year-to-year changes in the rate of *E. coli* O157:H7-positive raw ground beef samples were statistically significant, which was consistent in samples obtained from both federally inspected establishments and retail outlets *(30).*

4. ASSISTING INDUSTRY IN IMPLEMENTING HACCP

To help implementing the PR/HACCP Final Rule, FSIS made expert advice and guidance available to smaller plants for the development, implementation, and evaluation of PR/HACCP programs. With this information and assistance, establishments' HACCP teams were then required to develop and assess the scientific basis for their decisions made under HACCP and food safety procedures related to microbiological hazard analysis and critical limits.

Once a HACCP plan was implemented at an establishment, FSIS could assess the operation based on plant adherence to its validated food safety system. Microbiological samples submitted by field inspection program personnel were then analyzed by the FSIS laboratories for verification of compliance with food safety standards. Specialized FSIS audit teams could also visit and review first hand the activities and documentation at federally inspected establishments.

These assessments provided a technical review of the corrective action(s) implemented or proposed to eliminate the problem occurring in the establishment. Various specialized teams were formed as evaluation needs were determined.

4.1. Technical Assessment Groups

FSIS' first assessment teams, Technical Assessment Groups (TAG), were convened to respond to establishment food safety problems by conducting document reviews. An ad hoc TAG gathered scientific and technical information necessary to assist field inspection personnel in making sound and appropriate decisions in the application of policy in unique field settings. They also evaluated establishments' microbiological testing programs and their proposed process corrective actions or reprocessing procedures involving contamination incidents or temperature deviations. Whereas these proved helpful, the assessments of TAG were considered secondary to an on-site visit and their use was deemphasized in place of other assessment groups.

4.2. In-Depth Verification Teams

When an establishment had an on-going problem with process control, multiple *Salmonella* set failures, or did not provide sufficient information, FSIS activated an In-Depth Verification (IDV) team. An IDV review was "an assessment as to whether an establishment is carrying out activities that meet the requirements of the PR/HACCP Final Rule" *(31).* IDV reviews were designed to be either targeted for cause or random. The targeted reviews were performed by FSIS when:

(a) The establishment failed to meet PR/HACCP *Salmonella* testing performance standards on two consecutive sets;
(b) Persistent problems were identified by in-house inspection program personnel;
(c) The agency needed to decide if it would institute proceedings for withdrawal of inspection;
(d) Specific information was needed in order to determine regulatory compliance;
(e) The establishment had repeated positive pathogen test results in RTE product, repeatedly been implicated in illnesses, or involved in recalls.

IDV reviews supplemented the verification tools used by in-plant inspection program personnel and examined the technical and scientific merit of an establishment's HACCP system in a more rigorous and integrated manner. The IDV team wrote an evaluative report documenting its findings which was provided to both the inspection force and the establishment to assist in their decision-making processes for corrective actions. FSIS often used this report to assist field personnel as a scientific basis for taking subsequent regulatory action.

4.2.1. IDV Findings

FSIS recently examined IDV contributions to pathogen reduction in the US food supply *(32)*. Through 2003, 77 IDVs were held in response to failure of PR/HACCP *Salmonella* testing ($n=60$), presence of *Lm* in RTE foods ($n=9$), presence of *E. coli* O157:H7 in raw ground beef ($n=4$), and others (metal contamination, sanitation failures, zero tolerance failures for fecal contamination, and undercooked product) ($n=4$). IDVs for *Salmonella* "B" set failures were held in 3.2% of all establishments subject to PR/HACCP testing *(32)*. Table 2a and 2b shows a breakdown of where IDVs were performed. Of these, 16 IDVs took place in large plants, 34 in small plants, and 10 in very small plants (Table 2a.). The greatest number of IDVs for second set *Salmonella* failures was conducted in ground beef ($n=19$), market hog ($n=17$), and broiler ($n=11$) establishments (Table 2b).

PR/HACCP *Salmonella* data revealed that following an IDV, establishments were more likely to pass subsequent *Salmonella* testing (Fig. 3). Therefore, IDVs had likely a positive impact on food safety by identifying regulatory non-compliance, and processing conditions and practices that put products at risk for contamination.

4.3. Intensified Verification Testing

FSIS also specifically addressed RTE product contamination. *Lm* is, of particular, concern to the agency because when *Lm* is found in RTE products, it means *Lm* has generally stemmed from post-process contamination of exposed product. Unlike *Salmonella* and *E. coli*, *Lm* is an environmental organism usually found in soil and water. The expectation of finding *Lm* was not related to feces and, therefore, required a different type of assessment. The Intensified Verification Testing (IVT) concept was initiated to address *Lm* and an IVT team was deployed specifically to take *Lm* samples at establishments with a history of repeat *Lm*-positive test results. An IVT was used to focus resources at establishments producing RTE meat and poultry products.

At the establishment, the team of microbiologists took multiple samples, including:

- Product: The team collected RTE product at random intervals during the IVT. Product was collected in its final, packaged form after the establishment's pre-shipment review.
- Product contact surface sites: Those areas that the product may contact during production, such as conveyors, knives, slicers, packagers, tables, chutes, racks, etc.

Table 2a
Relative Number of In-Depth Verification (IDVs) Reviews
by Establishment Size

Establishment size	Total no. of establishments	No. of IDVs[a]/(%)[b]
Large	234	16 (6.8)
Small	773	34 (4.4)
Very small	834	10 (1.2)
Not recorded	15	0 (0)
Total	**1856**	**60 (3.2)**

From ref. *(31)*.
[a]Number of establishments of that size subjected to an IDV.
[b]Percentage of establishments of that size subjected to an IDV.

Table 2b
In-Depth Verification (IDVs) Reviews Held for Second Set
Salmonella **Failure, by Commodity, 2000–2003**

Commodity type	Total no. of establishments	Number of IDVs[a]/(%)[b]
Ground beef	1227	19 (1.5)
Market hog	315	17 (5.4)
Broilers	217	11 (5.1)
Cow/bull	131	8 (6.1)
Ground turkey	43	3 (7.0)
Steer/heifer	126	1 (0.8)
Ground chicken	24	1 (4.2)
Total	**1856**	**60 (3.2)**

From ref. *(31)*.
[a]Number of establishments producing the product subjected to an IDV.
[b]Percentage of establishments producing the product subjected to an IDV.

- Indirect contact sites: Areas that are near product contact areas, but indirectly or intermittently contact product. Indirect contact sites may include employee utensils, tools, and equipment, sides of tables or equipment, light switches, cart handles, etc.
- Non-contact sites: Areas that do not come into direct contact with product, but may contribute to contamination during production. Non-contact sites may include floors, walls, drains, overhead cooling units, and doors.

Based on the sample test results from any of the above surfaces and the IVT report, the need for regulatory action was determined *(33)*.

4.4. Food Safety Assessments

Since the complexity of food safety programs grew, the types of assessments also continued to evolve at FSIS. In order to improve efficiency in on-site plant reviews, FSIS created a new classification of inspection program personnel called Enforcement, Investigations, and Analysis Officers (EIAOs). EIAOs superseded TAG teams and IDVs. EIAOs are intensely trained in microbiological sampling, food technology, and HACCP principles. EIAOs and other agency personnel close to the establishments could perform the fundamentals of IDVs and IVTs, which are currently called Food

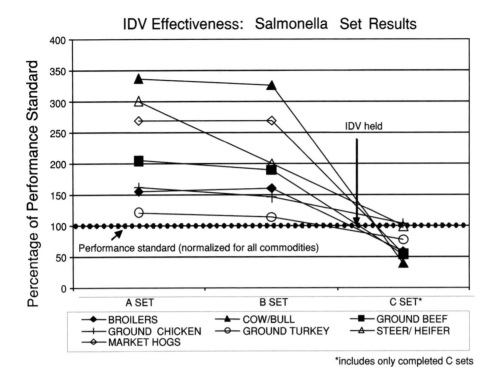

Fig. 3. Mean *Salmonella* set results by commodity, for all establishments where an IDV has been held, 2000–2003 *(28)*.

Safety Assessments (FSAs). Headquarters microbiological staff continue to offer technical help to the EIAOs.

FSAs are a tool used by the agency to address reoccurring problems in establishments. As industry's compliance with regulations continues to improve and the numbers of food-borne illnesses continue to decrease, FSIS will reevaluate its program to make it more effective and efficient.

5. EFFECTIVENESS OF THE IMPLEMENTATION OF HACCP

The Centers for Disease Control and Prevention reported in 2002 and again in 2004 that HACCP implementation was an important factor in the overall decline in bacterial foodborne illnesses *(34,35)*. From 1996 to 2001, there was an overall 15% reduction of *Salmonella,* a 21% reduction of *E. coli* O157:H7, a 27% reduction of *Campylobacter,* and a 35% reduction of *Listeria* attributable to the implementation of HACCP among other factors (Fig. 4) *(34)*. Between 2001 and 2003, reductions continued to 17% for *Salmonella,* 28% for *Campylobacter,* and 42% for *E. coli* O157:H7, while *Listeria* remained the same at 35% (Fig. 4) *(35)*. Illnesses caused by *Salmonella* Typhimurium (typically associated with meat and poultry) decreased by 38%. Dissemination of these data assisted in gaining worldwide recognition of HACCP as an effective food safety tool for food manufacturing establishments and the concept has spread to restaurants and other institutions in food handling. There has also been a trend toward global expansion of the use of HACCP principles.

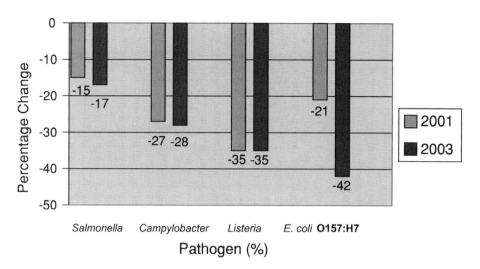

Fig. 4. The success of HACCP – the decline in bacterial foodborne illnesses reported to FoodNet from 1996 to 2001 and 1996 to 2003 *(34,35).*

6. GLOBAL FOOD SAFETY: THE WIDESPREAD INFLUENCE OF HACCP

The implementation of PR/HACCP by the United States has far reaching international food safety implications. Other countries are looking at successful models to adopt for reducing foodborne illnesses. Industrialization, urbanization, and rising wealth have revolutionized food production and the food supply, resulting in mass production and distribution, plus a proliferation of food service establishments and outlets. The globalization of food and feed trade, facilitated by international agreements aimed at promoting liberalization of trade, offers many benefits but also presents new risks. All foods, such as fish, milk, juices, fruits, and vegetables in addition to meat and poultry products, are major trade commodities, and can be a vehicle for worldwide transmission of infectious diseases *(36–39)*. There becomes a need for countries to have equivalent food safety systems in order to trade with one another and ensure food safety for the importing country.

6.1. Regulating International Food Safety through HACCP

To address international food safety concerns, the Food and Agriculture Organization and the World Health Organization created the Codex Alimentarius Commission (Codex) in 1963. Codex is an international inter-governmental body that develops science-based food safety and commodity standards, guidelines, and recommendations to promote the health and economic interests of consumers, whereas encouraging fair international trade in food. The *Recommended International Code of Hygienic Practice – General Principles of Food Hygiene*, third revision (1997) endorsed HACCP as a guiding principle of food safety by adding a HACCP annex with the Codex recommendations on food hygiene *(39)*.

6.2. Role of HACCP in Food Safety Harmonization

Progress in HACCP implementation varies from country to country. Many countries (e.g., the United States, European Union Member States, Canada, Australia, and

New Zealand) have mandated the use of science-based HACCP system requirements for particular sectors of their domestic food industries *(40–43)*.

In other countries that are beginning to develop exporting capabilities, lack of expertise and resources for training are major impediments to the domestic implementation of HACCP, with most progress made with food produced for international trade. An improved understanding and appropriate resources to support HACCP are needed by many developing countries. Basic hygienic controls need to be implemented by the industry before the HACCP system can be effectively implemented globally *(44,45)*.

Under the Agreement on the Application of Sanitary and Phytosanitary Measures (SPS Agreement) *(39)*, World Trade Organization member countries agree to facilitate the provision of technical assistance to achieve international food safety standards to all other members. Improvements in the capacity of participating nations include the implementation of effective HACCP systems which would be globally beneficial from both public health and economic standpoints *(46,47)*.

7. FUTURE OF HACCP AND PATHOGEN REDUCTION INITIATIVES AT FSIS

FSIS relies on its partnerships with other public health agencies, such as the CDC to assist with measuring the public health impact of its pathogen-reduction programs. Recent data from the Foodborne Disease Active Surveillance Network (FoodNet) indicate that the FSIS regulatory programs have been effective in reducing the incidence of disease from certain foodborne pathogens *(48)*. The FoodNet program, an active surveillance program for laboratory confirmed cases of 10 human foodborne diseases tracks the incidence in 10 sites. The program uses this data to monitor national trends in foodborne disease. Since 1996, CDC has annually published a FoodNet report analyzing trends in foodborne illnesses.

The 2003 FoodNet report stated that the changes in incidences of infections occurred concurrently with implementation of HACCP. In addition, the report pointed out that the decline in human infections from *Salmonella* mirrored declines in *Salmonella* that FSIS found in meat and poultry products. In the report, CDC recommended that additional targeted efforts be made to reduce the prevalence of pathogens in animal reservoirs by focusing on attribution; a goal with which FSIS concurs.

Ongoing baseline studies of the prevalence and numbers of pathogens on products will provide FSIS with new information for developing improved sampling programs and agency risk management initiatives, such as performance standards and other regulatory options.

7.1. Salmonella *PR/HACCP Program Data Analysis, 2004–2005*

Since 2003, the agency analyzed a total of 54,750 non-targeted "A" set samples in calendar year 2004 and a total of 40,714 non-targeted samples in CY 2005 for the seven product categories. Of these for CY 2004, 2052 (3.7%) and for CY 2005, 2322 (5.7%) samples were positive for *Salmonella (48,49)*. Calendar year 2005 was the first year since HACCP implementation that there had not been an overall decrease in the percentage of positive samples when results are weighed against the proportion of samples collected for each category in 2001.

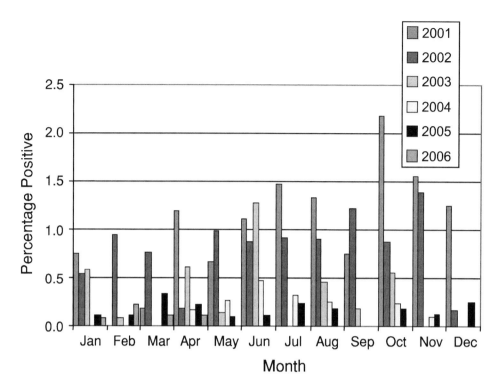

Fig. 5. The percentage of *E. coli* O157:H7 positives by month and year.

Out of 1004 "A" sets completed in CY 2004 and 755 "A" sets completed in CY 2005, 956 (95.2%) and 706 (93.5%) met the product-specific performance standard for 2004 and 2005, respectively. Compliance with product-specific performance standards ranged from 40 to 100% when the data were stratified by product class and establishment size *(48,49)*.

For CY 2004, the percentage of positive "A" set samples decreased for all three beef categories. The non-targeted testing program did not find a single positive in 1993 beef carcass samples (both steer/heifer and cow/bull) from large establishments. In CY 2004, the percentage of positive samples for market hogs increased from the CY 2003 level of 2.5% up to 3.1% after four consecutive years of declining levels *(48,49)*.

In CY 2005, the percentage of positive samples for all product classes was lower than the baseline prevalence rate determined prior to PR/HACCP implementation. In the history of the non-targeted regulatory sampling program, *Salmonella*-positive rates never exceeded these baseline rates with the exception of market hogs in CY 1999. When CY 2005 product-specific rates were further stratified by establishment size, broilers produced by very small establishments exceeded the baseline prevalence estimate for the first time since CY 2001 *(48,49)*.

7.2. E. coli *O157:H7 Data Analysis, 2003–2006*

Analysis of regulatory data collected showed that the percentage positives of *E. coli* O157:H7 for 2004 to April 2006 sustained monthly trends below 0.5% positive (Fig. 5). A cumulative trend of the same results showed lowered numbers in 2003 from previous

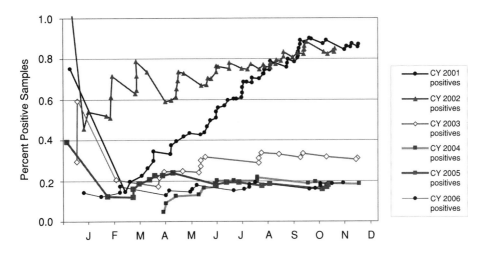

Fig. 6. Cumulative percentage positives of *E. coli* O157:H7 by month and year, 2001–2006.

years and further lowering and stability in the numbers of positives to 0.2% positive or below (Fig. 6) which meets the 2010 Healthy People target *(50)*.

7.3. Listeria Monocytogenes *Data Analysis, 2003–2005*

FSIS issued Directive 10,240.4 in October 2003 which defined two new sampling projects for CY 2004. These projects were identified as All Ready-to-Eat (ALLRTE) and Ready-to-Eat Risk1 (RTERISK1). Under the project ALLRTE, inspection program personnel were instructed to collect, at random, a RTE product that fit the previous FSIS definitions of targeted or low-targeted products. Most of the remaining, the samples would be scheduled under RTERISK1. For CY 2005, the ALLRTE project had 18 positive *Lm* results in 2806 samples, a positive rate of 0.64%, up slightly from the rate of 0.55 in 2004, but below the CY 2003 the overall percentage of 0.76% positive, or the CY 2002 percentage of 1.03% positive. For CY 2005, the RTERISK1 project had 39 positive *Lm* results in 6072 samples, a positive rate of 0.64%, a decrease from the positive rate of 1.01% in CY 2004, when a similar number of RTERISK1 samples were analyzed (5915). Similarly, the *Lm* project had a positive rate of 0.72% from 7089 samples in 2005 *(51)*.

From 2001 to 2004, FSIS analyzed samples from 5143 sliced, diced, or shredded products and recorded 91 positive results (1.77% positive). In CY 2005, there were only 26 *Lm* positives in 3855 samples (0.67% positive), a substantial decrease from the 4-yr average of 1.77% *(51)*.

7.4. Programmatic Updates in FSIS Since 2003

The introduction of risk-based sampling into the testing programs has helped to modify implementation of HACCP and further reduce bacterial contamination. Risk-based programs relate a hazard to public health outcome, such as illnesses. Risk analyses have guided the program development to make changes in agency functions to further potentially reduce the public health risks to consumers.

7.4.1. Baseline Studies

The newer baseline studies will provide FSIS and the regulated industry with data generated to determine the prevalence of foodborne pathogens in FSIS-regulated foods for public health evaluation by determining the quantitative levels of selected foodborne pathogens and microorganisms that serve as indicators of process control (e.g., *Campylobacter*, generic *E. coli*, *Salmonella*, *Enterobacteriaceae*, coliforms, and plate counts of aerobic microorganisms). These data will enable the Agency and industry to develop risk-based verification sampling programs, target interventions and effectively work toward reducing the risk of foodborne pathogens in FSIS-regulated products.

7.4.2. Risk-based Rules for Scheduling Salmonella *Set Samples for Raw Products*

FSIS is in the process of issuing new procedures for responding to how well establishments control *Salmonella* in raw products. The new procedures further strengthen FSIS' scientific and systematic approach to food safety and to the enforcement of current regulations. To guide FSIS resources, each plant performance is characterized into one of the three categories relative to the degree of potential *Salmonella* contamination. Category 1 meets 50% below the standard or better. The new category 2 still meets the standard, but is above 50% of the set performance limit. Category 3 fails the standard. Additional testing will also include public health trend surveillance and identification of specific *Salmonella* serotypes of the greatest human health concern. Follow-up actions include agency reaction to the presence of these serotypes, modifying scheduling frequency of sample sets, and conducting FSAs.

7.4.3. HACCP-based Initiatives in Regulating RTE Products

Unlike in raw products where the goal is to reduce pathogens and other hazards, the goal is to eliminate hazards in RTE products by achieving a targeted zero tolerance for pathogens, such as *Salmonella* and *Lm*. Integrating HACCP with the development of risk-based models offers a means to examine the food continuum from farm to table *(52)*. In December 2000, FSIS discontinued its RTE testing program based on product categories and introduced HACCP-driven testing based on processing categories as identified in 9 CFR 417.2. The product categories were identified based on factors affecting the probability of a product becoming contaminated with *Lm* during post-lethality exposure or factors that could relate to the effectiveness of the lethality step.

In 2005, FSIS implemented the first HACCP RTE verification project in which RTE establishments were identified based on their risk profile under FSIS Directive 10,240.4, Revision 1, 03/15/06. The establishment is targeted for sampling from a list of establishments identified with a particular risk ranking for *Lm*. The rankings are based on a number of factors including the RTE alternative(s) used by the establishment based on the type and number of interventions employed, the volume of production for post-lethality exposed products, and the sample test results from previous testing for *Lm*.

All RTE samples are analyzed for *Salmonella* and *Lm*. A few specific products containing beef, such as cooked beef patties and dry fermented sausages, are also analyzed for *E. coli* O157:H7.

There have been substantial percentage reductions in the pathogens in the FSIS-regulated RTE products *(53)*. From 2001 to 2004, out of the 5143 sliced, diced, or shredded products analyzed by FSIS, 91 were positive for *Lm* (1.77%). CY 2005 recorded only 26 *Lm* positives in 3855 samples (0.67%), a substantial decrease from the 4-yr average.

In addition, the FSIS RTE verification testing has consistently found very low levels of *Salmonella* in RTE products. During 2001 and 2002, there were 23 *Salmonella* positives in 14,121 samples (0.16%). The percentage of positive samples has been noticeably lower during the past 3 yr. During 2003 to 2005, only 10 *Salmonella* positives were found out of the 13,343 tested samples (0.07%). Finally, all of the 7137 RTE regulatory samples tested for *E. coli* O157:H7 between 1994 and 2005, were found negative *(52)*.

8. CONCLUSION

PR/HACCP provided FSIS and US processors with a regulatory framework to address both continuing and emerging public health hazards with the ability to reassess food hazards likely to occur. PR/HACCP is now a key preventative tool in protecting against foodborne illness. The use of CCPs and established critical limits in combination with monitoring programs has resulted in a significant reduction in the numbers of foodborne pathogens. FSIS' PR/HACCP program has likely contributed to reducing the incidence of foodborne illness due to *Salmonella, E. coli* O157:H7, *Lm,* and other pathogens in meat and poultry products. FSIS was able to incorporate additional public health safety programs into HACCP such as the specified risk material (SRM) removal, such as spinal cord, brain, dorsal root ganglia as related to bovine spongiform encephalopathy into the overall verification process. Industry application of HACCP principles and their success in achieving process control and the resultant reduction in foodborne pathogens, underscores the merit of FSIS' emphasis on a scientific approach to pathogen control.

FSIS expects future advances in pathogen reduction to continue to evolve as HACCP continues and advances in science are incorporated into the tools and programs aimed at preventing contamination of product and foodborne disease.

REFERENCES

1. Mead, P. S., Slutsker, L., Dietz, V., et al. (1999) Food-related illness and death in the United States. *Emerging Inf. Dis.* **5**(5). Available at http://www.cdc.gov/ncidod/EID/vol5no5/mead.htm (accessed on 3/25/07).
2. FSIS Mission Statement. Available at http://www.fsis.usda.gov/About_FSIS/index.asp (accessed on 3/25/07).
3. Meat News. (2004) *Annual Meat Report.* Available at http://www.meatnews.com/index.cfm?fuseaction=PArticle&artNum=7076 (accessed on 3/25/07).
4. USDA National Agriculture Statistics Service. (2004) Poultry slaughter, *2003 Annual Summary.* Available from http://usda.mannlib.cornell.edu/reports/nass/poultry/ppy-bban/pslaan04.pdf
5. USDA National Agricultural Statistics Service. (2004) Poultry– production and value, *2003 Summary.* Available from http://usda.mannlib.cornell.edu/usda/nass/PoulProdVa//2000s/2004/PoulProdVa-04-29-2004.pdf
6. Smith De Waal, C., Hicks, G., Barlow, K., Alderton, L., and Vegosen, L. (2006) Foods associated with foodborne illness outbreaks from 1990–2003. *Food Prot. Trends*, **25**(7), 466–473.
7. Pathogen reduction/hazard analysis and critical control points systems, final rule. *Federal Register* (1996), 38,805–38,989, July 25. Available from http://frwebgate.access.gpo.gov/cgi-bin/getpage.cgi?dbname=1996_register&position=all&page=38805
8. National Academies of Science. (1985) *An Evaluation of the Role of Microbiological Criteria for Foods and Food Ingredients.* NAS, National Research Council, National Academy Press, Washington, DC.

9. USDA/FSIS/NACMCF. (1989) *HACCP Principles for Food Production NACMCF.* USDA-FSIS Information Office, Washington, DC.

10. Fortin, N. D. (2003) The hang-up with HACCP: the resistance to translating science into food safety law. *Food Drug Law J.* **10,** 565–593.

11. Brown, M. H., Gill, C. O., Hollingsworth, J., et al. (2000) The role of microbiological testing in systems for assuring the safety of beef. *Int. J. Food Microbiol.* **62,** 7–16.

12. USDA/FSIS/NACMCF. (1989) *Hazard Analysis and Critical Control Point System.* USDA, Washington, DC.

13. Code of Federal Regulations, Title 9, Sections 310.25 and 318.94.

14. Baseline data is available at http://www.fsis.usda.gov/Science/Baseline_Data/index.asp (accessed on 3/25/07).

15. USDA/FSIS. (2000) *HACCP Implementation: Salmonella Compliance Test Results* January 26, 1998 to January 24, 2000. Available at http://www.fsis.usda.gov/OPHS/haccp/salcomp.pdf (accessed on 3/25/07).

16. USDA/FSIS. (1998) *HACCP-based Inspection Models Project In-plant Slaughter.* USDA, Washington, DC. Available at http://www.fsis.usda.gov/OA/haccp/himp.htm

17. American Meat Institute. (1994) *HACCP: The Hazard Analysis and Critical Control Point System in the Meat and Poultry Industry.* American Meat Institute Foundation, Washington, DC.

18. USDA/FSIS (1997a). *Guidelines for Escherichia coli Testing for Process Control Verification in Cattle and Swine Slaughter Establishments.* OPPDE *E.coli*-1. Available from http://www.fsis.gov/PDF/Guideline_for_Ecoli_Testing_Cattle_Swine–Estab.Pdf

19. USDA/FSIS (1997b). *Guidelines for Escherichia coli testing for process control verification in poultry slaughter establishments.* OPPDE *E.coli*-2. Available from http://fsis.usda.gov/PDF/Guideline_ Ecoli_Testing_Slaughter_Estab.Pdf

20. USDA/FSIS. (2004) *HACCP Overview: Slaughter Inspection Training.* USDA/FSIS, Washington, DC. Available at http://www.fsis.usda.gov/PDF/SIT_HACCPoverview.pdf

21. Grigg, N. P. (1998) Statistical process control in UK food production: an overview. *Br. Food J.* **100,** 371–379.

22. Bauman, H. E. (1992) Introduction to HACCP. In *HACCP Principles and Applications* (Pierson, M. D. and Corlett Jr., D. A., eds.), Van Nostrand Reinhold, New York, NY, pp. 1–5.

23. Stevenson, K. and Bernard, D. (eds.) (1999) HACCP. *A Systematic Approach to Food Safety.* The Food Processors Institute, Washington, DC.

24. DHEW. (1971) *Proceedings of the 1971 National Conference on Food Protection.* U.S. Department of Health, Education, and Welfare, Public Health Service, Washington, DC.

25. USDA/FSIS/NACMCF. (2004) *Response to the Questions Posed by FSIS Regarding Performance Standards with Particular Reference to Broilers (Young Chickens).* Available at http://www.fsis.usda.gov/OPHS/NACMCF/2004/NACMCF_broiler_4_13_ 04.pdf (accessed on 3/25/07).

26. USDA/FSIS/NACMCF. (2004) *Response to the Questions Posed by FSIS Regarding Performance Standards with Particular Reference to Raw Ground Chicken.* Available at http://www.fsis.usda.gov/ophs/nacmcf/2004/NACMCF_Ground_Chicken_082704.pdf (accessed on 3/25/07).

27. USDA/FSIS/NACMCF. (2004) *Response to the Questions Posed by FSIS Regarding Performance Standards with Particular Reference to Raw Ground Turkey.* Available at http://www.fsis.usda.gov/ophs/nacmcf/2004/NACMCF_Ground_Turkey_082704.pdf (accessed on 3/25/07).

28. Eblen, D. R. and Barlow, K. E. (2004) FSIS *Salmonella* PR/HACCP data 1998–2003: a summary, Abstract P035. In: *Abstract Book, International Association for Food Protection 91st Annual Meeting*, Phoenix, AZ, August 8–11, 2004.

29. USDA/FSIS. (2002) *New Measures to Address E. coli O157:H7 Contamination.* Available at http://www.fsis.usda.gov/OA/background/ec0902.pdf (accessed on 3/25/07).

30. Naugle, A. L., Holt, K. G., Levine, P., and Eckel, R. (2005) Food safety and inspection service regulatory testing program for *Escherichia coli* O157:H7 in raw ground beef. *J. Food Prot.* **68**(3), 462–468.

31. USDA/FSIS. (2001) *Conducting Targeted In-Depth Verification Reviews*, Dir. 5500.1. Available at http://www.fsis.usda.gov/OPPDE/rdad/FSISDirectives/5500.1.htm (accessed on 3/25/07).

32. Barlow, K. E. and Eblen, D. R. (2004) USDA FSIS in-depth verification (IDV) reviews, 2000–2003. Abstract P034, In: *Abstract Book, International Association for Food Protection 91st Annual Meeting*, Phoenix, AZ, August 8–11, 2004.

33. USDA/FSIS. (2003) *Control of Listeria monocytogenes in Ready-to-Eat Meat and Poultry Products*, GAO-03-1002R, July 21, 2003. Available at http://www.fsis.usda.gov/OPPDE/rdad/FRPubs/97-013F.htm (accessed on 3/25/07).

34. Centers for Disease Control and Prevention. (2002) Preliminary FoodNet data on the incidence of infection with pathogens transmitted commonly through food — selected sites, United States, 2001. *Morb. Mortal. Wkly Rep.* **51**(15), 325–329. Available at http://www.cdc.gov/mmwr/preview/mmwrhtml/mm5115a3.htm (accessed on 3/25/07).

35. Vugia, D., Cronquist, A., Hadler, J., et al. (2003) Preliminary FoodNet data on the incidence of infection with pathogens transmitted commonly through food — selected sites, United States, 2003. *Morb. Mortal. Wkly Rep.* **53**(16), 338–343. Available at http://www.cdc.gov/mmwr/preview/mmwrhtml/mm5316a2.htm (accessed on 3/25/07).

36. Kaferstein, F. K., Motarjemi, Y., and Bettcher, D. W. (1997) Foodborne disease control: a transnational challenge. *Emerg. Infect. Dis.* **3**(4), 503–510.

37. Kaferstein, F. K. (2003) Actions to reverse the upward curve of foodborne illness. *Food Control* **14**, 101–109.

38. Kruse, H. (1999) Globalization of the food supply–food safety implications. Special regional requirements: future concerns. *Food Control* **10**, 315–320.

39. FAO/WHO. (1997) Recommended international code of hygienic practice: general principles of food hygiene. In: *Codex Alimentarius, General Requirements (Food Hygiene)*, vol. 1B, 2nd ed., pp. 1–26.

40. McEachern, V. and Mountjoy, K. (1999) The Canadian Food Inspection Agency's integrated inspection system. *Food Control* **10**, 311–314.

41. Peters, R. E. (1999) Developing and implementing HACCP certification in Australia. *Food Control* **10**, 307–309.

42. Lee, J. A. and Hathaway, S. C. (1999) Experiences with HACCP as a tool to assure the export of food. *Food Control* **10**, 321–323.

43. Reichenbach, H. (1999) International food safety and HACCP conference — opening speech. *Food Control* **10**, 235–237.

44. Marthi, B. (1999) Food safety challenges in developing countries: the Indian situation. *Food Control* **10**, 243–245.

45. Jirathana, P. (1998) Constraints experienced by developing countries in the development and application of HACCP. *Food Control* **9**, 97–100.

46. Orriss, G. D. and Whitehead, A. J. (2000) Hazard analysis and critical control point (HACCP) as a part of an overall quality assurance system in international food trade. *Food Control* **11**, 345–351.

47. Otsuki, T., Wilson, J. S., and Sewadeh, M. (2001) Saving two in a billion: quantifying the trade effect of European food safety standards on African exports. *Food Policy* **26**, 495–514.

48. USDA/FSIS. (2006) *Progress Report on* Salmonella *Testing of Raw Meat and Poultry Products, 1998–2004*. Available at http://www.fsis.usda.gov/PDF/Progress_Report_Salmonella_Testing_1998-2004.pdf (accessed on 3/25/07).

49. USDA/FSIS. (2006) *Progress Report on* Salmonella *Testing of Raw Meat and Poultry Products, 1998–2005*. Available at http://www.fsis.usda.gov/Science/Progress_Report_Salmonella_Testing/index.asp (accessed on 3/25/07).

50. U.S. Department of Health and Human Services. (2000) Healthy People 2010, vol. 1. In: *Objectives for Food Safety, Focus Area 10.1. Reduce Infections Caused by Key Foodborne Pathogens, Targets and Baseline*. Available at http://www.healthypeople.gov/document/html/objectives/10-01.htm (accessed on 3/25/07).
51. USDA/FSIS. (2006) *The FSIS Microbiological Testing Program for Ready-to-Eat (RTE) Meat and Poultry Products, 1990–2005*. Available at http://www.fsis.usda.gov/Science/Micro_Testing_RTE/index.asp#results03 (accessed on 3/25/07).
52. Buchanan, R. L. and Whiting, R. C. (1998) Risk assessment: a means for linking HACCP plans and public health. *J. Food Prot.* **61**(11)**,** 1531–1534.
53. USDA/FSIS. (2006) *The FSIS Microbiological Testing Program for Ready-to-Eat (RTE) Meat and poultry Products, 1990–2005*. Available at http://www.fsis.usda.gov/Science/Micro_Testing_RTE/index.asp (accessed on 3/25/07).

16

Use of Antimicrobials in Food Animal Production

Frank M. Aarestrup and Lars B. Jenser

Abstract

The usage of antimicrobial for therapy, metaphylactic, prophylactic and as growth promoters is described. Already in the end of the 1960'ties the usage of antimicrobial for other purposes than therapy promotion was questioned and with the increased consumption of antimicrobial in food animal production higher prevalence of antimicrobial resistant bacteria have been observed. Figures for usage are hard to obtained but published numbers indicate very different patterns of usage with room for improvement and prudent usage in several countries. High usage of antimicrobial have led to banning of the usage of growth promoters in the European Union while several compound are still used for growth promotion and therapy in United States. Resistant bacteria selected for could via the farm to fork the transmitted from the animal reservoir to humans causing reduced treatment possibilities and prolonged hospitalization. Indication of spread of antimicrobial resistance between the animal and human reservoir is given.

Key Words: Antimicrobials; growth promoters; antimicrobial usage; and epidemiology of antimicrobial resistance; interventions and ban of growth promoters.

1. INTRODUCTION

Antimicrobials are substances of natural, semisynthetic, or synthetic origin that kill or inhibit the growth of a microorganism but cause little or no damage to the host (1). This broader term than antibiotic, which refers to substances that are produced by microorganisms, will be used to include compounds with antibacterial effects used in modern animal production.

Antimicrobials have been used for treatment of animals and for growth promotion since the late 1940s. With the discovery of a large numbers of antimicrobials, these were introduced for routine therapeutic treatment of animals in the 1950s and soon after these compounds were shown to have a growth-promoting effect if fed to animals. One of the first-identified growth promoters, aureomycin, belongs to the tetracyclines (2–6). Antimicrobials have been commonly used for growth promotion since 1949 in the Unites States and since 1953 in the United Kingdom (7).

Modern agricultural production is very intensive with optimization of every step in the production and minimizing labor. Today most food animals in industrialized countries are reared in large groups on small areas with up to thousand animals living together and with an attempt to achieve quick weight gains.

As a consequence of this, a large number of substances with antimicrobial activity are used in modern food production. These include antimicrobials used for therapy, antimicrobial growth promoters, disinfectants, and metals. The most

From: *Infectious Disease: Foodborne Diseases*
Edited by: S. Simjee © Humana Press Inc., Totowa, NJ

commonly used ones are antimicrobials for therapy and growth promotion. In the following sections, a description of the usage of antimicrobials are given and the possible consequences hereof.

2. CLASSIFICATION OF DRUGS USED IN ANIMAL PRODUCTION

2.1. Therapeutic Antimicrobial

Several antimicrobials are used for the treatment of infections in animals. Among these are penicillins, cephalosporins, tetracyclines, chloramphenicols, aminoglycosides, spectinomycin, lincosamide, macrolides, nitrofurans, nitroimidazoles, sulfonamides, trimethoprim, polymyxin, and quinolones *(1)*. Most antimicrobials are used to treat enteric and pulmonary infections, skin and organ abscesses, and mastitis *(8)*.

2.2. Coccidiostats

Coccidiostats are substances, some with antimicrobial effects, used to prevent and treat coccidiosis in poultry. In Europe, antimicrobials used as coccidiostats must be approved by the European Union. Those currently authorized include decoquinate, diclazuril, halofuginone, robenidine, narazan, narazan/nicarbazine, lasalocid-sodium, and maduramicin-ammonium. They are mostly used in broilers, but also to some extent in turkeys and laying hens. In Denmark, mainly ionophores, such as salinomycin and monensin, are used as coccidiostats and the usage inclined to 25,493 kg active compounds during 1994–1999 but has since declined to a total of 11,133 kg active compounds in 2003 *(9)*.

2.3. Growth Promoters

Several classes of antimicrobials have been used for growth promotion. In the European Union, 11 different compounds from eight structural different classes were approved until 1995 (Table 1). Some of these compounds are structurally closely related to the antimicrobial used in the therapy of humans and animals (*see* Table 1), and some have been used both for growth promotion and therapy (tylosin). Usage of antimicrobials as growth promoters belonging to the same antimicrobial group as the therapeutic-used antimicrobials will diminish the potential of the therapeutic antimicrobial in human therapy because of the selection of antimicrobial resistance that can be transferred through the food chain to humans. In 1969, the Swann committee *(7)* suggested not to use antimicrobials that were used for therapy as growth promoters. Since 1998, the European Union has banned several compounds and, has phased out the use of antimicrobial growth promoters by 2005.

In the United States, compounds belonging to the same antimicrobial group, as previously were used as in Europe, are still used, whereas others like glycopeptides have never been used for growth promotion. Contradictory to the usage in Europe, several compounds, such as tetracycline, penicillin, and sulphonamides, are used both for growth promotion and for therapeutic treatment *(10)*. Antimicrobials used are evaluated for safety to human health and are classified as critically important, highly important, important, or not important according to a published guidance *(11,12)*. Several organizations have developed principles for prudent usage of antimicrobials *(13,14)* and organizations such as the Alliance for the Prudent Usage of Antibiotics

Table 1
Antimicrobial Used for Growth Promotion in Europe and United States

Antimicrobial group	Antimicrobial growth promoter	United States[a]	Europe	Related to antibiotic used in human treatment
Polypeptides	Bacitracin	In use (swine, poultry)	Banned (1999)	Bacitracin[b]
Flavofosfolipid	Flavomycin/ Bambermycin	In use (broilers)	Banned (2006)	None
Glycopeptides	Avoparcin	Not used	Banned (2006)	Vancomycin, Teicoplanin
Ionophores	Monensin	Not used	Banned (2006)	None
	Salinomycin	Not used	Banned (2006)	None
Macrolides	Tylosin	In use (swine)	Banned (1999)	Macrolides (erythromycin)
	Spiramycin	Not used	Banned (1999)	Macrolides (erythromycin)
Oligosaccharides	Avilamycin	Not used	Banned (2006)	Evernimicin[c]
Quinoxalines	Carbadox	In use (swine)	Until 1999[d]	None
	Olaquindox	Not used	Until 1999[d]	None
Streptogramins	Virginiamycin	In use (broilers)	Banned (1999)	Quinupristin/ Dalfopristin, Pristinamycin
Sulfonamides	Sulfathiazole	In use[d] (swine)	Not used	Sulfonamides
Tetracyclines	Tetracyclines	In use (swine)[e]	Not used	Tetracyclines
Penicillin	Penicillin	In use (swine)[e]	Not used	Penicillin
Pleuromuttilin	Tiamulin	In use (swine)	Prophylactic usage	None

[a]Adopted from GAO-04-490, April 2004 *(10)*.
[b]Skin infections.
[c]Redrawn before released for human treatment due to side effects.
[d]Redrawn due to carcinogenic effects.
[e]Used in chlortetracycline/penicillin/sulfathiazole combinations.

(APUA) and Union of Concerned Scientists (UCS) have created an awareness of the potential problem in using antimicrobials for growth promotion that could reduce the potential of therapeutic antimicrobials.

3. REASONS FOR USAGE AND CONSUMPTION OF ANTIMICROBIALS

3.1. Usage

In modern food animal production, antimicrobials are normally used in one of the four different ways. (1) *Therapy:* treatment of infections in clinical-sick animals, preferably with a bacteriological diagnostic. (2) *Metaphylactics:* treatment of clinical-healthy animals belonging to the same flock or pen as animals with clinical signs. In this way,

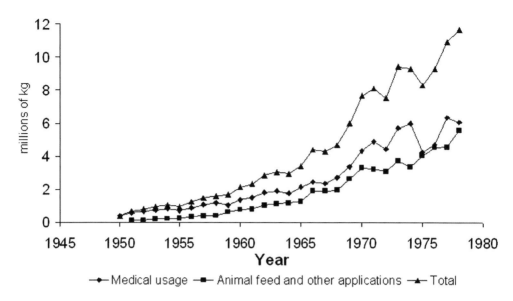

Fig. 1. Antimicrobial production from 1950 to 1978 (in millions of kg) in the United States *(15)*.

infections may be treated before they become clinically visible and the entire treatment period may thereby be shortened. In addition, this can, because of the modern production systems, often be the only way to treat large broiler flocks with water medication. (3) *Prophylactics:* treatment of healthy animals in a period where they are stressed to prevent disease. Examples include medicated early weaning. This use of antimicrobials can be a sign of management problems and, in most countries, is not considered legal or imprudent. (4) *Growth promotion:* inclusion of antimicrobials continuously in animal feed to improve growth. The antimicrobials are used in subtherapeutic concentrations. How this beneficial effect is achieved is not well established and this usage has been seriously questioned in several countries in recent years. The European Union has banned the usage of specific antimicrobials for growth promotion, and the quantities of antimicrobials used for growth promotion have been reduced.

3.2. Consumption

It is difficult to obtain solid data on the consumption of antimicrobials in the production of food animals. Exact figures are very rare and only estimates are available for a few countries.

In the United States, the consumption of antimicrobials increased tremendously throughout the 1950s to 1970s (Fig. 1). In 1951, a total of 110 mt were produced for additional to animal feed and other application, whereas 580 mt were produced for medical use in humans and animals *(15)*. In 1978, 5580 mt were produced as feed additives, whereas 6080 mt were produced for treatment of humans and animals. Thus, an increase of 50 and 10 times, respectively, for growth promotion and treatment was observed. Recently, the UCS *(16)* estimated the total usage of antimicrobials in food animal production in United States to be 11,150 mt, whereas the usage for treatment of humans was estimated to be 1361 mt.

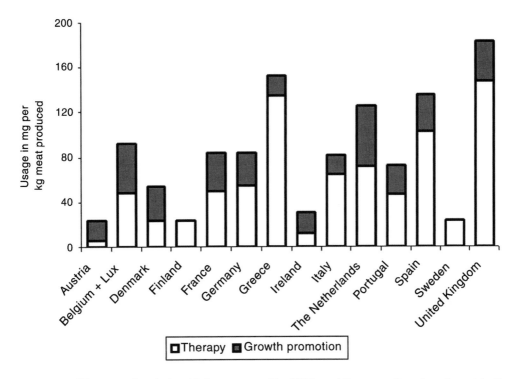

Fig. 2. Milligrams of antimicrobial agents used in 1997 per kilogram of produced meat in the different countries in the European Union (EMEA).

In the United Kingdom, the estimated usage of antimicrobials in 1996 was 650 mt for therapy and 100 mt for growth promotion *(17)*. For human treatment, approx 470 mt were used in 1997 *(17)*.

In Denmark, Norway, and Sweden, monitoring programs estimate the usage of antimicrobials in production animals. In Sweden, estimates of antimicrobials used for therapy as coccidiostats and for feed medication are given and have been monitored yearly since 1998 *(18)*. In Norway, estimates of antimicrobials used for therapy and for growth promotion are given and have been monitored since 1995 *(19)*. Finally, in Denmark, estimates of antimicrobials used for therapy, growth promotion, and as coccidiostats are given and have been monitored yearly since 1996 *(9)*. In Denmark, based on the VETSTAT program, usage can be monitored down to farm level.

The European Agency for the Evaluation of Medical Products *(20)* has estimated the amount of antimicrobials used for treatment and growth promotion for food animals in the different EU countries in 1997. The estimate for the production of animals from 1996 was included for comparison. In Fig. 2, the usage of antimicrobials to produce 1 kg of meat in different EU countries is presented. Even though problems exist in validating the estimated figures and production methods are different in individual countries, major differences in the amount of antimicrobials used were identified for the production of the same amount of meat. This provides room for major reductions in some countries.

4. BANNING ANTIMICROBIALS USED IN ANIMAL PRODUCTION

An association between usage of antimicrobials for growth promotion and occurrence of resistance was established before 1969 *(7)*. The Swann report recommended that antimicrobials could be used for growth promotion under the limitations that the used antimicrobials had no economical values otherwise in production, had little or no application as therapeutic agents in humans or animals, and would not impair the efficacy of a prescribed therapeutic antibiotic or antibiotics through the development of resistant bacteria.

5. EFFECT OF BANNING AND OTHER INTERVENTION OF USAGE OF ANTIMICROBIALS

In Sweden, antimicrobials used for growth promotion were banned in 1986. All antimicrobials for veterinary use were then classified as medicine and were available only by veterinary prescription *(21)*. It was also required that antimicrobial feed additives given to poultry should be proven not to increase colonization and shedding of *Salmonella.* No negative clinical or economical effect of the ban was detected in the Swedish slaughter pig production, whereas in poultry production the usage of antimicrobials for growth promotion had a verified protective effect against necrotic enteritis. Introducing better hygiene and management routines circumvented the effect of banning in the use of antimicrobials for growth promotion. In the piglet production, significant problems with weaning diarrhea were observed, resulting in an increase in the usage of antimicrobials for therapy. The termination of antimicrobials used for growth promotion resulted in a 35% reduction in the overall consumption of antimicrobials for animal production, and by other means of intervention, the used amount of antimicrobials for animals was reduced by a total of 50% *(21)*. From the last year (1985) when the antimicrobial was used for growth promotion and until 1999, more than a 50% reduction was archived in the overall usage of antimicrobials for food animal production in Sweden *(22)*. Similar results have been obtained in Norway *(23)* and Finland *(24)*.

In Denmark, larger quantities of antimicrobials were used until 1998 in animal feed for growth promotion than for therapy *(25,26)*. As provided by Danish regulation, all sales of veterinary medicines must take place by prescription from a veterinarian. The usage of antimicrobials were increased from 1986 to 1994 (Fig. 3). This correlated with a simultaneous increase in pig production, but the increased production could not only be justified by this increase in antimicrobial usage and could not be related to significant animal health problems.

In the mid-1990s, large amounts of tetracycline were used prophylactically. From 1995, a new regulation removed the economical incitement for the veterinarians to sell antimicrobials to the farmers resulting in a decrease in the consumption of antimicrobials. In 1995, the Danish authorities observed an increase in the use of antimicrobials for treatment of animals. Furthermore, the use of antimicrobials for growth promotion came under increased scrutiny because of quantities used and possible co-selection for resistance to therapeutic antimicrobials. The glycopeptide avoparcin was banned in Denmark in 1995 based on the selection of resistance to glycopeptide-resistant enterococci that, through the food chain, could be transferred to humans. For human treatment, the glycopeptide vancomycin was used as a last resort against methicillin-resistant

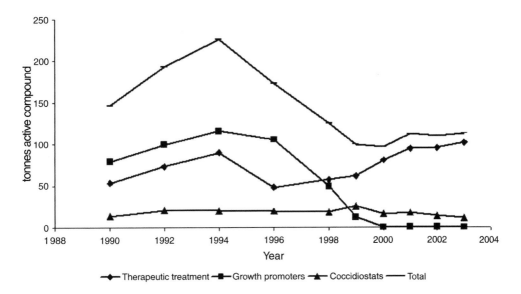

Fig. 3. Estimate consumption of antimicrobials in Denmark from 1900 to 2003 *(9)*.

Staphylococcus aureus and multi-resistant Gram-positive bacteria. During 1996 and 1997, the total consumption of antimicrobials was increased reflecting the increase in the food animal production. In 1997, avoparcin was banned in all EU countries based on a precautionary action.

In January 1998, the streptogramin virginiamycin was banned in Denmark based on crossresistance to Quinupristin–Dalfopristin (Synercid), an antimicrobial released for human treatment *(27)*. In December 1998, the European Commission decided to ban the use of bacitracin (a polypeptide), spiramycin (a macrolide), tylosin (a macrolide), and virginiamycin (a streptogramin) for growth promotion starting on July 1, 1999.

Furthermore, the Danish food animal industries decided to voluntarily stop all usage of antimicrobial for growth promotion from the end of 1999. Consequently, the usage of growth promoters has decreased significantly during 1998 and 1999 and has now been terminated. These initiatives follow recommendations by the World Health Organization *(28)*. By banning the use of growth promoters, the overall usage of antimicrobials for production animals was reduced from 205,448 kg active compounds in 1994 to 101,900 kg active compounds in 2003. Following the voluntary ban on antimicrobials as growth promoters, the amount of antimicrobials used for therapy increased 178% from 1998 to 2003, mostly based on usage of macrolides, tetracycline, and penicillin *(9)*. No negative effect of termination in the usage of antimicrobials as growth promoters on poultry products was observed *(29)*, whereas the removal of antimicrobials as growth promoters from weaned pigs resulted in an increase of antimicrobial consumption for therapy *(30)*.

6. RESISTANCE EPIDEMIOLOGY: SPREAD OF RESISTANCE FROM ANIMALS TO HUMANS

Spread of antimicrobial resistance from animals to humans has mainly been documented for zoonotic bacteria. A zoonosis is an infection or infectious disease that are transmissible under normal condition from vertebrate animals to humans *(31)*. Well-known foodborne zoonotic agents are *Salmonella, Campylobacter, Yersinia, Listeria,* and enterohemorrhagic *Escherichia coli*. Several studies have shown that zoonotic bacteria will acquire resistance among food animals, after which they transfer to and cause infections in humans.

One of the most pronounced examples in recent years is the appearance of fluoro-quinolone resistance among food animals subsequently followed by spread of resistant zoonotic bacteria to humans. Fluoroquinolones are, in several countries, the drug of choice for treatment of gastrointestinal infections in humans, and an emergence of resistance among zoonotic organisms such as *Salmonella* and *Campylobacter* is a matter of increasing concern.

The first observations came from the Netherlands where water medication with the fluoroquinolone enrofloxacin in the poultry production was followed by an emergence of fluoroquinolone-resistant *Campylobacter* species among both poultry and humans *(32)*. Since then, several studies have documented an increase in the occurrence of resistance to fluoroquinolones among *Campylobacter* from food animals and humans following the introduction of fluoroquinolones for treatment of infections in food animals *(33)*.

In Germany, an increase in the occurrence of fluoroquinolone resistance among *Salmonella* Typhimurium DT204c was observed after the introduction of enrofloxacin for veterinary use in 1989 *(34)*. Most recently in the United Kingdom, substantial increases in resistance to fluoroquinolone in *Salmonella* Hadar and *Salmonella* Virchow, and in multiresistant *Salmonella* Typhimurium DT104 have followed the licensing for veterinary usage of the fluoroquinolone enrofloxacin in 1993 and danofloxacin in 1996 *(35)*.

Resistant genes are transmissible and can be transferred both intra- and interspecies. Bacteria of animal origin may act as reservoirs for resistant genes that can be transferred to the human reservoir. However, because antimicrobials belonging to same classes are used in veterinary and human medicine, it may be very difficult to determine the direction of transfer. Thus, when studying the spread of antimicrobial resistance from animals to humans, the best cases are often found while introducing new antimicrobials for the usage only in animals. The usages of other nonstructural related antimicrobials or even nonantimicrobials such as metals and disinfectants can make it difficult to establish a clear linkage between prevalence of antimicrobial resistance and usage of specific antimicrobials.

A number of observations of horizontal spread of resistance from bacteria in food animals to bacteria in humans have been reported.

The streptothricin antimicrobial nourseothricin was introduced in animal husbandry for growth promotion in the former German Democratic Republic in 1983. No similar compounds have been used prior to the introduction, and resistance was only observed at very low frequencies. After the introduction, *E. coli* isolates with transferable resistant plasmids emerged among pigs *(36)*. This plasmid was subsequently found in isolates from the pigs, farmers and their families, and was furthermore found in *E. coli* isolates of the gut flora or causing urinary tract infections among humans living in the same geographic region *(37)*. Streptothricin resistance has also been found in *Shigella* isolates *(38)* and among *Campylobacter* isolates *(39)* as well as in staphylococci linked to aminoglycoside resistance *(40)* and enterococci here genetically linked to aminoglycoside and macrolides *(41)*.

After the introduction of the aminoglycoside apramycin for veterinary use in the beginning of the 1980s, resistance to apramycin emerged among *E. coli* isolates found in cattle and pigs in France and in the United Kingdom *(42,43)*. Apramycin has never been used for the treatment of infections in humans. The resistant gene (AAC(3)IV) encoding apramycin resistance co-selects for tobramycin, gentamicin, kanamycin, and neomycin resistance, but presence of this gene was first observed after the introduction of apramycin usage *(44)*. This gene and similar resistant plasmids were subsequently found in *S. enterica* from animals and human clinical isolates of *E. coli, S. enterica,* and *Klebsiella pneumoniae (45–51)*. These observations strongly indicate that this resistant gene primarily emerged among food animals selected by the usage of apramycin and then spread horizontal to bacteria of human origin where the usage of gentamicin for human treatment selected for its presence.

The glycopeptide avoparcin has been used for several decades as a growth promoter in Europe. A high prevalence of glycopeptide resistance was found among enterococci isolated from production animals in several European countries *(52–54)*. Unique identical isolates were identified among isolates of human and animal origin when standard typing methods were used *(55,56)*, whereas typing of strains indicated different clones in the human and animal reservoirs *(57,58)*. Genetic studies showed that predominantly the *van*A gene cluster encoded glycopeptides resistance among bacteria isolates from animals *(59–61)*. These studies together with genetic characterization of glycopeptide resistant isolates from humans *(62)* detected genetic variations in the *van*A gene cluster that could be used for determining ways of transmission. Especially prevalence of a single base-pair variant in the *van*X gene of the *van*A gene cluster in the different animal and human reservoirs indicated that glycopeptide resistance had spread from animals to humans *(63)*.

Studies from Germany *(64)*, the Netherlands *(65)*, and Belgium *(66)* have indicated that banning of the growth promoter avoparcin has reduced the presence of glycopeptide resistant among enterococci of human origin. Especially in the study from Belgium *(66)*, in which the prevalence of *Van*A, the dominant resistant determinant in animals, was reduced among patients after the ban of avoparcin.

7. SUMMARY AND CONCLUSIONS

Any usage of antimicrobials, even in subtherapeutic doses, will select for antimicrobial resistance. Studies have shown that humans and animals are not distinct reservoirs, but that antimicrobial-resistant bacteria and antimicrobial-resistant genes are exchanged between the two reservoirs. The frequency by which this happens is difficult to determine, and different environmental factors could select for different clones in the human and animal reservoirs. Transfer of resistance between the reservoirs can happen even if the resistant bacteria is only transient, as when consuming animal products, hence the prevalence of resistant bacteria in production animals and their products should be reduced as much as possible.

REFERENCES

1. Prescott, J. F. (2000) Antimicrobial drug resistance and its epidemiology. In: *Antimicrobial Therapy in Veterinary Medicine* (Prescott, J. F., Baggot, J. D., and Walker, R. D., eds.), Iowa State University Press, Ames, pp. 27–49.

2. Carpenter, L. E. (1950) Effect of aureomycin on the growth of weaning pigs. *Arch. Biochem.* **27**, 469–471.
3. Cunha, T. J., Burnside, J. E., Meadows, G. B., et al. (1950) Effect of APF supplement on efficiency of feed utilization for pig. *J. Anim. Sci.* **9**, 615–618.
4. Jukes, T. H., Stocstad, E. L. R., Taylor, R. R., Cunha, T. J., Edwards, H. W., and Meadows, G. B. (1950) Growth-promoting effect of aureomycin on pigs. *Arch. Biochem.* **26**, 324–325.
5. Stokstad, E. L. R. and Jukes, T. H. (1950) Further observations on the "Animal Protein Factor". *Proc. Soc. Exp. Biol. Med.* **73**, 523–528.
6. Whitehill, A. R., Oleson, J. J., and Hutchings, B. L. (1950) Stimulatory effect of aureomycin on the growth of chicks. *Proc. Soc. Exp. Biol. Med.* **74**, 11–13.
7. Swann, M. M. (1969) *Joint Committee on the Use of Antibiotics in Animal Husbandry and Veterinary Medicine.* HMSO, London.
8. Teuber, M. (2001) Veterinary use and antibiotic resistance. *Curr. Opin. Microbiol.* **4**, 493–499.
9. DANMAP. (2003) Use of antimicrobial agents and occurrence of antimicrobial resistance in bacteria from food animals, food and humans in Denmark. Danish Zoonoses Center, Sørborg, ISSN: 1600-2032. Available at www.dfvf.dk.
10. GAO. (2004) Antibiotic resistance. Federal Agencies. Need to better focus efforts to address risk to humans from antibiotic use in animals. United States General Accounting Office, Washington, DC.
11. Committee on Drug use in Food Animals, Panel on Animal Health, Food Safety and Public Health, Board of Agriculture, National Research Council, Food and Nutrition Board, Institute of Medicine. (1999) *The Use of Drugs in Food Animals: Benefits and Risks*, National Academic Press, Washington, DC.
12. FDA. (2003) Guidance for industry: evaluating the safety of antimicrobial new animal drugs with regard to their microbiological effect on bacteria of human health concern. #152. Center for Veterinary Medicine, Food and Drug Administration, Rockville, MD.
13. American Veterinary Medical Association. (2002) Position on Antimicrobials in Livestock feed. Available at http://avma.org/issues/policy/jtua_feeds.asp (accessed on 06 February 2007).
14. American Association of Swine Practitioners (2000) AASP basic guidelines of judicious therapeutic use of antimicrobials in pork production. *J. Swine Health Product.* **8**, 90–93.
15. Black, W. D. (1984) The use of antimicrobial drugs in agriculture. *Can. J. Physiol. Pharmacol.* **62**, 1044–1048.
16. Mellon, M., Benbrook, C., and Benbrook, K. L. (2001) *Hogging it. Estimates of Antimicrobial Abuse in Livestock.* Union of Concerned Scientists, UCS Publications, Cambridge, MA.
17. Harvey, J. and Mason, L. (1998) *The Use and Misuse of Antibiotics in UK Agriculture.* Soil Association, Bristol, UK.
18. Swarm. (2002) Swedish veterinary antimicrobial resistance monitoring. ISSN: 1650-6332. National Veterinary Institute, SE-751 89 Uppsala, Sweden.
19. NORM/NORM-Vet. (2002) Consumption of antimicrobial agents and occurrence of antimicrobial resistance in Norway. Tromsø/Oslo. ISSN: 1502-2037. Available at http://www.vetinst.no/Arkiv/Zoonosesenteret/NORM_NORM_VET_2002.pdf (accessed on 06 February 2007).
20. EMEA. (1999) Antibiotic resistance in the European Union associated with therapeutic use of veterinary medicines. Report and qualitative risk assessment by the Committee for Veterinary Medical Products. The European Agency for the evaluation of Medical products, July 14, 1999.
21. Wierup, M. (2001) The experience of reducing antibiotics used in animal production in the Nordic countries. *Int. J. Antimicrob. Agents* **18**, 287–290.
22. Wierup, M. (2001) The Swedish experience of the 1986 ban of antimicrobial growth promoters, with special reference to animal health, disease prevention, productivity and usage of antimicrobials. *Microb. Drug Res.* **7**, 183–190.

23. Grave, K. and Rönning, M. (2000) Prescribing patterns of veterinary antibacterial drugs in Norway during 1995–1999. *Nor. Veterinaertidskrift* **112**, 23–28.

24. Anonymous. (2000) Bacterial resistance to antimicrobial agents in Finland. FINRES 1999. Ministry of Agriculture and Forestry, Ministry of Social Affairs and health, http://www. mmm.fi/el/julk/finres99en.html (accessed on 06 February 2007).

25. Aarestrup, F. M., Bager, F., Madsen, M., Jensen, N. E., Meyling, A., and Wegener, H. C. (1998) Surveillance of antimicrobial resistance in bacteria isolated from food animals to antimicrobial growth promoters and related therapeutic agents in Denmark. *APMIS* **106**, 606–622.

26. Aarestrup, F. M., Bager, F., Jensen, N. E., Madsen, M., Meyling, A., and Wegener, H. C. (1998) Resistance to antimicrobial agents used for animal therapy in pathogenic-, zoonotic- and indicator bacteria isolated from different food animals in Denmark: a baseline study for the Danish Integrated Antimicrobial Resistance Monitoring Programme (DANMAP). *APMIS* **106**, 745–770.

27. Johnson, A. P. and Livermore, D. M. (1999) Quinupristin/dalfopristin, a new addition to the antimicrobial arsenal. *Lancet* **354**, 2012–2013.

28. World Health Organization. (2000) WHO global principles for the containment of anti- microbial resistance in animals intended for food. Report of a WHO consultation, June 5–9, Geneva, Switzerland.

29. Emborg, H. D., Ersbøll, A. K., Heuer, O. E., and Wegener, H. C. (2001) The effect of discontinuing the use of antimicrobial growth promoters on productivity in the Danish broiler production. *Prevent. Vet. Med.* **50**, 53–70.

30. Larsen, P. B. (2004) *Working Papers from the International Symposium: "Beyond Antimicrobial Growth Promoters in Food Animal Production"*, DIAS Report Animal Husbandry, Foulum, Denmark, **57**, 67–72.

31. Acha, P. N. and Szyfres, B. (1994) *Zoonoses and Communicable Diseases Common to Man and Animals*, 2nd edn, Pan American Health Organisation, Pan American Sanitary Bureau, Regional Office of the World Health Organization, Washington, DC.

32. Endtz, H. P., Ruijs, G. J., van Klingeren, B., Jansen, W. H., van der Reyden, T., and Mouton, R. P. (1991) Quinolone resistance in Campylobacter isolated from man and poultry following the introduction of fluoroquinolones in veterinary medicine. *J. Antimicrob. Chemother.* **27**, 199–208.

33. Engberg, J., Aarestrup, F. M., Smidt, P. G., Nachamkin, I., and Taylor, D. E. (2001) Quinolone and macrolide resistance in *Campylobacter jejuni* and *coli*: a review of mechanisms and trends over time of resistance profiles in human isolates. *Emerg. Infect. Dis.* **7**, 24–34.

34. Helmuth, R. (2000) Antibiotic resistance in *Salmonella*. In: Salmonella *in Domestic Animals* (Wray, C. and Wray, A., eds.), CAB International, Wallingford, UK, 89–106.

35. Threlfall, E. J., Ward, L. R., Skinner, J. A., and Rowe, B. (1997) Increase in multiple antibiotic resistance in nontyphoidal salmonellas from humans in England and wales: a comparison of data for 1994 and 1996. *Microb. Drug Resist.* **3**, 263–266.

36. Tschäpe, H., Tietze, E., Prager, R., Voigt, W., Wolter, E., and Seltmann, G. (1984) Plasmid borne streptothricin resistance in Gram-negative bacteria. *Plasmid* **12**, 189–196.

37. Hummel, R., Tschäpe, H., and Witte, W. (1986) Spread of plasmid-mediated nourseothricin resistance due to antibiotic use in animal husbandry. *J. Basic Microbiol.* **26**, 461–466.

38. Witte, W. (1998) Medical consequences of antibiotic use in agriculture. *Science* **279**, 996–997.

39. Bottcher, I. and Jacob, J. (1992) The occurrence of high-level streptothricin resistance in thermotolerant campylobacters isolated from the slurry of swine and the environment. *Int. J. Med. Microbiol. Virol. Parasitol. Infect. Dis.* **277**, 467–473.

40. Derbise, A., Aubert, S., and El Solh, N. (1997) Mapping the region carrying the three contiguous resistance genes *aadE*, *sat4* and *aphA-3* in the genomes of staphylococci. *Antimicrob. Agents Chemother.* **41**, 1024–1032.

41. Werner, G., Hildebrandt, B., and Witte, W. (2003) Linkage of *erm*(B) and *aadE-sat4-aphA-3* in multiple-resistant *Enterococcus faecium* isolates of different ecological origins. *Microb. Drug Resist.* **9,** S9–S16.

42. Chaslus-Dancla, E. and Lafont, J. P. (1985) Resistance to gentamicin and apramycin in *Escherichia coli* from calves in France. *Vet. Rec.* **117,** 90–91.

43. Wray, C., Hedges, R. W., Shannon, K. P., and Bradley, D. E. (1986) Apramycin and gentamicin resistance in *Escherichia coli* and salmonellas isolated from farm animals. *J. Hyg. Camb.* **97,** 445–456.

44. Hedges, R. W. and Shannon, K. P. (1984) Resistance to apramycin in *Escherichia coli* isolated from animals: detection of a novel aminoglycoside-modifying enzyme. *J. Gen. Microbiol.* **130,** 473–482.

45. Chaslus-Dancla, E., Martel, J. L., Carlier, C., Lafont, J. P., and Courvalin, P. (1986) Emergence of aminoglycoside 3-N-acetyltransferase IV in *Escherichia coli* and *Salmonella typhimurium* isolated from animals in France. *Antimicrob. Agents Chemother.* **29,** 239–243.

46. Chaslus-Dancla, E., Pohl, P., Meurisse, M., Marin, M., and Lafont, J. P. (1991) High genetic homology between plasmids of human and animal origins conferring resistance to the aminoglycosides gentamicin and apramycin. *Antimicrob. Agents Chemother.* **35,** 590–593.

47. Hunter, J. E., Shelley, J. C., Walton, J. R., Hart, C. A., and Bennett, M. (1992) Apramycin resistance plasmids in *Escherichia coli*: possible transfer to *Salmonella typhimurium* in calves. *Epidemiol. Infect.* **108,** 271–278.

48. Hunter, J. E., Hart, C. A., Shelley, J. C., Walton, J. R., and Bennett, M. (1993) Human isolates of apramycin-resistant *Escherichia coli* which contain the genes for the AAC(3)IV enzyme. *Epidemiol. Infect.* **110,** 253–259.

49. Johnson, A. P., Burns, L., Woodford, N., et al. (1994) Gentamicin resistance in clinical isolates of *Escherichia coli* encoded by genes of veterinary origin. *J. Med. Microbiol.* **40,** 221–226.

50. Pohl, P., Glupczynski, Y., Marin, M., Van Robaeys, G., Lintermans, P., and Couturier, M. (1993) Replicon typing characterization of plasmids encoding resistance to gentamicin and apramycin in *Escherichia coli* and *Salmonella typhimurium* isolated from human and animal sources in Belgium. *Epidemiol. Infect.* **111,** 229–238.

51. Threlfall, E. J., Rowe, B., Ferguson, J. L., and Ward, L. R. (1986) Characterization of plasmids conferring resistance to gentamicin and apramycin in strains of *Salmonella typhimurium* phage type 204c isolated in Britain. *J. Hyg. Camb.* **97,** 419–426.

52. Aarestrup, F. M. (1995) Occurrence of glycopeptide resistance among *Enterococcus faecium* isolates from conventional and ecological farms. *Microb. Drug Resist.* **1,** 255–257.

53. Klare, I., Heier, H., Claus, H., Reissbrodt, R., and Witte, W. (1995) *van*A-mediated high-level glycopeptide resistance in *Enterococcus faecium* from animal husbandry. *FEMS Microbiol. Lett.* **125,** 165–171.

54. Kruuse, H., Johansen, B. K., Rorvik, L. M., and Schaller, G. (1999) The use of avoparcin as a growth promoter and the occurence of vancomycin-resistant *Enterococcus* species in Norwegian poultry and swine production. *Microb. Drug Resist.* **5,** 135–139.

55. Van den Bogaard, A. E., Jensen, L. B., and Stobberingh, E. E. (1997) Vancomycin-resistant enterococci in turkeys and farmers. *N. Engl. Med.* **337,** 1558,1559.

56. Jensen, L. B., Hammerum, A. M., Poulsen, R. L., and West, H. (1999) Vancomycin-resistant *Enterococcus faecium* strains with highly similar pulsed-field gel electrophoresis patterns containing similar Tn*1546*-like elements isolated from a hospitalized patient and pigs in Denmark. *Antimicrob. Agents Chemother.* **43,** 724, 725.

57. Willems, R. J., Top, J., van den Braak, N., et al. (2000) Host specificity of vancomycin-resistant *Enterococcus faecium*. *J. Infect. Dis.* **182,** 816–823.

58. Bruinsma, N., Willems, R. J., van den Bogaard, A. E., et al. (2002) Different levels of genetic homogeneity in vancomycin-resistant and susceptible *Enterococcus faecium* isolates

from different human and animal sources analysed by amplified-fragment length polymorphism. *Antimicrob. Agents Chemother.* **46,** 2779–2783.

59. Jensen, L. B., Ahrens, P., Dons, L., Jones, R. N., Hammerum, A. M., and Aarestrup, F. M. (1998) Molecular analysis of Tn*1546* in *Enterococcus faecium* isolated from animals and humans. *J. Clin. Microbiol.* **36,** 437–442.

60. Simonsen, G. S., Haaheim, H., Dahl, K. H., et al. (1998) Transmission of *Van*A-type vancomycin-resistant enterococci and *van*A resistance elements between chicken and human at avoparcin exposed farms. *Microb. Drug Resist.* **4,** 313–318.

61. Descheemaeker, P. R., Chapelle, S., Devriese, L. A., Butaye, P., Vandamme, P., and Goossens, H. (1999) Comparison of glycopeptide-resistant *Enterococcus faecium* isolates and glycopeptide resistance genes of human and animal origin. *Antimicrob. Agents Chemother.* **43,** 2032–2037.

62. Woodford, N., Adebiyi, A. M., Palepou, M. F., and Cockson, B. D. (1998) Diversity of *Van*A glycopeptide resistance elements in enterococci from human and nonhuman sources. *Antimicrb. Agents Chemother.* **42,** 502–508.

63. Jensen, L. B. (1998) Differences in the occurrence of two base-pair variants of Tn*1546* from vancomycin resistant enterococci from humans, pigs and poultry. *Antimicrob. Agents Chemother.* **42,** 2463–2464.

64. Klare, I., Badstubner, D., Konstabel, C., Bohme, G., Claus, H., and Witte, W. (1999) Decreased incidence of *Van*A-type vancomycin-resistant enterococci isolated from poultry meat and from fecal samples of humans in the community after discontinuation of avoparcin usage in animal husbandry. *Microb. Drug Res.* **5,** 45–52.

65. Van den Bogaard, A. E., Bruinsma, N., and Stobberingh, E. E. (2000) The effect of banning avoparcin on VRE carriage in the Netherlands. *J. Antimicrob. Chemother.* **45,** 146–148.

66. Ieven, M., Vercauteren, E., Lammens, C., Ursi, D., and Gossens, H. (2001) Significant decrease of GRE colonization rate in hospitalized patients after avoparcin ban in animals. Abstract of the 41st Interscience Conference on Antimicrobial Agents and Chemotherapy, Chicago, IL. Abstract LB-8.

17

Alternatives to Antimicrobials*

Toni L. Poole, Todd R. Callaway, and David J. Nisbet

Abstract

The emergence of multidrug resistant pathogens has stimulated a need to find alternatives to antimicrobials. Currently, no single treatment is available that can eliminate the need for antimicrobials; particularly for immunocompromised individuals. Prudent use to protocols have been called for to decrease the consumption of antimicrobials. This includes the use to antimicrobials for individuals clinically diagnosed with bacterial infections and excludes antimicrobial treatment for viral infections, disease prophylaxis, and growth promotion (1). Most clinicians and scientists agree that unnecessary use of antimicrobials should be eliminated; few agree on what constitutes unnecessary use. Modern medicine and modern food animal production practices have contributed to the current problem, and more than the cessation of antimicrobial use for prophylaxis and growth promotion is necessary to reduce the incidence of multidrug resistant pathogens in hospitals and the environment.

There are countless preharvest protocols in food animal production for disease prophylaxis and many more are currently under investigation. Potential strategies that could be incorporated with the current management practices include: new diagnostic procedures, vaccination and treatment-based new technologies, competitive exclusion, and the use of probiotics. New treatment options are also under study that include: bacteriophage therapy and compounds directed at new bacterial metabolic targets (e.g., programmed cell death pathways). The combined application of preharvest prevention and treatment strategies has the potential to greatly reduce the amount of antimicrobials currently in use.

1. INTRODUCTION

In the last 50 yr, there have been tremendous advances in human and veterinary medicine. These advances have been accompanied by new challenges. In both human and animal populations, immune status and viral diseases predispose the host to secondary bacterial infections by opportunistic pathogens. A significant proportion of bacterial infections that require treatment are due to opportunistic pathogens. Demographic studies of human populations have shown a dramatic increase in immunocompromised individuals over the last 50 years (2,3). In large scale poultry and livestock production facilities, viral infections and stress play a significant role in the immune status of animals, and the eventual need for antimicrobial treatment for opportunistic infections. These and other factors have contributed to the increased use of antimicrobials. Alternative treatments are needed for the immunocompromised human and animal populations if we are to reduce substantially the consumption of antimicrobials. For example, the elimination

---

Disclaimer: Mention of a trade name, proprietary product, or specific equipment does not constitute a guarantee or warranty by the US Department of Agriculture and does not imply its approval to the exclusion of other products that may be suitable.

From: *Infectious Disease: Foodborne Diseases*
Edited by: S. Simjee © Humana Press Inc., Totowa, NJ

of posttransplantation immunosuppression would have a significant impact on the need for antimicrobials in human medicine.

Possibly the single most challenging issue in poultry and livestock production is preventing the spread of infectious diseases through entire herds or flocks. Ideally, sick animals would be identified on an individual basis and quarantined; thus, preventing exposure to the infectious agent with the rest of the flock or herd. That strategy works best in swine and poultry production facilities that use an 'all-in, all-out' method of rearing; however, disease rollover still occurs. Part of the problem lies in the inability to detect infected individuals. Many viral and bacterial pathogens result in subclinical infections, or are shed prior to the onset of clinical symptoms. Human pathogens that reside persistently or transiently in the gastrointestinal tract of food animals may have little or no affect on the animal's health and may be shed sporadically *(4)*. Fecal shedding from man and animals should be considered a major facilitator of bacterial dissemination. Fecal shedding of *E. coli* O157:H7 by beef cattle has been correlated with carcass contamination *(5)*. Control of fecal shedding represents a significant control point for bacterial dissemination. Now and in the future, preharvest management interventions will be aimed at reducing pathogen load, pathogen exposure, and promoting the immune defenses of poultry and livestock.

2. DISEASE PREVENTION

Disease prevention in food animals has and will continue to require a multifocal approach involving vaccination and farm-management practices. There is a need for new treatments that can replace the current dependence on antibiotics in both veterinary and human medicine. New vaccines, immunomodulatory drugs, and competitive exclusion (CE) technologies represent three areas of intense investigation that may be used alone or in combination to prevent diseases that in the past have been met with insurmountable obstacles.

2.1. Vaccines

The development of vaccines to prevent infectious diseases predates the discovery of antimicrobials by over 100 yr. Unfortunately, vaccination has not resulted in the eradication of infectious disease, and many diseases thought conquered are reemerging. Technological advances have provided new approaches to vaccine design. Even so, conventional live-attenuated and killed vaccines are still the most cost-efficient and efficacious means of immunization. Live-attenuated vaccines capable of replication elicit the most protective immune response; however, they also carry risks. Whole-killed vaccines are considerably safer. Generally, they produce an antibody response, but not a significant mucosal or cell-mediated immune response. An antibody response alone is often not sufficient to provide protection. Therefore, the current challenge is to develop vaccines that produce a protective immune response similar to live vaccines, but have the safety of killed vaccines.

There are literally hundreds of vaccines approved for use in domestic animals *(6,7)*, which include live-attenuated, killed, DNA, marker, and subunit vaccines. Ideally, a vaccine should be safe for the individual; effective after a single dose; elicit a humoral, cellular, and mucosal immune response that is protective against all field-strains; and confer life-time protection *(7)*. In food animal production in which the vaccination involves entire herds or flocks, a vaccine must also be stable at room temperature, easy to administer, and low in cost. Many vaccines are made more efficacious by the use of adjuvants; for a review on adjuvants in veterinary medicine *see* ref. *(8,9)*.

Many new technologies, including newer vaccines, require changes in management practices that introduce significant additional expense to the producer and overdue stress to the animals. These technologies are considered impractical. Products that are defined as recombinant or genetically modified are controversial and are not likely to gain widespread use in the immediate future.

2.1.1. Live-Attenuated Vaccines

Historically, modified live vaccines (viral or bacterial) were produced by serial passage or other growth conditions that resulted in a less pathogenic strain. Very little is known about the process of attenuation at the molecular level, and it is likely to differ for individual pathogens. Ever expanding molecular technologies are beginning to explain the mechanism of attenuation for some pathogens *(10)* and this should in turn further vaccine development.

2.1.2. Subunit Vaccines

Subunit vaccines consist of proteins produced by the pathogen that are believed to confer a protective immune response in a natural infection (e.g., inactivated toxin proteins, bacterial capsular polysaccharides, and viral proteins). The vaccine can consist of one or more proteins that may be purified from cell cultures or cloned and expressed from a protein expression vector. Subunit vaccines were developed in the belief that they would elicit a protective immune-response while eliminating the potential for pathogenesis and an inflammatory response, thus providing greater safety than a live-attenuated vaccine. The predicted advantages of using a protein expression vector are quantity, purity, and cost. Protein expression vectors have been engineered to over express the protein of interest and may encode a tag for easy purification. However, there are numerous problems associated with overproduction of proteins. In *E. coli*, it is not uncommon for proteins to be sequestered in inclusion bodies. If this occurs, the proteins must be denatured and refolded. This is a process may not yield the intended results. There are several types of protein expression systems available and each has its own list of problems.

2.1.3. DNA Vaccines

DNA vaccines represent a novel delivery system for provision of antigenic proteins. The gene of interest is cloned into a protein expression vector that can be transcribed and translated into the protein of choice. In this case, the DNA vector (e.g., plasmid) is injected directly into the host where it is taken up by the host cells. Transcription and translation into proteins occurs within the host cells. This differs from subunit vaccines in which the protein is generated and purified in a laboratory. Plasmids used as vaccines are designed such that they are unable to replicate themselves *(11)*.

2.2. Immunomodulators

Immunomodulators include cytokines, pharmaceuticals, probiotics, nutraceuticals, and medicinal plant products *(12)*. They may be used as adjuvants to enhance the immune response to a vaccine, or in some cases as a treatment for disease. It is important that products in these categories elicit an appropriate immune response without inducing immune-mediated tissue damage. Both the innate and adaptive arms of the host's immune system respond to pathogens, and over the last few years, a great deal is known about the molecular function and interrelatedness to these two branches of the immune response. The innate immune system responds to infection first by initiating a rather nonspecific defense response. The adaptive immune response is mediated by a large repertoire of antigen specific T and B lymphocytes that undergo gene rearrangement, and thus specifically recognize a broad spectrum of foreign antigens that subsequently provide memory of these antigens *(13,14)*. Signal transduction pathways are integral components of both innate and adaptive immune responses and are involved in linking both the systems *(15)*. Intensive research into the molecular mechanisms of these signal transduction pathways continues to provide information on how individual proteins are involved in infection, inflammatory disease, autoimmune disease, allergy, and cancer. This knowledge has already led to new treatments for sepsis and rheumatoid arthritis and may have the greatest potential for new pharmaceuticals that in some cases may relieve our reliance on antimicrobials. Although these new technologies have a great potential for treating a wide variety of diseases, therapies that treat inflammatory diseases carry the risk of reducing the hosts ability to respond to infectious diseases.

2.2.1. Innate Immunity

A significant breakthrough in understanding how the innate immune system recognizes pathogens has resulted from the discovery of pattern recognition receptors (PRR) and their associated signal-transduction pathways. Toll-like receptors (TLRs) are a major class of the PRR family, and are the human homologue of Toll receptors first described in *Drosophila (16)*. To date, 11 human TLRs and 10 murine TLRs have been identified *(14,17–20)*. TLRs recognize conserved components of infectious agents, termed pathogen-associated molecular patterns (PAMPs). The structural component TLRs recognize, are conserved, and are essential for the survival of the pathogen. PAMP binding to a TLR alone, as a heterodimer with other TLRs, or in combination with other cofactor receptors triggers signal transduction pathways that initiate inflammatory and cell-mediated immune responses. Although a limited number of TLRs may recognize a number of PAMPs, only a small number of the ligands are currently known; for a review of TLRs and their ligands *see* ref. *(15)*. TLR2 complexes recognize bacterial lipoteichoic acid, lipoproteins, and peptidoglycan. TLR3 recognizes double-stranded RNA that leads to the recognition of double-stranded RNA viruses *(20,21)*. Recognition of double-stranded RNA provides specificity to viruses because single-stranded and double-stranded viruses have double-stranded intermediates during replication. RNA viruses also require RNA-dependent RNA polymerases for replication. These polymerases are virus specific and could theoretically be a PAMP. For a review of TLR recognition specificities to viruses *see* ref. *(20)*. TLR4 recognizes Gram-negative bacterial lipopolysaccharide and TLR9 recognizes unmethylated CpG DNA, a motif specific to bacterial DNA *(22)*. Specificity is an important issue because in many cases broad spectrum treatments have detrimental affects on the host. For example, broad spectrum antibiotic treatment often results in diarrhea.

2.3. Competitive Exclusion Technology

In modern food-animal production facilities, it is often difficult to prevent early exposure of newborns to human enteropathogens *(23–25)*. Newborn animals may not have contact with their mothers or natural environmental factors such as soil that contribute to the establishment of healthy flora *(26)*. This problem was recognized over 20 yr ago, and in an effort to counterbalance this deficiency, day-of-hatch chicks were orally provided normal flora from the cecal contents of a healthy adult chicken *(27)*. This technology referred to as competitive exclusion (CE) *(27)* has proven to be an effective method for prevention of animal and human pathogens that colonize the food-animal gastrointestinal tract *(27–32)*. In chickens, CE provides the most efficient protection against entero-pathogens when provided immediately after birth *(33,34)*. In neonatal pigs raised off sow, a CE culture was shown to decrease fecal shedding and reduce mortality from challenge with enterotoxigenic *E. coli (31)*. The mechanisms that provide protection are thought to be the same as those that are involved in maintaining niche selection pressure, discussed below *(34,35)*.

Nisbet et al. *(32)* further developed the technology of CE by establishing an anaerobic continuous-flow culture of chicken cecal bacteria (CF3). Because the bacteria in CF3 were obtained from a healthy adult chicken that had never been treated with antibiotics, the bacteria were niche-adapted to the chicken intestinal tract prior to in vitro culture *(36)*. This may have contributed to their ability to maintain complex species diversity as well as stable cell concentrations within the continuous-flow system. The continuous-flow system provides a means for reproducible application of CE technology.

2.3.1. Niche Adaptation

In nature, nonpathogenic bacteria from the environment colonize the human and animal gastrointestinal tract shortly after birth. Synergistic and antagonistic interactions between the host and the microflora entering the gastrointestinal tract lead to a succession of bacterial species until a stable consortium results *(37–39)*. The bacteria that persist in the gastrointestinal tract must tolerate or adapt to the conditions present in the niche during the initial colonization. Our knowledge regarding the complex composition of the gastrointestinal population is less than complete, because many of the resident species are difficult to culture. Modern culture-independent molecular methods used to identify bacterial strains have greatly expanded our knowledge of species diversity, but are not as useful for understanding ecological relationships among microflora in their natural environment *(40)*. The diversity of gastrointestinal microflora differs among animal species, among individual animals, and along the length of the gastrointestinal tract itself *(37–39,41–43)*.

Many of the bacterial species present in the gastrointestinal tract produce substances that are inhibitory to other bacterial species. These substances include bacteriocins, organic acids, and hydrogen peroxide *(35,38,44,45)*. The bacteria that produce these substances are generally immune to their toxic effects. Many of the inhabitants of the gastrointestinal tract are intrinsically tolerant to the low pH that results from the produc-tion of organic acids. Susceptible species must adapt to the hostile local environment if they are to persist in the niche. Factors such as attachment-site specificity, cell invasion, or residence in mucus may provide protection from the hostile environment. In some cases, acquisition of specific genes that encode resistance or immunity proteins may confer resistance against specific inhibitors. Niche-specific selection pressure is main-tained by the continuous production of inhibitory substances by the resident microflora *(38,46,47)*. Coordination of nutrient utilization and host-specific factors also contribute to the selection process that ultimately determines the species profile of the gastro-intestinal microflora *(39)*. It is likely that many more mechanisms exist that allow species survival in hostile environments.

Autochthonous bacteria are the species considered to have evolved with the host and are considered normal flora to the host *(26,37)*. Allochthonous species are the bacteria nonindigenous to a particular host species. However, this is a fine line, and the factors that disrupt the natural balance may allow the establishment of allochthonous species *(26,37)*. This is an important distinction while considering issues of food safety and antimicrobial use. Antimicrobials not only reshape the profile of established bacteria already present in the gastrointestinal tract, but also disrupt the inhibitory stringency of the niche allowing other species, such as *Salmonella*, to persist that might normally be transient *(48–51)*. Disruption of the normal flora also increases susceptibility to intestinal disease by microorganisms that are pathogenic to food animals *(26,52)*.

There is an important distinction between CF3, and the most probiotic and CE cultures described in the scientific literature. CF3 went through a natural selection process in the chicken gastrointestinal tract. Most other experimental cultures were selected and combined by scientists. Attempts in our laboratory to generate stable, long-term, diverse continuous-flow cultures by hand-selecting and combining bacterial isolates have failed. In our experience, defined cultures of scientist-selected isolates have only been successful in continuous-flow culture if all of the isolates originated from one animal *(25)*. Even then, combining them in the correct proportion may yield varying results *(31,35)*.

Because the CF3 culture naturally supports *Enterococcus faecium*, *Enterococcus faecalis*, *Enterococcus avium*, and *Enterococcus gallinarum*, a derivative continuous-flow culture of CF3, termed CCF, was used to study antibiotic resistance gene transfer among enterococci. Because enterococci predominate in CCF, it had been assumed that CCF culture conditions would easily support exogenous enterococci *(50)*. This was not the case, as all enterococci added to the CCF culture were rapidly eliminated. Moreover, CCF displaced a continuous-flow monoculture of vancomycin-resistant *E. faecium* (VRE) *(50)*. CCF treatment of day-of-hatch chicks almost completely prevented VRE colonization when challenged 48 h posttreatment *(43)*.

The question has been put forth, "If exogenous *E. faecium* cannot survive in CCF, how do the *E. faecium* naturally present in CCF survive?" We suspect that coadaptation is a critical factor and *E. faecium* adapted to inhibitory factors in the chick cecum in a sequential order during the initial microbial colonization. After the niche had stabilized, it may have been more difficult for exogenous bacteria to adapt to the cocktail of inhibitory factors they were exposed to simultaneously.

From the studies previously described, we capitalized on the ability of CCF to eliminate exogenous VRE and developed a model to assess the ecological interactions of a mixed anaerobic culture treated with antibiotics of various activities *(51)*. Parameters used to predict the efficacy of CCF included bacterial cell counts, volatile fatty acid (VFA) production, and rate of VRE clearance.

The most interesting result was produced by CCF cultures treated with 10 µg/mL vancomycin. VFA levels in these cultures suggested that it would be unable to inhibit the VRE challenge organism. However, it eliminated VRE in 9 d as compared to 7 d for CCF control cultures. In contrast, VRE was able to persist in CCF treated with 40 µg/mL vancomycin. Treatment with 40 µg/mL vancomycin eliminated all of the endogenous enterococci in CCF cultures including *E. faecalis*, whereas *E. faecalis* remained in the

10 µg/mL vancomycin-treated cultures. This was significant because we had previously shown that CCF *E. faecalis* produces a bacteriocin-like substance that is inhibitory to *E. faecium*. To reexamine the effect of low-level vancomycin treatment on CCF and factor out the possibility of *E. faecium* inhibitory specificity, we substituted *E. coli* as the posttreatment challenge organism. In this experiment, CCF was unable to eliminate *E. coli*. These results suggest that CCF possesses a certain level of inhibitory specificity against *E. faecium*. Once again it seems likely that endogenous CCF *E. faecium* have adapted, by an unknown mechanism, to the conditions produced by CCF.

Studies with the CCF culture may have explained a significant role for niche adaptation in the establishment and stability of gastrointestinal microbial ecology *(48,50,51)*. This may have relevance to the ecology of antimicrobial resistance. Instead of cycling antimicrobials and allowing bacteria to consecutively adapt, it may be preferable to hit them with multifaceted inhibitory cocktails as may be found in the gastrointestinal niche. The multifocal approach could be considered analogous to the treatment of HIV. This approach is believed to decrease the consecutive selection of resistant mutants by targeting multiple viral enzymes at one time *(53)*.

Niche-selection pressure may explain why most probiotic preparations fed to animals are not maintained after oral administration was discontinued *(54)*. Wagner et al. *(55)* have developed an assay to assess the efficacy of CE cultures. This assay is based on short-term mixed-batch cultures. They compared PREEMPT, an FDA approved product, derived from CF3, to a similar batch culture derived by selecting similar species purchased from ATCC. The use of a short-term culture of selected ATCC isolates provides no data on the survivability of these isolates in a long-term culture. Our experience suggests that 29 ATCC isolates would be reduced to two or three isolates in a week's time in a continuous-flow culture and would have no value as a CE treatment. Although selected probiotic cultures may be strongly inhibitory to enteropathogens in vitro, they are of little value if they cannot become stable inhabitants of the gastrointestinal tract. The health benefits of probiotic cultures may be greatly increased if cultures are derived from niche-adapted mixed cultures of bacteria.

2.4. Targeting Specific Metabolic Pathways

Enterobacteriaceae, *E. coli* and *Salmonella*, have the capacity to respire under anaerobic conditions by converting nitrate to nitrite *(56)*. This reaction catalyzed by nitrate reductase. Nitrate reductase also uses chlorate as a substrate converting chlorate into the cytotoxic compound, chlorite *(56)*. The accumulation of chlorite in the cytoplasm results in cell death. For this reason, chlorate is selectively lethal to bacterial species that express nitrate reductase.

The addition of chlorate to drinking water and diets of food animals has been shown to reduce populations of enteric pathogens in poultry and livestock, including ruminant species *(57–60)*. Due to toxicity issues, long-term administration of chlorate is not recommended. The strategy currently under consideration is the treatment of poultry and livestock for a short period prior to slaughter to reduce the pathogen load at slaughter *(61)*.

3. DISEASE TREATMENTS

New disease complexes continue to emerge; whereas, infectious diseases once thought eradicated are reemerging because of the acquisition of multidrug resistance. There are a very few new antimicrobials in development because there is little economic incentive for pharmaceutical companies to produce drugs of last resort that will only be used by a small percentage of the population. However, as clinical failures increase the demand for new treatments will outweigh the obstacles.

3.1. New Antimicrobials

Oxazolidinones represent a fairly new class of antimicrobial for the treatment of Gram-positive infections. Oxazolidinone development was considered in the 1980s, but was halted because of toxicity issues *(62,63)*. Linezolid (Zyvox, Pfizer) is currently the only member of the oxazolidinone class that has been approved for use in the United States *(64)*. Oxazolidinones bind to the 50S subunit of the bacterial ribosome and are believed to prevent the formation of the ribosomal initiation complex; thus, inhibiting translation initiation *(65,66)*. Crossresistance with other protein synthesis inhibitors is not known to occur *(67)*. The only known mechanism of resistance against oxazolidinones is the target modification of the 50S subunit. Presently, this is the result of mutation and is not conferred by a mobile DNA element *(66)*.

Unlike the oxazolidinones, most new antimicrobials are derivatives of the established antimicrobial classes. Because the mechanism of action is the same for most members of an antimicrobial class, resistant isolates present in the environment are often crossresistant to the new drugs. For example, Virginamycin-resistant bacteria found in agricultural venues are crossresistant to quinupristin-dalfopristin (Synercid, Monarch Pharmaceuticals) that was approved for human use in 1999.

3.2. Novel Treatment Strategies

Advances in molecular biology and immunology continue to provide new targets for drug development. A new target that may hold promise makes use of our relatively new knowledge of programmed cell death (PCD) systems. The MazEF suicide module of *E. coli* is currently the prototypic PCD system. The mechanism of action relies on stability differences between MazF, a stable toxin, and MazE, a labile antitoxin. Normally the proteins are coexpressed and the binding of MazE to MazF inactivates the toxin. The MazEF complex also negatively regulates the transcription of the *mazEF* genes. When the labile antitoxin is degraded and not replaced, the MazF triggers cell death. Antimicrobials that inhibit transcription or translation also trigger *mazEF* PCD *(68)*. Many antimicrobials that inhibit protein synthesis are phenotypically bacteriostatic because protein synthesis may resume once the administration of the drug ceases. However, a bactericidal phenotype may be observed if the same antimicrobial compound induces *mazEF* cell-mediated death *(69)*. Thymine starvation has also been reported to trigger *mazEF* cell-mediated death *(70)*. Knowledge of PCD and other metabolic systems should provide new avenues for drug development in this century.

3.3. Bacteriophage

Bacteriophages are viruses that infect bacteria by injecting their genetic material (e.g., DNA or RNA) into a bacterial cell. They fall into two general categories, lytic and temperate. Lytic phages infect the bacterial cell, replicate, and cause cell lysis; thus, killing the bacterial cell. Temperate (lysogenic) phages can replicate and lyse the host cell, or they can integrate into the host's genome becoming a prophage prior to replication and lysis *(71)*. The process of integration into the genome can lead to the acquisition of genetic material from the host. Antimicrobial resistance and virulence genes are believed to have been horizontally transferred between bacterial species in this manner *(71–74)*. For this reason it is important to use only lytic phages for the development of antibacterial therapeutics. Medicinal uses for bacteriophage therapy include: gastrointestinal, systemic, and cutaneous infections. A number of potential preharvest and postharvest uses are currently under investigation for the biological control of foodborne pathogens *(75–79)*.

The use of bacteriophages for clinical treatment of bacterial infections predates the discovery of antimicrobials. However, there were a number of problems associated with the initial attempts of bacteriophage therapy *(80)*. The lack of phage biology knowledge, narrow target specificity, purity of preparations, and proper storage considerations were not well understood in the 1930s and 1940s. This led to unpredictable results and clinical failures. When antimicrobials became readily available, bacteriophage research for clinical use ceased in the United States and Western Europe *(80–82)*. Now that multidrug resistant pathogens have become prevalent, and there is a renewed interest in the use of bacteriophage as a clinical treatment for bacterial infections. Advances in biochemistry and medicine have solved many of the early obstacles associated with bacteriophage therapy. For example, the problem of bacterial endo- and exotoxin contamination has been eliminated by modern purification technology. The narrow host range specificity of bacteriophage is desirable in the gastrointestinal tract, because the ability to target specific pathogens leaves the normal flora largely undisturbed. In some cases, the host range is so narrow that phage may only infect a single bacterial strain *(81)*. One approach to overcome this problem has been the use of multiphage preparations that target a number of common clinical strains that are usually encountered *(83,84)*. However, narrow host-range specificity has also been used to the advantage of scientists for use as a diagnostic tool. Bacteriophages specific to *Mycobacterium tuberculosis*, *Escherichia coli* O157:H7, *Listeria monocytogenes*, *Salmonella* spp., and *Staphlylococcus aureus* have been modified to carry a reporter gene *(85–88)*. If the phage-specific bacterial host is present for phage transduction to occur, the reporter gene product should be detectable; thus, confirming the presences of the pathogen. This may eventually be useful for the early detection of tuberculosis *(88)*.

Bacteriophage are ubiquitous in the environment and are commonly isolated from the gastrointestinal tracts of food animals *(89–95)*. Bacteriophage therapy has been successfully used in vivo to reduce pathogenesis by enteric pathogens in mice, calves, piglets, sheep, poultry, and fish *(96–101)*. Because the gastrointestinal tract is a reservoir for many foodborne and opportunistic pathogens, bacteriophage therapy may be a suitable biological control strategy to reduce pathogen load. Although bacteria may develop phage resistance, this is less of a concern because a single dose is preferable to multiple doses.

Over all there are many potential advantages with regard to bacteriophage therapy. Bacteriophage themselves are not toxic to the host and can be administered in low doses owing to the self-perpetuating nature of the infection. They are also self-limiting when no more target organism is present to infect. The host can develop an immune response to bacteriophage; however, only one dose is usually necessary to achieve the results.

4. CONCLUSIONS

To maintain a safe food supply, a broad range of preharvest and postharvest management practices have been implemented, and many more are currently under investigation. Even so, it has become clear that commensal and pathogenic microorganisms exhibit a genetic fluidity that will necessitate continual development of intervention strategies as new pathogens emerge and previously known pathogens reemerge.

We cannot eliminate the genetic reservoir that provides antimicrobial resistance genes to bacteria, but we may be able to make vast improvements over the current situation. To curtail the emergence of multidrug resistant bacteria several approaches need to be pursued. New management strategies must be used that reduce clonal expansion and dissemination of resistant bacterial isolates into the environment. These may include nutritional supplements that promote growth without affecting the bacterial flora in the gut. Such supplements should not select for resistance, nor should they increase fecal shedding of enteric pathogens. Combination therapy should also be considered for clinical therapeutics, so that consecutive resistance is not generated against antimicrobials. This approach has been successful in reducing the emergence of resistant HIV mutants *(53)*, and appears to be similar to the mechanism used in nature to competitively exclude enteric pathogens from the gut. Strategies that maintain competition in the gut may reduce the ability of pathogens to disseminate.

Finally, advances in innate and adaptive immunology, vaccine technology, and molecular biology will continue to provide new treatments that assure greater specificity against pathogens, while reducing toxicity to the individual. Treatments that reduce the susceptibility of humans and animals to opportunistic pathogens, even immunocompromised patients, may soon be on the horizon.

REFERENCES

1. van den Bogaard, A. E. and Stobberingh, E. E. (1999) Antibiotic usage in animals. *Drugs* **58,** 589–607.
2. Altekruse, S. F., Swerdlow, D. B. L., and Wells, S. J. (1998) Factors in the emergence of food borne diseases. *Vet. Clin. North Am. Food Anim. Pract.* **14,** 1–15.
3. Morris, G. J. J. and Potter, M. (1997) Emergence of new pathogens as a function of changes in host susceptibility. *Emerg. Inf. Dis.* **3,** 435–441.
4. Callaway, T. R., Anderson, R. C., Edrington, T. S., et al. (2004) Pre-harvest supplementation strategies to reduce carriage and shedding of food-borne pathogens. *Anim. Health Res. Rev.* **5,** 35–47.
5. Elder, R. O., Keen, J. E., Siragusa, G. R., Barkocy-Gallagher, G. A., Koohmaraie, M., and Lagreid, W. W. (2000) Correlation of enterohemorrhagic *Escherichia coli* O157:H7. *Proc. Natl Acad. Sci. USA* **97,** 2999–3003.
6. Sharma, J. M. (1999) Introduction to poultry vaccines and immunity. *Adv. Vet. Med.* **41,** 481–488.
7. Babiuk, L. A. (1999) Broadening the approaches to developing more effective vaccines. *Vaccine* **17,** 1587–1595.
8. Bowersock, T. L. and Martin, S. (1999) Vaccine delivery to animals. *Advan. Drug Deliv. Rev.* **38,** 167–194.
9. Bowersock, T. L. (2002) Evolving importance of biologics and novel delivery systems in the face of microbial resistance. *AAPS Pharm. Sci.* **4,** 1–17.
10. Badgett, M. R., Auer, A., Carmichael, L. E., Parish, C. R., and Bull, J. J. (2002) Evolutionary dynamics of viral attenuation. *J. Virol.* **76,** 10,524–10,529.

11. Donnelly, J. J., Ulmer, J. B., Shiver, J. W., and Liu, M. A. (1997) DNA vaccines. *Annu. Rev. Immunol.* **15,** 617–648.

12. Blecha, F. (2001) Immunodulators for prevention and treatment of infectious diseases in food-producing animals. *Vet. Clin. North Am. Food Anim. Pract.* **17(3),** 621–633.

13. O'Neill, L. A. J. (2003) Therapeutic targeting of Toll-like receptors for inflammatory and infectious diseases. *Curr. Opin. Pharm.* **3,** 396–403.

14. Janeway, C. A. Jr. and Medzhitov, R. (2002) Innate immune recognition. *Annu. Rev. Immunol.* **20,** 197–216.

15. Werling, D. and Jungi, T. W. (2003) Toll-like receptors linking innate and adaptive immune response. *Vet. Immunol. Immunopath.* **91,** 1–12.

16. Stein, D., Roth, S., Vogelsang, E., and Nusslein-Volhard, C. (1991) The polarity of the dorsoventral axis in the *Drosophila* embryo is defined by and extracellular receptor signal. *Cell* **65,** 725–735.

17. Medzhitov, R., Preston-Hurlburt, P. P., and Janeway, C. A. Jr. (1997) A human homologue of the *Drosophila* Toll protein signals activation of adaptive immunity. *Nature* **388,** 394–397.

18. Kiyoshi, T., Kaisho, T., and Akira, S. (2003) Toll-like receptors. *Annu. Rev. Immunol.* **21,** 335–376.

19. Takeuchi, O., Kawai, T., Sanjo, H., et al. (1999) TLR6: a novel member of an expanding Toll-like receptor family. *Gene* **231,** 59–65.

20. Boehme, K. W. and Compton, T. (2004) Innate sensing of viruses by Toll-like receptors. *J. Virol.* **78,** 7867–7873.

21. Kobayashi, K. S. and Flavell, R. A. (2004) Shielding the double-edged sword: negative regulation of the innate immune system. *J. Leukoc. Biol.* **75,** 428–433.

22. Krieg, A. M. (2002) CpG motifs in bacterial DNA and their immune effects. *Annu. Rev. Immunol.* **20,** 709–760.

23. Fuller, R. (1989) Probiotics in man and animals. *J. Appl. Bacteriol.* **66,** 365–378.

24. Jensen, B. B. (1998) The impact of feed additives on the microbial ecology of the gut in young pigs. *J. Anim. Feed Sci.* **7,** 45–64.

25. Gustafson, R. H. and Bowen, R. E. (1997) Antibiotic use in agriculture. *J. Appl. Microbiol.* **83,** 531–541.

26. Klaenhammer, T. R. (2001) *Probiotics and Prebiotics.* American Society for Microbiology, Washingtion, DC.

27. Nurmi, E. and Rantala, M. (1973) New aspects of salmonella infection in broiler production. *Nature* **241,** 210–211.

28. Anderson, R. C., Stanker, L. H., Young, C. R., et al. (1999) Effect of competitive exclusion treatment on colonization of early-weaned pigs by *Salmonella* serovar cholerasuis. *Swine Health Prod.* **12,** 155–160.

29. Snoeyenbos, G. H., Weinak, O. M., and Smyser, C. F. (1978) Protecting chicks and poultry from salmonellae by oral administration of "normal" gut microflora. *Avian Dis.* **22,** 273–287.

30. Barnes, E. M., Impey, C. S., and Stevens, B. J. H. (1979) Factors affecting the incidence and anti-*Salmonella* activity of the anaerobic cecal flora of the chick. *J. Hygiene* **82,** 263–283.

31. Genovese, K. J., Anderson, R. C., Harvey, R. B., and Nisbet, D. J. (2000) Competitive exclusion treatment reduces the mortality and fecal shedding associated with enterotoxigenic *Escherichia coli* infection in nursery-raised neonatal pigs. *Can. J. Vet. Res.* **64,** 204–207.

32. Nisbet, D. J., Corrier, D. E., Ricke, S. C., Hume, M. E., Byrd, J. A. I., and Deloach, J. R. (1996) Maintenance of biological efficacy in chicks of a cecal competitive-exclusion culture against *Salmonella* by continuous-flow fermentation. *J. Food Prot.* **59,** 1279–1283.

33. Corrier, D. E., Byrd J. A. II, Hume, M. E., and Nisbet, D. J. (1998) Effect of simultaneous or delayed competitive exclusion treatment on the spread of *Salmonella* in chicks. *Appl. Poult. Sci.* **7,** 132–137.

34. Stavric, S., Gleeson, T. M., Blanchfield, B., and Pivnick, H. (1987) Role of adhering microflora in competitive exclusion of *Salmonella* from young chicks. *J. Food Prot.* **50,** 928–932.

35. Nisbet, D. J., Corrier, D. E., Ricke, S. C., Hume, M. E., Byrd J. A. II, and DeLoach, J. R. (1996) Caecal propionic acid as a biological indicator of the early establishment of a microbial ecosystem inhibitory to *Salmonella* in the chicks. *Anaerobe* **2,** 345–350.

36. Corrier, D. E., Nisbet, D. J., Scanlan, C. M., Hollister, A. G., and DeLoach, J. R. (1995) Control of *Salmonella typhimurium* colonization in broiler chicks with a continuous-flow culture. *Poult. Sci.* **74,** 916–924.

37. Savage, D. C. (1977) Microbial ecology of the gastrointestinal tract. *Annu. Rev. Microbiol.* **31,** 107–133.

38. Brook, I. (1999) Bacterial interference. *Crit. Rev. Microbiol.* **25,** 155–172.

39. Hooper, L. V., Xu, J., Falk, P. G., Midtvedt, T., and Gordon, J. I. (1999) A molecular sensor that allows a gut commensal to control its nutrient foundation in a competitive ecosystem. *Proc. Natl Acad. Sci. USA* **96,** 9833–9838.

40. Hugenholtz, P., Goebel, B. M., and Pace, N. R. (1998) Impact of culture-independent studies on the emerging phylogenetic view of bacterial diversity. *J. Bacteriol.* **180,** 4765–4774.

41. Apajalahti, J. H. A., Kettunen, A., Bedford, M. R., and Holben, W. E. (2001) Percent G + C profiling accurately reveals diet related difference in the gastrointestinal microbial community of broiler chickens. *Appl. Environ. Microbiol.* **67,** 5656–5667.

42. Gong, J., Forster, R. J., Yu, H., et al. (2002) Diversity and phylogenetic analysis of bacteria in the mucosa of chicken ceca and comparison with bacteria in the cecal lumen. *FEMS Microbiol. Lett.* **208,** 1–7.

43. Knarreborg, A., Simon, M. A., Enberg, R. M., Jensen, B. B., and Tannock, G. W. (2002) Effects of dietary fat source and subtherapeutic levels of antibiotic on bacterial community in the ileum of broiler chickens at various ages. *Appl. Environ. Microbiol.* **68,** 5918–5924.

44. Hinton, A. and Hume, M. E. (1995) Antibacterial activity of the metabolic by-products of a *Veillonella* sp. and *Bacteroides fragilis*. *Anaerobe* **1,** 121–127.

45. Riley, M. A. and Gordon, D. M. (1999) The ecological role of bacteriocins in bacterial competition. *Trends Microbiol.* **7,** 129–133.

46. Lan, R. and Treeves, R. R. (2000) Intraspecies variation in bacterial genomes: the need for a species genome concept. *Trends Microbiol.* **8,** 396–401.

47. Reeves, P. R. (1992) Variation in O-antigens, nich-specific selection and bacterial populations. *FEMS Microbiol. Lett.* **100,** 509–516.

48. Poole, T. L., Genovese, K. J., Knape, K. D., Callaway, T. R., Bischoff, K. M., and Nisbet, D. J. (2003) Effect of subtherapeutic concentrations of tylosin on the inhibitory stringency of a mixed anaerobe continuous-flow culture of chicken microflora against *Escherichia coli* 0157:H7. *J. Appl. Microbiol.* **94,** 73–79.

49. Poole, T. L. (2001) Persistence of a vancomycin-resistant *Enterococcus faecium* in an anaerobic continuous-flow culture of porcine microflora in the presence of subtherapeutic concentrations of vancomycin. *Microb. Drug Resist.* **7,** 343–348.

50. Poole, T. L., Genovese, K. J., Anderson, T. J., Bischoff, K. M., and Nisbet, D. J. (2001) Inhibition of a vancomycin-resistant enterocci by an anaerobic continuous flow culture of chicken microflora. *Microb. Ecol. Health Dis.* **13,** 246–253.

51. Poole, T. L., Genovese, K. J., Callaway, T. R., Bischoff, K. M., Donskey, C. J., and Nisbet, D. J. (2004) Competitive exclusion of a glycopeptide resistant *Enterococcus faecium* GRE in the presence of vancomycin, but not equivalent concentrations of tylosin or gentamicin. *Poult. Sci.* **83,** 1099–1105.

52. Collins, F. M. and Cart, P. B. (1978) Growth of salmonellae in orally infected germ-free mice. *Infect. Immun.* **21,** 41–47.

53. Kaufman, G. R. and Cooper, D. A. (2000) Antiretroviral therapy of HIV-1 infection: established treatment strategies and new therapeutic options. *Curr. Opin. Microbiol.* **3,** 508–514.

54. Netherwood, T., Bowden, R., Harrison, P., O'Donnell, A. G., Parker, D. S., and Gilbert, H. J. (1999) Gene transfer in the gastrointestinal tract. *Appl. Environ. Microbiol.* **65**, 5139–5141.

55. Wagner, D. R., Holland, M., and Cerniglia, C. E. (2002) An in vitro assay to evaluate competitive exclusion products for poultry. *J. Food Prot.* **65**, 746–751.

56. Stewart, V. J. (1988) Nitrate respiration in relation to facultative metabolism in enterobactera. *Microbiol. Rev.* **52**, 190–232.

57. Anderson, R. C., Buckley, S. A., Callaway, T. R., et al. (2001) Effect of sodium chlorate on *Salmonella typhimurium* concentrations in the pig gut. *J. Food Prot.* **64**, 255–259.

58. Callaway, T. R., Anderson, R. C., Genovese, K. J., et al. (2001) Sodium chlorate supplementation reduces *E. coli* O157:H7 populations in cattle. *J. Anim. Sci.* **80**, 1683–1689.

59. Edrington, T. S., Callaway, T. R., Anderson, R. C., et al. (2003) Reduction of *E. coli* O157:H7 populations in sheep by supplementation of an experimental sodium chlorate product. *Small Ruminant Res.* **49**, 173–181.

60. Anderson, R. C., Callaway, T. R., Buckley, S. A., et al. (2001) Effect of oral sodium chlorate administration on *Escherichia coli* O157:H7 in the gut of experimentally infected pigs. *Int. J. Food Microbiol.* **71**, 125–130.

61. Anderson, R. C., Carr, M. A., Miller, R. K., et al. (2005) Effects of experimental chlorate preparations as feed and water supplements on *Escherichia coli* colonization and contamination of beef cattle and carcasses. *Food Microbiol.* **22**, 439–447.

62. Eliopoulos, G. M. (2003) Quinupristin-dalfopristin and linezolid: evidence and opinion. *Clin. Infect. Dis.* **36**, 473–481.

63. Bozdogan, B. and Applebaum, P. C. (2004) Oxazolidinones: activity, mode of actin, and mechanism of resistance. *Int. J. Antimicrob. Agents* **23**, 113–119.

64. Copra, I. (1998) Research and development of antibacterial agents. *Curr. Opin. Microb.* **1**, 495–501.

65. Zhou, C. C. (2002) [1]H Nuclear magnetic resonance study of oxazolidinone binding to bacterial ribosomes. *Antimicrob. Agents Chem.* **46**, 625–629.

66. Murray, R. W., Schaad, R. D., Zurenko, G. E., and Marotti, K. R. (1998) Ribosome from an oxazolidinone-resistant mutant confer resistance to eperezolid in a *Staphylococcus aureus* cell-free transcription-translation assay. *Antimicrob. Agents Chemother.* **42**, 947–950.

67. Fines, M. and Leclercq, R. (2000) Activity of linezolid against Gram-positive cocci possessing genes conferring resistance to protein synthesis inhibitors. *J. Antimicrob. Chem.* **45**, 797–802.

68. Engelberg-Kulka, H., Boaz, S., Reches, M., Amitai, S., and Hazan, R. (2004) Bacterial programmed cell death systems as targets for antibiotics. *Trends Microbiol.* **12(2)**, 66–71.

69. Sat, B., Hazan, R., Fisher, T., Khaner, H., Glaser, G., and Engelberg-Kulka, H. (2001) Programmed cell death in *Escherichia coli*: some antibiotics can trigger *mazEF* lethality. *J. Bacteriol.* **183**, 2041–2045.

70. Sat, B., Reches, M., and Engelberg-Kulka, H. (2003) The *Escherichia coli mazEF* suicide module mediates thymineless death. *J. Bacteriol.* **185**, 1803–1807.

71. Canchaya, C., Fournous, G., Chibani-Chennoufi, S., Dillmann, M., and Brussow, H. (2003) Phage as agents of lateral gene transfer. *Curr. Opin. Microb.* **6**, 417–424.

72. Huang, A., Friesen, J., and Brunton, J. L. (1987) Characterization of a bacteriophage that carries the genes for production of shiga-like toxin 1 in *Escherichia coli*. *J. Bacteriol.* **169**, 4308–4312.

73. O'Brien, A. D., Newland, J. W., Miller, S. F., Holmes, R. K., Smith, H. W., and Formal, S. B. (1985) Shiga-like toxin-converting phage from *Escherichia coli* strains that cause hemorrhagic colitis or infantile diarrhea. *Science* **226**, 694–696.

74. Newland, J. W., Strockbine, N. A., Miller, S. F., O'Brien, A. D., and Holmes, R. K. (1985) Cloning of shiga-like toxin structural genes from a toxin converting phage of *Escherichia coli*. *Science* **230**, 178–181.

75. Leverentz, B., Conway, W. S., Alavidze, Z., et al. (2001) Examination of bacteriophage as a biocontrol method for *Salmonella* on fresh-cut fruit: a model study. *J. Food Prot.* **64,** 1116–1121.
76. Leverentz, B., Conway, W. S., Camp, J. M., et al. (2003) Biocontrol of *Listeria monocytogenes* on fresh-cut produce by treatment with lytic bacteriophages and a bacteriocin. *Appl. Environ. Microbiol.* **69,** 4519–4526.
77. Huff, W. E., Huff, G. R., Rath, N. C., Balog, J. M., and Donoghue, A. M. (2003) Bacteriophage treatment of a severe *Escherichia coli* respiratory infection in broiler chickens. *Avian Dis.* **47,** 1399–1405.
78. Modi, R., Hirvi, A., Hill, A., and Griffiths, M. W. (2001) Effect of phage on survival of *Salmonella enteritidis. J. Food Prot.* **64,** 927–933.
79. Goode, D., Allen, V. M., and Barrow, P. A. (2003) Reduction of experimental *Salmonella* and *Campylobacter* contamination of chicken skin by application of lytic bacteriophages. *Appl. Environ. Microbiol.* **69,** 5032–5036.
80. Barrow, P. A. and Soothill, J. S. (1997) Bacteriophage therapy and prophylaxis: rediscovery and renewed assessment of potential. *Trends Microbiol.* **5,** 268–271.
81. Summers, W. C. (2001) Bacteriophage therapy. *Annu. Rev. Microbiol.* **55,** 437–451.
82. Sulakvelidze, A., Alavidze, Z., and Morris, G. J. J. (2001) Bacteriophage therapy. *Antimicrob. Agents Chem.* **45,** 649–659.
83. Biswas, B., Adhya, S., Washart, P., et al. (2002) Bacteriophage therapy rescues mice bacteremic from a clinical isolate of vancomycin-resistant *Enterococcus faecium. Infect. Immun.* **70,** 204–210.
84. Goodridge, L., Gallaccio, A., and Griffiths, M. W. (2003) Morphological, host range, and genetic characterization of two coliphages. *Appl. Environ. Microbiol.* **69,** 5364–5371.
85. Chen, J. and Griffiths, M. W. (1996) *Salmonella* detection in eggs using *lux*$^{+}$ bacteriophages. *J. Food Prot.* **59,** 908–914.
86. Gaeng, S., Scherer, S., Neve, H., and Lowssner, J. (2000) Gene cloning and expression and secretion of *Listeria monocytogenes* bacteriophage-lytic enzymes in *Lactococcus lactis. Appl. Environ. Microbiol.* **66,** 2951–2958.
87. Waddell, T. E. and Poppe, C. (2000) Construction of mini-Tn*10luxABcam/Ptac*-ATS and its use for developing a bacteriophage that transduces bioluminescence to *Escherichia coli* O157:H7. *FEMS Microbiol. Lett.* **182,** 285–289.
88. Hazbon, M. H., Guarin, N., Ferro, B. D., et al. (2003) Photographic and luminometric detection of luciferase reporter phages for drug susceptibility testing of clinical *Mycobacterium tuberculosis* isolates. *J. Clin. Microbiol.* **41,** 4865–4869.
89. Adams, J. C., Gazaway, J. A., Brailsford, M. D., Hartman, P. A., and Jacobson, N. L. (1966) Isolation of bacteriophages from the bovine rumen. *Experientia* **22,** 717–718.
90. Rogers, C. G. and Sarles, W. B. (1963) Characterization of *Enterococcus* bacteriophages from the small intestine of the rat. *J. Bacteriol.* **85,** 1378–1385.
91. Hoogenraad, N. J., Hird, F. J. R., Holmes, I., and Miller, N. F. (1967) Bacteriophages in rumen contents of sheep. *J. Gen. Virol.* **1,** 575–576.
92. Brailsford, M. D. and Hartman, P. A. (1968) Characterization of *Streptococcus durans* bacteriophages. *Can. J. Microbiol.* **14,** 397–402.
93. Orpin, C. G. and Munn, E. A. (1973) The occurrence of bacteriophages in the rumen and their influence on rumen bacterial populations. *Experientia* **30,** 1018–1020.
94. Iverson, W. G. and Millis, N. F. (1976) Characterization of *Streptococcus bovis* bacteriophages. *Can. J. Microbiol.* **22,** 847–852.
95. Klieve, A. V. and Bauchop, T. (1988) Morphological diversity of ruminal bacteriophages from sheep and cattle. *Appl. Environ. Microbiol.* **54,** 1637–1641.
96. Park, S. C., Shimanmura, I., Fukunaga, M., Mori, K., and Nakai, T. (2000) Isolation of bacteriophages specific to a fish pathogen, *Pseudomonas plecoglossicida*, as a candidate for disease control. *Appl. Environ. Microbiol.* **66,** 1416–1422.

97. Smith, D. L. and Huggins, M. B. (1987) Effectiveness of phages in treating experimental *E. coli* in calves, piglets and lambs. *J. Gen. Microbiol.* **129,** 2659–2675.

98. Smith, D. L. and Huggins, M. B. (1987) The control of experimental *E. coli* diarrhea in calves by means of bacteriophage. *J. Gen. Microbiol.* **133,** 1111–1126.

99. Smith, D. L. and Huggins, M. B. (1987) Factors influencing the survival and multiplication of bacteriophages in calves and in their environment. *J. Gen. Microbiol.* **133,** 1127–1135.

100. Smith, H. W. and Huggins, M. B. (1982) Successful treatment of experimental *E. coli* infections in mice using phages: its general superiority over antibiotics. *J. Gen. Microbiol.* **128,** 307–318.

101. Nakai, T., Sugimoto, R., Park, K.-H., et al. (1999) Protective effects of bacteriophage on experimental *Lactococus garvieae* infection in yellowtail. *Dis. Aquat. Org.* **37,** 33–41.

18

Microbial Risk Assessment

Carl M. Schroeder, Elke Jensen, Marianne D. Miliotis, Sherri B. Dennis, and Kara M. Morgan

Abstract

Microbial risk assessment (MRA) is used to evaluate foodborne hazards, the likelihood of exposure to those hazards, and the resulting public-health impact. It is generally recognized to consist of four parts: hazard identification, hazard characterization, exposure assessment, and risk characterization. Model predictions generated by MRAs are most often expressed as the estimated likelihood of foodborne illness and/or number of deaths in a given population for a given period. MRA is used increasingly to inform decision-making aimed both at managing human health risks from foodborne pathogens and at devising standards for promoting safe and fair international food trade. This chapter discusses how MRA fits into the larger context of risk analysis, describes in detail the process of MRA, reviews examples of recently completed MRAs, and suggests steps towards further improvement of the MRA process.

1. INTRODUCTION

1.1. Historical Context

During the late 1960s and early 1970s various naturally occurring and commercial chemicals, including environmental pollutants, were identified as carcinogens (*see* ref. *[1]*). The Federal agencies responsible for regulating these and other hazardous substances quickly became the center of a whirlwind of controversy surrounding the scientific validity, or alleged lack thereof, of the rationale underpinning their regulatory decision-making. Consequently calls emerged for substantive changes in both the institutional processes for evaluating health risk data and the way in which the results were translated into regulatory policy *(2)*. Based largely on these developments a US Congressional directive issued in the early 1980s gave rise to formation of the National Research Council's (NRC) *Committee on the Institutional Means for Assessments of Risks to Public Health*. The overall goal of the committee was to define ways to "strengthen the reliability and objectivity of scientific assessment that forms the basis for federal regulatory policies applicable to carcinogens and other public health hazards" *(3)*. The results of the committee's work were summarized and published in 1983 as an NRC report entitled *Risk Assessment in the Federal Government: Managing the Process (4)*. It is this report, commonly referred to as "The Red Book" (based on the color of its cover), which has emerged as the seminal title in risk assessment, and which, based on its general acceptance by the scientific community and continued practical use, can in many ways be used to mark the beginning of a formalized concept of risk assessment.

From: *Infectious Disease: Foodborne Diseases*
Edited by: S. Simjee © Humana Press Inc., Totowa, NJ

Recent years have witnessed adoption of the risk assessment process for characterizing risks from pathogenic microorganisms. In 1994, the General Agreement on Tariffs and Trade (GATT) established the Agreement on the Application of Sanitary and Phytosanitary Measures (the SPS Agreement) and the Agreement on Technical Barriers to Trade (the TBT Agreement). These two agreements are intended to (1) insure that no member state is prevented from adopting and implementing measures to protect human, animal, and plant health and (2) facilitate international free trade of foods by mandating that public-health protections established by member states are based on sound science, rather than used as *de facto* nontariff trade barriers (*see* ref. *[5]*). Establishment of guidelines and procedures by which member countries conduct risk assessments for developing food safety regulation fall under the auspices of the Food and Agricultural Organization/World Health Organization (FAO/WHO) Food Standard Programme's *Codex Alimentarius* Commission (Codex) *(6,7)*. Today microbial risk assessments (MRAs) are a standard component, *inter alia*, in the effort to protect public health and facilitate free trade.

1.2. The Bigger Picture: Risk Assessment As It Relates to Risk Analysis

Before proceeding to a more detailed description of the risk assessment process, it is important first to establish a broader picture of the role of risk assessment in decision-making. Risk assessment is one component of risk analysis, the process for gathering information, doing analysis, and making decisions about risks. Risk analysis should include a planning phase, the risk assessment, the structuring of the risk management decisions, and the strategy for risk communication.

For even the most straightforward risk analysis, many different types of research may be needed to inform the problem, including, for example, epidemiology, toxicology, microbiology, human behavior, and engineering. Because of its multidisciplinary character, using a team of individuals with different types of expertise is a very effective approach to conducting risk analysis. Also, because it links in to several organizational levels (scientific/technical staff to do the technical work, communications staff to do the risk communication, and management to make decisions), it is critical for risk analysis to be a team-driven effort.

1.2.1. Planning

The formulation of the problem, or the planning phase, is critical to achieving the highest quality results from the risk assessment. Planning should include all members of the team to ensure that everyone has input into the formulation of the questions that the research will answer, and that the degree of uncertainty in these answers is acceptable. The planning phase should include discussion of criteria for the quality of the analysis, which will drive the data quality needs for the project. For example, if only a rough estimate is needed and a high degree of uncertainty is acceptable, the quality of the input data may not be as critical as when a high degree of certainty is needed, in which case the quality of the data and the quality of any models used will need to be developed to assure a high degree of certainty. The planning phase should also coordinate the needs of the risk managers (by defining the questions that need answers) with the ability of the analysts (taking into account the available resources, including expertise, staff, time, data, and models). Risk communicators should also be involved in the planning to ensure that the strategy for communicating about these risks is taken into account from the start. This approach leads to an efficient and effective strategy for making decisions about risk.

1.2.2. Assessment

A risk assessment is the process that develops the estimate of the likelihood and severity of a particular outcome, given a well-defined scenario. This measure of likelihood and severity is called the risk estimate. Risk assessment is the analysis step in which data are gathered and assessed, models are chosen or developed, and results are derived. A risk assessment could be done on the back-of-an-envelope in 10 min, or it could be documented in a 500-page document with a CD of accompanying spreadsheets and models. The differences between such approaches would be the degree of confidence or uncertainty in the estimates. Again this should match the needs of the risk analysis, as defined in the planning stage. According to the National Academies of Science *(4)*, risk assessment includes four steps: hazard identification, hazard characterization, exposure assessment, and risk characterization. This framework has been adopted by Codex and is described in section 3 below.

1.2.3. Management

Risk management is the process of making decisions about risks, informed by the results of the risk assessment. It includes determining whether the risk calculated in the risk assessment is acceptable or unacceptable, given the accompanying level of uncertainty. In the case of unacceptable risk, risk management is most likely to involve the development of strategies for reducing exposure, and may also include strategies for additional research or data collection to reduce the uncertainty in the risk estimate. In the case of acceptable risk, the risk management strategy may involve efforts to ensure the exposure stays at or below a certain level.

Risk communication is often part of a risk management strategy. The National Research Council of the National Academies *(8)* has described it as "an interactive process of exchange of information and opinion among individuals, groups and institutions. It involves multiple messages about the nature of risk and other messages, not strictly about risk, that express concerns, opinions, or reactions to risk messages or to legal and institutional arrangements for risk management." (p. 21). Further information on the risk analysis process may be found at http://www.who.int/foodsafety/micro/riskanalysis/en/. Meanwhile, having reviewed the overarching framework of risk analysis, and having understood where risk assessment fits into this process, we shall now turn our attention to MRA.

2. EXPLORING MICROBIAL RISK ASSESSMENT

2.1. Expertise Required to Conduct a Quantitative MRA

MRA typically requires access to a full range of professional expertise and information. This includes not only identification of the type, extent, quality, and availability of data, but also of key organizations with responsibilities for, or capacities to support, microbiological risk assessment. In our experience, input and critical evaluation of experts in a number of scientific disciplines, including microbiology, chemistry, epidemiology, medicine, mathematics, statistics, toxicology, and food science, have typically proven necessary for conducting useful MRAs. It is critically important that before undertaking a quantitative MRA appropriately trained personnel be identified and brought onboard.

2.2. Range of Microbial Risk Assessments

MRAs fall into two broad categories: qualitative and quantitative. In the former, risk is described as the likelihood of illness (high vs low) whereas in the latter, it is expressed as the predicted number of illnesses. Because of their relative complexity, and because they are typically the preferable of the two formats for informing decision-making, we shall focus on quantitative MRA.

Quantitative MRAs provide numerical expressions of risk and indication of the attendant uncertainties *(6)*. Quantitative MRA modeling, a relatively new approach in the field of microbial risk, uses probability models to evaluate the likelihood of adverse human health effects from exposure to pathogenic microorganisms. Predictions are presented as distributions, rather than as point estimates. This approach offers several advantages, one of which is that the variability of a contributing factor may be found to have substantial significance, which can be difficult to assess properly by point estimates alone. Distributions can further be used to reflect the presence of uncertainty in data interpretation. Another advantage is that quantitative MRA modeling facilitates the identification of factors most important in determining the magnitude of a risk. Lastly, quantitative models are flexible; inputs and model components can be changed readily as new data become available. Nearly all of today's quantitative MRAs are conducted using Monte Carlo sampling – a random sampling of individual probability distributions to produce hundreds or thousands of outcome scenarios. Although further discussion of Monte Carlo sampling is outside the scope of this chapter, the reader is recommended to consult the text by Vose *(9)*.

MRAs can be further delineated into one of at least four specific types: (1) pathogen–commodity product pathway assessments, (2) relative risk rankings of food commodities, (3) geographical assessments, and (4) risk–risk assessments *(10)*.

2.2.1. Product Pathway Risk Assessments

The purpose of conducting a product pathway risk assessment is to identify factors in the farm-to-table continuum that are likely to affect public health related to ingestion of a particular pathogen via consumption of a particular food *(10)*. An example of this type of assessment is the USDA *Escherichia coli* O157:H7 in ground beef risk assessment (described in section 5 below). In conducting a risk assessment of this type, the risk assessment team is tasked with modeling the growth/decline of bacteria throughout the farm-to-table continuum. The strength of the farm-to-table risk assessment process is that it allows risk assessors to examine the predicted effects of various mitigation options, including those related to animal production, food processing, and consumer behavior, on the risk of foodborne illnesses.

2.2.2. Relative Risk Rankings

The primary purpose of relative risk rankings is to identify food commodities that pose the greatest risk of illness from a particular foodborne pathogen. An example of this type of assessment is the DHHS/USDA *Listeria monocytogenes* risk assessment (described in section 5 below). Relative risk rankings are particularly useful in instances where epidemiologic data suggest several foods may be of concern for a particular pathogen. Results of risk rankings help to focus efforts and resources aimed at reducing foodborne illnesses.

2.2.3. Geographical Assessments

Geographical assessments, as the name suggest, are designed to elucidate both the pathways that are likely to lead to exposure to pathogens and the factors the either limit or allow the risk to occur. They are effective in identifying (1) the risk of introduction of disease agents through food animals or animal products (e.g., intentionally, as in bioterrorism, or unintentionally) and (2) mitigations to reduce consumer exposure to foodborne pathogens. The USDA/Harvard University risk assessment for bovine spongiform encephalopathy (BSE) is an example of this type of assessment.

2.2.4. Risk–Risk Assessments

Risk–risk assessments examine a trade off of one risk for another in situations in which reducing the risk of one hazard likely increases the risk of another. An example of this type of assessment is determination of the impact on public health by treating drinking water or water used in processing/manufacturing with a chemical, say chlorine, to eliminate a microbial pathogen, *Mycobacterium* spp., for instance. The risk from chlorine exposure would be expected to inversely correlate to that from ingesting mycobacteria.

3. MICROBIAL RISK ASSESSMENT PROCESS

The Codex Committee on Food Hygiene *(6)* and the National Advisory Committee on the Microbiological Criteria for Foods *(11)* have proposed a framework for conducting MRAs. It includes four steps: (1) hazard identification (the identification of biological, chemical, and physical agents capable of causing adverse health effects and which may be present in a particular food or group of foods); (2) exposure assessment (the qualitative and/or quantitative evaluation of the likely intake of biological, chemical, and physical agents via food as well as exposures from other sources if relevant); (3) hazard characterization/dose–response (the qualitative and/or quantitative evaluation of the nature of the adverse health effects associated with the hazard); and (4) risk characterization (the integration of the hazard identification, hazard characterization, and exposure assessment determinations to provide qualitative or quantitative estimates of the likelihood and severity of the adverse effects which could occur in a given population).

3.1. Hazard Identification

For microbial agents, hazard identification identifies the microorganisms or the microbial toxins of concern *(6)*, i.e., those for which an association between disease and the presence of a pathogen in a food is recognized and acknowledged *(12)*. Hazard identification is predominately a qualitative process. Information on hazards can be obtained from scientific literature, various databases, and through expert elicitation. Pertinent information may be found in data from basic research and clinical studies, and from epidemiological studies and surveillance.

3.2. Exposure Assessment

Exposure assessment estimates the level of microbiological pathogens or microbiological toxins, and the likelihood of their occurrence in foods at the time of consumption.

It includes an assessment of the extent of actual or anticipated human exposure to the hazard. Levels of microorganisms can be dynamic and, while they may be kept low, such as by adhering to proper time/temperature controls during food processing, can substantially increase with food mishandling *(6)*. The exposure assessment, therefore, usually describes the pathway from production to consumption. Scenarios can be constructed to predict the range of possible exposures. Factors to be considered include the frequency of contamination of foods by the pathogenic agent and its level in those foods over time. Qualitatively, foods can be categorized according to the likelihood of whether they will be contaminated at their source; whether they can support the growth of the pathogen of concern; whether there is substantial potential for abusive handling; or whether they will undergo a heating process or other lethality step.

In turn, the presence, growth, survival, and death of microbial pathogens in foods are influenced by several factors. These may include the (1) characteristics of the pathogenic microorganism, (2) microbiological ecology of the food, (3) initial bacterial contamination of the raw food material, (4) level of sanitation and process controls and the methods of processing, (5) packaging, distribution, and storage of the foods, and (6) food preparation steps such as cooking and hot-holding. As a result, predictive microbiology can be a useful tool in performing the exposure assessment. The recently released ComBase database (available at http://wyndmoor.arserrc.gov/combase/default.aspx), which estimates the behavior of microorganisms in response to environmental conditions, is a helpful tool to formulate the quantitative aspects of exposure assessment.

Other factors to be considered in the exposure assessment include patterns of consumption (including both the portion size and frequency of consumption), which is typically related to socio-economic and cultural backgrounds, ethnicity, seasonality, consumer age differences (population demographics), regional differences, consumer preferences and behavior, and the role of the food handler as a source of contamination. The potential impact of abusive environmental time and temperature relationships should also be considered in assessing the degree of exposure to a hazardous microbe. For instance, human exposure levels to a foodborne pathogen can rapidly increase by a million-fold within even a relatively short period of temperature abuse, whereas heating food immediately prior to consumption can reduce pathogen levels considerably.

3.3. Hazard Characterization

The primary component of the hazard characterization is the dose–response relationship, defined as the determination of the relationship between the magnitude of exposure to a chemical, biological, or physical agent and the severity and/or frequency of associated adverse health effects *(6)*. Hazard characterization thus provides a qualitative or quantitative description of the severity and duration of adverse effects that may result from ingestion of a foodborne microorganism or its toxin. The dose–response relationship can be derived using data from clinical feeding trials, epidemiological (outbreak or surveillance) investigations *(13)*, or both *(14)*. For instance, a combination of epidemiologic and food survey data was used by Buchanan et al. *(15)* to estimate a dose–response relationship for *L. monocytogenes* illness from cheese. Extrapolation of animal studies for determining dose–response may also be used to derive the dose–response, such as in the HHS/USDA *L. monocytogenes* risk assessment *(16)*. Generally speaking, data from controlled human clinical feeding trials are preferable to those from outbreak or surveillance investigations, which in turn are preferable to animal surrogate models.

Several important factors relating to the microorganism of concern must be borne in mind when performing the hazard characterization *(17)*. These include the facts that: (1) microorganisms can replicate; (2) the virulence and infectivity of microorganisms can change depending on their interaction with their host and environment; (3) genetic material can be transferred between microorganisms leading to the transfer of antibiotic resistance and virulence factors; (4) microorganisms can be spread through secondary and tertiary transmissions; (5) microorganisms can persist in certain individuals leading to continued shedding and further chance of spreading infection; and (6) although the likelihood of disease increases with increasing numbers of pathogenic microorganisms consumed, there is potential for low levels of microorganisms to cause disease.

Similarly, aspects regarding the microbial ecology of foods must be carefully thought-out prior to performing the hazard characterization. For instance, food attributes such as high fat or salt content (and food matrix effects in general) are important considerations in assessing the potential for microbial virulence. As an example, studies with *V. cholerae* O1 indicate that a food matrix, such as cooked rice, provides buffering capacity and may have substantive impact on dose–response relationships *(18)*. Other studies have shown that high intake of milk fat inhibits intestinal colonization of *Listeria* but not of *Salmonella* in rats *(19)*.

Lastly, it is important at the hazard characterization stage of the risk assessment to remember that microbial virulence may also be affected by several host factors, such as (1) increased host susceptibility because of breakdowns of physiological barriers; (2) host susceptibility characteristics, such as age and immune status; and (3) population characteristics, such as access to medical care. The FAO/WHO guidelines for conducting hazard characterizations are available at http://www.who.int/foodsafety/publications/ micro/en/pathogen.pdf.

3.4. Risk Characterization

Risk characterization brings together the qualitative and/or quantitative information of the hazard characterization and exposure assessment to provide an estimate of risk for a given population. This estimate can be assessed by comparison with independent epidemiological data that relate hazards to disease prevalence. The degree of confidence in the final estimation of risk will depend on the variability, uncertainty, and assumptions identified in the previous steps of MRA. Differentiation of uncertainty and variability is important in subsequent selections of risk management options. Uncertainty is associated with the data themselves, and with the choice of model. Data uncertainties include those that might arise in the evaluation and extrapolation of information obtained from epidemiological, microbiological, and laboratory animal studies. Uncertainties also arise whenever attempts are made to use data concerning the occurrence of certain phenomena obtained under one set of conditions to make estimations or predictions about the phenomena likely to occur under other sets of conditions for which the data are not available.

An important component of risk characterization is sensitivity analysis. This is done to determine the parameters that contribute most to the total uncertainty of the risk assessment output. A common type of sensitivity analysis is the use of "what if?" scenarios. By using this methodology, the risk assessor can examine the relative impact of changing different parameters in the model. For example, one might find in the conceptual model that a change in the prevalence of pathogen at the farm has a dramatic effect on risk, but that a change in the consumer behavior has little or no effect on predictions of risk. In this case, the results of the sensitivity analysis would indicate that the greatest amount of resources should be focused on reducing the on-farm prevalence of the pathogen. Sensitivity analysis may also be used to identify the crucial data gaps (*see* section 4.2 below), those parameters that drive the risk estimates, but for which there is little available data.

The end result or output of the risk characterization is the estimate or prediction of illness associated with a particular microorganism, given the uncertainty and variability inherent in the assumptions and data. The effect of control measures, if evaluated in the risk assessment, is also described in the risk characterization.

4. OUTCOMES OF RISK ASSESSMENTS

The principal outcome of a quantitative MRA is a prediction of the numbers of illnesses, the severity of illness outcome, or both. Risk of illness outcomes from quantitative MRAs are typically reported either on a per-serving basis or as population estimates. The latter are useful when focusing on specific groups of interest, such as children, the elderly, or immune-compromised individuals.

Predictions from quantitative MRAs are most often given as distributions, in which case the best estimate is usually associated with the distribution's mean or median. Distributions show a range of outcomes associated with foodborne illnesses where the upper and lower confidence bounds (typically reported as the 5th and 95th percentile confidence intervals) give a feel for overall estimate uncertainty and variability (Fig. 1). Clearly defining confidence bounds for quantitative MRA results helps to underscore that they are the best estimates, and not absolutes.

4.1. How Outcomes From Risk Assessments Are Used

Usually, one tries to start by characterizing the "baseline" model estimates by comparing the predicted number of illnesses to epidemiologically based estimates. It is then possible to examine the impact of different assumptions used in the model. An important feature of quantitative MRA is the ability to examine the impact of interventions at points along the farm-to-fork continuum on reducing foodborne infections and to identify at which point(s) the impact is greatest. This is useful in setting risk-based performance standards and in informing the development of verification sampling protocols. Outcomes of MRAs are also used to explore the impact of different risk management options. For example, a risk manager can see the impact of setting a performance standard, such as requiring a certain level of lethality to be achieved by the processors.

4.2. Identification of Research Needs (Data Gaps)

As discussed above, data gaps are identified as a result of sensitivity analysis. This is one of the most important exercises in a risk assessment in that it provides critical information that may be used to prioritize and facilitate future research. A good risk assessment report is the one that clearly and concisely identifies research needed to be undertaken to strengthen the future assessments. A classical example of data gaps commonly encountered in MRAs is that resulting from the relatively meager information available with which to support quantitative modeling of dose–response relationships for many, if not most, microbial pathogens *(20).*

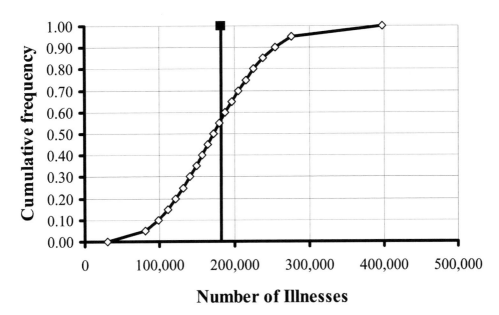

Number of Illnesses

Fig. 1. Predictions from quantitative MRAs are often given as distributions, as in the hypothetical case shown (for the number of illnesses caused by bacterium Y). The best estimate is usually associated with the mean or median of the distribution. In this case the median estimate was 182,000 illnesses, as indicated by the filled box and solid vertical line. The open diamonds and attached line indicate the range of estimate uncertainty, extending from 81,000 illnesses at the 5th percentile to 276,500 illnesses at the 95th percentile. By presenting results in this manner, risk assessors make it easier for risk analysts to get a feel for overall estimate uncertainty. Clearly defining confidence bounds also helps to underscore the fact that the results are simply best estimates, and not absolutes.

5. EXAMPLES OF QUANTITATIVE MICROBIAL RISK ASSESSMENTS

5.1. L. monocytogenes *Risk Assessment of DHHS/USDA*

The *L. monocytogenes* risk assessment (LMRA), commissioned in response to a presidential request for federal agencies to develop control plans to reduce listeriosis by 50% by the year 2005 *(21)*, was a joint effort led by Food and Drug Administration's Center for Food Safety and Applied Nutrition (FDA/CFSAN) in collaboration with USDA's Food Safety and Inspection Service (FSIS), and with consultation with the US Centers for Disease Control and Prevention (CDC). The purpose the assessment was to identify which foods should receive the most regulatory attention in an effort to improve public health.

5.1.1. Scope of the LMRA

The disease endpoint used in the LMRA was that of patient fatality. The number of listeriosis cases was estimated from the predicted number of deaths using a correction factor based on CDC's Foodborne Disease Active Surveillance Network (FoodNet) data. The LMRA provided analyses and models that (1) estimated the potential level of exposure of three age-based population groups (perinatal, intermediate-age, and elderly) and the total United States population to *L. monocytogenes* contaminated foods for 23 food categories, and (2) related this exposure to public health consequences.

5.1.2. Description of the LMRA Process

Once the decision was made to conduct a risk assessment, a formal announcement detailing the purpose and scope of the project and requesting submission of relevant data, was published in the Federal Register. Next, peer review of key assumptions, data, and modeling approaches used in the assessment were sought from the National Advisory Committee on the Microbiological Criteria of Foods, and during public meetings with agency stakeholders. Scientific experts, both internal and external to the agencies, then reviewed the draft risk assessment document and model, whereupon it was further revised. Following agency approval and clearance, the draft risk assessment was issued for public comment in 2001. Public comments, in addition to newly available data and modeling techniques, were reviewed by FDA and USDA/FSIS, and the model revised accordingly. The updated risk assessment report and the model were released in October 2003. The risk assessment is available electronically at http://www.foodsafety.gov.

5.1.3. Results and Conclusions From LMRA

The LMRA models provided an estimated rate of fatal infection from *L. monocytogenes* on an individual serving basis for a particular food category. It also provided an estimated number of fatal infections per year in the United States for each food category, the "per annum risk." Because the per annum risk was derived from the "per-serving" risk, there was generally a higher degree of uncertainty associated with this risk. The results in Table 1 show that the risk of listeriosis from consumption of different ready-to-eat foods was predicted to vary greatly among the various food categories.

A statistical technique referred to as "cluster analysis" was used to group the LMRA simulation outputs for each food category, accounting for the differences in the median predications as well as the upper and lower bounds of the predicted risks. When performed at the 90% confidence level, the per-serving predictions fell into four clusters and the per-annum predictions into five clusters for the 23 food categories. These clusters were used, in turn, to develop a two-dimensional matrix of per-serving and per-annum groups (Fig. 2). Risk managers used the resulting five cluster risk designations to develop different approaches to controlling listeriosis based on the relative risk and characteristics of specific foods.

Although the LMRA purposely did not look into the pathways for the manufacture of individual foods, the models developed can be used to estimate the likely impact of control strategies by changing one or more input parameters and measuring the change in the model outputs. In the case of the LMRA, scenarios were run to allow comparison of the baseline calculations (shown in Table 1) to new situations that might arise as a result of potential risk-reduction strategies.

5.1.4. Impact of the LMRA

The scientific evaluations and mathematical models developed for LMRA provided a systematic assessment of the scientific knowledge needed to assist not only in reviewing the effectiveness of current policies, programs, and practices, but also in identifying new strategies to minimize the public health impact of foodborne *L. monocytogenes*. Moreover, the assessment provided a foundation to assist future evaluations of the potential effectiveness of new strategies for controlling foodborne listeriosis.

Table 1
Predicted Median Cases of Listeriosis for the Total US Population

Food categories	Predicted median cases of listeriosis	
	Per-serving basis	Per-annum basis
Deli meats	7.7×10^{-8}	1598.7
Frankfurters, not reheated	6.5×10^{-8}	30.5
Pate and meat spreads	3.2×10^{-8}	3.8
Unpasteurized fluid milk	7.1×10^{-9}	3.1
Smoked seafood	6.2×10^{-9}	1.3
Cooked ready-to-eat crustaceans	5.1×10^{-9}	2.8
High fat and other dairy products	2.7×10^{-9}	56.4
Soft unripened cheese	1.8×10^{-9}	7.7
Pasteurized fluid milk	1.0×10^{-9}	90.8
Fresh soft cheese	1.7×10^{-10}	<0.1
Frankfurters, reheated	6.3×10^{-11}	0.4
Preserved fish	2.3×10^{-11}	<0.1
Raw seafood	2.0×10^{-11}	<0.1
Fruits	1.9×10^{-11}	0.9
Dry/semi-dry fermented sausages	1.7×10^{-11}	<0.1
Semi-soft cheese	6.5×10^{-12}	<0.1
Soft ripened cheese	5.1×10^{-12}	<0.1
Vegetables	2.8×10^{-12}	0.2
Deli-type salads	5.6×10^{-13}	<0.1
Ice cream and other frozen dairy products	4.9×10^{-14}	<0.1
Processed cheese	4.2×10^{-14}	<0.1
Cultured milk products	3.2×10^{-14}	<0.1
Hard cheese	4.5×10^{-15}	<0.1

Adapted from the HHSD/USDA *Listeria monocytogenes* risk assessment (*see* http://www.foodsafety.gov).

5.2. E. coli O157:H7 Risk Assessment of USDA

The *E. coli* O157:H7 risk assessment (ECRA) was conducted by the USDA FSIS in consultation with the FDA, CDC, NACMF, and various professionals from federal and state governments, industry, and academia. The ECRA was initiated in response to identification of *E. coli* O157:H7 in cattle and ground beef, and in light of the serious public health impact of O157:H7 foodborne outbreaks. Two of the main purposes in conducting the ECRA were to (1) comprehensively evaluate the risk of illness from O157:H7 in ground beef and (2) identify those points in the farm-to-table chain where mitigations to reduce illness are likely to be most effective.

5.2.1. Scope of ECRA

The ECRA was farm-to-table in scope. It was a "baseline" risk assessment that was designed to reflect the full range of practices, behaviors, and conditions throughout animal production, slaughter, processing, transportation, storage, preparation, and consumption (Fig. 3). The ECRA was confined to examine the risk of illness from O157:H7 in ground beef servings; exposures from produce, water, cross-contamination, and others were not included in the risk assessment model. Output estimates were

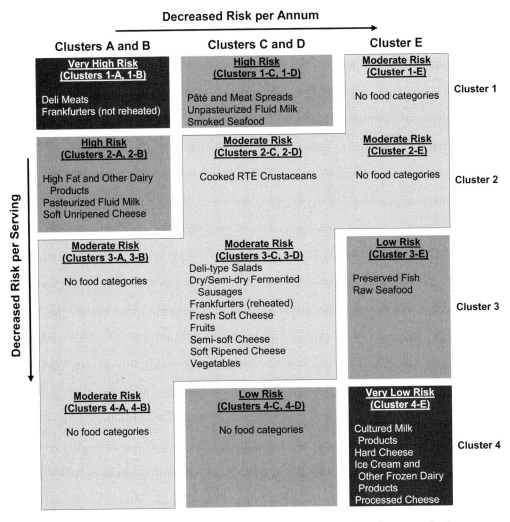

Fig. 2. Two-dimensional matrix of food categories based on cluster analysis.

generated for a variety of risk assessment endpoints, including risks of (1) illness, (2) hospitalization, (3) hemolytic uremic syndrome (HUS), and (4) death.

5.2.2. Description of ECRA Process

In August 1998, USDA FSIS announced plans to develop a risk assessment for human illness from *E. coli* O157:H7 in ground beef. Accordingly, an announcement was published in the Federal Register and a call for data was made. Scientific information available as of July 2001 was then compiled, analyzed, and integrated into the risk assessment framework. Scientists from inside and outside the agency reviewed the ECRA throughout its construction, and their feedback was used continually to update the risk assessment model. Following agency clearance, the ECRA was released for public comment. At this time the National Academies of Science (NAS) were contracted to assemble a panel of experts and conduct a peer-review of the draft ECRA. Based on public comments and the NAS peer review, the ECRA is currently under revision. Once

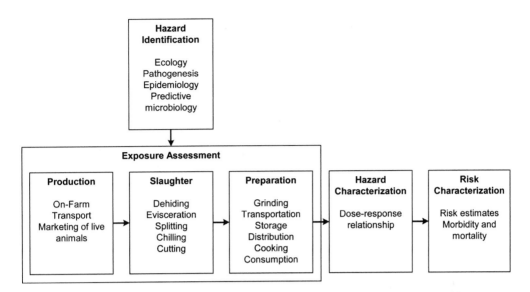

Fig. 3. Farm-to-table risk assessment model for *E. coli* O157:H7 in ground beef risk assessment. The assessment reflected, to the extent practicable, a full range of current practices, behaviors, and conditions in the farm-to-table continuum (production, slaughter, processing, transportation, storage, preparation, and consumption).

revision is completed, the updated ECRA will be publicly released. The draft report may be found at http://www.usda.fsis.gov. The NAS review of the ECRA is available in print *(22)* or online at http://www.nationalacademies.org.

5.2.3. Results and Conclusions From ECRA

Less than 0.007–0.018% (depending upon time of year) of cooked ground beef products were predicted to contain *E. coli* O157:H7. The median probability of illness for the general population of the United States from consuming O157:H7 in ground beef was predicted to be approx 1 illness per million servings (Table 2). The predicted risk of illness was strongly dependent on the time of year during which the ground beef is consumed (about one illness in every 600,000 servings consumed during June through September compared to about 1 in every 1.6 million servings during October through May). Children under the age of five were predicted to have a risk of illness roughly 2.5 times that of the general population. Research needs identified in conducting the ECRA included investigations regarding (1) the effect of carcass chilling on reducing (or increasing) *E. coli* O157:H7; (2) the influence of food matrix effects, competitive microbiota, and environmental conditions on the maximum density of O157:H7 in ground beef; and (3) retail and consumer storage, cooking, and consumption patterns of ground beef vis-à-vis seasons; i.e., winter, spring, summer, and fall.

5.2.4. Impact of ECRA

Similar to the LMRA, the scientific evaluations and mathematical models developed for the ECRA yielded an assessment of the scientific knowledge to assist in reviewing the effectiveness of current policies, programs, and practices, and in identifying strategies to reduce the public health impact of *E. coli* O157:H7 in ground beef. For instance, the ECRA was used to determine the effect of various mitigations in the slaughter house

Table 2
Risk of Illness From *E. coli* O157:H7 in Ground Beef for the US Population Using Median Exposure and Dose–Response Distributions

Log of *E. coli* O157:H7 per-serving	No. of *E. coli* O157:H7 per-serving	Probability of exposure (Ex)	Probability of illness given exposure (DR)	Risk of illness (Ex × DR)
0.0	1	5.5×10^{-05}	1.7×10^{-04}	9.5×10^{-09}
1.0	10	6.1×10^{-06}	1.7×10^{-03}	1.0×10^{-08}
2.0	100	7.7×10^{-07}	1.6×10^{-02}	1.3×10^{-08}
3.0	1,000	4.3×10^{-07}	1.2×10^{-01}	5.0×10^{-08}
4.0	10,000	2.7×10^{-07}	3.6×10^{-01}	9.7×10^{-08}
5.0	100,000	1.8×10^{-07}	5.8×10^{-01}	1.0×10^{-07}
6.0	1,000,000	1.2×10^{-07}	7.3×10^{-01}	8.9×10^{-08}
7.0	10,000,000	7.4×10^{-08}	8.2×10^{-01}	6.1×10^{-08}
8.0	100,000,000	3.8×10^{-08}	8.9×10^{-01}	3.4×10^{-08}
9.0	1,000,000,000	1.4×10^{-08}	9.3×10^{-01}	1.3×10^{-08}
10.0	10,000,000,000	8.5×10^{-10}	9.5×10^{-01}	8.1×10^{-10}
11.0	100,000,000,000	2.0×10^{-12}	9.7×10^{-01}	1.9×10^{-12}
Population risk of illness from *E. coli* O157:H7 per-serving				9.6×10^{-07}

The annual risk of illness from *E. coli* O157:H7 in ground beef for the general US population is nearly one illness in 1 million servings of ground beef (9.6×10^{-7}). Adapted from ref. *(23)*.

on the risk of illness from *E. coli* O157:H7. Results from the ECRA also provided the basis for the subsequent ruling that *E. coli* O157:H7 is reasonably likely to occur in ground beef. Further impacts are likely to be realized following the completion of the revised ECRA.

5.3. Other Food Safety Risk Assessments

A variety of food safety risk assessments have been performed by various organizations, including government and academic institutions. Table 3 highlights some important MRAs performed to date. For more information, *see* http://www.foodriskclearinghouse. umd.edu/risk_assessments.cfm and http://www.foodrisk.org/risk_assessments.cfm.

6. CONCLUSIONS

Risk assessment is a tool that provides a method for structuring disparate information into an organized, systematic framework. It is a science-based approach, but developing a risk assessment is not a purely technical task. Interpreting the science and fitting different types of data generated from different sources to answer a question (which in most instances the data were not generated to answer) takes more than just technical knowledge. To ensure a high-quality risk assessment is developed, it takes an impressive array of expertise, practical experience, perspective, and most importantly, careful planning.

But, as with most things, there is room for improvement. What can be done to strengthen the MRA process? Here are a few suggestions:

(1) A common criticism is that it is very simple to "slant" a risk assessment by excluding or including certain data, or by cherry-picking the models used. Therefore, it is important to establish well-defined criteria for selecting data and models before the assessment begins.

Table 3
Summary of Some Recently Completed Food Safety Microbial Risk Assessments

Pathogen/ commodity of interest	Source (year of publication)	Summary	Ref.
Bacillus cereus in pasteurized milk	National Institute of Public Health and the Environment, the Netherlands (1997)	When stored at ≤7°C for not more than 7 d (conditions for storing pasteurized milk in the Netherlands), the number of *B. cereus* cells present in milk at the time of expiration was predicted not to exceed 10^5 cfu/mL. Additional data on the dose of *B. cereus* likely to cause human illness are needed	(24)
Campylobacter spp. In broiler chickens and *Vibrio* spp. in seafood	Food and Agricultural Organization/World Health Organization (2001)	A good example of how risk assessment can be applied to pathogen–commodity combinations. Several key data gaps were identified regarding the risk of human illness from thermophilic campylobacters in broiler chickens, including, most importantly, a general lack of "systematic and fundamental" investigations to determine those steps in the production-to-consumption continuum that may lead to human infection subsequent to chicken consumption. Interventions such as thorough washing and safe preparation of seafood were identified as key mitigations for reducing risk of illness from vibrios in seafood. A valuable discussion pertaining to the potential for risk assessments to be conducted by developing countries can be found here	(25)
Campylobacter spp. in chickens	Institute of Food Safety and Toxicology, the Danish Veterinary and Food Administration (2003)	Results of this risk assessment, based on data from Danish surveillance programs, suggested that the incidence of campylobacteriosis from chicken consumption would be reduced approx 30 times by either (1) introducing a two-log reduction reduction of the number of *Campylobacter* on chicken carcasses, (2) reducing the prevalence of *Campylobacter* in flocks by approx 30 times, or (3) improving kitchen hygiene by approx 30 times. Persons 18–29 yr of age were predicted to be at the greatest risk for campylobacteriosis from chicken	(26)

(Continued)

Table 3 (*Continued*)

Pathogen/ commodity of interest	Source (year of publication)	Summary	Ref.
Fluoroquinolone-resistant *Campylobacter* spp. in chicken	Center for Veterinary Medicine, the US Food and Drug Administration (2001)	This assessment provided quantitative estimates for the human health risk posed by fluoroquinolone-resistant campylobacters in chicken. Estimates were given for the number of persons in the United States who contracted fluoroquinolone-resistant campylobacter infections from chicken and received fluoroquinolones	(27)
Escherichia coli in ground beef	Health Canada (1998)	The model predicted a probability of HUS 3.7×10^{-6} and a probability of mortality 1.9×10^{-7} per meal for the very young. The average probability of illness was predicted to be reduced by 80%, if a hypothetical mitigation strategy directed at reducing microbial growth during retail storage via reduced storage temperature was introduced	(28)
Escherichia coli in steak tartare	Netherlands National Institute of Public Health and the Environment (2001)	The results this farm-to-table risk assessment suggested the prevalence of raw tartare patties contaminated with Shiga toxin-producing *E. coli* O157 (0.3%) in the Netherlands is low, but that the incidence rate of diarrheal illness (8 per 100,000 person yr) is high. Intervention at the farm or during slaughter was predicted more efficient for reducing STEC O157 health risks than intervention by consumers	(29)
Listeria monocytogenes in deli meats	Food Safety and Inspection Service, US Department of Agriculture (2003)	Food contact surfaces found positive for *Listeria* spp. greatly increased the likelihood of finding ready-to-eat (RTE) deli meat product lots positive for *L. monocytogenes*. Multiple interventions, including increased inspection and sanitation of food contact surface areas, use of growth inhibitors, and product reformulation, were predicted to be more effective in reducing the risk of illness from *L. monocytogenes* than any single intervention	(30)
Listeria monocytogenes in soft cheese	Epidemiology and Animal Health Management Laboratory, Alfort	The average number of expected cases of listeriosis per year in France was estimated at 57 for a high-risk	(31)

(*Continued*)

Table 3 (*Continued*)

Pathogen/ commodity of interest	Source (year of publication)	Summary	Ref.
	Veterinary School, Maisons-Alfort, France (1998)	subpopulation and one for a low-risk healthy subpopulation. Reducing the frequency of environmental milk contamination and eliminating *L. monocytogenes* mastitis decreased substantially the expected incidence of listeriosis	
Listeria monocytogenes in selected categories of RTE foods	Department of Health and Human Services, the US Department of Agriculture (2003)	Results supported the findings of epidemiological investigations that certain foods are more likely to be the vehicles for *Listeria monocytogenes* and quantified the magnitude of the differences in the predicted risk of listeriosis on a per-serving and per-annum basis for different RTE foods. The exposure models and 'what-if' scenarios identified five factors that affect consumer exposure: (1) amounts and frequency of consumption of a RTE food; (2) frequency and levels of *Listeria* in food; (3) potential of the food to support growth of *Listeria* during refrigeration storage; (4) refrigeration temperature; and (5) duration of refrigeration	*(16)*
Salmonella Enteritidis in pasteurized liquid eggs	Center for Food Safety and Nutrition, the US Food and Drug Administration/Food Safety and Inspection Service, the US Department of Agriculture (1997)	Results revealed that inadequate pasteurization temperatures and/or storage of eggs at abusive temperatures between the farm and egg-breaker plant likely result in a hazardous liquid egg product. Pasteurization at proper temperatures, however, was predicted to provide sufficient consumer protection from (1) a high-incidence of *S.* Enteritidis-infected birds and (2) temperature abuse during egg storage	*(32)*
Salmonella Enteritidis in shell eggs	Food Safety and Inspection Service, US Department of Agriculture (2005)	Rapid cooling of eggs after lay and pasteurization of eggs were predicted to be effective mitigations for reducing eggborne salmonellosis	*(35)*
Salmonella Typhimurium DT104 in dry-cured pork sausages	Danish Bacon and Meat Council, the National Committee for Pig Production, and the Danish Meat Research	The findings of this assessment predicted that when *Salmonella* are present in raw pork, they are usually in low numbers, and that during processing any *Salmonella* present will be reduced to	*(33)*

(*Continued*)

Table 3 (*Continued*)

Pathogen/ commodity of interest	Source (year of publication)	Summary	Ref.
	Institute (2002)	approx 2–3 log-units. Dry-cured sausages, either produced within or imported to Denmark, will infrequently contain *S.* Typhimurium DT104; in those cases in which they occur, the number of bacterial cells was predicted to be low (1–4 cells per serving)	
Vibrio parahaemolyticus in raw molluscan shellfish	US Food and Drug Administration (2004)	Water temperature at the time of harvest was found to be the driving force influencing the initial levels of *V. parahaemolyticus* in oysters. There was a significant reduction in the probability of illness when the oysters were cooled immediately after harvest. Other postharvest practices, especially intervention measures, were found to greatly influence the levels of pathogenic bacteria in the resultant number of illness	*(14)*

For further information and a more detailed list *see* Forsythe *(5)* and http://www.foodrisk.org/risk_assessments.cfm.

(2) Methods for communicating the uncertainty in data and models used to develop the risk assessment are crude at best, and are easily overlooked by senior management decision-makers. It is critical to have a system that provides the time and expertise to ensure high quality, balanced assessment. When time or other resources are short, the quality of the risk assessment is likely to be adversely impacted. However, detecting such problems is difficult without a thorough peer-review.

(3) There are many types of foodborne illnesses and, despite the enthusiasm for risk assessment, it is not feasible to perform MRAs on the entire gamut of foodborne pathogen/food commodity combinations. Prioritization is therefore essential. It is possible to further improve upon the efficient use of resources, including staff expertise and time, finances, and opportunities for collaboration/leveraging.

(4) Finally, risk assessors are the most important resource for conducting MRAs. Education efforts, including attempts to spark interest in students early in their careers, further development of university graduate programs in risk assessment, and continuing education programs to foster professional development are necessary to expand the pool of well-qualified risk assessors.

The above suggestions for improvement notwithstanding, MRA has been refined to the point where it is all but universally recognized as a useful decision-making tool in the absence of complete or reliable data. For this tool to be used most efficiently and effectively, careful consideration and planning must be brought to bear throughout the MRA process. Results from MRAs should not be used as the sole basis for policy or regulatory decisions. Rather, they should be used to complement other information, such as epidemiologic and surveillance data, financial considerations, and consumer feedback, in which case they bolster sound decision-making. Used in this manner, quantitative MRA proves a valuable tool for curtailing foodborne infections, and ultimately, protecting the public health.

ACKNOWLEDGMENTS

Joseph Rodricks provided valuable insight into the risk assessment process, particularly regarding its historical context. Janell Kause, Robert Buchanan, Hanne Rosenquist, Bjarke Bak Christensen, David Goldblatt, and Katie Pritchard graciously shared their time and expertise in reviewing various drafts of the manuscript. Any errors are ours alone.

REFERENCES

1. Rodricks, J. V. (1992) *Calculated Risks: The Toxicity and Human Health Risks of Chemicals in Our Environment*. Cambridge University Press, Cambridge, UK.
2. Rodricks, J. V. (2003) What happened to the Red Book's second most important recommendation? *Human Ecol. Risk Assess.* **9,** 1169–1180.
3. Press, F. (1983) Letter from the Chairman, pp. iii–iv. In (National Research Council) *Risk Assessment in the Federal Government: Managing the Process.* Committee on the Institutional Means for Assessment of Risks to Public Health, Commission on Life Sciences, National Academy Press, Washington, DC, 191 pp.,
4. National Research Council. (1983) *Risk Assessment in the Federal Government: Managing the Process.* Committee on the Institutional Means for Assessment of Risks to Public Health, Commission on Life Sciences, National Academy Press, Washington, DC, 191 pp.,
5. Forsythe, S. J. (2002) The Microbiological Risk Assessment of Food. Blackwell, Oxford, UK.
6. Food and Agricultural Organization of the United Nations. (1999) Principles and guidelines for the conduct of microbiol risk assessment. *In* Codex Alimentairus Food Hygiene Basic Texts, 2nd edition, pp. 53–62. FAO, Rome.
7. Food and Agricultural Organization of the United Nations. (2000) Report of the Joint FAO/WHO expert consultation on risk assessment of microbiological hazards in foods. FAO Headquarters, Rome, 17–21 July 2000. *FAO Food and Nutrition Paper, 71.*
8. Committee on Risk Perception and Communication, NRC. (1989) *Improving Risk Communication*, The National Academies Press, Washington, DC.
9. Vose, D. (2000) *Risk Analysis: A Quantitative Guide*, John Wiley & Sons, Chichester, UK.
10. Wachsmuth, K., Kause, J., Maczka, C., Ebel, E., Schlosser, E., and Anderson, S. (2003) *Microbial Risk Assessment as a Tool for Guiding Food Safety*. Marcel Dekker, New York.
11. National Advisory Committee on Microbiological Criteria for Foods (NACMCF). (1998) Principles of risk assessment for illness caused by foodborne biological agents. *J. Food Prot.* **61,** 1071–1074.
12. Lammerding, A. M. and Paoli, G. M. (1997) Quantitative risk assessment: an emerging tool for emerging foodborne pathogens. *Emerg. Infect. Dis.* **3,** 483–487.
13. Kasuga, F., Hirota, M., Wada, M., et al. (2006) Outbreak data for *Salmonella* dose response. Proceedings of the First International Conference on Microbial Risk Assessment, College Park, MD. *J. Food Prot.* (In press).
14. Food and Drug Administration. (2001) Public health impact of *Vibrio parahaemolyticus* in raw molluscan shellfish. Available at http://vm.cfsan.fda.gov/~dms/vprisk.html (accessed on 1/8/07).
15. Buchanan, R. L., Damert, W. G., Whiting, R. C., and van Schothorst, M. (1997) Use of epidemiologic and food survey data to estimate a purposefully conservative dose–response relationship for *Listeria monocytogenes* levels and incidence of listeriosis. *J. Food Prot.* **60,** 918–922.

16. HHS/USDA. (2003) Quantitative assessment of relative risk to public health from food-borne *Listeria monocytogenes* among selected categories of ready-to-eat foods. Available at http://www.foodsafety.gov/~dms/lmr2-toc.html (accessed on 1/8/07).

17. Dennis, S. B., Miliotis, M. D., and Buchanan, R. L. (2002) Hazard characterization/dose–response assessment. In: *Microbial Risk Assessment in Food Processing* (Brown, M. and Stringer, M., eds.), Woodhead, Cambridge, UK, pp. 83–86.

18. Levine, M. M., Black, R. E., Clements, M. L., Nalin, D. R., Cisneros, L., and Finkelstein, R. A. (1981) Volunteer studies in development of vaccines against cholera and enterotoxigenic *Escherichia coli*: a review. In *Acute Enteric Infections in Children: New Prospects for Treatment and Prevention* (Holme, T., Holmgren, J., Merson, M. H., and Molby, R., eds.), Elsevier/North-Holland, Amsterdam, pp. 443–459.

19. Sprong, R. C., Hulstein, M. F., and Van der Meer, R. (1999) High intake of milk fat inhibits intestinal colonization of *Listeria* but not of *Salmonella* in rats. *J. Nutr.* **129,** 1382–1389.

20. Coleman, M. and Marks, H. (1998) Topics in dose–response modeling. *J. Food Prot.* **61,** 1550–1559.

21. President's Council on Food Safety. (1999) Egg safety from production to consumption: an action plan to eliminate *Salmonella* Enteritidis illnesses due to eggs. Partnership of the US Department of Agriculture, Department of Health and Human Services, Environmental Protection Agency and the Department of Commerce, Washington, DC.

22. Committee on the Review of the USDA *E. coli* O157:H7 Farm-to-Table Process Risk Assessment. (2002) *Escherichia coli* O157:H7 in Ground Beef: Review of a Draft Risk Assessment. The National Academies Press, Washington, DC.

23. Food Safety and Inspection Service. (2000) Draft Risk Assessment of the Public Health Impact of *Escherichia coli* O157:H7 in Ground Beef. The US Department of Agriculture, Washington, DC.

24. Notermans, S., Dufrenne, J., Teunis, P., Beumer, R., Te Giffel, M., and Peeters Weem, P. (1997) A risk assessment study of *Bacillus cereus* present in pasteurized milk. *Food Microbiol.* **14,** 143–151.

25. FAO/WHO. (2001) Hazard identification, exposure assessment and hazard characterization of *Campylobacter* spp. in broiler chickens and *Vibrio* spp. in seafood—Joint FAO, Rome/WHO Expert Consultation, Geneva, Switzerland, 23–27 July, 2001.

26. Rosenquist, H., Nielsen, N. L., Sommer, H. M., Norrung, B., and Christensen, B. B. (2003) Quantitative risk assessment of human campylobacteriosis associated with thermophilic *Campylobacter* species in chickens. *Int. J. Food Microbiol.* **83,** 87–103.

27. Center for Veterinary Medicine, Rockville, Maryland, US Food and Drug Administration. (2001) *Risk Assessment on the Human Health Impact of Fluoroquinolone Resistant Campylobacter Associated with the Consumption of Chicken.*

28. Cassin, M. H., Lammerding, A. M., Todd, E. C., Ross, W., and McColl, R. S. (1998) Quantitative risk assessment for *Escherichia coli* O157:H7 in ground beef hamburgers. *Int. J. Food Microbiol.* **41,** 21–44.

29. Netherlands National Institute of Public Health and the Environment. (2001) Risk assessment of Shiga-toxin producing *Escherichia coli* O157 in steak tartare in the Netherlands. RIVM, Bilthoven. Report 257851003.

30. Gallagher, D. L., Ebel, E. D., and Kause, J. R. (2003) FSIS risk assessment for *Listeria monocytogenes* in deli meats. Food Safety and Inspection Service, US Department of Agriculture, Washington, DC.

31. Bemrah, N., Sanaa, M., Cassin, M. H., Griffiths, M. W., and Cerf, O. (1998) Quantitative risk assessment of human listeriosis from consumption of soft cheese made from raw milk. *Prev. Vet. Med.* **37,** 129–145.

32. Whiting, R. C. and Buchanan, R. L. (1997) Development of a quantitative risk assessment model for *Salmonella enteritidis* in pasteurized liquid eggs. *Int. J. Food Microbiol.* **36,** 111–125.

33. Alban, L., Olsen, A. M., Nielsen, B., Sorensen, R., and Jessen, B. (2002) Qualitative and quantitative risk assessment for human salmonellosis due to multi-resistant *Salmonella* Typhimurium DT104 from consumption of Danish dry-cured pork sausages. *Prev. Vet. Med.* **52,** 251–265.

34. Lindqvist, R., Sylven, S., and Vagsholm, I. (2002) Quantitative microbial risk assessment exemplified by Staphylococcus aureus in unripened cheese made from raw milk. *Int. J. Food Microbiol.* **78,** 155–170.

35. Schroeder, C. M., Latimer, H. K., Schlosser, W. D., et al. (2006) Overview and summary of the Food Safety and Inspection Service risk assessment for *Salmonella* Enteritidis in shell eggs, October 2005. *Foodborne Pathogens and Diease.* **3,** 403–412.

19

Food Irradiation and Other Sanitation Procedures

Donald W. Thayer

Abstract

   Radiation pasteurization of food can be used as a terminal intervention step in HACCP programs to protect the public from foodborne pathogens that may be very difficult to control by any other method. The appropriate radiation doses and the effects of environmental factors such as temperature, atmosphere, and water activity required to control the following foodborne pathogens have been determined. *Bacillus cereus, Campylobacter jejuni, Clostridium botulinum, Clostridium perfringens, Cyclospora cayetanensis, Escherichia coli* O157:H7, *Listeria monocytogenes, Salmonella* spp., *Staphylococcus aureus, Toxoplasma gondii, Vibrio cholerae, Vibrio vulnificus, Vibrio parahaemolyticus,* and *Yersinia enterocolitica* are typical of the foodborne microorganisms that can be inactivated by food irradiation. The endospore-forming bacteria are considerably more resistant to ionizing radiation than are the nonspore formers; however, even these will be reduced in numbers by pasteurization doses. Radiation and thermal processing were demonstrated to interact, producing a greater inactivation of salmonellae on poultry meat than would be predicted from the individual processes. Salmonellae did not multiply at significantly greater rates on irradiated meat.

1. INTRODUCTION

   The term food irradiation refers to the treatment of food with ionizing radiation to achieve a beneficial effect. Almost from the time of the discovery of radioactivity itself, scientists noted that pathogens could be inactivated by irradiation. Minck *(1)* suggested in 1896 that X-rays might be therapeutic in controlling human pathogens. Green *(2)* discovered that radiation from radium would inactivate *Staphylococcus aureus, Vibrio cholerae,* and *Bacillus anthracis.* A patent was issued in 1918 to Gillett *(3)* for a device containing 16 X-ray tubes that could produce X-rays to inactivate trichinae in pork. Schwartz *(4)* discovered that encysted trichinae could be inactivated with X-rays. He did not, as some have suggested, either apply for or receive a patent for the process.

1.1. Factors Influencing Effectiveness of Treatment

   The effectiveness of ionizing irradiation to inactivate food-spoilage organisms and foodborne pathogens depends on a number of factors, some of which are intrinsic to the organism of interest and the food product upon which it is found and some of which are processing factors, such as the temperature and atmosphere at the time of irradiation. In general, the radiation resistance of microorganisms is indirectly related to their complexity and size, as follows in the order of most to least sensitive: protozoa > log-phase bacterial cell > stationary-phase cell > bacterial spore > virus > prion. The radiolytic products of water, such as hydrogen peroxide, are themselves very toxic to microorganisms; thus,

From: *Infectious Disease: Foodborne Diseases*
Edited by: S. Simjee © Humana Press Inc., Totowa, NJ

a foodborne pathogen will be more sensitive to ionizing radiation in foods with a high percentage of water than in very dry foods. The complexity of the medium in or on which the microorganism is present also significantly affects sensitivity to ionizing radiation because of competition for free radicals. If antioxidants are present in the foodstuff, they may protect the microorganism against the radiation.

1.2. Regulation of Irradiated Foods in the United States

In the United States, the US Food and Drug Administration limits the radiation sources to the following: γ-rays from sealed units of the radionuclides cobalt-60 or cesium-137, electrons generated from machine sources at energies not to exceed 10 MeV, and X-rays generated from the machine sources at energies not to exceed 7.5 MeV *(5)*. The sources of radiation never touch the food and are incapable of generating any radioactivity in the food itself. A food may be irradiated to: (1) inhibit sprouting in tubers and bulbs, (2) alter growth and maturation inhibition of fresh foods, (3) disinfest foods of arthropod pests, (4) inactivate foodborne pathogens and spoilage organisms, and (5) sterilize foods. The inactivation of pathogens and spoilage organisms by ionizing radiation is influenced by the temperature and atmosphere during the process of irradiation as well as the food's chemical composition. Irradiated foods have been demonstrated to be wholesome and nutritious. As with any other food-processing technology, maintenance of good organoleptic properties of foods requires the proper application of the technology and good food-science. The wholesomeness of irradiated foods is beyond the scope of this article and has been the subject of many studies. The reader is urged to consult reviews and contemporary research on the subject of nutritional and toxicological safety of irradiated foods *(6–10)*. Discussions and descriptions of the technology for the irradiation of foods may be found in several literature *(11–13)*. As a sanitation practice, the inactivation of foodborne pathogens and spoilage organisms is discussed in this chapter.

1.3. Definitions

The SI unit of absorbed radiation dose is the gray (Gy) and is equivalent to the absorption of 1 J/kg. The older special unit for absorbed radiation dose was the rad defined as follows: 1 rad = 10^{-2} J/kg, 100 rad = 1 kGy.

1.4. Limitations

The limitations imposed by US regulations for the irradiation of foods *(5,14,15)* can serve as a guideline, but not a limit, for our discussion. The following are the purposes for which foods currently may be irradiated in the United States *(15)* as stated in 21CFR179.26.

1.4.1. Trichinella spiralis

"For control of *Trichinella spiralis* in pork carcasses or fresh, nonheat processed cuts of pork carcasses." The minimum dose is 0.3 kGy and not to exceed 1 kGy.

1.4.2. Growth and Maturation Inhibition

"For growth and maturation inhibition of fresh foods." The maximum dose must not exceed 1 kGy.

1.4.3. Disinfestation of Arthropod Pests

"For disinfestation of arthropod pests from foods." The maximum dose must not exceed 1 kGy.

1.4.4. Dehydrated Enzyme Preparations

"For microbial disinfection of dry or dehydrated enzyme preparations (including immobilized enzymes)." The maximum dose must not exceed 10 kGy (1 Mrad).

1.4.5. Dry or Dehydrated Spice

"For microbial disinfection of the following dry or dehydrated aromatic vegetable substances when used as ingredients in small amounts solely for flavoring or aroma: culinary herbs, seeds, spices, vegetable seasonings that are used to impart flavor but that are not either represented as, or appear to be, a vegetable that is eaten for its own sake, and blends of these aromatic vegetable substances. Turmeric and paprika may also be irradiated when they are to be used as color additives. The blends may contain sodium chloride and minor amounts of dry food ingredients ordinarily used in such blends." The maximum dose must not exceed 30 kGy (3 Mrad).

1.4.6. Poultry

"For control of foodborne pathogens in fresh or frozen, uncooked poultry; any packed products that are: (1) whole carcasses or disjointed portions of such carcasses that are "ready-to-cook poultry" within the meaning of 9CFR 381.1 *(16)*, or (2) mechanically separated poultry product (a finely comminuted ingredient produced by the mechanical deboning of poultry carcasses or parts of carcasses). The maximum dose shall not exceed 3 kGy (300 krad) and any packaging used shall not exclude oxygen" *(15)*.

1.4.7. Space-Flight, Shelf-Stable Meats

"For the sterilization of frozen, packaged meats used solely in the National Aeronautics and Space Administration space flight programs. Minimum dose: 44 kGy (4.4 Mrad). Packaging materials used need not comply with Section 179.25(c) provided that their use is otherwise permitted by applicable regulations in parts 174–186 of this chapter" *(15)*.

1.4.8. Meat

"For control of foodborne pathogens in, and extension of the shelf life of, refrigerated or frozen, uncooked meat products within the meaning of 9CFR301.2(rr), meat by-products within the meaning of 9CFR301.2(tt), or meat food-products within the meaning of 9CFR301.2(uu), with or without nonfluid seasoning, that are otherwise composed solely of intact or ground meat, meat by-products, or both meat and meat by-products. Dose not to exceed 4.5 kGy maximum for refrigerated products, and not to exceed 7.0 kGy maximum for frozen products"*(15)*.

1.4.9. Fresh Shell Eggs

"For control of *Salmonella* in fresh shell eggs; not to exceed 3 kGy."

1.4.10. Seeds

"For control of microbial pathogens on seeds for sprouting; not to exceed 8 kGy."

2. *SALMONELLA*

Treatment with ionizing radiation is an effective method for the reduction or elimination of contaminating *Salmonella* serovars from foods, and their inactivation has been studied since 1904 *(2)*.

2.1. Eradication or Control of **Salmonella** *and Extension of Shelf Life of Poultry Meat*

The elimination of salmonella by ionizing irradiation of fresh or frozen poultry carcasses and mechanically deboned meat has been the subject of several dozen studies since the initial work on the high-dose extension of the shelf life of chicken meat at the Massachusetts Institute of Technology by Proctor et al. *(17)*. The irradiation of poultry meat was reviewed by Thayer *(18)*. Kiss and Farkas *(19)* found that a dose of 2–5 kGy of γ radiation administered to eviscerated chicken almost completely eliminated salmonellae and extended the shelf life 2–3 times of carcasses stored at 0–4°C. Lescano et al. *(20)* found that the chicken breasts irradiated at 2.5 kGy were free of salmonellae and were organically acceptable for up to 22 d. Electron-beam irradiation was found to be effective for the control of foodborne pathogens and had little effect on the sensory properties of boneless chicken breast *(21)*.

2.2. Effect of Irradiation Temperature on Radiation Resistance

Licciardello *(22)* discovered that the radiation sensitivity of *Salmonella* Typhimurium increased as the irradiation temperature was increased from 0 to 54.4°C. The cells were markedly more sensitive at irradiation temperatures above 43.3°C. Previte et al. *(23)* discovered that the radiation D-value for five strains of *S.* Typhimurium irradiated at 4°C on autoclaved chicken varied from 0.052 to 0.068 Mrad (0.52–0.68 kGy). They also discovered that the D-value increased to 2.93 kGy when the samples were irradiated at −80°C. Licciardello et al. *(24)* observed an increase in the radiation resistance of salmonellae on chicken at subfreezing temperatures. Working with broilers artificially contaminated on the skin with *S.* Panama, Mulder *(25)* irradiated them at −18°C and obtained a D-value of 1.29 kGy. Mulder et al. *(26)* treated 240 naturally contaminated broilers with 2.5 kGy of γ radiation at −20°C and found that the treatment reduced the *S.* Panama contamination by 2.5 log-cycles. When freshly eviscerated broilers, either with or without salt treatment, were irradiated at a dose of 2.5 kGy and stored at 1.6°C, *Salmonella* spp. were eliminated and the broilers had a shelf life of 15 d *(27)*. Mulder compared the effectiveness of irradiating broilers before or after freezing and then stored them for up to 3 mo at −18°C *(28)*. Salmonellae were not detected after storage for 1 mo in those irradiated before freezing and after 3 mo in those irradiated after freezing. Klinger et al. *(29)* were unable to find salmonellae on broilers irradiated to 2 kGy, though they were easy to find on untreated samples. Hanis et al. *(30)* discovered that chicken carcasses artificially inoculated with 10^6 CFU/g of *S.* Typhimurium were free of salmonellae by a dose of 10 kGy at either −15 or +10°C, but not by treatment at 5 kGy. Thayer and Boyd *(31)* investigated the effects of temperature from −20 to +20°C and γ radiation dose from 0 to 3.6 kGy in sterile, mechanically deboned chicken meat and developed equations from the results that will predict the log-reduction of *S.* Typhimurium at a given temperature and radiation dose. They also found that *S.* Typhimurium was more resistant to γ radiation when vacuum packaged than when air was present during irradiation *(32)*.

2.3. Effect of Irradiation Atmosphere on Radiation Resistance

In the 1950s, oxygen was identified by basic researchers of radiation biology as increasing mutation frequency and lethality of ionizing radiation for cells *(33–35)*. These observations also had implications for changes in the sensory quality of irradiated foods. Investigators discovered that the decimal reduction dose for *Salmonella* spp. on anaerobic chicken meat was approx 2.8 times greater than the observed value when oxygen was present *(24)*. Licciardello et al. *(24)* estimated that a dose of 3 kGy would result in a contamination rate of only one in 80,000 poultry carcasses treated. Thayer et al. *(36)* compared the γ radiation resistance of six serovars of *Salmonella* when irradiated at 5°C in buffer, brain heart infusion broth, and mechanically deboned chicken meat while vacuum packed or with air in the package. In each case, the measured radiation resistance of the serovars was significantly greater in chicken than in either broth or buffer, illustrating the necessity for obtaining D-values for the pathogens in the actual food products. They found no significant differences because of the presence of air in the package during irradiation.

2.4. Serovar

Significant differences in the γ radiation D-values were observed by Thayer et al. *(36)* for the serovars: *S.* Newport, 0.38 ± 0.03 kGy; *S.* Arizonae, 0.44 ± 0.06 kGy; *S.* Typhimurium, 0.51 ± 0.03 kGy; *S.* Anatum, 0.52 ± 0.11 kGy; *S.* Dublin, 0.53 ± 0.11 kGy; and *S.* Enteritidis, 0.77 ± 0.10 kGy.

2.5. Competition and Growth Rates of Survivors on Irradiated Product

A concern about the irradiation of any food was that by reducing the normal flora on the product, a surviving pathogen might multiply much more rapidly than normal. Licciardello et al. *(24)* discovered that the surviving *S.* Typhimurium did not multiply more rapidly on irradiated chicken meat. Thayer and Boyd *(32)* used response–surface methodology to study the effects of γ radiation from 0 to 3.60 kGy, temperature from −20 to +20°C, and atmosphere of either air or vacuum pack on the survival of a streptomycin-resistant isolate of *S.* Typhimurium on mechanically deboned chicken meat and chicken drumsticks. Significant effects for temperature and radiation dose, but not for atmosphere, were found. The predictive equations for the survival of this pathogen were not markedly different from those derived for sterile, mechanically deboned chicken, indicating that the residual indigenous microflora did not significantly alter the radiation resistance. Thayer et al. *(37)* discovered that the γ-injured *S.* Typhimurium cells on mechanically deboned chicken meat were much more sensitive to heat than the nonirradiated cells, which implies any cells surviving the irradiation process were unlikely to survive cooking. This increased sensitivity of the salmonellae to γ radiation was retained during refrigerated storage of the irradiated chicken.

Szczawińska et al. *(38)* investigated the effect of irradiation on the ability of salmonellae to compete by irradiating mechanically deboned chicken meat with 0, 1.25, or 2.50 kGy of γ radiation at 5°C, challenging each lot of meat with an inoculation of approx 10^5 CFU/g of nonirradiated salmonellae (*S.* Dublin, *S.* Enteritidis, or *S.* Typhimurium) and then storing the inoculated samples at 5, 10, or 20°C. The assumption was that the study would mimic poultry meat that had become contaminated following irradiation. The final populations of *S.* Dublin and *S.* Typhimurium were only slightly greater in the irradiated meat after incubation at 10 or 20°C; however, there was no apparent difference between the populations in the meat irradiated to 1.25 kGy vs in the meat irradiated to 2.50 kGy. Because the residual populations of indigenous microflora were greatly reduced in the 2.5 kGy vs the 1.25 kGy irradiated chicken meat, one would have expected a significant difference in the population of salmonellae after incubation if competition were a significant factor.

2.6. Virulence

Because ionizing radiation was known to be mutagenic, the possibility of increased or decreased virulence of surviving pathogens was the subject of several investigations. Previte *(39)* discovered, in 1968, that the toxicity of *S.* Typhimurium endotoxin decreased progressively when exposed to 10, 50, or 20 kGy of γ radiation. A dose of 50 kGy inactivated approx 50% of the lethal lipopolysaccharide. Ley et al. *(40)* followed salmonellae through six cycles of irradiation on frozen meat and discovered that there were no changes in their normal taxonomic characteristics, though they became slightly more sensitive to irradiation. There was no increase in the resistance of *S.* Typhimurium surviving 1, 5, or 10 treatments with 5 kGy to tetracycline, chloramphenicol, or polymixin *(41)* nor were there changes in virulence. When a lesser dose of 2.3 kGy was used, there were increases in both antibiotic and radiation resistance after 10 cycles. It is hard to conceive how, in the normal food-production process, with shipment of the food to the consumer for cooking and consumption, pathogens would actually be subjected to recycling. Maxcy *(42)* concluded that at doses low enough to retain good sensory properties, any pathogenic survivors would be weakened and present no unique problem of acquired resistance through recycling.

2.7. Dose Rate

It has been stated by many authors that dose rate might influence the actual dose of ionizing radiation required to inactivate various pathogens. This is especially of concern when comparing inactivation doses for electron-beam radiation to those for γ-rays. There is support for concern because the very high-dose rates generated by electron-beam irradiation resulted in greater retention of thiamin in radiation-sterilized chicken meat than in γ radiation-sterilized chicken meat *(8)*. The D-values for *Listeria monocytogenes*, *Staphylococcus aureus*, *Escherichia coli* O157:H7, *S.* Typhimurium, *Yersinia enterocolitica*, *Vibrio parahaemolyticus*, and *Campylobacter jejuni* were not significantly different when derived from exposures to γ radiation at the rates of 0.78, 2.6, and 22 kGy/h *(43)*.

2.8. Water

The literature is replete with observations of increased sensitivity of salmonellae and other microorganisms to ionizing radiation in aqueous suspension *(25,36,43)*. Thayer et al. *(44)* found that the resistance to γ radiation of *S.* Typhimurium in chicken meat increased when the amount of water was decreased or when NaCl was added to the meat, decreasing the water activity. However, when the water activity was decreased by the addition of sucrose, no such increase in radiation resistance was observed, implying that it was the amount of water in the suspending medium and not its water activity that was the controlling factor.

2.9. Substrate

Those involved with the application of ionizing radiation for the control of foodborne pathogens on foods quickly found that the observed radiation resistance was often very different when the organisms were on a food product rather than in an aqueous suspension. To a large extent, we can attribute such variances to the chemical composition of the food and its ability to react with the ionizing radiation as a scavenger for free radicals allowing interaction of the radiation with the microorganism. We already know from our earlier discussion that different serovars of *Salmonella* may vary in their resistance to ionizing radiation; however, the D-value for *S.* Typhimurium was 0.37 kGy on ground beef *(45)* and 0.44 kGy on minced chicken *(46)*. A mixture of *Salmonella* serovars had a D-value of 0.54 kGy *(47)* on sprouts, but a D-value of 0.97 kGy on alfalfa seeds *(48)*. Thayer et al. *(49)* postulated and tested the concept that many of the differences in radiation resistance attributed to the food substrate might be because of the variations in primary factors such as oxygen tension, pH, irradiation temperature, bacterial growth stage, amount of water, food additives, cultural conditions, and the methods used to enumerate the number of survivors. Using identical cultural conditions, Thayer et al. *(49)* found that the D-value for a mixture of *Salmonella* serovars *S.* Dublin 15480, *S.* Enteritidis 13076, *S.* Newport 6962, *S.* Senftenberg 8400, and *S.* Typhimurium 14028 did not differ significantly on ground beef, lamb, turkey breast, and turkey leg meats. The mean D-value for this mixture of *Salmonella* serovars was 0.70 ± 0.04 kGy at 5°C. The D-value for this mixture was, however, significantly lower on pork meat (0.51 ± 0.03 kGy) than on the other meats. In a separate study of the radiation resistance of this same mixture of *S.* serovars on bison, ostrich, alligator, and caiman meats, the D-values obtained with all of the meats were not significantly different from each other, and averaged 0.53 ± 0.02 kGy, not statistically different from that observed for pork *(50)*. The authors concluded that one could expect to obtain similar control of foodborne pathogens on edible meats and poultry products.

Clavero et al. *(51)* discovered that the D-values for a mixture of *Salmonella* serovars were not significantly different in low (8–14%) and high fat (27–28%) raw ground beef, averaging 0.64 kGy when irradiated at 2–5°C.

2.10. Salmonella *Control in Eggs and Egg Products*

There was a very early recognition that irradiation might be a good method to control *Salmonella* in liquid whole egg used for the production of dried egg powder *(52)*. Proctor et al. *(52)* found that the taste panels could not distinguish between scrambled eggs prepared from spray dried irradiated and nonirradiated liquid whole egg; *Salmonella* could be completely eliminated. Irradiation of either shell eggs or of liquid whole egg was less feasible because of the high lipid-content of the yolk and because of the adverse changes in the albumen fraction of shell eggs *(53–57)*. Kijowski et al. *(58)* irradiated frozen whole egg and obtained D-values of 0.39 and 0.52 kGy for *Salmonella* spp. and *E. coli*, respectively. No detrimental effects of irradiation were noted at doses up to 2.5 kGy. In most foods *E. coli* is more sensitive to radiation than is *Salmonella*.

Apparently, natural antibacterial factors in the egg influence the radiation resistance of *Salmonella*. Matić et al. *(59)* discovered that the same degree of control by a 2.4-kGy irradiation dose of *Salmonella* in egg powder could be achieved by 1 kGy when the irradiated product was stored for 3 wk. Kijowski et al. *(58)* discovered that frozen liquid whole egg could be irradiated to a dose of 2.5 kGy without detrimental functional, chemical, and sensory effects and reduced the probability of survival of *Salmonella* by 6 logs. The irradiation resistances of five *S.* Enteritidis isolates were compared on and within shell eggs *(60)*. The D-values ranged from 0.32 to 0.41 kGy, indicating that a dose of 1.5 kGy should reduce counts by 4 logs. Investigators irradiated liquid whole egg at 60°C and obtained greater rates of *Salmonella* inactivation than with either heat or radiation alone *(61)*.

2.11. Salmonella *Control on Fruits, Vegetables, and Produce*

Salmonella contamination of produce is a continuing source of foodborne disease, in part, because it is frequently consumed raw *(62)*. The subject of irradiation of fruits, vegetables, and produce has been reviewed, though few studies included pathogens *(63–70)*. Hagenmaier and Baker *(71)* observed that even a very low radiation dose of 0.19 kGy increased the shelf life of fresh-cut iceberg lettuce and reduced its microbial population. The radiation D-values for the *Salmonella* spp. isolated from meat or vegetables on alfalfa sprouts were 0.54 and 0.46 kGy, respectively *(47)*. A radiation dose of 2 kGy increased the shelf life of alfalfa sprouts by 10 d, and irradiation of the seeds at doses up to 2 kGy did not unacceptably decrease the yield of alfalfa sprouts per given weight of seed *(72)*. Rajkowski et al. *(73)* discovered that the shelf life of the broccoli sprout could be extended 10 d by a radiation dose of 2 kGy. The D-values for vegetable isolates of *Salmonella* were 1.10 kGy when they were present on the seeds. A radiation dose of 2 kGy significantly reduced both sprout yield and germination of the broccoli seeds. Thayer et al. *(48)* discovered that when *Salmonella* isolates from either meat or vegetables were present on alfalfa seeds, the average D-value for the inactivation of *Salmonella* was 0.97 ± 0.03 kGy, much higher than the D-values for the same isolates on meat and poultry. The higher than expected resistance to γ radiation was not because of the low amounts of water (~12%) in the alfalfa seeds. Significantly, a 2-kGy radiation dose should inactivate 2 logs of *Salmonella* on alfalfa seeds. A natural contaminant of alfalfa seeds that had been responsible for several outbreaks of salmonellosis, *S.* Mbandaka, had a D-value of 0.81 ± 0.02 kGy on alfalfa seeds *(74)*. An absorbed dose of 4 kGy but not 3 kGy eliminated *S.* Mbandaka from the seed. This dose was much greater than expected because the organism could only be detected by an enrichment culture, and the investigators concluded after several studies that the maximum contamination level per seed and not contamination level of the lot of seed determined the radiation dose required for inactivation *(73)*. *Salmonella* are more sensitive to γ irradiation on lettuce than on alfalfa sprouts, as Niemira *(75)* compared the sensitivities of a mixture of *S.* Anatum and *S.* Stanley on Boston, green leaf, iceberg, and red leaf lettuce and obtained D-values of 0.24 ± 0.01, 0.31 ± 02, 0.25 ± 0.01, and 0.23 ± 0.01 kGy, respectively.

2.12. Salmonella *Control in Juice*

Juices were recognized as possible sources of foodborne disease, and irradiation was investigated as a method for pasteurization or sterilization *(76–87)*. In spite of the linkage of *Salmonella* contamination of juices *(88–90)* to foodborne disease, relatively few studies have explicitly examined its control in juice by irradiation. The radiation resistance of *S*. Enteritidis varied only slightly in five different commercial orange juices from 0.35 to 0.37 kGy *(91)*. The γ radiation D-values at 2°C in undiluted orange juice for *S*. Anatum, *S*. Newport, *S*. Infantis, and *S*. Stanley were 0.48, 0.48, 0.35, and 0.38 kGy, respectively *(92)*; variations in the amount of pulp did not alter the radiation resistance.

3. CAMPYLOBACTER

Campylobacteriosis in 1996 accounts for 46% of the laboratory confirmed cases of bacterial gastroenteritis reported to the US Centers for Disease Control and Prevention *(93)*. Poultry, meat, and raw milk are the major sources of human infection with *Campylobacter jejuni (93)*. Tarkowski et al. *(45)* obtained D-values of 0.15, 0.14, and 0.16 kGy for three strains of *C. jejuni* in ground beef. Lambert and Maxcy *(94)* determined that the D-values for log-phase cells of *C. jejuni* irradiated in ground beef were 0.32, 0.16, and 0.17 kGy at temperatures of −30 ± 10, 0–5, and 30 ± 10°C, respectively. When irradiated in ground turkey meat the results were very similar, 0.29, 19, and 0.16 kGy at temperatures of −30 ± 10, 0–5, and 30 ± 10°C, respectively. However, only two replications, and in some cases only two dose-levels, were investigated. The authors' results did not indicate that growth stage affected the radiation resistance of the cells in ground beef. Yogasundram et al. *(95)* discovered that a contamination level of 10^3 of *C. jejuni* on chicken drumsticks could be reduced by 1 log by a dose of 0.5 kGy; a 1 kGy treatment completely eliminated it. Clavero et al. *(51)* measured D-values for *C. je*juni of 0.24 ± 0.02 and 0.18 ± 0.005 kGy in low-fat (8.2–13.9%) raw ground beef at −17 to −15°C or at 3–5°C, respectively. The D-values for *C. jejuni* were 0.21 ± 0.02 and 0.20 ± 0.02 kGy in high-fat (26.8–27.1%) raw ground beef at −17 to −15°C or at 3–5°C, respectively. Dion et al. *(43)* discovered that a 28-fold difference in γ-ray dose rate did not alter the D-value for inactivation of *C. jejuni* in chicken breast meat. Patterson *(96)* compared the γ-radiation sensitivity of three strains of *C. jejuni* and one strain each of *C. fetus* and *C. lari* on poultry meat. The D-values ranged from 0.12 to 0.25 kGy. The author suggested that the radiation doses sufficient to eliminate *Salmonella* or *Listeria* would also eliminate *Campylobacter* spp. The D-values for inactivation of *Arcobacter butzleri* (0.27 kGy) and *C. jejuni* (0.19 kGy) in vacuum packaged pork were determined with electron-beam irradiation *(97)*. A 1.5 kGy dose of ionizing radiation would be expected, therefore, to eliminate 5 logs of *A. butzleri* and 7 logs of *C. jejuni*. The available evidence indicates that food irradiation offers an excellent opportunity to control this pathogen on food.

4. ESCHERICHIA COLI

Escherichia coli in early studies was viewed primarily as a tool in the study of genetics and radiation biology *(98,99)*. These studies and others, however, produced information for the practical application of ionizing radiation for the treatment of foods, such as the effects of temperature during irradiation *(100)* and the effect of oxygen *(33)*. Fram et al. *(101)* demonstrated that the percentage of a given species killed by a specific dose of X-rays was the same regardless of the concentration of cells in the initial suspension and also that the inactivation followed first-order kinetics. Stapleton *(102)* discovered that there were variations in the radiation resistance of *E. coli* that were dependent on the growth cycle of the bacterium.

4.1. Poultry

Mulder *(25)* investigated the inactivation of *Escherichia coli* K12 by irradiation on inoculated deep-frozen chicken broiler carcasses. The poultry was also inoculated with *S. panama*. At an irradiation temperature of −18°C and a γ-ray dose rate of 1.5 kGy/h, the D-value for the inactivation of *E. coli* K12 on skin was 0.58 kGy. Patterson *(46)* inoculated sterile chicken mince with an unidentified strain of *E. coli* and obtained a D-value of 0.39 kGy in air or 0.27 in vacuum. This result is unusual in that the D-value would be expected to be lower in the presence of oxygen.

4.2. Beef

In 1993, Thayer and Boyd *(103)* reported the results of an investigation of the sensitivity of *E. coli* O157:H7 ATCC 43895 when irradiated on sterile, mechanically deboned chicken meat or on sterile ground beef. At an irradiation temperature of 0°C in either vacuum-packed hamburger or mechanically deboned chicken meat, stationary phase cells of *E. coli* O157:H7 had a D-value of 0.27 ± 0.03 kGy. When the irradiation temperature was decreased to −5°C, the D-value increased to 0.44 ± 0.03 kGy in mechanically deboned chicken meat. The authors found a very significant effect of irradiation temperature on the inactivation of *E. coli* O157:H7, indicating that a dose of 1.5 kGy should eliminate 5.36 logs at 5°C but only 2.64 logs at −20°C. These authors found in a challenge study that a dose of 0.75 kGy reduced an average inoculum of $10^{4.8}$ CFU/g of *E. coli* O157:H7 in ground beef to less than 10 CFU/g, and survivors were not detected in ground beef irradiated to an absorbed dose of 1.5 kGy even after the samples were temperature abused at 35°C for 20 h. Thayer et al. *(49)* found that D-values of a cocktail of *E. coli* O157:H7 did not differ when irradiated in ground beef, pork, lamb, or turkey under identical conditions. Thayer and Boyd *(104)* determined D-values for the inactivation of *E. coli* O157:H7 in ground beef when irradiated at temperatures from −76 to +20°C; the results confirmed their earlier studies and provided D-values for temperatures that might be encountered under commercial conditions of irradiation.

Clavero et al. *(51)*, using a five strain mixture of *E. coli* O157 and irradiating the samples of inoculated raw ground beef in a commercial irradiator, observed D-values of 0.24 and 0.31 kGy at irradiation temperatures of 3–5°C and −17 to −15°C. The D-values were not affected by the amount of fat in the meat. Ito and Harsojo *(105)* found D-values of 0.26 and 0.46 kGy for the inactivation of strain IID959 in ground beef by γ irradiation at fresh and frozen conditions, respectively.

4.3. Juice

Contamination with *E. coli* O157:H7 is not limited to meats *(62,63)* and research has established that ionizing irradiation can be an effective treatment for its elimination from fruit juices, produce, and seeds used for the production of food sprouts. Buchanan et al. *(106)* discovered that the D-values for three isolates of *E. coli* O157:H7 in apple juice increased when the cells were acid adapted or when there was a high level of suspended solids in the cider. They concluded that a dose of 1.8 kGy should be sufficient to achieve the 5D inactivation recommended by the National Advisory Committee for

Microbiological Criteria *(90)*. Fan and Thayer *(107)* found that a dose of 3.55 kGy required for a 5D reduction of *Salmonella* in apple juice resulted in increased production of malondialdehyde after storage; however, malondialdehyde production was reduced when the product was flushed with nitrogen and frozen before irradiation. Fan et al. *(108)* determined that a 5D ionizing radiation dose at 5°C, 3.55 kGy, for the control of *Salmonella* would result in a 16.6% loss of the total ascorbic acid in orange juice. However, these losses were also reduced by flushing with nitrogen and freezing the product before irradiation.

4.4. Food Sprouts and Seeds

Rajkowski and Thayer *(47)* discovered that the D-values 0.26–0.34 kGy at an irradiation temperature of 19°C for the inactivation of *E. coli* O157:H7 on inoculated alfalfa, broccoli, and radish sprouts were very similar to those observed for its inactivation on meat. The total ascorbic acid in irradiated alfalfa sprouts was measured at 1, 7, and 14 d of storage following radiation doses of 0–2.57 kGy at 5°C and was found to be slightly higher in the sprouts irradiated at the maximum dose than in the nonirradiated controls *(109)*.

Unfortunately, with the exception of very large commercial producers, most sprouts are grown by relatively small firms with insufficient product for irradiation to be a practical step in their HACCP plan. (Sprout producers are considered to be food processors, not farmers, by the US Food and Drug Administration.) This fact makes it a better alternative to treat the seed rather than the sprouts; this also keeps the pathogens out of the food-processing facility. However, there is an obvious limitation to irradiation of seeds in that a too rigorous treatment may decrease sprouting or the yield of sprouts per gram of seed. There is no single answer to the problem as seeds very widely in their sensitivity to ionizing radiation *(63)*. Thayer et al. *(48)* found that the D-values for the inactivation of disease outbreak isolates of *E. coli* O157:H7 and *Salmonella* on alfalfa seeds were 0.60 ± 0.01 and 0.97 ± 0.03 kGy, respectively. A 2-kGy dose for alfalfa seed, recommended by Rajkowski and Thayer *(72)*, should produce a 3.3-log reduction of *E. coli* O157:H7 without a significant reduction in the yield of sprouts. Rajkowski et al. *(73)* discovered that the D-value for inactivation of *E. coli* O157:H7 when present on broccoli seeds was 1.11 ± 0.12 kGy and that 2 kGy was the maximum dose that allowed good productivity of sprouts. Alfalfa sprouts grown from irradiated seeds consistently were found to have higher quantities of antioxidants, including ascorbic acid, than those grown from nonirradiated seeds, even after the sprouts were stored for 21 d at 7°C *(110)*. Bari et al. *(111)* compared chemical and irradiation treatments for killing *E. coli* O157:H7 on alfalfa, radish and mung bean seeds and concluded that only an irradiation dose of 2.0 kGy in combination with dry heat completely eliminated it from alfalfa seeds and mung beans, but a dose of 2.5 kGy was required to eliminate the pathogen from radish seeds.

4.5. Produce

Outbreaks of disease have been linked to fresh cut lettuce contaminated with *E. coli* O157:H7 *(62)*. Niemira et al. *(112)* investigated the radiation sensitivity of an outbreak isolate of *E. coli* O157:H7 on four types of lettuce and found significant differences depending upon the lettuce type. The D-values were 0.12 ± 0.004 and 0.14 ± 0.003 kGy on the surface of red leaf lettuce and either iceberg or Boston lettuce, respectively.

The D-values were significantly higher when homogenates of the lettuce were used; the D-value in homogenates of red leaf lettuce was 0.34 ± 0.01 kGy. Chlorination of iceberg lettuce with 200 ng/mL and then irradiating it to 0.55 kGy provided a 5.4-log reduction in *E. coli* O157:H7 *(113)*.

5. *LISTERIA*

Several foodborne outbreaks of listeriosis prompted concern about the presence of *L. monocytogenes* not only in milk and cheese, but also, in or on poultry, meat, and ready-to-eat foods and for the potential of its control through irradiation *(114–116)*. Huhtanen et al. *(117)* discovered that the mean D-value for inactivation of seven isolates of *L. monocytogenes* on mechanically deboned chicken meat was 0.46 ± 0.05 kGy at 2–4°C. El Shenawy et al. *(118)* found that D-values for three isolates of *L mono-cytogenes* ranged from 0.51 to 1.0 kGy when present in minced beef and γ-irradiated at 20–21°C. The Scott A strain was least resistant to irradiation. Patterson et al. *(119)* found D-values of 0.42–0.55 kGy for four strains of *L. monocytogenes* on sterile chicken mince. It is unknown if the author refrigerated the samples in any way during the irradiation treatment. Farag et al. *(120)* obtained almost identical D-values with γ- and 10 MeV-irradiation. Lewis and Corry *(121)* irradiated 32 of 64 matched chicken carcasses to an absorbed dose of 2.5 kGy. The number of carcasses contaminated with *L. monocytogenes* was lower in the irradiated birds. *L. innocua* was found on 44% of the nonirradiated chickens but on none of those that was irradiated. Tarte et al. *(122)* found that the D-values for *L. monocytogenes* NADC 2045 Scott A, *L. monocytogenes* NADC 2783, *L. monocytogenes* ATCC 15313, *L. ivanovi*, and *L. innocua* in ground pork were 0.447, 0.424, 0.445, 0.372, and 0.638 kGy, respectively. Thayer et al. *(49)* found that the D-values for a mixture of four pathogenic isolates of *L. monocytogenes* averaged 0.47 kGy and were not significantly different when irradiated at 5°C on vacuum-packed beef, pork, lamb, turkey breast, and turkey leg meat.

5.1. Temperature

Monk et al. *(123)* irradiated low- and high-fat frozen and refrigerated ground beef patties inoculated with *L. monocytogenes* using a commercial radiation source and found that the D-values ranged from 0.507 to 0.610 kGy. Neither the amount of fat nor the irradiation temperature influenced the D-value. Andrews et al. *(124)* also did not observe increased resistance to radiation by *L. monocytogenes* when it was frozen suspended in tryptic soy broth; the opposite effect was found. The D-values at −80, 4, and 20°C were 0.43, 0.58, and 0.62 kGy, respectively. Thayer and Boyd *(125)* examined the radiation resistance of *L. monocytogenes* in ground beef at 5°C intervals for irradiation temperatures from −20 to +5°C using a temperature-controlled irradiator and found that radiation resistance increased as the irradiation temperature decreased. In a manner resembling the Arrhenius plot of reaction rate vs temperature, a straight line was obtained when the logs of the D-values were plotted vs the reciprocal of the absolute temperature from −5 to −20°C. The D-values were 0.445, 0.453, 0.772, 0.854, 1.006, and 1.298 kGy at irradiation temperatures of 5, 0, −5, −10, −15, and −20°C, respectively. Survival of *L. monocytogenes* followed a predictable pattern when irradiated at temperatures of −60 to +20°C. There was no statistically significant increase in survival as the temperature decreased from 20 to 0°C, but survival increased sharply from 0 to −20°C and less so from −20 to −30°C, at which point the curve leveled off. This type of effect would be predicted for a decrease in secondary reactions as the temperature reached the point at which the migration of free radicals became severely inhibited by the ice structure. Kamat and Nair *(126)* found significantly increased survival of *L. monocytogenes* when irradiated at cryogenic temperatures (dry ice).

5.2. Synergy Between Heat and Irradiation

Grant and Patterson *(127)* found that the D-value for *L. monocytogenes* cells on minced roast beef heated at 65°C was 34.0–53.0 s. When the samples were preirradiated to a dose of 0.8 kGy the D-value was 15.3–16.8 s. The decreased resistance to heat persisted for up to 2 wk during the storage of the irradiated beef at 2–3°C. The results suggest that any *L. monocytogenes* present in irradiated foods that are intended to be reheated prior to consumption would be more easily killed.

5.3. Milk and Cheese

In an interesting application it was found by Bougle and Stahl *(128)* that a fully pathogenic strain of *L. monocytogenes* inoculated at 10^4/g, though not at 10^5/g, into raw milk camembert cheese could be completely destroyed by a dose of 2.6 kGy of ionizing radiation without altering the organoleptic properties. An irradiated camembert was marketed in France. Hashisaka et al. *(129,130)* determined that the D-values for *L. monocytogenes* at −78°C were 1.4 and 2.0 kGy in mozzarella cheese and ice cream, respectively. This study investigated the production of low bacterial products for immunosuppressed patients and unfortunately found that many milk products had off flavors as a result of sterilization treatments *(131)*. Ennahar et al. *(132)* proposed using low energy electrons to irradiate the rind of soft and red smear cheeses to eliminate *L. monocytogenes* which is associated primarily with the rind of the cheese. A dose of 3.0 kGy eliminated 10^5 CFU/g of *L. monocytogenes* and avoided changes in sensory properties of the cheeses.

5.4. Modified Atmosphere Packaging

Grant and Patterson *(133)* irradiated two isolates of *L. monocytogenes* in minced pork packed in either oxygen permeable packaging or under a modified atmosphere of 25% CO_2 and 75% N_2. They obtained D-values of 0.65 and 0.57 kGy in air and 0.71 and 0.60 kGy in the modified atmosphere. Thayer and Boyd *(134)* found that the recovery and multiplication of radiation-injured *L. monocytogenes* cells on radiation-sterilized ground raw turkey was inhibited by modified atmospheres containing 40.5 or 64% CO_2 and nitrogen during storage at a mild abuse temperature of 7°C for 28 d.

5.5. Ready to Eat Precooked Foods

Grant and Patterson *(135)* determined the D-values for *L. monocytogenes* on individual components of a chilled "ready meal" and found some significant variations. The D-values were 0.53, 0.56, 0.56, 0.64, and 0.65 kGy on mashed potato, gravy, cauliflower, beef, and roast potato, respectively. A *sous-vide* treatment of vacuum and cooking chicken to an internal temperature of 65.6°C had little effect on the survival of *L. monocytogenes*; but following a combination of *sous-vide* with an irradiation dose of 2.9 kGy, the pathogen was undetectable during 8 wk of storage at 2°C. Grant et al. *(136)* found that irradiation extends the lag phase of *L. monocytogenes*. This work was extended by Patterson et al. *(119)* who found that in cooked poultry meat at 6°C, the lag phase for *L. monocytogenes* was only 1 d in unirradiated samples but 18 d in samples that had received a dose of 2.5 kGy. Thayer et al. *(137)* found that the D-values for the inactivation of *L. monocytogenes* on air-packed raw and cooked ground turkey meat were significantly different, 0.70 ± 0.04 and 0.60 ± 0.02 kGy, respectively. *L. monocytogenes* cells surviving a low dose of 1 or 2 kGy multiplied more rapidly on cooked turkey than on raw turkey during 21 d of storage at 7°C.

Clardy et al. *(138)* investigated the potential for irradiation control of *L. monocytogenes* that might contaminate commercial ham and cheese sandwiches that are sold frozen. The investigators inoculated the frozen ham slice with *L. monocytogenes*, resealed the packages, and refroze the sandwiches before determining the D-value. The sensory effects of irradiation were determined with uninoculated sandwiches. At an irradiation temperature of −40°C, the D-values ranged from 0.71 to 0.81 kGy. The sensory panelists were able to identify the irradiated sandwiches but were divided on whether or not the irradiation had adversely affected the quality.

Frankfurters have been recalled on several occasions because of contamination with *L. monocytogenes*. Irradiation offers an exceptionally good method for the decontamination of packaged frankfurters. Sommers and Thayer *(139)* determined the D-values for the inactivation of *L. monocytogenes* on the surface of several types and brands of commercial frankfurters and found that D-values ranged from 0.49 to 0.71 kGy with an average of 0.61 kGy. Product formulation significantly affected the radiation resistance; the higher D-values tended to be associated with mixed meat and poultry frankfurters. Stadium franks have a bright red color because they have been surface-treated with sodium erythorbate. Because erythorbate is an antioxidant, there was a concern that its presence might protect *L. monocytogenes* during irradiation. Sommers et al. *(140)* found that though radiation resistance increased when cells were suspended in 0.1% solutions, no such increase occurred when cells were on the surface of treated frankfurters. Sommers and Fan *(141)* discovered that the D-value for the inactivation of *L. monocytogenes* when it was surface inoculated onto slices of beef bologna was not influenced by the antioxidant values of the meat, which increased with increasing dextrose concentrations from 0 to 8%. The D-values ranged from 0.59 ± 0.02 to 0.61 ± 0.04 kGy. Sommers and Fan *(142)* discovered that adding sodium diacetate to bologna emulsion decreased the D-value for the inactivation of *L. monocytogenes*. Sommers et al. *(143)* found that the D-value for *L. monocytogenes* decreased when sodium diacetate and potassium lactate were added to the bologna mixture. Sodium diacetate and potassium lactate mixtures inhibit the multiplication of *L. monocytogenes*. Sommers et al. *(144)* discovered that the D-value for *L. monocytogenes* was inversely dependent to the concentration of citric acid in which the frankfurters were dipped prior to packaging. The D-values were 0.61, 0.60, 0.54, and 0.53 kGy when frankfurters were dipped in 0, 1, 5, or 10% citric acid, respectively.

5.6. Produce

Contaminated produce has been responsible for the dissemination of *L. monocytogenes* to man, and produce serves as an excellent medium for its multiplication *(62)*. Irradiation is an appropriate method to eliminate this organism from produce and to increase shelf life significantly. Farkas et al. *(145)* investigated the inactivation of *L. monocytogenes* on chilled cut bell peppers and carrots and found that a very low dose of 1 kGy eliminated the pathogen and extended the shelf life. These authors also found that the pathogen rapidly multiplied on unirradiated bell pepper. Because frozen vegetables sometimes get included in salad mixes without prior cooking, they can be a possible source of foodborne pathogens. Niemira et al. *(146)* irradiated inoculated broccoli, corn, lima beans, and peas at subfreezing temperatures and determined that the D-values for the inactivation of *L. monocytogenes* at −5°C ranged from 0.50 kGy for broccoli to 0.60 kGy for corn. Significant effects were noted for temperature and type of vegetable. Niemira *(75)* discovered that the D-value for the inactivation of *L. monocytogenes* at 4°C was approx 0.19 kGy on Boston, green leaf, iceberg, and red leaf lettuce, indicating that a 5-log reduction could be achieved by a dose of 1 kGy. Niemira et al. *(147)* determined that the D-value for *L. monocytogenes* was 0.21 ± 0.01 kGy at 2°C when the pathogen was inoculated onto endive. It was determined that a 4-log reduction could be achieved with little or no impact on the product's texture or color.

6. VIBRIO

The possibility of inactivating *Vibrio cholerae* by irradiation was recognized and tested by Green in 1904 *(2)*. Among the many pathogenic *Vibrio* species is *V. vulnificus*, which has proven to be resistant to technologies that have eliminated other vibrios from shell fish *(148)*. Matches and Liston *(149)* determined the sensitivity of 27 *Vibrio parahaemolyticus* isolates in homogenized fish and crab meat to γ radiation. In crab meat, reductions varied at 24°C from 2.8 logs at 1.0 kGy to 5.4 logs at 0.25 kGy. Campanini et al. *(150)* determined that the D-values for *V. parahaemolyticus* and *V. cholerae* in pH 7 phosphate buffer were 0.10 and 0.11 kGy, respectively. Bandekar et al. *(151)* discovered that the D-value for the inactivation of *Vibrio parahaemolyticus* on frozen shrimp was 0.10 kGy. Ito et al. *(152)* found that the dose necessary to eliminate *V. parahaemolyticus* from frozen shrimp was 1.5–2.0 kGy. Hau et al. *(153)* found that the D-value for *V. cholerae* on grass prawns was 0.11 kGy. Sang et al. *(154)* determined that radiation doses of 1.0 and 0.5 kGy eliminated *V. cholerae* from frog legs at −20 and 4°C, respectively. Ama et al. *(155)* found that irradiating fresh oyster at 40°C more effectively inactivated *Vibrio vulnificus* than either heating or irradiation at refrigeration temperatures. Kwon and Byun *(156)* discovered that *V. parahaemolyticus* was more sensitive to radiation following a heat treatment. de Moraes et al. *(157)* found that the D-values for *V. cholerae* in oysters ranged from 0.173 to 0.235 kGy. Jakabi et al. *(158)* found that a dose of 1.0 kGy was sufficient for a 5- to 6-log reduction of *V. parahaemolyticus* in oysters. The results clearly indicate that irradiation is an effective sanitation treatment for the elimination of *Vibrio* species from food.

7. YERSINIA

El Zawahry and Rowley *(159)* determined that the D-values for *Yersinia enterocolitica* in ground beef were 0.195 and 0.388 kGy at irradiation temperatures of 25 and −30°C, respectively. Tarkowski et al. *(45)* determined that the D-values for *Y. enterocolitica* were 0.10, 0.16, and 0.21 kGy for serotypes 0:3, 0:5/27, and 0:9, respectively. Grant and Patterson *(133)* determined the D_{10}-values for two isolates of *Y. enterocolitica* on pork packed with air or under a modified atmosphere consisting of 25% CO_2 : 75% N_2. The

D_{10}-values ranged from 0.164 to 0.204 kGy on nonselective media. The presence of the modified atmosphere increased the radiation resistance of one of the isolates significantly from 0.164 to 0.176 kGy. The radiation resistances of log- and stationary-phase *Y. enterocolitica* cells were not significantly different *(43)*. *Y. enterocolitica* inoculated at 10^5 onto the surface of steak or ground beef was undetectable following a radiation dose of 2.0 kGy *(160)*. Kamat et al. *(161)* found that the D-value for *Y. enterocolitica* was 0.25 kGy in a homogenate of 10% raw pork meat. These authors found that while *Y. enterocolitica* could be eliminated by an irradiation dose of 4–3 kGy from salami and cooked ham that had been inoculated at 10^6 CFU/g, the pathogen was not eliminated from raw pork. However, lower numbers, $<10^3$ CFU/g, were completely eliminated by a dose of 1 kGy, even at –40°C. Shenoy et al. *(162)* found that heat shocking *Y. enterocolitica* cells did not increase their resistance to radiation. Sommers et al. *(163)* determined that the D-value for a mixture of four *Y. enterocolitica* strains inoculated into ground pork increased when the radiation temperature decreased from 0.19 to 0.55 kGy at irradiation temperatures of +5 and –76°C, respectively. Sommers and Bhaduri *(164)* discovered that any *Y. enterocolitica* that might survive a low dose irradiation treatment would have a high probability of being less virulent because of the loss of the virulence plasmid as the result of the treatment. Sommers and Novak *(165)* discovered that the presence or absence of the virulence plasmid in *Y. enterocolitica* did not affect its resistance to ionizing radiation.

8. *STAPHYLOCOCCUS*

Staphylococcus pyogenes aureus was among the bacteria first found to be sensitive to radiation from radium by Green in 1904, and Chambers and Russ in 1912 *(2,166)*. Baker *(167)* quantified the response of *Staphylococcus aureus* to the β-rays from radium and observed that when the log concentration was plotted vs the exposure time in hours, the inactivation was linear or first order. Fram et al. *(101)* found that *S. aureus* was more resistant to X-rays than *Aerobacter aerogenes*, *E. coli*, *Serratia marcescens*, *Pseudomonas aeruginosa*, and *P. fluorescens*. The percentage of bacteria of a given species that were killed by a specified dose was the same regardless of the concentration of bacteria in the irradiated suspension. Patterson *(46)* did not find any significant differences in the D-values, averaging 0.40 kGy at 10°C, for *S. aureus* inoculated onto sterile chicken mince and irradiated in air, CO_2, vacuum, or N_2. Licciardello et al. *(168)* found that the D-value for the inactivation of *S. aureus* in minced clam and mussel meat was 100 krad (0.10 kGy). Unfortunately, the irradiation temperature was not stated, making it difficult to assess the significance of the data. Grant and Patterson *(135)* measured the D-values for *S. aureus* NCTC 10655 when irradiated at a temperature of 0–3°C in the components of a roast beef meal and obtained D-values of 0.39, 0.36, 0.43, 0.39, and 0.42 kGy in roast beef, gravy, cauliflower, roast potato, and mashed potato, respectively. Hau et al. *(153)* measured the D-value for *S. aureus* ATCC 10832 on the surface of frozen grass prawns and obtained a value of 0.29 kGy at –10 ± 2°C.

Thayer and Boyd *(169)* measured D-values of 0.27 ± 0.02 and 0.36 ± 0.01 kGy for the inactivation of log- and stationary-phase cells of *S. aureus* ATCC 13565 inoculated into sterile, mechanically deboned chicken meat and irradiated at a temperature of 0°C. Results obtained with air- and vacuum-packed samples were not significantly different. A higher D-value of 0.47 ± 0.01 kGy was obtained for a mixture of four strains (ATCC

13565, B121, B124, and B176) of *S. aureus* inoculated into sterile, mechanically ground chicken meat, vacuum packed, and irradiated at 0°C. These authors *(169)* conducted a challenge study in which vacuum-packaged, sterile, mechanically-deboned chicken meat was inoculated with approx $10^{3.9}$ CFU of *S. aureus* and irradiated samples at doses from 0 to 3.0 kGy at 0°C. One set of samples was analyzed immediately after irradiation and a second set was temperature abused at 35°C for 20 h before analysis. In the non-temperature abused samples, *S. aureus* was not detected in samples that had been treated with 1.50 kGy or higher. In samples that were temperature abused, nonirradiated samples and those treated with 0.75 kGy were positive for *S. aureus*; however, toxin was detected only in the nonirradiated samples. Neither viable *S. aureus* nor toxin was found in samples that received 1.50 kGy or greater treatment. The authors found significantly increased survival of *S. aureus* in frozen samples. Grant et al. *(170)* discovered that following a 2-kGy treatment, *S. aureus* was not detected during storage at 15°C for 7 d in roast beef and gravy that had been inoculated with approx 10^2 CFU/g. With a greater inoculum 2 kGy produced a 3–4 log-reduction in the population and a significant delay before toxin was detectable under abuse conditions.

Thayer et al. *(49)* tested the hypothesis that radiation resistances of *S. aureus* were specific for the meat on which the bacteria were irradiated. Using identical conditions of preparation and irradiation and a mixture of ATCC 25923 and 13565, they found that there were only minor variations in the D-values. The D-values for the inactivation of *S. aureus* inoculated onto sterile meat and irradiated at 5°C were 0.46 ± 0.02, 0.40 ± 0.03, 0.43 ± 0.02, 0.45 ± 0.03, and 0.46 ± 0.05 kGy for beef, lamb, pork, turkey breast, and turkey leg meats, respectively. Thayer et al. *(50)* conducted a similar study with exotic meats and found D-values of 0.40 ± 0.02, 0.34 ± 0.02, 0.36 ± 0.01, and 0.38 ± 0.01 kGy for the inactivation of *S. aureus* ATCC 25923 and ATCC 13565 on ground bison, ostrich, alligator, and caiman meats, respectively. Thayer and Boyd *(104)* determined the D-value for *S. aureus* on ground beef at several irradiation temperatures and found significantly increased values when the product was irradiated under frozen conditions, e.g., 0.51 ± 0.02 and 0.88 ± 0.05 kGy at 0 and −20°C, respectively. Lamb et al. *(171)* determined the D-value for the inactivation of *S. aureus* on ready to eat ham and cheese sandwiches to be 0.63 kGy. The authors concluded that irradiation was a viable method for the preservation of such products.

9. BACILLUS CEREUS

The spores of *Bacillus cereus* are much more resistant to ionizing radiation than are their vegetative cells, requiring care in designing and evaluating experiments and in applying the technology, especially if the aim is to produce a shelf-stable food product. On the other hand, ionizing irradiation of dry spices, herbs, vegetable seasonings, and enzymes to eliminate pathogens and spoilage organisms is undoubtedly the largest commercial use of the technology. The reader is referred for detailed discussion and irradiation practice to the following: Farkas *(172)*, Kiss and Farkas *(173)*, and IGFI Document 5 *(174)*.

Basic research on the radiation resistance of the spores of *Bacillus* species dates back to the work of Green in 1904 *(2)*. Yamazaki et al. *(175)* discovered that the medium in which the spores were produced influenced their radiation resistance and that some of the observed variations were associated with variations in Mn and Ca in the media. Berg and

Grecz *(176)* discovered that radiation resistance of *Bacillus cereus* spores was directly related to the amount of dipicolinic acid (DPA) in the spore. Farkas and Roberts *(177)* discovered that *B. cereus* spores were more sensitive to sodium chloride when they were irradiated and then heated. Ma and Maxcy *(178)* discovered that the spores of *B. cereus* were more resistant to radiation at ambient temperature than at −30°C. Kamat and Lewis *(179)* found that the D-value for spores of *B. cereus* BIS-59 was 4 kGy and that for vegetative cells was 0.30 kGy. They did not observe a decreased D-value for DPA depleted spores. Kamat and Pradhan *(180)* found that DPA did not seem to influence the sensitivity of *B. cereus* spores to γ radiation. Hashisaka et al. *(130)* discovered that the 12-D value for the inactivation of *B. cereus* spores inoculated into ice cream or yoghurt was 43–50 kGy at −78°C. Grant and Patterson *(135)* found that the D-value for the inactivation of *B. cereus* vegetative cells in the components of a chilled ready-meal of roast beef and mashed potato was 0.126–0.288 kGy. Grant et al. *(170)* found that toxin production was delayed by a radiation dose of 2 kGy in a roast beef ready-meal inoculated with ~10^2 cells/g when temperature abused at 22°C. It is unclear in the latter study if the authors were certain that no spores were present in the inoculum. Thayer and Boyd *(181)* discovered that the D-values for *B. cereus* ATCC 33018 log-phase cells, stationary-phase cells, and spores were 0.18, 0.43, and 2.56 kGy, respectively, on mechanically deboned chicken meat when irradiated at a temperature of 5°C. The radiation resistance of vegetative cells was greater at −20 than at +20°C; however, the effect of temperature on the radiation resistance of spores was small. The radiation resistance of a mixture of the spores of six strains of *B. cereus* was 2.78 in ground beef, ground pork loin, and beef gravy.

10. *CLOSTRIDIUM*

In general the spores of *Clostridium* are very resistant to ionizing radiation, and a very high dose is required to establish a 12D inactivation of *C. botulinum*. When we are interested in the inactivation of *C. botulinum*, we are usually working with vacuum-packed products that are stored at room temperature on which there is a potential to produce botulinum toxin. Because of this potential, the recommended minimum dose for the sterilization of meat, poultry, and fish products is 45 kGy and is usually administered to vacuum-packed, enzyme-inactivated products while they are deeply frozen *(182)*. Combination treatments may be used, but the sum of the treatments must always be equivalent to canning. The microbiological and food technological conditions required to produce irradiated shelf-stable meat and poultry products was reviewed in ref. *(183)*.

Anellis et al. *(184)* established that the 12D-value for the inactivation of *C. botulinum* type A or B spores in enzyme-inactivated beef to be 43 kGy, *in vacuo*, at −30 ± 10°C. Anellis et al. *(185)* established that the 12D dose for enzyme-inactivated pork was 42.7 kGy with a shoulder of 1.1 kGy. Anellis et al. *(185)* also determined the 12D dose for enzyme-inactivated chicken to be 42.7 kGy, with a shoulder of 5.1 kGy.

11. FOODBORNE PARASITES

The incidence of foodborne disease caused by parasitic helminths and protozoa remain a significant problem worldwide *(186)*. The process of food irradiation could make a significant contribution in the control of such diseases *(186,187)*.

Though there is no explicit regulatory approval of food irradiation for the control or inactivation of protozoa in the United States, there are several studies suggesting that many could be inactivated on foods by low-dose treatment with ionizing radiation. *Entamoeba histolytica* cysts are inactivated by 0.25 kGy *(188)*. *Toxoplasma gondii* cysts are killed by ionizing radiation doses of 0.5 kGy or above *(189–193)*. Dubey et al. *(193)* demonstrated that ionizing irradiation (0.5 kGy) killed *T. gondii* oocysts on raspberries as a model for the inactivation of other coccidian parasites such as *Cyclospora* or *Cryptosporidium* on fruits and vegetables.

Trichinella spiralis infested pork can be made safe for consumption by low-dose irradiation. Brake et al. *(194)* discovered that a γ-radiation dose of 0.15 kGy prevented larval development in the intestine though it did not kill the first-stage larvae outright. The data from this study was used by the FDA in approving irradiation to a dose of 0.30 kGy to control *Trichinella spiralis* in pork carcasses or fresh, nonheat processed cuts of pork carcasses.

REFERENCES

1. Minck, F. (1896) Zur Frage über die Einwirkung der Röntgen'schen Strahlen auf Bakterien und ihre eventuelle therapeutische Verwendbarkeit (To the problem on the action of X-rays on bacteria and their possible therapeutic application). *Munch. Med.Wochenschr.* **5**, 101–102.
2. Green, A. B. (1904) A note on the action of radium on microorganisms. *Proc. R. Soc. London* **B73**, 375–381.
3. Gillett, D. C. (1918) Apparatus for preserving organic materials by the use of X-rays. US Patent 1,275,417.
4. Schwartz, B. (1921) Effects of X-rays on trichinae. *J. Agric. Res.* **20**, 845–854.
5. Department of Health and Human Services, Food and Drug Administration. (2004) 21 CFR Part 179 irradiation in the production, processing and handling of food: final rule. *Fed. Reg.* **69**, 76,844–76,847.
6. Diehl, J. F. (1995) *Safety of Irradiated Foods*, 2nd edn, Marcel Dekker, New York.
7. Joint FAO/IAEA/WHO Study Group. (1999) *High-Dose Irradiation: Wholesomeness of Food Irradiated with Doses Above 10* kGy. World Health Organization, Geneva.
8. Thayer, D. W. (1990) Food irradiation: benefits and concerns. *J. Food Qual.* **13**, 147–169.
9. Thayer, D. W., Christopher, J. P., Campbell, L. A., et al. (1987) Toxicology studies of irradiation-sterilized chicken. *J. Food Prot.* **50**, 278–288.
10. Thayer, D. W., Josephson, E. S., Brynjolfsson, A., and Giddings, G. G. (1996) Radiation pasteurization of food. *CAST Issue Paper No. 7*, Council for Agricultural Science and Technology, Ames, IA, pp. 1–10.
11. Kerr, W. (2003) Pulsed X-ray treatments of foods. *Encyclopedia of Agricultural, Food and Biological Engineering*, Marcel Dekker, New York, pp. 819–821.
12. Thayer, D. W. (2003) Ionizing irradiation, treatment of food. *Encyclopedia of Agricultural, Food, and Biological Engineering*, Marcel Dekker, New York, pp. 536–539.
13. Wilkinson, V. M. and Gould, G. W. (1996) *Food Irradiation: A Reference Guide.* Butterworth-Heinemann, Oxford.
14. USDA FSIS. (2003) 9CFR424 Code of Federal Regulations. Title 9 Animals and Animal Products, Chapter III Food Safety Inspection Service, Department of Agriculture, Part 424 Preparation and processing operations, Sections 424.21 and 424.22, US Government Printing Office: Washington, DC.
15. USDA FDA. (2003) 21CFR179 Code of Federal Regulations. Title 21 Food and Drugs, Chapter I Food and Drug Administration, Department of Health and Human Services, Part 179—Irradiation in the production, processing and handling of food, Sections 179.21, 179.25, 179.26, and 179.45, US Government Printing Office, Washington, DC.
16. USDA FSIS. (2003) 9CFR381 Code of Federal Regulations. Title 9 Animals and Animal Products, Chapter III Food Safety Inspection Service, Department of Agriculture, Part 381

Poultry products inspection regulations, Section 381.1, US Government Printing Office, Washington, DC.

17. Proctor, B. E., Nickerson, J. T. R., and Licciardello, J. J. (1956) Cathode ray irradiation of chicken meat for the extension of shelf life. *Food Res.* **21,** 11–20.

18. Thayer, D. W. (2000) Irradiation of poultry meat. Proceedings of the 35th National Meeting on Poultry Health and Processing, October 18–20, 2000, Ocean City, MD. pp. 87–93.

19. Kiss, I. and Farkas, J. (1972) Radurization of whole eviscerated chicken carcass. *Acta Aliment. Acad. Sci. Hung.* **1,** 73–86.

20. Lescano, G., Narvaiz, P., Kairiyama, E., and Kaupert, N. (1991) Effect of chicken breast irradiation on microbiological, chemical and organoleptic quality. *Lebensm. Wiss. Technol.* **24,** 130–134.

21. Lewis, S. J., Velasquez, A., Cuppett, S. L., and McKee, S. R. (2002) Effect of electron beam irradiation on poultry meat safety and quality. *Poult. Sci.* **81,** 896–903.

22. Licciardello, J. J. (1964) Effect of temperature on radiosensitivity of *Salmonella typhimurium*. *J. Food Sci.* **29,** 469–474.

23. Previte, J. J., Chang, Y., and El-Bisi, H. M. (1970) Effects of radiation pasteurization on *Salmonella*. I. Parameters affecting survival and recovery from chicken. *Can. J. Microbiol.* **16,** 465–471.

24. Licciardello, J. J., Nickerson, J. T. R., and Goldblith, S. A. (1970) Inactivation of *Salmonella* in poultry with gamma radiation. *Poult. Sci.* **49,** 663–675.

25. Mulder, R. W. A. W. (1976) Radiation inactivation of *Salmonella panama* and *Escherichia coli* K 12 present on deep-frozen broiler carcasses. *Eur. J. Appl. Microbiol.* **3,** 63–69.

26. Mulder, R. W. A. W., Notermans, S., and Kampelmacher, E. H. (1977) Inactivation of salmonellae on chilled and deep frozen broiler carcasses by irradiation. *J. Appl. Bacteriol.* **42,** 179–185.

27. Kahan, R. S. and Howker, J. J. (1978) Low-dose irradiation of fresh, non-frozen chicken and other preservation methods for shelf-life extension and for improving its public-health quality. *Food Preservation by Irradiation*, Vol. II, International Atomic Energy Agency, Vienna, pp. 221–242.

28. Mulder, R. W. A. W. (1982) The use of low temperatures and radiation to destroy *Enterobacteriaceae* and salmonellae in broiler carcasses. *J. Food Technol.* **17,** 461–466.

29. Klinger, I., Fuchs, V., Basker, D., Juven, B. J., Lapidot, M., and Eisenberg, E. (1986) Irradiation of broiler chicken meat. *Isr. J. Vet. Med.* **42,** 181–192.

30. Hanis, T., Jelen, P., Klir, P., Mñuková, M., Pérez, B., and Pesek, M. (1989) Poultry meat irradiation—effect of temperature on chemical changes and inactivation of microorganisms. *J. Food Prot.* **52,** 26–29.

31. Thayer, D. W. and Boyd, G. (1991) Effect of ionizing radiation dose, temperature, and atmosphere on the survival of *Salmonella typhimurium* in sterile, mechanically deboned chicken meat. *Poult. Sci.* **70,** 381–388.

32. Thayer, D. W. and Boyd, G. (1991) Survival of *Salmonella typhimurium* ATCC 14028 on the surface of chicken legs or in mechanically deboned chicken meat gamma irradiated in air or vacuum at temperatures of −20 to +20 degree. *Poult. Sci.* **70,** 1026–1033.

33. Anderson, E. H. (1951) The effect of oxygen on mutation induction by X-rays. *Proc. Natl Acad. Sci. USA* **37,** 340–349.

34. Anderson, R. S. and Turkowitz, H. (1941) The experimental modification of the sensitivity of yeast to roentgen rays. *Am. J. Roentgenol. Radium Therapy Nucl. Med.* **46,** 537–542.

35. Hollaender, A. and Stapleton, G. E. (1953) New aspects of the oxygen concentration effect in X-ray inactivation of bacterial suspensions. *Fed. Proc.* **12,** 70.

36. Thayer, D. W., Boyd, G., Muller, W. S., Lipson, C. A., Hayne, W. C., and Baer, S. H. (1990) Radiation resistance of *Salmonella*. *J. Ind. Microbiol.* **5,** 383–390.

37. Thayer, D. W., Songprasertchai, S., and Boyd, G. (1991) Effects of heat and ionizing radiation on *Salmonella typhimurium* in mechanically deboned chicken meat. *J. Food Prot.* **54,** 718–724.

38. Szczawińska, M. E., Thayer, D. W., and Phillips, J. G. (1991) Fate of unirradiated *Salmonella* in irradiated mechanically deboned chicken meat. *Int. J. Food Microbiol.* **14,** 313–324.

39. Previte, J. J. (1968) Immunogenicity of irradiated *Salmonella typhimurium* cells and endotoxin. *J. Bacteriol.* **95,** 2165–2170.

40. Ley, F. J., Kennedy, T. S., Kawashima, K., Roberts, D., and Hobbs, B. C. (1970) The use of gamma radiation for the elimination of *Salmonella* from frozen meat. *J. Hyg. Camb.* **68,** 293–311.

41. Previte, J. J., Chang, Y., Scrutchfield, W., El-Bisi, H. M. (1971) Effects of radiation pasteurization on *Salmonella*. II. Influence of repeated radiation-growth cycles on virulence and resistance to radiation and antibiotics. *Can. J. Microbiol.* **17,** 105–110.

42. Maxcy, R. B. (1983) Significance of residual organisms in foods after substerilizing doses of gamma radiation a review. *J. Food Saf.* **5,** 203–211.

43. Dion, P., Charbonneau, R., and Thibault, C. (1994) Effect of ionizing dose rate on the radioresistance of some food pathogenic bacteria. *Can. J. Microbiol.* **40,** 369–374.

44. Thayer, D. W., Boyd, G., Fox J. B. Jr., and Lakritz, L. (1995) Effects of NaCl, sucrose, and water content on the survival of *Salmonella typhimurium* on irradiated pork and chicken. *J. Food Prot.* **58,** 490–496.

45. Tarkowski, J. A., Stoffer, S. C. C., Beumer, R. R., and Kampelmacher, E. H. (1984) Low dose gamma irradiation of raw meat. I. Bacteriological and sensory quality effects in artificially contaminated samples. *Int. J. Food Microbiol.* **1,** 13–23.

46. Patterson, M. (1988) Sensitivity of bacteria to irradiation on poultry meat under various atmospheres. *Lett. Appl. Microbiol.* **7,** 55–58.

47. Rajkowski, K. T. and Thayer, D. W. (2000) Reduction of *Salmonella* spp. and strains of *Escherichia coli* O157:H7 by gamma radiation of inoculated sprouts. *J. Food Prot.* **63,** 871–875.

48. Thayer, D. W., Rajkowski, K. T., Boyd, G., Cooke, P. H., and Soroka, D. S. (2003) Inactivation of *Escherichia coli* O157:H7 and *Salmonella* by gamma irradiation of alfalfa seed intended for production of food sprouts. *J. Food Prot.* **66,** 175–181.

49. Thayer, D. W., Boyd, G., Fox, J. B. Jr., Lakritz, L., and Hampson, J. W. (1995) Variations in radiation sensitivity of foodborne pathogens associated with the suspending meat. *J. Food Sci.* **60,** 63–67.

50. Thayer, D. W., Boyd, G., Fox, J. B. Jr., and Lakritz, L. (1997) Elimination by gamma irradiation of *Salmonella* spp. and strains of *Staphylococcus aureus* inoculated in bison, ostrich, alligator, and caiman meat. *J. Food Prot.* **60,** 756–760.

51. Clavero, M. R. S., Monk, J. D., Beuchat, L. R., Doyle, M. P., and Brackett, R. E. (1994) Inactivation of *Escherichia coli* O157:H7, Salmonellae, and *Campylobacter jejuni* in raw ground beef by gamma irradiation. *Appl. Environ. Microbiol.* **60,** 2069–2075.

52. Proctor, B. E., Joslin, R. P., Nickerson, J. T. R., and Lockhart, E. E. (1953) Elimination of *Salmonella* in whole egg powder by cathode ray irradiation of egg magma prior to drying. *Food Technol.* **7,** 291–296.

53. Ball, H. R. and Gardner, F. A. (1968) Physical and functional properties of gamma-irradiated liquid egg white. *Poult. Sci.* **47,** 1481–1487.

54. Katušin-Ražem, B., Ražem, D., Matic, S., Mihoković, V., Kostromin-Šooš, N., and Milanović, N. (1989) Chemical and organoleptic properties of irradiated dried whole egg and egg yolk. *J. Food Prot.* **52,** 781–786.

55. Ma, C.-Y., Sahasrabudhe, M. R., Poste, L. M., Harwalkar, V. R., Chambers, J. R., and O'Hara, K. P. J. (1990) Gamma irradiation of shell eggs. Internal and sensory quality,

physicochemical characteristics, and functional properties. *Can. Inst. Food Sci. Technol. J.* **23,** 226–232.

56. Katušin-Ražem, B., Mihaljević, B., and Ražem, D. (1992) Radiation-induced oxidative chemical changes in dehydrated egg products. *J. Agric. Food Chem.* **40,** 662–668.

57. Ma, C. Y., Harwalkar, V. R., Poste, L. M., and Sahasrabudhe, M. R. (1993) Effect of gamma irradiation on the physicochemical and functional properties of frozen liquid egg products. *Food Res. Int.* **26,** 247–254.

58. Kijowski, J., Lesnierowski, G., Zabielski, J., Fiszer, W., and Magnuski, T. (1994) Chapter 28 Radiation pasteurization of frozen whole egg. In: *Egg Uses and Processing Technologies New Developments* (Sim, J. S. and Nakai, S., ed.), CAB International, Oxon, UK, pp. 340–348.

59. Matić, S., Mihoković, V., Katušin-Ražem, B., and Ražem, D. (1990) The eradication of *Salmonella* in egg powder by gamma irradiation. *J. Food Prot.* **53,** 111–114.

60. Serrano, L. E., Murano, E. A., Shenoy, K., and Olson, D. G. (1997) D Values of *Salmonella enteritidis* isolates and quality attributes of shell eggs and liquid whole eggs treated with irradiation. *Poult. Sci.* **76,** 202–206.

61. Schaffner, D., Hamdy, M. K., Toledo, R. T., and Tift, M. L. (1989) *Salmonella* inactivation in liquid whole egg by thermoradiation. *J. Food Sci.* **54,** 902–905.

62. Beuchat, L. R. (1996) Pathogenic microorganisms associated with fresh produce. *J. Food Prot.* **59,** 204–216.

63. Thayer, D. W. and Rajkowski, K. T. (1999) Developments in irradiation of fresh fruits and vegetables. *Food Technol.* **53,** 62–65.

64. Thomas, P. (1984) Radiation preservation of foods of plant origin. I. Potatoes and other tuber crops. *CRC Crit. Rev. Food Sci. Nutr.* **19,** 327–379.

65. Thomas, P. (1984) Radiation preservation of food of plant origin. II. Onions and other bulb crops. *CRC Crit. Rev. Food Sci. Nutr.* **21,** 95–136.

66. Thomas, P. (1986) Radiation of foods of plant origin. III. Tropical fruits: bananas, mangoes, and papayas. *CRC Crit. Rev. Food Sci. Nutr.* **23,** 147–205.

67. Thomas, P. (1986) Radiation of foods of plant origin. IV. Subtropical fruits: citrus, grapes and avocados. *CRC Crit Rev. Food Sci. Nutr.* **24,** 53–89.

68. Thomas, P. (1986) Radiation preservation of foods of plant origin. V. Temperate fruits: pome fruits, stone fruits, and berries. *CRC Crit. Rev. Food Sci. Nutr.* **24,** 357–400.

69. Thomas, P. (1988) Radiation preservation of foods of plant origin. VI. Mushrooms, tomatoes, minor fruits and vegetables, dried fruits, and nuts. *CRC Crit. Rev. Food Sci. Nutr.* **26,** 313–357.

70. Willemont, C., Marcotte, M., and Deschenes, L. (1996) Ionizing irradiation for preservation of fruits. In: *Processing Fruits: Science and Technology. Vol. 1. Biology, Principles and Applications* (Somogyi, L. P. and Ramaswarmy, H., eds.), Chapter 9, Tecnomic, Lancaster, PA, pp. 221–260.

71. Hagenmaier, R. D. and Baker, R. A. (1997) Low-dose irradiation of cut iceberg lettuce in modified atmosphere packaging. *J. Agric. Food Chem.* **45,** 2864–2868.

72. Rajkowski, K. T. and Thayer, D. W. (2001) Alfalfa seed germination and yield ratio and alfalfa sprout microbial keeping quality following irradiation of seeds and sprouts. *J. Food Prot.* **64,** 1988–1995.

73. Rajkowski, K. T., Boyd, G., and Thayer, D. W. (2003) Irradiation *D*-values for *Escherichia coli* O157:H7 and *Salmonella* sp. on inoculated broccoli seeds and effects of irradiation on broccoli sprout keeping quality and seed viability. *J. Food Prot.* **66,** 760–766.

74. Thayer, D. W., Boyd, G., and Fett, W. F. (2003) λ-Radiation decontamination of alfalfa seeds naturally contaminated with *Salmonella* Mbandaka. *J. Food Sci.* **68,** 1777–1781.

75. Niemira, B. A. (2003) Radiation sensitivity and recoverability of *Listeria monocytogenes* and *Salmonella* on 4 lettuce types. *J. Food Sci.* **68,** 2784–2787.

76. Bregvadze, U. D. (1963) Preserving fruit juices by the combined effects of gamma irradiation and sorbic acid. *Navy Fiz. Metody Obrabotki Pishch. Produktov, Kiev, Sb.* **1963,** 199–205.

77. Chachin, K. and Ogata, K. (1969) Changes in chemical constituents and quality of some juices irradiated with the sterilizing dose levels of gamma rays. *Food Irradiat. (Shokuhin Shosha)* **4,** 85–90.

78. Chuaqui-Offermanns, N. and McDougall, T. (1991) Effects of heat, radiation and their combination. *Rad. Phys. Chem.* **38,** 425–427.

79. Fetter, F., Stehlik, G., Kovacs, J., and Weiss, S. (1969) Das Flavourverhalten einiger Gamma-bestrahlter Fruchtsaefte. (Flavor changes in gamma-irradiated fruit juices.) *Mitteilungen: Rebe, Wein, Obstbau und Fruechteverwertung* **19,** 140–151.

80. Gagnon, M., Julien, J. P., and Riel, R. R. (1968) Irradiation of apple juice. I. Effects of gamma rays on the rate of filtration and the viscosity. *Can. Inst. Food Tech. J.* **1,** 117–122.

81. Gasco, L., Barrera, R., and Cruz, F. (1970) Aroma modification in irradiated fruit juices. Evolution of carbonyl compounds and volatile alcohols in apple juice treated with various doses of gamma-rays. *Rev. Agro. Tecnol. Aliment.* **10,** 105–116.

82. Kaupert, N. L., Lescano, H. G., and Kotliar, N. (1981) Conservation of apple and pear juice concentrates. Synergic effect of heat and radiation. In: *Combination Processes in Food Irradiation*, International Atomic Energy Agency, Vienna, pp. 205–216.

83. Kiss, I. and Farkas, J. (1968) Radiation sterilization of freeze-concentrated apple juice. *Elelmiszertudomany* **2,** 67–75.

84. Kiss, I. and Farkas, J. (1970) Ueber die Wirkung der ionisierende Strahlung auf den gefrierkonzentrierten Apfelsaft. (The effect of ionizing radiation on freeze-concentrated apple juice.) *Mitteilungen: Rebe, Wein, Obstbau und Fruechteverwertung* **20,** 296–304.

85. Kiss, I. and Farkas, J. (1973) Combination of cryscentration with ionizing irradiation for the preservation of fruit-juice. *Kiserl. Kozl. Budap.* **63(E),** 3–18.

86. Obara, T., Shimotsuura, A., Shimazu, F., and Watanabe, W. (1958) Preservation of vegetable foods irradiated with radioactive rays. I. Influence of γ-ray irradiation upon components contained in citrus juice. II. Influence of γ-ray irradiation on vitamin C contained in citrus juice under various conditions. *Radioisotopes (Tokyo)* **7,** 127–132; 133–140.

87. Proctor, B. E. and O'Meara, J. P. (1951) Effect of high-voltage cathode rays on ascorbic acid. *Ind. Eng. Chem.* **43,** 718–721.

88. Cook, K. A. (1998) Outbreak of *Salmonella* Serotype Hartford infections associated with unpasteurized orange juice. *JAMA* **280,** 1504–1509.

89. Tauxe, R., Kruse, H., Hedberg, C., Potter, M., Madden, J., and Wachsmuth, K. (1997) Microbial hazards and emerging issues associated with produce: a preliminary report to the National Advisory Committee on Microbiological Criteria for Foods. *J. Food Prot.* **60,** 1400–1408.

90. National Advisory Committee on Microbiological Criteria for Food. (1997) Recommendations for controlling the transmission of pathogenic microorganisms in juices. Food Safety and Inspection Service, US Department of Agriculture, Washington, DC.

91. Niemira, B. A. (2001) Citrus juice composition does not influence radiation sensitivity of *Salmonella* Enteritidis. *J. Food Prot.* **64,** 869–872.

92. Niemira, B. A., Sommers, C. H., and Boyd, G. (2001) Irradiation inactivation of four *Salmonella* serotypes in orange juices with various turbidities. *J. Food Prot.* **64,** 614–617 (Erratum **64,** 872).

93. Altekruse, S. F., Stern, N. J., Fields, P. I., and Swerdlow, D. L. (1999) *Campylobacter jejuni* — An emerging foodborne pathogen. *Emerg. Inf. Dis.* **5,** 28–35.

94. Lambert, J. D. and Maxcy, R. B. (1984) Effect of gamma radiation on *Campylobacter jejuni*. *J. Food Sci.* **49,** 665–667, 674.

95. Yogasundram, K., Shane, S. M., Grodner, R. M., Lambremont, E. N., and Smith, R. E. (1987) Decontamination of *Campylobacter jejuni* on chicken drumsticks using chemicals and radiation. *Vet. Res. Commun.* **11,** 31–40.

96. Patterson, M. F. (1995) Sensitivity of *Campylobacter* spp. to irradiation in poultry meat. *Lett. Appl. Microbiol.* **20,** 338–340.

97. Collins, C. I., Murano, E. A., and Wesley, I. V. (1996) Survival of *Arcobacter butzleri* and *Campylobacter jejuni* after irradiation treatment in vacuum-packaged ground pork. *J. Food Prot.* **59,** 1164–1166.

98. Roepke, R. R. and Mercer, F. E. (1947) Lethal and sublethal effects of X-rays on *Escherichia coli* as related to the yield of biochemical mutants. *J. Bacteriol.* **54,** 731–743.

99. Billen, D., Stapleton, G. E., and Hollaender, A. (1953) The effect of X-irradiation on the respiration of *Escherichia coli*. *J. Bacteriol.* **65,** 131–135.

100. Anderson, E. H. and Billen, D. (1955) The effect of temperature on X-ray induced mutability in *Escherichia coli*. *J. Bacteriol.* **70,** 35–43.

101. Fram, H., Proctor, B. E., and Dunn, C. G. (1950) Effects of X-rays produced at 50 kilowatts on different species of bacteria. *J. Bacteriol.* **60,** 263–267.

102. Stapleton, G. E. (1955) Variations in the sensitivity of *Escherichia coli* to ionizing radiations during the growth cycle. *J. Bacteriol.* **70,** 357–362.

103. Thayer, D. W. and Boyd, G. (1993) Elimination of *Escherichia coli* O157:H7 in meats by gamma irradiation. *Appl. Environ. Microbiol.* **59,** 1030–1034.

104. Thayer, D. W. and Boyd, G. (2001) Effect of irradiation temp on inactivation of *Escherichia coli* O157:H7 and *Staphylococcus aureus*. *J. Food Prot.* **64,** 1624–1626.

105. Ito, H. H. (1998) Irradiation effect of *Escherichia coli* O157:H7 in meats. *Food Irradiat. Jpn.* **33,** 29–32.

106. Buchanan, R. L., Edelson, S. G., Snipes, K., and Boyd, G. (1998) Inactivation of *Escherichia coli* O157:H7 in apple juice by irradiation. *Appl. Environ. Microbiol.* **64,** 4533–4535.

107. Fan, X. and Thayer, D. W. (2002) γ-Radiation influences browning, antioxidant activity, and malondialdehyde level of apple juice. *J. Agric. Food Chem.* **50,** 710–715.

108. Fan, X. T., Thayer, D. W., and Handel, A. P. (2002) Nutritional quality of irradiated orange juice. *J. Food Proc. Pres.* **26,** 195–211.

109. Fan, X. and Thayer, D. W. (2001) Quality of irradiated alfalfa sprouts. *J. Food Prot.* **64,** 1574–1578.

110. Fan, X., Rajkowski, K., and Thayer, D. W. (2003) Quality of alfalfa sprouts grown from irradiated seeds. *J. Food Qual.* **26,** 165–176.

111. Bari, M. L., Nazuka, E., Sabina, Y., Todoriki, S., and Isshiki, K. (2003) Chemical and irradiation treatments for killing *Escherichia coli* O157:H7 on alfalfa, radish, and Mung bean seeds. *J. Food Prot.* **66,** 767–774.

112. Niemira, B. A., Sommers, C. H., and Fan, X. T. (2002) Suspending lettuce type influences recoverability and radiation sensitivity of *Escherichia coli* O157:H7. *J. Food Prot.* **65,** 1388–1393.

113. Foley, D. M., Dufour, A., Rodriguez, L., Caporaso, F., and Prakash, A. (2002) Reduction of *Escherichia coli* O157:H7 in shredded iceberg lettuce by chlorination and gamma irradiation. *Rad. Phys. Chem.* **63,** 391–396.

114. Glass, K. A. and Doyle, M. P. (1989) Fate of *Listeria monocytogenes* in processed meat products during refrigerated storage. *Appl. Environ. Microbiol.* **55,** 1565–1569.

115. Pini, P. N. and Gilbert, R. J. (1988) The occurrence in the UK of *Listeria* species in raw chickens and soft cheese. *Int. J. Food Microbiol.* **6,** 317–326.

116. WHO. (1988) Foodborne listeriosis. Report of a World Health Organization Informal Working Group. *WHO/EFE/FOS* 88.5, February 15–19, 1988, Geneva.

117. Huhtanen, C. N., Jenkins, R. K., and Thayer, D. W. (1989) Gamma radiation sensitivity of *Listeria monocytogenes*. *J. Food Prot.* **52,** 610–613.

118. El Shenawy, M. A., Yousef, A. E., and Marth, E. H. (1989) Radiation sensitivity of *Listeria monocytogenes* in broth or in raw ground beef. *Lebens. Wissensch. Technol.* **22,** 387–390.

119. Patterson, M. F., Damoglou, A. P., and Buick, R. K. (1993) Effects of irradiation dose and storage temperature on the growth of *Listeria monocytogenes* on poultry meat. *Food Microbiol.* **10,** 197–203.

120. Farag, M. D. E.-D. H., Shamsuzzaman, K., and Borsa, J. (1990) Radiation sensitivity of *Listeria monocytogenes* in phosphate buffer, trypticase soy broth, and poultry feed. *J. Food Prot.* **53,** 648–651.

121. Lewis, S. J. and Corry, J. E. L. (1991) Survey of incidence of *Listeria monocytogenes* and other *Listeria* spp. in experimentally irradiated and in matched unirradiated raw chickens. *Int. J. Food Microbiol.* **12,** 257–262.

122. Tarte, R., Murano, E. A., and Olson, D. G. (1996) Survival and injury of *Listeria monocytogenes, Listeria innocua* and *Listeria ivanovii* in ground pork following electron beam irradiation. *J. Food Prot.* **59,** 596–600.

123. Monk, J. D., Clavero, M. R. S., Beuchat, L. R., Doyle, M. P., and Brackett, R. E. (1994) Irradiation inactivation of *Listeria monocytogenes* and *Staphylococcus aureus* in low- and high-fat, frozen and refrigerated ground beef. *J. Food Prot.* **57,** 969–974.

124. Andrews, L. S., Marshall, D. L., and Grodner, R. M. (1995) Radiosensitivity of *Listeria monocytogenes* at various temperatures and cell concentrations. *J. Food Prot.* **58,** 748–751.

125. Thayer, D. W. and Boyd, G. (1995) Radiation sensitivity of *Listeria monocytogenes* on beef as affected by temperature. *J. Food Sci.* **60,** 237–240.

126. Kamat, A. S. and Nair, M. P. (1995) Gamma irradiation as a means to eliminate *Listeria monocytogenes* from frozen chicken meat. *J. Sci. Food Agric.* **69,** 415–422.

127. Grant, I. R. and Patterson, M. F. (1995) Combined effect of gamma radiation and heating on the destruction of *Listeria monocytogenes* and *Salmonella typhimurium* in cook-chill roast beef and gravy. *Int. J. Food Microbiol.* **27,** 117–128.

128. Bougle, D. L. and Stahl, V. (1994) Survival of *Listeria monocytogenes* after irradiation treatment of Camembert cheeses made from raw milk. *J. Food Prot.* **57,** 811–813.

129. Hashisaka, A. E., Weagant, S. D., and Dong, F. M. (1989) Survival of *Listeria monocytogenes* in mozzarella cheese and ice cream exposed to gamma irradiation. *J. Food Prot.* **52,** 490–492.

130. Hashisaka, A. E., Matches, J. R., Batters, Y., Hungate, F. P., and Dong, F. M. (1990) Effects of gamma-irradiation at −78°C on microbial populations in dairy products. *J. Food Sci.* **55,** 1284–1289.

131. Hashisaka, A. E., Einstein, M. A., Rasco, B. A., Hungate, F. P., and Dong, F. M. (1990) Sensory analysis of dairy products irradiated with cobalt-60 at −78°C. *J. Food Sci.* **55,** 404–408, 412.

132. Ennahar, S., Kuntz, F., Strasser, A., Bergaentzle, M., Hasselmann, C., and Stahl, V. (1994) Elimination of *Listeria monocytogenes* in soft and red smear cheeses by irradiation with low energy electrons. *Int. J. Food Sci. Tech.* **29,** 395–403.

133. Grant, I. R. and Patterson, M. F. (1991) Effect of irradiation and modified atmosphere packaging on the microbiological safety of minced pork stored under temperature abuse conditions. *Int. J. Food Sci. Tech.* **26,** 521–533.

134. Thayer, D. W. and Boyd, G. (1999) Irradiation and modified atmosphere packaging for the control of *Listeria monocytogenes* on turkey meat. *J. Food Prot.* **62,** 1136–1142.

135. Grant, I. R. and Patterson, M. F. (1992) Sensitivity of foodborne pathogens to irradiation in the components of a chilled ready meal. *Food Microbiol.* **9,** 95–103.

136. Grant, I. R., Nixon, C. R., and Patterson, M. F. (1993) Comparison of the growth of *Listeria monocytogenes* in unirradiated and irradiated cook-chill roast beef and gravy at refrigeration temperatures. *Lett. Appl. Microbiol.* **17,** 55–57.

137. Thayer, D. W., Boyd, G., Kim, A., Fox J. B. Jr., and Farrell, H. M. (1998) Fate of gamma-irradiated *Listeria monocytogenes* during refrigerated storage on raw or cooked turkey breast meat. *J. Food Prot.* **61,** 979–987.

138. Clardy, S., Foley, D. M., Caporaso, F., Calicchia, M. L., and Prakash, A. (2002) Effect of gamma irradiation on *Listeria monocytogenes* in frozen, artificially contaminated sandwiches. *J. Food Prot.* **65,** 1740–1744.

139. Sommers, C. H. and Thayer, D. W. (2000) Survival of surface-inoculated *Listeria mono-cytogenes* on commercially available frankfurters following gamma irradiation. *J. Food Sci.* **20,** 27–137.

140. Sommers, C. H., Handel, A. P., and Niemira, B. A. (2002) Radiation resistance of *Listeria monocytogenes* in the presence or absence of sodium erythorbate. *J. Food Sci.* **67,** 2266–2270.

141. Sommers, C. H. and Fan, X. T. (2002) Antioxidant power, lipid oxidation, color, and viability of *Listeria monocytogenes* in beef bologna treated with gamma radiation and containing various levels of glucose. *J. Food Prot.* **65,** 1750–1755.

142. Sommers, C. and Fan, X. (2003) Gamma irradiation of fine-emulsion sausage containing sodium diacetate. *J. Food Prot.* **66,** 819–824.

143. Sommers, C., Fan, X. T., Niemira, B. A., and Sokorai, K. (2003) Radiation (gamma) resistance and postirradiation growth of *Listeria monocytogenes* suspended in beef bologna containing sodium diacetate and potassium lactate. *J. Food Prot.* **66,** 2051–2056.

144. Sommers, C. H., Fan, X. T., Handel, A. P., and Sokorai, K. B. (2003) Effect of citric acid on the radiation resistance of *Listeria monocytogenes* and frankfurter quality factors. *Meat Sci.* **63,** 407–415.

145. Farkas, J., Saray, T., Mohacsi-Farkas, C., Horti, K., and Andrassy, E. (1997) Effects of low-dose gamma radiation on shelf-life and microbiological safety of pre-cut-prepared vegetables. *Adv. Food Sci.* **19,** 111–119.

146. Niemira, B. A., Fan, X. T., and Sommers, C. H. (2002) Irradiation temperature influences product quality factors of frozen vegetables and radiation sensitivity of inoculated *Listeria monocytogenes*. *J. Food Prot.* **65,** 1406–1410.

147. Niemira, B. A., Fan, X., Sokorai, K. J. B., and Sommers, C. H. (2003) Ionizing radiation sensitivity of *Listeria monocytogenes* ATCC 49594 and *Listeria innocua* ATCC 51742 inoculated on endive (Cichoriumendiva). *J. Food Prot.* **66,** 993–998.

148. Tamplin, M. L. and Capers, G. M. (1992) Persistence of *Vibrio vulnificus* in tissues of Gulf Coast oysters, *Crassostrea virginica*, exposed to seawater disinfected with UV light. *Appl. Environ. Microbiol.* **58,** 1506–1510.

149. Matches, J. R. and Liston, J. (1971) Radiation destruction of *Vibrio parahaemolyticus*. *J. Food Sci.* **36,** 339–340.

150. Campanini, M., Zupan, J., Pani, L., and Vicini, E. (1974) Resistance of *Vibrio para-haemolyticus* and *V. cholerae* to unfavourable conditions. *Ind. Conserv.* **49,** 170–172.

151. Bandekar, J. R., Chander, R., and Nerkar, D. P. (1987) Radiation control of *Vibrio para-haemolyticus* in shrimp. *J. Food Saf.* **8,** 83–88.

152. Ito, H., Sangthong, N., and Ishigaki, I. (1988) Effect of gamma-irradiation on frozen shrimps: inactivation of microorganisms and shelf-life extension of defrosted shrimps. *Food Irradiat. (Shokuhin Shosha)* **23,** 72–76.

153. Hau, L. B., Liew, M. H., and Yeh, L. T. (1992) Preservation of grass prawns by ionizing radiation. *J. Food Prot.* **55,** 198–202.

154. Sang, F. C., Hugh-Jones, M. E., and Hagstad, H. V. (1987) Viability of *Vibrio cholerae* 01 on frog legs under frozen and refrigerated conditions and low dose radiation treatment. *J. Food Prot.* **50,** 662–664.

155. Ama, A. A., Hamdy, M. K., and Toledo, R. T. (1994) Effects of heating, pH and thermo-radiation on inactivation of *Vibrio vulnificus*. *Food Microbiol.* **11,** 215–227.

156. Kwon, O. H. and Byun, M. W. (1996) The combined effect of heat and gamma irradiation on the inactivation of selected microorganisms associated with food hygiene. *J. Korean Soc. Food Sci. Nutr.* **25,** 804–809.

157. de Moraes, I. R., Del Mastro, N. L., Jakabi, M., and Gelli, D. S. (2000) Radiosensitivity of *Vibrio cholerae* O1 incorporated in oysters, to (60)CO. *Rev Saude Publ.* **34,** 29–32 (in Portuguese).

158. Jakabi, M., Gelli, D. S., Torre, J. C., et al. (2003) Inactivation by ionizing radiation of *Salmonella enteritidis, Salmonella infantis*, and *Vibrio parahaemolyticus* in oysters (*Crassostrea brasiliana*). *J Food Prot.* **66,** 1025–1029.

159. El Zawahry, Y. A. and Rowley, D. B. (1979) Radiation resistance and injury of *Yersinia enterocolitica*. *Appl. Environ. Microbiol.* **37**, 50–54.
160. Fu, A.-H., Sebranek, J. G., and Murano, E. A. (1995) Survival of *Listeria monocytogenes*, *Yersinia enterocolitica* and *Escherichia coli* O157:H7 and quality changes after irradiation of beef steaks and ground beef. *J. Food Sci.* **60**, 972–977.
161. Kamat, A. S., Khare, S., Doctor, T., and Nair, P. M. (1997) Control of *Yersinia enterocolitica* in raw pork and pork products by γ-irradiation. *Int. J. Food Microbiol.* **36**, 69–76.
162. Shenoy, K., Murano, E. A., and Olson, D. G. (1998) Survival of heat-shocked *Yersinia enterocolitica* after irradiation in ground pork. *Int. J. Food Microbiol.* **39**, 133–137.
163. Sommers, C. H., Niemira, B. A., Tunick, M., and Boyd, G. (2002) Effect of temperature on the radiation resistance of virulent *Yersinia enterocolitica*. *Meat Sci.* **61**, 323–328.
164. Sommers, C. H. and Bhaduri, S. (2001) Loss of crystal violet binding activity in stationary phase *Yersinia enterocolitica* following gamma irradiation. *Food Microbiol.* **18**, 367–374.
165. Sommers, C. H. and Novak, J. S. (2002) Radiation resistance of virulence plasmid-containing and plasmid-less *Yersinia enterocolitica*. (Research Note) *J. Food Prot.* **65**, 556–559.
166. Chambers, H. and Russ, S. (1912) The bactericidal action of radium emanation. *Proc. R. Soc. London* **B5**, 198–212.
167. Baker, S. L. (1935) A quantitative comparison of the effects of the beta rays of radium on the agent of the Rous sarcoma, on the bacteriophage, on tetanus toxin and on certain bacteria, antibodies and ferments. *Br. J. Exp. Pathol.* **16**, 148–155.
168. Licciardello, J. J., D'Entremont, D. L., and Lundstrom, R. C. (1989) Radio-resistance of some bacterial pathogens in soft-shell clams (*Mya arenaria*) and mussels (*Mytilus edulis*). *J. Food Prot.* **52**, 407–411.
169. Thayer, D. W. and Boyd, G. (1992) Gamma ray processing to destroy *Staphylococcus aureus* in mechanically deboned chicken meat. *J. Food Sci.* **57**, 848–851.
170. Grant, I. R., Nixon, C. R., and Patterson, M. F. (1993) Effect of low-dose irradiation on growth of and toxin production by *Staphylococcus aureus* and *Bacillus cereus* in roast beef and gravy. *Int. J. Food Microbiol.* **18**, 25–36.
171. Lamb, J. L., Gogley, J. M., Thompson, M. J., Solis, D. R., and Sen, S. (2002) Effect of low-dose gamma irradiation on *Staphylococcus aureus* and product packaging in ready-to-eat ham and cheese sandwiches. *J. Food Prot.* **65**, 1800–1805.
172. Farkas, J. (1983) Radurization and radicidation: spices. In: *Preservation of Food By Ionizing Radiation* (Josephson, E. S. and Peterson, M. S. eds.), CRC, Boca Raton, FL.
173. Kiss, I. and Farkas, J. (1988) Irradiation as a method for decontamination of spices. *Food Rev. Int.* **4**, 77–92.
174. IGFI. (1991) Code of good irradiation practice for the control of pathogens and other microflora in spices, herbs and other vegetable seasonings. *Document No. 5*, International Consultative Group on Food Irradiation, Vienna, pp. 1–12.
175. Yamazaki, K., Ito, N., Sato, K., and Oka, M. (1968) Effects of growth media composition on radioresistance of bacterial spores. *Food Irradiat. (Shokuhin Shosha)* **3**, 13–19.
176. Berg, P. E. and Grecz, N. (1970) Relationship of dipicolinic acid content in spores of *Bacillus cereus* T to ultraviolet and gamma radiation resistance. *J. Bacteriol.* **103**, 517–519.
177. Farkas, J. and Roberts, T. A. (1976) The effect of sodium chloride, gamma irradiation and/or heating on germination and development of spores of *Bacillus cereus* T in single germinants and complex media. *Acta Aliment.* **5**, 289–302.
178. Ma, K. and Maxcy, R. B. (1981) Factors influencing radiation resistance of vegetative bacteria and spores associated with radappertization of meat. *J. Food Sci.* **46**, 612–616.
179. Kamat, A. S. and Lewis, N. F. (1982) Influence of heat and radiation on the germinability and viability of *B. cereus* BIS-59 spores. *Indian J. Microbiol.* **23**, 198–202.
180. Kamat, A. S. and Pradhan, D. S. (1987) Involvement of calcium and dipicolinic acid in the resistance of *Bacillus cereus* BIS-59 spores to u.v. and gamma radiations. *Int. J. Radiat. Biol.* **51**, 7–18.

181. Thayer, D. W. and Boyd, G. (1994) Control of enterotoxic *Bacillus cereus* on poultry or red meats and in beef gravy by gamma irradiation. *J. Food Prot.* **57,** 758–764.
182. IAEA. (1995) Shelf-stable foods through irradiation processing. *IAEA-TECDOC-843.* International Atomic Energy Agency, Vienna.
183. Thayer, D. W. (2001) Development of irradiated shelf-stable meat and poultry products. In: *Food Irradiation: Principles and Applications* (Molins, R. A., ed.), Chapter 13, Wiley Interscience, New York, pp. 329–345.
184. Anellis, A., Shattuck, E., Rowley, D. B., Ross, E. W. Jr., Whaley, D. N., and Dowell, V. R. Jr. (1975) Low-temperature irradiation of beef and methods for evaluation of a radappertization process. *Appl. Microbiol.* **30,** 811–820.
185. Anellis, A., Shattuck, E., Morin, M., et al. (1977) Cryogenic gamma irradiation of prototype pork and chicken and antagonistic effect between *Clostridium botulinum* types A and B. *Appl. Environ. Microbiol.* **34,** 823–831.
186. Roberts, T., Murrell, K. D., and Marks, S. (1994) Economic losses caused by foodborne parasitic diseases. *Parasitol. Today* **10,** 419–423.
187. Loaharanu, P. and Murrell, D. (1994) A role for irradiation in the control of foodborne parasites. *Food Sci. Technol.* **5,** 190–195.
188. Schneider, C. R. (1960) Radiosensitivity of *Entamoeba histolytica* cysts. *Exp. Parasitol.* **9,** 87–91.
189. Dubey, J. P., Brake, R. J., Murrell, K. D., and Fayer, R. (1986). Effect of irradiation on the viability of *Toxoplasma gondii* cysts in tissues of mice and pigs. *Am. J. Vet. Res.* **47,** 518–522.
190 Dubey, J. P. and Thayer, D. W. (1994) Killing of *Toxoplasma gondii* tissue cysts by irradiation under defined conditions. *J. Parasitol.* **80,** 764–767.
191. Dubey, J. P., Jenkins, M. C., and Thayer, D. W. (1996) Irradiation killing of *Toxoplasma gondii* oocysts. *J. Eukaryot. Microbiol* **45,** S 123.
192. Dubey, J. P., Jenkins, M. C., Thayer, D. W., Kwok, O. C. H., and Shen, S. K. (1996). Killing of *Toxoplasma gondii* oocysts by irradiation and protective immunity induced by vaccination with irradiated oocysts. *J. Parasitol.* **82,** 724–727.
193. Dubey, J. P., Thayer, D. W., Speer, C. A., and Shen, S. K. (1998) Effect of gamma irradiation on unsporulated and sporulated *Toxoplasma gondii* oocysts. *Int. J. Parasitol.* **28,** 369–375.
194. Brake, R. J., Murrell, K. D., Ray, E. E., Thomas, J. D., Muggenburg, B. A., and Sivinski, J. S. (1985) Destruction of *Trichinella spiralis* by low-dose irradiation of infected pork. *J. Food Saf.* **7,** 127–143.

20

Molecular Techniques of Detection and Discrimination of Foodborne Pathogens and Their Toxins

Steven L. Foley and Kathie Grant

1. INTRODUCTION

The ability to detect foodborne pathogens and their toxins is important in reducing the role that these pathogens play in human disease. In order to ensure the safety of food and thereby to protect the public health, there is an ever-increasing demand from the food industry, the food consumer, and public-health authorities for more rapid methods to detect foodborne pathogens. Along with pathogen detection, the ability to distinguish between individual pathogens and their products is essential for identifying and understanding the sources of pathogens implicated in diseases. Taken together, discriminatory identification and typing methods provide valuable information that allow scientists to further understand how pathogens pass through the food chain and contribute to infection. To help understand the methods available for the detection and discrimination of foodborne pathogens, this chapter provides coverage of methods for rapid pathogen detection, the detection of foodborne toxins, and an examination of methods that can be used to distinguish among foodborne pathogens.

2. RAPID METHODS FOR FOOD PATHOGEN DETECTION

Traditional microbiological detection and identification methods for foodborne pathogens are well known for being both time consuming and laborious to perform, and are increasingly seen as being unable to meet the demands for rapid food testing. A rapid method is generally characterized as a test giving quicker results than the standard accepted method of isolation and biochemical and/or serological identification *(1)*. The demand for speedy results combined with major advances in a range of technologies has led to a vast array of rapid methods being developed and investigated over the past two decades. The use of such methods, either in a laboratory or for on-site testing of food production premises, depends not only on the time taken to get results but on a range of other factors such as robustness, sensitivity, specificity, accuracy, reliability, standardization, evaluation, throughput, ease of use, potential for automation, as well as cost, convenience, validation, and the throughput of the end user. The major advances in rapid methods have, in general, been in immunological and nucleic acid-based detection systems, although significant developments have been made and are being made in

From: *Infectious Disease: Foodborne Diseases*
Edited by: S. Simjee © Humana Press Inc., Totowa, NJ

other areas including impedance and conductance, bacteriophages, biosensors, microscopy as well as in miniaturized, automated biochemical detection kits. Many of the methods, as well as being used on their own for rapid detection, may also be used in combination, for example, immunomagnetic separation followed by bioluminescent detection. This has led to an almost exponential increase in the number of rapid methods being reported for detecting foodborne pathogens.

2.1. Application to Foods

Because of interference from the food matrix, almost all new rapid methods for food pathogen detection are performed following enrichment culture. Although this overcomes many of the problems associated with their application directly to food samples, it still necessitates a time-consuming incubation step, typically requiring at least 8–24 h. A number of separation and concentration techniques to isolate pathogens or their nucleic acids from foods have been investigated in order to overcome this lengthy step. Major difficulties in detecting pathogens directly in foods arise from the diverse range of food matrices that may be encountered, the inhibitory nature of the food constituents to various detection assays, and the necessity to ensure the absence of microorganisms in 25 g of food *(1)*. A further problem is the need of the ability to detect viable organisms and not those that have been destroyed by food processing. All of these points must be addressed in a simple, robust, rapid, generic method, and one that is amenable to automation, before the direct detection of pathogens in foods becomes a reality.

A variety of techniques have been investigated to overcome cultural enrichment, including centrifugation, filtration, aqueous polymer two-phase separation systems, ultrasound, immunomagnetic, bacteriophage, and lectin separation as well as dielectrophoresis *(2)*. A number have been shown to improve the sensitivity of the rapid detection methods compared to the previous method being applied directly to foods. Although, as yet, none has been able to deliver 100% separation of bacteria from foods, many have improved the sensitivity of several rapid-detection methods when carried out without enrichment culture. There is a great interest in this area at present, as the development of a simple generic method for food sample preparation to facilitate the direct use of rapid methods would be a huge benefit to all concerned with the microbiological safety of foods.

Although enrichment culture may be a lengthy step, its application before the use of a rapid-detection method provides a solution to those problems that are a challenge to the direct detection of pathogens in foods. Enrichment culture itself is an amplification technique, which allows small numbers of microorganisms to be detected and to grow only viable organisms. Another advantage may be in enabling the repair of stressed or injured cells that may be able to cause disease but would not be detectable by direct methods. Use of enrichment culture also dilutes the food material, thus diluting out inhibitors and can result in microorganisms being in a more similar, simpler matrix.

As increasing numbers of rapid tests for foodborne pathogens become commercially available, it is important that they are evaluated fully both by the user laboratory and also collectively conforming to approved official standards in order to ensure that the tests are as good as they claim and to evaluate how well they perform compared to the existing standard accepted detection method. However, at present, it is difficult to keep up with the sheer numbers of tests that are coming on to the market.

2.2. Immunological Assays

Immunological methods, in particular, have had a major impact on the development of rapid methods for foodborne pathogens and constitute the largest proportion of commercially available rapid tests for food pathogens. Immunoassays come in a variety of formats all of which involve the specific binding of antibody to an antigen of the food pathogen. Latex agglutination is one of the simplest examples, in which the antibody-coated colored latex particles or colloidal gold particles specifically cross-link with antigens of the pathogen to give visible clumping, thereby enabling rapid identification of pure cultures isolated from foods. There are many commercial assays of this type for a range of food pathogens including *Campylobacter*, *Escherichia coli* O157:H7, *Listeria monocytogenes*, *Salmonella* and *Shigella*, *Staphylococcus aureus*.

The most common form of immunoassay used for food pathogen detection is the enzyme-linked immunosorbent assay (ELISA). Standard assays are based on an antibody sandwich technique in which an antigen on the food pathogen in an enrichment culture bind to a specific antibody coated onto a solid support, typically a 96-well microtiter plate. Unwanted material is then removed by washing and detection of the pathogen is carried out by the addition of a second specific antibody coupled to an easily assayed enzyme *(3)*. Enzymes commonly used are alkaline phosphatase, β-galactosidase or horseradish peroxidase, which can be easily detected using colorimetric substrates. Alternative detection systems to enzymes are also in use with the sandwich assay format and these include fluorescence, or chemiluminescence, impedance, and polymerase chain reaction (PCR).

There is a huge range of commercial solid-phase assays available for a wide variety of food pathogens. The elaborate list is available at two authoritative references: Association of Official Analytical Chemists (AOAC) International website (https://ecam.commer.net/aboutecam.asp) which lists methods that have been evaluated to officially accepted standards and also produces a list of nonofficially validated commercial assays; and the United States Food and Drug Administration (FDA) website (http://www.cfsan.fda.gov/~ebam/bam-toc.html) which also lists a range of commercially available tests. It is important to note that owing to the rapid development of such tests no single list is exhaustive and the ones cited therein certainly do not claim to be.

The standard ELISA format allows for high sample throughput but the individual steps of the ELISA procedure, such as adding sample, washing, adding antibody complexes, adding detection reagents, etc., are fairly labor intensive and led to the automation in part or all of the ELISA processes. One example of a fully automated ELISA is BioMerieux's Vidas System in which the entire procedure take between 45 min to 2 h to complete after addition of an overnight enrichment broth. At present Vidas assays are available for *Listeria*, *L. monocytogenes*, *Salmonella*, *E. coli*, and *Campylobacter*. This particular company also markets an immunoconcentration kit for *Salmonella* and *E. coli* O157, which may help to shorten the preenrichment step. A range of commercial ELISAs for the detection of several foodborne pathogens is available *(4)*.

Although ELISAs are most frequently used in a 96-well format, other solid supports including dipsticks, paddles, magnetic particles, and nylon or nitrocellulose membranes can also be used and may be more convenient, particularly for lower throughputs. A significant step forward in the immunodetection field has been the development of lateral flow assays, which are both simple and extremely rapid to perform, typically taking

10–15 min. Although they are sandwich-based assays (in a membrane strip format), no washing steps or further manipulations are required once the overnight enrichment sample has been applied. Target bacteria, if present, react with specific antibody that is labeled with either colored latex particles, colloidal gold, or with a fluorescent marker. This antigen–antibody complex migrates laterally by capillary action to a second antibody specific for the target organism. This effectively captures the first antibody complex and a colored or fluorescent line forms owing to the label on the first antibody. However, in the absence of target bacteria no initial antigen–antibody complex forms and the first antibody migrates past the second specific antibody to react with a further antibody specific to the first to give a colored or fluorescent line. This line appears in a different area of the strip and acts as a control indication of the system if working properly *(5)*. Assays using fluorescently labeled antibodies for pathogen detection require a suitable device for detecting fluorescence. Again, many commercial companies are marketing lateral flow assays as rapid detection tests for foodborne pathogen, and examples of these can be found on the AOAC and FDA websites.

Because of interference from the food matrix or debris, ELISA for foodborne pathogens generally require the food sample to undergo an overnight incubation step before assay, to ensure that the target organism has reached a detectable level. This has led to numerous immunoassays being used in combination with separation and concentration methods to eliminate or shorten the preincubation time, for example, the use of immunoseparation to concentrate specific target bacteria from a short enrichment incubation then coupled with a specific ELISA to identify the pathogen. Antigen–antibody reactions have successfully provided a rapid means of detecting foodborne pathogens and recent developments have seen dramatic improvements in their automation and sensitivity. Their flexibility lends itself to their use in combination with other technologies. Exciting further developments for rapid direct detection of pathogens in foods are likely to occur.

2.3. Nucleic Acid Technologies

Major advances in nucleic acid-based food pathogen detection assays have occurred from the 1980s onward. This type of technology affords a high level of specificity as it depends on detecting specific nucleic-acid sequences in the target organism by hybridizing them to a short synthetic oligonucleotide complementary to the specific nucleic-acid sequence in the target organism. Several different types of assays including nucleic-acid hybridization, amplification, microarrays, and chips have either been developed or are being developed for use as rapid methods to detect food pathogens. One reason why nucleic-acid detection is attractive is that many conventional microbiological tests, including antigen–antibody detection, rely on the expression of phenotypic characteristics and these may vary depending on the assay or test conditions. The advantage of nucleic acids as a target for detection is its unambiguous nature; a bacterial culture has a mutation rate of approx 1 in 100 million cells *(114)*.

Simple hybridization assays for detecting the range of foodborne pathogens have been available over a decade and some have gained AOAC approval *(6)*. These assays generally consist of a labeled DNA probe, which hybridizes to a specific ribosomal RNA (rRNA) sequence in the target bacterial cell. rRNA is present at a much higher copy number than DNA and offers an increased sensitivity when used as the detection target for hybridization. Such assays are now available in simple to use 96-well formats enabling a high sample throughput. Although hybridization assays can be performed rapidly, typically 1–1.5 h, they do require the use of overnight enrichment cultures.

The advent of nucleic-acid amplification techniques heralded a revolution in rapid diagnostics. The theoretical ability to amplify a single copy of a nucleic-acid target offered the opportunity to increase the sensitivity of rapid detection tests and resolve the requirement for enrichment culture. However, in practice familiar problems of food matrix interference and inhibition together with the use of small sample volumes for nucleic-acid amplification assays has meant that techniques are not, in general, performed directly on foods but after cultural enrichment.

2.3.1. PCR Methods

PCR is the most well-known and established nucleic-acid amplification technique for detecting pathogenic microorganisms. In this process a specific region of nucleic acid in the target organism is amplified up to a billionfold during a rapid three-step cycling process reaching easily detectable levels. Cell lysis is usually required to release the nucleic acid from the microorganism under detection and then, following denaturation of the target sequence, two short synthetic oligonucleotides or primers specifically anneal to complementary sequences on opposite strands of the DNA flanking the region of interest. The primers are then extended, using the DNA strand on which they have annealed to as a template. The thermophilic enzyme Taq polymerize catalyzes the addition of complementary bases. Each cycle of denaturation, annealing, and extension results in a doubling in the amount of specific DNA of interest and, as all DNA produced at the end of a cycle then goes on to act as a template in the next cycle; there is an exponential increase in the amount of specific DNA generated. In conventional or block-based PCR, the accumulated DNA products are either visualized as a band on an ethidium bromide-stained agarose gel electrophoresis are identified on their base pair size *(7)*. However, this method of detection does not confirm the PCR product is the specific gene fragment of interest only that it approximately has the expected number of base pairs. Further characterization such as through hybridization of PCR products to a specific probe or sequencing of PCR products would confirm they were the specific DNA region of interest. Some of the PCR detection methods based on the use of DNA probes can be performed in 96-well formats. For example, in a PCR ELISA system, the PCR-probe hybrid is captured in a 96-well plate and detected colorimetrically using a labeled antibody. There are many PCR assays that have been described for food pathogens such as *Salmonella*, *E. coli* O157, *L. monocytogenes*, *Campylobacter*, and several others are commercially available (FDA, AOAC). PCR offers a rapid, specific, and sensitive method for the detection of food pathogens following enrichment, but strict measures, including the use of separate areas for performing different stages of the PCR process and the use of appropriate controls, must be in place to ensure that either samples or reagents do not become contaminated with amplified material.

PCR assays can also be multiplexed to detect more than one specific target in one reaction, thus offering the potential of detecting several food pathogens in a single procedure. Primers for amplifying several different targets can be combined in one reaction and following PCR-specific amplicons detected on size difference by agarose gel electrophoresis or by using specific hybridization probes. Successful multiplex PCR assays depend on judicious primer design, to avoid primer dimer formation, and careful optimization of the PCR assay in order to prevent individual reactions having an adverse affect on one another. There are numerous examples available in the literature on the use of multiplex PCR assays for the detection of foodborne pathogens (8–10).

2.3.2. Real-Time PCR Methods

The development of real-time PCR has been a major advance in nucleic-acid amplification technology. PCR amplicons are detected when PCR is occurring using fluorescently labeled molecules. In this way post-PCR manipulations such as agarose gel electrophoresis are eliminated and the whole procedure once setup is a closed tube system, helping to minimize potential contamination from amplicons. A further major advantage of real-time PCR is that amplicon detection occurs during the exponential phase of amplification; this enables accurate quantification of the initial starting nucleic acid and can be related to the number of microorganisms in a sample (1). This offers great potential for quantitating the number of pathogens directly in foods. There are a number of different formats of real-time PCR, which depend on the method used for amplicon detection. The simplest method is independent of the target sequence relying on the incorporation of a double-stranded (ds) DNA-binding dye, SYBR Green I (9). Amplicon accumulation is monitored at the end of each PCR cycle by an increase in fluorescence owing to the binding of SYBR Green I. The amplicon can be detected more specifically at the end of the amplification process by melt curve analysis. Each amplicon has a sequence-dependent temperature at which its double strands denature. This temperature is detected by monitoring the decrease in fluorescence when increasing the temperature of the amplicon from 65 to 94°C. A commercialized example of this type of assay is the BAX assay (Qualicon), which received AOAC approval for screening for *Salmonella*, *E. coli* O157, and *L. moncytogenes* from enrichment cultures (11,12).

The common alternative methods of amplicon detection are sequence specific and depend on the use of DNA probes and fluorescence resonance energy transfer (FRET). FRET is the transfer of resonance energy from one fluorophore to another owing to their close proximity and their overlapping emission and absorption spectra (13). The effect of FRET is to alter the emission spectra of the fluorophores to which the resonance energy has been transferred. The use of DNA probes confers both increased specificity and sensitivity to the PCR assay. There are a range of different probe formats including hydrolysis probes, hybridization probes, molecular beacons, molecular scorpions, and protein nucleic-acid probes. Hydrolysis probes are the basis for TaqMan-based real-time PCR assays (ABI) and depend on the 5′′ thermonuclease-dependent activity of Taq polymerase (14). A DNA probe complementary to a central region of 100–150 bp amplicon and with an annealing temperature 10°C above that of the PCR primers, is labeled at its 3′-end with a fluorophore that quenches the fluorescence of a reporter fluorophore attached to the 5′-end resulting from FRET. This probe hybridizes to the appropriate strand of the amplicon before the primers bind at the annealing temperature. The primers bind at 60°C and Taq polymerase begins extending the 3′-end but on reaching the bound probe it cleaves the 5′-end of the probe, associated with the exonuclease activity, and thus releases the reporter fluorophore from the quencher. FRET is no longer occurring, the emission spectrum is altered and there is a resultant increase in fluorescence at a specific wavelength. The advantages of TaqMan-based assays are that they are all designed to run under the same PCR cycling conditions and are performed in a 96-well microtiter format. This allows for high throughput and enables assays for different targets to run at the same time on one microtiter plate. Many assays for a variety of food pathogens have been described using TaqMan-based assays and several companies are marketing such assays for the detection of food pathogens *(15–17)*.

The hybridization probe format uses two probes that bind within a few base pairs of each other to one strand of the amplicon. The upstream probe is labeled with an acceptor fluorophore at its 3′-end whereas the downstream probe is labeled at its 5′-end with a fluorophore that acts as the donor for resonance energy transfer. Following denaturation the two probes bind to the amplicon at a temperature approx 10°C above the annealing temperature of the PCR reaction. This brings the probes into sufficiently close proximity to one another for resonance energy transfer to occur from the donor fluorophore to the acceptor, resulting in an increase in fluorescence being detected from the acceptor fluorophore. Assays for a range of food pathogens have been described that use the hybridization probe format for detecting real-time PCR products. Commercial assays are available for the detection of *Salmonella*, *E. coli* O157, and *L. monocytogenes* within 1 h following enrichment culture using the Lightcycler real-time PCR platform (Roche Diagnostics, Indianapolis, IN). With this particular PCR, thermal cycler reactions take place in glass capillaries that are heated and cooled very rapidly by air, enabling the PCR process to be accomplished very rapidly *(1,9,18,19)*.

Amplicons can also be detected using molecular beacons. These are hairpin-shaped oligonucleotides with a loop sequence complementary to the PCR amplicon, flanked by arm sequences that are complementary to each other *(20)*. The 5′- and 3′-end of the oligonucleotide are fluorescently labeled with reporter and donor fluorophores, respectively, and in the absence of target sequence the hairpin loop structure is favored in which the reporter fluorophore is quenched by the donor. However, in the presence of target or amplicon, owing to greater stability, the loop of the hairpin preferentially binds to its complementary sequence, resulting in the separation of the reporter from the quencher and a concomitant increase in fluorescence at a particular wavelength.

With molecular scorpions the probe element is physically coupled to one of the PCR primers *(21)*. This primer/probe has a reporter fluorophore at its 5′-end and a quencher fluorophore located on the 3′-end of a separate complementary oligonucleotide. The scorpion also possesses a blocker to prevent Taq polymerase copying the probe region. When the scorpion anneals to target DNA sequence, it is extended by Taq polymerase; denaturation then occurs which separates newly extended scorpion primer and target DNA as well as the oligonucleotide plus quencher fluorophore from the probe region of the scorpion. As the temperature reaches that of annealing, the probe region of the scorpion binds intramolecularly to the newly extended target region. Fluorescence is detectable, as the 5′ reporter fluorophore is no longer being quenched. Any unextended primer is quenched through hybridization with the quencher oligonucleotide.

Sequence-specific amplicon detection can also be achieved through the use of peptide nucleic-acid (PNA) probes which are believed to have enhanced sensitivity and specificity for detecting amplicons owing to a more stable hybridization with the DNA target *(22)*. With these probes the negatively charged sugar phosphate backbone of DNA is replaced by an achiral, uncharged pseudopeptide backbone composed of repeated *N*-(2-aminoethyl) glycine units *(22,23)*. Individual nucleotide bases attached to each of the units enables the PNA to hybridize to target nucleic acid. PNA probes can either be dual labeled with reporter and quencher fluorophores or may be labeled singly with a reporter dye attached to the PNA probe that has low fluorescence when unbound to its target DNA, but is highly fluorescent when bound.

Real-time PCR assays can also be multiplexed to detect more than one target gene or pathogen. Individual amplified products can be detected either by melt curve analysis, owing to differences in amplicon length and/or guanine–cytosine (GC) content, or by oligonucleotide probes through the use of different fluorescent labels. Many of the newer real-time machines have four optical channels enabling the fluorescence of up to four different fluorophores to be distinguished. However, because of the complicated kinetics of four different reactions taking place in a single tube, most real-time technologies, at present, tend to advocate duplexing. A number of multiplex real-time PCR assays for food pathogens have been described in the literature *(9,24–27)*.

Many real-time PCR assays have been developed for foodborne pathogens affording the rapid, sensitive, specific detection of a range of pathogens following enrichment culture. The advantages of these assays together with their ease of use and amenability to automation make them highly attractive for use in combination with rapid generic methods for the separation and concentration of bacteria or nucleic acids from foods in order to overcome the lengthy cultural enrichment step. It is possible that the research and developments in this area will grow and lead to rapid, specific, sensitive detection assays that can be performed directly on food samples in the near future.

Although PCR is the most well-known, a number of other nucleic-acid amplification systems have been reported for the detection of foodborne pathogens including nucleic-acid sequence-based assays (NASBA) and strand displacement amplification (SDA). Both methods, unlike PCR, are isothermal taking place at a single temperature, which can obviate the need for a thermal cycler. With NASBA the target nucleic acid specifically detected is RNA and this has the potential for the direct detection of viable pathogens in food samples *(28)*. Most PCR assays for the detection of bacterial food pathogens amplify and detect DNA. However, DNA is known to be relatively stable and can persist in food samples after heat processing has rendered bacteria nonviable. Thus, the DNA amplified directly from a food sample may not be a reliable indicator of the presence of viable pathogenic bacteria. Messenger RNA (mRNA), on the other hand, is much less stable and has a shorter half-life than DNA and in general, mRNA is thought to be a more suitable indicator of bacterial viability *(28)*. In reality stability can depend on the particular transcript and the physiological state of the cell, and thus the mRNA target for detection must be selected judiciously *(28,29)*. Bacterial target mRNA can be detected by PCR using the enzyme reverse transcriptase, however, the procedure necessitates the complete removal of DNA from the sample and the use of several sample controls to ensure that any amplification detected is associated with mRNA and not with DNA. With NASBA, copy DNA (cDNA) is synthesized using a specific primer to the target mRNA and the enzyme avian myeloblastosis virus–reverse transcriptase (AMV-RT). The RNA portion of the resulting RNA–cDNA duplex is then degraded using RNase H and second strand synthesis of the cDNA occurs using a second specific primer and the DNA polymerase activity of AMV-RT. This ds-DNA is then transcribed by T7 RNA polymerase to produce thousands of RNA transcripts. The entire procedure occurs at 41°C and at this temperature genomic DNA from the target microorganism remains double-stranded and unable to act as a target for amplification *(28)*. Although less well-developed than PCR there are a number of reports in the literature on NASBA assays, including a few using real-time methodologies, to detect mRNA from foodborne pathogens *(28,30,31)*. Further reports on the use of this technology for food-pathogen detection are anticipated with much interest.

SDA is also an isothermal nucleic-acid amplification technique. SDA, however, occurs in two phases. There is an initial target generation phase in which the modified primers and a polymerase enzyme generate copies of the specific target sequence containing a restriction site at either end. This is followed by an amplification phase in which a restriction enzyme nicks one strand of the modified target and thus initiates transcription and subsequent displacement of the downstream DNA strand. Each displaced strand continues to participate in the target generation step, and so, there is a continual process of nicking, extension, strand displacement, and amplification resulting in exponential amplification of the original target DNA. SDA combined with fluorescence polarization as a detection method has been investigated to rapidly identify *E. coli* O157:H7 *(32)*.

2.3.3. mRNA and Cell Viability

Most nucleic-acid amplification methods for detecting bacterial food pathogens including PCR and SDA are based on the amplification of target DNA. However, DNA is relatively stable and can be amplified from bacterial cells that have been autoclaved. In addition to foods, it is important to detect pathogens that pose a risk to human health and not those that have been rendered harmless by food processing. To address this problem a number of investigators have used mRNA as the target in PCR assays for bacterial food pathogens. In comparison to DNA mRNA is relatively short-lived typically having a half-life of minutes as opposed to hours or days; and thus, the detection of mRNA as a product of active gene expression is believed to be an indicator of cell viability. By using reverse transcriptase PCR (RT-PCR) a cDNA of the target mRNA is synthesized before amplification; this cDNA then continues to be used in a standard PCR assay. Although RT-PCR and real-time RT-PCR have been used for the detection of bacterial food pathogens *(26)*, there are several difficulties with the methodology that prevent it being widely adopted as a rapid food pathogen detection method. Isolating mRNA from bacterial cells can be problematic as it is highly susceptible to digestion by nucleases. This can have a direct impact on the sensitivity of the assay. Another requirement is the need to remove any DNA to prevent it acting as a target during PCR. To ensure that amplification is because of the presence of mRNA alone necessitates the use of several more control reactions than an assay where DNA is the target. These additional factors can make RT-PCR technically challenging to perform, particularly on large numbers of samples. However, mRNA is the direct target for the NASBA technique described in the section on alternative amplification techniques. There are increasing numbers of reports in the literature of its use to determine the presence of viable microorganisms *(33–35)*. The use of NASBA in combination with fluorescently labeled molecular beacon probes has been used to investigate gene expression in *Bacillus* species during growth in milk *(31)*. Such an approach offers the potential for real-time detection of viable bacteria cells *(26)*.

2.4. Microarrays

There has been a variety of interest in recent years in the use of DNA microarray technology as a rapid method for identifying and characterizing pathogens *(26)*. In essence, DNA microarrays consist of multiple-specific oligonucleotides or PCR probes spotted mechanically on to a glass microchip in a lattice-type configuration. Target nucleic acid, which may be either PCR products, genomic DNA, or RNA, cDNA is then applied to the microarray and hybridization detected by a fluorescent label incorporated directly into the target nucleic acid. The fluorescence pattern is then recorded and analyzed using a scanner *(36)*. Potentially thousands of oligonucleotides or probes can be spotted on to a single microarray slide offering the opportunity not only of detecting a broad range of pathogens simultaneously with high specificity, but also of providing detailed genetic characterization or genotyping of specific foodborne pathogens. Several microarrays have been developed for food pathogens. One such array simultaneously detects and genotypes enterohemorrhagic *E. coli* (EHEC) in chicken rinsate *(37)*, whereas another based on the *gyrB* gene has been used to rapidly detect and identify *E. coli*, *Salmonella*, and *Shigella (38)*. Different species of *Listeria* have also been discriminated by the use of a microarray based on six virulence determinant genes *(39)*. Microarrays for broad-range bacterial detection are often based on 16S rRNA gene sequences enabling universal primers to amplify 16S rDNA and using specific 16S rDNA probes on the microarray for hybridization to identify different bacterial species. One example of this is a 16S rDNA-based microarray developed to detect 40 bacterial species that predominate in the human gastrointestinal tract. This enabled the rapid detection of these species in DNA directly extracted from fecal samples from 11 individuals *(40)*. The latest technological developments are leading to the production of more cost-effective microarrays and are beginning to incorporate the use of three-dimensional arrays to facilitate extremely rapid hybridization and real-time detection *(41)*. Such progress is paving the way for the wider use of microarrays for rapid detection of pathogens and particularly for their application for food microbiology.

3. TOXIN DETECTION

There is a wide variety of microorganisms that are able to produce toxins. In this section, we discuss the detection of the toxins produced by *S. aureus*, *Vibrio cholerae*, *Clostridium botulinum*, *C. perfringens*, *Bacillus cereus*, and *E. coli*. A number of different types of methods have been developed for the detection of toxin genes and their toxic products. These methods include the detection of toxin genes by amplification methods and hybridization probing. The toxins themselves have been detected using immunological assays, such as ELISA and agglutination tests, and by bioassays such as mouse neutralization testing and cytotoxicity assays in tissue culture.

3.1. Nucleic Acid Detection

PCR methods are useful for amplifying specific genes that encode bacterial toxins. The presence or absence of a gene provides insights into the toxin-producing ability of an organism. PCR methods for toxin detection have been developed for a number of bacterial species including *V. cholera*, *E. coli*, and staphylococcal isolates to name a few. Many of the PCR-based methods utilize PCR to detect different toxin or virulence genes that allow the bacteria to become pathogens. Table 1 contains a list of PCR primers that have been employed for the detection of microbial toxin genes. For example, PCR can be used to detect cholera toxin genes (*ctxA* and *ctxB*) which are present in strains of *V. cholera* O1 that are known to cause epidemic cholera. The genes code for

Table 1
PCR Primers for the Detection of Bacterial Toxin Genes

Species	Toxin	Gene	Primer sequence (5′→3′)	Ref.
Bacillus cereus	Hemolysin BL	*hblA*	AAGCAATGGAATACAATGGG AGAATCTAAATCATGCCACTGC	*(45)*
		hblB	AAGCAATGGAATACAATGGG AATATGTCCCAGTACACCCG	*(45)*
		hblC	GATAC(T/C)AATGTGGCAACTGC TTGAGACTGCTCG(T/C)TAGTTG	*(45)*
		hblD	ACCGGTAACACTATTCATGC AGATCCATATGCTTAGATGC	*(45)*
	Nonhemolytic enterotoxin	*nheA*	GTTAGGATCACAATCACCGC ACGAATGTAATTTGAGTCGC	*(45)*
		nheB	TTTAGTAGTGGATCTGTACGC TTAATGTTCGTTAATCCTGC	*(45)*
		nheC	TGGATTCCAAGATGTAAGG ATTACGACTTCTGCTTGTGC	*(45)*
	Enterotoxin T	*bceT*	CGTATCGGTCGTTCACTCGG GTTGATTTTCCGTAGCCTGGG	*(45)*
	Cytotoxin K	*cytK*	ACAGATATCGG(G/T)CAAAATGC GAACTC(G/C)(A/T)AACTGGGTTGGA	*(45)*
Clostridium botulinum	Type A toxin	*BoNT(A)*	AGCTACGGAGGCAGCTATGTT CGTATTTCCAAAGCTGAAAAGG	*(43)*
	Type B toxin	*BoNT(B)*	CAGGAGAAGTGGAGCGAAAA CTTGCGCCTTTGTTTTCTTG	*(43)*
	Type E toxin	*BoNT(E)*	CCAAGATTTTCATCCGCCTA GCTATTGATCCAAAACGGTGA	*(43)*
	Type F toxin	*BoNT(F)*	CGGCTTCATTAGAGAACGGA TAACTCCCCTAGCCCCGTAT	*(43)*
Clostridium perfringens	Enterotoxin	*cpe*	TTGTTAATACTTTAAGGATATGTATCC TCCATCACCTAAGGACTG	*(106)*
E. coli	Shiga toxin 1	*stx1*	AGTCGTACGGGGATGCAGATAAAT CCGGACACATAGAAGGAAACTCAT	*(48)*
	Shiga toxin 2	*stx2*	TTCCGGAATGCAAATCAGTC CGATACTCCGGAAGCACATTG	*(48)*
Staphylococcus aureus	Enterotoxin A	*entA*	CCTTTGGAAACGGTTAAAACG TCTGAACCTTCCCATCAAAAAC	*(46)*
	Enterotoxin B	*entB*	TCGCATCAAACTGACAAACG GCAGGTACTCTATAAGTGCCTGC	*(46)*
	Enterotoxin C	*entC*	CTCAAGAACTAGACATAAAAGCTAGG TCAAAATCGGATTAACATTATCC	*(46)*
	Enterotoxin D	*entD*	CTAGTTTGGTAATATCTCCTTTAAACG TTAATGCTATATCTTATAGGGTAAACATC	*(46)*
	Enterotoxin E	*entE*	CAGTACCTATAGATAAAGTTAAAACAAGC TAACTTACCGTGGACCCTTC	*(46)*
Vibrio cholerae	Cholera toxin	*ctxA*	CGGGCAGATTCTAGACCTCCTG CGATGATCTTGGAGCATTCCCAC	*(107)*

a A–B polypeptide toxin, in which the B subunit (*ctxB*) allows for binding of the toxin in the intestinal epithelium, and the A subunit (*ctxA*) activates adenylate cyclase activity leading to the secretion of large amounts of fluid characteristic of cholera *(42)*.

Other PCR targets include the amplification of the botulinum neurotoxin (BoNT) gene from the different strains of *C. botulinum*. There are seven types of *C. botulinum* (A–G) that are distinguished by the BoNT produced by the strains. Strains A, B, E, and F produce neurotoxins, which are toxic to humans *(43)*. *C. perfringens* also produces a toxin that affects humans. The *C. perfringens* enterotoxin (CPE) leads to the observed food poisoning symptoms, including abdominal pain and diarrhea *(44)*. *B. cereus* strains are also able to produce a series of enterotoxins that are made up of multiprotein complexes. A series of PCR primer pairs have been designed to identify these proteins that make up the function *B. cereus* enterotoxins (Table 1) *(45)*. *S. aureus* strains are capable of producing multiple toxins (enterotoxins) that are associated with food poisoning as well as toxic shock syndrome toxins and exfoliative toxins *(46)*. PCR primer pairs used for the detection of enterotoxins A–E are provided in Table 1.

Toxin genes are also the targets of many *E. coli* differentiation strategies. Enterotoxigenic *E. coli* (ETEC) are an important cause of diarrhea in humans and animals. These organisms are known to produce heat-labile (LT) and/or heat-stable (ST) enterotoxins. Enterohemorrhagic *E. coli* (EHEC) strains are known to contain shiga toxin (Stx) genes *(47,48)*. In addition to PCR, a number of gene-specific hybridization probes have been designed and utilized for the detection of toxin genes in food and stool samples. The procedure for hybridization detection is described earlier in this chapter. Probes have been designed for *C. botulinum* neurotoxins *(49)*, CPE *(44)*, *E. coli* Shiga toxins *(48)*, staphylococcal exotoxins *(46)*, and cholera toxins *(50)*.

3.2. Immunological and Biochemical Methods

Many biological toxin detection methods rely on the presence of immunological reactions to detect toxins. One set of tests that is commonly employed are ELISAs. The procedure is described earlier in the chapter. ELISA tests for toxins have been generated for staphylococcal enterotoxins A, B, C, and E and found to have detections levels of less than 0.5 µg/100 g in ground beef *(51)*. ELISA tests have also been employed for the detection of botulinum toxins. The sensitivity of detection for botulinum toxins obtained by standard ELISA testing is somewhat less than that obtained by mouse bioassays; however, signal amplification methods have been used to increase the sensitivity to near that of the bioassays *(52)*. ELISA tests have also been developed for the detection of enterotoxins produced by *E. coli*. Monoclonal antibodies with specificity to Shiga-like toxins have been used to detect the presence of Shiga-like toxin I and II from *E. coli (53)*.

Additional immunological-based methods include latex agglutination techniques such as reverse passive latex agglutination (RPLA). As mentioned above, latex agglutination utilizes antibody-coated colored latex particles or colloidal gold particles that cross-link with the toxin to give visible clumping to identify toxins in foods. With RPLA, toxins are soluble, so positive reactions are indicated by the formation of a diffuse lattice pattern *(4)*. There are commercial tests available using RPLA for toxins including those of *B. cereus*, *C. perfringens*, Shiga toxin-producing *E. coli*, *S. aureus*, and *V. cholerae*. The FDA-BAM has an extensive listing of immunological-based methods for rapid detection of bacterial toxins on its website http://www.cfsan.fda.gov/~ebam/bam-a1. html (accessed on 10 November 2006).

Gel diffusion assays are another set of immunologic methods that have been used in toxin detection. Gel diffusion methods include electroimmunodiffusion, micro-Ouchterlony slide test, Oudin or single-diffusion tube test, and the microslide double diffusion test *(54)*. These methods rely on diffusion of the toxin and detecting antibody in a semisolid gel (often agarose) matrix. Interaction of the toxin and antibody results in a visible precipitation band that indicating the presence of the toxin *(55)*. Gel diffusion assays have been used in the detection of toxins from *C. botulinum*, *C. perfringens*, and *S. aureus (54)*.

Biochemical detection of toxin-producing microorganisms typically utilizes chemical reactions to detect unique characteristics of the toxins or toxin-producing microorganisms. One example is the thermostable nuclease test, which has been used for the detection of enterotoxin-producing *S. aureus* strains. It appears that *S. aureus* strains that produce enterotoxins typically produce higher levels of the heat-stable DNAse than other organisms. As the level of enterotoxin appears to rise, so does the nuclease, which is more easily detected chemically than then toxin. Because the enzyme, like the enterotoxin, is heat stable, the assay can be used detect toxin on heat-treated foods *(54)*.

3.3. Bioassays

Animal bioassays for toxin detection typically involve the injection of the toxin into an animal, or a part of an animal, and examine the animal or its tissue for toxin-induced lesions or death. Mice lethality tests are typically used for the detection of BoNT. The test involves isolating the potential toxin from the food source, dividing the sample, inactivating the potential toxin in half of the sample and injecting it intraperitoneally into one mouse and injecting the noninactivated sample into a second mouse. The mice are observed for up to 3 d to check for symptoms of botulism or death. The mouse with the inactivated sample serves as a control and should not show significant pathology or die *(56)*. Mouse lethality assays have also been used in the detection of toxins from *C. perfringens (57)*.

Other animal bioassays involve testing in only a portion of the animal. One such group of tests is the ligated loop tests. These tests assay for the ability of toxins to elicit fluid release from the host enteric tissue leading to fluid accumulation in the intestine. To test for the accumulation of fluid, a portion of the small intestines in the test animal (most commonly a rabbit) is exposed and sections of the intestine tied off with suture. The suspension carrying the potential toxin is injected into the lumen of the intestine and animal is sewn up. After an incubation period, the animal is sacrificed and the intestines checked for the accumulation of fluid, which would indicate potential enterotoxicity. The ligated loop method has been used in the study of *B. cereus*, *C. perfringens*, *E. coli*, and *Vibrio* enterotoxins *(54)*. Other tests have utilized injecting suspensions into the skin of test animals and examining for changes in the tissue that would indicate the presence of toxins in the test sample.

In addition to in vivo assays using live animals, a number of cell culture assays have been developed to assess the presence of pathogens and their toxins in food test samples. Often, culture cells will be induced to change shape when they are exposed to certain toxins, these morphological changes can be visualized and the presence of the toxin detected. Cholera toxin and *E. coli* enterotoxins have been found to distort the shape of Chinese hamster ovary cells and CPE has been shown to disrupt the plating efficiency in cultured Vero cells *(58,59)*.

4. PHENOTYPIC METHODS FOR PATHOGEN DISCRIMINATION

Traditionally, bacterial pathogens have been identified by the examination of phenotypic characteristics of organisms. The methods of characterization used include biochemical profiles based on metabolic activity combined with methods such as serotyping and phage typing. These and other phenotypic-typing techniques utilize the expression products of particular genes that may be present in the different strains of bacteria to separate organisms. A number of phenotypic-typing methods have passed the test of time and have remained in use for many decades. Because of their importance for the detection and characterization of foodborne pathogens, some of the phenotypic techniques will be discussed in the following subsections.

4.1. Serotyping

A widely used method to differentiate among foodborne pathogens is serotyping. Serotyping uses differences in the somatic (O) and flagellar (H) surface antigens to separate strains into distinct serotypes *(60)*. For serotyping, the suspension of bacteria is mixed and incubated with a panel of antisera specific for a variety of O and H epitopes. Specific agglutination profiles are used to determine the serotype of the isolate being tested. Currently, there are over 2500 identified serotypes of *Salmonella*, with more types being added every year *(61)*. In the past, serotyping results were used to speciate *Salmonella* isolates, there was one species per serotype. In the current naming scheme, the Salmonellae are divided into two species, *S. enterica* and *S. bongori*, each with multiple serotypes *(62)*. Serotyping has been used to classify a number of other foodborne pathogens including *Campylobacter* and *E. coli (63)*.

4.2. Phage Typing

Phage typing is phenotypic method of discrimination that has been utilized for many decades to discriminate among closely related bacterial strains *(64)*. There are a number of phage typing schemes available for discriminating among foodborne pathogens including *Salmonella, E. coli,* and *Campylobacter (63,67,68)*. Phage typing has been shown to be useful in the description of pandemic clones such *S. enterica* serovar Typhimurium definitive type 104 (DT104), a relatively common cause severe gastrointestinal illness in humans and is typically resistant to multiple antibiotics and the pathogenic *S. enterica* serovar Enteriditis phage types 4 and 8 *(68)*. Additionally, phage typing has proven useful to separate *E. coli* O157:H7 strains associated with an outbreak from the strains not associated with an outbreak *(67)*.

Phage typing utilizes the selective ability of bacteriophages to infect certain strains of bacteria *(64)*. This differential ability of bacteriophages to infect bacteria is related to the phage receptors present on the surface of the bacteria *(65)*. If the bacterial cell harbors the appropriate receptor for a particular phage, the phage will infect the organism and lyse the bacterium. The phage type is assigned to the specific strain of bacteria based upon whether or not specific typing phages are able to infect and lyse the cells, which is evident through the formation of plaques on an otherwise confluent lawn of bacterial growth

(66). A noted problem with phage typing is that due to the limited number of available phages, many strains are untypeable *(69).* Also, because the typing procedure is technically demanding and requires the preservation of multiple biologically active phage stocks, it is typically performed by reference laboratories, which can limit its availability *(70).*

5. MOLECULAR METHODS FOR PATHOGEN DISCRIMINATION

5.1. Plasmid Typing

The examination of plasmids may be an important way to discriminate among potential foodborne pathogens. Bacteria can share genetic material with one another through the transfer of plasmids *(71).* Plasmid DNA is typically isolated using a procedure that limits the isolation of chromosomal DNA, when retaining the plasmid DNA *(74,75).* The isolated plasmid DNA is analyzed using agarose gel electrophoresis, and the number and size of the plasmids present provides a potential method to discriminate between strains *(76).* An alternative approach to using the size and number of plasmids to discriminate among isolates is to restrict the plasmid DNA with enzymes to create restriction fragment profiles that can be used to differentiate among the isolates *(76).* Restriction analysis of plasmids eliminates a critical drawback of traditional plasmid profiling to type bacteria. Because the topography of plasmids affects their electrophoretic mobility, a plasmid that is nicked or unwound will migrate at different rates. Multiple bands could be present for the same plasmid, thus making analysis more difficult *(71).* Additionally, some strains lack plasmids making it impossible to distinguish among isolates using plasmid profiling. Plasmids are also mobile elements, the gain or loss of plasmids can lead to problems trying to determine the relatedness of foodborne bacterial isolates *(77).*

5.2. Restriction Fragment Length Polymorphism Analysis (RFLP)

RFLP profiles are created following restriction enzyme digestion of bacterial DNA. The restriction patterns produced can be visualized following gel electrophoresis. Typically this method generates a very large number of DNA fragments making it difficult to distinguish band patterns with standard gel electrophoresis. When restriction fragmentation is combined with Southern blotting using probes for repetitive DNA target, RFLP can be an effective technique for differentiating among strains of foodborne pathogens. Ribotyping and insertion sequence (IS) RFLP typing are examples of this form of RFLP analysis. Additionally, restriction fragment analysis can be done without blotting through the use of a rare-cutting enzyme and separation by pulsed-field gel electrophoresis (PFGE) (discussed below).

IS typing has been widely used as an epidemiological tool for investigating foodborne pathogens. Many enteric organisms contain IS*200* sequences that can be utilized for RFLP blotting. IS*200* sequences are present in many *Salmonella*, as well as in some *E. coli* and *Yersinia*. These sequences are approx 700 bp in length and have been found randomly around the genome of enteric organisms *(78).* A drawback of IS*200* analysis is that some strains lack IS*200* sequences; thus, the technique would not be effective in differentiating these isolates *(79).* RFLP with IS blotting is widely being replaced by PCR amplification methods, such as IS*200* repetitive element-PCR (Rep-PCR), for typing foodborne pathogens.

Ribotyping utilizes the number and location of the rRNA gene sequences to probe the restricted cellular DNA *(80).* Enzymes commonly used to restrict foodborne pathogens for DNA include *Pst*I, *Pvu*II, *Hind*III, and *Eco*RI *(80,81,115)*. The restriction fragments profiles are hybridized with a probe that is homologous to the highly conserved regions of rRNA *(82)*. Differences in the DNA sequences in the regions flanking the rRNA lead to variability in the size of the junction fragments. This sequence variability produces distinct patterns that can be used to discriminate between related strains *(81,83)*. Ribotyping can be highly reproducible, especially when the technique is automated using a system where most of the procedure is robotically controlled. There are a few potential weaknesses in ribotyping, some bacterial groups have a limited number of rRNA genes leading to the identification of very few bands for significant analysis. Also, genome modification, such as methylation, can lead to problems with DNA restriction and poor band-pattern resolution *(84)*.

PFGE is an additional restriction-based typing method that is considered by many to be the "gold standard" molecular typing method for bacteria *(84)*. PFGE is used by public health monitoring systems such as PulseNet, a multiagency program to track the spread of foodborne pathogens and assist in determining sources of foodborne disease outbreaks *(85)*. RFLP typing by electrophoresis is often limited by the size of DNA fragments that can be efficiently separated. Standard agarose or acrylamide gel electrophoresis can only separate fragments up to about 50 kb in size. With PFGE, DNA fragments are separated under conditions where there is incremental switch of the polarity of the electric field in the running apparatus. This electrophoretic approach allows for the resolution of DNA fragments up to 800 kb in size *(86)*. When DNA is restricted with a rare-cutting restriction enzyme, PFGE provides a DNA "fingerprint" that reflects the DNA sequence of the entire bacterial genome *(84)*.

For PFGE, a bacterial suspension is prepared with an optimal cell concentration, mixed with molten agarose, and cast into plug molds. The embedded cells are treated with detergents and/or enzymes, such as sarcosine and proteinase K, to lyse the cells and release the DNA. The plug is thoroughly washed to remove cellular debris and treated with a rare-cutting restriction enzyme. Following enzyme treatment, the plugs are inserted into an agarose gel and the restriction fragments separated under conditions of switching polarity. Following electrophoresis, the pattern of DNA separation is visualized after staining with a fluorescent dye, such as ethidium bromide *(87)*. The banding pattern from one isolate can be compared with those of other isolates and information about the relatedness of the strains can be detected. Tenover and others developed a standardized criterion to determine the genetic relatedness of isolates based on their PFGE banding patterns *(88)*. Recently, additional standardized criteria have been developed for PFGE analysis of heterogeneous foodborne pathogens that allows for better resolution of pathogen relatedness *(116)*. This standardization of technique has allowed PFGE to become a widely accepted method for comparing the genetic identity of bacteria during a disease outbreak *(85)*. PFGE typing has demonstrated a high level of reproducibility for foodborne pathogens. Because all of the DNA extraction and restriction enzyme digestion occurs within the agarose plugs, the free DNA is not pipetted, thus the banding patterns generated are related to restriction enzyme digestion, not nonspecific mechanical shearing *(89)*. A potential drawback of PFGE is that it tends to be somewhat labor intensive, sometimes requiring a few days to perform the procedure and interpret the results *(90)*.

5.3. Amplified Fragment Length Polymorphisms (AFLP)

There are a number of methods that utilize PCR to differentiate among foodborne pathogens. One such method is AFLP. AFLP uses both restriction digest analysis and PCR amplification to genotype of bacterial strains. Discrimination among isolates is based on variability in restriction enzyme recognition sites within the genome. Bacterial DNA is digested with one or more enzymes and short polynucleotide linker sequences are ligated to the free DNA ends. The linker sequences serve as targets for PCR primers to anneal and facilitate amplification of the restriction fragments. Following amplification, fragments are separated by electrophoresis and the amplification profiles analyzed and compared to other isolates to distinguish among strains *(91)*. AFLP can be further automated through the use of fluorescently labeled PCR primers. Amplified fragments are separated and detected using an automated DNA sequencer.

To limit the number of PCR products amplified to a manageable number for analysis, PCR primers are designed to contain one to three additional nucleotides on their 3′-end. The additional bases of the primer potentially anneal with unknown sequence in the template strand. For each additional nucleotide added to the primer that could interact with the template, the number of amplicons is reduced by about a factor of 4, because of the requirement that the added bases need to be complimentary to bases in the unknown template sequence *(92)*. Following PCR, amplicons are separated with a DNA sequencing gel, and an elution profile is generated using the fluorescent intensity of the incorporated primers. The resulting profile intensities are compared to those of other isolates to determine the relationship of strains to one another. With the increased signal resulting from PCR amplification of the restriction fragments, only a small amount of DNA is required for analysis. The PCR primers are directed toward the linker DNA because the prior knowledge of the organism's DNA sequence is not required for analysis. The fairly random fragments generated represent a wide range of different locations throughout the bacterial genome, giving a relatively good coverage of the genome to identify differences *(93)*. AFLP is a technique that is well suited for detecting differences among closely related strains *(91)*.

5.4. Random Amplified Polymorphic DNA PCR and Arbitrarily Primed PCR

Two additional PCR-based typing methods are, namely, random amplified polymorphic DNA PCR (RAPD-PCR) and arbitrarily primed PCR (AP-PCR), which are similar methods to PCR subtyping. These similar techniques use random PCR primers to amplify DNA sequences. When the random primers bind the DNA template in close proximity to one another, they amplify the region of the genome between the primers creating variably sized amplification products. For AP-PCR, the primers are usually 20–34 bases, whereas for RAPD-PCR the primers are approx 6–10 bp in length *(94,95)*. Annealing of the random primers is done at relatively low temperature (lower stringency conditions) to facilitate primer binding to genomic sequences with incomplete homology leading to PCR amplification. With AP-PCR, the first few amplification cycles use the lower annealing temperature, which is followed by a higher temperature and stringency for the remainder amplification cycles *(94)*. Following amplification, the PCR products are subjected to agarose gel electrophoresis and the amplicon banding patterns from isolates are compared to one another to differentiate the strains of the bacteria.

Use of methods such as AP-PCR and RAPD-PCR has many benefits. The typing results are obtained in a short amount of time, only a small amount of template DNA is needed and a prior knowledge of the DNA sequence is not required because of the use of the random primers. If increased separation among strains is required, multiple sets of primers can be used to increase discrimination *(96)*. As with many PCR-based typing methods, there is often difficulty in reproducing the same result from one time to the next and from one operator to the next. This difficulty is likely related to the low stringency conditions used to amplify the products, small differences in reaction conditions, and/or reagent concentrations can alter the banding patterns produced leading to difficulty in interpreting and comparing results *(90)*.

5.5. Repetitive Element PCR (Rep-PCR)

In most species of foodborne pathogens, there are repeated DNA sequences spread throughout their genome. These repetitive elements can serve as targets for PCR primers that facilitate the amplification of genomic DNA between the repetitive elements. In order for PCR to work, two of the elements must be in close proximity to allow amplification *(97)*. Because the regions that flank the repeated elements are variable in size, different-sized amplicons will be formed creating a profile that is viewed following gel electrophoresis. Differences in the size of the amplified fragments produce different banding patterns that are compared to one another *(98)*. A number of repeated sequences have been used as targets for Rep-PCR analysis including enterobacterial repetitive intergenic consensus PCR (ERIC-PCR) and repetitive extragenic palindromic PCR (REP-PCR). ERIC sequences are conserved sequences present in enteric bacteria, whereas REP elements are relatively short sequences of DNA that have highly conserved regions of palindromic DNA that serve as good PCR primer targets *(97,99)*. Additionally, insertion sequences such as IS*200*, serve as target sequence for primers in a type of Rep-PCR *(79)*.

Like AP-PCR and RAPD-PCR, Rep-PCR returns typing results in a short time with good discrimination *(90)*. An additional benefit is that the techniques are fairly easy to perform and can be adapted to run on an automated DNA sequencer. Automation of the techniques helps to facilitate analysis of the amplification profiles and increase reproducibility of the results, which can be a problem in manual Rep-PCR *(100)*. Rep-PCR results can be stored in database libraries to facilitate comparison of profiles among researchers and public health officials *(101)*. A weakness of Rep-PCR typing is that certain strains lack a sufficient number of repetitive elements to generate a discriminatory profile leading to difficulty in making relational inferences about isolates *(90)*.

6. FUTURE DIRECTIONS AND CONCLUSIONS

The trend in pathogen detection is the continuing to move toward rapid methods for the detection and characterization of pathogens. This movement is facilitated by the desire to reduce the amount of time it takes to go from the collection of a sample to pathogen identification and characterization to food safety intervention. Ideally results will be ready in hours instead of days or weeks. A number of techniques have recently been developed with utility for use in identification and characterization, including immunological assays, real-time PCR assays, DNA microarrays, multilocus sequence typing (MLST), and multilocus variable number tandem repeat analysis (MLVA). Real-time

PCR and DNA microarray assays are continuously developed that allows for very rapid identification and differentiations of pathogens. These microarrays contain a large number of oligonucleotides that can be used to screen for the presence or absence of multiple virulence or other discriminatory genes that can be used to identify and assess the potential pathogenicity of the organisms *(36)*. Similarly, real-time PCR could be used to screen isolates for a more limited set of genes to separate isolates. To increase PCR efficiency, multiple primer sets can be combined in multiplex PCR assays to reduce the number of reactions required for analysis. An added benefit of gene detection from a clinical standpoint is that specific genes of clinical importance, such as those for toxins and antimicrobial resistance, can be detected *(102)*. As the number of pathogen genomes that have been sequenced continues to rise, the design of microarrays and real-time PCR assays should only improve to provide better identification and characterization of bacterial foodborne pathogens.

Another trend in molecular typing is the advancement of sequence-based methodologies. One example of this trend is MLST, the foundation of this set of techniques is that in different strains of bacteria there will be variability in the sequence of particular genes. This variability would be associated with mutation or recombination events that lead to genetic divergence of different strains that can be used to determine the relatedness of bacteria. Typically, multiple housekeeping genes (genes required for basic cellular functions) are sequenced. Whereas generally conserved, housekeeping genes usually have sufficient variability to develop distinct alleles for the different strains *(103)*. The alleles from each of the genes sequenced are looked at as a group and the strain is assigned a specific sequence type. The genetic relatedness of the strains is determined by comparing differences in allele profiles that make up the sequence type. In cases where all alleles are the same, the sequence types are identical and the two strains are defined as clonal *(104)*. An advantage of MLST is that the data is portable, Internet-based MLST databases have been set up for some of the foodborne pathogens to facilitate the rapid exchange of MLST data *(104)*. Table 2 lists many of the genes that have been used for typing different foodborne pathogens. Advances in gene selection and the establishment of additional online databases will help facilitate wider acceptance of sequence typing, especially if the cost and difficulty of DNA sequencing continues to decline.

Bacterial genome sequencing has spawned additional sequence-based typing methods. One such method is MLVA, a technique that utilizes copy number differences in repeated sequences located throughout the bacterial genome. Differences in the copy numbers of the repeated elements at the variable loci are used to differentiate isolates. By examining multiple repeat regions a discriminatory pattern can often be determined. The technique has shown promise with some of the strains that appear homogeneous with PFGE, such as *E. coli* O157:H7 isolates. MLVA is able to distinguish among many of these isolates and will prevent nonclonal strains from being called identical *(105)*.

As is shown in this chapter, there are a wide variety of methods that can be used for the detection and discrimination of foodborne pathogens and their toxins. The continuing development of rapid methods of detection should help to minimize the impact of foodborne pathogens on human health. One of the areas of need in the detection of foodborne pathogens is in the area of development of culture-independent methods detection and discrimination. Removing the need for overnight enrichment for viable pathogens would allow for much more rapid methods that is valuable in determining whether a

Table 2
Gene Sequence Targets for MLST

Bacterial species	Genes	Ref.
Campylobacter jejuni	*aspA, glnA, gltA, pgm, tkt, uncA*	*(108)*
Escherichia coli O157:H7	*arcA, aroE, dnaE, mdh, gnd, gapA, pgm, espA, ompA, eaeA, fliC, hlyA, uidA*	*(105,109)*
Listeria monocytogenes	*abcZ, dat, ldh, sod, cat, dapE, pgm, bglA, lhkA*	*(110)*
Salmonella enterica	*glnA, manB, pduF,* 16sRNA gene, *pefB, hilA, fimH, aroC, dnaN, hemD, hisD, purC, sucA, thrA*	*(111,117, 118)*
Staphylococcus aureus	*arcC, aroE, glpF, gmk, pta, tpi, yqiL*	*(112)*
Vibrio cholerae	*ctxA, ctxB, tcpA, gyrB, pgm, recA*	*(113)*

food source is contaminated or not. When pathogens are detected there are a number of methods available to characterize the organisms responsible for foodborne illnesses. The various discrimination methods have strengths and weakness that make choosing the proper technique important. The use of rapid, reliable, and sensitive methods allow for the development of intervention strategies that could help to limit the spread of foodborne pathogens.

REFERENCES

1. Ellingson, J. L., Anderson, J. L., Carlson, S. A., and Sharma, V. K. (2004) Twelve hour real-time PCR technique for the sensitive and specific detection of *Salmonella* in raw and ready-to-eat meat products. *Mol. Cell Probes.* **18,** 51–57.
2. Benoit, P. W. and Donahue, D. W. (2003) Methods for rapid separation and concentration of bacteria in food that bypass time-consuming cultural enrichment. *J. Food Prot.* **66,** 1935–1948.
3. Robinson, M., Gustad, T. R., and Meinhardt, S. (1997) Non-specific binding of mouse IgG1 to *Heligmosomoides polygyrus*: parasite homogenate can affinity purify mouse monoclonal antibodies. *Parasit. Immunol.* **114,** 79–84.
4. Doyle, M. P., Beuchat, L. R., and Montville, T.J. (2001) *Food Microbiology Fundamentals and Frontiers*, ASM, Washington, DC.
5. Aldus, C. F., van Amerongen, A., Ariens, R. M., Peck, M. W., Wichers, J. H., and Wyatt, G. M. (2003) Principles of some novel rapid dipstick methods for detection and character-ization of verotoxigenic *Escherichia coli*. *J. Appl. Microbiol.* **95,** 380–389.
6. Curiale, M. S., Sons, T., Fanning, L., et al. (1994) Deoxyribonucleic acid hybridization method for the detection of Listeria in dairy products, seafoods, and meats: collaborative study. *J. AOAC Int.* **77,** 602–617.
7. de Boer, E. and Beumer, R.R. (1999) Methodology for detection and typing of foodborne microorganisms. *Int. J. Food Microbiol.* **50,** 119–130.
8. Hirose, K., Itoh, K., Nakajima, H., et al. (2002) Selective amplification of tyv (rfbE), prt (rfbS), viaB, and fliC genes by multiplex PCR for identification of *Salmonella enterica* serovars Typhi and Paratyphi A. *J. Clin. Microbiol.* **40,** 633–636.
9. Jothikumar, N., Wang, X., and Griffiths, M. W. (2003) Real-time multiplex SYBR green I-based PCR assay for simultaneous detection of *Salmonella* serovars and *Listeria monocytogenes*. *J. Food Prot.* **66,** 2141–2145.
10. Meng, J., Zhao, S., Doyle, M. P., Mitchell, S. E., and Kresovich, S. (1997) A multiplex PCR for identifying Shiga-like toxin-producing *Escherichia coli* O157:H7. *Lett. Appl. Microbiol.* **24,** 172–176.

11. Shearer, A. E., Strapp, C. M., and Joerger, R. D. (2001) Evaluation of a polymerase chain reaction-based system for detection of *Salmonella enteritidis, Escherichia coli* O157:H7, *Listeria* spp., and *Listeria monocytogenes* on fresh fruits and vegetables. *J. Food Prot.* **64,** 788–795.

12. Strapp, C. M., Shearer, A. E., and Joerger, R. D. (2003) Survey of retail alfalfa sprouts and mushrooms for the presence of *Escherichia coli* O157:H7, *Salmonella*, and Listeria with BAX, and evaluation of this polymerase chain reaction-based system with experimentally contaminated samples. *J. Food Prot.* **66,** 182–187.

13. Didenko, V. V. (2001) DNA probes using fluorescence resonance energy transfer (FRET): designs and applications. *Biotechniques.* **31,** 1106–1121.

14. Batt, C. A. (1997) Molecular diagnostics for dairy-borne pathogens. *J. Dairy Sci.* **80,** 220–229.

15. Fratamico, P. M. (2003) Comparison of culture, polymerase chain reaction (PCR), TaqMan Salmonella, and Transia Card Salmonella assays for detection of *Salmonella* spp. in naturally-contaminated ground chicken, ground turkey, and ground beef. *Mol. Cell Probes* **17,** 215–221.

16. Fratamico, P. M. and Bagi, L. K. (2001) Comparison of an immunochromatographic method and the TaqMan *E. coli* O157:H7 assay for detection of *Escherichia coli* O157:H7 in alfalfa sprout spent irrigation water and in sprouts after blanching. *J. Ind. Microbiol. Biotechnol.* **27,** 129–134.

17. Iijima, Y., Asako, N. T., Aihara, M., and Hayashi, K. (2004) Improvement in the detection rate of diarrhoeagenic bacteria in human stool specimens by a rapid real-time PCR assay. *J. Med. Microbiol.* **53,** 617–622.

18. Fitzmaurice, J., Glennon, M., Duffy, G., Sheridan, J. J., Carroll, C., and Maher, M. (2004) Application of real-time PCR and RT-PCR assays for the detection and quantitation of VT 1 and VT 2 toxin genes in *E. coli* O157:H7. *Mol. Cell Probes* **18,** 123–132.

19. Perelle, S., Josefsen, M., Hoorfar, J., Dilasser, F., Grout, J., and Fach, P. (2004) A LightCycler real-time PCR hybridization probe assay for detecting food-borne thermophilic Campylobacter. *Mol. Cell Probes* **18,** 321–327.

20. Chen, W., Martinez, G., and Mulchandani, A. (2000) Molecular beacons: a real-time polymerase chain reaction assay for detecting *Salmonella. Anal. Biochem.* **280,** 166–172.

21. Whitcombe, D., Theaker, J., Guy, S. P., Brown, T., and Little, S. (1999) Detection of PCR products using self-probing amplicons and fluorescence. *Nat. Biotechnol.* **17,** 804–807.

22. Stender, H., Fiandaca, M., Hyldig-Nielsen, J. J., and Coull, J. (2002) PNA for rapid microbiology. *J. Microbiol. Methods* **48,** 1–17.

23. Isacsson, J., Cao, H., Ohlsson, L., et al. (2000) Rapid and specific detection of PCR products using light-up probes. *Mol. Cell Probes* **14,** 321–328.

24. Beuret, C. (2004) Simultaneous detection of enteric viruses by multiplex real-time RT-PCR. *J. Virol. Methods.* **115,** 1–8.

25. Jinneman, K. C., Yoshitomi, K. J., and Weagant, S. D. (2003) Multiplex real-time PCR method to identify Shiga toxin genes stx1 and stx2 and *Escherichia coli* O157:H7/H-serotype. *Appl. Environ. Microbiol.* **69,** 6327–6333.

26. McKillip, J. L. and Drake, M. (2004) Real-time nucleic acid-based detection methods for pathogenic bacteria in food. *J. Food Prot.* **67,** 823–832.

27. Tamarapu, S., McKillip, J. L., and Drake, M. (2001) Development of a multiplex polymerase chain reaction assay for detection and differentiation of *Staphylococcus aureus* in dairy products. *J. Food Prot.* **64,** 664–668.

28. Cook, N. (2003) The use of NASBA for the detection of microbial pathogens in food and environmental samples. *J. Microbiol. Meth.* **53,** 165–174.

29. Norton, D. M. and Batt, C. A. (1999) Detection of viable *Listeria monocytogenes* with a 5′ nuclease PCR assay. *Appl. Environ. Microbiol.* **65,** 2122–2127.

30. D'Souza, D. H. and Jaykus, L. A. (2003) Nucleic acid sequence based amplification for the rapid and sensitive detection of *Salmonella enterica* from foods. *J. Appl. Microbiol.* **95,** 1343–1350.

31. Gore, H. M., Wakeman, C. A., Hull, R. M., and McKillip, J. L. (2003) Real-time molecular beacon NASBA reveals hblC expression from *Bacillus* spp. in milk. *Biochem. Biophys. Res. Commun.* **311,** 386–390.

32. Ge, B., Larkin, C., Ahn, S., et al. (2002) Identification of *Escherichia coli* O157:H7 and other enterohemorrhagic serotypes by EHEC-*hlyA* targeting, strand displacement amplification, and fluorescence polarization. *Mol. Cell Probes* **16,** 85–92.

33. Simpkins, S. A., Chan, A. B., Hays, J., Popping, B., and Cook, N. (2000) An RNA transcription-based amplification technique (NASBA) for the detection of viable *Salmonella enterica*. *Lett. Appl. Microbiol.* **30,** 75–79.

34. Uyttendaele, M., Schukkink, R., van Gemen, B., and Debevere, J. (1995) Development of NASBA, a nucleic acid amplification system, for identification of *Listeria monocytogenes* and comparison to ELISA and a modified FDA method. *Int. J. Food Microbiol.* **27,** 77–89.

35. Uyttendaele, M., Bastiaansen, A., and Debevere, J. (1997) Evaluation of the NASBA nucleic acid amplification system for assessment of the viability of *Campylobacter jejuni*. *Int. J. Food Microbiol.* **37,** 13–20.

36. Chizhikov, V., Rasooly, A., Chumakov, K., and Levy, D.D. (2001) Microarray analysis of microbial virulence factors. *Appl. Environ. Microbiol.* **67,** 3258–3263.

37. Call, D. R., Brockman, F. J., and Chandler, D. P. (2001) Detecting and genotyping *Escherichia coli* O157:H7 using multiplexed PCR and nucleic acid microarrays. *Int. J. Food Microbiol.* **67,** 71–80.

38. Kakinuma, K., Fukushima, M., and Kawaguchi, R. (2003) Detection and identification of *Escherichia coli*, *Shigella*, and *Salmonella* by microarrays using the gyrB gene. *Biotechnol. Bioeng.* **83,** 721–728.

39. Volokhov, D., Rasooly, A., Chumakov, K., and Chizhikov, V. (2002) Identification of Listeria species by microarray-based assay. *J. Clin. Microbiol.* **40,** 4720–4728.

40. Wang, R. F., Beggs, M. L., Erickson, B. D., and Cerniglia, C. E. (2004) DNA microarray analysis of predominant human intestinal bacteria in fecal samples. *Mol. Cell Probes* **18,** 223–234.

41. Bodrossy, L. and Sessitsch, A. (2004) Oligonucleotide microarrays in microbial diagnostics. *Curr. Opin. Microbiol.* **7,** 245–254.

42. Fields, P. I., Popovic, T., Wachsmuth, K., and Olsvik, O. (1992) Use of polymerase chain reaction for detection of toxigenic *Vibrio cholerae* O1 strains from the Latin American cholera epidemic. *J. Clin. Microbiol.* **30,** 2118–2121.

43. Lindstrom, M., Keto, R., Markkula, A., Nevas, M., Hielm, S., and Korkeala, H. (2001) Multiplex PCR assay for detection and identification of *Clostridium botulinum* types A, B, E, and F in food and fecal material. *Appl. Environ. Microbiol.* **67,** 5694–5699.

44. Damme-Jongsten, M., Rodhouse, J., Gilbert, R.J., and Notermans, S. (1990) Synthetic DNA probes for detection of enterotoxigenic *Clostridium perfringens* strains isolated from outbreaks of food poisoning. *J. Clin. Microbiol.* **28,** 131–133.

45. Guinebretiere, M. H., Broussolle, V., and Nguyen-The, C. (2002) Enterotoxigenic profiles of food-poisoning and food-borne *Bacillus cereus* strains. *J. Clin. Microbiol.* **40,** 3053–3056.

46. Becker, K., Roth, R., and Peters, G. (1998) Rapid and specific detection of toxigenic *Staphylococcus aureus*: use of two multiplex PCR enzyme immunoassays for amplification and hybridization of staphylococcal enterotoxin genes, exfoliative toxin genes, and toxic shock syndrome toxin 1 gene. *J. Clin. Microbiol.* **36,** 2548–2553.

47. Persing, D. H. (1993) *Diagnostic Molecular Microbiology Principles and Applications*. American Society for Microbiology, Washington, DC.

48. Pulz, M., Matussek, A., Monazahian, M., et al. (2003) Comparison of a shiga toxin enzyme-linked immunosorbent assay and two types of PCR for detection of shiga toxin-producing *Escherichia coli* in human stool specimens. *J. Clin. Microbiol.* **41,** 4671–4675.

49. Fach, P., Gibert, M., Griffais, R., Guillou, J. P., and Popoff, M. R. (1995) PCR and gene probe identification of botulinum neurotoxin A-, B-, E-, F-, and G-producing *Clostridium* spp. and evaluation in food samples. *Appl. Environ. Microbiol.* **61,** 389–392.

50. Takeda, T., Peina, Y., Ogawa, A., et al. (1991) Detection of heat-stable enterotoxin in a cholera toxin gene-positive strain of *Vibrio cholerae* O1. *FEMS Microbiol. Lett.* **64,** 23–27.

51. Notermans, S., Dufrenne, J., and Schothorst, M. (1978) Enzyme-linked immunosorbent assay for detection of *Clostridium botulinum* toxin type A. *Jpn. J. Med. Sci. Biol.* **31,** 81–85.

52. Doellgast, G. J., Triscott, M. X., Beard, G. A., et al. (1993) Sensitive enzyme-linked immunosorbent assay for detection of *Clostridium botulinum* neurotoxins A, B, and E using signal amplification via enzyme-linked coagulation assay. *J. Clin. Microbiol.* **31,** 2402–2409.

53. Downes, F. P., Green, J. H., Greene, K., Strockbine, N., Wells, J. G., and Wachsmuth, I. K. (1989) Development and evaluation of enzyme-linked immunosorbent assays for detection of shiga-like toxin I and shiga-like toxin II. *J. Clin. Microbiol.* **27,** 1292–1297.

54. Jay, J. M. (2000) *Modern Food Microbiology*, Aspen, Gaithersburg, MD.

55. Klein, J. and Horejsí, V. (1997) *Immunology*, Blackwell, Oxford.

56. Solomon, H. M. and Lilly, T. (2001) *Clostridium botulinum*. Bacteriological Analytical Manual Online. Center for Food Safety & Applied Nutrition, US Food and Drug Administration (accessed on 11 August 2004).

57. Stark, R. L. and Duncan, C. L. (1971) Biological characteristics of *Clostridium perfringens* type A enterotoxin. *Infect. Immun.* **4,** 89–96.

58. Guerrant, R. L., Brunton, L. L., Schnaitman, T. C., Rebhun, L. I., and Gilman. A. G. (1974) Cyclic adenosine monophosphate and alteration of Chinese hamster ovary cell morphology: a rapid, sensitive in vitro assay for the enterotoxins of *Vibrio cholerae* and *Escherichia coli. Infect. Immun.* **10,** 320–327.

59. McDonel, J. L. and McClane, B. A. (1981) Highly sensitive assay for *Clostridium perfringens* enterotoxin that uses inhibition of plating efficiency of Vero cells grown in culture. *J. Clin. Microbiol.* **13,** 940–946.

60. Voogt, N., Wannet, W. J., Nagelkerke, N. J., and Henken, A. M. (2002) Differences between national reference laboratories of the European community in their ability to serotype *Salmonella* species. *Eur. J. Clin. Microbiol. Infect. Dis.* **21,** 204–208.

61. Popoff, M. Y., Bockemuhl, J., Brenner, F. W., and Gheesling, L. L. (2001) Supplement 2000 (no. 44) to the Kauffmann-White scheme. *Res. Microbiol.* **152,** 907–909.

62. Brenner, F. W., Villar, R. G., Angulo, F. J., Tauxe, R., and Swaminathan, B. (2000) *Salmonella* nomenclature. *J. Clin. Microbiol.* **38,** 2465–2467.

63. Frost, J. A., Kramer, J. M., and Gillanders, S. A. (1999) Phage typing of *Campylobacter jejuni* and *Campylobacter coli* and its use as an adjunct to serotyping. *Epidemiol. Infect.* **123,** 47–55.

64. Schmieger, H. (1999) Molecular survey of the *Salmonella* phage typing system of Anderson. *J. Bacteriol.* **181,** 1630–1635.

65. Snyder, L. and Champness, W. (1997) *Molecular Genetics of Bacteria*. ASM Press, Washington, DC.

66. Hickman-Brenner, F. W., Stubbs, A. D., and Farmer, J. J. III. (1991) Phage typing of *Salmonella enteritidis* in the United States. *J. Clin. Microbiol.* **29,** 2817–2823.

67. Barrett, T. J., Lior, H., Green, J. H., et al. (1994) Laboratory investigation of a multistate food-borne outbreak of *Escherichia coli* O157:H7 by using pulsed-field gel electrophoresis and phage typing. *J. Clin. Microbiol.* **32,** 3013–3017.

68. Humphrey, T. (2001) *Salmonella typhimurium* definitive type 104. A multi-resistant *Salmonella. Int. J. Food Microbiol.* **67,** 173–186.

69. Amavisit, P., Markham, P. F., Lightfoot, D., Whithear, K. G., and Browning, G. F. (2001) Molecular epidemiology of *Salmonella* Heidelberg in an equine hospital. *Vet. Microbiol.* **80,** 85–98.

70. Arbeit, R. D. (1995) Laboratory procedures for the epidemiologic analysis of microorganisms. In: *Manual of Clinical Microbiology* (Murray, P. R., Baron, E. J., Pfaller, M. A., Tenover, F. C., and Yolken, R. H., eds.), ASM, Washington, DC, pp. 190–208.

71. Summers, D. K. (1996) *The Biology of Plasmids*, Blackwell, Oxford.

72. Johnson, T. J., Giddings, C. W., Horne, S. M., et al. (2002) Location of increased serum survival gene and selected virulence traits on a conjugative R plasmid in an avian *Escherichia coli* isolate. *Avian Dis.* **46,** 342–352.

73. Tenover, F. C. (2001) Development and spread of bacterial resistance to antimicrobial agents: an overview. *Clin. Infect. Dis.* **33(3),** S108–S115.

74. Birnboim, H. C. and Doly, J. (1979) A rapid alkaline extraction procedure for screening recombinant plasmid DNA. *Nucleic Acids Res.* **7,** 1513–1523.

75. Kado, C. I. and Liu, S. T. (1981) Rapid procedure for detection and isolation of large and small plasmids. *J. Bacteriol.* **145,** 1365–1373.

76. Nauerby, B., Pedersen, K., Dietz, H. H., and Madsen, M. (2000) Comparison of Danish isolates of *Salmonella enterica* serovar enteritidis PT9a and PT11 from hedgehogs (*Erinaceus europaeus*) and humans by plasmid profiling and pulsed-field gel electrophoresis. *J. Clin. Microbiol.* **38,** 3631–3635.

77. Kumao, T., Ba-Thein, W., and Hayashi, H. (2002) Molecular subtyping methods for detection of *Salmonella enterica* serovar Oranienburg outbreaks. *J. Clin. Microbiol.* **40,** 2057–2061.

78. Beuzon, C. R. and Casadesus J. (1997) Conserved structure of IS200 elements in *Salmonella. Nucleic Acids Res.* **25,** 1355–1361.

79. Millemann, Y., Gaubert, S., Remy, D., and Colmin, C. (2000) Evaluation of IS200-PCR and comparison with other molecular markers to trace *Salmonella enterica* subsp. enterica serotype typhimurium bovine isolates from farm to meat. *J. Clin. Microbiol.* **38,** 2204–2209.

80. Ling, J. M., Lo, N. W., Ho, Y. M., et al. (2000) Molecular methods for the epidemiological typing of *Salmonella enterica* serotype Typhi from Hong Kong and Vietnam. *J. Clin. Microbiol.* **38,** 292–300.

81. Gendel, S. M. and Ulaszek, J. (2000) Ribotype analysis of strain distribution in *Listeria monocytogenes. J. Food Prot.* **63,** 179–185.

82. Chisholm, S. A., Crichton, P. B., Knight, H. I., and Old, D. C. (1999) Molecular typing of *Salmonella* serotype Thompson strains isolated from human and animal sources. *Epidemiol. Infect.* **122,** 33–39.

83. Snipes, K. P., Hirsh, D. C., Kasten, R. W., et al. (1989) Use of an rRNA probe and restriction endonuclease analysis to fingerprint *Pasteurella multocida* isolated from turkeys and wildlife. *J. Clin. Microbiol.* **27,** 1847–1853.

84. Olive, D. M. and Bean, P. (1999) Principles and applications of methods for DNA-based typing of microbial organisms. *J. Clin. Microbiol.* **37,** 1661–1669.

85. Swaminathan, B., Barrett, T. J., Hunter, S. B., and Tauxe, R. V. (2001) PulseNet: the molecular subtyping network for foodborne bacterial disease surveillance, United States. *Emerg. Infect. Dis.* **7,** 382–389.

86. Schwartz, D. C. and Cantor, C. R. (1984) Separation of yeast chromosome-sized DNAs by pulsed field gradient gel electrophoresis. *Cell.* **37,** 67–75.

87. Gautom, R. K. (1997) Rapid pulsed-field gel electrophoresis protocol for typing of *Escherichia coli* O157:H7 and other gram-negative organisms in 1 day. *J. Clin. Microbiol.* **35,** 2977–2980.

88. Tenover, F. C., Arbeit, R. D., Goering, R. V., et al. (1995) Interpreting chromosomal DNA restriction patterns produced by pulsed-field gel electrophoresis: criteria for bacterial strain typing. *J. Clin. Microbiol.* **33,** 2233–2239.

89. Birren, B. and Lai, E. (1993) *Pulsed Field Gel Electrophoresis A Practical Guide*, Academic, San Diego.

90. Swaminathan, B. and Barrett, T. J. (1995) Amplification methods for epidemiologic investigations of infectious diesease. *J. Microbiol. Meth.* **2,** 129–139.
91. Mueller, U. G. and Wolfenbarger, L. L. (1999) AFLP genotyping and fingerprinting. *Trends Ecol. Evol.* **14,** 389–394.
92. Savelkoul, P. H., Aarts, H. J., de Haas, J., et al. (1999) Amplified-fragment length polymorphism analysis: the state of an art. *J. Clin. Microbiol.* **37,** 3083–3091.
93. Desai, M., Threlfall, E. J., and Stanley, J. (2001) Fluorescent amplified-fragment length polymorphism subtyping of the *Salmonella enterica* serovar enteritidis phage type 4 clone complex. *J. Clin. Microbiol.* **39,** 201–206.
94. Welsh, J. and McClelland, M. (1990) Fingerprinting genomes using PCR with arbitrary primers. *Nucleic Acids Res.* **18,** 7213–7218.
95. Williams, J. G., Kubelik, A. R., Livak, K. J., Rafalski, J. A., and Tingey, S. V. (1990) DNA polymorphisms amplified by arbitrary primers are useful as genetic markers. *Nucleic Acids Res.* **18,** 6531–6535.
96. Franklin, R. B., Taylor, D. R., and Mills, A. L. (1999) Characterization of microbial communities using randomly amplified polymorphic DNA (RAPD). *J. Microbiol. Meth.* **35,** 225–235.
97. Versalovic, J., Koeuth, T., and Lupski, J. R. (1991) Distribution of repetitive DNA sequences in eubacteria and application to fingerprinting of bacterial genomes. *Nucleic Acids Res.* **19,** 6823–6831.
98. Georghiou, P. R., Doggett, A. M., Kielhofner, M. A., et al. (1994) Molecular fingerprinting of *Legionella* species by repetitive element PCR. *J. Clin. Microbiol.* **32,** 2989–2994.
99. Hulton, C. S., Higgins, C. F., and Sharp, P. M. (1991) ERIC sequences: a novel family of repetitive elements in the genomes of *Escherichia coli, Salmonella typhimurium* and other enterobacteria. *Mol. Microbiol.* **5,** 825–834.
100. Del Vecchio, V. G., Petroziello, J. M., Gress, M. J., et al. (1995) Molecular genotyping of methicillin-resistant *Staphylococcus aureus* via fluorophore-enhanced repetitive-sequence PCR. *J. Clin. Microbiol.* **33,** 2141–2144.
101. Garaizar, J., Lopez-Molina, N., Laconcha, I., et al. (2000) Suitability of PCR fingerprinting, infrequent-restriction-site PCR, and pulsed-field gel electrophoresis, combined with computerized gel analysis, in library typing of *Salmonella enterica* serovar enteritidis. *Appl. Environ. Microbiol.* **66,** 5273–5281.
102. Gordon, D. M. (2001) Geographical structure and host specificity in bacteria and the implications for tracing the source of coliform contamination. *Microbiology* **147,** 1079–1085.
103. Maiden, M. C., Bygraves, J. A., Feil, E., et al. (1998) Multilocus sequence typing: a portable approach to the identification of clones within populations of pathogenic microorganisms. *Proc. Natl Acad. Sci. USA* **95,** 3140–3145.
104. Enright, M. C. and Spratt, B.G. (1999) Multilocus sequence typing. *Trends Microbiol.* **7,** 482–487.
105. Noller, A. C., McEllistrem, M. C., Stine, O. C., et al. (2003) Multilocus sequence typing reveals a lack of diversity among *Escherichia coli* O157:H7 isolates that are distinct by pulsed-field gel electrophoresis. *J. Clin. Microbiol.* **41,** 675–679.
106. Kokai-Kun, J. F., Songer, J. G., Czeczulin, J. R., Chen, F., and McClane, B. A. (1994) Comparison of Western immunoblots and gene detection assays for identification of potentially enterotoxigenic isolates of *Clostridium perfringens. J. Clin. Microbiol.* **32,** 2533–2539.
107. Singh, D. V., Isac, S. R., and Colwell, R. R. (2002) Development of a hexaplex PCR assay for rapid detection of virulence and regulatory genes in *Vibrio cholerae* and *Vibrio mimicus. J. Clin. Microbiol.* **40,** 4321–4324.
108. Dingle, K. E., Colles, F. M., Wareing, D. R., et al. (2001) Multilocus sequence typing system for *Campylobacter jejuni. J. Clin. Microbiol.* **39,** 14–23.

109. Foley, S. L., Simjee, S., Meng, J., White, D. G., McDermott, P. F., and Zhao, S. (2004) Evaluation of molecular typing methods for *Escherichia coli* O157:H7 isolates from cattle, food, and humans. *J. Food Prot.* **67,** 651–657.

110. Salcedo, C., Arreaza, L., Alcala, B., de la, F. L., and Vazquez, J. A. (2003) Development of a multilocus sequence typing method for analysis of *Listeria monocytogenes* clones. *J. Clin. Microbiol.* **41,** 757–762.

111. Kotetishvili, M., Stine, O. C., Kreger, A., Morris, J. G. Jr., and Sulakvelidze, A. (2002) Multilocus sequence typing for characterization of clinical and environmental *Salmonella* strains. *J. Clin. Microbiol.* **40,** 1626–1635.

112. Enright, M. C., Day, N. P., Davies, C. E., Peacock, S. J., and Spratt, B. G. (2000) Multilocus sequence typing for characterization of methicillin-resistant and methicillin-susceptible clones of *Staphylococcus aureus. J. Clin. Microbiol.* **38,** 1008–1015.

113. Kotetishvili, M., Stine, O. C., Chen, Y., et al. (2003) Multilocus sequence typing has better discriminatory ability for typing *Vibrio cholerae* than does pulsed-field gel electrophoresis and provides a measure of phylogenetic relatedness. *J. Clin. Microbiol.* **41,** 2191–2196.

114. Zhao, X. and Drlica, K. (2001) Restricting the selection of antibiotic-resistant mutants: a general strategy derived from fluoroquinolone studies. *Clin. Infect Dis.* **33,** S147–S156.

115. Wassenaar, T. M. and Newell, D. G. (2000) Genotyping of *Campylobacter* spp. *Appl. Environ. Microbiol.* **66,** 1–9.

116. Barrett, T. J., Gerner-Smidt, P., and Swaminathan, B. (2006) Interpretation of pulsed-field gel electrophoresis patterns in foodborne disease investigations and surveillance. *Foodborne Pathog. Dis.* **3,** 20–31.

117. Harbottle, H., White, D. G., McDermott, P. F., Walker, R. D., and Zhao, S. (2006) Comparison of multilocus sequence typing, pulsed-field gel electrophoresis, and antimicrobial susceptibility typing for characterization of *Salmonella enterica* serotype Newport isolates. *J. Clin. Microbiol.* **44,** 2449–2457.

118. Foley, S. L., White, D. G., McDermott, P. F., et al. (2006) Comparison of subtyping method for *Salmonella enterica* serovar Typhimurium from food animal sources. *J. Clin. Microbiol.* **44,** 3569–3577.

21

Future Directions in Food Safety*

Ross C. Beier and Suresh D. Pillai

Abstract

The recent success that the USDA Food Safety Inspection Service has had in 2003 and 2004 of reversing the steadily increasing trend in Class 1 recalls is welcomed. In agreement with those statistics are the FSIS microbiological results for *Escherichia coli* O157:H7 in raw ground beef, which also showed a decrease in 2003. But there is much work to be done in food safety and much more to achieve. It is imperative that while addressing food-safety issues, we should understand the role that the environmental microbiology, public health epidemiology, aerobiology, molecular microbial ecology, occupational health, industrial processes, municipal water quality, and animal health have on food safety. Although it is a difficult task, a concerted effort by industry, academic, and governmental researchers can accomplish the goal. Here we discuss the future directions and applications in the distribution and spread of foodborne hazards, methods for microbial detection and differentiation, intervention strategies for farm pathogen reduction, targeting waste at animal production sites, considerations on antimicrobial resistance, food-safety storage and preparation strategies, food irradiation, new and emerging food-safety hazards, and quantitative microbial food-safety risk assessment. Although this does not comprise an exhaustive list of food-safety issues, these are the areas that, we think, require considerable attention by researchers. Not only we need to strive to improve food safety through new strategies, processes, and applications, but we also need to be flexible and observant to readily handle the new and emerging food-safety problems, whether they are within our borders or global. At present, the United States has one of the safest food-safety systems in place. However, although this is not a time for complacency, our research endeavors should be designed to keep pace with the food-safety needs of the future.

1. INTRODUCTION

Thousands of people around the world die each year from contaminated food. Foods can be contaminated with microbial pathogens and toxins at all points of the food-production cycle from preharvest through postharvest as well as within the home. Pathogens that can be deadly to humans are found to reside within food animals. The US Centers for Disease Control and Prevention (CDC) reported that the outbreaks of foodborne diseases for which the etiology and transmission vehicle could be traced have resulted from foods of animal origin about 50% of the time *(1)*. The CDC estimates that foodborne disease causes approx 76 million illness cases in the United States each year. This results in approx 325,000 hospitalizations and 5000 deaths each year. The costs in terms of overall medical expenses and lost wage-productivity are estimated to be between

*Mention of a trade name, proprietary product, or specific equipment does not constitute a guarantee or warranty by the US Department of Agriculture and does not imply its approval to the exclusion of other products that may be suitable.

From: *Infectious Disease: Foodborne Diseases*
Edited by: S. Simjee © Humana Press Inc., Totowa, NJ

US$6.5 and 34.9 billion *(2,3)*. One of the main pathogens for concern, *Salmonella*, is an important cause for human and animal diseases worldwide *(4)*, and can be serious or fatal for the elderly and immunocompromised. The CDC estimates that each year in the United States over 1.3 million illnesses and 553 deaths are caused by foodborne transmission of *Salmonella*. The cost of medical care and lost productivity due to *Salmonella* infections in the United States were estimated at $2.3 billion per year in 1998 dollars *(5)*. Pathogenic *Escherichia coli* primarily cause one of three types of infections: enteric infections, urinary tract infections, or septicemic infections *(6)*. Among the enteric *E. coli*, bacteria capable of producing Shiga toxins are of particular concern. Shiga toxins are potent protein-synthesis inhibitors capable of killing cells in picogram quantities. *E. coli* O157:H7 is the most common Shiga toxin-producing *E. coli (7)*. It is estimated that in the United States 62,000 human illnesses and 52 deaths are caused by foodborne transmission of *E. coli* O157:H7 each year *(2)*.

Increased occurrence of pathogens that cause major foodborne disease outbreaks (*E. coli* O157:H7, *Salmonella*, and *Listeria monocytogenes*) are key concerns for food producers, food processors, governmental regulators, public health officials, and consumers worldwide. These challenges have led to increased global demand for actions to improve the safety of raw and manufactured foods *(8)*. The US Department of Agriculture–Food Safety Inspection Service (USDA–FSIS) has adopted an overall food-safety strategy to reduce the risk of foodborne illness associated with pathogens such as *Salmonella* spp., *Campylobacter* spp., *E. coli* O157:H7, *L. monocytogenes*, *Clostridium botulinum*, *Clostridium perfringens*, *Staphylococcus aureus*, *Aeromonas hydrophila*, and *Bacillus cereus (9)*. In 1996, the USDA–FSIS issued a final rule that requires the meat and poultry establishments to adopt a scientific Hazard Analysis Critical Control Point (HACCP) system *(10)*. HACCP is a mandatory food system used by all food companies in the European Union *(11)*, and it is internationally recognized as being a most effective means for producing safe food *(12–14)*.

In the years following the implementation of the HACCP program, the total number of Class 1, or high risk, recalls steadily increased (Fig. 1). Then, in 2003, FSIS outlined five goals to improve the health status of consumers. This document is called "Enhancing Public Health: Strategies for the Future" *(15)*. In this document "FSIS outlined a series of new and comprehensive scientific initiatives to better understand, predict, and prevent microbiological contamination of meat, poultry, and egg products, thereby improving health outcomes for American families" *(16)*. The addition of these initiatives along with other improvements in food safety has resulted in a downward trend in the number of Class 1 recalls during 2003 and in the first half of 2004 (Fig. 1). FSIS has been obtaining microbiological results for *E. coli* O157:H7 in raw ground beef products since 1994 *(17)*. Late in 1999, FSIS introduced new methods that further increased the sensitivity of the microbiological tests. Table 1 shows the overall results of the determinations for the years 2000–2003. The percent positives average around 0.8% from the years 2000–2002, but declined to 0.3% in 2003 *(17)*. These results correlate well with the observed recalls over this period. Another indicator that the FSIS scientific policies and programs have improved food safety is the data from late 2003 that show a 25% drop in the percentage of positive *L. monocytogenes* regulatory samples compared to those of the year 2002, and a 70% decline compared to the years prior to the implementation of the HACCP program *(16)*.

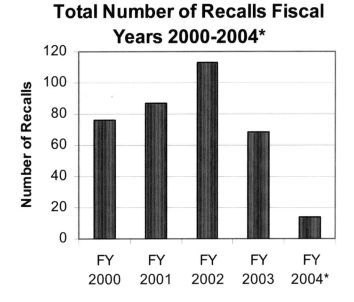

*Contains data collected through June 30, 2004

Fig. 1. Total number of recalls during the fiscal years 2000–2004 *(16)*.

Table 1
Microbiological Results of Raw Ground Beef Products
Analyzed for *E. coli* O157:H7, 2000–2003 *(17)*

Year	No. of samples	No. of positives	%Positives
2000	6375	55	0.86
2001	7010	59	0.84
2002	7025	55	0.78
2003	6584	20	0.30

These FSIS results are very promising and the contamination trend appears to be in the right direction. But let us make no mistake that there is plenty to do in food safety. The recent trends shown above by FSIS are just a beginning. We definitely want to see recall rates hovering around 0%, and contamination rates at zero. Are these rates achievable? Can these rates be maintained?

Many of the pathogens that can cause widespread disease outbreaks among the human population are transmitted through surface and groundwater and may ultimately end up on food, either through irrigation water, through contact or through contaminated processing water. Given the multiple routes through which humans can contaminate the environment as well as be exposed to microbial pathogens, it is impossible or illogical to separate food-safety principles from the environmental quality or from the public health principles (Fig. 2). It is also extremely critical to emphasize that to address food-safety issues holistically, a clear understanding of environmental microbiology,

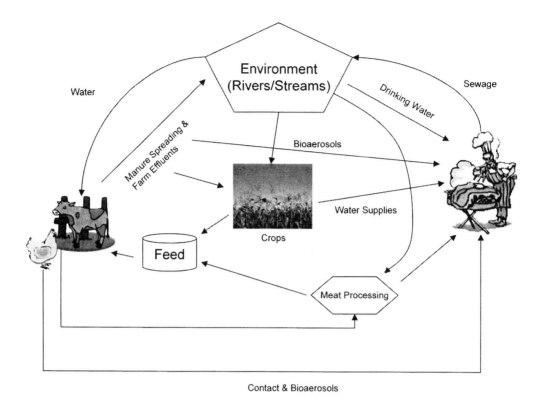

Fig. 2. Spread of microbial pathogens among food, environment, wastes, and humans.

public health epidemiology, aerobiology, molecular microbial ecology, occupational health, industrial processes, municipal water quality, and animal health are needed (Fig. 3). Otherwise, the solutions that are devised and the outcomes that are achieved can be seriously compromised. In the mean time however, we need to be adamant about learning more about host–pathogen interactions, distribution and spread of foodborne hazards, including handling the glut of manure buildup at animal production facilities, antimicrobial resistance involvement from prescription and on-farm abuse, and the abuse because of the incorporation of antibiotics in a myriad of household products available in the open market. We must also strive to improve and simplify verification tests, improve decontamination strategies and on-farm pathogen reduction strategies, be fully prepared to handle the new and emerging food-safety hazards, and be more analytical in addressing microbial food-safety issues. Last but not least we need to come up with a good plan to impart and implement food-safety storage and preparation strategies for the producer, retailer, food service operation, and the consumer. Some of these areas will be highlighted in this chapter. These areas as well as others have been discussed in detail in other recent publications *(18,19)*.

2. DISTRIBUTION AND SPREAD OF FOODBORNE HAZARDS

The scientific community, the regulatory community, and the general public need to realize that is impossible and incorrect to cubbyhole food safety separately from environmental quality and public health. As shown in Fig. 2, the organisms that are of concern

Fig. 3. Interrelationship of food safety with other disciplines.

to food safety find their origins in the environment, in farm animals, and in humans. Organisms are transmitted between humans, animals, the environment, and foods through air, water, soil, and equipment routes. Multiple disciplines and factors ultimately impact the safety and quality of foods (Fig. 3). Although the transmission of hepatitis A infection may be decreased by vaccinating food-service workers, there are other routes of hepatitis A transmission that are extremely difficult to control. The safety in food-service operations can depend highly on the incoming food ingredients. Enteric viruses such as hepatitis A (HAV) can get into the food chain through contaminated irrigation water or wash water. In 1988, 202 cases of HAV were reported in one outbreak, and upon investigation it was traced back to commercially distributed lettuce. Contamination in the lettuce was found to be prior to local distribution *(20)*. In 2003, approx 555 people were infected with HAV at one restaurant, resulting in three deaths, through contaminated green onions which was traced back to contamination in the distribution system or during growing, harvesting, packing, or cooling *(21)*. Pathogens such as *L. monocytogenes* find ecological niches (to survive) within food-processing facilities in locations such as within drains and on food-handling equipment. Thus, the design, the material, and the ability to clean and disinfect the equipment have significant impacts on our ability to control this pathogen.

In the future, improved analytical methods will lead to the identification of the environmental reservoirs of a number of pathogens that are relevant to food safety. Improved surveillance and tracking studies will lead to the identification of many of these pathogens that could be classified as emerging or re-emerging pathogens. Studies on the biotic and abiotic factors that influence the emergence of these pathogens will gain increased attention worldwide. The influence of global climate change on infectious disease trends will be another intensively studied area. The fate of newly identified pathogens in air, water, and soil would need to be explained to identify the environmental management strategies to control these pathogens. Improved waste-handling and waste-treatment systems for both municipal and animal wastes have to be developed to minimize or eliminate infections from pathogens such as SARS. Water treatment technologies incorporating UV disinfection, reverse osmosis, chlorine dioxide, and ozone need to be evaluated for specific applications on the farm and in postharvest processing applications.

3. METHODS FOR MICROBIAL DETECTION AND DIFFERENTIATION

The detection of specific microbial pathogens is expected to increase worldwide. In 1999, the United States food industry performed as many as 144 million microbiological tests *(22)*. These numbers were 23% higher than what was observed in the preceding year. It must be emphasized that though the number of microbiological tests were indeed significant, only approx 26% of these tests were pathogen-specific tests. Regulatory pressures on the food industry such as the "zero tolerance" of pathogens like *L. monocytogenes* and *E. coli* O157:H7 have been responsible for the increase in tests to detect these organisms *(23)*. Even though there are commercial diagnostic assays to detect many of the key foodborne bacterial pathogens, there are no commercial kits to detect enteric viruses in food. This is an unfortunate situation given that viruses account for over 65% of known foodborne illnesses in humans. A variety of food items such as fresh produce, frosted bakery products, infected food-handlers, and shellfish are known to have transmitted viral infections to the human population *(24–26)*. Methods are urgently needed to process the food samples to extract the viruses out of the samples, concentrate the viruses, and ultimately detect the viruses without interference from the sample matrix. Current methods to detect enteric viruses using tissue culture are extremely time-consuming, labor intensive, and are consequently expensive. Qualitative molecular methods that provide only a cursory positive/negative result for the presence of enteric viruses in food may have only limited applicability given the potential to detect noninfectious virus particles. There have been a few reports on the detection of infectious virus particles using molecular methods. These methods need to be rigorously tested in multiple laboratories and on different types of samples to prove their ultimate utility for the food industry.

Critical issues related to the use of molecular methods for pathogen detection in the food industry include choosing the appropriate sample volumes to be tested, sample concentration and purification procedures, and ultimately regulatory acceptance of the molecular methods. Molecular methods will find increasing applications especially for microbial risk-assessment studies. The USDA and the FDA must deal with the issue of re-emerging organisms and have the responsibility of identifying and prioritizing the critical microbial contaminants in foods. Using indicator organisms in lieu of specific pathogens can be fraught with limitations. Although spectacular breakthroughs have been achieved in terms of biosensors for pathogen detection, much work still remains to be done with respect to the first step in detecting pathogens, namely sample processing. The processing has to be extremely efficient at recovering small numbers of organisms on food samples but also has to be amenable to downstream applications such as molecular assays (i.e., PCR, quantitative PCR, or biosensors) without any interference from the sample matrix. PCR is ideally suited for the food industry. It may provide more effective product quality monitoring and prevent costly recalls. Even though molecular methods will not totally replace the conventional microbiological assays, the future molecular methods can allow for faster, more sensitive, and more characterization capabilities. This will entail an increased awareness of QA/QC as it pertains to molecular methods and a trained workforce that can keep up with the rapidly changing technologies.

Differentiation among foodborne pathogens is accomplished in two main ways. The traditional phenotypic methods are one way of identification and classification of bacteria. The second way uses genotypic or other molecular methods *(27)*. Many of the best discriminatory methods, like pulse-field gel electrophoresis (PFGE), take days to conduct. The CDC supported PulseNet which relies on PFGE fingerprinting is considered the golden standard for epidemiological tracking. In the future, methods to differentiate bacteria strains will be developed that are rapid, reproducible, and hopefully automated. The current 16S rDNA-based fingerprinting method (termed Riboprinting), though automated and sophisticated, is still far from being as discriminatory as PFGE. It is critical to improve and devise new discriminatory tools. Methods based on multilocus variable number tandem repeat analysis (MLVA) show promise. This method has been able to distinguish among *E. coli* O157:H7 isolates that appear homogeneous by PFGE *(28,29)*, and distinguished between *Bacillus anthracis* isolates *(30–32)*. It is the new methodologies, such as this one, that will help us to move forward into the future with more discriminatory results and rapid response time. In the final analysis, however, it will be the regulatory and liability pressures that will ultimately dictate whether or not, and to what extent the food industry will adopt molecular methods *(23)*.

4. ON-FARM PATHOGEN REDUCTION INTERVENTION STRATEGIES

Because the farm is the initial location for the foods of animal origin, it is on the farm where the improvement of food safety of animal origin needs to begin. New additional food-safety interventions should be developed to decrease pathogen contamination during livestock and poultry production. A number of products have been developed using the strategy known as competitive exclusion (CE). Competitive exclusion is the process of using microbial cultures to out compete pathogenic bacteria. A defined culture having 29 nonpathogenic bacteria was developed that decreased the prevalence of *Salmonella* in chicks *(33,34)*. Another CE product was developed using an undefined culture that decreased the prevalence of *Salmonella* in poultry *(35)*. Currently, a CE culture to be used in swine is under development. In vitro studies using a continuous flow culture in chemostats demonstrated that the culture designated RPCF decreased *Salmonella enterica* serovar Choleraesuis, *E. coli* F-18, and *E. coli* O157:H7 within 24 h post-inoculation, and *S. enterica* serovar Typhimurium was reduced by 48 h post-inoculation *(36)*. The RPCF culture reduced *Salmonella* serovar Choleraesuis colonization in early-weaned pigs *(37)*. The use of RPCF to protect suckling neonatal pigs against infection with enterotoxigenic *E. coli* resulted in significant reduction of *E. coli* compared to the controls *(38)*. During field trials, the RPCF culture reduced the disease associated with enterotoxigenic strains of *E. coli* in weaned pigs *(39)*. The RPCF culture has shown successful results in decreasing disease caused by *E. coli* in neonatal and weaned pigs. This disease can be fatal and the mortality rate can reach high levels *(40)*. Antibiotics have traditionally been used as a treatment of choice. But over time, *E. coli* have become more resistant to antibiotics and CE is a potential alternative treatment. When developed properly, it could become the method of choice for reducing the disease associated with enterotoxigenic strains of *E. coli*. More work in these and other animal species should be investigated using the CE strategy.

An important area for the use of intervention strategies to reduce pathogens is by using feed additives or treatments. Some of these strategies use heating, pelleting, chemical treatments, or a combination of treatments. The use of chlorate treatment is a very interesting strategy. The reduction of chlorate by nitrate reductase (NR) increases the death rates of *E. coli* and *Salmonella (41)*. *Salmonella* and *E. coli* possess respiratory NR activity *(42)* that can reduce chlorate to cytotoxic chlorite *(43)*, while most gastro-intestinal anaerobes lack NR and are not affected *(44)*. Nitrate adaptation in broilers produced a higher reduction of *Salmonella* Typhimurium following chlorate treatment *(45)*. Chlorate treatment via oral gavage of weaned pigs resulted in reduced cecal concentrations of *Salmonella (44)*, and reduced *E. coli* O157:H7 in the pig gut *(46)*. *E. coli* O157:H7 in sheep can also be reduced by chlorate supplementation *(47,48)*. This is an example of a unique chemical treatment that affects pathogens with little or no affect on other gastrointestinal anaerobes and is worthy of further pursuit.

New innovative approaches to successfully develop intervention strategies would be advantageous. Also, educational efforts are needed that focus on the producers of beef, dairy, pork, egg, and poultry describing the best intervention strategies to decrease pathogen contamination on the farm.

5. TARGETING WASTE AT ANIMAL PRODUCTION SITES

Recently there have been some improvements in wastewater discharge to surface waters *(49)*. It is also common to spread large amounts of animal waste on agricultural lands. Prior to spreading the animal waste, there usually is some form of storage. Temperature was targeted as being the single most effective environmental factor affecting pathogen survival in storage *(50)*. Another factor that influences the availability of pathogens in overland flow is their survival in soil. But there is still a lack of basic data on the levels of pathogens found in animal manures *(50)*. Given the diversity of microbial pathogens and the diversity of soil types and geographical locations, it is expected that data gaps still exist. However, future studies should try to integrate as many different parameters as possible into the study design.

The understanding of how microorganisms partition during overland-flow transport is just beginning to emerge. Existing pathogen transport models do not take into account the interactions between the pathogens and the soil and waste particles. Improvements are needed in pathogen transport models *(49)*. Although significant amounts of information on virus and bacterial transport in the subsurface are known, our knowledge of particle-assisted pathogen migration during runoff is still rudimentary. With a better understanding of overland-flow transport of pathogens, better management strategies can be developed and implemented. Solutions to handle the glut of manure should also include both ground-water and surface runoff water protection strategies toward contamination with bacterial load.

Surface runoff and ground-water contamination by pathogens from intense animal production and waste lagoons is an emerging problem. In 2004, it was reported that North Carolina had 4500 active and 1700 inactive swine waste lagoons *(51)*. Because of the problem with overflow and leakage into surface and ground water, North Carolina is no longer permitting waste lagoons *(51)*. The United States is faced with a manure glut because of the large numbers of animals that are produced *(19)*. The amount of manure was estimated in 1997 to be approx 1.36 billion tons *(19)*. Because of public health and environmental concerns it would be advantageous to develop new technologies to handle large amounts of high nitrogenous wastes that will result in marketable by-products, limit the amount of land required for waste management, limit the impact on surface and ground water, and ultimately also help with global warming.

Until new waste treatment technologies are developed and rigorously tested we should assume that animal wastes harbor harmful pathogens, and their contact with humans and crops should be limited to the extent possible. It is important to thoroughly understand the impact of animal wastes on public health and the environment. The municipal waste industry is currently facing enormous challenges related to the disposal of municipal sewage sludge (biosolids) close to human population centers because of poor management. We must determine the magnitude of the contamination from bacteria found in animal wastes on surface water, ground water, and soil (and the plants that are grown on this soil). To embrace this problem with a better understanding, it will require numerous studies that evaluate the bacteria in animal production facilities, in surface water, and at sites where waste leakage or water runoff is likely.

We only need to remind ourselves of the recent Walkerton, Ontario, tragedy to put in perspective how contamination from cattle manure can cause a large disaster *(52)*. This disaster occurred in a pristine Canadian small farming community, where more than 2300 people became ill and seven people died *(53)*. The tragedy was caused by contamination of a drinking water well by cattle manure containing *E. coli* O157:H7. The potential contamination of water wells should be at the top of the list as a critical focal point for development and implementation of prevention strategies. Another area of concern directly associated with water-well contamination is the use of contaminated water on truck crops. How about the process of recycling this water? What steps are needed to be sure that this water is pathogen-free (including *E. coli* O157:H7) when reused on truck crops?

6. ANTIMICROBIAL RESISTANCE CONSIDERATIONS

Because some antibiotics or classes of antibiotics are used to treat both human and animal illnesses, there is a large debate over the emergence of antibiotic-resistant pathogens. Some health professionals believe that antibiotic-resistant pathogens have emerged because of the therapeutic use of antibiotics to treat animal diseases and by the use of growth promoters in animal feeds *(54–56)*. However, others believe that over-prescription and abuse of antibiotics to treat human illnesses have caused the emergence of antibiotic-resistant pathogens *(57)*. Antimicrobial resistance has become a highly controversial topic in both animal husbandry and clinical medicine. Even though much is known about the role of gene transfer mechanisms in the development of antibiotic resistance, it is extremely important to clearly understand the molecular mechanisms that signal the involvement of horizontal gene transfer of antibiotic resistance genes among bacteria. Once the mechanisms and magnitude of resistance gene transfer are clearly understood and quantified, strategies can be sought to reduce the potential for dissemination of these genes. Also, the role of host signals in the development of antibiotic-resistant bacteria in humans and animals warrants further study.

Research reports have expressed concern that the use of biocides may contribute to the development of antibiotic resistance *(58)*. The use of antimicrobials such as triclosan in hand and dish soaps, deodorant soaps, hand creams, shower gels, surgical scrubs, facial cleaners, sanitizers, coatings and gloves may very well be the suspect because of the possibility of introducing antimicrobial resistance. It was thought that triclosan was a nonspecific biocide that disrupted cell membranes, thereby leaving the bacteria unable to proliferate *(59)*. However, recent research has shown that triclosan acts at specific targets *(60,61)*. The basis of triclosan's activity is to inhibit fatty-acid biosynthesis *(62)*. Triclosan is a substrate of a multidrug efflux pump in *Pseudomonas aeruginosa*, and triclosan will select for itself as well as for multidrug-resistance in bacteria *(63)*. Therefore, it is quite clear that triclosan should not be in every product possible, as it is now. Especially, it should not be used in surgical scrubs. In the future, chemicals like triclosan should not find their way into consumer products; it is expected that sooner or later triclosan will be removed from the glut of products on the market that now contain it. However, there is a high-dollar commercial investment in using triclosan or chemicals like it in consumer products. This allows the tag of "antibacterial" to be used, which is worth a lot to the companies that manufacture these products.

There has been an ongoing concern whether antibiotic and biocide resistance may be linked in some way *(64–66)*. Indeed a link has been shown *(63,67,68)*. Exposure to antibiotics or triclosan can select for multidrug-resistance by over-expression of identical multidrug efflux systems *(63)*. There is an urgent need for improved surveillance systems to monitor the development of biocide resistance in enteric pathogens. We must be mindful of using biocides in animal production that are also used in food-processing plants, food-preparation facilities, and in human clinical settings. We also must consider stopping the use of biocides like triclosan that show a link between the developments of cross-resistance to antibiotics.

The debate over the origin of antibiotic resistance will ultimately affect the success or failure of global marketplace opportunities. The success of the global marketplace will require truly common sense global food-safety standards and equivalent food-safety systems in place. Sperber stated that industry had responded more quickly on *L. monocytogenes* in 1986 and *E. coli* O157:H7 in 1994 because of the growing alliance between the food industry, trade associations, and research teams *(11)*. He commented, "you put these groups together and it is a powerful alliance" *(11)*. Further, he recommends that for a true global effort to produce food-safety rules that are based on common sense requires the addition and contribution of regulators to the alliance.

We need to be forward looking in terms of antimicrobial resistance considerations. Not only should we be improving detection methods, the understanding of resistance mechanisms, and the types of environmental or human-made pressures that increase resistance, but we need to be putting together the international cooperation needed to swiftly handle an emerging resistance problem anywhere in the world.

7. FOOD-SAFETY STORAGE AND PREPARATION STRATEGIES

A comparison of food-safety knowledge with home food-handling practices found that the level of knowledge of food-safety practices is much higher than the level of in-home application of food-handling practices *(69,70)*. Anderson and coworkers placed cameras in the kitchens of 100 middle-class families to directly observe their food handling and preparation behaviors *(70)*. The researchers evaluated washing hands, cleaning working surfaces and vegetables, modes of cross-contamination, and cooking and chilling behaviors. The results showed the researchers why foodborne illness is such a problem in American homes. However, in a survey that was conducted by the researchers, nearly all subjects were concerned about food safety *(70).*

The Council for Agricultural Science and Technology (CAST) discussed consumer education pointing out that the educational goal of increasing the application of safe food-handling and food-consumption practices will continue to target the consumer. However, after much educational effort, the gap between current food-safety education and the ideal remains large *(19).* Consumers are increasingly aware that foodborne illnesses are a problem, but they continue to blame others and are unaware of the role that they must play in food safety *(19,71).* McIntosh took a critical look at consumer food-handling behavior *(71).* He concluded that the consumers are aware of food-safety problems, and yet do not make improvements in their own food-handling behavior *(71).* McIntosh points out that perhaps an intervention having long-term consequences might be within the public school system, and to potentially use the health education class to introduce and discuss food safety *(71).* He notes that adolescents are in the process of evolving food habits that will carry over to adulthood *(71).*

We would like to make further suggestions regarding the use of public school system as a vehicle to embed food-safety strategies into every individual. It is our opinion that during the Junior or Senior year of High School a class be mandatory for every student on food-safety strategies and behavior, and food-safety handling practices for various foods. These handling behaviors should incorporate both a variety of pathogens as well as some common food toxicants. Therefore, the principles of handwashing, surface cleaning, eliminating cross-contamination, cooking, and chilling should all be demonstrated and practiced by the students. The students should also gain knowledge in this food-safety class of the safe ways to store and handle those foods that contain toxicants (e.g., potatoes, celery, parsley, limes, and sweet potatoes).

8. FOOD IRRADIATION

Foodborne illness is preventable. It can be prevented by improved food-production methods, improved food-processing technologies, and improved food preparation and consumption methods within homes. A number of food-processing technologies have been developed and employed in recent years. However, none of the current technologies have had the same level of promise, and unfortunately the level of criticism, as had food irradiation. Ionizing irradiation is one of the most extensively studied food-processing technologies. It was found in 1904 that ionizing radiation could destroy bacteria, and the technology was evaluated in 1921 to destroy trichinae in pork *(72).* Today, we have nationally and internationally approved irradiation protocols for a variety of food products including uncooked meat and poultry products.

Ionizing radiation is defined as a radiation that has enough energy to remove electrons from atoms thereby leading to the formation of ions. There are different types of ionizing radiation such as X-rays, gamma rays and beta rays (E-beam), depending on the source of the radiation. All types of ionizing radiation function in the same way, i.e., causing ionization of atoms in the food material by stripping off electrons. Over 40 different countries have approved the use of food irradiation. The United States has the most number of approvals for the use of irradiation on foods. However, the United States Food and Drug Administration (FDA) still considers irradiation as a "food additive." The inappropriateness of this classification is evident since other processes such as baking, frying, and boiling, which also cause chemical changes in the food are not considered as additives *(73)*.

Internationally, foods such as apples, strawberries, bananas, mangoes, onions, potatoes, spices and seasonings, meat, poultry, fish, frog legs, and grains have been irradiated for many years. There is a worldwide standard for food irradiation. The standard was adopted by the Codex Alimentarius Commission, a joint body of the Food and Agricultural Organization (FAO) of the United Nations and the World Health Organization *(73)*. Achieving low doses of irradiation is extremely important to eliminate minor traces of microbial contamination and retain the sensory qualities of specific food items. In conjunction with low-dose irradiation capabilities, dosimeters that can measure such low doses are urgently required. There is a need to develop irradiation protocols for achieving uniform dose distribution on uniquely shaped foods, such as apples, cantaloupes, tomatoes, and heterogeneously shaped packages. Standardized protocols are needed for dosimeter placement and product presentation to the E-beam to achieve minimal min : max ratios for efficient pathogen kill. Studies are also needed to identify the irradiation dose for different food-matrix properties that can eliminate viral pathogens on fruits and vegetables that are minimally processed and are highly vulnerable to fecal contamination *(74)*.

9. NEW AND EMERGING FOOD-SAFETY HAZARDS

During the early 1990s, there was a widespread concern over the possibility of developing resistance in pathogens to antimicrobial agents used in food-animal production. In 1997, a group from Denmark published an exhaustive study on the issue *(75)*. Seyforth and coworkers concluded that although *Salmonella* Typhimurium isolated from animals and humans showed antimicrobial resistance, multiple-resistance was most often acquired outside of Denmark *(75)*. Then Aarestrup and Wegener stated in a review that "There is an urgent need to implement strategies for prudent use of antibiotics in food animal production to prevent further increases in the occurrence of antimicrobial resistance in foodborne human pathogenic bacteria such as *Campylobacter* and *E. coli*" (*[76]*, *p. 639*). However, their concerns immediately became reality. A variant of *Salmonella* Typhimurium strain DT104, resistant to quinolones, was responsible for the death of two people in 1998 *(77)*. This bacterium was traced back to a single Danish swine herd that had been treated with quinolones *(78)*. This episode clearly documents the spread of zoonotic (the transmission of infections under normal conditions from animals to humans) bacteria from animals to humans. The threat of emerging resistant pathogenic bacteria is real. We must be focused with a close watch on the potential for this to happen anywhere in the world *(79)*.

9.1. Bovine Spongiform Encephalopathy Responsiveness and Preparedness

The emergence of bovine spongiform encephalopathy (BSE) is one example of a challenge to the food production safety system upon which the existing food-safety monitoring systems did not have an appreciable impact *(11)*. See a review on risks for human health from animal transmissible spongiform encephalopathies (TSEs) *(80)*. Since BSE is a threat to human and animal health, and fell outside of the existing food-safety health measures, firewalls were developed by the United States to prevent its introduction and amplification *(19)*. Three firewalls have been introduced, briefly: (1) importation of live ruminants and ruminant products are restricted from countries with BSE cases; (2) the USDA performs immunohistological exams of all brains from cattle condemned for central nervous system disorders; and (3) the FDA has prohibited feeding ruminant meat and bone meal to ruminants *(19)*.

Based on strain typing tests in mice, BSE also causes the human disease, variant Creutzfeldt Jakob disease (vCJD) *(81)*, and both BSE and vCJD had similar incubation times in prion gene replacement studies *(82,83)*. All of the different types of tests used to evaluate the similarities and differences between BSE and vCJD have clearly shown that BSE and vCJD are the same TSE strain *(80)*. While these and other tests for strain-typing showed that chronic wasting disease (CWD) in the family Cervidae, which includes mule deer, white-tailed deer, and Rocky Mountain elk, possesses different patterns from BSE or vCJD. Please see other reviews for further reading concerning CWD *(84,85)*.

Sheep have been given the same food as cattle in some areas, and this may result in the possibility of sheep also contracting BSE. There is a concern of possible trans-mission of vCJD to humans through BSE-infected sheep, which would be very difficult to distinguish in sheep from scrapie *(80)*. Therefore, methodology to test for BSE in sheep vs scrapie is needed.

Scrapie was recognized in the United States since 1947. It was a speculation that CWD was derived from scrapie. Observations from captive cervids provide evidence of lateral transmission of CWD, which is similar to scrapie, but the details concerning the transmission of CWD still remain to be determined. The CWD agent has been demonstrated to be in various tissues suggesting that the CWD agent may be shed through the alimentary tract, and shedding probably precedes the onset of the clinical disease in both deer and elk *(85)*. However, CWD is like BSE because of the long incubation periods and subtle early clinical signs. There have been no cases reported of human prion disease associated with CWD, whereas human exposure to the BSE agent has resulted in over one hundred deaths due to vCJD *(85)*. The prevalence of CWD has been steadily increasing, and therefore, we need to carefully assess the potential risk that CWD exposure may pose to humans. Recently, the sheep scrapie agent was used to experimentally induce spongiform encephalopathy in elk, and could not be distinguished from CWD of elk *(86)*.

Certainly, much remains to be learned about the transmission and epidemiology of BSE and CWD. Current improvements in CWD testing using enzyme-linked immunosorbent assay (ELISA) technology proved that the ELISA is an excellent rapid test for screening large number of samples and there is a tremendous improvement in CWD detection *(87)*, and similar improvements are needed in the detection of BSE. A much higher proportion of cattle need to be evaluated for BSE. ELISA methods have been developed in collaboration with the French Commissariat à L'Energie Atomique (CEA) for a very highly sensitive and specific ELISA for BSE. This test is available in many countries including the United Kingdom, France, Germany, Belgium, Luxembourg, the Netherlands, Norway, Sweden, Switzerland, Italy, and Spain *(88)*, but not in the United States.

9.2. Is there a Link Between Johne's and Crohn's Diseases?

There are concerns that the cause of Johne's disease in ruminants is also the cause of Crohn's disease in humans *(89,90)*. Johne's disease is a chronic inflammatory bowel disease of ruminants caused by *Mycobacterium avium* subsp. *paratuberculosis* (MapTb) *(91)*. It is estimated that Johne's disease costs the US dairy industry $200–250 million annually in reduced productivity *(92)*. Crohn's disease is also a chronic inflammatory gastrointestinal disease found in humans that can affect the whole digestive tract, but most commonly affects the lower portion of the small intestine (ileum), where it connects with the large intestine *(93)*. The cause of Crohn's disease is not known *(93)*. It is estimated that almost one million Americans live with Crohn's disease, and that up to 75% of those who live with Crohn's disease may require surgery at some point *(93)*.

Paratuberculosis in ruminants has a prolonged incubation period, resulting in most animals remaining subclinical *(90)*. It is these subclinical animals that can spread the disease to other animals. It is estimated that more than 20% of cow herds in the United States are infected with Johne's disease *(94)*. The survival, in some cases, of MapTb during pasteurization of raw milk *(19,95,96)*, the isolation of MapTb from wildlife, and the similar manifestation of Johne's and Crohn's disease point out why there is a concern. There is a need for more accurate diagnostic testing for MapTb *(90)*. However, the MapTb link between Johne's and Crohn's diseases is highly debated. The debate is primarily based on an inability to satisfy Koch's postulates, since the presence of MapTb has not been demonstrated in all cases of Crohn's disease *(90)*.

The knowledge of the genomic structure of this group of organisms is incomplete, and there is evidence suggesting that *M. avium* and *Map*Tb may represent only two forms of a continuum of complex *M. avium* isolates *(90)*. Recently developed animal models offer the opportunity to study specific interactions between the host/pathogen and potentially lead to improved diagnosis and therapeutic treatment *(90)*. This area would benefit from more researchers evaluating the causes of Crohn's disease, and more pointedly, improved and more accurate diagnostic testing.

10. QUANTITATIVE MICROBIAL FOOD-SAFETY RISK ASSESSMENT

In quantitative microbial risk assessment (QMRA), risk assessment is the first component of the risk analysis process, and it is followed by risk management and risk communication *(97,98)*. Mena and coworkers have discussed QMRA and its application for foodborne pathogens *(98)*. The goal of the risk analysis process is to lead to risk management decisions to better utilize intervention strategies or monitoring procedures. QMRAs are needed to identify the critical points within the food supply system at which additional interventions will have the greatest impact on decreasing the health hazards and improving the overall suitability of foods. Therefore, the result from QMRA can be folded right back into HACCP programs.

The use of QMRA in the food industry has the promise of being very powerful. However, there are many data gaps that must be addressed before QMRA can be fully utilized *(19)*. Also, QMRA is not simple, for it requires microbial modeling. A controversial area is the choice of the model to fit the data with *(98)*. QMRA provides the necessary approach to predict the public health significance of new food production and processing practices, the emergence of foodborne pathogens, and the use of particular critical points for intervention strategies or monitoring procedures *(98)*. With aggressive data gathering along with adoption of models proven to fit real world scenarios, QMRA will become the premier risk assessment tool in the food industry.

REFERENCES

1. USDHHS–CDC (US Department of Health and Human Services–Centers for Disease Control and Prevention). (1996) Surveillance for foodborne-disease outbreaks: United States, 1988–1992. Centers for Disease Control and Prevention Surveillance Summary. *Morb. Mort. Wkly Rep.* **45,** SS–5.
2. Mead, P. S., Slutsker, L., Dietz, V., et al. (1999) Food-related illness and death in the United States. *Emer. Infect. Dis.* **5,** 607–625.
3. Buzby, J. C. and Roberts, T. (1997) Economic costs and trade impacts of microbial foodborne illness. *World Health Stat. Q.* **50(1–2),** 57–66.
4. Lax, A. J., Barrow, P. A., Jones, P. W., and Wallis, T. S. (1995) Current perspectives in salmonellosis. *Br. Vet. J.* **151,** 351–377.
5. Frenzen, P. D., Riggs, T. L., Buzby, J. C., et al. (1999) *Salmonella* cost estimate updated using FoodNet data. *FoodReview* **22(2),** 10–15.
6. Waghela, S. D. (2004) Pathogenic *Escherichia coli*. In: *Preharvest and Postharvest Food Safety: Contemporary Issues and Future Directions* (Beier, R. C., Pillai, S. D., Phillips, T. D., and Ziprin, R. L., eds.), IFT/Blackwell, Ames, IA, pp. 13–25.
7. Crump, J. A., Braden, C. R., Dey, M. E., et al. (2003) Outbreaks of *Escherichia coli* O157 infections at multiple county agricultural fairs: a hazard of mixing cattle, concession stands and children. *Epidemiol. Infect.* **131,** 1055–1062.
8. Castell-Perez, M. E. and Moreira, R. G. (2004) Decontamination systems. In: *Preharvest and Postharvest Food Safety: Contemporary Issues and Future Directions* (Beier, R. C., Pillai, S. D., Phillips, T. D., and Ziprin, R. L., eds.), IFT/Blackwell, Ames, IA, pp. 337–347.
9. USDA–FSIS (US Department of Agriculture–Food Safety Inspection Service). (2002) Pathogen Reduction; Hazard Analysis and Critical Control Point (HACCP) Systems. *Title 9 Code of Federal Regulations*, Parts 304, 308, 310, 320, 327, 381, 416, and 417. Government Printing Office, Washington, DC.
10. Federal Register. (1996) Pathogen Reduction; Hazard Analysis and Critical Control Point (HACCP) Systems; Final Rule. United States Department of Agriculture–Food Safety and Inspection Service. *Title 9 CFR*, Parts 304, 308, 310, 320, 327, 381, 416, and 417. *Fed. Regist.* **61,** 38,805–38,989.
11. Sperber, W. H. (2004) Advancing the food safety agenda. *Food Saf. Mag.* **10(3),** 32, 34–36.
12. Stevenson, K. E. and Bernard, D. T. (1999) Introduction to hazard analysis and critical control point systems. In: *HACCP: A Systematic Approach to Food Safety*, 3rd edn, The Food Processors Institute, Washington, DC, pp. 1–4.
13. Mortimore, S. and Wallace, C. (2001) *Food Industry Briefing Series: HACCP*, Iowa State University Press/Blackwell Science, Ames, IA.
14. Keeton, J. T. and Harris, K. B. (2004) The hazard analysis and critical control point system and importance of verification procedures. In: *Preharvest and Postharvest Food Safety: Contemporary Issues and Future Directions* (Beier, R. C., Pillai, S. D., Phillips, T. D., and Ziprin, R. L., eds.), IFT Press/Blackwell, Ames, IA, pp. 257–269.
15. USDA–FSIS (US Department of Agriculture–Food Safety Inspection Service). (2003) Enhancing public health: strategies for the future. Available at http://www.fsis.usda.gov/Frame/FrameRedirect.asp?main=/oa/speeches/2003/em_sma.htm (accessed on 16 September 2004).
16. USDA–FSIS (US Department of Agriculture–Food Safety and Inspection Service). (2004) Fulfilling the vision: initiatives in protecting public health. Available at www.fsis.usda.gov (accessed on 31 August 2004), p. 5.
17. USDA–FSIS (US Department of Agriculture–Food Safety and Inspection Service). (2003) Microbiological results of raw ground beef products analyzed for *Escherichia coli* O157:H7. Available at http://www.fsis.usda.gov/Science/Ground_Beef_E.Coli_Testing_Results/index. asp (accessed on 13 September 2004).

18. Beier, R. C., Pillai, S. D., Phillips, T. D., and Ziprin, R. L. (eds.) (2004) *Preharvest and Postharvest Food Safety: Contemporary Issues and Future Directions*, IFT/Blackwell, Ames, IA.

19. CAST (Council for Agricultural Science and Technology). (2004) *Intervention Strategies for the Microbiological Safety of Foods of Animal Origin.* Issue Paper 25. Council for Agricultural Science and Technology, Ames, IA.

20. Rosenblum, L. S., Mirkin, I. R., Allen, D. T., et al. (1990) A multifocal outbreak of hepatitis A traced to commercially distributed lettuce. *Am. J. Public Health* **80,** 1076–1079.

21. Dato, V., Weltman, A., Waller, K., et al. (2003) Hepatitis A outbreak associated with green onions at a restaurant—Monaca, Pennsylvania, 2003. *Morb. Mort. Wkly Rep.* **52,** 1155–1157.

22. Strategic Consulting. (2000) *Pathogen Testing in the US Food Industry.* Strategic Consulting, Woodstock, VT.

23. Pillai, S. D. (2004) Molecular methods for microbial detection. In: *Preharvest and Postharvest Food Safety: Contemporary Issues and Future Directions* (Beier, R. C., Pillai, S. D., Phillips, T. D., and Ziprin, R. L., eds.), IFT/Blackwell, Ames, IA, pp. 289–302.

24. Richards, G. P. (2001) Enteric virus contamination of foods through industrial practices: a primer on intervention strategies. *J. Ind. Microbiol. Biotechnol.* **27,** 117–125.

25. Frankhauser, R. L., Monroe, S. S., Noel, J. S., et al. (2002) Epidemiologic and molecular trends of "Norwalk-like viruses" associated with outbreaks of gastroenteritis in the United States. *J. Infect. Dis.* **186,** 1–7.

26. Goyal, S. M. (2004) Viruses in food. In: *Preharvest and Postharvest Food Safety: Contemporary Issues and Future Directions* (Beier, R. C., Pillai, S. D., Phillips, T. D., and Ziprin, R. L., eds.), IFT/Blackwell, Ames, IA, pp. 101–117.

27. Foley, S. L. and Walker, R. D. (2004) Methods for differentiation among bacterial foodborne pathogens. In: *Preharvest and Postharvest Food Safety: Contemporary Issues and Future Directions* (Beier, R. C., Pillai, S. D., Phillips, T. D., and Ziprin, R. L., eds.), IFT/Blackwell, Ames, IA, pp. 303–316.

28. Keys, C., Kemper, S., and Keim, P. (2002) MLVA: a novel typing system for *E. coli* O157:H7. Abstracts of the American Society for Microbiology General Meeting, May 19–23, Salt Lake City, UT.

29. Keys, C., Jay, Z., Fleishman, A., et al. (2003) VNTR Mutations in *E. coli* O157:H7—rates, products and allelic effects. Abstracts of the American Society for Microbiology General Meeting, May 18–22, Washington, DC.

30. Keim, P., Smith, K. L., Keys, C., et al. (2001) Molecular investigation of the Aum Shinrikyo anthrax release in Kameido, Japan. *J. Clin. Microbiol.* **39,** 4566–4567.

31. Fouet, A., Smith, K. L., Keys, C., et al. (2002) Diversity among French *Bacillus anthracis* isolates. *J. Clin. Microbiol.* **40,** 4732–4734.

32. Takahashi, H., Keim, P., Kaufmann, A. F., et al. (2004) *Bacillus anthracis* incident, Kameido, Tokyo, 1993. *Emerg. Infect Dis.* **10,** 117–120.

33. Hume, M. E., Corrier, D. E., Nisbet, D. J., and DeLoach, J. R. (1996) Reduction of *Salmonella* crop and cecal colonization by a characterized competitive exclusion culture in broilers during growout. *J. Food Protect.* **59,** 688–693.

34. Nisbit, D. J., Corrier, D. E., and DeLoach, J. R. (1997) Probiotic for control of *Salmonella* in fowl produced by continuous culture of fecal/cecal material. US Patent No. 5,604,127.

35. Bailey, J. S., Cason, J. A., and Cox, N. A. (1998) Effect of *Salmonella* in young chicks on competitive exclusion treatment. *Poult. Sci.* **77,** 394–399.

36. Harvey, R. B., Droleskey, R. E., Hume, M. E., et al. (2002) In vitro inhibition of *Salmonella enterica* serovars Choleraesuis and Typhimurium, *Escherichia coli* F-18, and *Escherichia coli* O157:H7 by a porcine continuous-flow competitive exclusion culture. *Curr. Microbiol.* **45,** 226–229.

37. Anderson, R. C., Stanker, L. H., Young, C. R., et al. (1999) Effect of competitive exclusion treatment on colonization of early-weaned pigs by *Salmonella* serovar Choleraesuis. *Swine Health Prod.* **7,** 155–160.

38. Genovese, K. J., Harvey, R. B., Anderson, R. C., and Nisbet, D. J. (2001) Protection of suckling neonatal pigs against infection with an enterotoxigenic *Escherichia coli* expressing 987P fimbriae by the administration of a bacterial competitive exclusion culture. *Microb. Ecol. Health Dis.* **13,** 223–228.

39. Harvey, R. B., Ebert, R. C., Schmitt, C. S., et al. (2003). Use of a porcine-derived, defined culture of commensal bacteria as an alternative to antibiotics to control *E. coli* disease in weaned pigs: field trial results. Proceedings of the 9th International Symposium on Digestive Physiology in Pigs, Banff, AB, Canada, Vol. 2, pp. 72–74.

40. Genovese, K. J., Anderson, R. C., Harvey, R. B., and Nisbet, D. J. (2000) Competitive exclusion treatment reduces the mortality and fecal shedding associated with enterotoxigenic *Escherichia coli* infection in nursery-raised neonatal pigs. *Can. J. Vet. Res.* **64,** 204–207.

41. Tamási, G. and Lantos, Z. (1983) Influence of nitrate reductases on survival of *Escherichia coli* and *Salmonella enterititdis* in liquid manure in the presence and absence of chlorate. *Agric. Wastes* **6,** 91–97.

42. Brenner, D. J. (1984) Enterobacteriaceae. In: *Bergey's Manual of Systematic Bacteriology* (Krieg, N. R. and Holt, J. G., eds.), Vol. 1, Williams & Wilkins, Baltimore, MD, pp. 408–420.

43. Stewart, V. (1988) Nitrate respiration in relation to facultative metabolism in enterobacteria. *Microbiol. Rev.* **52,** 190–232.

44. Anderson, R. C., Buckley, S. A., Callaway, T. R., et al. (2001) Effect of sodium chlorate on *Salmonella* Typhimurium concentrations in the weaned pig gut. *J. Food Prot.* **64,** 255–258.

45. Jung, Y. S., Anderson, R. C., Byrd, J. A., et al. (2003) Reduction of *Salmonella* Typhimurium in experimentally challenged broilers by nitrate adaptation and chlorate supplementation in drinking water. *J. Food Prot.* **66,** 660–663.

46. Anderson, R. C., Callaway, T. R., Buckley, S. A., et al. (2001) Effect of oral sodium chlorate administration on *Escherichia coli* O157:H7 in the gut of experimentally infected pigs. *Int. J. Food Microbiol.* **71,** 125–130.

47. Callaway, T. R., Edrington, T. S., Anderson, R. C., et al. (2003) *Escherichia coli* O157:H7 populations in sheep can be reduced by chlorate supplementation. *J. Food Prot.* **66,** 194–199.

48. Edrington, T. S., Callaway, T. R., Anderson, R. C., et al. (2003) Reduction of *E. coli* O157:H7 populations in sheep by supplementation of an experimental sodium chlorate product. *Small Ruminant Res.* **49,** 173–181.

49. Tyrrel, S. F. and Quinton, J. N. (2003) Overland flow transport of pathogens from agricultural land receiving faecal wastes. *J. Appl. Microbiol.* **94,** 87S–93S.

50. Nicholson, F. A., Hutchinson, M. C., Smith, K. A., et al. (2000) *A Study on Farm Manure Application to Agricultural Land and an Assessment of the Risks of Pathogens Transfer into the Food Chain.* Project Number FS2526, Final Report to the Ministry of Agriculture, Fisheries and Food, London.

51. Humenik, F. J., Rice, J. M., Baird, C. L., and Koelsch, R. (2004) Environmentally superior technologies for swine waste management. *Water Sci. Technol.* **49(5–6),** 15–22.

52. Ali, S. H. (2004) A socio-ecological autopsy of the *E. coli* O157:H7 outbreak in Walkerton, ON, Canada. *Soc. Sci. Med.* **58,** 2601–2612.

53. Bruce-Grey-Owen Sound Health Unit. (2000) *The Investigative Report of the Walkerton Outbreak of Waterborne Gastroenteritis, May–June, 2000.* Released on 10 October 2000 during a public meeting, Walkerton, ON.

54. Kelley, T. R., Pancorbo, O. C., Merka, W. C., and Barnhart, H. M. (1998) Antibiotic resistance of bacterial litter isolates. *Poult. Sci.* **77,** 243–247.

55. Rajashekara, G., Haverly, E., Halvorson, D. A., et al. (2000) Multidrug-resistant *Salmonella typhimurium* DT104 in poultry. *J. Food Protect.* **63,** 155–161.
56. Teuber, M. (2001) Veterinary use and antibiotic resistance. *Curr. Opin. Microbiol.* **4,** 493–499.
57. Price, D. (2000) Real antibiotics issue cannot be overlooked. *Feedstuffs* **72,** 8, 18.
58. Levy, S. B. (2001) Antibacterial household products: cause for concern. *Emerg. Infect. Dis.* **7,** 512–515.
59. Regös, J., Zak, O., Solf, R., et al. (1979) Antimicrobial spectrum of triclosan, a broadspectrum antimicrobial agent for topical application. II. Comparison with some other antimicrobial agents. *Dermatologica* **158,** 72–79.
60. Heath, R. J., Li, J., Roland, G. E., and Rock, C. O. (2000) Inhibition of the *Staphylococcus aureus* NADPH-dependent enoyl–acyl carrier protein reductase by triclosan and hexachlorophene. *J. Biol. Chem.* **275,** 4654–4659.
61. Heath, R. J. and Rock, C. O. (2000) A triclosan-resistant bacterial enzyme. *Nature* **406,** 145–146.
62. Levy, C. W., Roujeinikova, A., Sedelnikova, S., et al. (1999) Molecular basis of triclosan activity. *Nature* **398,** 383–384.
63. Chuanchuen, R., Beinlich, K., Hoang, T. T., et al. (2001) Cross-resistance between triclosan and antibiotics in *Pseudomonas aeruginosa* is mediated by multidrug efflux pumps: exposure of a susceptible mutant strain to triclosan selects *nfxB* mutants overexpressing *MexCD-OprJ*. *Antimicrob. Agents Chemother.* **45,** 428–432.
64. Lambert, R. J. W., Joynson, J., and Forbes, B. (2001) The relationships and susceptibilities of some industrial, laboratory and clinical isolates of *Pseudomonas aeruginosa* to some antibiotics and biocides. *J. Appl. Microbiol.* **91,** 972–984.
65. White, D. G. and McDermott, P. F. (2001) Biocides, drug resistance and microbial evolution. *Curr. Opin. Microbiol.* **4,** 313–317.
66. Beier, R. C., Bischoff, K. M., and Poole, T. L. (2004) Disinfectants (biocides) used in animal production: antimicrobial resistance considerations. In: *Preharvest and Postharvest Food Safety: Contemporary Issues and Future Directions* (Beier, R. C., Pillai, S. D., Phillips, T. D., and Ziprin, R. L., eds.), IFT/Blackwell, Ames, IA, pp. 201–211.
67. Schweizer, H. P. (1998) Intrinsic resistance to inhibitors of fatty acid biosynthesis in *Pseudomonas aeruginosa* is due to efflux: application of a novel technique for generation of unmarked chromosomal mutations for the study of efflux systems. *Antimicrob. Agents Chemother.* **42,** 394–398.
68. Beier, R. C., Bischoff, K. M., Ziprin, R. L., et al. (2005) Chlorhexidine susceptibility, virulence factors and antibiotic resistance of beta-hemolytic *Escherichia coli* isolated from neonatal swine with diarrhea. *Bull. Environ. Contam. Toxicol.* **75(5),** 835–844.
69. Albrecht, J. (1995) Food safety knowledge and practices of consumers in the US. *J. Cons. Stud. Home Econ.* **19,** 103–118.
70. Anderson, J. B., Shuster, T. A., Gee, E., et al. (2001) A camera's view of consumer food handling and preparation practices. Safe Food Institute Online, Utah State University, Logan. Available at http://www.safefoodinstitute.org/finding. htm (accessed on 15 September 2004).
71. McIntosh, W. A. (2004) Food safety risk communication and consumer food-handling behavior. In: *Preharvest and Postharvest Food Safety: Contemporary Issues and Future Directions* (Beier, R. C., Pillai, S. D., Phillips, T. D., and Ziprin, R. L., eds.), IFT/Blackwell, Ames, IA, pp. 405–414.
72. Josephson, E. S. (1983) An historic review of food irradiation. *J. Food Saf.* **5,** 161–190.
73. Pillai, S. D. (2004) Food irradiation: a solution to combat worldwide food-borne illnesses. Proceedings of the International Congress of Bioprocessing in the Food Industry, July 11–13, Clermont-Ferrand, France.
74. Pillai, S. D. (2004) Food irradiation. In: *Preharvest and Postharvest Food Safety: Contemporary Issues and Future Directions* (Beier, R. C., Pillai, S. D., Phillips, T. D., and Ziprin, R. L., eds.), IFT/Blackwell, Ames, IA, pp. 375–387.

75. Seyfarth, A. M., Wegener, H. C., and Frimodt-Møller, N. (1997) Antimicrobial resistance in *Salmonella enterica* subsp. *enterica* serovar Typhimurium from humans and production animals. *J. Antimicrob. Chemother.* **40**, 67–75.

76. Aarestrup, F. M. and Wegener, H. C. (1999) The effects of antibiotic usage in food animals on the development of antimicrobial resistance of importance for humans in *Campylobacter* and *Escherichia coli*. *Microbes Infect.* **1**, 639–644.

77. Ferber, D. (2000) Superbugs on the hoof? *Science* **288**, 792–794.

78. Mølbak, K., Baggesen, D. L., Aarestrup, F. M., et al. (1999) An outbreak of multidrug-resistant, quinolone-resistant *Salmonella enterica* serotype Typhimurium DT104. *N. Engl. J. Med.* **341**, 1420–1425.

79. CIDRAP (Center for Infectious Disease Research & Policy). (2004) Links between human and animal disease surveillance growing. University of Minnesota, Minneapolis–St. Paul, MN. Available at http://www.cidrap.umn.edu/cidrap/content/bt/bioprep/news/sept2104vetspub_rev. html (accessed on 22 September 2004).

80. Schmerr, M. J. (2004) Do animal transmissible spongiform encephalopathies pose a risk for human health? In: *Preharvest and Postharvest Food Safety: Contemporary Issues and Future Directions* (Beier, R. C., Pillai, S. D., Phillips, T. D., and Ziprin, R. L., eds.), IFT/Blackwell, Ames, IA, pp. 173–187.

81. Bruce, M. E., Will, R. G., Ironside, J. W., et al. (1997) Transmissions to mice indicate that "new variant" CJD is caused by the BSE agent. *Nature* **389**, 498–501.

82. Hill, A. F., Desbruslais, M., Joiner, S., et al. (1997) The same prion strain causes vCJD and BSE. *Nature* **389**, 448–450.

83. Scott, M. R., Will, R., Ironside, J., et al. (1999) Compelling transgenetic evidence for transmission of bovine spongiform encephalopathy prions to humans. *Proc. Natl Acad. Sci. USA* **96**, 15,137–15,142.

84. APHIS (Animal and Plant Health Inspection Services). (2004) Chronic wasting disease. Available at http://www.aphis.usda.gov/vs/nahps/cwd/ (accessed on 23 September 2004).

85. CWDA (Chronic Wasting Disease Alliance). (2004) Chronic wasting disease: implications and challenges for wildlife managers. Available at http://www.cwd-info.org/index.php/fuseaction/about.overview (accessed on 23 September 2004).

86. Hamir, A. N., Miller, J. M., Cutlip, R. C., et al. (2004) Transmission of sheep scrapie to elk (*Cervus elaphus nelsoni*) by intracerebral inoculation: final outcome of the experiment. *J. Vet. Diagn. Invest.* **16**, 316–321.

87. Hibler, C. P., Wilson, K. L., Spraker, T. R., et al. (2003) Field validation and assessment of an enzyme-linked immunosorbent assay for detecting chronic wasting disease in mule deer (*Odocoileus hemionus*), white-tailed deer (*Odocoileus virginianus*), and Rocky Mountain elk (*Cervus elaphus nelsoni*). *J. Vet. Diagn. Invest.* **15**, 311–319.

88. Bio-Rad Laboratories. (2004) BSE testing. Available at http://www.bio-rad.com (Food/Animal/Environmental Testing→TSE Testing→BSE Testing; accessed on 28 October 2004).

89. CAST (Council for Agricultural Science and Technology). (2001) *Johne's Disease in Cattle*. Issue Paper No. 17. Council for Agricultural Science and Technology, Ames, IA.

90. Ficht, T. A., Adams, L. G., Khare, S., et al. (2004) Global analysis of the *Mycobacterium avium* subsp. *paratuberculosis* genome and model systems exploring host–agent interactions. In: *Preharvest and Postharvest Food Safety: Contemporary Issues and Future Directions* (Beier, R. C., Pillai, S. D., Phillips, T. D., and Ziprin, R. L., eds.), IFT/ Blackwell, Ames, IA, pp. 87–99.

91. Johne, H. A. and Frothigham, L. (1895) Ein eigenthumlicher fall von tuberkulose beim rind. *D. Z. Thmd.* **21**, 438–454.

92. Ott, S. L., Wells, S. J., and Wagner, B. A. (1999) Herd-level economic losses associated with Johne's disease on US dairy operations. *Prev. Vet. Med.* **40**, 179–192.

93. YourMedicalSource.com. (2004) What is Crohn's disease (CD)? Available at http://health. yahoo.com/Health/Centers/Digestive/71.html (accessed on 20 September 2004).

94. AgriNews. (2004) Johne's stalks dairy, beef. Rochester, MN. Available at http://webstar.postbulletin.com/agrinews/39098956185410.bsp (accessed on 22 September 2004).
95. Chiodini, R. J. and Hermon-Taylor, J. (1993) The thermal resistance of *Mycobacterium paratuberculosis* in raw milk under conditions simulating pasteurization. *J. Vet. Diagn. Invest.* **5,** 629–631.
96. Sung, N. and Collins, M. T. (1998) Thermal tolerance of *Mycobacterium paratuberculosis.* *Appl. Environ. Microbiol.* **64,** 999–1005.
97. NRC (National Research Council). (1994) *Science and Judgment in Risk Assessment.* National Academy, Washington, DC.
98. Mena, K. D., Rose, J. B., and Gerba, C. P. (2004) Addressing microbial food safety issues quantitatively: a risk assessment approach. In: *Preharvest and Postharvest Food Safety: Contemporary Issues and Future Directions* (Beier, R. C., Pillai, S. D., Phillips, T. D., and Ziprin, R. L., eds.), IFT/Blackwell, Ames, IA, pp. 415–426.

Index

Printed in the United States of America